THE PERFORMANCE AND DESIGN OF ALTERNATING CURRENT MACHINES

THE PERFORMANCE AND DESIGN OF ALTERNATING CURRENT MACHINES

TRANSFORMERS, THREE-PHASE INDUCTION MOTORS AND SYNCHRONOUS MACHINES

BY

M. G. SAY

Ph.D., M.Sc., A.C.G.I., D.I.C., M.I.E.E., F.R.S.E.

FORMERLY PROFESSOR OF ELECTRICAL ENGINEERING,
HERIOT-WATT COLLEGE, EDINBURGH

THIRD EDITION

CBS Publishers & Distributors Pvt. Ltd.

New Delhi • Bengaluru • Chennai • Kochi • Kolkata • Mumbai
Hyderabad • Nagpur • Patna • Pune • Vijayawada

ISBN: 81-239-1027-4

First Indian Edition: 1983
Reprint: 1998, 1999, 2000, 2002

This edition has been published in India by arrangement with
M/s Pitman, London

Published by **Satish Kumar Jain** and produced by **Varun Jain** for
CBS Publishers & Distributors Pvt. Ltd.,
4819/XI Prahlad Street, 24 Ansari Road, Daryaganj, New Delhi - 110002
delhi@cbspd.com, cbspubs@airtelmail.in • www.cbspd.com
Ph.: 23289259, 23266861, 23266867 • Fax: 011-23243014

Corporate Office: 204 FIE, Industrial Area, Patparganj, Delhi - 110 092
Ph: 49344934 • Fax: 011-49344935
E-mail: publishing@cbspd.com • publicity@cbspd.com

Branches:
• *Bengaluru:* 2975, 17th Cross, K.R. Road, Bansankari 2nd Stage,
Bengaluru - 70 • Ph: +91-80-26771678/79 • Fax: +91-80-26771680
E-mail: cbsbng@gmail.com, bangalore@cbspd.com
• *Chennai:* No. 7, Subbaraya Street, Shenoy Nagar, Chennai - 600030
Ph: +91-44-26681266, 26680620 • Fax: +91-44-42032115
E-mail: chennai@cbspd.com
• *Kochi:* Ashana House, 39/1904, A.M. Thomas Road, Valanjambalam,
Ernakulum, Kochi • Ph: +91-484-4059061-65
Fax: +91-484-4059065 • E-mail: cochin@cbspd.com
• *Kolkata:* 6-B, Ground Floor, Rameshwar Shaw Road, Kolkata - 700014
Ph: +91-33-22891126/7/8 • E-mail: kolkata@cbspd.com
• *Mumbai:* 83-C, Dr. E. Moses Road, Worli, Mumbai - 400018
Ph: +91-9833017933, 022-24902340/41 • E-mail: mumbai@cbspd.com

Representatives:

• Hyderabad: 0-9885175004	• Nagpur: 0-9021734563
• Patna: 0-9334159340	• Pune: 0-9623451994
• Jharkhand: 0-9811541605	• Uttarakhand: 0-9716462459

Printed at:
J.S. Offset Printers, Delhi

PREFACE
TO THE THIRD EDITION

THE original plan of combining performance with design has prove
to be an acceptable one, particularly as the emphasis is on the
performance rather than on the design. Most readers are user
rather than designers, and they will find as comprehensive a dis
cussion of a.c. machines, their basic theory of operation and applica
tion, as can be concentrated into a volume of this size. The con
siderable range of a.c. machines has made it necessary to limit the
scope to transformers, three-phase induction machines and syn
chronous machines. All other machines are to some degree specia
and of subsidiary importance.

The book is not intended as an instruction in commercial design
that is an art best known to practising designers. An adequate
study of turbo-alternator design, for example, would take a volume
in itself, at least one-half of which would be devoted to subject
matter of entirely mechanical concern.

In this third edition the opportunity has been taken of a complet
conversion to rationalized M.K.S. units. All formulae, whether
concerned with electrical, magnetic, mechanical or thermal quan
tities, are consistent among themselves in respect of units. Only in
numerical examples are decimal multiples or submultiples employed
if convenient, and where ratings are given in units of MVA. o
kVA. instead of the basic volt-ampere, the fact is stated.

The text has been largely re-fashioned to give emphasis to
essential unity of all electromagnetic machines, a unity that ha
been obscured by the technological accretions in which each clas
of machine has, in the last half-century, become separately embedded
Further, an introduction is now given to the general circuit theor
of machines, a powerful method of dealing with their behaviou
under transient conditions.

Thanks are again due to the author's good friends in the industry
whose help has been warm and unstinted. Mr. E. Openshaw Taylo.
B.Sc., D.I.C., Assistant Professor of Electrical Engineering at th
Heriot-Watt College, has rendered invaluable help and critica
advice during the new revision. He is also the author of a companio
volume *The Performance and Design of A.C. Commutator Motors.*

The author has had many letters with suggestions for improve
ment of the text, and will continue to welcome the comments of hi
readers

HERIOT-WATT COLLEGE,   M.G.S.
EDINBURGH

UNIT SYSTEM

BESIDES the usual "practical" volt (V.), ampere (A.), ohm (Ω.), mho ($\mho$.), henry (H.) and farad (F.), a self-consistent unit system with a rationalized M.K.S. basis is used throughout. The principal units are listed below, with their abbreviations:

Length, mass, time	metre, kilogram, second	m., kg., s.
Area, volume	sq. metre, cu. metre	m.^{2}, m.^{3}
Density	kilogram per cu. metre	kg./m.^{3}
Force	newton	N.
Work and energy	metre-newton = joule	m.-N., J.
Power	joule per sec. = watt	W.
Torque	newton-metre	N.-m.
Pressure	newton per sq. metre	N./m.^{2}
Angle	radian	rad.
Velocity: angular	radian per sec.	rad./s.
linear	metre per sec.	m./s.
Rotational speed	revolution per sec.	r.p.s.
Inertia	kilogram-sq. metre	kg.-m.^{2}
Temperature	degree Celsius (centigrade)	°C.
Thermal resistivity	watt per metre per sq. metre per deg. C.	thermal Ω.
Resistivity	ohm per metre per sq. metre	Ω-m.
Magnetic flux, linkage	weber, weber-turn	Wb. Wb.-T.
Magnetic flux-density	weber per sq. metre	Wb./m.^{2}
Magnetomotive force	ampere-turn	AT.
Magnetizing force	amp.-turn per metre	AT./m.
Reluctance	amp.-turn per weber	AT./Wb.
Permeance	weber per amp.-turn	Wb./AT.
Permeability (abs.)	weber per sq. meter per amp.-turn per metre	H./m.
Specific magnetic loading	weber per sq. metre	Wb./m.^{2}
Specific electric loading	amp.-cond. per metre	AC./m,
Current density	amp. per sq. metre	A./m.^{2}

CONVERSIONS—

Force: 1 N. = 1 J./m. = 0·102 kg. (force) = 0·225 lb. (force).
Energy: 1 J. = 0·738 ft.-lb. = 0·24 cal.
Power: 1 W. = 0·738 ft.-lb./s. = 1·34/1000 h.p.
Inertia: 1 kg.-m.^{2} = 0·738 slug-ft.^{2} = 23·7 lb.-ft.^{2}.
Torque: 1 N.-m. = 0·738 lb.-ft.

CONTENTS

poles—Leakage of non-salient poles—Armature leakage—Slot leakage—Overhang leakage—Zig-zag leakage—Differential leakage—Phase reactance of an induction motor—Phase reactance of a synchronous machine

INSETS

READING GUIDE

THE following schedule gives a course of introductory reading for those studying a.c. machinery for the first time. Only the simpler aspects of theory, performance and construction are included. The references are to Chapters and their numbered paragraphs:

Fundamental Principles	I: 1, 2, 4. II: 1–3.
Transformers	III: 1–6. IV: 1–4. V: 1, 5, 7, 9, 10.
General Machines	VIII: 1–5. IX: 1–3. X: 1–4, 8, 11.
Induction Machines	XII: 1–4. XIII: 1· 4, 8, 11, 14. XV: 1–3.
Synchronous Machines	XVII: 1–7. XVIII: 1–3, 8, 10. XX: 2, 3, 6.
Converting Machines	XXII: 1–3. XXIII: 1–4, 8, 9, 11.

LIST OF SYMBOLS

Capitals

A, B, C, D	quadripole parameters.
A	integral constant.
AT	ampere-turns.
B	flux density.
$\overline{B}$	specific magnetic loading.
C	capacitance.
C	number of coils.
D	damping-torque constant.
D	distance between core-axes.
D	stator bore.
E	e.m.f.
$\overline{E}$	average e.m.f.
F	magnetomotive force.
G	density.
G	output coefficient.
G	weight.
H	barometric height, mm.
H	inertia constant.
I	current.
J	current per conductor.
K	capacitance.
K	constant or factor.
L	core length.
L'	core length, contracted.
L	inductance.
L	length or distance dimension.

M	mutual inductance.
M	torque.
N	linkages.
N	number of phases or rings.
N	speed of rotation, r.p.m.
P	force.
P	power.
Q	reactive volt-amperes.
R	radius.
R	resistance.
S	cooling surface.
S	number of slots.
S	reluctance.
S	volt-ampere rating.
S	$= (1-s)$.
T	absolute temperature.
T	number of turns.
U	volume.
V	voltage.
W	weight, lb.
W	width of core.
X	reactance.
Y	admittance.
Y	pole-pitch.
Z	impedance.
Z	number of conductors.

Small Letters

a	cross-section.
a	h.c.f. of S and p.
a	number of parallel paths.
a	width of duct.
ac	amp.-conductors per unit length.
at	amp.-turns per unit length.
b	breadth.
b	component flux density.
b	conductor or coil width.
b	pole-arc.
c	capacitance.
c	cooling coefficient.
c	ratio, damping ratio.
c	spacing.
c_1	stator constant.
d	diameter.
d	diameter of bare wire.
d_1	diameter of insulated wire.
e	e.m.f.
f	factor, function.
f	frequency.
f	mechanical force.
g	gravitational constant.

g	slots per pole.
g'	slots per pole per phase.
h	height or depth.
i	current.
j	current per conductor.
j	$+ 90°$ rotation operator.
k	constant.
k	coupling coefficient.
k	unbalance coefficient.
l	length.
m	layers in slot.
m	mass.
m	number of coils in series.
m	number of starter steps.
n	harmonic order.
n	integer.
n	number of h.v. coils.
n	speed, r.p.s.
n_1	synchronous speed.
p	force, stress or pressure.
p	number of pole-pairs.
p	operator d/dt.
p	specific loss.

xix

SMALL LETTERS—(*contd.*)

r	radius.	w	width or dimension.
r	ratio.	x	fraction, multiplier.
r	resistance.	x	reactance.
r'	equivalent resistance.	x'	equivalent reactance.
s	slip.	y	admittance.
t	thickness.	y'	equivalent admittance.
t	time.	y	pitch.
t	turns-ratio.	y'	contracted pitch.
u	number of coil-sides per slot.	z	impedance.
u	velocity, or peripheral speed.	z'	equivalent impedance.
v	voltage.	z	number of conductors.

GREEK LETTERS

α	angle.	Λ	permeance.
α	attenuation coefficient.	μ	absolute permeability.
α	decay factor.	μ_0	magnetic space constant, $4\pi/10^7$.
α	eddy-current loss coefficient.	μ_r	relative permeability.
α	$= r_2/x_2$.	ν	cyclic speed-irregularity.
β	angle.	ν	magnetic leakage coefficient.
β	angular phase-pitch (elec.).	Φ	magnetic flux.
β	phase coefficient.	ϕ	component flux.
γ	angular slot-pitch (elec.).	ϕ	load phase-angle.
γ	specific cost.	ψ	angle between coil e.m.f.'s (elec.).
γ	starter coefficient.	ρ	resistivity.
δ	current-density.	ρ_i	thermal resistivity of insulation.
δ	deflection.	ρ	resistor step.
δ	divergence angle.	σ	phase-spread.
ε	base of natural logarithms.	σ	power or load angle.
ε	chording angle.	θ	angle.
ε	eccentricity.	θ	arctan (x/r).
ε	regulation.	θ	temperature-rise.
η	efficiency.	τ	time-constant.
λ	internal phase-angle.	ω	angular frequency $= 2\pi f$.
λ	permeance coefficient.	ω_r	angular velocity.
λ	wavelength.		

SUBSCRIPTS

a, b, c	phases A, B, C.	c	core.
a	acceleration.		critical.
	active component.		I^2R.
	alternating component.	d	conducted.
	armature.		direct.
	axial.		disturbing.
	duct.		duct.
ad	armature direct-axis.		eddy-loss.
aq	armature quadrature-axis.	e	coil-span.
av	average.		electrical.
b	brush.		e.m.f.
c	cage-rotor.	eq	equipotential.
	capacitor.	eu	e.m.f., unsaturated.
	coil.	f	field.
	commutator.		form-factor, load-factor.
	compensating.		friction.
	conductor.	g	gap.
	converter.	h	harmonic.
	copper.		hysteretic.

SUBSCRIPTS—(contd.)

i	current.		s	space.
	induction.			starting.
	initial.			steady-state.
	inlet.			synchronous, synchronizing.
	insulation.		sc	short-circuit.
	iron.		sci	ideal short-circuit.
l	leakage.		sd	direct-axis synchronous.
	limb.		sdu	direct-axis syn., unsaturated.
	line.		sl	shoe-leakage.
	load.		sq	quadrature-axis synchronous.
	loss.		t	tensile.
m	distribution-factor.			tooth.
	magnetizing.			torque.
	maximum.			total.
	mechanical.			transient.
	mutual.			turn.
mc	mean-conductor.		$l\frac{1}{3}$	one-third distance
mm	maximum mechanical.		v	voltage.
mt	mean-turn.		w	width.
n	normal-rated.			winding-factor.
	harmonic-order.			window.
o	opening.		x	leakage.
	overall.			reactance.
	overhang.		y	yoke.
p	pth conductor.		z	impedance.
	pole.			zigzag.
	power.		0	characteristic or surge.
	pressure.			initial.
	pulsation.			natural-frequency.
pd	direct-axis pulsational.			no-load.
ph	phase.			zero-phase-sequence.
pl	pole-leakage.		1	input.
pq	quadrature-axis pulsational.			primary.
q	quadrature-axis.			stator.
r	radial.			synchronous.
	radiated.			upper-limit.
	reactive component.		2	auxiliary.
	remanent.			lower-limit.
	resistance.			outer-cage.
	resultant.			output.
	retaining-ring.			rotor.
	rotational.			secondary.
rd	direct-axis rotational.		3	inner-cage.
rq	quadrature-axis rotational.			tertiary.
s	shoe.		$+$	positive-phase-sequence.
	slot.		$-$	negative-phase-sequence.

CHAPTER I

INTRODUCTORY

1. Alternating Current Machines. The synchronous generator, converter and motor, the transformer, and the induction motor, form the subjects of study in this book. They are of sufficient importance to account between them for almost all of the output of manufacturers of electrical machinery, apart from d.c. motors.

2. The Alternating Current System. The wide application of a.c. machines may be said to be due to

(a) The heteropolar arrangement.

(b) The transformer.

Electrical energy is particularly useful by reason of the ease with which it can be distributed and converted to other desired forms. At present, the chief source of electrical energy is the mechanical form, either direct (as in hydraulic storage reservoirs or rivers) or from chemical energy via heat. A convenient direct method of converting mechanical to electrical energy is by use of the mechanical force manifested when a current-carrying conductor is situated in a magnetic field. The movement of a conductor through a polar magnetic field having consecutive reversals of polarity results in the production of an induced e.m.f. which changes sign (i.e. direction) in accordance with the change in the magnetic polarity: in brief, an alternating e.m.f. Thus all ordinary *electromagnetic machines* are inherently *a.c. machines*.

The wider use of electricity has necessitated the development of systems of transmission and distribution. *Transmission systems* are required (a) to enable sources of natural energy (e.g. waterfalls or storage lakes or coal fields), far distant from the centres of energy demand (e.g. large works, industrial areas or cities), to be economically utilized; (b) to enable generating plant to be concentrated in a few large, favourably-situated stations; and (c) to permit of the interconnection of networks or distributing areas to ensure reliability, economy and continuity of supply.

Frequent reference is made to the companion volume, *The Performance and Design of Direct Current Machines*, by Dr. A. E. Clayton. For brevity, this is contracted to [A] in the text. Since references are valuable if they do not involve unnecessary searching, such references are confined chiefly to a few easily-obtained textbooks, as follows—

[A] *Performance and Design of D.C. Machines.* A. E. Clayton and N. N. Hancock (Pitman).

[B] *Electrical Engineering Design Manual.* M. G. Say (Chapman & Hall).

[C] *Performance and Design of A.C. Commutator Motors.* E. O. Taylor (Pitman).

I

A high voltage is desirable for transmitting large powers in order to decrease the I^2R losses and reduce the amount of conductor material. A very much lower voltage, on the other hand, is required for distribution, for various reasons connected with safety and convenience. The transformer makes this easily and economically possible.

Thus the fact that generation is inherently a production of alternating e.m.f.'s, coupled with the advantage of a.c. transformation, provides the basic reason for the widespread development of the a.c. system. The collateral development of a.c. motors and other methods of utilizing a.c. energy has been a natural step.

It must by no means be assumed that the a.c. system represents a finality. Apart from the superiority of d.c. motors in certain cases, particularly where speed-control is desired, there are signs that a high-voltage d.c. transmission system may evolve in the near future. However, while changes, small or great, in methods of electrical production and use may be confidently expected, it is probable that a.c. machines will remain important components of electrical generation, application and utilization.

3. Circuit Behaviour. A voltage v, applied to a circuit of series resistance R and inductance L, produces a current such that

$$v = Ri + L(di/dt)$$

at every instant. If v is a constant direct voltage V, the *steady-state* current is $I = V/R$, because $di/dt = 0$ in the absence of changes; but whenever the conditions are altered (e.g. by switching or by changing the parameters) the new steady state is attained through a *transient* condition. The instantaneous current can then be written $i = i_s + i_t$, with i_s the final steady-state current and i_t the temporary transient component. The time-form of i_t depends only on the parameters, and for the RL circuit considered is the exponential

$$i_t = i_0 \varepsilon^{-(R/L)t} = i_0 \exp(-R/L)t.$$

Fig. 1 (a) shows the steady-state and transient components and the total resultant current for an initiation transient when the drive-voltage is V. As the current is zero at the switching instant (and cannot rise thereafter faster than V/L), then i_t must cancel i_s initially: hence

$$i = i_s + i_t = i_s - i_s \exp(-R/L)t = (V/R)[1 - \exp(-R/L)t]$$

This is a well-known case of practical interest in the switching of d.c. field circuits.

If v is the sine voltage $v_m \sin \omega t$, the steady-state current is the sinewave $i_s = (v_m/Z) \sin(\omega t - \theta)$, where $Z = \sqrt{(R^2 + \omega^2 L^2)} = \sqrt{(R^2 + X^2)}$, and $\theta = \arctan(X/R)$. Suppose the voltage to be switched on at some instant in its cycle corresponding to $t = t_0$: the steady-state current for this instant would be $(v_m/Z) \sin(\omega t_0 - \theta)$. But the current is actually zero, so that again i_t initially cancels i_s.

Subsequently i_t decays exponentially as before. The current following circuit switching is therefore $i = i_s + i_t$, giving

$$i = (v_m/Z)\,[\sin\,(\omega t - \theta) - \sin\,(\omega t_0 - \theta)\,.\,\exp\,(-R/L)(t - t_0)].$$

In machine circuits R is often much smaller than X so that $Z \simeq X$ and $\theta \simeq 90°$. For these conditions, approximately,

$$i \simeq (v_m/X)[-\cos\,\omega t + \cos\,\omega t_0\,.\,\exp\,(-R/L)(t - t_0)].$$

If the voltage is switched on at a zero ($t_0 = 0$, π/ω, $2\pi/\omega$ . . .), the transient term has its greatest value v_m/X, and the resultant current

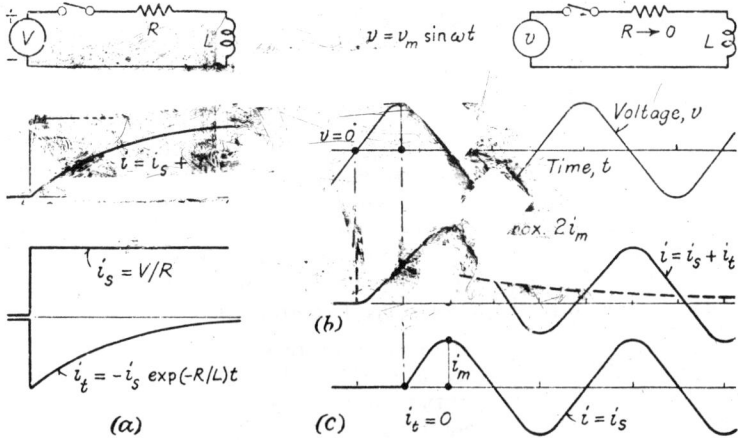

Fig. 1. Transient Phenomena in RL Circuit

starts from zero with complete asymmetry, Fig. 1 (b). In contrast, if the switch is closed on peak voltage, the resultant current attains steady state instantaneously without any transient, Fig. 1 (c). The current in (b) has an amplitude nearly double that in (c), an example of the *doubling effect*. Intermediate switching instants give partial asymmetry, with smaller transient components.

TRANSIENT AND STEADY STATES. It is usual to develop circuit theory on a steady-state basis, using complex algebraic treatment with complexor or "vector" diagrams. The consideration of transient conditions demands a return to more basic concepts.

STEADY-STATE CONVENTIONS. A complexor* voltage V, of magnitude V and drawn at an arbitrary angle α to a horizontal datum, can be variously described as

$$V = V\underline{/\alpha} = V_1 + jV_2 = V\,(\cos\alpha + j\sin\alpha) = V\,.\,\exp\,(j\alpha).$$

* Voltages and currents are not vectors in the true physical sense, and are here called "complexors." However, in deference to common usage the term "vector" is also occasionally used in the text.

Fig. 2A shows a circuit of series resistance R and inductive reactance $j\omega L = jX$. An r.m.s. voltage V is applied, and an r.m.s. current I flows such that $V = I(R + jX) = IZ$. The impedance operator $Z = Z/\theta = \sqrt{(R^2 + X^2)}$. $/\text{arc tan } (X/R)$: it does not stand for a sinusoidally-varying quantity. If $I = I/\beta$, then $I = V/Z$ and $\theta = \alpha - \beta$.

The voltage can be regarded as the complex sum of components IR and IX. The latter is the component opposing E_x (the e.m.f. of

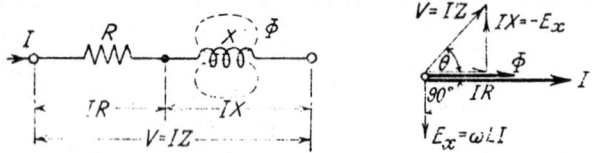

FIG. 2A.   VECTOR OR COMPLEXOR DIAGRAM

self-inductance in L) whose instantaneous value is $e_x = - L(di/dt)$. The magnetic flux Φ of the coil, proportional to the current if saturation effects can be ignored, is represented by a complexor in phase with the current. Then E_x lags on it by $\pi/2$. The r.m.s. flux could quite properly be employed, but the peak value is usual as a practical measure of the degree of magnetic saturation.

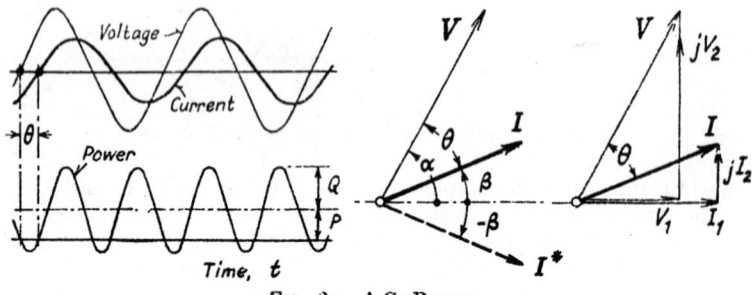

FIG. 2B.   A.C. POWER

POWER. The instantaneous power in a single-phase circuit pulsates at double supply frequency. In general it is asymmetric about the time axis so that it cannot be represented in a diagram containing V and I. The average of the power/time curve is the mean power dissipation P watts, about which the remaining alternating part has an amplitude Q vars (reactive volt-amperes), Fig. 2B.

If $V = V/\alpha$ and $I = I/\beta$, the product $VI = VI/\alpha + \beta$ is unrelated to the power P. But the power can be derived from the voltage V and the conjugate of I, namely $I^\star = I/-\beta$: for

$$VI^\star = V/\alpha \cdot I/-\beta = VI/\alpha - \beta = VI/\theta$$
$$= VI (\cos \theta + j \sin \theta) = P + jQ.$$

The average power is $P = VI \cos \theta$ and the reactive volt-amperes—or peak rate of energy storage in reactive circuit components—is $Q = VI \sin \theta$. With V leading I as in an inductive circuit, θ and Q are positive. Alternatively I could be taken with the conjugate of V to give $V^{\star}I = P - jQ$ for an inductive circuit.

With voltage and current $V/\underline{\alpha}$ and $I/\underline{\beta}$ it is thus only necessary to take the product $S = VI$ volt-amperes with the cosine and sine of the angular difference θ. Further, $P = S \cos \theta$ and $Q = S \sin \theta$, so that $S = \sqrt{(P^2 + Q^2)}$. If the voltage and current are known in the rectangular form $V = V_1 + jV_2$ and $I = I_1 + jI_2$, the process is to take pairs of co-phasal components for the power, and of quadrature components for the reactive volt-amperes:

$$P = V_1 I_1 + V_2 I_2 \text{ and } Q = V_2 I_1 - V_1 I_2.$$

The result for Q will be positive if V leads I.

4. Per-unit Values. Calculations and comparisons of machine performance are commonly expressed in *per-unit* values. Normal voltage is 1.0 p.u. (or 100 per cent) voltage, and full-load current is 1.0 p.u. current. Their product is 1.0 p.u. (i.e. full-load) rating. The advantage is that products of per-unit values require no adjustment, whereas with percentage values it is necessary to divide the product by 100.

Consider a 1 200-kVA., 3·3-kV., star-connected generator with a phase resistance of $0.18\ \Omega$. Unit phase voltage is 1.9 kV., and unit phase current 210 A. The full-load IR drop is $210 \times 0.18 = 38$ V., which is 0.02 p.u. (or 2 per cent) of 1.9 kV. The resistance is thus 0.02 p.u. For half-load current the drop would be 0.5 p.u. $\times$ 0.02 p.u. $= 0.01$ p.u. In percentage form it would be $50 \times 2 = 100$, requiring division by 100 to obtain the correct drop of 1 per cent of the phase voltage.

CHAPTER II
FUNDAMENTAL PRINCIPLES

1. The Electromagnetic Machine. At voltages that can be developed and used by normal means, the electrostatic forces are very weak. On the other hand, even with comparatively small currents a considerable mechanical force can be produced by electromagnetic means; in consequence the machines in common use as generators and motors are *electromagnetic* in type. Only in very high-voltage generators and transformers have the electrostatic field and its effects to be considered as regards their secondary influences on the operating characteristics, apart, of course, from the question of insulation.

2. Induction and Interaction. There are two related principles forming the foundations upon which are based all electromagnetic machines concerned in the conversion of electrical energy to or from mechanical energy. These are (*a*) the law of *induction* and (*b*) the law of *interaction*.

(1) LAW OF INDUCTION The essentials for the production of an electromotive force are electric and magnetic circuits, mutually interlinked. The summation of the products of webers of magnetic induction with complete turns of the circuit is termed the *total flux-linkage*: if it is made to change, an e.m.f. is induced in the electric circuit. This e.m.f. persists only while the change is taking place, and has a magnitude proportional to the rate of the change with time. The instantaneous e.m.f. is

$$e = -\ (dN/dt) \text{ volts}$$

with the linkage N in weber-turns. The e.m.f. has a direction such as to oppose the change. Thus if the electric circuit were closed on itself, and the number of line-linkages formed by it and some externally-produced magnetic field were reduced, then the e.m.f. induced would produce a current in the closed circuit, generating a self-magnetic field superimposed upon the external field and tending to make up the deficiency.

For engineering purposes the induction law is generally used in the simplified form—

$$e = -\ T_c(d\Phi/dt) \text{ volts} \quad . \quad . \quad . \quad . \quad . \quad (1)$$

Here T_c is the number of turns in the electric circuit, all of which are linked completely with all the Φ webers of induction of a given flux. For this purpose a gross flux may be resolved into (i) a *mutual* or working component, and (ii) a *leakage* component.

The electromagnetic method of producing an e.m.f. in a circuit (in order that the e.m.f. shall produce a current and thus enable electrical energy to be delivered) is therefore to provide a magnetic field linked with an electric circuit, and to change the number of line-linkages $N = T_c\Phi$. Considering for simplicity that the electric circuit comprises a coil of T_c turns, then the change of line-linkages may be accomplished in a variety of ways—

(i) Supposing the flux constant in value, the coil may move through the flux (relative motion of flux and coil);

(ii) Supposing the coil stationary with reference to the flux, the flux may vary in magnitude (flux pulsation);

(iii) Both changes may occur together; i.e. the coil may move through a varying flux.

In case (i) above the flux-cutting rule can be applied. The e.m.f. in a single conductor of length l m. can be calculated from the

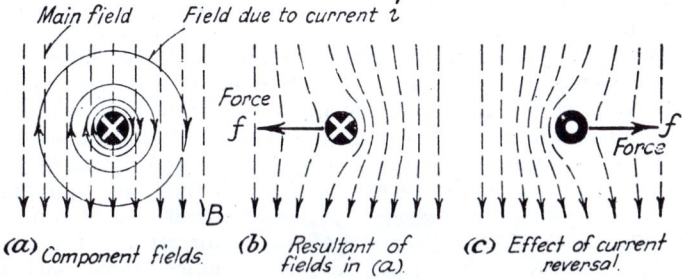

FIG. 3. ELECTROMAGNETIC INTERACTION

rate at which it cuts across a magnetic field of density B webers per m.^{2} when moving at u m. per sec. at right-angles to the direction of the flux—

$$e = Blu \text{ volts} \qquad . \qquad . \qquad . \qquad . \qquad (2)$$

This is referred to as the *e.m.f. of rotation*. It is always associated with the conversion of energy between the mechanical and electrical forms. The e.m.f. in a coil in case (ii) is found directly from eq. (1) as the *e.m.f. of pulsation* or *transformation*. No motion is involved and there is no energy conversion. For case (iii) both e.m.f.'s are produced: this case is treated generally in Chapter XXVI.

(2) LAW OF INTERACTION. When a conductor of length l m., carrying a current i amperes, lies in and perpendicular to the direction of a magnetic field of density B webers per m.^{2}, a mechanical force is developed on it of magnitude

$$f = Bli \text{ newtons} \qquad . \qquad . \qquad . \qquad . \qquad (3)$$

in a direction perpendicular to both current and field. In the diagram (a) of Fig. 3, B represents the density of an original magnetic field. The introduction of a conductor carrying a current brings at the same time a new magnetic field. The original field and the conductor field combine to form a resultant field (Fig. 3 (b)).

The field density actually existing round the conductor is not B. It is greater than B on one side and less on the other. The distortion is an essential feature in the production of mechanical force.

GENERATORS AND MOTORS. In a *generator*, an e.m.f. is produced by the movement of a coil in a magnetic field. The current produced by the e.m.f. interacts with the field to produce a mechanical force opposing the movement, and against which the essential movement has to be maintained. The electrical power ei is produced therefore from the mechanical power supplied.

In a *motor*, we may suppose a conductor or coil to lie in a magnetic field. If current is supplied to the coil, a mechanical force is manifested and due to this force the coil will move. Immediately that relative movement takes place between coil and field, however, an e.m.f. is induced, in opposition to the current. To maintain the current and the associated motor action, it is therefore necessary to apply to the coil, from an external source, a voltage sufficient to overcome the induced e.m.f. Thus the motor requires electrical power to produce a corresponding amount of mechanical power.

The directions of flux, current and movement in generator and motor action are given in Fig. 4. The coil is free to move about the

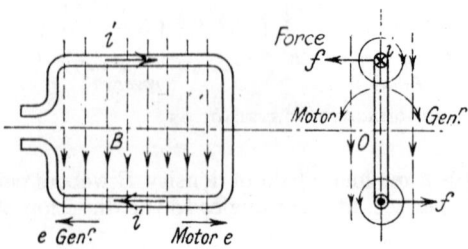

axis O. The component fields are shown, the direction of the mechanical force, and the directions of rotation for motor and generator action. The direction of the e.m.f. is such as to maintain the current in a generator and to oppose it in a motor. The action is reversible: i.e. the same

FIG. 4. ELEMENTARY ELECTROMAGNETIC MACHINE

arrangement may act either as generator or motor.

TRANSFORMERS. Forces are developed in the transformer but are not allowed to produce movement. Consequently there is no concern with mechanical power, and only transformer e.m.f.'s are generated.

3. **Classification.** The principles of § 2 are applied as in Fig. 5.

(a) *Rotary Machines.* Two magnetic elements Fig. 5 (a), one fixed (*stator*) and the other (*rotor*) capable of relative rotation, are separated by a narrow annular gap. The stator is usually the outer element for mechanical convenience. Each element carries one or more windings, and a mutual magnetic flux crosses the gap to link them. Rotor rotation results in e.m.f. induction and in electromechanical power conversion through interaction torques.

(*b*) *Transformers.* The *induction-regulator* transformer, in which the relative position of the elements has to be adjustable, is identical with (*a*). For the normal *static* transformer the gap is not necessary, so the flux is established in a closed magnetic circuit, Fig. 5 (*b*).

In spite of constructional variety, the electrical differences

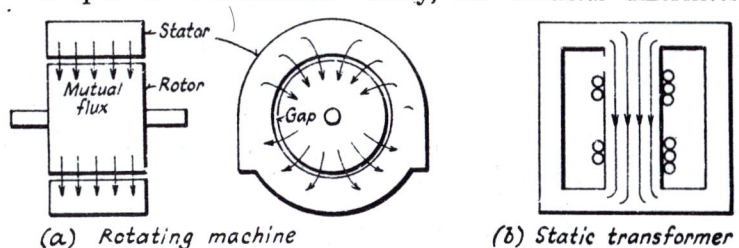

(*a*) *Rotating machine*   (*b*) *Static transformer*

FIG. 5. ELEMENTS OF ELECTROMAGNETIC MACHINES

between the several types of machine are only secondary, and result from—

(i) the kind of power system, d.c. or a.c., on which the machine is to work; and

(ii) the kind of connections made between the windings and the power system: i.e. tapped phase windings or commutators with brushes.

The following combinations are considered in detail in Chapter XXVI.

CLASSIFICATION OF ELECTROMAGNETIC MACHINES

Case	Flux	Coils	Connections	Type-name
A	Constant	Moving	Commutator	D.C. machine
B	Constant	Moving	Tappings	Synchronous machine
C	Pulsating	Fixed	Commutator	Transverter
D	Pulsating	Fixed	Tappings	Transformer, regulator
E	Pulsating	Moving	Commutator	A.C. commutator machine
F	Pulsating	Moving	Tappings	Induction machine

This book is devoted to cases B, D and F: none has a commutator. Static transformers are treated first, then a general discussion of rotating machines in Chapter VIII to precede a more detailed examination of cases F and B.

4. Three-phase Complexor Diagram. The working flux Φ in an a.c. machine results from a current I representing the combined m.m.f.'s of all windings linking the magnetic circuit. The e.m.f. E generated in a winding in which Φ varies sinusoidally lags 90° on Φ and I. The complexor relationship is the same as that of I, Φ and E_x in Fig. 2A.

For three-phase machines under steady, balanced conditions the complexor diagram is drawn with the voltage and current of phase A, but with the flux Φ_m common to all phases. Consider phase A on no load when the axis of Φ_m passes the phase-centre: this is the instant of zero linkages and of maximum induced e.m.f. E_a. It is reasonable to draw Φ_m along the pole-axis; and to preserve the time-quadrature relation, E_a must be drawn at right-angles (lagging) to Φ_m, as shown in Fig. 6. Thus for this instant E_a lies in the

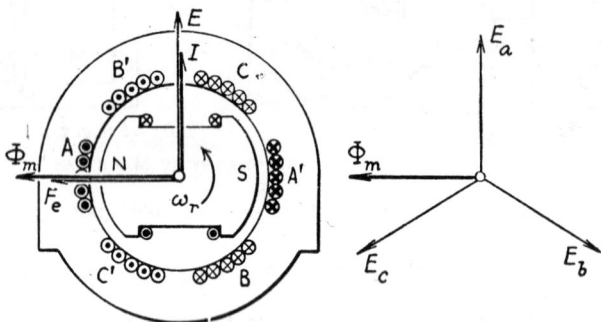

FIG. 6. CONVENTION FOR THREE-PHASE MACHINE

m.m.f. axis of phase A. Peak current I_a, if in phase with peak e.m.f. E_a, must then also be drawn in the axis of phase A. As shown by the dots and crosses, this will be the instantaneous m.m.f. axis of the whole armature. The conductors of phase A produce interaction torque with Φ_m corresponding to the power $E_a I_a$.

Thus the flux is coincident with the pole axis on no load, or with the axis of resultant m.m.f., on load. E_a lags 90° on Φ_m, and the current I_a has the direction of the axis of phase A.

SYMMETRICAL COMPONENTS. The e.m.f.'s generated by the machine in Fig. 6 have the *positive* phase-sequence ABC. With unbalanced loading, the asymmetric voltages and currents may be resolved into symmetrical components with *positive, negative* and *zero phase-sequence*. The behaviour of machines under conditions of unbalance can then be discussed in terms of these symmetrical components.

CHAPTER III

TRANSFORMERS: THEORY

1. Use of the Transformer. The part played by the transformer in making possible the a.c. distribution of electrical energy has been mentioned already. It has been said that almost the entire world production of electrical energy is transformed twice, thrice, or even four times before being utilized. The annual output of transformers in Britain is 10 000 000 kVA. The wide field covered is indicated by the following list—

SIZES AND USES OF TRANSFORMERS

Type	Application	Order of Size, kVA.
Power . .	Generator	60 000 upwards
	Main high-voltage transmission lines .	50 000 upwards
	Secondary transmission lines . .	10 000–50 000
	Bulk supply to large consumers . .	500–20 000
	Distribution	50– 2 000
	House	0·5– 25
Plant . .	Converter, rectifier, startor, furnace, testing units	Wide range
	Traction (locomotive)	500–3 000
Special unit .	Instrument transformers . .	Fractional
	Small mains, lamp, radio-feed, bell transformers	Fractional
	Welding, neon-sign	1–100

The physical basis of the transformer is *mutual induction* between two circuits linked by a common magnetic field. Small coreless transformers as used with high-frequency currents are more susceptible to treatment directly on this basis, while power transformers have characteristics that enable simplifying assumptions to be made, but which may cause the physical principles to be obscured. The general case illustrates more directly, perhaps, the physical principles involved.

2. Mutual Induction. An electric circuit carrying a current has associated with it, as a part of the electrical phenomenon of current flow, a *magnetic field* in its immediate neighbourhood. If the current in the circuit is alternating, then the magnetic field at any point in the surrounding medium will change in magnitude and direction in accordance with the changes of current with time.

If another circuit (the *secondary*) be in the vicinity of the first (the *primary*), it will link some of the magnetic flux produced by the primary (Fig. 7). With an alternating primary current (and therefore flux) the changing linkages will produce in the secondary an e.m.f.

$$e_{2M} = -\frac{dN_{2M}}{dt} \text{ volts}$$

where N_{2M} represents the linkages in the T_2 turns of the secondary winding with that part Φ_{1M} of the flux Φ_1 produced by the primary that links the secondary. If the second-

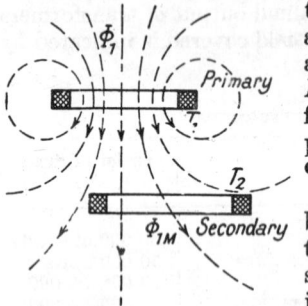

ary coil is suitably shaped and favour-ably placed relatively to the primary, $N_{2M} \simeq T_2\Phi_{1M}$: in general N_{2M} will differ from this simple product as it is not possible to secure that *all* the flux Φ_{1M} links *all* the turns T_2 completely.

The e.m.f. e_{2M} is said to be produced by reason of the *mutual induction* of the primary and secondary circuits. A similar effect will naturally take place if the respective roles of the two circuits

FIG. 7. MUTUAL INDUCTION   are interchanged, and it is shown in textbooks of electrical technology that the mutual inductance is the same irrespective of which circuit is primary and which secondary, in any given case. The coefficient of mutual inductance L_{12} in henrys may be defined as the e.m.f. in volts induced in one circuit when the current in the other is changed at the rate of 1 A. per sec.; or the energy in the common magnetic field in joules when each circuit carries 1 A.

The mutually induced e.m.f. in the secondary circuit will, if the circuit be closed through a load. circulate current in the load and dissipate energy therein. This energy can come only from the primary, to which the whole operation is due. Thus energy is being transferred from primary to secondary by means of the mutual magnetic field. This is important, and is the principle underlying the transformer effect. The process briefly is: the primary produces a pulsating magnetic field in which energy is stored and restored periodically. The e.m.f. e_{2M} and the current i_2 associated with it in the secondary circuit abstract energy from the common field and pass it on to the secondary load. If there is no secondary load the magnetic field energy passes into and out of the primary circuit as a continual pulsation of energy from electrical to and from magnetic form

The more closely the primary and secondary circuits are mutually linked, the more direct becomes the exchange of energy oetween them. If the two circuits link a common iron core, Fig 8, the effects are—

(*a*) A great increase in the total flux by virtue of the improved permeance of the magnetic circuit;

(*b*) A smaller magnetizing current (i.e. primary current with secondary open-circuited), since the increased flux per ampere induces more primary e.m.f.

(*c*) A much greater proportion of mutual to non-mutual or leakage flux: the latter has air paths whereas the former occupies the permeable iron core.

(*d*) The introduction of losses in the core, so that the field can no longer be established without loss.

The voltage applied to the primary is almost completely concerned in opposing the induced e.m.f. due to the mutual flux, so that if the primary voltage is constant, the mutual flux remains approximately constant regardless of the load connected across the secondary coil.

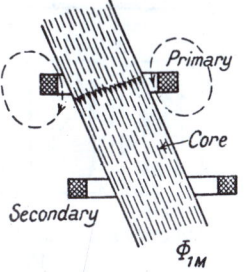

FIG. 8. INCREASE OF MUTUAL INDUCTANCE BY IRON CORE

LINKED ELECTRIC AND MAGNETIC CIRCUITS IN POWER TRANSFORMERS. The power transformer is required to pass electrical energy from one circuit to another, via the medium of the pulsating mutual magnetic field, as efficiently and economically as possible. Our knowledge of magnetic materials indicates the use of iron or steel for the conveyance of the flux with much greater ease than any other known material. The coils

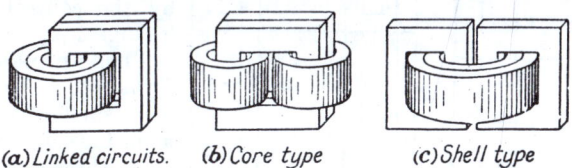

(*a*) Linked circuits.　(*b*) Core type　(*c*) Shell type

FIG. 9. LINKED ELECTRIC AND MAGNETIC CIRCUITS

are therefore made to embrace an iron core, which serves as a good conducting path for the mutual magnetic flux, ensuring that the flux links each coil fairly completely.

The elementary linked circuits are shown diagrammatically in Fig. 8. The use of an iron core permits of much greater freedom in the shape and arrangement of the primary and secondary coils, since the great majority of the flux will be conveyed by the core almost regardless of the relative positions of the two sets of coils—primary and secondary—that link it. In practice, two general forms are usual: these are obtained from the simple linked circuits (*a*) of Fig. 9 by splitting either the coils (*b*) or the core (*c*) to give the *core* and *shell* types, of which the elementary forms are shown for single-phase transformation. In core types, to avoid undue leakage flux, it is usual to have half the primary and half the secondary wind-

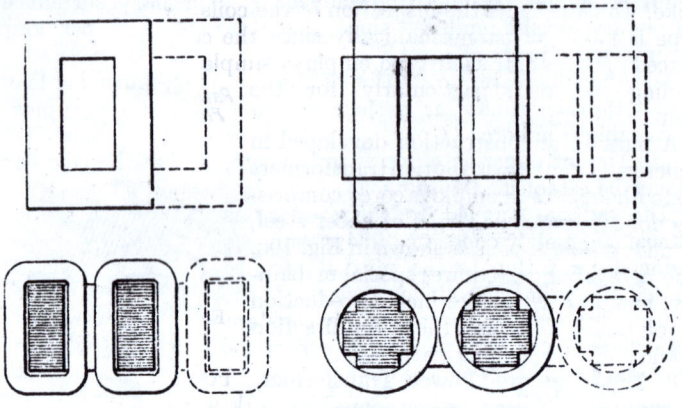

RECTANGULAR CORE TYPE   CIRCULAR CORE TYPE

FIG. 10A. CORE-TYPE TRANSFORMERS

ing side-by-side or concentrically on each limb; not primary on one limb and secondary on the other. The form of the transformer construction is determined by the constructional methods employed and by the control of the leakage flux.

Three-phase transformers are developed from single-phase types as in Figs. 10A and 10B. The three-phase shell arrangement is merely three single-phase transformers assembled together. The three-phase core type, on the other hand, embodies the principle that the sum of the fluxes in each phase in a given direction along the cores is zero, i.e. that the flux going up one limb can be returned down the other two. Thus only one-half of a

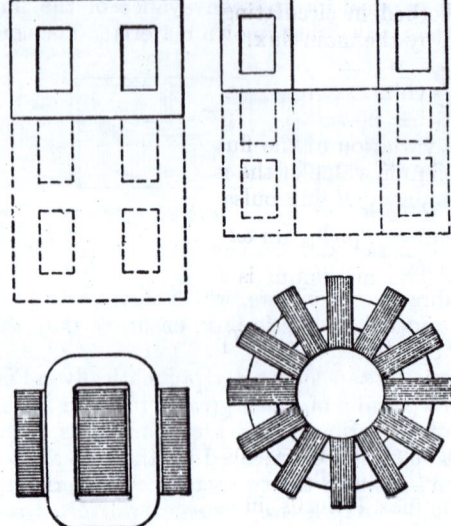

RECTANGULAR SHELL TYPE   CIRCULAR SHELL TYPE

FIG. 10B. SHELL-TYPE TRANSFORMERS

complete magnetic circuit is necessary for each phase. Each set of phase windings occupies one limb only.

The core type is more easily repaired on site, by removing the yoke, which permits the inspection of the coils and cores. The shell type is more robust mechanically since the coils are more readily braced. The radial shell type employs simple round coils, and the cooling is good, particularly for the iron.

A method of construction developed in America for small distribution transformers up to about 5 kVA. employs cores comprising long continuous strips of sheet steel, wound round the coils as shown in Fig. 10c. The core winding requires special machinery, but the advantages include reduction of joints and the use of the grain-direction of the steel for the flux-path.

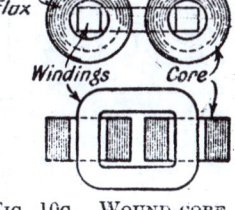

FIG. 10c. WOUND-CORE TRANSFORMER

3. **Theory of the Power Transformer.** Power transformers for normal purposes can be considered to work with an approximately constant *mutual* or *main* flux, Φ_m. Since the resistance of the windings is always small, and the leakage flux (not linking both windings and therefore not contributing to the transformer-action) is only a small fraction (e.g. 5 per cent) of the total, the resistance and leakage reactance voltage drops are small compared with the reactance due to the mutual flux. Approximately, therefore, the whole of the applied voltage V_1 is absorbed in circulating the primary current against the e.m.f. induced by the main flux. This e.m.f. is

$$e_1 = - T_1 \frac{d\Phi}{dt} \text{ volts} \quad . \quad . \quad . \quad . \quad (1)$$

Assuming sinusoidal time variation of the flux, let $\Phi = \Phi_m \cos \omega t$. where Φ_m is the time-maximum value of the mutual flux in webers; $\omega = 2\pi f$; and f is the frequency of flux pulsation. Then

$$e_1 = - T_1 (d/dt)(\Phi_m \cos \omega t) = T_1 \omega \Phi_m \sin \omega t \text{ volts} \quad . \quad (4)$$

the instantaneous value. The maximum is $e_{1m} = T_1 \omega \Phi_m$ and the r.m.s. value

$$E_1 = (1/\sqrt{2})\omega T_1 \Phi_m = (\sqrt{2})\pi f T_1 \Phi_m = 4.44 f T_1 \Phi_m \text{ volts} \quad (5)$$

In each turn, the e.m.f. is

$$E_t = 4.44 f \Phi_m \text{ volts} \quad . \quad . \quad . \quad . \quad (6)$$

whether the turn is part of the primary, secondary, or other winding, provided only that it links the flux Φ_m.

If the secondary winding has T_2 turns, in each of which an e.m.f. E_t is induced, the total e.m.f. is

$$E_2 = 4.44 f T_2 \Phi_m \text{ volts.} \quad . \quad . \quad . \quad . \quad (7)$$

Obviously the primary and secondary e.m.f.'s bear the same ratio as the turns, or

$$E_1/E_2 = T_1'/T_2 \qquad \qquad . \qquad . \qquad . \qquad . \qquad . \quad (8)$$

Further, considered with respect to the common flux which produces them, both e.m.f.'s are *in phase*. From eq. (4) it appears that the e.m.f.'s lag by 90° in time on the flux. The applied primary voltage V_1 opposes E_1, while E_2 provides the secondary output voltage V_2. Thus V_1 and V_2 are substantially in *phase-opposition*.

At normal loads the m.m.f. required to maintain the main flux is small compared with the m.m.f. of either current alone. Consequently the primary and secondary m.m.f.'s substantially balance each other, or $I_1T_1 \simeq I_2T_2$, whence

$$I_1/I_2 \simeq T_2/T_1 \qquad . \qquad . \qquad . \qquad . \qquad . \quad (9)$$

a relation inverse to that for the e.m.f.'s, eq. (8). The relative phase of the currents is that of *opposition*, since their m.m.f.'s oppose.

The action of the transformer can be summarized as follows: Let an alternating voltage V_1 be applied to a primary coil of T_1

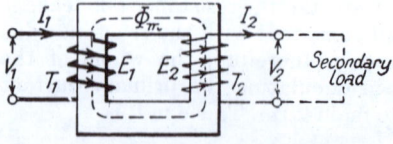

turns linking a suitable iron core (Fig. 11). A magnetizing reactive current then flows in the coil, establishing a flux Φ_m in the core (and small additional fluxes elsewhere, neglected for a first consideration). The magnitude of Φ_m

FIG. 11   ELEMENTARY TRANSFORMER

is such that it induces in the coil an e.m.f. E_1 of self-induction to counterbalance the applied voltage V_1 and establish electrical equilibrium. If there be a secondary coil of T_2 turns, linking the same core, then by mutual induction an e.m.f. E_2 is developed therein. Should a load (i.e. an impedance of some finite value) be connected to the second coil, a current I_2 will flow in the secondary circuit under the influence of the induced e.m.f. E_2. The secondary current will, by Lenz's law, tend to reduce the pulsating flux Φ_m, but this is prevented by an immediate and automatic adjustment of the primary current I_1, thereby maintaining the flux Φ_m at the value required to produce the e.m.f. of self-induction E_1. Any reduction of the flux would cause a diminution of E_1, leaving a voltage-difference between V_1 and E_1 which would be sufficient to increase the primary current and thereby re-establish the flux. Thus any current which flows in the secondary causes its counterpart to flow in the primary, it being a condition of working of the transformer that the flux Φ_m shall always be maintained at a value such that the voltage V_1 applied to the primary terminals shall be balanced by the induced e.m.f. E_1, neglecting drops. It is, therefore, evident that energy is conveyed from primary to secondary by the flux: the primary stores energy in the magnetic field, and an extraction of some of this for the secondary load is made up by the addition of energy from the primary, which consequently takes an increased current. Since the input to the transformer is $V_1 I_1$ = $E_1 I_1$ volt-amperes, and the output is $E_2 I_2$, then

$$E_1 I_1 = E_2 I_2$$

neglecting losses; whence

$$I_2/I_1 = E_1/E_2 = T_1/T_2 . \qquad . \qquad . \qquad . \qquad (10)$$

a relation which neglects also the magnetizing current demanded by the imperfection of the core. From eq. (10) it follows that $I_2 T_2 = I_1 T_1$ or that the m.m.f.'s of the primary and secondary windings are equal. They are also in magnetic opposition on account of the phase-opposition of I_1 and I_2.

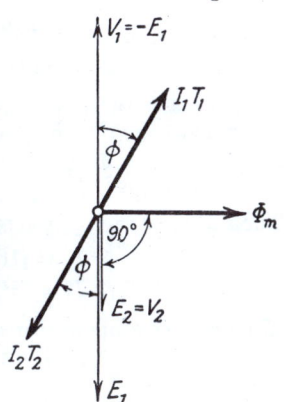

In an *ideal transformer*, there are no voltage drops in resistance or leakage reactance, the m.m.f. required to maintain the main flux is vanishingly small, and there is no core loss. In these circumstances, eq. (10) holds exactly, and the phase relations are the simple ones shown in Fig. 12. Here V_1, the applied primary voltage, is equal in magnitude, and in exact phase opposition, with E_1. $E_2 = V_2$, the secondary terminal voltage. According to the nature of the secondary load, the phase and magnitude of the secondary current I_2 will be determined. The m.m.f. is proportional to $I_2 T_2$. To this, the primary m.m.f. $I_1 T_1$ will be equal and opposite.

FIG. 12. COMPLEXOR DIAGRAM: IDEAL TRANSFORMER

The stability of the conditions represented by Fig. 12 can be understood from the following considerations. Since there are no resistance and leakage reactance drops, the e.m.f. E_1 must always be equal to the applied (constant) voltage V_1. The flux Φ_m inducing E_1 is thus constant. As, in the ideal transformer, it requires no m.m.f., there must be no resultant m.m.f. in the common magnetic circuit. The appearance of $I_2 T_2$ due to the secondary load current must thus be counterbalanced by the equal and opposite $I_1 T_1$: or, the rate $E_2 I_2$ of energy abstraction from the secondary must be provided for by the introduction of energy at an equal rate $E_1 I_1$ ($= V_1 I_1$) into the primary.

EXAMPLE. The core of a three-phase, 50-cycle, 11 000/550-V. mesh/star, 300-kVA., core-type transformer has a *gross* cross-section of 400 cm.^{2} Find (*a*) the number of h.v. and l.v. turns per phase, (*b*) the e.m.f. per turn, and (*c*) the full-load h.v. and l.v. phase currents. The core density should not exceed 1·3 Wb./m.^{2}

The phase-voltage ratio for mesh/star connection is 11 000/(550/$\sqrt{3}$) = 11 000/317 V. The *net* core section, i.e. allowing for core-plate insulation, is 0·9 × 400 = 360 cm.^{2} Assuming the induction density to be 1·3 Wb./m.^{2}, the e.m.f. per turn from eq. (6) is

$$E_t = 4{\cdot}44 . 50 . (360 \times 10^{-4} \times 1{\cdot}3) = 10{\cdot}4 \text{ V}.$$

The number of secondary turns is

$$T_2 = 317/10{\cdot}4 = 30{\cdot}5, \text{ say } \underline{31 \text{ turns,}}$$

since fractions of a turn are not possible. The adjustment to 31 turns, however, makes the e.m.f. per turn now

$$E_t = 317/31 = \underline{10{\cdot}2 \text{ V.,}}$$

and the actual core density

$$B = 1{\cdot}3 \,(10{\cdot}2/10{\cdot}4) = 1{\cdot}275 \text{ Wb./m.}^2$$

The primary turns are most directly obtained from the secondary turns and the voltage ratio:

$$T_1 = 31 \,(11\,000/317) = \underline{1\,075 \text{ turns.}}$$

The rating of each phase is 100 kVA., so that the currents are

$$I_1 = (100/11\,000)10^3 = 9{\cdot}1 \text{ A.,}$$

and

$$I_2 = (100/317)10^3 = 316 \text{ A.}$$

The primary ampere-turns are

$$I_1 T_1 = 9{\cdot}1 . 1\,075 = \underline{9\,800 \text{ A.T.}};$$

and for the secondary

$$I_2 T_2 = 316 . 31 = \underline{9\,800 \text{ A.T.}}$$

The magnetizing current would scarcely alter this equality: with a core-length of 40 cm. per phase and a density of 1·275 Wb./m.^{2}, the excitation required would be about 14 A.T. per cm., or a total of 14 × 40 = 560 A.T. peak value, i.e. 400 A.T. r.m.s. This represents a magnetizing current of about 4 per cent of full-load current.

4. The Complexor Diagram. Much of the behaviour and operating characteristics of the transformer can be derived from its "vector" diagram It is also convenient for introducing the modifications due to the divergence from the "ideal" state of the actual transformer.

MAGNETIZING CURRENT AND CORE LOSS. Although the iron core is highly permeable, it is not possible to generate a magnetic field in it without the application of at least a small m.m.f. Thus even when the secondary winding is on open circuit and giving no current, a small *magnetizing* current is needed to maintain the magnetic circuit or core in the magnetized state. The magnetizing m.m.f. is a function of the length, the net cross-sectional area, and the permeability of the iron path. The m.m.f. of the primary circuit on no load is of the order of 5 per cent of its m.m.f. on full load.

The magnetizing current (symbol I_{0r}) is to be considered as in time phase with its associated flux: a reasonable assumption, as when there is no magnetizing current and consequently no m.m.f., there will be no flux. (See Fig. 13.) For present purposes I_{cr} is considered to be sinusoidal, although this is not the case for the saturation values normally employed in practice.

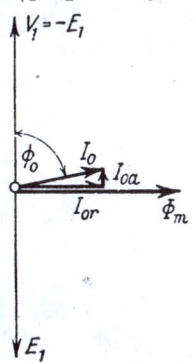

The pulsation of the flux in the core is productive of *core loss*, due to hysteresis and eddy-currents, as described in Chapter IX. The continuous loss of energy to the core, which produces heat eventually dissipated from the transformer, requires a continuous supply from the electrical source to which the primary is connected. Since the only way in which an electric circuit can furnish energy is for a current to flow under the influence of a voltage, there must be a current component I_{0a} which supplies the losses. The product $E_1 I_{0a}$ thus represents the power input to supply core loss. The voltage E_1 is concerned (rather than V_1) because this is the voltage directly associated with the magnetic flux.

On no load, the currents I_{0r} for magnetization and I_{0a} for losses will flow. The currents are in phase quadrature, since one is in phase with Φ_m while the other is in the direction of E_1, which in turn is in phase-quadrature to Φ_m. The components I_{0a} and I_{0r} have as resultant the *no-load current* I_0. so that

$$I_0 = \sqrt{(I_{0r}^2 + I_{0a}^2)} \qquad \qquad \qquad (11)$$

The reactive component of the no-load current, I_{0r}, cannot physically be dissociated from the active component I_{0a}, for the latter is an essential part of the magnetization of the core, accounting for the hysteresis and eddy-current effects which modify and complicate the phenomenon of magnetization in a ferrous metal.

The no-load current makes a phase-angle ϕ_0 of the order of arc cos 0·2 with the applied voltage V_1.

VOLTAGE DROPS IN RESISTANCE AND LEAKAGE REACTANCE. The practical transformer has coils of finite resistance. In large transformers the effect of the resistance in causing a voltage drop is nearly negligible, but the I^2R loss due to the same cause is of great importance in design, as producing the major portion of the load losses that must be dissipated by the cooling and ventilating system.

The resistance of the transformer windings is actually distributed uniformly, but may conveniently be conceived as concentrated. The drop across the resistance r_1 of the primary winding is then I_1r_1, and this has to be provided for by a component of the applied voltage under all conditions in order that the current can be made to circulate through the winding. Similar considerations apply to the secondary resistance voltage I_2r_2.

The leakage reactance is not so readily appreciated. Magnetic flux cannot be confined into a desired path so completely as an electric current. The greater proportion (i.e. the *mutual flux*) is

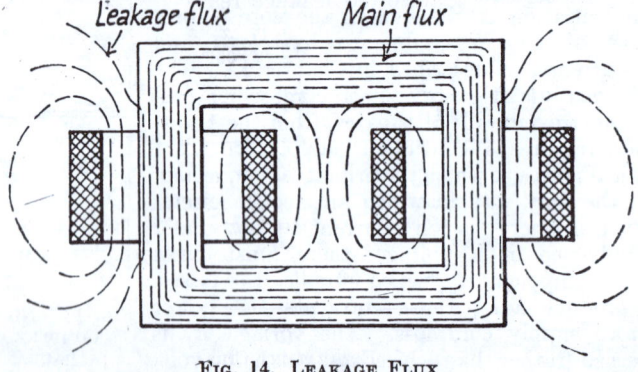

Leakage flux   Main flux

FIG. 14. LEAKAGE FLUX

produced in the core provided for it, but a small proportion (Fig. 14) called the *leakage flux* links one or other winding, but not both, so that it does not contribute to the transfer of energy from primary to secondary. On account of the leakage flux, however, both primary and secondary windings have *leakage reactance*, that is, each will become the seat of an e.m.f. of self-induction, of magnitude a small fraction (e.g. 3 per cent) of the e.m.f. due to the main flux. The terminal voltage V_1 applied to the primary must therefore have a component I_1x_1 to balance the primary leakage e.m.f. In the secondary, similarly, an e.m.f. of self-induction is developed, which can be considered as additional to E_2; alternatively, the reversed drop I_2x_2 can be introduced to take account of the leakage reactance.

The primary and secondary coils in Fig. 14 are shown on separate limbs, an arrangement that would result in an exceptionally large leakage. Leakage flux plots for the more usual winding arrangements are shown in Figs. 76 and 97. Physically the leakage flux is the result of the opposing ampere-turn distributions of both windings, and its subdivision into separate components, one for each winding, though arbitrary, is convenient.

Leakage between primary and secondary could be eliminated if the windings could be made to occupy the same space. This, of course, is physically impossible, but an approximation to it is achieved if the coils of primary and secondary are sectionalized and interleaved: such an arrangement leads to a marked reduction of the leakage reactance. If, on the other hand, the primary and secondary are kept separate and widely spaced, there will be much more room for leakage flux and the leakage reactance will be greater. It is thus possible to control the reactance within limits. The calculation of reactance is detailed in Chapter VII, §7.

THE EQUIVALENT CIRCUIT. The transformer shown diagrammatically in Fig. 15 (a) can be resolved into an equivalent circuit (b) in which the resistance and leakage reactance of primary and

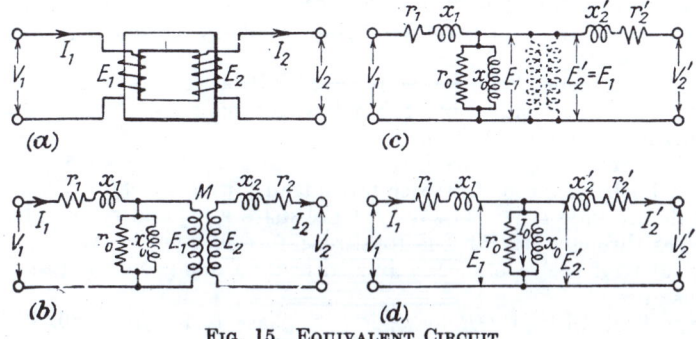

FIG. 15. EQUIVALENT CIRCUIT

secondary respectively are represented by the "lumped" r_1, x_1, r_2 and x_2, as if these were external to a transformer of which the windings were without resistance and leakage. Similarly a shunt circuit r_0 and x_0 can be introduced such that $E_1/r_0 = I_{0a}$ and $E_1/x_0 = I_{0r}$, the two quadrature components of the magnetizing current. The windings of the transformer are now "ideal," and represent the seat of the induced e.m.f.'s E_1 and E_2, which are related by the expression $E_1/E_2 = T_1/T_2$, the turn-ratio.

Suppose $T_1 = T_2$, then $E_1 = E_2$, and the two sides of the transformer may be joined in parallel (c), and the energy transmitted from primary to secondary without a transformer at all (d). The circuit, Fig. 15 (d), represents exactly the electrical characteristics of a transformer with unity turn-ratio: that is, the resistance and reactance voltages, no-load current, core and I^2R losses, are reproduced and give the same characteristics as the transformer.

An equivalent circuit is useful for calculations of regulation, parallel operation, etc. Since in the majority of cases the turn-ratio is not unity, it is necessary to imagine the actual secondary winding of T_2 turns replaced by an *equivalent* winding of T_1 turns. for which the I^2R loss and the *per-unit* or *percentage* reactance voltage must be the same as in the actual secondary. For this, the equivalent secondary must have resistance r_2' and leakage reactance x_2' such that

$$I_1{}^2 r_2' = I_2{}^2 r_2, \text{ or } r_2' = r_2 \frac{I_2{}^2}{I_1{}^2} \simeq r_2 \left(\frac{T_1}{T_2}\right)^2; \qquad . \qquad . \qquad . \quad (12)$$

and $\quad \dfrac{I_1 x_2'}{E_1} = \dfrac{I_2 x_2}{E_2}, \text{ or } x_2' = x_2 \dfrac{I_2}{I_1} \cdot \dfrac{E_1}{E_2} \simeq x_2 \left(\dfrac{T_1}{T_2}\right)^2. \qquad . \qquad . \quad (13)$

The current I_2' in the equivalent secondary is very nearly I_1, for in Fig. 15 (d) the shunt current I_0 is, at normal loads, only a small fraction of I_1.

The equivalent secondary is thus obtained from the actual secondary by multiplying its resistance and reactance by the square of the turn-ratio. The output voltage V_2' is (T_1/T_2) times as great as the actual secondary voltage V_2 on account of the change in turn-ratio to unity.

The argument above can be reversed if it be desired to consider a transformer as comprising the actual secondary winding and an equivalent primary of T_2 turns.

EXAMPLE. The full-load I^2R loss on the h.v. side of a 300-kVA., 11 000/550-V., mesh/star three-phase transformer is 1·86 kW.; and on the l.v. side it is 1·44 kW. (a) Calculate r_1, r_2 and r_2' for phase values throughout. (b) The total reactance is 4 per cent: find x_1, x_2 and x_2' if the reactance is divided in the same proportion as the resistance.

(a) The full-load I^2R loss per h.v. phase is 1·86/3 = 0·62 kW., and the current per phase is 9·1 A. Whence

$$I_1{}^2 r_1 = 0.62 \text{ kW., and } r_1 = 620/9.1^2 = 7.50 \ \Omega.$$

Similarly for the l.v. phase,

$$I_2{}^2 r_2 = 1.44/3 = 0.48 \text{ kW., } r_2 = 480/316^2 = 0.0048 \ \Omega.$$

The secondary equivalent resistance in primary terms is, from eq. (12),

$r_2' = r_2 (T_1/T_2)^2 = r_2 (V_1/V_2)^2$ where V_1 and V_2 are phase values, 11 000 V. and 317 V. respectively. Hence

$$r_2' = 0.0048 \ (11 \ 000/317)^2 = 5.77 \ \Omega.,$$

a figure comparable with r_1 as might be expected. The total I^2R loss is, per phase of h.v. and l.v. together,

$$I_1{}^2 r_1 + I_2{}^2 r_2 = I_1{}^2 r_1 + I_2{}'^2 r_2' = I_1{}^2 (r_1 + r_2') = I_1{}^2 R_1.$$

Evaluating this, the loss is

$$9.1^2 \ (7.50 + 5.77) = 1.1 \text{ kW. per phase}$$
$$= 3.3 \text{ kW. for all three phases.}$$

This agrees with the total I^2R loss given as 1·86 + 1·44 = 3·3 kW.

(*b*) The reactance is 4 per cent for the transformer, and will be the same for each phase. A 4 per cent, or 0·04 per-unit, reactance is such that, for full-load current

$$I_1X_1 = I_1(x_1 + x_2') = 4 \text{ per cent of } V_1 = 0.04V_1,$$

whence $\quad X_1 = 0.04 \cdot 11\,000/9\cdot1 = 48\cdot4\,\Omega.$

If x_1 and x_2' are related in the same proportion as r_1 and r_2' (a fairly reasonable assumption, but one that cannot be verified), then

$$x_1 = \underline{27\cdot4\,\Omega} \text{ and } x_2' = \underline{21\cdot0\,\Omega}.$$

From eq. (13)

$$x_2 = x_2'(T_2/T_1)^2 = x_2'(V_2/V_1)^2$$

whence $\quad x_2 = 21\cdot0(317/11\,000)^2 = \underline{0\cdot0174\,\Omega.}$

COMPLEXOR DIAGRAM OF THE TRANSFORMER ON LOAD. It would be inconvenient to draw the diagram of a transformer using the actual numerical values of current and voltage, particularly if the step-up or step-down ratio were large. The diagram is drawn for the equivalent transformer, making the voltages and currents of primary and secondary comparable.

In Fig. 16, then, the complexor Φ_m represents the main or mutual flux, considered to be of constant peak value: a close approximation to actuality in transformers of normal design. The e.m.f.'s induced in the primary and equivalent secondary windings are respectively E_1 and E_2', which lag by 90° on Φ_m, and are equal in magnitude by definition of equivalence. The current I_0 is that which, flowing in the T_1 turns of the primary, produces the requisite exciting ampere-turns, and which contains an active component to supply the core loss.

The load connected to the secondary terminals is assumed to be such as to permit the current I_2' to flow in the equivalent secondary circuit. The primary current must be such as to balance on the magnetic circuit the ampere-turns of the secondary, leaving as resultant the essential I_0. Thus the primary current will be the complexor sum of I_0 and $-I_2'$, indicated in Fig. 16 by I_1.

It is now possible to obtain the terminal voltage. On the primary side, the applied voltage V_1 must have the components: (*a*) $-E_1$ to neutralize E_1, the e.m.f. due to induction by Φ_m; (*b*) I_1r_1 to circulate the current through the resistance of the primary winding; and (*c*) I_1x_1, a component to overcome the e.m.f. E_{x1} of leakage reactance. E_{x1} is due to the pulsation of the primary leakage flux, which is in phase with (and due to) the primary current. Building up these three components, then V_1 is obtained as the primary applied voltage: it makes the angle ϕ_1 with the primary current.

On the secondary side it must be observed that the e.m.f. E_2'

resembles that of a generator, since the current I_2' in the secondary circuit is circulated by it, against the load impedance together with the resistance and leakage reactance of the secondary winding. Thus E_2' comprises three components : (a) the terminal voltage V_2', (b) the resistance voltage $I_2'r_2'$, and (c) the leakage reactance voltage $I_2'x_2'$, which opposes the e.m.f. E_{x2}' of secondary leakage.

The phase relationships of the secondary resistance and reactance voltages to the secondary current are naturally the same as those of

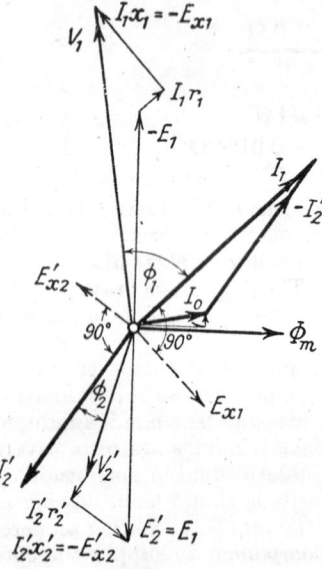

the primary drops and current. The drops are such as to reduce V_1 to E_1 and E_2' to V_2'.

The secondary terminal phase-difference ϕ_2 between V_2' and I_2' is settled by the impedance of the secondary load. The angle was presumed known in drawing the diagram, and inspection of Fig. 16 will show that, since V_1 and not Φ_m (and E_1) is known to start with, the diagram would involve geometrical construction if drawn strictly from measurable values.

For a power transformer of more than a few kVA., the magnitudes of the voltage drops and the no-load current in Fig. 16 are exaggerated. On full load, the resistance and reactance voltages of each winding are generally of the order of $\frac{1}{2}$ and $2\frac{1}{2}$ per cent ($0\cdot005$ and $0\cdot025$ p.u.) re-

FIG. 16. COMPLEXOR DIAGRAM
OF TRANSFORMER ON LOAD

spectively, and the no-load current about 5 per cent ($0\cdot05$ p.u.) of the full-load current. For miniature transformers of a few volt-amperes, considerable increases in these figures may be expected.

5. Efficiency. The transformer is not called upon to convert electrical energy into mechanical energy or *vice versa*, and consequently has no moving parts. The losses are confined to—

Core Loss, due to the pulsation of the magnetic flux in the iron producing eddy-current and hysteresis losses (Chapter IX, §2);

I^2R *Loss*, due to the heating of the conductors by the passage of the current;

Stray Loss, due to stray magnetic fields causing eddy currents in the conductors or in surrounding metal (e.g. the tank);

Dielectric Loss in the insulating materials, particularly in the oil and the solid insulation of high-voltage transformers.

NO-LOAD LOSS. On no load the secondary circuit is open, and the primary current is I_0 only. The I^2R loss due to this is in most cases quite negligible : e.g. the I^2R loss at full load may be of the order of 1 per cent of the rated capacity; since the no-load current is only about one-twentieth of the full-load current, the I^2R loss due to it is only (with these figures) $\frac{1}{400}$ of 1 per cent.

The power input on no load is consequently concerned with the core and dielectric loss, the latter being negligible except for very high voltage transformers for testing. The induced e.m.f. E_1 is almost equal numerically to the applied voltage V_1, and the flux is determined by eq. (5),

$$\Phi_m = E_1/4\cdot44fT_1 \text{ webers,}$$

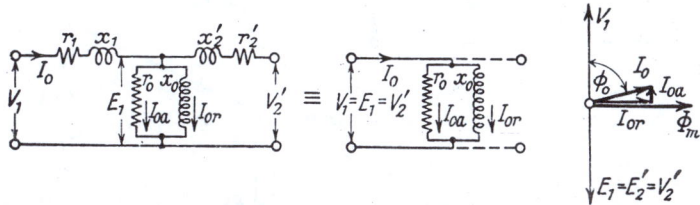

FIG. 17. EQUIVALENT CIRCUIT AND COMPLEXOR DIAGRAM: NO LOAD

where $E_1 \simeq V_1$. Conditions on no load are represented in the equivalent circuit of Fig. 17, in which r_1 and x_1 can be neglected in comparison with r_0 and x_0. The total no-load current is $I_0 = V_1\sqrt{[(1/r_0)^2 + (1/x_0)^2]}$, and the active component $I_{0a} = V_1/r_0$ is concerned with the core loss due to the pulsation of Φ_m. The core-loss power $V_1 I_{0a}$ on no load is consequently measurable by a no-load test. (See Chapter VI, §6.)

SHORT-CIRCUIT LOSS. Consider the complexor diagram in Fig. 18. Suppose the voltage V_1 be reduced to a small fraction of normal value (for safety) and the secondary terminals be short-circuited. A current will circulate in the secondary winding, but the terminal voltage V_2' will vanish by reason of the short circuit. E_2' will be concerned solely in neutralizing the small impedance voltage $I_2'z_2'$, (since V_1, and therefore E_1 and Φ_m, are small). On the primary side the current I_1 is roughly equal and in phase opposition to I_2', while I_0 almost vanishes when the flux is so reduced. The applied voltage has components I_1r_1, I_1x_1 and $-E_1 = -E_2' = -I_2'z_2'$. The conditions are shown in Fig. 18.

The applied voltage on short circuit is seen to be, very nearly

$$V_1 = I_1 z_1 + (-E_1) = I_1 z_1 + (-I_2'z_2')$$
$$\simeq I_1(z_1 + z_2') = I_1 Z_1$$

since $I_1 \simeq -I_2$. Reference to the equivalent circuit of Fig. 15 (d), assuming the secondary terminals short-circuited and the shunt circuit $x_0 r_0$ absent, shows that the simple circuit remaining has a voltage V_1 applied to an impedance z_1 and z_2' in series, so that

$$V_1 = I_1(z_1 + z_2')$$

which is the same as the previous complexor expression.

If V_1 be adjusted so that I_1 has its full load value (requiring the applied voltage to be 0·05-0·1 p.u. of rated value), the e.m.f.'s E_1 and E_2' are very small (0·025-0·05 p.u. of normal no-load values

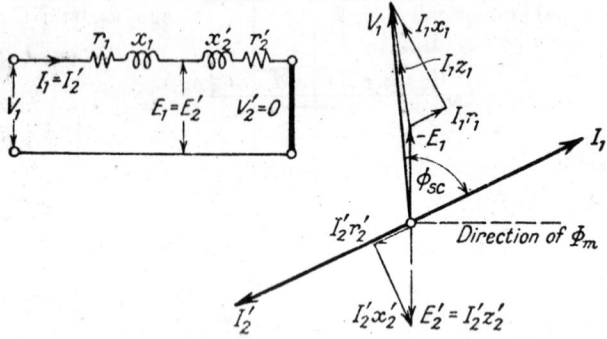

Fig. 18. Equivalent Circuit and Complexor Diagram:
Short Circuit

and the flux is proportionally reduced. The core loss, which is proportional approximately to the square of the flux-density, is consequently negligible, and so is the magnetizing current. The omission of the shunt part of the equivalent circuit is therefore justified for a consideration of short-circuit conditions. The power input on short-circuit is absorbed in heating the coils: i.e. the I^2R loss of both primary and secondary windings together. In any measurement of short-circuit power input, the stray loss is included, because with normal currents circulating in the windings, the leakage fluxes are also normal (or nearly so), and the stray eddy losses produced by them will be included in the power input.

EFFICIENCY ON LOAD. With transformers of normal design, the flux varies only a few per cent between no-load and full-load conditions. Consequently it is permissible to regard the core loss as constant, regardless of load. Let this loss be P_i.

If the short-circuit loss in I^2R with full-load S kVA. be P_c, the loss for any other load (neglecting magnetizing current) will be $x^2 P_c$, where x is the per-unit load considered. Thus at half full-load, the I^2R loss will be $\frac{1}{4}P_c$; at one-tenth of full load, $\frac{1}{100}P_c$; and so on.

The total loss at any load xS kilovolt-amperes at power factor $\cos \phi$ is $P_i + x^2 P_c$, and the efficiency is

$$\eta = \frac{xS \cdot \cos \phi}{xS \cdot \cos \phi + P_i + x^2 P_c} = 1 - \frac{P_i + x^2 P_c}{xS \cdot \cos \phi + P_i + x^2 P_c} \quad (14)$$

The second form is better for the purposes of calculation, since transformer efficiencies are very high and it is easier to figure the divergence from unity.

In a case such as this, where the loss comprises a constant part and a part varying as the square of the load, differentiation shows that maximum efficiency occurs when the variable loss is equal to the constant loss, i.e. $x^2 P_c = P_i$. (See Chapter VII, §4.)

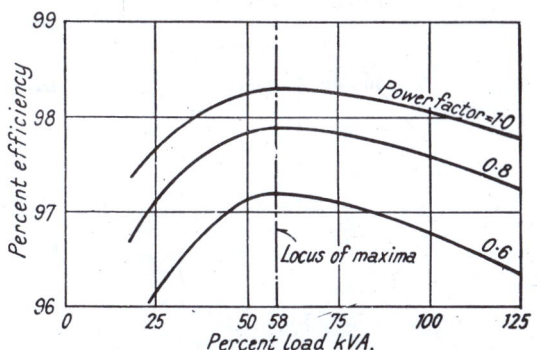

FIG. 19. EFFICIENCY CURVES

Maximum efficiency therefore occurs at the fraction x of the full-load rating S where

$$x = \sqrt{(P_i/P_c)} \text{ per-unit} \qquad . \qquad (15)$$

Generally $x < 1$, so that maximum efficiency occurs below full load. The choice of x is a matter for the designer, and depends on the service for which the transformer is to be designed. Thus with $P_c = 3P_i$, the load for maximum efficiency is $\sqrt{(1/3)} = 0·58$ of full load.

Eq. (14) shows that the efficiency is dependent on the power factor $\cos \phi$ of the load. Greatest efficiency is obtained naturally when the load is non-reactive. If the load is purely reactive, or approximates to this (as in cable testing), the efficiency may be very low. This emphasizes the fact that a.c. machines are built to produce voltage and current, and their size depends on the voltage and current values demanded, not upon the phase relationship between them. Fig. 19 is a typical efficiency characteristic. It is drawn for a transformer with 0·005 p.u. core loss, and 0·015 p.u. full-load I^2R loss, for loads of power factor unity, 0·8 and 0·6 respectively, and is constructed from calculations based on eq. (14), as follows—

EXAMPLE. A 300 kVA. transformer has a core loss of 1·5 kW. and a full-load I^2R loss of 4·5 kW. Calculate its efficiency for $\frac{1}{4}$, $\frac{1}{2}$, $\frac{3}{4}$, 1, and $1\frac{1}{4}$ times full-load output at power factors of (a) unity. (b) 0·8, and (c) 0·6 respectively.

The core loss is $1.5/300 = 0.005$ p.u., and the I^2R loss is $4.5/300 = 0.015$ p.u. on full load. Calculations may be made directly in per-unit values. The full-load kVA. rating is 1.0 p.u., and the full-load kW. is $1.0 \cos \phi$ p.u.

Load, p.u.		0·25	0·5	0·75	1·0	1·25
Core loss, p.u. . .		0·005	0·005	0·005	0·005	0·005
I^2R loss, p.u. . .		0·0009	0·0037	0·0084	0·015	0·0234
Total loss, p.u. . .		0·0059	0·0087	0·0134	0·020	0·0284
Output, p.u.	p.f. = 1·0	0·25	0·50	0·75	1·00	1·25
	p.f. = 0·8	0·20	0·40	0·60	0·80	1·00
	p.f. = 0·6	0·15	0·30	0·45	0·60	0·75
Effici-ency, p.u.	p.f. = 1·0	0·9767	0·9828	0·9824	0·9804	0·9777
	p.f. = 0·8	0·9710	0·9786	0·9781	0·9756	0·9723
	p.f. = 0·6	0·9618	0·9717	0·9710	0·9677	0·9634

The per-unit efficiency is calculated from the expression

$$1 - \frac{\text{total loss p.u.}}{\text{output p.u.} + \text{total loss p.u.}}$$

The efficiency is plotted as a percentage in Fig. 19.

6. Regulation. The regulation of a transformer refers to the change of secondary terminal voltage between no-load and load conditions: it is usually quoted as a per cent or per-unit value for full-load at given power factor.

On no load, $V_1 \simeq E_1 = E_2' = V_2'$ numerically, as will be seen from the diagram in Fig. 17. Thus the voltage at the secondary terminals on no load is obtained from V_2' (or V_1) by multiplying by the turn-ratio T_2/T_1.

On load, the drop $I_1 z_1$ is subtracted from V_1 to obtain $E_1 = E_2'$, then a further drop $I_2 z_2'$ is deducted to obtain V_2'. Neglecting the no-load current, $I_2' = I_1$, and both drops in the form $I_1(z_1 + z_2')$ may be deducted directly from V_1 to obtain V_2'. This process is illustrated by the equivalent circuit in Fig. 20 (a), which neglects the shunt circuit carrying I_0, and in which it is obvious that $V_2' = V_1 - I_1(z_1 + z_2')$.

Thus the *regulation* is given by the numerical difference $V_1 - V_2'$. The complexor diagrams, Fig. 20 (b), are drawn to show the conditions obtaining in the equivalent circuit, and in these diagrams the regulation is greatly exaggerated. In all normal transformers the drop $I_1(z_1 + z_2')$ for full-load current is only a small fraction of V_1. A diagram drawn more nearly in proportion is given in Fig. 21 (a).

Assuming that the angle θ between V_2' and V_1 is negligible, the numerical voltage difference between V_1 and V_2' can be written

$$V_1 - V_2' = I_1(r_1 + r_2') \cos \phi + I_1 (x_1 + x_2') \sin \phi$$

$$= I_1 R_1 \cos \phi + I_1 X_1 \sin \phi,$$

where $R_1 = (r_1 + r_2')$ is the *total resistance in primary terms.*
and $X_1 = (x_1 + x_2')$ is the *total reactance in primary terms.* $\left. \right\}$ (16)

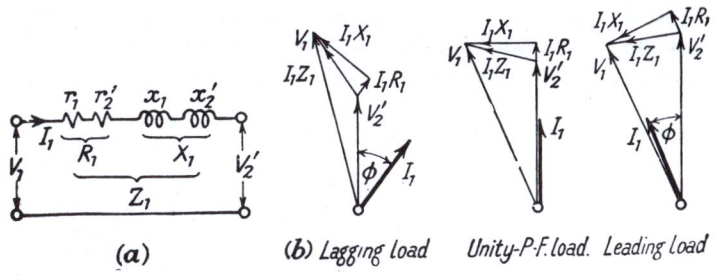

(a)   (b) *Lagging load*   *Unity-P-F. load.*   *Leading load*

FIG. 20. REGULATION

The per-unit regulation, for full-load rated output S and full-load current I_1, is

$$\varepsilon = \frac{I_1 R_1 \cos \phi + I_1 X_1 \sin \phi}{V_1} = \frac{I_1 R_1}{V_1} \cos \phi + \frac{I_1 X_1}{V_1} \sin \phi.$$

In the first of these terms

$$\frac{I_1 R_1}{V_1} = \frac{I_1{}^2 R_1}{V_1 I_1} = \frac{P_c}{S} = \varepsilon_r = \begin{cases} \textit{per-unit } I^2 R \textit{ loss or} \\ \textit{per-unit resistance} \end{cases} \qquad . \qquad (17)$$

In the second term

$$\frac{I_1 X_1}{V_1} = \varepsilon_x = \textit{per-unit reactance} \quad . \qquad . \qquad . \qquad . \qquad (18)$$

The *per-unit regulation* is therefore

$$\varepsilon = \varepsilon_r \cos \phi + \varepsilon_x \sin \phi \qquad . \qquad . \qquad . \qquad . \qquad (19)$$

and the angle ϕ refers to the load phase-angle at the secondary terminals.

If the regulation is large it is necessary to relinquish the assumption that V_1 and V_2' in Fig. 21 (a) are parallel. The revised expression is

$$\varepsilon = \varepsilon_r \cos \phi + \varepsilon_x \sin \phi + \tfrac{1}{2}(\varepsilon_x \cos \phi - \varepsilon_r \sin \phi)^2 \qquad . \qquad . \qquad (20)$$

The expression is obtained trigonometrically from the construction in Fig. 21 (b). If the secondary output is I_2 at V_2 and phase angle

ϕ, the regulation given by eq. (19) is $(\overline{BC} + \overline{CD})/\overline{OA} = \overline{BD}/\overline{OA}$. The true regulation is $\overline{BA}/\overline{OA}$, so that it is necessary to make the small addition $\overline{DA}/\overline{OA}$. This term is

$$\frac{\overline{DA}}{\overline{OA}} \simeq \frac{\overline{FD}^2}{2\overline{OA}^2} = \frac{(\overline{FH} - \overline{DH})^2}{2\overline{OA}^2} = \frac{(I_2 X_2 \cos\phi - I_2 R_2 \sin\phi)^2}{2V_1'^2},$$

whence eq. (20). In Fig. 21, (a) is drawn in primary and (b) in secondary terms: both naturally lead to the same result.

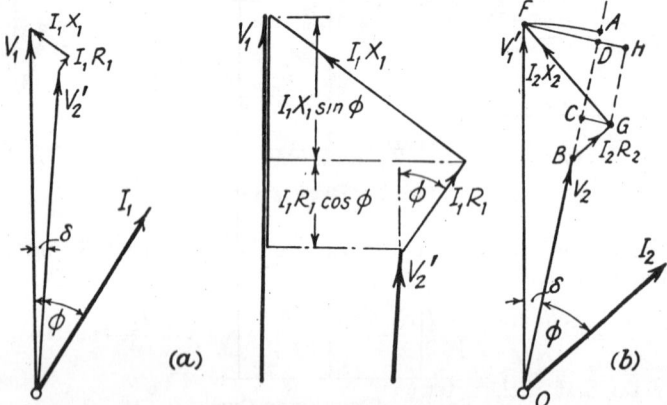

FIG. 21. CALCULATION OF REGULATION

Eq. (19) indicates that the regulation on full load varies with the power factor at the secondary terminals. It will be a maximum when $\phi = $ arc tan $(\varepsilon_x/\varepsilon_r)$, as can be seen from Fig. 21 (a), where the greatest difference between V_1 and V_2' will occur when the angle ϕ coincides with the internal angle arc tan $(X_1/R_1) = $ arc tan $(\varepsilon_x/\varepsilon_r)$ of the total impedance. The regulation will be zero when $\varepsilon_r \cos\phi + \varepsilon_x \sin\phi = 0$, i.e. when tan $\phi = -(\varepsilon_r/\varepsilon_x)$, giving $\phi = -$ arc tan $(\varepsilon_r/\varepsilon_x)$ corresponding to a negative (leading) angle. At leading power factors below this the regulation will be negative, i.e. the secondary terminal voltage will *rise* between no load and full load. The full-load regulation at various power-factors is shown for a typical case in Fig. 22; while Fig. 20 provides a vectorial explanation, the impedance drop $I_1(z_1 + z_2') = I_1 Z_1$ being exaggerated for clarity.

The numerical values ε_r and ε_x are readily calculable from the short-circuit test, as described in Chapter VI, §7.

Regulation is a numeric, not a complex quantity, so that the regulation produced by two impedances connected in series can be combined algebraically. Thus we could write

$$\varepsilon = \varepsilon_{12} = \varepsilon_1 + \varepsilon_2$$

where ε_1 and ε_2 are the parts of the regulation associated with the individual impedances. In the present case the latter represent the primary and equivalent secondary leakage impedances in the

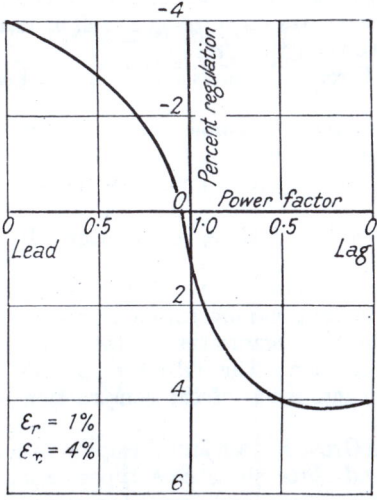

FIG. 22. REGULATION CURVE

equivalent circuit of a two-winding transformer. There is a clear distinction between r_1 and r_2', but the division in $X_1 = x_1 + x_2'$ is arbitrary. However, if the windings are regarded as having distinct leakage reactances, each can be credited with individual leakage impedances and, consequently, per-unit regulations ε_1 and ε_2. This takes care of all mutual effects and is extremely convenient where the transformer has more than two windings. Such cases occur where two or more loads are fed at different voltages from the same primary, or where the method of tap-changing introduces additional load circuits. The method of using per-unit impedances for individual windings is exemplified in Chapter V, § 11.

CHAPTER IV

TRANSFORMERS: CONSTRUCTION

1. Constructional Parts. The transformer is a comparatively simple structure, since there are no rotating parts, or bearings. The chief elements of the construction are—

(1) *Magnetic Circuits*, comprising limbs, yokes, and clamping structures.

(2) *Electric Circuits*, the primary, secondary and (if any) tertiary windings, formers, insulation and bracing devices.

(3) *Terminals*, tappings and tapping switches, terminal insulators and leads.

(4) *Tank*, oil, cooling devices, conservators, dryers and ancillary apparatus.

Improvements are continually being made in construction, and the practice of different manufacturers depends considerably on the size of unit made, the organization of the factory, and the individuality of the designers. The substance of this chapter is to be taken only as an indication of the construction of modern transformers.

The practice in Great Britain and Europe is to concentrate mainly on the single- and three-phase core types. Some shell types are built for single-phase use, and somewhat rarely for three phases. A few special constructions are sometimes employed. Attention here will be directed chiefly to the single-phase core and shell and the three-phase core types of construction.

2. Core Construction. Special alloy steel of high resistance and low hysteresis loss is used almost exclusively in transformer cores, and all electrical-steel manufacturers have suitable grades. (See Chapter IX.) Induction densities up to $1 \cdot 35$–$1 \cdot 55$ Wb./m.^{2} are possible, the limit for 50 c/s being the loss and the magnetizing current.

As the flux in the cores is a pulsating one, the magnetic circuit must be laminated and the separate laminations insulated in order to retain the advantages of subdivision. Paper, japan, varnish china clay or phosphate may be used. The last-named is able to withstand the annealing temperatures of cold-rolled steels, so that it can be applied to the whole sheet before cutting and annealing.

Burring of the edges of the plates may cause a considerable increase in core loss by providing paths for eddy currents should the sharp edges cut through the insulation and establish contact between adjacent plates. Burrs are removed before core assembly. Silicon alloy steels are hard, and cause wear of the punching tools, so that the removal of burrs needs special attention.

It is found that the magnetic properties of transformer sheet steels vary in accordance with the direction of the grain produced by rolling. Sheets are therefore cut as far as possible along the grain which is the direction in which the material has a higher permeability.

It must not be forgotten that lamination and insulation of core plates reduce the effective or net core area, for which due allowance must be made. In building the core, considerable pressures are used to minimize air gaps between the plates which would constitute avoidable losses of area and might contribute to noisy operation.

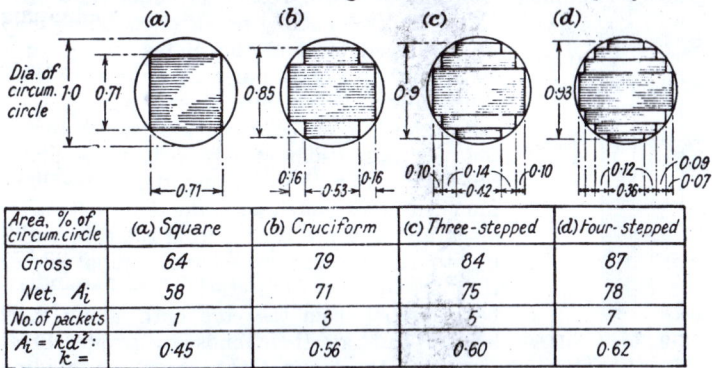

Area, % of circum.circle	(a) Square	(b) Cruciform	(c) Three-stepped	(d) Four-stepped
Gross	64	79	84	87
Net, A_i	58	71	75	78
No. of packets	1	3	5	7
$A_i = k d^2$: $k =$	0·45	0·56	0·60	0·62

FIG. 23. SECTIONS OF CORE-TYPE TRANSFORMER LIMBS

The reduction of core sectional area due to the presence of paper, surface oxide, etc., is of the order of 10 per cent.

CORE SECTIONS. As has been seen, iron losses make it imperative to laminate transformer cores. The problem of building up these thin sheets into a mechanically strong core is largely a matter for the draughtsman.

With core-type transformers in small sizes, the simple rectangular limb can be used with either circular or rectangular coils. As the size of the transformer increases, it becomes wasteful to employ rectangular coils, and circular coils are usually preferred. For this purpose the limbs can be square, as in Fig. 23 (a), where the circle represents the inner circumference of the tubular former carrying the coils. Clearly a considerable amount of useful space is wasted, the length of the circumference of the circumscribing circle being large in comparison with the cross-section of the limb. A very common improvement is to employ cruciform limb sections, as in Fig. 23 (b). This demands at least two sizes of plate. With large transformers further core-stepping may be introduced to reduce the length of mean turn and the consequent I^2R loss. Any saving due to core-stepping must, of course, be balanced against the cost of extra labour in shearing several new plate sizes and the reduction in cooling space between core and coil. A typical core section with three steps is shown in Fig. 23 (c). The figures indicate the best proportions for the core packets, and the gross area expressed as a fraction of the area of the circumscribing circle. The three-stepped limb is commonly employed and even more steps may be used for very large transformers. In this case, the section is generally too massive to be built solidly : cooling ducts are left between the packets.

Cores for shell-type transformers are usually of simple rectangular cross-section, the coils being also rectangular (Fig. 10B).

ASSEMBLY. A number of methods are available for clamping stacks of stampings together to form cores. In small sizes (e.g. below 50 kVA.) string or cotton webbing may be employed to bind the plates. Very large cores may also be clamped by strong binding.

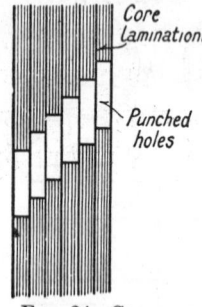

FIG. 24. COOLING DUCTS

Cores may also be clamped between iron frames (after the fashion of miniature transformers), but the most usual way is to bolt the plates together. For this purpose the plates have punched holes which accommodate bolts after assembly, and are used during the core building to register the successive laminations as they are added to the stack. The bolts must be insulated from the core both along their length and at their ends to avoid short-circuiting the laminations, thereby providing eddy current paths. For the same reason it is usual to avoid a double row of core-clamping bolts, for if the insulation of one bolt in each row becomes impaired, a considerable area of the core is surrounded by what amounts to a short-circuited turn, and excessive losses may occur locally. The slight tendency of the plates to "fan" out at their edges, due to central clamping, increases the space between adjacent laminations and provides a safeguard against electrical contact at the sharp edges.

When core-clamping bolts are employed, stiffening or *flitch* plates are used to give the built-up core more rigidity and to prevent bulging between bolts. Stiffener plates are insulated from the cores and are discontinuous at joints to obviate any tendency for the flux to use them as a conducting path in parallel with the laminations.

For small- and medium-sized oil-immersed transformers the dissipation of the core losses (as heat) is simple, as the surface of the laminations is large compared with their volume and losses. Very large cores, on the other hand, have a relatively small surface/volume ratio, so that the cooling surface must in some way be augmented by additional ducts. There are two ways of arranging ducts—either parallel or perpendicular to the direction of the laminations. The first is easy, the second requires special punching. Unfortunately the first method does not present to the oil any additional plate edges. Heat flows twenty times more readily along the laminations to the edges than from plate to plate across the intervening insulation, which has naturally a low heat conductivity. One method of contriving the exposure of a greater surface of plate-edge is shown in Fig. 24, where the upward orientation of the duct facilitates the passage of hot oil, but avoids splitting the core in two.

In any but the smaller sizes it is impracticable to cut complete plates, i.e. complete magnetic circuits. If this were attempted, the wastage of sheet would be great, and the difficulty of inserting the

FIG. 25. CORE OF THREE-PHASE CORE-TYPE TRANSFORMER
(*Bruce Peebles*)

coils—which must interlink the core—almost insuperable, since each turn would have to be separately threaded through. It therefore becomes necessary to make the coils separately, and to place them on the cores or to build up the plates through them. This necessitates one or more joints in the magnetic circuit. Although joints introduce gaps in the continuity of the circuit, the plates can be suitably arranged to reduce the effect of the gaps on the magnetic conductivity of the joints. As the core is built up, the joints in one layer are arranged at places other than those at which the joints in the previous layer occurred, so that the joint in one layer is covered by the iron of the preceding and succeeding plates. For example, a single-phase shell-type transformer of small output may have a core composed of T- and U-shaped stampings. Successive layers are reversed and so arranged that no two joints fall together.

In assembling such a transformer, the coils are made and finished, and the core plates inserted on each side until the complete core is built up, after which it is clamped.

For large shell-type transformers, where the cores are riveted or bolted along their length, the plates are assembled on pins which ensure correct registering. After assembly, the pins are withdrawn,

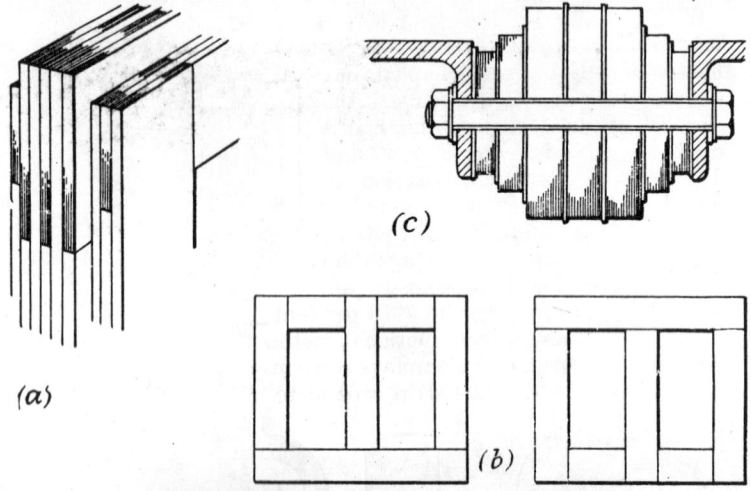

FIG. 26. CORE CONSTRUCTION OF CORE-TYPE TRANSFORMER

leaving the holes ready for receiving the insulated core-clamping bolts. Fig. 29 shows a shell-type transformer under construction.

For core-types the clamps and flitch plates are also arranged in a jig with upright pins, on which the core plates are threaded. The joints between core legs and yokes are invariably interleaved, Fig. 26 (*a*). It is usual to build with laminations in threes or fours to shorten the building time and reduce the chance of buckling the plates. Successive sets of plates are arranged as in (*b*) for inter-leaving. Sheets of pressboard may be inserted at intervals in a thick core, (*c*). Small earth clips make electrical connection between packets. For grain-oriented steel intricate mitred joints have been developed.

With the largest cores, the limbs may be built with a central axial slot, and with spacers between the laminations at intervals in the stack, to facilitate cooling.

After the core has been built, the second set of clamps and flitch plates is added and the core tightened. The yoke is removed to admit the coils after the whole core has been stood upright: it is then replaced.

The yoke of a core-type transformer is subdivided in a similar way to the limb, and the relative areas of the several packets bear the same relation, to avoid the flux changing from one portion to another at its passage between limb and yoke: such interchange would be productive of eddy-current losses. Sometimes (in core-type transformers) the yokes are made, as a whole, of about 20 per cent greater area than the limbs, thereby reducing the iron loss in parts which do not involve an increase in the length of copper.

With very large three-phase core types, a limiting factor in the design is the loading gauge of the road or railway route along which the transformer must be hauled. It becomes necessary to reduce the overall height. A common method is to use a *five-limbed core* (Figs. 27 and 41), which needs a cross-section in the yokes less than that required in the usual three-limbed construction; it may be about 30 per cent less than that of the limbs. The core losses are, however, generally larger by 5–10 per cent.

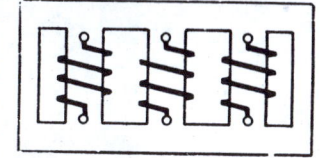

Fig. 27. Five-limbed Core

A development in constructional methods for small (e.g. rural-distribution) transformers employs cores made from lengths of cold-rolled grain-oriented steel strip, wound to shape and impregnated.

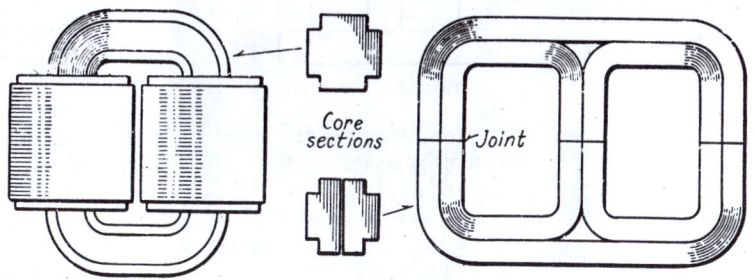

Single-phase wound core

Core sections

Joint

Three-phase "C" core

Fig. 28. Small Cores for Grain-oriented Steel

Then either (i) the core is cut across to make a pair of "C" cores, threaded into coils and clamped; or (ii) it is left uncut and the coils are wound on it by a special winding machine. In the former case construction is easy, but the second method yields minimum core loss and magnetizing current. Fig. 28 illustrates the method. The three-phase type shown on the right can be produced by method (i) only.

CONSTRUCTIONAL FRAMEWORK. Considerable use is made of channel- and angle-section rolled steel in the framework of core-type transformers, as is shown in the illustrations of this chapter. A typical construction is to clamp the top and bottom yokes between channel sections, held firmly by tie-bolts. The bottom pair of channels has cross channels as feet. The upper pair carries clamps for the high- and low-voltage connections.

3. **Windings.** In addition to the classification as *circular* or *rect-*

FIG. 29. SINGLE-PHASE SHELL-TYPE TRANSFORMER IN COURSE OF ASSEMBLY
(*Metropolitan-Vickers*)

ingular, transformer coils can be either *concentric* or *sandwiched.* The terms are almost self-explanatory. In Fig. 30 (*a*) a single-phase core-type transformer with cylindrical coils is shown (a very common arrangement), and in Fig. 30 (*b*) a single-phase shell-type with sandwich coils. The latter are used almost invariably with shell-type transformers. In Fig. 30 the letters *L* and *H* refer to the low- and high-voltage windings respectively. On account of the easier insulation facilities, the low-voltage winding is placed nearer to the core in the case of core-type and on the outside positions in the case of shell-type transformers. The insulation spaces between low- and high-voltage coils also serve to facilitate cooling.

Cylindrical concentric helix windings, commonly employed for core-type transformers, can often be built up (generally with axial spacing strips to improve oil circulation between the coil and the tube) on bakelite tubes, which facilitate erection, and form a strong foundation for winding the coils. Wherever possible, simple helical coils are used, preferably in a single layer. Usually the voltage of the low-voltage side is sufficiently small to permit of this,

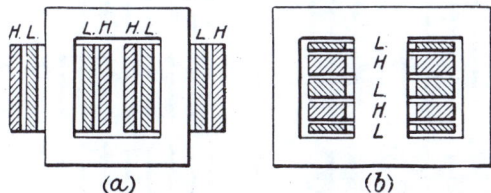

(a)   (b)

Fɪɢ. 30. Cᴏɴᴄᴇɴᴛʀɪᴄ ᴀɴᴅ Sᴀɴᴅᴡɪᴄʜ Wɪɴᴅɪɴɢs

and frequently a helical winding in one or two layers can be used for the high-voltage winding. Where this is not suitable, the coil must be sectionalized in order to reduce the voltage between layers. In this way it becomes unnecessary to put insulation between successive layers over and above that on the wires themselves. With a *sectionalized* winding the voltage per section is of the order 1 000 V. or less, but it is possible to reach 5 000 to 6 000 V. per coil, unsectionalized. The chief difficulty in the making of large concentric coils is the handling of several hundred pounds of copper in a single coil. Care has to be taken to wind the coils tightly and to keep them perfectly circular. For insulation between high- and low-voltage windings bakelite or elephantide tubes may be used. They can be stressed up to about 20 kV. per cm. radially, the oil in the duct being regarded as an additional margin.

Cross-over coils are suitable for currents not exceeding about 20 A. They are used for h.v. windings in comparatively small transformers, and comprise wires of small circular section with double cotton covering. The coil is wound on a former with several layers of several turns per layer, tape being interleaved axially to give greater rigidity to the coil. The coil ends (one from inside and one from outside) are joined to other similar coils in series, spaced with blocks of insulating material to allow of free oil circulation.

Disc coils are made up of a number of flat sections, comprising layers wound spirally from inside outwards as shown in Fig. 31. Generally, rectangular wire is employed, wound on the flat side, so that each disc is mechanically strong. Sectional or continuous disc coils are commonly used. Every turn being in contact with the oil, the cooling is good.

Sandwich windings, commonly employed for shell-type transformers, allow of easy control over the reactance. The nearer two coils are together on the same magnetic axis, the greater is the proportion of mutual flux and the less is the leakage flux. If it were

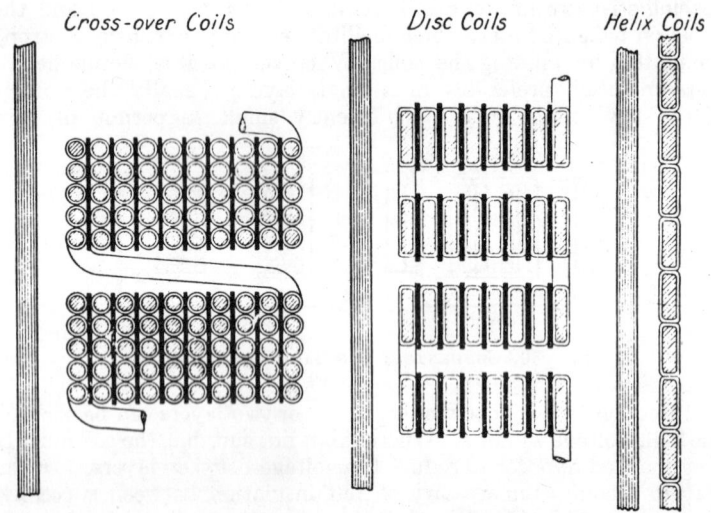

FIG. 31. TRANSFORMER COILS

possible to accommodate the two coils in the same space, the whole flux would link both windings, and there would be no leakage flux. Leakage can be reduced by subdividing the low- and high-voltage coils. Each high-voltage section lies between two low-voltage sections. The end low-voltage sections contain half the turns of the normal low-voltage sections. In order to balance the magnetomotive forces of adjacent sections, each normal section, whether high or low voltage, carries the same number of ampere-turns. The higher the degree of subdivision, the smaller is the reactance.

Cross-over of stranded conductors. Conductors of large cross-section are not employed, as being too stiff to handle, and leading to excessive I^2R loss. The leakage flux pulsates over the cross-section of the windings, and may induce eddy e.m.f.'s which produce circulating currents and additional losses (often referred to as *stray* losses). The conductor must consequently be subdivided for the same reason as cores are laminated. A 7·5 mm. square conductor might be approaching the upper limi of size for a 50-c/s transformer. If a larger section is needed, insulated strands in parallel must be used, and balance between all strands attained by transposing their relative positions within the coil. The arrangement of coils and calculation of parasitic eddy currents is referred to in more detail in Chapter X, §9.

INSULATION. The insulation between the h.v. and l.v. windings, and between l.v. winding and core, comprises bakelite-paper

FIG. 32. CORE AND WINDINGS OF THREE-PHASE 200-KVA., 50-CYCLE 6 600/440-V. CORE-TYPE TRANSFORMER
(Bruce Peebles)

cylinders or elephantide wrap. Fig. 36, which is typical, shows the construction, assembly and insulation of cylinder and disc coils, bakelite cylinders, hardwood spacers and end packing. Core details are also shown. There are two helical coils in series (Nos. 1 and 2) for the l.v. winding, composed of conductors with six mutually insulated strands.

The insulation of the conductors may be of paper, cotton or glass tape, glass tape being used for air-insulated transformers. The paper is wrapped round the conductor in a suitable machine, preferably without overlap of adjacent turns. Paper is not flexible, and a "half-lap" wrapping would cause it to buckle. The wrapped conductor is lashed with cotton strands wound openly, to give some mechanical protection.

Paper insulation usually necessitates the use of round coils, while the cross-over of the several strands in a conductor must be properly shaped, and not merely twisted.

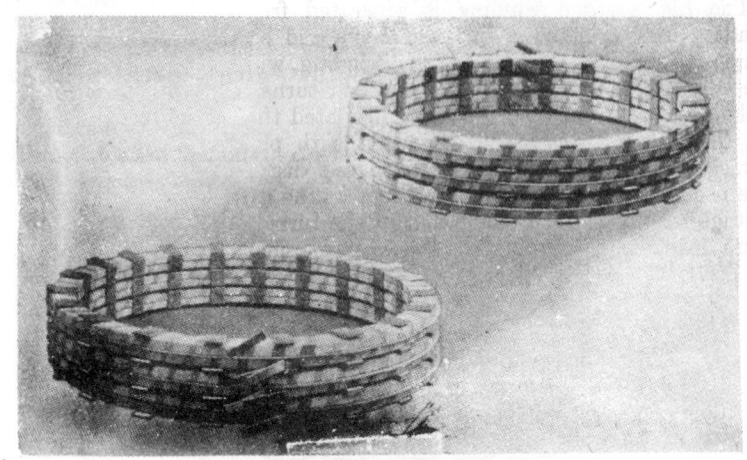

FIG. 33. DISC COILS
(Bruce Peebles)

FIG. 34. HELIX AND DISC COILS ON SINGLE-PHASE, 2 000-KVA.
50-CYCLE, 22/10·5-KV. TRANSFORMER
(Bruce Peebles)

The high-voltage winding is separated from the low-voltage winding in Fig. 36 by a series of ducts and bakelite cylinders or barrels. Details of the high-voltage winding, which is sectionalized, are shown. It will be seen that the end turns, i.e. those turns in coils 3, 4, and 5, are more heavily insulated than Nos. 6 and 7 to 23. The reason for this is connected with phenomena occurring during switching operations, or with line disturbances. Owing to strain on the insulation between turns at the line end of the high-voltage winding, about 5 per cent of the turns are reinforced with extra insulating material. In the example shown in Fig. 36 the

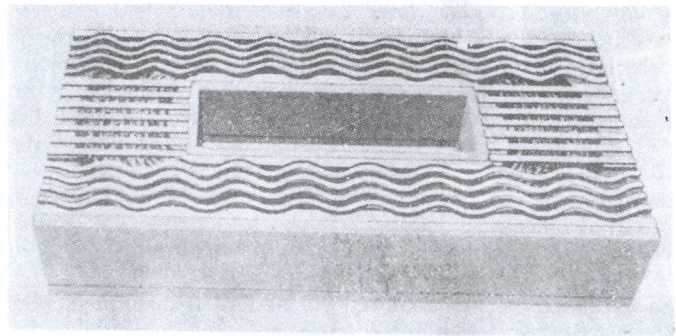

FIG. 35. SANDWICH COIL FOR SHELL-TYPE TRANSFORMER
(Metropolitan Vickers)

reinforcement is graded down to the normal insulation in coils 7 to 23.

For large h.v. transformers the end-turn reinforcement is a matter of careful design. Merely increasing the insulation thickness may result in markedly raising the impact thereon of impulse voltages. The winding insulation has to be co-ordinated with the means adopted for controlling the distribution of impulse electric stresses, Fig. 37 (a).

Unimpregnated paper insulation readily absorbs oil when the transformer is inserted in its tank. Small thin-wire coils may be varnished for adhesion. A difficulty, however, is introduced by the shrinking of insulation after a few months of service, resulting in a loosening of the windings. This predisposes to movement of the coils and breakdown on sudden short-circuit. The coils can be pre-shrunk under pressure before assembly to simulate service conditions. They may then be assembled and tightened up without danger of undue further shrinkage.

The permittivity of transformer oil is about 2·2; of bakelite cylinders about 4·4. The electric stress is therefore twice as great on the oil in the annular ducts as on the bakelite cylinders. In transformers for very high line voltages—e.g. 150–220 kV.—the radial width of the oil ducts is determined by the electric stress and has to be made much wider than would be necessary from considerations of cooling. On such transformers the ducts may be entirely filled with insulating paper, bent over at the coil ends to form an earth barrier. The paper, oil-saturated, has a high dielectric strength, so that the h.v.-l.v. spacing can be reduced. The greater freedom of choice of winding separation permits of better control over the leakage flux. See Fig. 37 (b).

The multi-layer h.v. barrel winding in Fig. 37 (c) is a special helix arrangement for transformers of exceptionally high line voltage. The innermost layer is wound over a neutral shield. Succeeding layers are shortened, giving the additional clearance to the yokes appropriate to the voltage of the layer to earth. The outermost layer, enclosed by a static shield, is connected to the line terminal, and the successive layers joined in series to give the electric stress distribution effect of a capacitor bushing to surge voltages applied to the line terminal. The line and neutral shields may comprise close-wound helices of conducting strip material applied over the appropriate insulated paper layers: the strips are connected electrically, but arranged so that they do not form a short-circuited turn.

Tappings are usually required on modern transformers. Details of these are given in Chapter V, §8.

LEADS AND TERMINALS. The connections to the windings are copper rods or bars, insulated wholly or in part, and taken to the bus-bars directly in the case of air-cooled transformers, or to the insulator bushings on the tank top in the case of oil-cooled transformers. The shape and size of the conductors are of importance in very high voltage systems, not on account of the current-carrying capacity, but because of dielectric stresses, corona, etc., at sharp bends and corners with such voltages.

BUSHINGS. Up to voltages of about 33 kV., ordinary porcelain insulators can be used, which do not require special comment. Above this voltage the use of condenser and of oil-filled terminal bushings, or, for certain cases, a combination of the two, has to be considered. Of course, any conductor can be effectively insulated by air provided that it is at a sufficient distance from other conducting bodies and sufficiently proportioned to prevent corona phenomena. Such conditions are naturally unobtainable with transformers where the conductor has to be taken through the cover of the containing tank, although common enough with overhead transmission lines.

The *oil-filled* bushing consists of a hollow porcelain cylinder of special shape (Fig. 38) with a conductor (usually a hollow tube) through its centre. The space between the conductor and the porcelain is filled with oil, the dielectric strength of which is greater

result is a series of capacitors formed by the conductor and the first tin-foil layer, the first and second tin-foil layers . . . and so on. The capacitance of the capacitors is controlled by their length and the radial separation of their tin-foil plates. Fig. 38 illustrates this. If the thickness of bakelized paper separating successive tin-foil layers is kept constant, and the capacitances of the capacitors are kept constant by successively reducing the length of the tin-foil layers proceeding outward, then the voltage across each capacitor will be the same, giving a practically uniform dielectric stress throughout the radial depth of the insulator. By arranging the dielectric stress to come within the limits of the material used, a bushing can be built to withstand any desired voltage. In Fig. 38 the short stepped end is oil-immersed beneath the tank cover, the smooth long end projecting outwards. For use in outdoor substations, the bushing is covered by a porcelain *rain-shed*, the annular space between the rain-shed and the bushing proper being filled up with bitumen. The rain-sheds are corrugated circumferentially to accord with the estimated electric field distribution and to provide a long leakage path.

The oil-immersed ends of h.v. bushings may be of *re-entrant* form, reducing the immersed length and permitting a more uniform distribution of the axial and radial electric stress components.

4. **Cooling.** Consider a transformer with k times the linear dimensions of another smaller but otherwise similar unit. Its core and conductor areas are k^2 times greater and its rating (with the same flux and current densities) increases k^4 times. The losses increase by the factor k^3 but the surface area is multiplied only by the factor k^2. Thus the loss per unit area to be dissipated is increased k times. Large transformers are therefore more difficult to cool than small ones, and require more elaborate methods.

The cooling of transformers differs from that of rotary machinery in that there is no inherent relative rotation to assist in the circulation of ventilating air. Luckily the losses are comparatively small, and the problem of cooling (which is essentially a problem of preserving the insulation—solid and liquid—from deterioration) can in most cases be solved by reliance on natural self-ventilation. The various methods are—

Simple Cooling.

AN: Natural cooling by atmospheric circulation, without any special devices. The transformer core and coils are open all round to the air. This method is confined to very small units of a few kVA. at low voltages.

AB: In this case the cooling is improved by an air blast, directed by suitable trunking and produced by a fan.

ON: The great majority of transformers are oil-immersed with natural cooling, i.e. the heat developed in the cores and coils is passed to the oil and thence to the tank walls, from which it is dissipated. The advantages over air cooling include freedom from the possibility of dust clogging the cooling ducts, or of moisture affecting the insulation, and the design for higher voltages is greatly improved.

OB: In this method the cooling of an ON-type transformer is improved by air blast over the outside of the tank.

OFN: The oil is circulated by pump to natural air coolers.

OFB: For large transformers artificial cooling may be used. The OFB method comprises a forced circulation of the oil to a *refrigerator*, where it is cooled by air-blast.

OW: An oil-immersed transformer of this type is cooled by the circulation of water in cooling tubes situated at the top of the tank but below oil-level.

OFW: Similar to OFB, except that the refrigerator employs water instead of air blast for cooling the oil, which is circulated by pump from the transformer to the cooler.

Mixed Cooling.

ON/OB: As ON, but with alternative additional air-blast cooling.

ON/OFN, ON/OFB, ON/OFW, ON/OB/OFB, ON/OW/OFW: Alternative cooling conditions in accordance with the methods indicated.

A transformer may have two or three ratings when more than one method of cooling is provided. For an ON/OB arrangement these ratings are approximately in the ratio 1/1·5; for ON/OB/OFB in the ratio 1/1·5/2.

NATURAL OIL COOLING. The diagram in Fig. 39 is drawn to indicate on the left the thermal flow of oil in a transformer tank. The oil in the ducts, and at the surfaces of the coils and cores, takes up heat by conduction, and rises, cool oil from the bottom of the tank rising to take its place. A continuous circulation of oil is completed by the heated oil flowing to the tank sides (where cooling to the ambient air occurs) and falling again to the bottom of the tank. Oil has a large coefficient of volume expansion with increase of temperature, and a substantial circulation is readily obtained so long as the cooling ducts in the cores and coils are not unduly restricted.

Fig. 39 also shows on the right a curve typical of approximate temperature distribution, the figures quoted being rises in degrees centigrade. On full load with continuous operation, the greatest temperature-rise will probably be in the coils. The maximum oil temperature may be about 10° less than the coil figure, and the mean oil temperature another 15° less.

The best dissipator of external heat is a plain blacked tank. But to dissipate the loss in a large transformer a plain tank would have an excessively large surface area and cubic capacity, and would require a great quantity of oil. Both space and oil are expensive.

Artificial means for increasing the surface area without increasing the cubical contents have, therefore, been developed. These comprise special tank constructions such as

(a) Fins, welded vertically to the tank sides;
(b) Corrugations;
(c) Round- or elliptical-section tubes;
(d) Auxiliary radiator tanks.

These are illustrated in Fig. 40. Little need be said of (a) or (b), the former being not very effective and the latter rather difficult in construction, although formerly used in Europe. Method (c) is

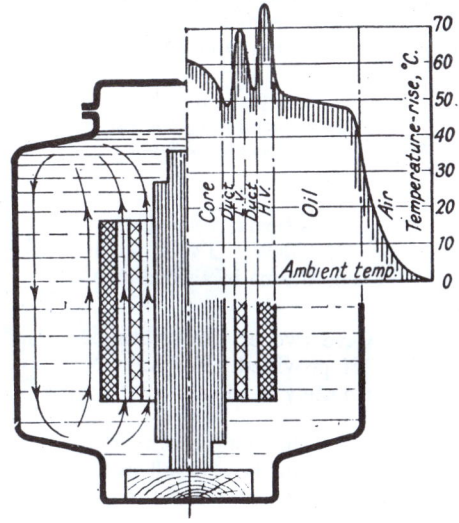

Fig. 39. Oil-circulation and Temperature Distribution

extremely common for a wide range of sizes, while (d) is used when there is insufficient room to accommodate all the tubes required by a large transformer.

The tubed tank provides considerable cooling surface, and the tubes being connected with the tank at the top and bottom only provide a head sufficient to generate a syphoning action, which improves the oil circulation quite apart from enhanced cooling. The tubes may sometimes be "gilled," i.e. wound with a strip-on-edge metal helix, to increase cooling surface and the eddying air flow which more effectively removes heat. The baffling action, however, also tends to restrict the total air flow so that the net gain is not commensurate with the added cooling surface.

For the largest sizes the *radiator* type of cooling is used, where separate radiator tanks with fins or corrugations, connected at top and bottom to the main tank, dissipate the heat by oil circulation. One such arrangement is shown diagrammatically in Fig. 40 (*d*).

The limit of output with oil-insulated, self-cooled transformers is reached when the tank becomes too large and costly. Another limit which obtains in some cases is the railway loading gauge, which precludes the transport of transformers and tanks exceeding a certain

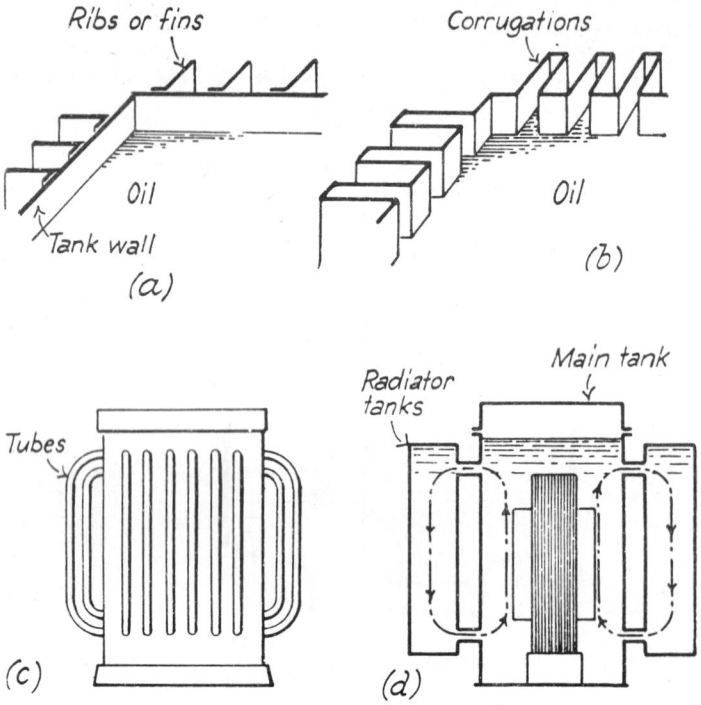

FIG. 40. COOLING METHODS

size. In carriage by road the available routes with their grades, bridges, etc., may decide the type of transformer and tank which can be used. For larger units, transport in parts must be resorted to, with erection on site. A tank dissipating about 50 kW is regarded by some manufacturers as the limit for self-cooling.

FORCED OIL COOLING. When forced cooling becomes necessary in large high-voltage oil-immersed transformers, the choice of the method of cooling will depend largely upon the conditions obtaining at the site. Air-blast cooling can be used, a hollow-walled tank being provided for the transformer and oil, the cooling air being blown through the hollow space. The heat removed from the inner walls of the tank can be raised to five or six times that dissipated naturally, so that very large transformers can be cooled in this way.

A cheap method of forced cooling where a natural head of water is obtainable is the use of a cooling coil, consisting of tubes through which cold water is circulated, inserted in the top of the tank (Fig. 43). This method has, however, the disadvantage that it introduces

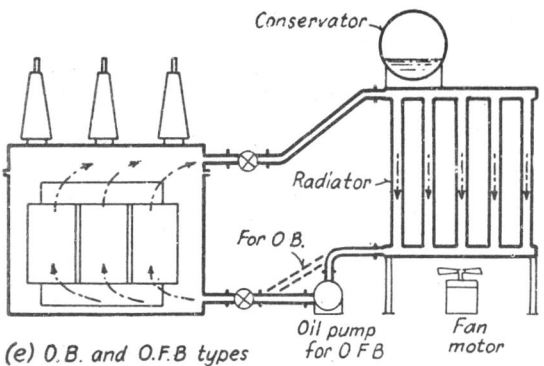

(e) O.B. and O.F.B types

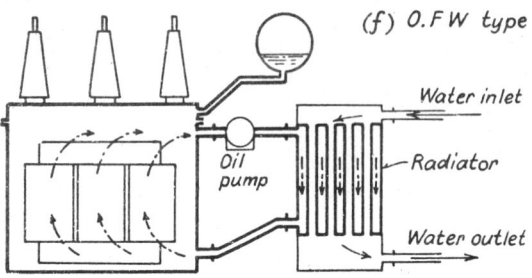

(f) O.F W type

Fig. 40 (contd.) Cooling Methods

into the tank a system containing water under a head greater than that of the oil. Any leakage will, therefore, be from the water to the oil, so that there is a risk of contaminating the oil and reducing its insulating value. Fins are placed on the copper cooling tubes to assist in the conduction of heat from the oil, since heat passes three times as rapidly from the copper to the cooling water as from the oil to the copper tubing. The inlet and outlet pipes are lagged to avoid water from the ambient air condensing on them and getting into the oil.

For large installations the best cooling system appears to be that in which oil is circulated by pump from the top of the transformer tank to a cooling plant, returning when cold to the bottom of the tank. When the cooling medium is water, this has the advantage that the oil can be arranged to work at a higher static head than the water, so that any leakage will be in the direction of oil to water. The system is suitable for application to banks of transformers, but for reliability not more than, say, three tanks should be connected in one cooling pump circuit.

Fig. 40 (e) and (f) shows diagrammatically the usual methods of cooling employed where separate radiators are necessary. The oil-circulation pump in (e) is incorporated only if the natural thermal head is insufficient to generate an adequate oil flow.

Until recently all large units employed oil-circulating systems, but considerable advances have been made towards increasing the size of self-cooled units by special radiators. It is possible to build entirely self-cooled units up to 40 000 kVA., with the advantage of eliminating breakdown risks due to auxiliary pumping equipment. The addition of an air-blast system to circulate cooling air over the radiators permits the increase of size to about 75 000 kVA. Although an auxiliary fan is involved, the transformer is still capable of half-load operation should the air blast fail. A temperature device can be used to bring the fan into action when the oil temperature reaches a desired limit; this improves the overall efficiency at small loads. An arrangement of this type is illustrated in Fig. 41.

Tank-less, air-insulated transformers have been built up to 1 500 kVA., but larger sizes require forced air circulation.

INTERNAL COOLING. The heating of the coils depends on their thermal conductivity, which is itself a function of (a) the thickness of the winding, and (b) the external insulation.

A coil design, which allows the copper heat to flow radially outwards with little cross insulation in the path of the flow, leads to economical rating in that a high current density can be employed for a given temperature rise without sacrifice of efficiency. The *strip-on-edge* winding, consisting of a single layer of copper of rectangular section wound on edge on a bakelite cylinder with one edge bare and in contact with oil, dissipates heat most effectively. In some designs the flow of heat can be so much improved that the transformer output entails a larger size of tank.

With cores, *ageing* is not to be feared when modern steels are used and correctly handled, but heating and cooling, with the accompanying expansion and contraction, lead sometimes to a loosening of the core construction. Owing to the laminated nature of the cores, and the presence (on the surfaces of the plates) of oxide films and paper, varnish, etc., the flow of heat in cores is almost wholly towards the edges. On account of the rather greater exposure of iron to the oil in shell-type transformers, these are better than the core-type as regards the cooling of the iron. On the other hand, the exposed coils of the core type will cool more readily than those of the shell type.

Fig. 41. Core and Windings (*above*) and Complete Unit (*below*) of ON/OFB 22½/45-MVA., 132/33-kV. Transformer

Forced oil circulation and blast fan started automatically by thermostat switch

(*Crompton-Parkinson*)

TANKS. Small tanks are constructed from welded sheet steel, and larger ones from plain boiler plate. The lids may be of cast iron, a waterproof gasket being used at the joint. The fittings include thermometer pockets, drain cock, rollers or wheels for moving the transformer into position, eye-bolts for lifting, conservators and breathers. Cooling tubes are welded in, but separate radiators are individually welded and afterwards bolted on.

Conservators are required to take up the expansion and contraction

FIG. 42. 2 000-KVA., 20/6·6-KV., 50-CYCLE TRANSFORMER IN TANK

With Conservator, Breather, Tapping Switches, Relief Valve, Dial Thermometer and Swivel Rollers

(*Bruce Peebles*)

of the oil with changes of temperature in service without allowing the oil to come in contact with the air, from which it is liable to take up moisture. The conservator may consist of an airtight cylindrical metal drum supported on the transformer lid or on a neighbouring wall, or of a flexible flat corrugated disc drum. The tank is filled when cold, and the expansion is taken up in the conservator. The displacement of air due to change of oil volume takes place through a *breather* containing calcium chloride or silica gel, which extracts the moisture from the air.

Some tank details are illustrated in Figs. 42 and 43.

TRANSFORMER OIL. Oil in transformer construction serves the double purpose of cooling and insulating. For use in transformer tanks, oil has to fulfil certain specifications and must be carefully selected. All oils are good insulators, but animal oils are either too viscous or tend to form fatty acids, which attack fibrous materials (e.g. cotton) and so are unsuitable for transformers. Vegetable oils (chiefly resinous) are apt to be inconsistent in quality and, like animal oils, tend to form destructive, fatty acids. Of the mineral oils, which alone are suitable for electrical purposes, some have a bituminous and others a paraffin base. The crude oil, as tapped, is distilled, producing a range of volatile spirits and oils ranging from the very light to the heavy, and ending with semi-solids like petroleum jelly, paraffin wax, or bitumen.

In the choice of an oil for transformer use the following characteristics have to be considered.

Viscosity. This determines the rate of cooling, and varies with the temperature. A high viscosity is an obvious disadvantage because of the sluggish flow through small apertures which it entails.

Insulating Property. It is usually unnecessary to trouble about the insulating properties of an oil, since it is always sufficiently good. A more important matter, however, is the reduction of the dielectric strength due to the presence of moisture, which must be rigorously avoided. A very small quantity of water in oil greatly lowers its value as an insulator, while the presence of dust and small fibres tends to paths of low resistivity.

Flash Point. The temperature at which the vapour above an oil surface ignites spontaneously is termed the flash point. A flash point of not less than about 160° C. is usually demanded for reasons of safety.

Fire Point. The temperature at which an oil will ignite and continue burning is about 25 per cent above the flash point, or about 200° C.

Purity. The oil must not contain impurities such as sulphur and its compounds. Sulphur when present causes corrosion of metal parts, and accelerates the production of sludge.

Sludging. This is the most important characteristic. Sludging means the slow formation of semi-solid hydrocarbons, sometimes of an acidic nature, which are deposited on windings and tank walls. The formation of sludge is due to heat and oxidation. In its turn it makes the whole transformer hotter, thus aggravating the trouble, which may proceed until the cooling ducts are blocked and the transformer becomes unusable owing to overheating. Experience shows that sludge is formed more quickly in the presence of bright copper surfaces. The chief remedy available is to use oil which remains without sludge formation after long periods of heating in the presence of oxygen, and to employ expansion chambers to restrict the contact of hot oil with the surrounding air.

Acidity. Among the products of oxidation of transformer oil are CO_2, volatile water-soluble organic acids, and water. These in combination can attack and corrode iron and other metals. The provision of breathers not only prevents the ingress of damp air, but helps on out-breathing to absorb any moisture produced by oxidation of the oil. Oil conservators are desirable to avoid the condensation of water-soluble acids on the under surface of the tank lid from which acidic droplets may fall back into the oil.

INHIBITED OIL. The deterioration of oil during its working life can be retarded by the use of anti-oxidants, particularly oxidation "inhibitors." The latter, which are usually of the phenolic or amino type, convert chain-forming molecules in the oil into inactive molecules, being gradually consumed in the process. Inhibitors greatly prolong the phase in the service life of the oil which precedes the onset of deterioration and during which the acid and sludge formations are substantially zero.

SYNTHETIC TRANSFORMER OIL. This has been developed to avoid the risk of fire and explosion, present always with normal mineral oils. *Chlorinated diphenyl*, a synthetic oil suitable for transformers, is chemically stable, non-oxidizing, rather volatile, and heavier than water. Its dielectric strength is higher than that of mineral oil, and moisture has a smaller tendency to migrate through it. The permittivity is 4·5, compared with about 2·5. This high figure is roughly the same as the permittivity of the solid insulating material used in a transformer, so that the distribution of electric stress will differ markedly from that when mineral oil is used, the stress in the oil being relieved at the expense of the solid insulation. The oil is a powerful solvent of most varnishes, gums, binders and paints, which must consequently be barred from transformers designed for synthetic oil cooling. When decomposed by electric arc, hydrogen chloride gas is the chief product: this may combine with water to form hydrochloric acid.

TEMPERATURE RISE. The temperature rises permitted in the British Standard Specification for power and lighting transformers are tabulated on p. 129.

CHAPTER V

TRANSFORMERS: OPERATION

1. **Equivalent Circuits.** Within normal limits of load and rating, and under normal working conditions, the transformer behaves as a simple voltage-changer, with an efficiency approximating closely to unity. The secondary effects of a transformer on a circuit are not usually important in normal operation, but may be considerable during faults or transient conditions. In consequence, a transformer may have to be considered from a variety of quite different viewpoints under its many conditions of operation. In dealing with the practical solution of transformer problems, much labour can be saved by simplifying the conditions down to include those characteristics that are essential, and neglecting those whose influence is sufficiently small. As has already been shown, the equivalent circuit Fig 15 (d) for normal conditions of voltage, frequency and load may itself be simplified for no-load conditions to a simple shunt admittance or impedance, Fig. 17; or to a simple series impedance, Fig. 18, for short-circuit conditions. The "complete" and so-called "equivalent" circuit is, however, itself a simplification. The transformer is a highly complex electromagnetic device in which electric as well as magnetic fields play a part, and for certain problems, particularly those involving voltage surges, a more comprehensive picture must be visualized, such as that of Fig. 74. The nature of the problem determines the complexity of the equivalent circuit to be chosen. In the lists below is a brief summary of the chief problems appropriate to each viewpoint—

Series-parallel Impedance, Fig. 15 (d): losses and efficiency; connections.

Series Impedance, Fig. 18: regulation, tap-changing, short-circuit fault behaviour, I^2R loss, reactance, parallel operation.

Shunt Impedance, Fig. 17: core loss, magnetizing current and kVA., switching-in transients, harmonic currents.

Distributed Inductance-capacitance Network, Fig. 73: over-voltage transients, resonance, grading of insulation.

Practical Network, Fig. 44. Strictly, the turns-ratio of a transformer differs from the no-load voltage-ratio because of the leakage flux. The divergence is greater, the lower the rating of the transformer. Again, it may be desired (say for a network analyser) to represent the effect of a transformer in a network when its turns-ratio is not known. However, *any* transformer can be accurately represented by the equivalent circuit* of Fig. 44, constructed from

---

* Morris, *Proc. I.E.E.*, 97 (II), pp. 17 and 735 (1950).

direct measurements. The chain-dotted line in (a) represents a loss-free transformer which introduces the complex no-load voltage-ratio. The parameters are Z_m, a complex impedor found from the primary input voltage and current with the secondary open-circuited; Z_s, an impedance measured between the secondary terminals with the primary short-circuited; and t_v, the complex primary/secondary open-circuit voltage-ratio. The circuit, Fig. 44 (a), then follows from the Helmholtz-Thévenin theorem. An equivalent approach, shown in (b), is to invoke the theorem of passive quadripoles in which

$$V_1 = AV_2 + BI_2 \quad \text{and} \quad I_1 = CV_2 + DI_2$$

where the quadripole parameters $ABCD$ can all be found from open-circuit and short-circuit tests.

2. **Installation.** The great majority of modern transformers are oil-immersed, even if of comparatively low voltage, and it is usual

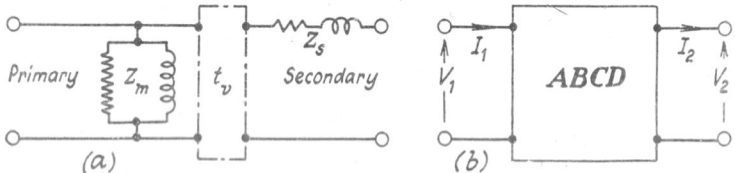

Fig. 44. "Practical" Equivalent Circuit

to dispatch them filled with oil, to avoid the absorption of moisture by the insulation. After inspection and adjustment of voltage tappings, the core is replaced in the tank, and the oil level made up to the marked depth. Should the transformer need "drying out," it is well wrapped with sacking or similar material to serve as lagging, and subjected to a short-circuit run, with alcohol thermometers arranged to take the temperature of the top layer of oil: a reading of 90° C. is high enough. At intervals the current is interrupted to take readings of insulation resistance, which drops at first, and then should rise to, say, 20 MΩ.

3. **Noise.** The rapid growth of domestic electrification and the ubiquity of even large power transformers for local distribution, make noise produced by such apparatus a problem of social consequence. Under no-load conditions, the "hum" developed by energized power transformers originates in the core, where the laminations tend to vibrate by magnetic forces. The noise is transmitted through the oil to the tank sides and thence to the surroundings. The essential factors in noise production are consequently: (a) magnetostriction, i.e. the very small extension, with corresponding reduction of cross-section, of sheet-steel strips when magnetized*; (b) the degree of mechanical vibration developed by the laminations, depending upon the tightness of clamping, size, gauge, associated

* E.g. extension of 0·000 12 per cent for a flux density of 1 Wb./m².

structural parts, etc.; (c) the mechanical vibration of the tank walls; and (d) the damping.

This problem has been studied experimentally.* It is established that magnetostriction is the first cause of transformer hum, but much of the noise depends on the natural frequency of vibration of the mechanical parts. Any constructional change which removes this frequency outside the audible range will be *ipso facto* an improvement. Transformer radiators and plain unreinforced tank walls are

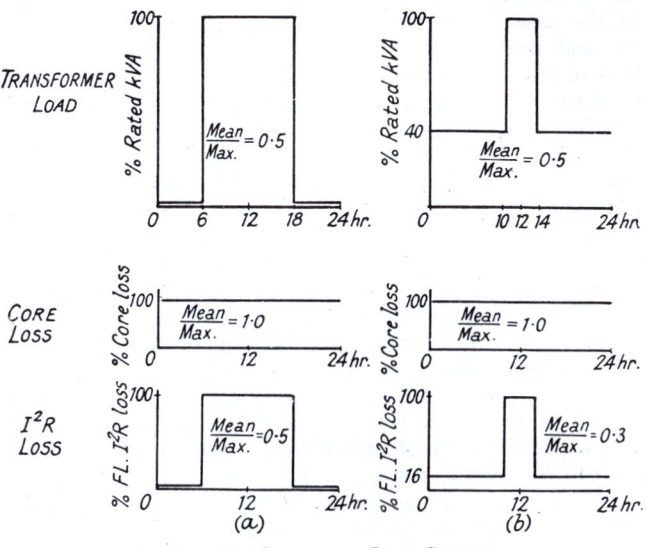

FIG. 45. LOAD AND LOSS CURVES

not usually troublesome as their natural frequencies lie as a rule below 100 cycles per sec. Stiffening and clamping may do more harm than good if the effect is to raise the natural vibratory period into the audible range.

In general, the total noise emission may be reduced by (a) preventing core-plate vibration, which necessitates the use of a lower flux-density and attention to constructional features such as clamping bolts, proportions, and dimensions of the "steps" in plate width, tightness of clamping and uniformity of plates; (b) sound-insulating the transformer from the tank by cushions, padding, or oil-barriers; (c) preventing vibration of the tank walls by suitable design of tank and stiffeners; and (d) sound-insulating the tank from the ground or surrounding air. There is no complete solution to the problem.

* Norris, *Engineer*, 155, pp. 446–8 (1923); Churcher and King, *J.I.E.E.*, 87, pp. 539–69 (1940); Grunder, *Bull. Oerl.*, 34, pp. 23–30 (1954).

4. Effect of Load-factor on Losses. A transformer design is affected by the kind of service it is called upon to perform. Factors such as standard frame-sizes demanded by standardization, requirements of impedance, tappings, limiting space, etc., tend to enforce a particular design, but a factor of considerable importance is (or should be) the distribution of the losses as between the core and the copper. The *load factors* of the I^2R and core losses are very different: for the former depends upon the square of the load kVA. whereas the latter is substantially constant. A minimum total of annual losses in the case of a transformer in continuous service is thus desirable, and requires attention to design and a pre-knowledge of the type of load for effective solution.

A transformer that is in circuit every day for twenty-four hours has a core-loss load factor of 100 per cent, irrespective of the load delivered. The I^2R losses, on the other hand, will vary over a wide range even for the same *average* load, since those losses are proportional to the square of the load. Fig. 45 shows two daily load curves of a hypothetical type. Each has the same *average* value, i.e. the number of kVAh. delivered is the same in each case. The root-mean-square load, upon which the I^2R losses over the twenty-four hours depend, is 50 per cent in case (*a*), and 30 per cent in case (*b*), but the load factors of the load, i.e. mean/maximum, are in both cases 50 per cent.

It is clear that, to obtain minimum total kWh. losses over a given period (say one year), the ratio of full-load I^2R to core loss will have to be adjusted in accordance with the type of load-curve. The load-factor of the load is not a reliable guide, because a variety of curves can be found, having the same load factor, or ratio mean load to maximum load, but different mean-square loads, and therefore different I^2R losses. An indication, based on the examination of the load curves of many typical power networks is as follows—

Load-factor of transformer, % .	30	40	50	60	70	80
Load-factor of I^2R loss, % .	16	25	35	47	60	72

Thus the losses in a transformer working on this type of load-curve with a load-factor of 50 per cent, may be expected to show over a period a total amount of lost energy equal to 100 per cent of the core loss and about 35 per cent of the full-load I^2R loss continuously. This suggests that a transformer in which the full-load I^2R loss is 100/35 or nearly three times the core loss will be of greatest economy on the assumption that the cost of wasted energy is the same for both losses. This is not necessarily the case, for the core loss is the greater when the transformer output is small, whereas at times of heavy load or overload the I^2R losses preponderate : and this is the time when energy is most valuable to the system. If account can be taken of this (not an easy matter to assess financially), a reduction of the I^2R loss may well prove economically favourable.

It is evident that the choice of a transformer on economic grounds is a very complex matter. Briefly, it will be necessary to consider capital cost, charges on capital, life, depreciation, load factor, loss ratio, energy costs, and type of tariff: a formidable list. An empirical estimate of the annual cost of losses is

$$[P_i + P_c(0\cdot3k_f + 0\cdot7k_f{}^2)]\gamma h,$$

for a transformer with core loss P_i and full-load I^2R loss P_c kW., and load factor k_f during the time h hours of connection per annum, with energy cost γ per kWh.

A loss ratio of full load I^2R to core loss of 3 is quite common. A higher ratio than this usually begins to involve excessive reactance and an increase in total full-load loss, since it is not easy to *reduce* the core loss economically by reducing the flux-density: the available method is to shorten the core and to use cold-rolled grain-oriented steel.

5. **Connections for Transformers.** With windings that can be connected in star, mesh, zigzag, or vee; with primary, secondary, tertiary, or auto connections; and transformers in single units or in banks of three, it is clear that the variety of connections is very great. No attempt will be made here to describe them completely. In many cases the characteristics, advantages, and drawbacks of a given type of connection can be estimated from the vector diagram of the primary and secondary e.m.f.'s.

A vector diagram can be constructed on the following general principles—

(a) The voltages of corresponding primary and secondary windings on the same limb (i.e. the input or applied primary voltage, and the developed secondary output voltage) are in phase opposition and the two induced e.m.f.'s are in phase (see Fig. 16).

(b) The e.m.f.'s induced in the three phases are equal, balanced, displaced mutually by one-third period in time, and have a definite sequence.

NOMENCLATURE. Transformer terminals conforming to B.S. 171: 1936 are brought out in rows, the h.v. on one side and the l.v. on the other, and are lettered from left to right facing the h.v. side. The h.v. terminals have capital letters (e.g. ABC); and l.v. terminals corresponding have small letters (e.g. abc). Tertiary windings, where provided, are lettered with capitals enclosed in circles. Neutral terminals precede line terminals. Each winding has two ends designated by the subscript numbers 1, 2; or if there are intermediary tappings, these are numbered in order of their separation from end 1. Thus an h.v. winding on phase A with four tappings would be numbered A_1, A_2, A_3 . . . A_6, with A_1 and A_6 forming the phase terminals.

If the induced e.m.f. in an h.v. phase A_1A_2 be in the direction A_1 to A_2 at a given instant, then the induced e.m.f. in the corresponding

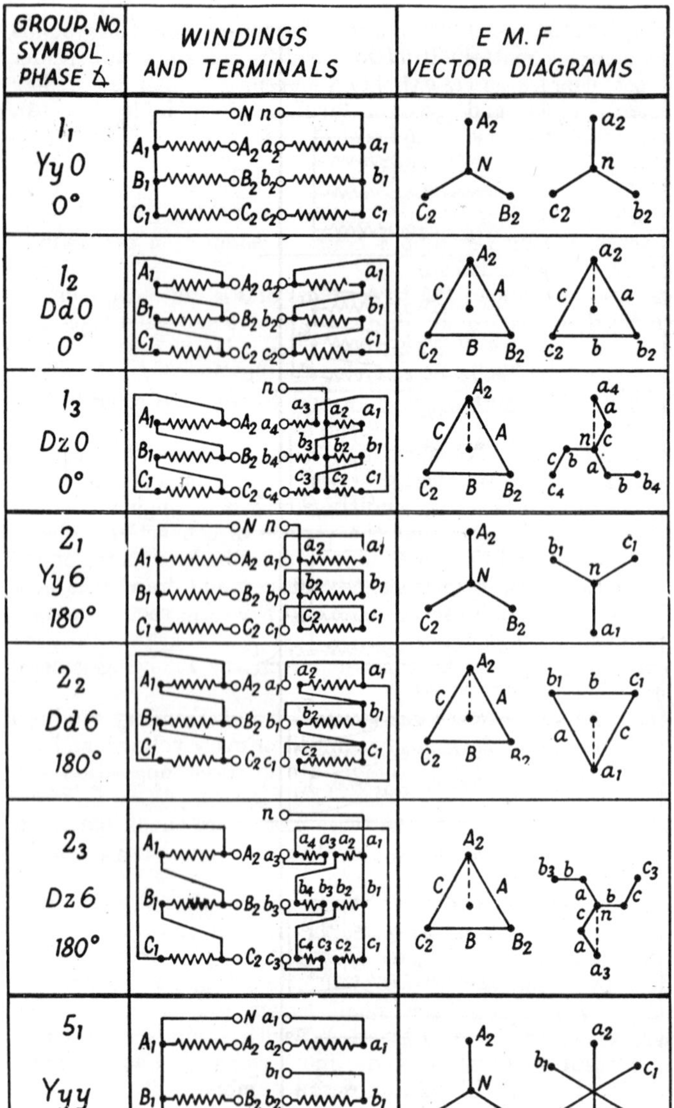

GROUP, NO. SYMBOL PHASE △	WINDINGS AND TERMINALS	E M.F VECTOR DIAGRAMS
1_1 Yy 0 $0°$		
1_2 Dd 0 $0°$		
1_3 Dz 0 $0°$		
2_1 Yy 6 $180°$		
2_2 Dd 6 $180°$		
2_3 Dz 6 $180°$		
5_1 Yyy		

Fig. 46. Standard Connections

GROUP, No. SYMBOL PHASE △	WINDINGS AND TERMINALS	E. M. F. VECTOR DIAGRAMS
3_1 Dy1 $-30°$		
3_2 Yd1 $-30°$		
3_3 Yz1 $-30°$		
4_1 Dy11 $+30°$		
4_2 Yd11 $+30°$		
4_3 Yz11 $+30°$		
3/2–Ph SCOTT		

FOR THREE-PHASE TRANSFORMERS

l.v. phase at the same instant will be from a_1 to $a_{\textrm{x}}$. The vector diagrams in Fig. 46 represent *induced* e.m.f.'s (not applied voltages) for a number of methods of connection.

Polyphase transformers are allotted symbols giving the type of phase-connection and the angle of *advance* turned through in passing from the vector representing the h.v. e.m.f. to that representing the l.v. e.m.f. at the corresponding terminal. The angle may be indicated by a clockface hour figure, the h.v. vector being 12 o'clock (zero) and the corresponding l.v. vector being represented by the hour hand. Thus, "Yzd 11" represents a (h.v. star/l.v. zigzag/tertiary delta)-connected 3-phase transformer, with the l.v. (secondary) e.m.f. vector in a given phase-combination at "11 o'clock," i.e. + 30° in advance of the 12 o'clock position of the h.v. e.m.f.

The groups into which all possible three-phase transformer connections are classified are—

Group 1: Zero phase displacement (Yy0, Dd0, Dz0).
Group 2: 180° phase displacement (Yy6, Dd6, Dz6).
Group 3: 30° lag phase displacement (Dy1, Yd1, Yz1).
Group 4: 30° lead phase displacement (Dy11, Yd11, Yz11).

The principal features of a few of the more common connections are noted below. The references to harmonics are explained in §12.

STAR/STAR (Yy0 or Yy6). This is the most economical connection for small, high-voltage transformers, as the number of turns per phase and the amount of insulation is a minimum. The possibility of utilizing both star points for a fourth wire may be useful.

Third-harmonic voltages are absent from the line voltage; unless there is a fourth wire no third-harmonic currents will flow (§12). If the transformer is worked at normal flux density, however, the neutral potentials will oscillate, while the third-harmonic phase-voltages may be high for shell-type three-phase units. The connection is most satisfactory with the three-phase core type: for other types the provision of a tertiary winding stabilizes the neutral conditions.

DELTA/DELTA (Dd0 or Dd6). This is an economical connection for large, low-voltage transformers in which the insulation problem is not urgent, as it increases the number of turns per phase and reduces the necessary sectional area of conductors. Large unbalance of load can be met without difficulty, while the closed mesh serves to damp out third-harmonic voltages. It is possible to operate the transformer on 58 per cent of its normal rating in vee connection should one of the phases develop a fault. This, however, is not usually practicable with three-phase units. The absence of a star-point may be disadvantageous.

STAR/DELTA (Dy or Yd). The star/delta arrangement is very common for power-supply transformers. It has the advantage of a star-point for mixed loading, and a delta winding to carry third-harmonic currents which stabilize the star-point potential. If the h.v. winding is the star-connected side, there is some saving in cost of insulation (see Star/Star). A delta-connected h.v. winding is almost universal, however, when it is desired to work motors and lighting from a four-wire l.v. side.

ZIGZAG/STAR (Yzl or Yz11). The interconnection between phases effects a reduction of third-harmonic voltages and at the same time permits of unbalanced loading; on account of the type of connection, however, the zigzag has to be confined to a fairly low-voltage winding. Since the phase voltages are composed on the zigzag side of two half-voltages with a phase difference of 60°, 15 per cent more turns are required for a given total voltage per phase compared with a normal phase connection, which may necessitate an increase in the frame size over that normally used for the rating. The zigzag/star connection has been employed where delta connections were mechanically weak (on account of large numbers of turns and small copper sections) in high-voltage transformers; also for rectifiers.

THREE-PHASE/SIX-PHASE (Yyy). One of the commonest of these is the star/double-star connection, used generally for the supply of six-ring synchronous converters, in which case the mesh-connected armature of the converter provides a low-impedance path for third-harmonic voltages. It is not, however, desirable to introduce harmonics into converter armatures if they can be avoided, on account of the possible effect on commutation.

GENERAL REMARKS ON THREE-PHASE CONNECTIONS. In three-phase working, which is becoming universal, a choice is possible between a three-phase unit and the bank of three single-phase units. A three-phase unit will cost about 15 per cent less than a bank, and will occupy considerably less space; this is reflected in power and substation building costs. There is no difference in reliability, but, as regards spare plant, it is cheaper to carry a single-phase than a three-phase unit if only one installation is concerned. Where there are several sets, this is less important. Single-phase banks are preferred in mines on account of the easier transport underground.

The possibility of connecting single-phase units in open delta allows of the supply being maintained up to 58 per cent of nominal full load if one of a bank develops a fault and is cut out of circuit: a practice more favoured in America than in Europe.

The choice between star and mesh connection merits separate consideration in each case. Star connection is cheaper, since mesh connection needs more turns and more insulation. The difference is small, however, at voltages below ~1 kV. With very high voltages a saving of 10 per cent may be effected, mainly on account of the insulation. An advantage of the star-connected winding with earthed neutral is that the maximum voltage to the core (frame or earth) is limited to 58 per cent of the line voltage, whereas with a delta-connected winding the earthing of one line (due to fault) increases the maximum voltage between windings and core to the full line voltage. Technically the mesh-connected primary is essential where the l.v. secondary is a star-connected four-wire supply to mixed three-phase and single-phase loads.

6. **Three-phase to Two-phase Connections.** These may be needed to supply two-phase furnaces; to interlink two- and three-phase systems; to supply three-phase apparatus from a two-phase source, and *vice versa*. The commonest connection for this purpose is the so-called *Scott* arrangement, but there are a few other methods.

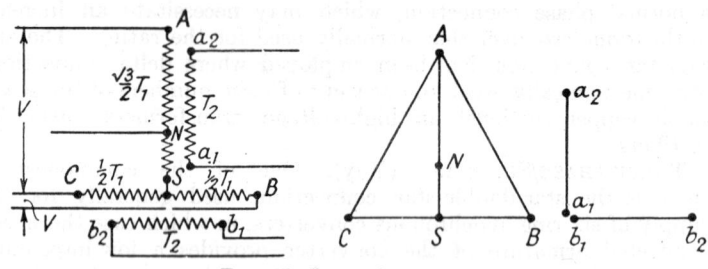

FIG. 47. SCOTT CONNECTION

The Scott connection requires two transformers of different ratings, but for interchangeability and the provision of spares, the transformers may be identical, both being provided with suitable tappings.

SCOTT THREE/TWO-PHASE CONNECTION (Fig. 47). The Scott connection is used to link two- and three-phase supply systems, with power flow in either direction. Consider in Fig. 47 the primary of a single-phase transformer connected between terminals BC of a three-phase supply. The voltage of the line being V, there will be a voltage V between B and C, and between A and B or C; while between A and the mid-point of the phase at S, will be a voltage $V(\sqrt{3})/2$. If the primary of a second transformer having $(\sqrt{3})/2$ $(= 0.87)$ times the turns of the first, be connected across AS, the voltage of S will be unaltered, and further across each turn of the second transformer there will be a voltage exactly equal to that per turn of the first. Thus, if the two secondaries have equal numbers of turns, they will produce secondary terminal voltages equal in magnitude but in phase quadrature.

The neutral point of the three-phase side can be located on the second or *teaser* transformer. The neutral must have a voltage $V/\sqrt{3}$ to A, and since the voltage A to S is $V(\sqrt{3})/2$, the neutral point will be $V[(\sqrt{3})/2 - 1/(\sqrt{3})]$ from S, or $0.288V$: that is, a number of turns above S equivalent to 29 per cent of the primary turns in the first or *main* transformer. The relative numbers of turns are shown in Fig. 47. Since 0·288 is one-third of 0·866, the neutral point divides the teaser winding AS in the ratio 1 : 2.

Let equal currents I_2 at unity p.f. be taken from the secondaries, Fig. 48 (a). Neglecting magnetizing current, the primary ampere-turns must balance the secondary ampere-turns in each transformer. The current in the teaser primary will be $2I_2T_2/(\sqrt{3})T_1$ $= 1·15I_2 . T_2/T_1$. In the main transformer the primary balancing current will be $I_2 . T_2/T_1$. The current in the teaser primary is in phase with the star voltage NA. The total current in the main primary is the resultant of two components; the first, $I_2 . T_2/T_1$ to

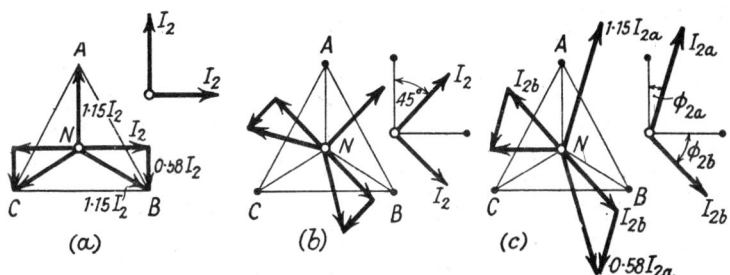

balance the secondary current; and the second, one-half of the teaser primary current in either direction from S, since the main transformer forms a return path for the teaser primary current which divides into halves at the mid-point tapping S. Thus the currents in the lines B and C are those obtained vectorially in Fig. 48 (a), in which, for simplicity, it is assumed that $T_1 = T_2$. Since the halves of the teaser current flow in opposite directions from S in the main primary, they have no magnetic effect on the core and play no part in balancing the main secondary ampere-turns. The geometry of the primary currents, with the rectangular components $I_2 . T_2/T_1$ and $0·58I_2 . T_2/T_1$, shows that the resultant currents in the B and C phases are in phase respectively with the star voltages NB and NC, and are equal to the teaser primary current. Thus the three-phase side is balanced for a balanced two-phase load of unity power-factor.

Fig. 48 (b), drawn for a balanced two-phase load at a power factor of 0·71 lagging, shows that the three-phase loading is again balanced, and that if the load is balanced on one side it will be balanced on the other. Under these conditions the main transformer rating is 15 per cent greater than that of the teaser, since the current is the same and the voltage 15 per cent in excess.

Fig. 48 (c) illustrates an unbalanced load on the secondary (two-phase) side, both currents and power factors being different. The geometrical construction is obvious from what has already been said about (a) and (b).

A Scott arrangement is possible with auto-transformers.

LE BLANC THREE/TWO-PHASE CONNECTION. The three-phase side of a Le Blanc transformer consists of three windings connected in star or delta and wound on the limbs of an ordinary three-limb core. If the two-phase voltage is to be V in each phase, the limbs ABC, Fig. 49 (a), are provided respectively with windings a_1 of voltage

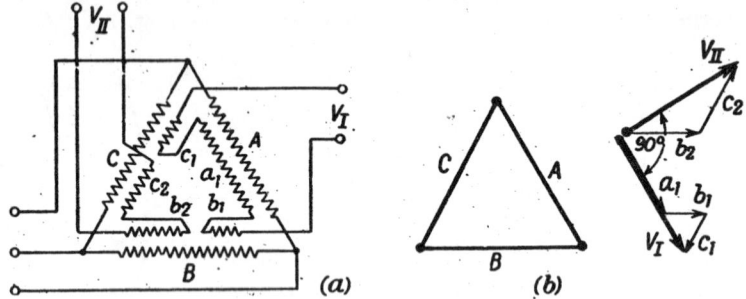

FIG. 49.   LE BLANC CONNECTION

$\frac{2}{3}V$; b_1 of $\frac{1}{3}V$ and b_2 of $V/\sqrt{3}$; and c_1 of $\frac{1}{3}V$ and c_2 of voltage $V/\sqrt{3}$. The windings when grouped give phase I of the two-phase supply with a_1, b_1 and c_1 in series, and phase II with b_2 and c_2 in series. Fig. 49 (b) shows the vector summation, resulting in e.m.f.'s of equal magnitude in phase quadrature. The use of the windings b_1 and c_1 in phase I is to obtain a balance of m.m.f.'s round the three interlinked magnetic circuits and to give balanced loading on the three-phase side when the two-phase output is balanced. The Le Blanc connection has the advantage of using a standard three-phase transformer core.

THREE/ONE-PHASE CONNECTION. The supply of one single-phase load from a three-wire three-phase system is governed by the principle that the input on the three-phase side is essentially single-phase in character: that is, the *algebraic* sum of the currents in two of the lines is equal to the current in the third. This principle assumes the absence of auxiliary energy-storage devices.[*] The actual division of the single-phase current in the three lines of the three-phase side can be made to assume any proportions between $1:1:0$ and $1:\frac{1}{2}:\frac{1}{2}$. Fig. 50 shows a series of possible connections, in which the single-phase load is assumed purely resistive and the load voltage is taken as equal to the voltage of the three-phase lines. (a) A direct application of the load across a pair of lines is simplest.

---

[*] The kinetic energy of a 3-ph./1-ph. motor-generator would give complete 3-phase balance. A static balancing circuit can be devised by use of a capacitor and an inductor: see Langley Morris, *J.I.E.E.*, 93 (II), p. 341 (1946).

(b) This is one phase of a Scott arrangement, using the teaser transformer: the step-up ratio of the teaser to the single-phase output winding is $2/\sqrt{3} = 1.15$, so that one three-phase line carries more current than the load. (c) An open-delta connection. (d) A combination of the two-phase Scott connection: the voltage of each part of the output connection must be $1/\sqrt{2} = 0.71$ of the required one-phase voltage.

It follows from the principles of symmetrical phase-sequence

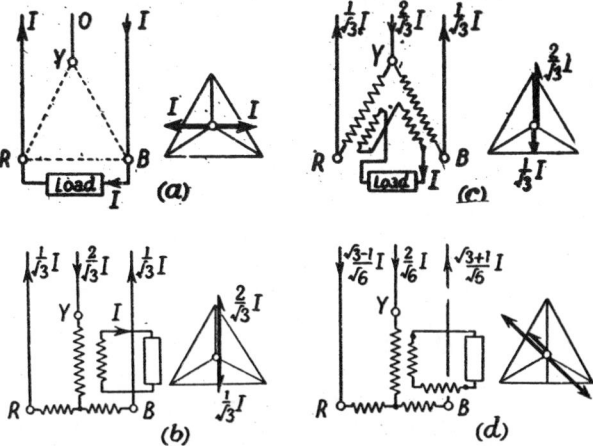

FIG. 50 THREE- TO SINGLE-PHASE CONVERSION

components that the three-phase supply comprises equal positive- and negative-sequence components on account of its "single-phase" nature already referred to. The positive-sequence currents convey the one-phase power and reactive volt-amperes, while the negative-sequence currents are ineffective with a positive-sequence voltage, and simply double the line loss. Thus the use of elaborate connection arrangements is not justifiable.

7. Auto-transformers. An auto-transformer has windings common to primary and secondary, so that the input and output circuits are electrically connected as one continuous winding per phase. In the common part of the winding, the input and output currents are superposed.

The principal application of the auto-transformer is in cases where separation of the primary and secondary circuits is not essential, and the voltage ratio is not great. Such applications include boosters, static balancers, and induction-motor startors. The advantage gained is a considerable saving in conductor material, cores, and losses.

Neglecting magnetizing current, let the single-phase auto-transformer in Fig. 51 (a) have T_1 primary turns tapped at T_2 for a lower-voltage secondary output. If the current delivered by the secondary be I_2 at voltage V_2, the corresponding input is I_1 at V_1, where $V_1/V_2 = I_2/I_1 = T_1/T_2$, neglecting losses. The ampereturns $I_2 T_2$ will oppose $I_1 T_1$ as in the normal transformer, implying that the common part of the winding carries the current $I_2 - I_1$. Consequently a smaller conductor section is needed, and assuming the same current density throughout the winding, the ratio

$$\text{Conductor material} \frac{\text{in auto-transformer}}{\text{in normal transformer}}$$

$$= \frac{I_1(T_1 - T_2) + (I_2 - I_1)T_2}{I_1 T_1 + I_2 T_2}$$

$$= 1 - \frac{2}{(T_1/T_2) + (I_2/I_1)} = 1 - (V_2/V_1) \qquad (21)$$

From eq. (21) it is seen that the more nearly V_2 approaches V_1, the greater is the saving. If $V_2 = V_1$, the auto-transformer, according to eq. (21), requires no conductor material, for actually no transformer is needed. The saving in material and cost is, of course, less than that indicated, on account of the fact that the core is nearly as large as with the normal transformer, and the cost and efficiency of transformers are not directly proportional to their kVA. ratings. For a voltage ratio of 2 : 1, approximately 50 per cent saving in copper would be effected, and the cost might be 65–70 per cent of that of a normal transformer of the same rating. Multi-ratio auto-transformers could be graded with a suitable conductor area for each section according to the kind of rating—constant output kVA., or constant output current, or constant load impedance. Such multiple grading is rather too complicated for it to be much used in actual manufacture.

Auto-transformers are not generally employed when the voltage-ratio exceeds 3 : 1, except for motor-starting duty, as the disadvantages preponderate. The latter are due to the direct electrical connection between the two sides, by reason of which disturbances on one system are liable to affect the other more seriously than when (as in the normal two-winding arrangement) there is only a magnetic link. The reactance is inherently low on account of (a) the "coalescence" of the primary and secondary windings and (b) the smaller currents, so that the conditions on short circuit are more severe. While it is quite possible to arrange for greater leakage by special means, such a step detracts from the advantage of using the auto-transformer at all.

EQUIVALENT CIRCUIT. The voltage V_1 applied to the primary of an auto-transformer opposes the induced e.m.f. E_1 and supplies the drops in resistance and leakage reactance. Taking values as marked in in Fig. 51 (a) and neglecting magnetizing current, then approximately

$$V_1 = -E_1 - I_1(r_1 \cos \phi + x_1 \sin \phi) + (I_2 - I_1)(r_2 \cos \phi + x_2 \sin \phi)$$

for a secondary (and primary) load phase-angle ϕ. The impedance drop term for the common part of the winding is *added* and not subtracted because the net current therein is opposed to the direction of I_1. The secondary e.m.f. is $E_2 = E_1/t$, where $t = T_1/T_2$, the primary/secondary turn-ratio. Across the secondary output terminals is the voltage

$$V_2 = E_2 - (I_2 - I_1)(r_2 \cos \phi + x_2 \sin \phi).$$

These equations are illustrated by the vector diagram, Fig. 51 (*b*).

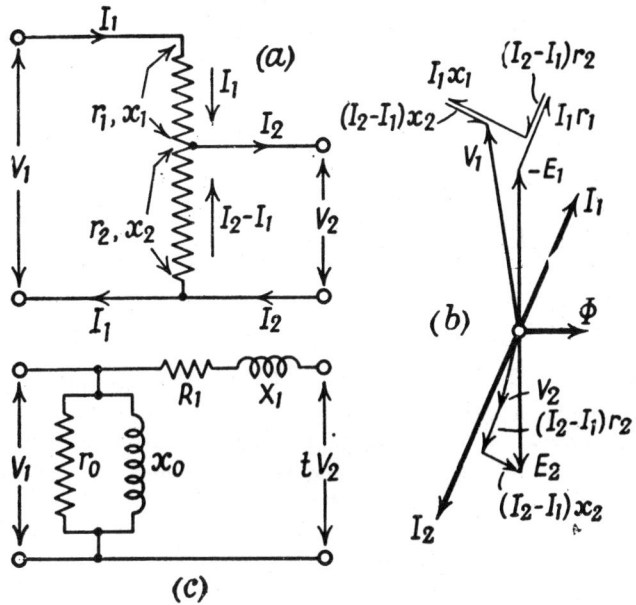

FIG. 51. AUTO-TRANSFORMER

Combining the expressions for V_1 and V_2, with $E_1 = tE_2$ and $I_2 = tI_1$, then

$$V_1 = tV_2 + I_1[\{r_1 + (t-1)^2 r_2\} \cos \phi + \{x_1 + (t-1)^2 x_2\} \sin \phi].$$

As for a normal two-winding transformer, p. 29, this can be written

$$V_1 = tV_2 + I_1(R_1 \cos \phi + X_1 \sin \phi),$$

where $R_1 = r_1 + (t-1)^2 r_2$ and $X_1 = x_1 + (t-1)^2 x_2$ are the equivalent resistance and reactance to primary terms. These compare with $R_1 = r_1 + t^2 r_2$ and $X_1 = x_1 + t^2 x_2$ for the two-winding transformer, the values, however, being in that case for each complete winding. A test with the secondary winding short-circuited gives $Z_1 = R_1 + jX_1$ as the ratio of the primary applied voltage and input current. It is of interest to observe that an auto-transformer may be tested for impedance, and the results calculated, exactly as if it were a two-winding transformer.

The resistance and reactance equivalents for the seconaary side are obtained from

$$R_2 = R_1/t^2 = r_2 + r_1/t^2 \text{ and } X_2 = X_1/t^2 = x_2 + x_1/t^2.$$

The complete equivalent circuit to primary terms is shown in Fig. 51 (c). The values r_0 and x_0 are obtained from an open-circuit test.

THREE-PHASE AUTO-CONNECTION. This is readily obtained.

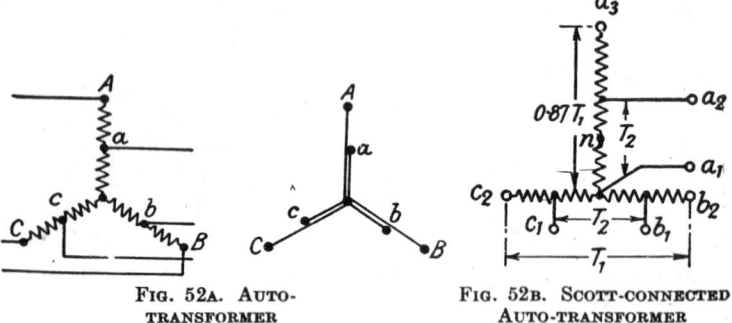

FIG. 52A. AUTO-TRANSFORMER

FIG. 52B. SCOTT-CONNECTED AUTO-TRANSFORMER

Fig. 52A shows a typical method. To avoid undue fault voltages it is usual to employ a star connection.

A commonly used three-phase connection is the Scott three-phase/two-phase auto arrangement in Fig. 52B. By changing the tapped turns T_2 (and extending the windings if necessary) any desired voltage-ratio can be obtained. It is not, however, permissible to interconnect the two secondary phases into a quarter-phase system.

Small- and medium-sized auto-transformers are very economical and are satisfactory for a variety of purposes such as supplying small loads (fans, sewing-machine motors) at a voltage lower than that of the supply. Large units are used to connect grid systems (such as 275- and 132-kV. networks).

Auto-transformer startors for induction motors with cage rotors

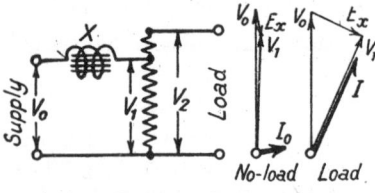

FIG. 53A. VOLTAGE REGULATION BY AUTO-TRANSFORMER AND REACTOR

consist of a three-phase star-connected auto-transformer with one, two, or more tappings, a double-throw switch, and (usually) overcurrent and undervoltage release devices. Such equipment is required to function for a few seconds while the motor is running up to normal speed, so that it is permissible to design it for high current- and flux-densities to save cost.

SIMPLE VOLTAGE REGULATION BY AUTO-TRANSFORMER AND REACTOR. By the simple arrangement shown diagrammatically in Fig. 53A, a static voltage regulator to compensate for feeder line drop can be obtained with loads of power factor approaching unity. On no load, the magnetizing current I_0 of the auto-transformer induces the e.m.f. E_x in the reactor, so that the voltage V_0 is reduced to V_1 at the transformer primary terminals. The output voltage will be $V_2 = V_1 . T_2/T_1$. If a unity-power-factor load-current I is taken, the e.m.f. in the reactor (which lags by 90° on the

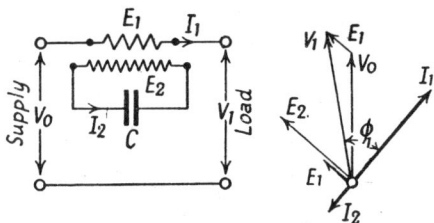

FIG. 53B. CAPACITOR BOOSTER

current) is such as to raise the voltage V_1 to a value greater than its no-load value, thus increasing the output voltage. A rise of about 6 per cent can be obtained.

CAPACITOR BOOSTER. Line drop on a distributor may be compensated by means of a series-connected transformer whose secondary is loaded on to a bank of condensers, as shown for one phase in Fig. 53B. Let the load current be I_1 and the supply voltage be V_0. The secondary current will be I_2, in phase-opposition to I_1 if magnetizing current be neglected. The secondary voltage E_2 must lag I_2 by 90° on account of the capacitive load. The corresponding primary e.m.f. is E_1. The voltage on the load side of the transformer is the vector addition of V_0 and E_1, giving V_1. So long as the load phase-angle ϕ_1 is lagging, $V_1 > V_0$. Further, the supply phase-angle has been reduced. The response of the booster is immediate, but it is clearly unable to boost high-power-factor loads, and would reduce the voltage on a load having a leading power factor.

8. **Voltage Regulation and Tap-changing.** A control of voltage in a supply network is essential for several reasons, including—

(a) The adjustment of consumers' voltage within the statutory limits;

(b) The control of the kilowatt and kilovar flow over a line interconnecting two generating stations;

(c) Seasonal (5–10%), daily (3–5%) or short-period (1–2%) regulation of the voltage of various parts of a network in accordance with cyclic changes of load distribution.

Much of this regulation is done by altering the ratio of transformation by tapping the windings to alter the number of turns. This process is termed *tap-changing*. It may be done when the transformers are out of circuit or (particularly with short-period and daily load cycles requiring voltage adjustment) when on load. The latter condition is more complicated and calls for special equipment known as *on-load tap-changing gear*.

The essentials of voltage control are indicated schematically in Fig. 54. At (*a*) is shown a simple tapping switch giving a voltage range on one or other side of the transformer by adding or subtracting turns, thus giving simple up-and-down voltage regulation.

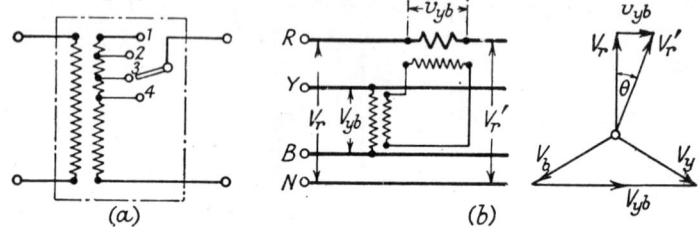

FIG. 54. METHODS OF VOLTAGE CONTROL

At (*b*) is a method of obtaining a control of phase-angle between the voltages of the input and output sides, by means of a special regulating transformer connected to a series or booster transformer. The equipment for one-phase boost is shown : it would be usual to provide equipment for performing the same operation for the other phases. Thus in the diagram line R has injected into it a proportion of the voltage V_{yb} between the lines Y and B, and marked v_{yb} in the vector diagram. The output from the arrangement, V'_r, differs from the input voltage V_r by the angle θ, which depends on the magnitude of v_{yb}. The latter voltage may be varied by tappings, to give a corresponding control of θ.

TAPPINGS. The tapped part of the windings may be at the phase ends, at the neutral point, or in the middle of the phases. The choice is largely a constructional question. Thus the advantage of having the tappings near the ends of the phases is that the number of bushing insulators is reduced, which may be useful where cover space is limited. The insulation between the several phase parts, when tappings are made at the neutral ends, is small. This has economic advantages when the voltage of the transformer is high. Where a comparatively large voltage ratio change is required, the tappings should be made at the centres of the phase windings, for with such an arrangement the magnetic dissymmetry is least and the reactance is not too much affected. The arrangement cannot, however, be used on l.v. windings placed next to the core.

When the voltage of a tapped winding is high, i.e. the number of turns is large, there is no difficulty in obtaining voltage variation within close percentage limits. With l.v. windings, of few turns,

on the other hand, the voltage per turn may be a large percentage of
the total voltage, and there is a severe limitation imposed on the
choice of voltages. It is not possible to tap anything other than an
integral number of turns. Thus a 1 000-kVA. transformer may
have a turn-e.m.f. of about 13 V. A 260-V. winding would have
twenty turns and it would not be possible to have voltage tappings
any closer than 5 per cent. Where convenient, it is therefore usual
to tap the h.v. side. A further advantage of this practice in a step-
down transformer is that, at times of light load when the secondary

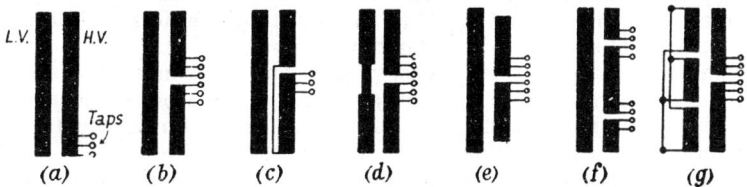

FIG. 55. METHODS OF TAPPING AND BALANCING

voltage has to be lowest, the maximum number of h.v. turns is
included on the primary side, thus reducing the e.m.f. per turn, the
flux-density and the core loss.

Certain transformers have reinforced insulation on the end turns,
to withstand surge voltages. It is essential that these be not tapped,
or that the reinforcement be carried far enough beyond the lowest
tap. In h.v. transformers, reinforcement may be deemed necessary
for the central taps, on account of the possibility of surge voltages
building up at these points.

An effect of great importance is that of the unbalanced axial
forces developed under fault conditions when tappings have produced magnetic asymmetry between the primary and secondary
windings. (See p. 108.) The windings have to be arranged to limit
the amount of unbalance. Some methods* of locating the tappings
are shown in Fig. 55. Tappings at one end of a winding (a) are
permissible in small transformers. In larger units they are usually
placed centrally, both for (b) delta- and (c) star-connected windings.
Linear AT unbalance is mitigated by "thinning down" the l.v.
winding (d), or by balancing parts of the winding more evenly
(e and f). The untapped side may be split into several parts (g)
corresponding to the tapped side, and connected in parallel: the
distribution of current in the parallel paths will be such as to give
almost complete balance. For very large tapping ranges, 10 per cent
and more, a special tapping coil may be used Its top and bottom
halves may be connected in parallel, or each of its tapped-out
sections may be distributed uniformly over the whole winding

* Billig, *J.I.E.E.*, 93 (II), p. 227 (1946).

length. In either case the distribution of the *electric* field due to a steep-fronted surge is affected (p. 105).

Other effects of the use of tappings include a change of reactance; alteration of I^2R and core losses; and increased difficulties of arranging for the parallel operation of dissimilar transformers under all conditions of load and voltage. Certain cases have to be specially considered; such as, for example, the single-phase core-type transformer, in which the tapped turns should be divided equally between the two limbs.

OFF-CIRCUIT TAP-CHANGING. This adjustment is obtained by tapping the respective windings as required, and taking connections therefrom to some position near the top of the transformer. Reconnection is effected by lifting off the tank cover, or by using hand holes in the cover. Alternatively the required reconnection can be made by carrying the tapping leads through the cover for changing by hand or by manually-operated switch. Such arrangements are suitable for seasonal alterations. Two usual types are—

(a) *Vertical tapping switch*, comprising an insulated bar for each phase carrying knife contacts, to be raised or lowered by handwheel and gearing to make connection with fixed contacts arranged on the outer coils of the phases.

(b) *Faceplate switch*, an arrangement similar in idea to the common faceplate types of motor starter. The spindles of the moving arms are geared to a rod with a handwheel external to the tank. The faceplate switches may be mounted on the top yoke, or near the centres of the limbs.

ON-LOAD TAP-CHANGING. For daily and short-period variations in voltage ratio, it must be possible to change the transformer tappings without interrupting the load. There are four methods of arranging for this.

(a) *Parallel Windings*. If it is possible to wind the tapped winding with two parallel circuits, these can be used alternately to avoid breaking the combined circuit. Thus in Fig. 56 (a), the transformer will be operating normally on tap 1 when 1a and 1b are closed. To change to tap 2, the circuit-breaker A (which is generally fitted externally) is opened. The load current is now carried exclusively by one of the parallel circuits. 1a is opened, then 2a closed, and A reclosed. The two circuits are again in parallel but with a difference in voltage. A circulating current will be superimposed on the main current, having a value determined by the resistance and reactance of the circulating circuit. The reactance is controlled by spacing the windings.

The next step is to open circuit-breaker B, open 1b and close 2b, then reclose B. Both branches are now working in parallel on tap 2.

During the transition period, when only one of the branches is carrying the load current, the greater impedance drop will cause a momentary fall in the terminal voltage. The number of tappings required is double that of a simple winding, and the provision of adequate reactance between the parallel branches may cause some complication in the design. On the other hand, most power transformers *require* parallel conductors, for conductors of sufficient section to carry the whole phase current might well be too large from the viewpoint of eddy currents.

(*b*) *Preventive Reactor*. The tapping switch comprises a double

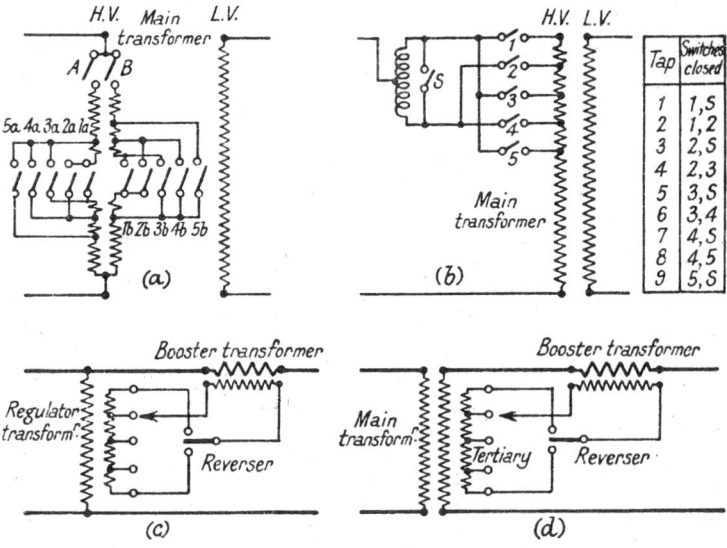

FIG. 56. METHODS OF TAP-CHANGING

contact joined through a reactor, centre-tapped. Normally both contacts are connected to the same tap, the current dividing between them and circulating in opposite directions through the reactor, thus neutralizing its reactance. During transition, two tappings are bridged across the reactor, which limits the short-circuit current flowing between the tappings. The method is sometimes employed with a resistance, but without a centre tap. The arrangement then resembles a battery end-switch (Fig. 57).

An alternative method is shown in Fig. 56 (*b*), where the reactor may be operated with either side connected to a tapping, or with both the mid-point of the reactor taking up a medium voltage between those of the tappings to which it is connected. The number of taps is consequently reduced. Since in this case the reactor is always in circuit, either partly or wholly, it adds to the loss and may be noisy. On certain taps (when only one side of the reactor is in use) some economy is obtained by short-circuiting it. Although shown at the end of the h.v. winding, the taps may be (and usually are) in the middle of the winding.

(c) *Regulator Transformer.* If it is not desirable to incorporate tap-changing gear in the main transformer, a regulating transformer can be used as in Fig. 56 (c). The rating of such a transformer is only a fraction (equal to the percentage boost) of that of the main transformer. The tap-changing gear is quite separate from the main transformer and is smaller in current-carrying capacity. Against

FIG. 57. CORE, WINDINGS AND ON-LOAD TAP CHANGER OF 50-MVA, 150/52·5-kV TRANSFORMER
(*Metropolitan-Vickers*)

this advantage is the cost of providing the additional unit and floor space.

(d) *Tertiary Winding.* A method analogous to (c) above is to use a tertiary winding on the main transformer instead of a separate regulator unit, Fig. 56 (d). In both (c) and (d), the booster transformer must not be open-circuited, as otherwise its core may be very highly saturated.

The cost of tap-changing equipment lies in (1) the mechanical arrangements for effecting the change; (2) the relay devices to prevent incorrect operation or to ensure that transformers in parallel are all operated together; (3) the electrical operation of the gear; and (4) the additional insulation required for tappings, switches, etc. The booster transformer system is the most expensive, but has the considerable advantage of employing a standard main transformer, and tap-devices dealing only with a small power at low voltage.

PRINCIPLES OF CONTROL. Tap changing automatically is initiated by a voltage-sensitive relay or *governor*. The greater its sensitivity the more marked will be the response to fluctuations of mean voltage, which causes operation more often than is necessary. Further, the sensitivity must be related to the voltage added to or subtracted from the output by one tap or step. Too insensitive a governor would result in the voltage being too low for heavy loads (i.e. for a low input voltage) and too high for light loads. A suitable compromise is to make the governor sensitivity equal to the voltage value of one regulator step.

It is uneconomical to allow a change to be initiated by a voltage fluctuation which may have passed by the time it is corrected, so

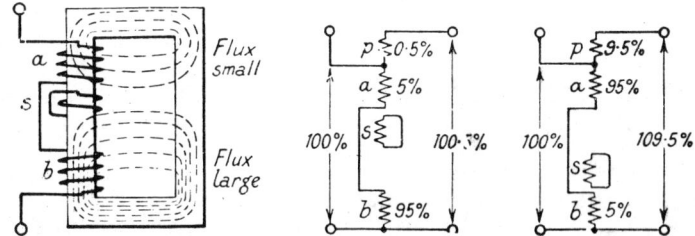

FIG. 58. MOVING-COIL VOLTAGE REGULATOR

FIG. 59. MOVING-COIL VOLTAGE REGULATOR: AUTO ARRANGEMENT

that a time-delay device is desirable. A typical delay is about half a minute.

9. **Transformer-type Voltage Regulators.** Smooth voltage-variation over a desired range is useful for the speed control of some types of motor, dimming of stage lighting, electro-chemistry and laboratory schemes.

MOVING-COIL REGULATOR. This comprises a special transformer with a movable short-circuited coil.

A two-limbed core, Fig. 58, is provided with windings *a* and *b* in opposition at the top and bottom respectively of one limb.* An isolated short-circuited coil *s* is free to be moved up and down the limb. If a voltage be applied to coil *a* alone, the current depends on the impedance of *a*, which is considerably affected by the position of *s*. The closer *s* is to *a*, the greater its short-circuiting effect (i.e. the greater its mutual inductance with *a*) and the lower the effective impedance of *a*.

If a voltage be applied to *a* and *b* in series, the voltage distribution between them is again affected by the position of *s*. The nearer *s* is to one of the coils, the lower is the voltage across it, and the

* In practice each coil occupies nearly one-half of the depth of the limb.

higher the voltage across the other, assuming the applied voltage to be constant. With suitable design, the voltages across a and b can be varied between 5 per cent and 95 per cent of the total applied voltage by altering the relative position of s. The coil s acts by its induced current to increase or decrease the leakage flux and therefore the total reactance.

A variety of arrangements is possible. Fig. 59 shows an auto arrangement in which an auxiliary winding p is used, with one-tenth of the turns of a or b. Coils a and p act together as a normal transformer, so that with s at the top, and 5 per cent of the applied voltage across a, there will be 0·5 per cent in p and the output voltage will be 100·5 per cent of the input. With s at the bottom position, the voltage of a rises to, say, 95 per cent, and that of p to 9·5 per cent. The output voltage is consequently 109·5 per cent, and a smooth variation between the extremes is possible.

Fig. 60 illustrates a different, two-winding arrangement. The turn-ratio of a and p is 6·43, and of b and q is 22·5. With s at the top and 5 per cent of the applied voltage on a, the output voltage is 5/6·43 + 95/22·5 = 5 per cent. With s at the bottom, the output voltage becomes 95/6·43 + 5/22·5 = 15 per cent. The output voltage could be varied down to zero if the direction of p or q were reversed and the turn-ratios suitably readjusted.

A unit of the type described is shown in Fig. 61. The parts are plainly visible. The moving coil is carried on a square-threaded shaft, which may be readily hand-controlled as the mechanical forces involved are small. The arrangement is simple, noiseless and without phase displacement. The losses and magnetizing current are of the same order as those of a corresponding induction regulator.

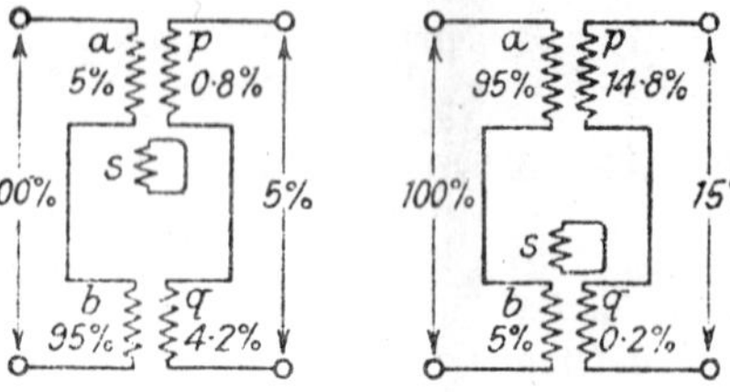

Fig. 60. Moving-coil Voltage Regulator

SLIDING-CONTACT REGULATORS. These, exemplified by the "Variac" and "Breco" designs, utilize a two-winding or auto arrangement in which the regulatable winding is bared and swept by a contact brush or roller. The voltage steps are reduced to the e.m.f. of a single turn, making the adjustment approximate to smooth and gradual control.

10. **Parallel Operation.** In common with other generating plant (and, viewed from the secondary side, the transformer is in a sense a generator), parallel operation of transformers is frequently necessary. When operating two or more transformers in parallel, their satisfactory performance requires that they have—

(a) the same voltage-ratio:

FIG. 61. AUTOMATIC THREE-PHASE MOVING-COIL REGULATOR GIVING 11½ PER CENT BOOST IN A 50-kVA., 400-V., 50-CYCLE CIRCUIT
(Ferranti)

(*b*) the same per-unit (or percentage) impedance;

(*c*) the same polarity;

(*d*) the same phase-sequence and zero relative phase-displacement.
Of these, (*c*) and (*d*) are absolutely essential, and (*a*) must be satisfied
to a close degree. There is more latitude with (*b*), but the more
nearly it is true, the better will be the load-division between the
several transformers.

(*a*) VOLTAGE-RATIO. An equal voltage-ratio (which, in high-
ratio cases is not necessarily achieved by equality of *turn*-ratio,
unless the designs are identical) is necessary to avoid no-load circulating current. Consider two transformers, connected in parallel
on their primary sides only. If voltage readings on the open secondaries do not show identical values, there will be circulating currents
between the secondaries (and therefore between the primaries also)
when the secondary terminals are connected in parallel. The impedance of transformers is small, so that a small percentage voltage-
difference may be sufficient to circulate a considerable current and
cause additional I^2R loss. When the secondaries are loaded the circulating current will tend to produce unequal loading conditions, and
it may be impossible to take the combined full-load output from
the parallel-connected group without one of the transformers
becoming excessively hot. (See also (*b*), following.)

(*b*) IMPEDANCE. The impedances of two transformers may differ
in magnitude and in quality (i.e. ratio of resistance to reactance),
and it is necessary to distinguish between *per-unit* and *numerical*
impedance. Consider two transformers of ratings in the ratio
2 : 1. To carry double the current, the former must have half the
impedance of the latter for the same regulation. The regulation
must, however, be the same for parallel operation, this condition
being enforced by the parallel connection. Hence the currents
carried by two transformers are proportional to their ratings if their
numerical or ohmic impedances are inversely proportional to those
ratings, and their *per-unit* impedances are identical. A transformer with an impedance of z per-unit requires z per-unit of normal
terminal voltage to circulate full load current on short circuit.
See p. 28.

A difference in the quality of the per-unit impedance results in
a divergence of phase angle of the two currents, so that one transformer will be working with a higher, and the other with a lower,
power factor than that of the combined output.

Case I. Equal Voltage Ratios. In this case the voltage ratios on
no load are presumed equal, and the voltages coincident in phase.
Unless the magnetizing currents are very different (expressed as a
fraction of the full-load current) the assumption of negligible phase-
difference is a close approximation to fact. Under these conditions
both sides of two transformers can be connected in parallel, and no
current will circulate between them on no load.

Neglecting magnetizing admittance, the case of two single-phase transformers in parallel is described by the equivalent circuits of Fig. 62. At (a) the two equivalent circuits are connected in parallel, and simplified to (b), in which the through connections are made common. The essence of (b) is, however, (c), in which it is clear that the case reduces to the consideration of two impedances in parallel. Let

$Z_1, Z_2 =$ the impedances of the two transformers;
$I_1, I_2 =$ their currents;
$V =$ common terminal voltage;
$I =$ total combined current.

These complexor quantities can be referred to either primary or secondary sides throughout, consistently.

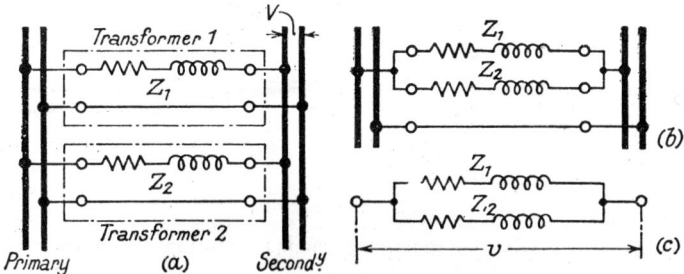

FIG. 62. TRANSFORMERS IN PARALLEL

The voltage drop v across the impedances in parallel is common to the two, so that

$$v = I_1 Z_1 = I_2 Z_2 = I Z_{12}. \qquad . \qquad . \qquad . \qquad (22)$$

where Z_{12} is the impedance of Z_1 and Z_2 in parallel combination, viz.

$$1/Z_{12} = 1/Z_1 + 1/Z_2 \text{ whence } Z_{12} = Z_1 Z_2/(Z_1 + Z_2) \qquad (23)$$

From eq. (22)

$$I_1 = I Z_{12}/Z_1 = I Z_2/(Z_1 + Z_2)$$
and $$I_2 = I Z_{12}/Z_2 = I Z_1/(Z_1 + Z_2).$$

Multiplying both sides by the common terminal voltage V,

$$VI_1 = VI \frac{Z_2}{Z_1 + Z_2} \text{ and } VI_2 = VI \frac{Z_1}{Z_1 + Z_2}.$$

Writing $VI \cdot 10^{-3} = S$, the combined load kVA., then the kVA. carried by each transformer is

$$S_1 = S \frac{Z_2}{Z_1 + Z_2} \text{ and } S_2 = S \frac{Z_1}{Z_1 + Z_2}. \qquad (24)$$

These expressions are complex, so that S_1 and S_2 are obtained in *magnitude* and *phase-angle* (i.e. S_1 and S_2 correspond to currents I_1 and I_2 having a calculable angular phase-difference from the common voltage V).

The problem may be solved graphically, although with more labour. If the impedance triangles of the two transformers are as given in Fig. 63 (a), it is clear that transformer 1 will carry the greater current since it has the lesser impedance. The currents I_1 and I_2 will, in fact, be related by the numerical expression

$$I_1/I_2 = Z_2/Z_1.$$

The drops I_1Z_1 and I_2Z_2 must be equal and in phase, since the transformers are connected in parallel on both primary and secondary

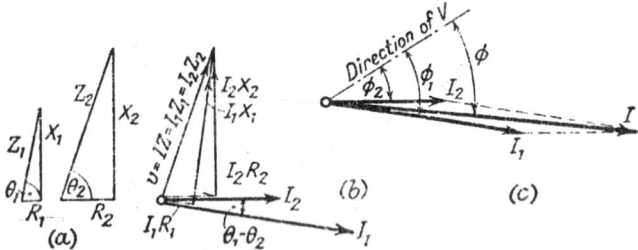

FIG. 63. TRANSFORMERS IN PARALLEL: VECTOR DIAGRAM

sides. The currents must lag on the voltage drop v by $\theta_1 = \arctan (X_1/R_1)$ and $\theta_2 = \arctan (X_2/R_2)$ respectively, as indicated in Fig. 63 (b). This indicates the method of solving the problem: draw I_1 and I_2 with an angular phase-difference of $(\theta_1 - \theta_2)$ and of magnitude, to some scale, inversely proportional to the respective impedances (Fig. 63 (c)). Sum I_1 and I_2 vectorially, giving I. According to the terms of the problem, the phase-angle and magnitude of I will be known from the conditions of loading, so that the angle ϕ to the terminal voltage V will be known. Inserting this, the transformer currents now are known in magnitude and phase with respect to V.

Stated briefly, if two transformers have equal impedances (i.e. equal resistance and equal reactance, not merely a *numerical* equality of impedance), they will share a load equally. If they have the same *per-unit* impedance, they will share a load in proportion to their respective ratings. It is therefore possible to operate any two transformers in parallel, regardless of their relative size, so long as the per-unit impedances are strictly equal in magnitude and phase-angle.

EXAMPLE. A 500-kVA. transformer with 0·01 p.u. resistance and 0·05 p.u. reactance is connected in parallel with a 250-kVA. transformer with 0·015 p.u. resistance and 0·04 p.u. reactance. The secondary voltage of each transformer is 400 V. on no load. Find how they share a load of 750 kVA. at power factor 0·8 lagging.

It is not necessary to find the ohmic values of the resistances and reactances, as from eq. (24) only the impedance ratios are required. The p.u. values given, however, refer to different ratings in the two cases, and must be adjusted to the same basic kVA. Since x per-unit reactance in a 250-kVA. transformer is equivalent (in ohms) to $2x$ in a 500-kVA. transformer, it is sufficient to take

$$Z_1 = 0.01 + j0.05 = 0.051/\underline{78.7°},$$

and $\qquad Z_2 = 2(0.015 + j0.04) = (0.03 + j0.08) = 0.085/\underline{69.4°}.$

Further

$$Z_1 + Z_2 = (0.04 + j0.13) = 0.136/\underline{72.9°}.$$

The total kVA. load is,

$$S = 750/\underline{-36.9°} \text{ kVA}.,$$

where $36.9°$ is the angle whose cosine is 0.8: the negative sign indicates a lag of the current on the voltage. From eq. (24)

$$S_1 = 750/\underline{-36.9°} . 0.085/\underline{69.4°}/0.136/\underline{72.9°} = 471/\underline{-40.4°} \text{ kVA}.$$

$$= 471 \text{ kVA at power factor } 0.762 \text{ lagging};$$

and similarly

$$S_2 = 750/\underline{-36.9°} . 0.051/\underline{78.7°}/0.136/\underline{72.9°} = 281/\underline{-31.1°} \text{ kVA}.$$

$$= 281 \text{ kVA. at power factor } 0.856 \text{ lagging}.$$

As a check, the kW. and kVAr. components should sum to the corresponding components of the total load, which comprises $750 \times 0.8 = 600$ kW. and $750 \times 0.6 = 450$ kVAr. For tne first transformer,

$$P_1 = 471 . 0.762 = 359 \text{ kW}.; \quad Q_1 = 471 . 0.648 = 305 \text{ kVAr}.;$$

and for the second,

$$P_2 = 281 . 0.856 = 241 \text{ kW}.; \quad Q_2 = 281 . 0.517 = 145 \text{ kVAr}.;$$

the active and reactive components being calculated from the kVA. by multiplying it by the cosine and sine respectively of the angle of lag. $P_1 + P_2 = 600$ kW., and $Q_1 + Q_2 = 450$ kVAr., thus confirming the solutions for S_1 and S_2.

Fig. 63 shows the graphical solution. The method of construction has already been described. As found above, $\theta_1 = 78.7°$, $\theta_2 = 69.4°$, $\theta_1 - \theta_2 = 9.3°$, and $\phi = -36.9°$. The currents can actually be scaled in kVA. at the common voltage. Note that the transformers are not loaded in proportion to their ratings.

Case II. Unequal Voltage Ratios. The general case, in which the voltage ratios of the transformers are unequal (i.e. their no-load secondary terminal voltages are unequal), is rather troublesome to solve graphically; the complex algebraic method, on the other hand, leads directly to a solution. Using the symbols of the previous case, with the addition of

E_1, E_2 = no load secondary e.m.f.'s;
    Z = load impedance at secondary terminals;

then

$$v = IZ_{12} = (I_1 + I_2)Z_{12}$$

as before. Also the e.m.f.'s of the transformers will be equal to the total drops in their respective circuits,

$$\left. \begin{array}{l} E_1 = I_1Z_1 + (I_1 + I_2)Z, \\ \text{and} \quad E_2 = I_2Z_2 + (I_1 + I_2)Z\,; \end{array} \right\} \qquad \cdots \cdots \quad (25)$$

whence $E_1 - E_2 = I_1Z_1 - I_2Z_2$ as might be expected. From this expression

$$I_1 = \frac{(E_1 - E_2) + I_2Z_2}{Z_1}$$

giving I_1 in terms of I_2. Substituting this in eq. (25) for E_2, gives

$$E_2 = I_2Z_2 + \left[\frac{(E_1 - E_2) + I_2Z_2}{Z_1} + I_2 \right]Z,$$

whence

$$\left. \begin{array}{l} I_2 = \dfrac{E_2Z_1 - (E_1 - E_2)Z}{Z_1Z_2 + Z(Z_1 + Z_2)} \\[2mm] \text{and by symmetry} \\[2mm] I_1 = \dfrac{E_1Z_2 + (E_1 - E_2)Z}{Z_1Z_2 + Z(Z_1 + Z_2)} \end{array} \right\} \qquad \cdots \cdots \quad (26)$$

Z is the secondary load impedance, and for convenience E_1 and E_2, Z_1 and Z_2 are respectively the open-circuit terminal secondary voltages and the transformer impedances referred to the secondary sides. The expressions in eq. (26) then give the secondary currents. The primary currents are obtained by multiplication by the turn-ratio and the addition (if not negligible) of the magnetizing currents. Usually E_1 and E_2 can be presumed to have the same phase, but eq. (26) could be used to demonstrate the results of connecting in parallel two transformers in which, by reason of some difference of internal connection, there is a phase divergence: e.g. the connection in parallel of a star/star and a star/delta transformer.

It is clear from eq. (25) that, on no load, there will be a vector circulating current between the two transformers, of amount

$$I_1 = -I_2 = (E_1 - E_2)/(Z_1 + Z_2)$$

a result obtainable from eq. (26) by writing $Z = \infty$. On short-circuit, the expected result from eq. (26) is

$$I_1 = E_1/Z_1 \text{ and } I_2 = E_2/Z_2$$

EXAMPLE. Solve the previous example for (a) the cross-current in the secondaries on no load, and (b) the sharing of 750 kVA. at

0·8 lagging, when the open-circui: secondary voltages are 405 V.
and 415 V. respectively.

It is more convenient to work with actual ohmic impedances in
this case. An assumption must be made for the output voltage on
load, but there will be but little error if it is assumed to be 400 V

Calculating the impedances referred to the secondary:

$$I_1R_1 = 0·01 \text{ p.u. of } 400 \text{ V.}, \quad \therefore R_1 = 0·01 \frac{400}{1\,250} = 0·0032 \ \Omega.$$

$$I_1X_1 = 0·05 \text{ p.u. of } 400 \text{ V.}, \quad \therefore X_1 = 5 \times 0·0032 = 0·0160 \ \Omega.$$

In a similar way, or by proportion,

$$R_2 = 0·0096 \ \Omega., \ X_2 = 0·0256 \ \Omega.$$

whence $\quad Z_1 = 0·0032 + j0·0160 = 0·0163\underline{/78·5°},$

and $\quad\quad Z_2 = 0·0096 + j0·0256 = 0·0275\underline{/69·4°},$

and $\quad Z_1 + Z_2 = 0·0128 + j0·0416 = 0·0436\underline{/72·9°}.$

With a load of 750 kVA. at power factor 0·8 lagging, the load impedance must be such that

$$VI . 10^{-3} = (V^2/Z)10^{-3} = 750\underline{/-36·9°},$$

so that $\quad Z = 400^2/(750.10^3\underline{/-36·9°})$
$$= 0·214\underline{/36·9°} = (0·171 + j0·128) \ \Omega.$$

(a) From eq. (26) for no load,
$$I_1 = - I_2 = \frac{405 - 415}{0·0436\underline{/72·9°}} = - 230 \text{ A.},$$

at power factor cos 72·9° = 0·294 lagging. This current is nearly
0·4 p.u. of the full-load rating of the smaller transformer.

(b) From eq. (26)
$$I_1 = \frac{405 . 0·0275\underline{/69·4°} - 10 . 0·214\underline{/36·9°}}{0·0163\underline{/78·5°} \cdot 0·0275\underline{/69·4°} + 0·214\underline{/36·9°} \cdot 0·0436\underline{/72·9°}}$$
$$= 970\underline{/-35·0°} \text{ A.}$$

Similarly
$$I_2 = \frac{415 . 0·0163\underline{/78·5°} + 10 . 0·214\underline{/36·9°}}{0·0163\underline{/78·5°} . 0·0275\underline{/69·4°} + 0·214\underline{/36·9°} . 0·0436\underline{/72·9°}}$$
$$= 875\underline{/-42·6°} \text{ A.}$$

The corresponding kVA. are—
$$S_1 = 400 . 970 . 10^{-3}\underline{/-35·0°}$$
$$= 388 \text{ kVA. at p.f. } 0·82 \text{ lagging,}$$

and $S_2 = 400 . 875 . 10^{-3}/\underline{-42.6°}$

$\underline{\underline{= 350 \text{ kVA. at p.f. } 0.736 \text{ lagging.}}}$

The total output power amounts to

$388 . 0.82 + 350 . 0.736 = 318 + 258 = 576 \text{ kW.},$

which is 4 per cent less than the 600 kW. required by the load on account of the assumption of the value of output voltage in order to assess Z. Note that the smaller transformer is overloaded by nearly 40 per cent.

A solution for two or more transformers operating in parallel is obtained very simply by application of the *parallel-generator theorem*, given in detail in Chapter XVII. For the example (b) above the short-circuit currents are

$$I_{1sc} = \frac{E_1}{Z_1} = \frac{405/0°}{0.0163/78.5°} = 4\,960 - j24\,360 \text{ A},$$

and $$I_{2sc} = \frac{E_2}{Z_2} = \frac{415/0°}{0.0275/69.4°} = 5\,310 - j14\,120 \text{ A};$$

so that the total short-circuit current is

$$I_{sc} = I_{1sc} + I_{2sc} = 10\,270 - j38\,480 = 39\,840/\underline{-75.1°} \text{ A}.$$

Taking the same load impedance estimate as before,

$$\frac{1}{Z} = \frac{1}{0.214/36.9°} + \frac{1}{0.0163/78.5°} + \frac{1}{0.0275/69.4°} = \frac{1}{0.0099/73.4°}.$$

The common terminal voltage becomes

$V = I_{sc} Z = 39\,840/\underline{-75.1°} . 0.0099/73.4° = 395/\underline{-1.7°} \text{ V}.$

The current output of the first transformer is

$$I_1 = \frac{E_1 - V}{Z_1} = \frac{405/0° - 395/-1.7°}{0.0163/78.5°} = \frac{15.4/49.5°}{0.0163/78.5°}$$

$$= 945/\underline{-29.0°} = 826 - j457 \text{ A}.$$

For the second,

$$I_2 = \frac{E_2 - V}{Z_2} = \frac{415/0° - 395/-1.7°}{0.0275/69.4°} = \frac{23.2/30.3°}{0.0275/69.4°}$$

$$= 845/\underline{-39.2°} = 655 - j534 \text{ A}.$$

The load current is

$$I = I_1 + I_2 = 1\,481 - j991 = 1\,785/\underline{-33.8°} \text{ A},$$

and the load kVA.

$S = VI . 10^{-3} = 395/\underline{-1.7°} . 1\,785/\underline{-33.8°} . 10^{-3} = 705/\underline{-32.1°}$

$\underline{\underline{= 600 \text{ kW. and } 375 \text{ kVAr. lagging.}}}$

The low kVAr. value is due to the overestimate of the load impedance.

(c) POLARITY. This can be either right or wrong. If wrong it results in a dead short-circuit. See Chapter VI, §3.

(d) PHASE-SEQUENCE. The question of phase-sequence, and the associated question of angular phase-difference due to dissimilarity of connection, occurs only with polyphase transformers. To start with, only a few of the many possible connections can be worked in parallel without producing excessive circulating current. Thus the secondary voltages of star/star and star/mesh transformers will differ in phase by 30°, as indicated in Fig. 46. So long as both primary and secondary voltages correspond between two transformers, then *aliter aequa* they can be connected in parallel. Any mixture of internal connections can be worked out if it is remembered that the primary and secondary coils on any one limb have induced e.m.f.'s that are in time-phase. The several connections produce various magnitudes and phases of secondary voltage: the magnitude can be adjusted for the purposes of parallel operation by suitable choice of turn-ratio, but the phase divergence cannot be compensated. Thus two sets of connections giving secondary voltages with a phase-displacement cannot be used for transformers intended for parallel operation.

The phase-sequence, or the order in which the phases reach their maximum positive voltages, must be identical for two paralleled transformers: otherwise during the cycle each pair of phases will be short-circuited. If three secondary terminals $a_1b_1c_1$ are to be paralleled with $a_2b_2c_2$, the polarity, ratio, etc., being correct, then a_1 may be connected to a_2. If this results in a voltage across b_1b_2 or c_1c_2, the sequence of one transformer is reversed, as would be shown by zero voltage across b_1c_2 and b_2c_1.

The following are typical of the connections for which, from the view point of phase-sequence and angular divergence, transformers can be operated in parallel—

Transformer 1: Yy   Yd   Yd
Transformer 2: Dd   Dy   Yz

Thus all transformers in the same group, Fig. 46, may from this standpoint be connected in parallel. Further, transformers having a + 30° angle may be paralleled with those having a − 30° angle by reversing the phase-sequence of both primary and secondary terminals.

11. **Tertiary Windings.** Transformers may be constructed with tertiary windings (i.e. windings additional to the normal primary and secondary) for any of the following reasons—

(a) For an additional load which for some reason must be kept insulated from that of the secondary;

(b) To supply phase-compensating devices, such as condensers, operated at some voltage not equal to that of the primary or secondary or with some different connection (e.g. mesh).

(c) In star/star-connected transformers, to allow sufficient earth-fault current to flow for operation of protective gear, to suppress harmonic voltages, and to limit voltage unbalance when the main load is asymmetrical; in each case the tertiary winding is delta-connected;

(d) As a voltage coil in a testing transformer;

(e) To load large split-winding generators;

(f) To interconnect three supply systems operating at different voltages.

Tertiary windings are frequently delta-connected: consequently, when faults and short-circuits occur on the primary or secondary sides (particularly between lines and earth), considerable unbalance of phase voltage may be produced, compensated by large tertiary circulating currents. The reactance of the winding must be such as to limit the circulating current to that which can be carried by the copper, otherwise the tertiary winding may overheat under fault conditions.

PRIMARY CURRENT. With given secondary and tertiary load currents I_2 and I_3, and the numbers of turns T_1, T_2 and T_3, the primary current is easily found by adding vectorially the secondary, tertiary and magnetizing ampere-turns or equivalent currents. Fig. 64A is drawn for the following example.

EXAMPLE. A 6 600/400/110-V., star/star/mesh, three-phase transformer has a magnetizing current of 5·5 A., and balanced three-phase loads of 1 000 kVA. at 0·8 lagging on the secondary, and 200 kVA. at 0·5 leading on the tertiary. Find the primary current and power factor.

Working throughout in equivalent primary currents,

$$I_0 = -j5·5.$$

$$-I_2' = [1\ 000 . 10^3/(\sqrt{3}) . 6\ 600]\ /\!\!-\!\text{arc cos } 0·8$$

$$= 87·5/\!\!-\!36·9° = (70 - j52·5)\ \text{A.}$$

$$-I_3' = [200 . 10^3/(\sqrt{3}) . 6\ 600]\ /\!\!+\!\text{arc cos } 0·5$$

$$= 17·5/\!\!+\!60° = (8·8 + j15·1)\ \text{A.}$$

The total primary current is therefore

$$I_1 = -j5·5 + 70 - j52·5 + (8·8 + j15·1)\ \text{A.}$$

$$= 78·8 - j42·9 = 89·7/\!\!-\!28·6°\ \text{A.}$$

The primary load is

$$S_1 = (\sqrt{3}) . 6\ 600 . 89·7 . 10^{-3} = 1\ 025\ \text{kVA.}$$ at power factor $\cos 28·6° = 0·878$, i.e. a kilowatt loading of $1\ 025 . 0·878 = 900\ \text{kW.}$, corresponding to 800 kW. in the secondary plus 100 kW. in the tertiary. Losses are neglected.

EQUIVALENT CIRCUIT: REGULATION. Properly interpreted, a three-circuit transformer may be represented by the equivalent

single-phase connection in Fig 64B, assuming each winding to /have
resistance and leakage reactance. All values are reduced to a
common-voltage and a common-rating
basis, for since it is no longer possible
to state that secondary output is equal
(nearly) to primary input, the simple
secondary impedance equivalents used
in the normal two-circuit transformer
cannot be applied. It is, of course,
still true that input is equal to output,
but the subdivision of the output between secondary and tertiary windings
is arbitrary. The equivalent resistance and reactance is consequently
expressed for each circuit on a basis
of the *same kVA.* at the *same voltage.*
Considering the primary and secondary working as a two-circuit
transformer, the total resistance voltage Ir_{12} is the sum of the drops

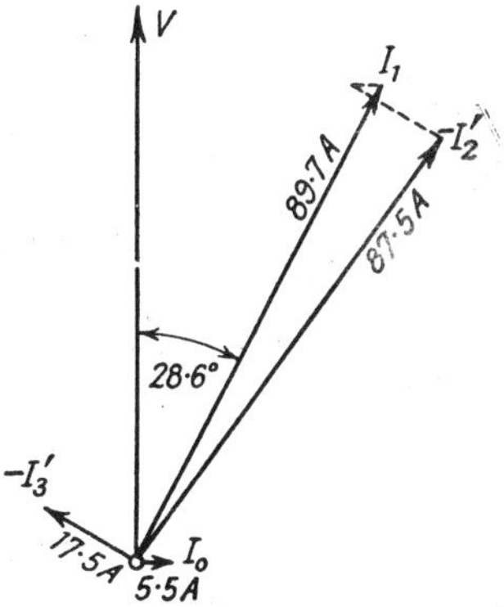

Fig. 64A. Vector Diagram
Three-winding Transformer

Ir_1 and Ir_2 in the two windings separately. A similar statement is
true for any pair of windings, and can be applied to the per-unit
reactance voltage as well. Thus[*]

$$Ir_{12} = Ir_1 + Ir_2, \qquad Ix_{12} = Ix_1 + Ix_2,$$
$$Ir_{23} = Ir_2 + Ir_3, \qquad Ix_{23} = Ix_2 + Ix_3,$$
$$Ir_{13} = Ir_1 + Ir_3, \qquad Ix_{13} = Ix_1 + Ix_3,$$

for the possible combinations of pairs of circuits taken together as
two-circuit transformers. Using these relations,

$$r_1 = \tfrac{1}{2}(r_{12} + r_{13} - r_{23}) = \tfrac{1}{2}\Sigma r - r_{23}.$$

The other corresponding relations are similar, whence

$$r_1 = \tfrac{1}{2}\Sigma r - r_{23}, \qquad x_1 = \tfrac{1}{2}\Sigma x - x_{23},$$
$$r_2 = \tfrac{1}{2}\Sigma r - r_{13}, \qquad x_2 = \tfrac{1}{2}\Sigma x - x_{13},$$
$$r_3 = \tfrac{1}{2}\Sigma r - r_{12}, \qquad x_3 = \tfrac{1}{2}\Sigma x - x_{12}.$$

Thus from a series of normal two-winding short-circuit tests, these
expressions give, for the assigned voltage and rating, the ohmic
values r_1, r_2, r_3 and x_1, x_2, x_3 in Fig. 64B, taking completely into
account all mutual effects.

Further, suppose the secondary in Fig. 64B is short-circuited, the
tertiary being left on open circuit. The total voltage applied to the
primary will be the sum of the primary and secondary impedance
voltages. A voltmeter across the tertiary output terminals is connected in effect across the junction of primary and secondary:
it therefore indicates the secondary impedance voltage (after

[*] Blume, Camilli, Boyajian and Montsinger. *Transformer Engineering*
(Wiley).

adjustment to the common voltage basis). In general, the voltage across one circuit when a second supplies short-circuit kVA. to the third, is the impedance voltage of the short-circuited circuit.

The *regulation* is best considered from the individual contribution

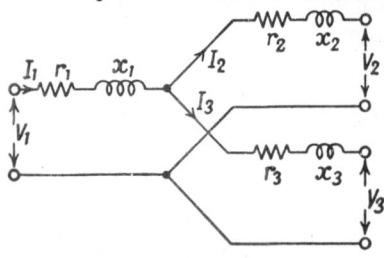

of each winding, the total regulation of two windings in combination being the algebraic (not complex) sum of the regulations of each. The regulation of a winding is that given by eq. (19) or (20), adjusted for the ratio k = (loading/basic rating). Thus for the primary alone,

FIG. 64B. EQUIVALENT CIRCUIT OF THREE-CIRCUIT TRANSFORMER

$$\varepsilon_1 = k(\varepsilon_{r1} \cos \phi_1 + \varepsilon_{x1} \sin \phi_1)$$

after eq. (19), with the additional term from eq. (20) if necessary.

The summation of the regulations of two circuits together is added if power flows from one to the other, and subtracted if not. For a transformer with primary fed and both secondary and tertiary loaded, the regulations become

$$\varepsilon_{12} = \varepsilon_1 + \varepsilon_2, \text{ and } \varepsilon_{13} = \varepsilon_1 + \varepsilon_3;$$

also,

$$\varepsilon_{23} = \varepsilon_2 - \varepsilon_3, \text{ and } \varepsilon_{32} = \varepsilon_3 - \varepsilon_2.$$

The primary load is the combination of secondary and tertiary loads, and its regulation ε_1 must be calculated on this basis.

EXAMPLE. Let the transformer of the previous example, p. 90, have respective primary, secondary and tertiary per-unit resistance voltages, Ir/V, of $\varepsilon_{r1} = 0.0053$, $\varepsilon_{r2} = 0.0063$ and $\varepsilon_{r3} = 0.0083$; and per-unit reactance voltages of $\varepsilon_{x1} = 0.0265$, $\varepsilon_{x2} = 0.025$, $\varepsilon_{x3} = 0.0332$. Find the regulations for the loading conditions given on a basis of 1 000 kVA.

The values of impedance voltages given are derived from a test such as that described above. The previous example (Fig. 64A) shows that the primary loading is 1 025 kVA. at a lower factor of 0.878 lagging. The regulation contributed by it is

$$\varepsilon_1 = k(\varepsilon_{r1} \cos \phi_1 + \varepsilon_{x1} \sin \phi_1)$$
$$= 1.025(0.0053 \cdot 0.88 + 0.0265 \cdot 0.48) = 0.0177 \text{ p.u.}$$

For the secondary, $k = 1\,000/1\,000 = 1$, and

$$\varepsilon_2 = 1(0.0063 \cdot 0.8 + 0.025 \cdot 0.6) = 0.02 \text{ p.u.}$$

For the tertiary, $k = 200/1\,000 = 0.2$, and

$$\varepsilon_3 = 0.2(0.0083 \cdot 0.5 - 0.0332 \cdot 0.87) = -0.0049 \text{ p.u.}$$

The primary-secondary regulation is

$$\varepsilon_{12} = \varepsilon_1 + \varepsilon_2 = 0.0177 + 0.02 = 0.0377 \text{ p.u.}$$

The primary-tertiary regulation is

$$\varepsilon_{13} = \varepsilon_1 + \varepsilon_3 = 0.0177 - 0.0049 = 0.0128 \text{ p.u.}$$

The regulation between secondary and tertiary is

$$\varepsilon_{23} = \varepsilon_2 - \varepsilon_3 = 0.02 - (-0.0049) = 0.0249 \text{ p.u.}$$

In the reverse direction the regulation $\varepsilon_{32} = -0.0249$ p.u.

TERTIARY WINDINGS IN STAR/STAR TRANSFORMERS. Star/star transformers comprising single-phase units, or three-phase *shell-* or *five-limb-core*-type units, suffer from the disadvantages (*a*) that they cannot readily supply unbalanced loads between line and neutral, and (*b*) that their phase voltages may be distorted by third-harmonic e.m.f.'s. These arise because their phase magnetic circuits are not interlinked, and out-of-balance (zero-phase-sequence) currents have an iron path for the production of z.p.s. flux. The z.p.s. impedance is therefore high. The flow of earth-fault current is severely restricted, and may be insufficient for the satisfactory operation of protective gear. The zero-sequence impedance to out-of-balance or earth-fault currents must include the very high impedance (0.5–5 p.u.) between the secondary winding of the loaded or faulted phase and the primary windings of the other two phases. Abnormal voltages therefore occur when z.p.s. current flows. The output load or fault current concerns one phase only, whereas the corresponding input load current has, in the absence of a primary neutral, to be conducted through both of the other phases, which act as reactors. The voltage of the loaded phase is reduced, that of the others being raised. By use of a delta-connected tertiary winding, induced currents are caused to circulate in it, apportioning the load more evenly over the three phases. The action is, in brief, to interlink magnetically the separate magnetic circuits.

The mesh-connected tertiary provides a path for the third-harmonic currents required for magnetizing to the comparatively high saturation flux densities now employed. The circulation of the third-harmonic currents then provides, with the primary current, the excitation necessary to secure a substantially sinusoidal flux and e.m.f.

The two disadvantages mentioned are to some degree mitigated in three-limb core-type transformers, because zero-sequence flux is forced out of the core limbs, leading to a lower z.p.s. impedance. It is nevertheless usual here also to provide a mesh-connected tertiary.

RATING OF TERTIARY WINDINGS. This depends upon the intended use. If to provide an additional load, the winding is designed and calculated on the same basis as the primary and secondary. When employed to balance loads and to control short-circuit currents, the tertiary winding carries current only for short periods, and its rating depends chiefly on its heat-capacity. In practice, whatever the intended function of the tertiary, its conductor section is generally determined by the fault conditions.

12. **Harmonics in Transformers.** The use of high induction densities in the cores of power transformers is imposed by the requirements of an economical design and the reduction of dead weight: Unfortunately this practice introduces a series of troubles due to the saturation of the magnetic circuit and the departure from rectilinearity of the flux/current relation, or B/at curve.

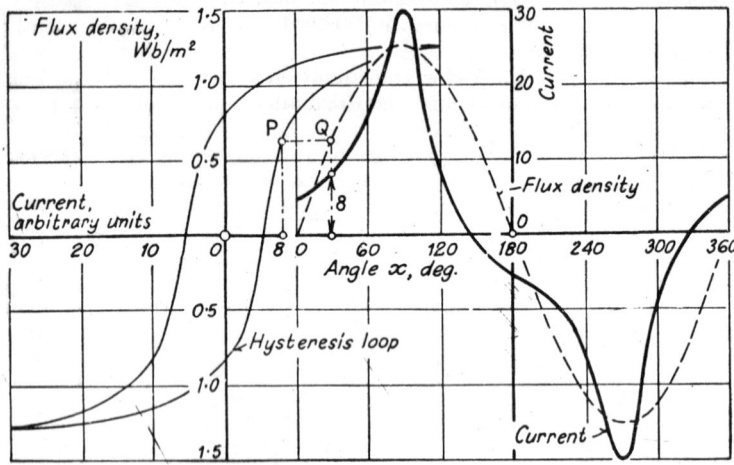

FIG. 65. MAGNETIZING CURRENT WAVE-FORM

Fig. 65 illustrates a method of plotting the magnetizing current variation to a base of time, assuming a sinusoidal flux-density

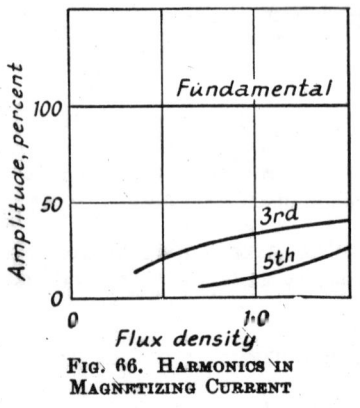

FIG. 66. HARMONICS IN MAGNETIZING CURRENT

(corresponding to a sinusoidal e.m.f.) and a corresponding B/at curve for the core material, including the effect of hysteresis. Thus an ascending flux-density corresponding to P on the hysteresis loop and to Q on the sinusoidal wave, requires 8 units of magnetizing current. The current curve plotted in this manner is seen to be far from sinusoidal; it comprises a fundamental term and a series of odd harmonics. The analysis of this typical case gives, in terms of the fundamental amplitude of 100 per cent,

$$100 \sin (x + 18°) - 39 \sin (3x + 7°) + 18 \sin (5x + 9°)$$
$$- 8 \sin (7x + 10°).$$

The displacement of 18° shows that the current has a component 100 sin 18° in phase with the component of terminal voltage — E_1, which is in phase quadrature to Φ_m and B_m. The component 100 cos 18° corresponds to the magnetizing current I_{0r} of the complexor diagram, Fig. 13. The component 100 sin 18° is, however, only a part of the current I_{0a}, since no account has been taken here of the eddy-current loss.

The essential point under consideration is the presence in the current of harmonics, chiefly the third and fifth, which become of

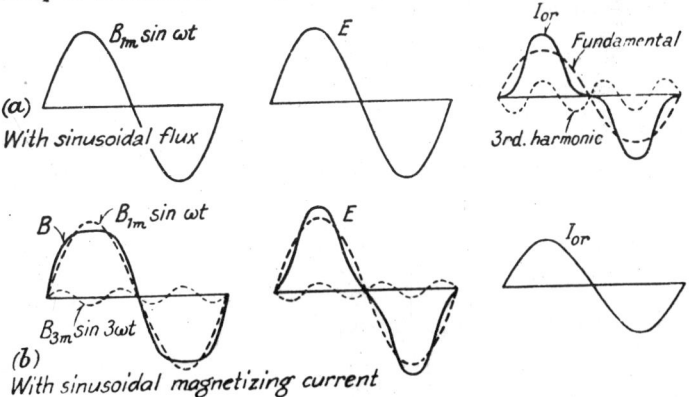

FIG. 67. FLUX, E.M.F., AND MAGNETIZING CURRENT HARMONICS

increasing amplitude as the maximum induction density is raised Fig. 66 gives typical percentage harmonic magnetizing curren' amplitudes. The fifth and seventh may be more important thar the third harmonic, since mesh connection does not suppress them

Thus a sinusoidal flux wave (required by a sinusoidal applied voltage) demands a magnetizing current with an harmonic content But a supply of strictly sinusoidal voltage *cannot* supply an harmoni current. If a sinusoidal magnetizing current is furnished, howevel the flux wave will fail to reach its sinusoidal peak value and wi become flat-topped. The e.m.f. induced by it will then be peaky with third (and other) harmonics. For, neglecting the shift produce by hysteresis, a flat-topped flux-density wave-form may be writte as

$$B = B_{1m} \sin \omega t + B_{3m} \sin 3\omega t,$$

omitting higher harmonics. The e.m.f. induced by such a density

$$e = - A_i \frac{dB}{dt} \text{ volts per turn,}$$

when the flux-density extends over the net core area A_i: when

$$e = [\omega A_i B_{1m} \cos \omega t + 3\omega A_i B_{3m} \cos 3\omega t]$$
$$= \omega A_i [B_{1m} \cos \omega t + 3B_{3m} \cos 3\omega t] \text{ volts.}$$

This expression represents a wave-form of a peaky nature (Fig. 67 (*b*)). Conversely, a peaky flux wave produces a flat-topped e.m.f. Also a 10 per cent third harmonic flux produces a 30 per cent harmonic e.m.f., so that a third-harmonic e.m.f. is readily introduced by a small departure of the flux from the sinusoidal.

Fig. 67 (*b*) summarizes these conclusions. A more searching discussion for cases of very high normal saturation is given later (p. 100).

The conditions investigated below presuppose a sinusoidal applied voltage, and impedance drops small enough to necessitate the induced e.m.f. being also sinusoidal between supply terminals.

(A). SINGLE-PHASE TRANSFORMERS. Considering the third harmonic component of the magnetizing current of a single-phase transformer, with sinusoidal impressed voltage and flux, it is clear that the supply voltage cannot itself give rise to any third harmonic current. To circulate this current, therefore, a triple-frequency e.m.f. must be generated in the transformer by a triple-frequency flux. The e.m.f. must be sufficient to circulate the third harmonic current through the transformer and the supply on the primary side, and through the transformer and load on the secondary side. (Since the load impedance is usually much greater than that of the supply, the third harmonic currents flow principally in the primary circuit.) The third harmonic flux is combined with the fundamental flux in such a way that the resultant flux has a wave-form intermediate between a sinusoid and that wave-form produced by a sinusoidal current, the exact degree of compromise being determined by the impedance of the circuits to third harmonic currents. If the supply impedance is negligibly small compared with the third harmonic impedance Z_3 of the transformer, then the third harmonic e.m.f. E_3 will be balanced by the corresponding impedance drop I_3Z_3, so that no harmonic voltage will appear in the line.

If the impedance to third harmonics could be made entirely negligible, then only a vanishingly small third harmonic e.m.f. would be needed to circulate a magnetizing current additive to the fundamental magnetizing current so as to permit a substantially sinusoidal flux.

The harmonic currents set up I^2R losses, and it is not immediately evident how these are supplied from the sinusoidal supply, which by definition cannot directly furnish harmonic power. Briefly, the explanation lies in the hypothesis that the fundamental power conveyed to the core is there subdivided, part being used as core loss for the non-sinusoidal flux, the remainder as harmonic I^2R loss by transformer action.

The remarks above with reference to the third harmonic are applicable without alteration to harmonics of other orders.

(B). THREE-PHASE BANKS OF SINGLE-PHASE TRANSFORMERS. In considering three-phase transformers it is essential to distinguish between the cases where the phases are magnetically separate and where they are magnetically (as well as electrically) interlinked. Where three-phase banks of single-phase transformers are used, the magnetic circuits are obviously separate, and each core must itself produce the flux demanded by the conditions of its electrical connections, which then determine the flow of harmonic currents.

In the following descriptions, strictly sinusoidal terminal voltages are assumed, and the loads are balanced. The supply and load are considered to be star-connected.

(a) **Dd** *Connection.* In a system of balanced three-phase voltages, the fundamentals, and the fifth, seventh, eleventh, thirteenth, and $(3m \pm 1)$th harmonics, all produce voltages displaced by $120°$ mutually. The third, ninth, $3m$th harmonics, on the other hand, have voltages in the three phases that are co-phasal in time—i.e all reach their positive and negative maxima together :[*] m takes all integral values greater than zero. The mesh connecting of the three phases forms a closed path for the harmonics of order $3m$ (often called *triplen* harmonics for brevity), and such harmonic e.m.f.'s will circulate corresponding harmonic currents in the mesh, so that each phase harmonic e.m.f. is absorbed by its own IZ drop: there is consequently no triplen harmonic voltage manifested at the line terminals. The impedance to harmonic currents is usually small (although it may be greater than the fundamental impedance of the transformer), and very small e.m.f.'s are sufficient to circulate considerable harmonic currents.

The action in the Dd case is then as follows. The supply voltage

---

[*] Let the voltage of phase I of a symmetrical three-phase system be

$$v_I = v_1 \sin (\omega t + \alpha_1) + v_3 \sin (3\omega t + \alpha_3) + v_5 \sin (5\omega t + \alpha_5) + \ldots$$

The voltage of phase II will be

$$v_{II} = v_1 \sin (\omega t - 120° + \alpha_1) + v_3 \sin (\overline{3\omega t - 120°} + \alpha_3)$$
$$+ v_5 \sin (5\omega t - 120° + \alpha_5) + \ldots$$

and of phase III will be

$$v_{III} = v_1 \sin (\omega t - 240° + \alpha_1) + v_3 \sin (\overline{3\omega t - 240°} + \alpha_3)$$
$$+ v_5 \sin (\overline{5\omega t - 240°} + \alpha_5) + \ldots$$

where v_1, v_3, . . . are the amplitudes of the fundamental and third . . . harmonic terms. Considering the third harmonics, these are,

in phase I, $v_3 \sin (3\omega t + \alpha_3)$;

in phase II, $v_3 \sin (\overline{3\omega t - 120°} + \alpha_3) = v_3 \sin (3\omega t + \alpha_3)$;

and in phase III, $v_3 \sin (\overline{3\omega t - 240°} + \alpha_3) = v_3 \sin (3\omega t + \alpha_3)$

Thus the third harmonics are *in time phase* with each other in all three phases. The same applies to the ninth, fifteenth, twenty-first, . . . harmonics, i.e. all *triplens* or odd multiples of 3.

The fifth, seventh, . . . and all other harmonics, form voltages displaced $120°$ in time mutually, with either the same or opposite phase sequence compared with that of the fundamentals.

—being sinusoidal—provides only a fundamental sinusoidal magnetizing current. Such a current will produce a non-sinusoidal
flat-topped flux curve, since there is no peaky current to overcome
the saturation at high flux densities. This flat-topped flux wave
induces a peaky e.m.f. wave (Fig. 67 (b)), which will not balance
the applied fundamental sine voltage, but will leave as resultant a
third harmonic component (and also higher harmonics). The third
harmonic components now circulate a triplen current in the mesh,
which is additive to the sinusoidal supply current, and builds up

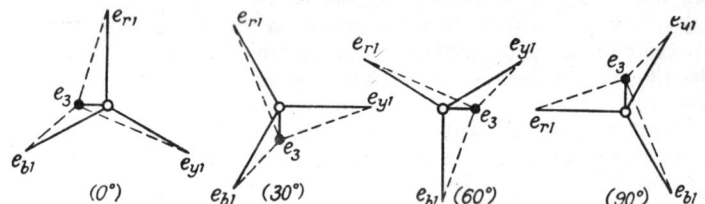

FIG. 68. OSCILLATION OF NEUTRAL

.he *total* current to the wave-form required—like that of Fig. 68 (a)
If the triplen mesh impedance is low, a very small departure of the
flux from a perfect sine is sufficient to produce a large third harmonic
magnetizing current. Thus the conditions are established whereby
the primary voltage and primary current are sine waves, the flux
is very nearly a sine wave, and the necessary triplen magnetizing
current component is circulated round the closed mesh.

Harmonics of other than triplen orders operate in the same manner as described under (A) above, p. 96.

(b) Yd *and* Dy *Connection without Neutral.* So long as there is
no neutral wire, either of these connections operates substantially
as in (a), but since there is a mesh connection on only one side, the
triplen mesh impedance will be greater, and the compensation of the
magnetizing current will demand a greater divergence of the flux
wave from a true sinusoid.

(c) Yy *Connection without Neutrals.* As already described, the
third harmonic voltages in a balanced three-phase system are equal
and co-phasal in time in all three legs of the system. With respect
to the star point of a star connection, therefore, the triplen e m f.'s
are directed all outwards, or all inwards, together at any given
instant. Between any two line terminals there are two legs, the
fundamentals of which together produce a line voltage $\sqrt{3}$ times
as much ; but the triplen harmonic e.m.f.'s will cancel each other
out. Thus under no circumstances can any triplen currents flow,
the input magnetizing currents remaining sinusoidal apart from the
fifth and seventh harmonics. The flux wave in each transformer
departs strongly from a sinusoid, becoming flat-topped after the

manner of Fig. 67 (*b*). The e.m.f. induced is consequently peaked. The balance between applied voltage and induced e.m.f. between line terminals is maintained since the triplen e.m.f.'s balance out, but the effect on the star-point voltage is to make it *oscillate*. The complexor conditions are shown in Fig. 68. The voltages e_{r1}, e_{y1}, e_{b1} represent the fundamental terms, mutually displaced by 120°. There is a third harmonic e.m.f. e_3, co-phasal in all legs, additional to the fundamental, and the resultant voltages across the legs are shown dotted. It is to be noted in the diagrams, which are drawn for four consecutive instants displaced by one-twelfth period, that the e_3 complexor rotates thrice as fast as the fundamental system.

The fifth, seventh, and other non-triplen harmonics are themselves balanced systems with a 120° phase-displacement, and appear across the lines to cause a self-suppressing circulating current.

(*d*) Yy *Connection with Neutral Wires.* The presence of a neutral connection renders the conditions for each of the three transformers similar to those described under (A) above. The neutral connection carries the three triplen harmonic compensating currents, which are all in phase: the current in the neutral is thus three times the triplen current of one transformer.

(*e*) *Other* Yd *Connections.* All the variants of the possible connections will fall under one or other of the subheadings (*a*) to (*d*) already dealt with. A mesh connection always provides a closed path for triplen currents, while a neutral conductor on the supply side (i.e. connected to the supply neutral, or earthed when there is another earthed neutral on the system) permits triplen currents to flow. A neutral conductor is not operative on the secondary side on open circuit.

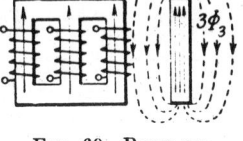

(*f*) *Tertiary* D *Winding.* If it be desired to effect some suppression of triplen harmonics without dispensing with the advantages of star connection on both primary and secondary sides, a solution is found in the provision of *tertiary* windings. one for each phase, connected in mesh. This may be provided exclusively for harmonic currents, or may have other functions (see § 11). If either or both of the primary and secondary have neutral connections, these will compete with the tertiary winding for some of the harmonic current, the division of which depends on the relative impedances of the alternative paths.

FIG. 69. PATH OF THIRD-HARMONIC FLUXES IN CORE-TYPE TRANSFORMER

(C) THREE-PHASE TRANSFORMER UNITS. The conclusions in (B) above can be applied directly to the shell-type three-phase transformer, in which the magnetic circuits of the separate phases are complete in themselves and do not interact. With the very common three-limbed core type, however, the phases are *magnetically* interlinked (Fig. 104 shows the interlinkage for fundamental sinusoidal fluxes and also for non-triplen flux harmonics). Any triplen flux harmonics that exist will be directed either all upwards or all downwards in the limbs together at any instant. The return paths of these fluxes must therefore lie outside the cores (Fig. 69) through the air or oil and the walls of the tank (if any). These paths are of high reluctance, so that there is a strong tendency to retain a sinusoidal flux and e.m.f. Occasionally the third harmonic fluxes have been found to cause losses in the tank walls. They can be further suppressed by surrounding the whole transformer by a copper ring, the induced currents in which damp down the values of the third harmonic fluxes.

In five-limbed transformers the end limbs provide a return path for triplen flux harmonics (Fig. 27).

EFFECTS OF TRANSFORMER HARMONICS. Briefly the effects of harmonic currents are—

(a) Additional I^2R loss due to circulating currents;

(b) Increased core loss;

(c) Interference magnetically with communication circuits and protective gear.

Harmonic voltages, depending on their magnitude and frequency may cause—

(a) Increased dielectric stresses;

(b) Electrostatic interference with communication circuits;

(c) Resonance between the inductance of the transformer windings and the capacitance of a feeder to which they are connected.

Third and other triplen harmonics, as described, are suppressed by delta-connected windings in the case of three-phase transformers.

Other harmonics may be substantially suppressed by connecting across the terminals of the transformer resonating shunts tuned to the frequencies which it is required to eliminate.

HARMONIC CURRENT COMPENSATION. When the flux-density in a transformer core is increased beyond 1·3 Wb./m.², the harmonic content of the magnetizing current—particularly the third (if it can flow), fifth and seventh—advances rapidly. It is, however, possible to secure almost sinusoidal magnetizing currents at densities of as much as 1·6 Wb./m.² by the use of special harmonic-compensating arrangements.

A well-saturated core-type star/star insulated-neutral transformer may have a magnetizing-current wave-form like that in Fig. 70 (a), exhibiting a strong fifth harmonic: the third is, of course, precluded by the connection, and further the flux is forced to be nearly sinusoidal by the high reluctance offered to third harmonic fluxes (Fig. 69). If the transformer is now provided with a five-limbed core (Fig. 70 (e)), the path for third harmonic fluxes is provided by the end limbs and the flux wave-form consequently flattens (Fig. 67 (b)). The wave-form of the magnetizing current, Fig. 70 (b), shows that the fifth harmonic has a reversed sign, and in addition there is a small reduction of the fundamental. Comparing wave-forms (*a*) and (*b*) it would appear that if the end-limbs of the transformer core have suitable intermediate reluctance, they would result in zero fifth harmonic in the magnetizing current. This is actually the case, as shown by the current oscillogram (*c*). The method is applicable to star/star transformers with insulated neutrals and to star/zigzag transformers. It is not applicable to delta connections.

The parallel connection of two transformers, one star/star and the other delta/delta, results in a suppression of fifth and seventh harmonic currents in the combined input magnetizing current since these are equal and in phase opposition in the two transformers.*

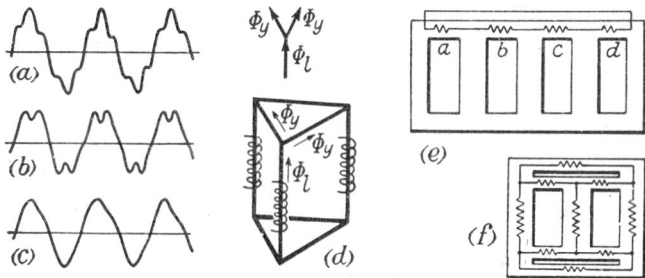

FIG. 70. HARMONIC CURRENT COMPENSATION

The same effect may be obtained in the one transformer by combining one core section having a (magnetic) star-connection with a second delta-connected section, to form one magnetic unit. If a *symmetrical* three-phase core-type transformer is considered, Fig. 70 (*d*), it will be seen that the limb flux Φ_l splits into two yoke fluxes Φ_y differing in phase by 30° each with respect to the limb flux. Now each flux requires, as well as a fundamental, a fifth and a seventh harmonic current for excitation. The yoke fifth and seventh harmonics will be displaced in phase with reference to the limb harmonics by $\pm 5 \times 30° = 150°$ and $\pm 7 \times 30° = 210°$ respectively. The two yokes together, by symmetry, take a combined fifth and seventh harmonic excitation in phase opposition to that of the limb to which they are magnetically connected. If the reluctances are suitably chosen, the combined exciting current for

* The magnetizing current in one line of a star-connected transformer is that for one phase only: in the case of a mesh-connection it is the vector difference of two phase currents, respectively in advance by 30° and 150° on the star phase current, and smaller in the ratio $1/\sqrt{3}$. The total line current for the two transformers together results in a fundamental of double that of the star transformer alone; the resultant 5th and 7th harmonic currents for the mesh transformer are in phase-opposition to those for the star transformer as regards the line, and so cancel each other. Thus the 5th's and 7th's circulate between only the transformers themselves and disappear from the lines.

limb and yokes together can be almost devoid of fifth and seventh harmonic current, even for high saturation densities. The latter, however, lead to considerable leakage unless the harmonic yoke excitation is provided by an appropriate winding, leaving the limbs to be harmonically excited by a winding thereon. If one winding is delta-connected, provision is then made for third-harmonic currents as well.

The magnetic star-mesh connection is inherent in a five-limbed core, Fig. 70 (e). Suppose the core to be folded so that the outer unwound limbs coalesce: since the flux in these limbs is always in anti-phase, the flux in yoke a would form a continuation of that in d, and the coalesced outer limbs would be redundant. In this condition the transformer resembles the symmetrical arrangement of (d). The magnetic star-mesh is obtainable in a three-limbed core, (f), by slitting the yokes. The yoke fluxes in (e) and (f) must form, in the several sections, equal components at 120°. This needs the careful balancing of the section reluctances, but an easier way is to wind round the yoke sections a mesh-connected winding to force the fluxes into the required vector-relationship in spite of any difference in reluctance. In (e) there is, in addition, a star-connected winding (besides the main input and output windings) interconnected with the mesh winding on the yokes. This aids the production between the limbs and core of the fifth and seventh harmonic currents.

13. Transients. It is customary to consider electrical phenomena in circuits in a *steady state*, that is, after the disappearance of the initial disturbances which occur between switching on and the attainment of steady conditions. The initial *transient* conditions may be of very short duration and of limited interest. On the other hand, in certain cases the transient conditions are vitally important, for momentary voltages or currents may arise of values large enough to cause damage by the breakdown of insulation, by the production of excessive mechanical stresses, or by intense localized heating.

The transient conditions of interest in transformer operation may conveniently be divided into current and voltage phenomena. The former concerns chiefly the switching on of transformers, while the latter is largely concerned with the safety of the insulation on the h.v. side when subjected to voltage surges of very steep wave-front.

SWITCHING-IN PHENOMENA. When the primary side of a transformer is switched on to normal voltage with the secondary side open-circuited, it acts exactly like a simple inductive reactor. At every instant the voltage applied must be balanced by the e.m.f. induced by the flux generated in the core by the magnetizing current, and by the small drops in resistance and leakage reactance. To a first approximation, it is substantially true that the e.m.f. induced must balance the voltage applied.

The precise value of the voltage at the instant at which the switch is closed may be anything between zero and its peak value. Further it may be rising or falling and may have either polarity at this instant.

Assume first that the core is initially demagnetized and that the switch closes at the *peak value* of the voltage, which obtains twice

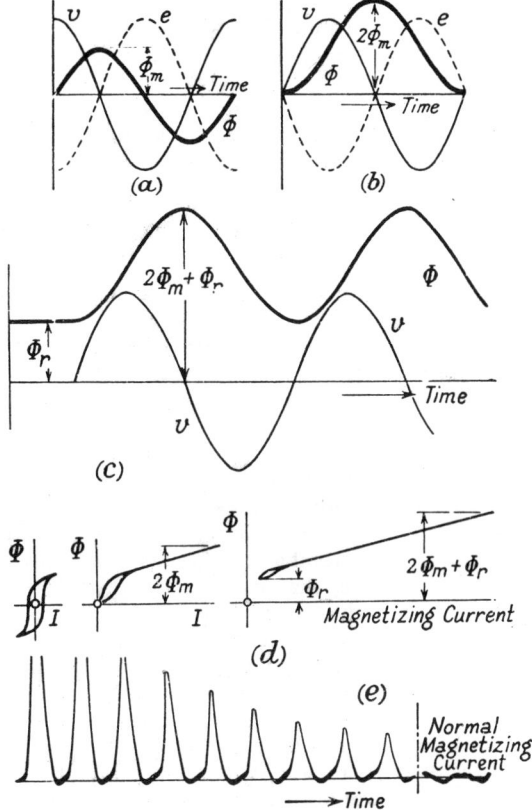

Fig. 71. Switching-in Transients

per cycle. The e.m.f. must be at once established: it requires a flux having a *maximum rate of change*, so that the flux starts to grow from zero value in the required direction, as also does the magnetizing current. These conditions, which obtain immediately after the closing of the switch, are, however, precisely the same as the final steady conditions, so that the energizing of the transformer proceeds without any transient (apart from a negligible capacitance effect: see p. 106). Thus when the transformer is switched on at the instant that the voltage applied passes through a peak value, the voltage, flux and e.m.f. waves immediately assume the normal relationship for an inductive circuit, illustrated in Fig. 71 (a).

Assume now that the switch closes at the instant that the applied voltage passes through a zero value, Fig. 71 (b). Throughout the first half-period the voltage is positive, and consequently the e.m.f. negative (i.e. in phase opposition) so that the flux for this half-period must continually increase. It begins at zero value with an increasing slope or rate-of-change and finally reaches *double* its normal peak value at the end of the first half-cycle. The same rate-of-change sequence takes place normally from a negative peak of flux to a positive peak, whereas since now the flux is initially zero the same total change (of twice the flux peak value) results in the flux attaining twice its normal maximum. This is referred to as the *doubling effect*.

If there were no resistance, and no losses, the conditions just described would continue indefinitely. The asymmetrical loss due to I^2R and core heating, however, rapidly reduces the flux waveform to symmetry about the time axis.

The doubling effect is not dangerous *per se*. With the high saturations in use with modern transformers, however, the magnetizing current required for producing twice normal flux-density may be very great indeed. For a density of 1·4 Wb./m.², an excitation of about 30 A.T. per cm. is required; for 2·1 Wb./m.², about 1 500 A.T. per cm.; or 50 times as much for 50 per cent increase in flux density. If the normal magnetizing current is 5 per cent of normal rated current, then the doubling effect may raise it initially to several times rated current. The double flux-density will not actually be reached, for with overcurrents the drops in resistance and reactance may become of considerable importance, and cannot be neglected in comparison with the e.m.f. Consequently the e.m.f. is not so large, and the flux does not attain its double peak. The effects of the initial current surge may nevertheless present the designer with problems of intense mechanical stress to solve.

Clearly, from the discussion of the limiting cases above, there will be a variety of possible circumstances under which a transformer can be suddenly energized: the switch may, in fact, close on a voltage of anything between peak value and zero. The severity of the surge of magnetizing current will correspond.

It is possible for remanent magnetism in the core to intensify the surge. Thus if the voltage is switched on at a zero in such a direction as to augment an existing remanent flux Φ_r, the resultant wave-form of the flux will be that given in Fig. 71 (c). The sinusoid of flux variation has still the double-amplitude $2\Phi_m$, but is entirely offset from the zero and rises to the peak $2\Phi_m + \Phi_r$. It rapidly tends towards symmetry on account of I^2R losses.

Hysteresis loops in terms of flux Φ and magnetizing current I for the three cases (a), (b) and (c), are shown in Fig. 71 (d). The first is normal and symmetrical; the second is unidirectional and the area of the loops is actually reduced in spite of the high saturation. The third curve shows the flux/current relation for case (c): supersaturated conditions predominate and the hysteresis loss is negligible. The peak current is, however, very high. Its value can be estimated approximately on the basis of the inductance of the primary as an air-cored coil, since the iron is so saturated that its permeability is little more than unity. The initial high current peaks decay rapidly, because the rate of decay is proportional to R/L, with R large (for losses) and L small. Thereafter, the distortion of the magnetizing current persists for several seconds, as the flux density falls nearer to normal, raising the inductance markedly. The switching-in current oscillogram, (e), appears to have extended zero-pauses. This is due to the great variation with saturation: the negative half-cycles are extremely small. Normal magnetizing current (Fig. 65) is scarcely observable.

Particularly large switching currents may occur if the secondary winding is short-circuited when the primary is energized, for the secondary equivalent short-circuit current is carried by the primary in addition to its own switching transient.

VOLTAGE SURGES. Transformers connected on one side to an overhead transmission line may be subjected to overvoltages produced in the line by switching, faults, atmospheric conditions or lightning discharges. These overvoltages, or surges, travel along the line at a speed approaching that of light, and with a wave-front more or less steep. Such a surge may have only a few micro-

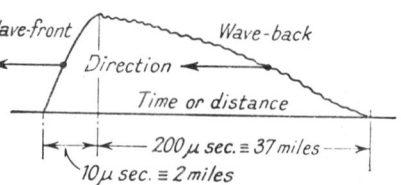

FIG. 72. SURGE VOLTAGE

seconds of time between its initial and peak values, and the maximum voltage may be anything up to that at which the transmission-line insulators flash over to earth. Fig. 72 shows a typical surge-voltage wave-form with a base of time and corresponding length of overhead transmission line.

The behaviour of a transformer to steep-fronted surges* is largely determined by the capacitances in and between successive h.v. coils, and of these coils to earth (i.e. to the l.v. winding, the core, and the tank wall). To a rapidly-changing voltage a capacitor offers a low impedance: thus the electric field, as represented by its equivalent capacitances (which are too small to effect normal-frequency operation), becomes a predominant factor in determining the

* See Allibone, McKenzie and Perry, "Effects of Impulse Voltages on Transformer Windings," *J.I.E.E.* 80, pp. 117, 342, 587 (1937); Norris, "The Lightning Strength of Power Transformers," *J.I.E.E.* 95 (II), p. 389 (1948); Taylor, *Power System Transients* (Newnes).

voltage distribution when the voltage has a high time-rate of change. A surge of 350 kV., rising uniformly to a peak in 10 μ-sec., has a rate of rise of 35 × 10⁶ kV./sec.: this is 1 000 times as great as the normal peak rate of voltage change in a 132-kV. 50-cycle star-connected transformer, and the importance of stray capacitance is multiplied by the same order.

Fig. 73 (*a*) gives schematically the arrangement of a h.v. winding with its associated inter-coil capacitances *c* and coil-to-earth capacitances *C*. With no earth capacitances *C*, any voltage impressed across AB would be uniformly-distributed if the "lumping" effect of individual h.v. coils be ignored. The unavoidable presence of the earth capacitances *C*, which are large compared with the series sum of the *c* capacitances, is such as to reduce the voltage steeply over the first few *c* capacitors for a steep wave front. They offer a path to earth which excludes the part of the h.v. winding remote

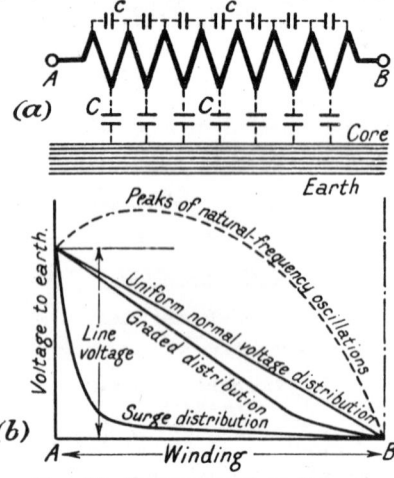

(a)

(b)

FIG. 73. DISTRIBUTION OF SURGE VOLTAGE

from the line, so that the drop across the first few coils is very high. Because the surge voltage itself may be several times the working line voltage, the first few coils may be subjected to an electric stress of e.g. 200 times normal. Thus the initial distribution of a surge voltage may approximate to that so marked in Fig. 73 (*b*).

After the initial incidence of the surge, the first few earth capacitances are charged, and the remainder are comparatively uncharged. This condition is unstable, and while the line-terminal voltage is being maintained by the wave-tail, the interval voltage distribution will tend to become uniform. The consequent interchange of stored energy between capacitors through coil-inductances generates complex oscillations at a variety of natural frequencies. The dotted curve in Fig. 73 (*b*) indicates a typical distribution of oscillatory voltage peaks. When these have become sufficiently attenuated, and if the surge voltage at the line terminal is maintained, the distribution sinks back to the uniform.

A more detailed approach to the surge problem analyses a steep-fronted surge into a Fourier series of component sines impressed on one end of a ladder network, Fig. 74 (*a*). At the critical frequency corresponding to Lc the series elements of the ladder are in rejector resonance. *Above* this frequency the component sines are presented with the equivalent circuit (*b*), resulting in the initial surge distribution indicated already in Fig. 73 (*b*). *Below* critical frequency the component sines encounter the equivalent circuit of Fig. 74 (*c*), and travel through the h.v. winding as through a low-pass filter. Both *c'* and *L'* depend on the frequency, and affect the velocity and attenuation of the travelling-wave components differently: this modifies the wave-front as it passes through the winding, as shown in (*d*). At the end of the winding reflection takes place in accordance

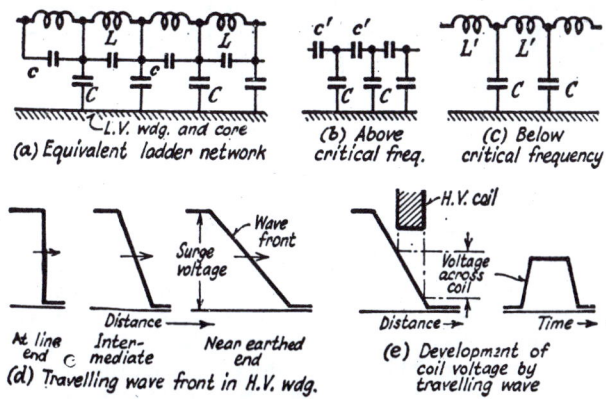

(a) Equivalent ladder network   (b) Above critical freq.   (c) Below critical frequency

(d) Travelling wave front in H.V. wdg.   (e) Development of coil voltage by travelling wave

FIG. 74. TRAVELLING WAVES IN H.V. WINDING

with the termination (e.g. earth or open-circuit). The *voltage to earth* existing at any point is then the resultant of the initial and travelling-wave voltage distributions at the instant considered.

The *voltage across an individual coil* is easily seen from Fig. 74 (*e*) to have a pulse form due to the passage of the travelling wave. Similar considerations lead to the form of the voltage between successive coils. The voltage distribution is affected by the capacitances and therefore by the design of the insulation, and it is essential to co-ordinate the insulation with the effect that it will have on the voltage distribution.

SURGE PROTECTION. The avoidance of breakdown under surges includes (*a*) the provision of reinforced end-turn insulation; (*b*) the use of protective devices; and (*c*) special types of stress-grading construction.

Method (*a*) has been employed in practice for many years, but is now suspect in view of greater knowledge of surge behaviour. End-turn reinforcement may actually make conditions worse by reducing *c* compared with *C* and intensifying the surge voltage build-up across the turns and coils at the line end.

External or internal devices such as *surge-absorbers* comprise series inductors at the line end of the h.v. winding to reduce the steepness of the wave-front, the losses absorbing surge energy and reducing the voltage reaching the h.v. winding.

Method (c) aims to produce for all conditions a uniformity of voltage distribution. The two basic types are—

1. *Shields.* If the coil-to-earth capacitances C can be neutralized by providing new capacitances C_1 to line (on precisely the same

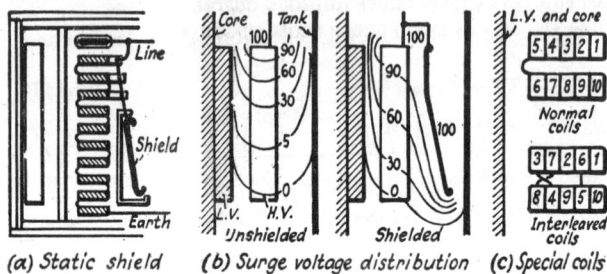

(a) Static shield   (b) Surge voltage distribution   (c) Special coils

FIG. 75. SURGE VOLTAGE GRADING

principle as is used for grading transmission-line suspension-insulator strings), then the coil-to-earth currents through C will be directly provided by line-to-coil currents through C_1, leaving the inter-coil currents through c all equal to produce a uniform distribution over the winding. This may be done by mounting, outside and around the winding, insulated metal shields connected to line, Fig. 75 (a). The effect on the electric field is shown diagrammatically in (b), the figures indicating equipotentials corresponding to a surge of value 100, for an unshielded and for a shielded transformer.

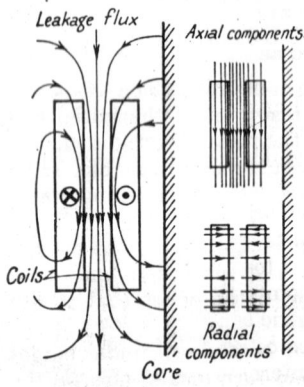

FIG. 76. COMPONENTS OF LEAKAGE FLUX

2. *Interleaved Turns.* Two normal disc-type coils are shown in Fig. 75 (c). If the order of turn-connection is altered to increase the voltage between adjacent turns, their capacitance current is increased so that the effective series coil-capacitance c is raised. This tends to counter the effect of the earth capacitances, for it is obvious that, if $c \gg C$, the voltage distribution under surge conditions must approach uniformity.

14. **Mechanical Stresses.** When a transformer is loaded, the primary and secondary ampere-turns are in magnetic opposition with reference to the core, but act cumulatively with respect to the space between them (F.3. 76). The effect of this is twofold: the magnetically-excited state of the inter-coil space gives rise to—

(a) Leakage flux: i.e. flux linking one or other winding;

(*b*) Mutual forces between windings.*

The mechanical repulsive force with normal load currents is low compared with the strength of the coils. Under fault conditions, particularly those of dead short-circuit, the forces produced may be increased several hundreds of times. Thus a transformer with a 0·05-p.u. reactance will carry a fault current initially 20 to 40 (or even occasionally more) times full-load current, with disruptive forces consequently 400 to 1 600 times those produced by full-load current.

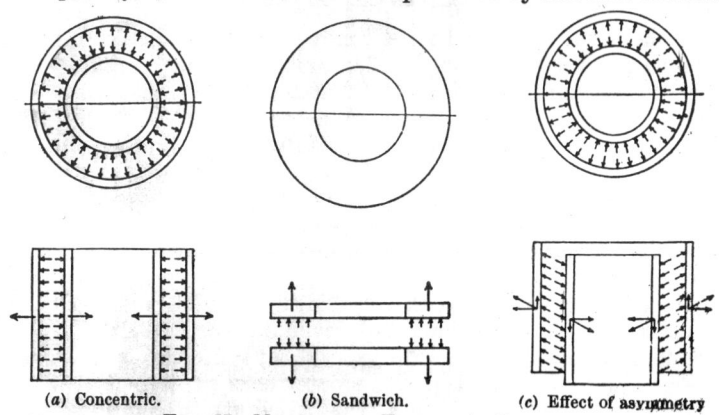

(*a*) Concentric.   (*b*) Sandwich.   (*c*) Effect of asymmetry

FIG. 77. MECHANICAL FORCES ON COILS

The forces produced may be considered as capable of analysis into the components—

(1) Radial, tending to burst outer and crush inner windings;
(2) Axial compression; and
(3) Unbalanced axial forces due to asymmetry.

RADIAL FORCE. Any electric circuit when carrying current develops mechanical forces, tending to enlarge its area. For a coil, this implies a tendency to become circular, provided that there is no magnetic asymmetry. The mechanical force is developed by the coil current interacting with the axial components of its own flux.

Considering a transformer with concentric coils, the flux initiating the mechanical forces occupies the space *between* the coils. Consequently the outer coil is subjected to internal pressure tending to burst it, but the inner coil is subjected to external pressure, and tends to collapse on to the core. A circular coil section is preferable in the latter as well as in the former case, as being the strongest shape mechanically for withstanding these pressures. Fig. 77 (*a*) illustrates the action of the hoop-stress and counter forces.

In shell-type transformers the coils are rectangular, but braced by enclosure within the core for a considerable part of their length.

* Norris, "Mechanical Strength of Power Transformers in Service," *Proc.I.E.E.*, 104(A), p. 289 (1957).

For an instantaneous current i in the T turns of one winding, the flux density in the annular duct between the cylindrical primary and secondary coils of a core-type transformer (Fig. 99 (a)) is $B_a = \mu_0 iT/L_c$. Its mean value in the windings is $\frac{1}{2}B_a$. The mean radial force per circumferential m. length and per m. of axial depth of coil is therefore $p = \frac{1}{2}\mu_0(iT/L_c)(iT/L_c)$. For the whole area L_cL_{mt} of the cylindrical face of a coil

$$p_r = \frac{1}{2}\mu_0(iT)^2(L_{mt}/L_c) \text{ newtons.}$$

On the worst fault conditions, i may reach twice the symmetrical peak short-circuit current, i.e. $2/\varepsilon$ times full-load current, where ε is the per-unit impedance. In a 100-kVA. transformer with 0·04 p.u. impedance the radial force may reach 40 tons: in a 100-MVA. transformer with 0·1 p.u. impedance it may approach 3 000 tons.

AXIAL COMPRESSION. So far, reference has been made only to stresses produced by the axial leakage flux. The radial components of the flux, which cross the windings chiefly at the ends (Fig. 76), will give rise to axial compressive forces tending to squeeze the windings together in the middle. With a symmetrical arrangement of windings these stresses are unimportant, even on short circuit.

Sandwich coils produce axial compressive stresses as illustrated in Fig. 77 (b). In shell types, the outer coils experience repulsion which is taken up by the core in the buried length, but those parts of the coils exterior to the core will have to be braced.

STRESSES DUE TO ASYMMETRY. Hitherto only symmetrical cases have been considered: i.e. windings symmetrically placed with respect to each other and of the same length. Ideal symmetry is never actually attained in practice, while the demand for wide tapping ranges (involving some parts of the various windings being out of circuit) makes complete symmetry quite impossible.

Considering concentric windings, Fig. 77 (c), if one is displaced relatively to the other, the currents give rise to unbalanced axial forces tending to increase the asymmetry. These stresses are transferred directly to the end supports and clamping devices, and increase with the amount of difference in length.

Thus in Fig. 78 (b) and (c), end effects will obtain, although considerably less than in case (a). It should be observed that the forces are so directed as to tend to increase the degree of asymmetry of the geometrical arrangement of the coils.

Cases (a) and (c) in Fig. 78 are typical respectively of end and centre-tapped coils. The mechanical superiority of (c) over (a) is obvious, but at the same time the use of tappings at all is an added complexity in the mechanical design. Internal faults, involving short-circuits between parts of a winding or tapping connections, may produce very large currents in restricted parts, and will increase asymmetry and mechanical stresses in consequence.

The self-compression effect is related to the radial force p_r in the ratio a/L_c. Additional forces due to asymmetry can be roughly estimated from the cross-flux developed by the out-of-balance m.m.f. $k(iT)$, Fig. 79. The windings are deemed to be enclosed between iron surfaces separated by a distance $2x = 2(a + b_1 + b_2)$. The cross-flux density has the maximum value $\mu_0 k(iT)/2x$. For the case shown the axial force will be that produced by the mean cross-flux density (one-half the maximum) acting on T turns, i.e.

$$p_a = \tfrac{1}{2}\mu_0 k(iT/2x) . L_{mt} . (iT)$$
$$= \tfrac{1}{2}\mu_0 k(iT)^2 L_{mt}/2(a + b_1 + b_2) \text{ newtons.}$$

The stress can become very large if the asymmetry is marked.

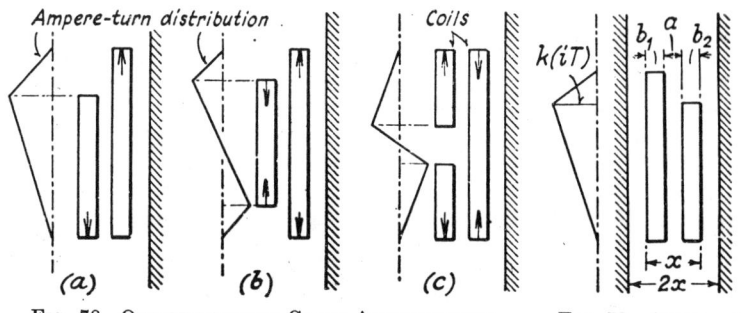

FIG. 78. OUT-OF-BALANCE CROSS AMPERE-TURNS   FIG. 79. AXIAL OUT-OF-BALANCE

BRACING. In so far as the stresses are inwards axially, they are taken by the windings themselves. Inward radial stresses are passed to the formers, packing pieces and cores. Outward axial stresses must be withstood by the end insulation. A well-constructed transformer will have a suitable choice of conductor dimensions and interturn insulation, and coils well supported and braced, with the compressive stresses kept in view. The end supports are not easily arranged to give good insulation and at the same time great mechanical strength, so that as far as possible the need for the latter quality must be avoided by maintaining symmetry, or by a suitable distribution of the unpreventable asymmetry caused by tappings, connections, etc., as indicated in Fig. 55.

15. **Low-voltage Transformers.** The use of the electric arc furnace has become increasingly common in recent years, particularly in the chemical and metallurgical industries. Such furnaces require large current at low and variable voltage, and demand special considerations in transformer design. Where possible, a shell design is used, as the windings on the low-voltage side can be sectionalized with greater uniformity of impedance. A three-phase furnace requiring

a single turn per secondary phase may, however, be of the core type. Similar considerations apply to electro-chemical loads.

The usual parallel connection of the secondary coils (on account of the large currents concerned) demands an even distribution of the leakage reactance on both primary and secondary sides: otherwise circulating currents may flow between parallel-connected sections.

The secondary is frequently constructed of sheet-type single turns, kept thin to avoid excessive eddy-currents. The turns are stamped to shape from copper sheet. The phases are led out with the connections arranged for minimum leakage. The reactance of leads may be seriously large, particularly if the transformer is some distance from the furnace.

Voltage regulation is provided by tapping the high-voltage side, to avoid complicated heavy-current connections. The tappings must be arranged symmetrically to avoid unbalanced mechanical forces in case of a short circuit—a not infrequent occurrence.

16. **High-voltage Testing Transformers.** At present h.v. testing technique requires four kinds of h.v. supply—

(a) Industrial-frequency alternating voltages up to 1 000–2 000 kV., or less;

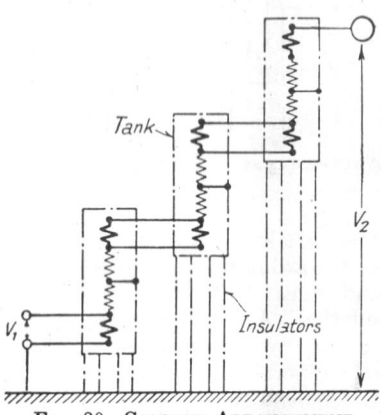

FIG. 80. CASCADE ARRANGEMENT OF H.V. TRANSFORMERS

(b) Constant direct voltages;

(c) High-frequency alternating voltages up to about 1 000 kV. at about 100–200 kc/s.;

(d) Surge or impulse voltages up to about 1 000 kV. or more, and of duration a few micro- or milli-seconds.

Type (a) is most common, being applied to routine tests of materials, machines and apparatus, and for use in connection with Schering bridge measurements. Type (b) is employed almost exclusively for the testing of cables on site (where portability is important), for the production of X-rays, and feeding the anodes of high-speed cathode-ray oscillographs. High-frequency generators (c) are required for testing at radio frequencies, and for investigations on porcelain insulators. Type (d) is required for the experimental investigation of transient disturbances on transmission lines due to lightning, switching, etc. (see § 13), particularly surge-voltages of steep wave-front.

All these supplies involve the use of transformers.

INDUSTRIAL-FREQUENCY TESTING TRANSFORMERS. The design and construction of h.v. testing transformers depends on the service

FIG. 81. MILLION-VOLT TESTING TRANSFORMER AT THE NATIONAL
PHYSICAL LABORATORY

(*Ferranti*)

conditions, which will usually be such as to involve intermittent use, with h.v. discharges on the output side amounting to a partial short circuit. It is possible to produce more than half a million volts (r.m.s.) from a single unit, but the problems of construction and transport may necessitate a cascade connection at voltages exceeding

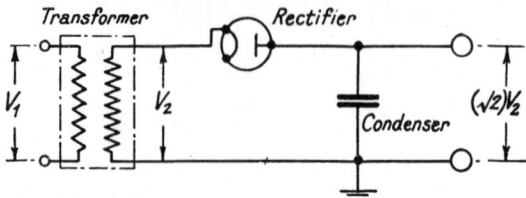

FIG. 82. HALF-WAVE RECTIFIER FOR HIGH DIRECT VOLTAGES

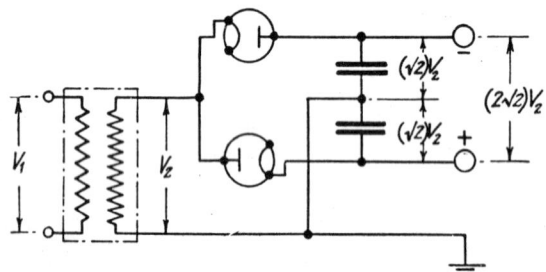

FIG. 83. FULL-WAVE RECTIFIER FOR HIGH DIRECT VOLTAGES

this figure. A typical arrangement is shown diagrammatically in Fig. 80. The mid-points of the transformers are connected to the tanks, and each end of the winding is suitably tapped and insulated to permit of one end being used for magnetizing and primary load current, the other end for the magnetizing and load current of the next transformer. All transformer tanks have to be insulated from earth (e.g. by mounting on bakelized paper cylinders), but the winding insulation has only to suffice for one-half of the voltage of the unit to the core. The three transformers of Fig. 80, if designed each for 333 kV. and insulated for 167 kV. to core, would in cascade produce a million volts to earth. Fig. 81 shows a typical three-unit h.v. testing transformer.

The cascade connection simplifies the insulation of the transformers, but the small portions of the windings which carry both magnetizing and load currents on the input sides give considerable reactance, and compensating windings may be necessary to maintain the voltage on load.

When a h.v. transformer is on open circuit, the secondary (h.v.)

side may have sufficient capacitance to cause a charging current to circulate, the counterpart of which has to be supplied to the primary together with the magnetizing current. It is quite possible for the

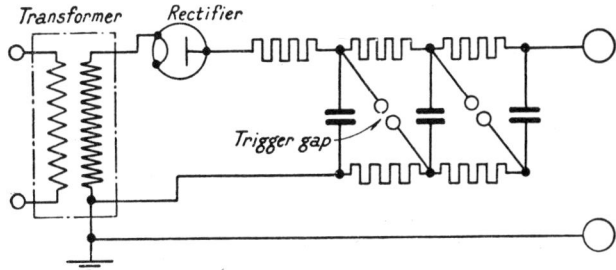

FIG. 84. SURGE GENERATOR

charging current to exceed the magnetizing current, so that the no-load current appears to be in leading phase quadrature with the applied primary voltage.

H.V. transformers are usually of the core type, since the insulation of the windings presents a simpler mechanical and electrostatic problem, and bakelized paper cylinders are used to a considerable extent to build up the concentric windings.

TRANSFORMERS FOR OTHER TESTS. In the three further h.v. supplies for testing listed on p. 112, the transformer is an essential feature, but in these cases its output is rectified, usually by means of h.v. thermionic diodes. The production of *high direct voltages*

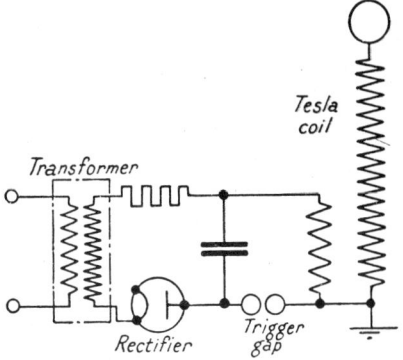

FIG. 85. HIGH-FREQUENCY GENERATOR

is illustrated in Figs. 82 and 83, which show half- and full-wave rectification, the capacitors being employed to maintain the output voltage (which has the open circuit values given in terms of the transformer secondary r.m.s. voltage). Since the rectifiers are at a high voltage to earth, their cathode supplies must be insulated, or (more usually) h.v. filament transformers used with secondaries well insulated from the primaries.

A similar arrangement (Fig. 84) is necessary for the *surge generator* in which a bank of h.v. capacitors is charged through a rectifier from the secondary of a h.v. transformer. When the voltage across the capacitors has risen to a pre-set value, the trigger gaps spark over together, connecting the capacitors momentarily in series and raising the voltage of the output terminals to a high value. If there is a circuit (not necessarily a metallic circuit, and usually an insulator) across the output terminals, the energy of the capacitors is released therein. A typical test specimen would be an artificial overhead line, or cable.

For the generation of *high-frequency* voltages the transformer (Fig. 85) charges a capacitor through a resistance and rectifier. When the voltage developed across the capacitor is sufficient, the trigger gap breaks down, and connects the primary of a Tesla coil in series with the capacitor, which then initiates a h.v. oscillatory resonant discharge. The oscillations are transformed up in the secondary of the Tesla coil. The latter may comprise two concentric coreless coils of spaced bar-conductor. The frequency developed depends upon the inductance and capacitance of the primary side, the degree of coupling, and the load on the secondary of the Tesla transformer.

17. **Transformer Protection: Buchholz System.** While it is not intended here to deal with protective devices for transformers in general (such as voltage-balance, core-balance and similar differential systems operating electrically), the Buchholz protection deserves attention, as being intimately concerned with the structure of the transformer and the non-electrical effects of fault conditions in it.

The Buchholz system is applicable to oil-immersed transformers— the great majority—and depends on the fact that transformer breakdowns are always preceded by the more or less violent generation of gas. A broken joint, for example, produces a local arc, and vaporizes the oil in the vicinity. An earth fault has the same result. Sudden short-circuits rapidly increase the temperature of the windings, particularly the inner layers, and pocketed oil is there vaporized. Discharges due to insulation weakness, e.g. by deterioration of the oil, will also cause oil dissociation accompanied by the generation of gas. Core faults, such as short circuits due to faulty core-clamp insulation, produce local heating and generate gas.

This generation of oil vapour or gas is utilized to actuate a relay which in turn signals the fault and cuts the transformer out of circuit. The relay is a hydraulic device, arranged in the pipe-line between the transformer tank and the separate oil conservator, Fig. 80 (a): the relay is shown in greater detail at (b). The vessel a is normally full of oil. It contains two floats, b_1 and b_2, which are hinged so as to be pressed by their buoyancy against two stops. If gas bubbles are generated in the transformer due to a fault, they will rise and traverse the pipe-line towards the conservator, and will be trapped in the upper part of the relay chamber, thereby displacing the oil and lowering the float b_1. This sinks and eventually closes

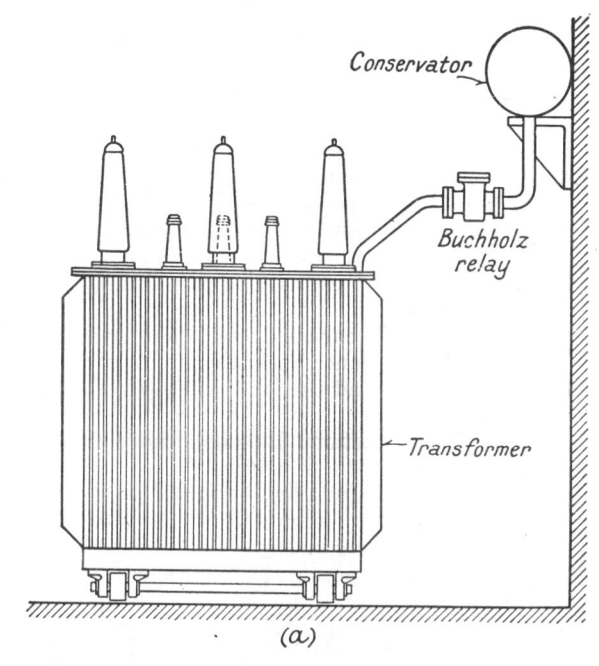

Conservator

Buchholz
relay

Transformer

(a)

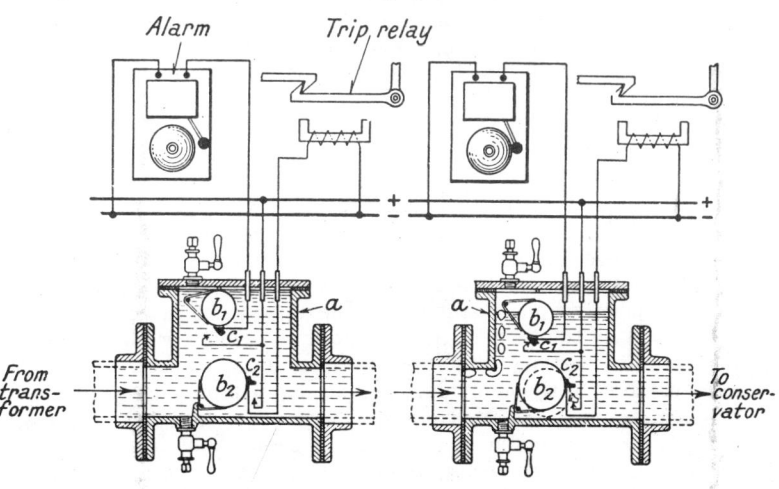

Alarm   Trip relay

+   −   +

From
trans-
former

To
conser-
vator

a b_1 c_1 c_2 b_2 a

(b)

FIG. 86. BUCHHOLZ PROTECTION

an external mercury contact c_1 (shown in Fig. 86 (b), however, as an inside contact, for simplicity), which operates an alarm.

A small window in the wall of the vessel shows the amount of gas trapped and its colour. From the rate of increase of gas an estimate

Colour	Cause
White　.　.　.	Destroyed paper
Yellow　.　.　.	Damaged wood
Black or grey　. 　.	Dissociated oil

can be made of the severity and continuance of the fault, while from the colour a diagnosis of the type of fault is possible. (See Table above.) A more definite diagnosis may be made by sampling the gas withdrawn at the stop-cock.

If the rate of generation of gas is small, the lower float b_2 is unaffected. When the fault becomes dangerous and the gas production violent, the sudden displacement of oil along the pipe-line tips the float b_2 and causes a second mercury contact c_2 to be momentarily closed, making the trip-coil circuit and operating the main switches on both h.v. and l.v. sides.

Gas is not produced until the local temperature exceeds about 150° C. Thus momentary overloads do not affect the relay unless the transformer is already hot. The normal to-and-fro movement of the oil produced by the cycles of heating and cooling in service is insufficient to cause relay operation. With new transformers, especially those in tanks provided with cooling tubes, there may be a slow release of occluded air which collects subsequently in the relay chamber.

CLASSIFICATION OF TRANSFORMER PROTECTION. The Table* on page 119 summarizes the causes of possible disturbance in transformer operation, with types of protection available.

18. **Impedance of Three-Phase Transformers to Phase-Sequence Component Currents.** The impedance of a transformer is independent of the phase-sequence of the applied voltage. Consequently its impedance to negative phase-sequence currents is the same as to those of positive phase-sequence. The impedance is that usually understood by the term in two-circuit transformers, while in transformers with tertiary windings the impedance concerned is that corresponding to the equivalent circuit Fig. 61D on page 92.

In determining the impedance to zero-phase-sequence currents, account must be taken of the winding connections, earthing and also in some cases the constructional type. For there to be a path for zero-phase-sequence currents implies a fault to earth and the flow of balancing currents in both circuits of the transformer.

---

* Szwander, *Met.-Vickers Gaz.*, 22. p. 349 (1944).

Origin of disturbance	Transformer features concerned	Type of protection	Kind of operation	Remarks
OVERCURRENTS **A.** *Overloads* exceeding permissible value	Deterioration of insulating material	(a) Winding temperature supervision	Alarm, isolation or temperature indication	Allows full use of transformer overload capacity.
		(b) Oil temperature supervision		Overload protection insufficient. Winding temperatures may be excessive
		(c) Time-delayed over-current fuses or relays	Isolation	Full overload capacity not used. Isolation too rapid on high and too slow on low overloads
B. *External Faults* 1. Large temporary currents	Mechanical	System protection	Isolation of faulty network	Transformer impedance should be large to limit fault current
2. Lower sustained fault currents	Thermal	Transformer overload		
C. *Internal Faults* 1. Large fault currents	Mechanical	Buchholz	Alarm and isolation	Gas ormation anticipates development of more serious faults
2. Small fault currents		Fuses or relays		Fault current must increase to operate. High setting necessary for discrimination
	Thermal	Restricted earth-fault		Full discrimination. Response to earth-faults only. Instantaneous. Each winding requires protection
		Balance		Full discrimination. Set tings independent of external factors
OVERVOLTAGES **A.** *External* 1. Atmospheric	End-turn reinforcement	System co-ordination		
	H.v. screening	Non-linear resistors	Lowers surge crest voltage	Service continually maintained
	Insulation co-ordination	Protective gaps		Disturbs service continuity
		Cables from overhead lines	Lowers steepness	
2. Switching and faults	Insulation in general	Neutral-earthing of system	Lowers magnitude of disturbance	Surge protection not effective
B. *Internal* 1. Peak voltage due to secondary third harmonic	Choice of suitable connections	Delta tertiary	Eliminates third harm. voltages	Three-phase three-limb core-type transformers not troublesome
2. Resonance of third harmonic with system capacitance	System insulation			

Fig. 87(A) sets out the connections and zero-sequence conditions for the more usual types of two-circuit transformers, neglecting resistance. The reactance x_{12} refers to $x_1 + x_2'$ or $x_1' + x_2$.

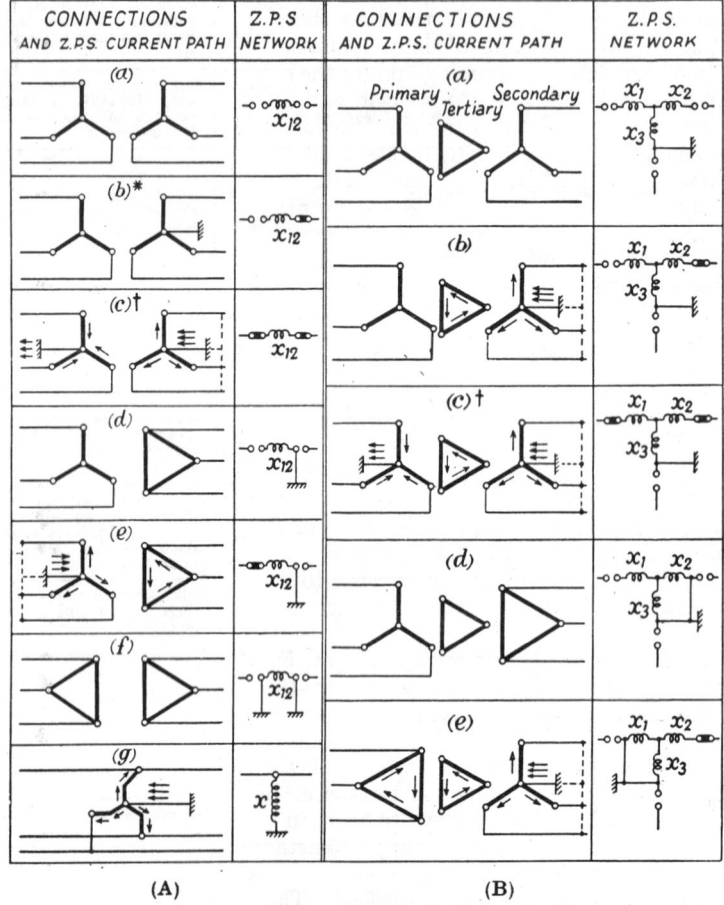

(A)   (B)

FIG. 87. ZERO-PHASE SEQUENCE CONNECTIONS
A. Two-circuit transformers
B. Three-circuit transformers
* Not applicable to 3-ph. core types.
† Assuming primary current path complete.

(a) *Yy (both neutrals isolated).* There being no earth circuit, an infinite impedance is offered to z.p.s. currents.

(b) *Yy (secondary neutral earthed).* An earth fault on the primary side will not result in zero z.p.s. current. The same is nearly true for an earth fault on the secondary side where the transformer comprises an arrangement without interlinked magnetic flux (i.e. a three-phase shell type or three single-phase units) because there is no path for the primary equivalent currents. In the case of a three-phase core-type unit, the z.p.s. fluxes produced by secondary z.p.s. currents can find a high-reluctance path like that shown in Fig. 69 for triplen harmonics, and for the same reason. The resulting z.p.s. reactance may be 30 per cent and upwards.

(c) *Yy (both neutrals earthed)*. Provided that the primary circuit is complete for z.p.s. currents, a secondary earth fault will produce z.p.s. components whose counterparts flow in the primary: and similarly for an earth fault in the primary circuit. The reactance per phase to z.p.s. currents is three times the reactance of the three-phase windings in parallel, i.e. the reactance of one phase. It is therefore identical with the reactance to p.p.s. and n.p.s. components. With three-phase core-type transformers the effect described under (b) above may reduce the reactance to z.p.s. components by about one-tenth.

(d) *Yd (neutral isolated)*. No z.p.s. currents can flow to or from the primary or secondary circuits, but can circulate within the delta.

(e) *Yd (neutral earthed)*. For an earth fault in the secondary circuit the reactance to z.p.s. is infinite. For a primary circuit earth fault compensating currents can flow in the delta secondary. The impedance to z.p.s. is as for (c) above.

(f) *Dd*. There can be no currents to or from either side to the associated circuit, and there will be no delta circulating z.p.s. currents.

(g) *Z earthing transformer*. To z.p.s. currents the two windings on each limb have cancelling ampere-turns, and the impedance thereto is that due to leakage. For p.p.s. or n.p.s. currents, neglecting magnetizing current, the connection offers infinite impedance.

Fig. 87(B) summarizes the more usual cases of three-winding transformers with delta-connected tertiaries. The symbols x_{12}, x_{23} and x_{13} have the significance given to them in §11, page 91.

(a) *Yy (both neutrals isolated)*. Reactance to z.p.s. currents is infinite.

(b) *Yy (secondary neutral earthed)*. There is no path for z.p.s. currents for a fault in the primary circuit. For a secondary circuit earth fault compensating currents can flow in the tertiary delta, the reactance being x_{23}.

(c) *Yy (both neutrals earthed)*. Z.p.s. currents can flow with a fault on either primary or secondary circuit because the tertiary can carry compensating currents. Currents can also flow in the unfaulted primary or secondary provided that the z.p.s. circuit is complete, modifying the total z.p.s. reactance in accordance with the additional parallel component.

(d) *Yd* (*neutral isolated*). There are no paths for z.p.s. currents.

(e) *Dy* (*neutral earthed*). Z.p.s. currents can flow when an earth fault occurs on the secondary circuit, with compensating currents in both primary and tertiary. The reactance to z.p.s. currents is $x_2 + (x_1 x_3 / x_{13})$. For faults in the primary circuit the z.p.s. reactance is infinite.

CHAPTER VÍ

TRANSFORMERS: TESTING

1. Objects of Tests. A transformer may be subjected to a range of
tests for a variety of purposes, including—
  (a) Routine tests after manufacture;
  (b) Acceptance tests, heat runs, etc.;
  (c) Specialized investigations on particular details of design, performance or operation.
It is beyond the scope of this book to consider (c), and attention
will be directed briefly to the testing for losses, performance,
reactance, etc. The theory and explanation of many of the tests
are contained in previous chapters.*

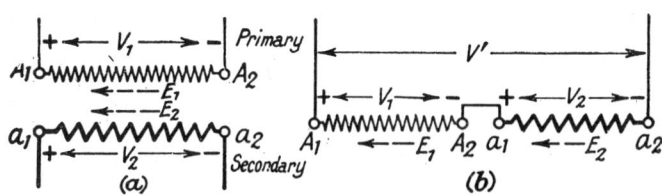

Fig. 88. Polarity Test

2. Phasing Out. In manufacture, where all connections are
traceable, the phases on primary and secondary sides may readily
be grouped. If not, all phases are short-circuited except a primary
and a supposedly corresponding secondary. A small direct current
is circulated in the primary and a voltmeter is connected across the
secondary. A momentary deflection of the voltmeter when the
primary current is made and broken confirms that the two windings
concerned belong to the same phase.

For this test all phase-ends should be separate. Difficulty may be
experienced with internally connected zigzag (interconnected star)
arrangements in three-phase windings.

3. Polarity. According to the specification No. 171: 1936 of the
British Standards Institution, terminals are distinguished by suffix
numbers in such a way that the same sequence of numbers represents
the same direction of *induced e.m.f.* in both primary and secondary
windings at any instant. Thus in Fig. 88 (a), if at any instant the
e.m.f. E_1 acts from A_2 to A_1 in the h.v. winding, E_2 will act from a_2
to a_1 in the l.v. winding: i.e. if at any instant A_1 is positive and A_2

---

* For further details of tests, see *Electrical Engineering Laboratory Manual:
Machines*, by Parker Smith (Oxford).

123

negative with respect to the applied voltage V_1, then the terminal voltage V_2 in the secondary l.v. winding will be positive at a_1 and negative at a_2.

If the two windings are connected in series by joining A_2 to a_1 (Fig. 88 (b)), and an alternating voltage V' applied across the free terminals A and a_2, then the markings are correct if V_1 across A_1A_2 is less than V'.

4. **D.C. Resistance.** Any suitable method may be used : preferably

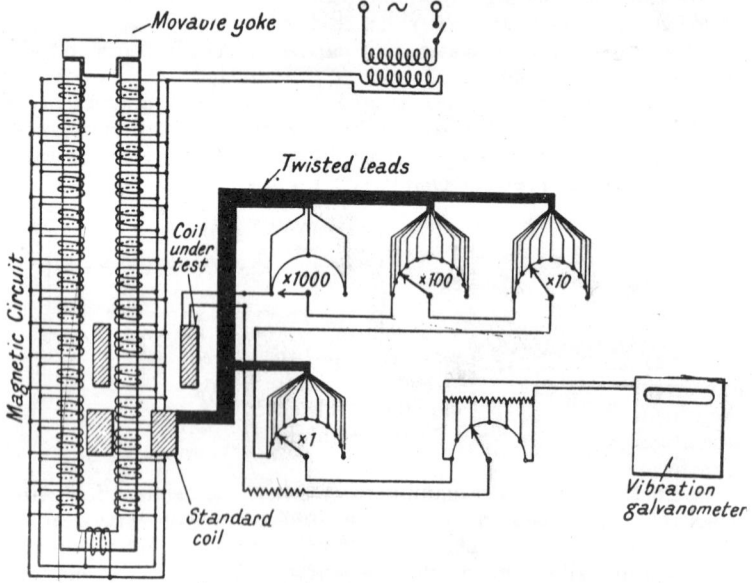

FIG. 89. COIL TESTER

a low-resistance bridge for large low-voltage transformers, in which the l.v. winding resistance may be very small. The measurement is made at known temperature, and may be used to check the design or to estimate eddy-current loss ratio.

5. **Voltage Ratio.** A direct measurement of voltage ratio by voltmeter is feasible if the voltages are reasonably low. High-voltage transformers with large ratios present difficulties in this respect, since it is necessary to measure the two voltages with entirely different instruments, the relation between them being subject to calibration error.

The voltage ratio can be found by reference to a standard transformer of the same nominal ratio. The h.v. sides are connected in parallel to a supply, and a voltmeter connected to read the voltage difference between the test and standard l.v. windings.

Most manufacturers employ bridge methods, for checking coils before assembly. The connection diagram in Fig. 89 refers to an arrangement comprising a magnetic circuit of U shape with closing yoke, energized from a suitable normal-frequency a.c. supply. The exciting coil is split into a large number of parallel parts in order to preserve a uniform induction density along the magnetic circuit. A standard coil of known turns, and the coil under test, are placed on the magnetic circuit, and the e.m.f.'s balanced on a bridge arrangement containing dials to which tappings are made to the sections of the standard coil, down to a single turn. By adjusting

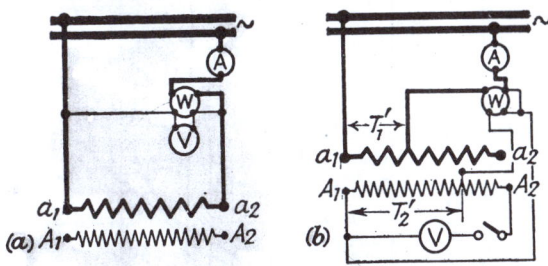

FIG. 90. NO-LOAD TEST

the dials, the number of standard tapped turns is made equal to the number of turns on the test coil, as exhibited by zero deflection on the tuned vibration galvanometer. Coils checked in this way before assembly may be safely used with certainty of accurate ratio in the transformer when built.

Assembled transformers may be checked on a *ratiometer*, which is essentially a potential divider excited from the same supply as the transformer under test, and subdivided so as to read the l.v. voltage in terms of the h.v. Balance is obtained by connecting the ratiometer tapping to the l.v. winding through an ammeter and adjusting the former until the current is zero.

6. **Magnetizing Current and Core Loss.** This test is made on a transformer complete with tank and oil, since it is performed at normal rated voltage and dielectric stress. One of the windings is left on open circuit and the other connected to a supply at normal voltage and frequency (Fig. 90 (*a*)). The l.v. winding is generally employed, as on the h.v. side the voltage may be difficult to manage and the no-load current rather small. Alternatively, if there are suitable l.v. and h.v. tappings, the method of Fig. 90 (*b*) may be used, where the voltage is increased by tapping the h.v. winding, and the current by tapping the l.v. winding. A h.v. voltmeter is necessary in order to indicate when the voltage is normal. The readings of the wattmeter have to be multiplied by the ratio T_1'/T_2'.

In either case readings of current and power at normal voltage and frequency are taken. The current comprises the magnetizing and core loss components I_{0r} and I_{0a} respectively, while the power indicated includes the core loss, the I^2R loss (which latter is usually negligible), and dielectric loss.

A test made at constant normal frequency and variable voltage V is instructive. After correcting if necessary for I^2R loss, P_0 represents chiefly core loss P_i, except with transformers for very high voltages, in which the dielectric loss may assume importance.

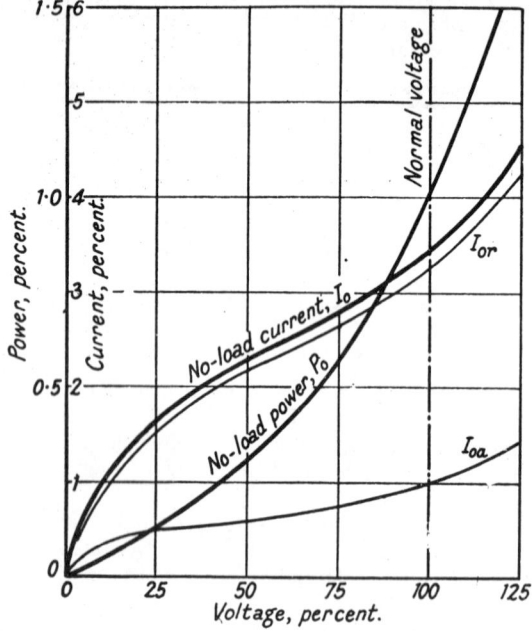

FIG. 91. NO-LOAD CURVES

The active current component is $I_{0a} = P_0/V$, the reactive (magnetizing) component is $I_{0r} = \sqrt{(I_0^2 - I_{0a}^2)}$. A plot of these gives the general shapes shown in Fig. 91. Since, at constant frequency, the flux density B_m is directly proportional to V, and the core loss roughly proportional to B_m^2, then the power/voltage curve is parabolic to a first approximation, while the active current $I_{0a} = P_0/V$ is proportional to V, roughly. The magnetizing current I_{0r} rises sharply at first at low induction densities, until the iron reaches its maximum permeability. Thereafter it rises less steeply until, before normal voltage is reached, saturation begins and the magnetizing current again starts to rise more steeply. Fig. 91 has ordinates of power in per cent of normal rated kVA. at unity power factor, and of current as per cent of rated current; it is typical of transformers of medium size (e.g. 100–1 000 kVA.).

CORE-LOSS VOLTMETER. Hysteresis accounts for 75 or 80 per cent of the core loss of modern transformers, and eddy-currents the remainder. The hysteresis loss depends only on a power of the maximum induction density B_m, so long as there is no more than one peak of density per half-cycle, but the eddy-current loss is a function of the wave-form of B, since it depends on induced e.m.f.'s. The r.m.s. value of the e.m.f. is not, however, very sensitive to harmonics.

The core loss should be measured with a sinusoidal applied voltage. To avoid the error due to changes in the peak value of the test voltage that may occur without change of r.m.s. value when the voltage is not sinusoidal, an obvious precaution is to measure the *mean* applied voltage, to which is proportional B_m and 70–80 per cent of the core loss. This follows from the relation $B_m \propto \int e \,.\, dt$ over a half-cycle. Thus when the wave-form of the applied voltage is in doubt, a nearly true measure of the core loss can be obtained by measuring the *mean* voltage applied, and adjusting it to $2\sqrt{2}/\pi$ times the rated r.m.s. value. For this, it is only necessary to employ a rectifier type of voltmeter arranged to indicate the peak value. A thermionic-valve or barrier-layer voltmeter is of use.

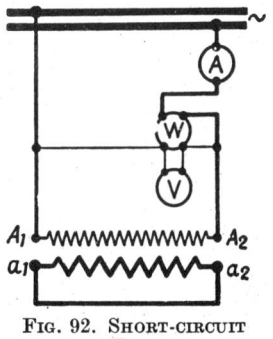

FIG. 92. SHORT-CIRCUIT TEST

7. **Impedance and I^2R Loss.** This test may be carried out with the transformer out of its tank, as the voltages are very low. One of the windings is short-circuited, and the other (usually the h.v. winding to reduce the current and increase the voltage to be measured) connected through a voltmeter, ammeter and wattmeter, duplicated where necessary for three-phase transformers, to a supply of normal frequency but only a small fraction of normal rated voltage (Fig. 92). The short-circuit connections must be carefully arranged to have as low a resistance as possible, since the l.v. windings of large transformers may have very low resistances.

A voltage V_{sc} sufficient to circulate full-rated current in primary and secondary windings (usually of the order of 5 per cent rated voltage) permits the measurement of the full load I^2R loss P_c. The wattmeter measurement includes a small core loss, usually less than $\frac{1}{2}$ per cent of that on normal voltage, so that a correction is rarely necessary.

The reactance, impedance, and regulation are calculable from the short-circuit test. Let

I = rated current per phase;

V = rated voltage per phase;

p = short-circuit watts per phase corresponding to I;

P_c = watts per phase corresponding to I in short-circuit test and corrected to 75° C.;

V_{sc} = volts per phase used in short-circuit test for current I;

E_x = reactance e.m.f. per phase (primary and secondary equivalent) at current I.

Then $E_x = \sqrt{[V_{sc}^2 - (p/I)^2]}$; $\varepsilon_x = E_x/V$; $\varepsilon_r = P_c/VI$. . (27)

The regulation is calculated by the expression of eqs. (19) *et seq.* (Chapter III, § 6.)

8. Efficiency. In practical testing it is unusual to perform a load test for either efficiency or temperature-rise: the efficiency is too high to measure accurately by any method other than that of loss-summation, and a load test for heating is uneconomical.

Using the definitions of B.S. No. 269: 1927, the losses to be accounted in the estimation of efficiency are—

Fixed loss: (a) core loss and (b) dielectric loss; measured together in a no-load test as P_i (corrected if necessary for no-load I^2R).

Direct load loss: I^2R loss in (c) primary and (d) secondary windings, computed for a mean conductor temperature of 75° C.

Stray load loss: the stray loss includes eddy-current losses in (e) the conductors and (f) other parts (e.g. tank walls, constructional parts, etc.). Losses (c), (d) and (e) are given by the corrected short-circuit test as P_c. Expressing the loss

$$\Sigma p = (a) + (b) + (c) + (d) + (e) \simeq P_i + P_c,$$

then the efficiency is $\eta = 1 - \dfrac{\Sigma p}{P + \Sigma p}$

for any load P. See also Chapter III, § 7. The loss (f) is generally small enough to be ignored.

EXAMPLE. A 300-kVA., 11 000/440-V., mesh/star, three-phase, 50-cycle transformer gave the following test results—

Open-circuit test on l.v. side at normal voltage and frequency: input 1·3 kW., 21·1 A.

Short-circuit test on h.v. side with voltage of 600 V.: input 2·80 kW., 15·0 A.; temperature of copper, 30° C. Calculate the per-unit resistance, reactance, and impedance; and the efficiency and regulation for full load at power factor 0·8 lagging.

The open-circuit test gives $P_i = 1·3$ kW.

The full-load l.v. line current is

$$I_2 = 300 . 10^3/[(\sqrt{3}) . 440] = 393 \text{ A.}$$

Thus the no-load current percentage is 100 . 21·1/393 = 5·4

The full-load h.v. line current is

$$I_1 = 300 . 10^3/[(\sqrt{3}) . 11\ 000] = 15·73 \text{ A.}$$

The loss with 15·0 A. is 2·80 kW. and with full-load current will be $p = 2·80\ (15·73/15·0)^2 = 3·08$ kW., at 30° C. = 1·03 kW. per ph. with $15·73/\sqrt{3} = 9·1$ A. per ph. The loss at 75° C. will be

$$3·08\ \frac{234·5 + 75}{234·5 + 30} = 3·6 \text{ kW., or } P_c = 3·6 \text{ kW.}$$

The per-unit resistance is
$$P_c/Q = 3\cdot6/300 = 0\cdot012 \text{ p.u.}$$
The voltage on short circuit for full-load current is $600 \times 15\cdot73/15$
$= 630$ V. From eq. (27)
$$E_x = \sqrt{[630^2 - (1\ 030/9\cdot1)^2]} = 620 \text{ V.};$$
$$\varepsilon_x = 620/11\ 000 = 0\cdot056 \text{ p.u.}$$
$$\varepsilon_r = 3\cdot6/300 = 0\cdot012 \text{ p.u.}$$
The per-unit impedance is therefore $\sqrt{(0\cdot012^2 + 0\cdot056^2)}$
$$= 0\cdot058 \text{ p.u.}$$
For full load at power factor 0·8 lagging the output is $0\cdot8 . 300$
$= 240$ kW. The losses total $1\cdot3 + 3\cdot6 = 4\cdot9$ kW. The efficiency is
$$100\left(1 - \frac{4\cdot9}{244\cdot9}\right) = 98\cdot00\%.$$
The regulation is, from eq. (19)
$$\varepsilon = 0\cdot012 . 0\cdot8 + 0\cdot056 . 0\cdot6 = 0\cdot043 \text{ p.u.}$$
for the same load and power factor.

9. **Temperature-rise.** This is measured during the performance of a *heat-run*, intended to reproduce as far as is practicable the service conditions of continuous rated load, to observe that the temperature-rises of the various parts conform to specification. The limits allowed by B.S. 171 : 1936 are—

Temperature-rise of windings by resistance,* except for windings of resistance less than 0·01 Ω, and of core and oil by thermometer—

Type	Windings Class A	Windings Class B	Oil	Core
AN, AB	55° C.	75° C.	—	As for adjacent windings
ON, OB, OW	60° C.	—	50° C.	
OFN, OFB	65° C.	—	50° C.	
OFW	70° C.	—	50° C.	

The temperature-rise limits quoted are with reference to standard ambient air or inlet water temperatures. The values adopted are: 25° C. for water; and 40° C. peak, with 35° C. as 24-hour average, for air. Transformers complying with the specification are suitable for operation continuously, 24 hours a day indefinitely, at *continuous maximum rating*. Some of the considerations involved in the thermal rating of transformers are discussed in Chapter VII.

*From the expression $R\theta_1 = R\theta_2 \dfrac{234\cdot5 + \theta_1}{234\cdot5 + \theta_2}$ where θ is in degrees centigrade.

It is noteworthy that higher limits are allowed for Class B than for Class A insulation. (See Chapter X, § 10.)

Methods of carrying out a heat-run economically, i.e. without loading the transformer normally, are—

(a) Back-to-back connection;
(b) Delta/delta connection;
(c) Equivalent open- or short-circuit run.

BACK-TO-BACK CONNECTION. The connection is shown in Fig. 93 (a) and (b). Consider the single-phase arrangement, which is

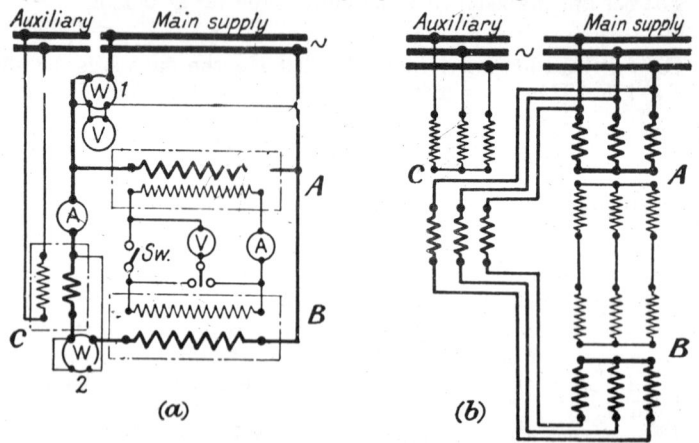

FIG. 93. BACK-TO-BACK TESTS

more obvious. Two identical transformers *A* and *B* (Fig. 93 (a)), connected in parallel on both sides to a supply of normal voltage and frequency so that their e.m.f.'s are in opposition round the closed circuits formed by the two primaries and the two secondaries. Disregarding the effect of the auxiliary transformer, then wattmeter W_1 will read the total core loss of the two transformers together. A voltmeter across the switch *Sw* should read zero, indicating complete equality of secondary voltages and phase opposition, and when it is closed, wattmeter W_1 should be unaffected.

A circulating e.m.f. is introduced into the primary circuit by means of the secondary of an auxiliary transformer *C*, energized from a suitable a.c. supply of any convenient frequency: it need not be the same as that of the main supply. The e.m.f. circulates a current, which has its counterpart in a current of equal ampere-turn value in the secondary. The e.m.f. necessary is the product of the circulating current and the impedance in primary terms. The e.m.f. and the circulating current are associated with wattmeter W_2, which consequently records the I^2R loss. The current is adjusted by control of the voltage applied to the primary side of the auxiliary transformer. By adjusting for a full-load value of the circulating current, the full-load I^2R loss is expended in the windings of the transformers under test, and at the same time the cores are energized to normal induction density. For a heat-run the wattmeters are not necessary: only a voltmeter to measure primary applied voltage, and a secondary ammeter to indicate circulating current.

If the two transformers are provided with suitable tappings, the auxiliary transformer may be dispensed with, and the circulating current produced by unbalance of the equality of e.m.f.'s when the secondaries are connected in parallel with slightly different numbers of turns.

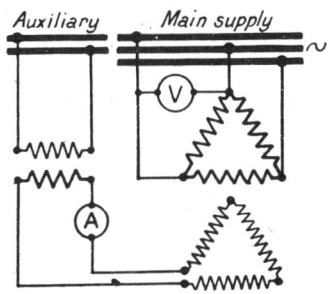

FIG. 94. HEAT-RUN

The scheme of connections for testing two identical three-phase transformers, shown in Fig. 93 (b), corresponds to the case of single-phase transformers. A tapping method may be used as an alternative where suitable taps are available.

Since the I^2R losses are provided by a circulating current, while the core losses are supplied by a current which divides into two paths in parallel, it will usually be found that one transformer is cooler than the other in a back-to-back test if the auxiliary and main supplies have the same frequency. This occurs because the phase-angles between the circulating and core-loss currents differ in the two transformer windings: thus if, in one transformer, the circulating current is nearly in phase with the magnetizing current, it will be in phase-opposition to it in the other, and the net current will be smaller, reducing the I^2R loss. A difference of temperature-rise from this cause is avoidable if the circulating current has a lower frequency: this also reduces the kVA. required.

DELTA/DELTA CONNECTION. The essential method is indicated in Fig. 94, in which the primary side is excited at normal voltage and frequency, while the secondary, connected in "open-delta," is supplied with a circulating current from an auxiliary single-phase supply of any convenient frequency. The method achieves the same result as the back-to-back connection without requiring two identical transformers. It is possible to apply the method whatever the normal internal connections if temporary alterations can be made where necessary.

EQUIVALENT SHORT-CIRCUIT RUN. Where the I^2R loss on full-load current is considerably greater than the core loss, as is usual in power transformers, an approximate test of temperature-rise may be made by a short-circuit test at a current somewhat greater than normal; so that the I^2R loss actually occurring is equal to the sum of the core and normal I^2R losses. The temperature-rise and resistance are estimated, the current adjusted to a value such that the I^2R loss when hot will be equal to the sum of the normal I^2R (hot) and core losses, and the test carried out, with adjustment of the exact value of the current near the end of the test to

$$I\sqrt{[1 + (P_i/P_c)]}$$

where I is the full-load rated current, P_i the normal core loss, and P_c the normal I^2R loss.

EQUIVALENT NO-LOAD RUN. Occasionally, as with high-frequency transformers, the core loss may be large compared with the full-load I^2R. In this case an approximation to full-load heating can be obtained by an open-circuit test at a suitable over-voltage.

10. **Insulation Tests.** "Megger" tests for insulation resistance are carried out between windings, between copper and core, and between the core-clamping bolts and the core.

A high-voltage test is made immediately after the heat-run on a new transformer. It may be a test by *direct* application of a voltage from a suitable source (e.g. a testing transformer) or an *induced-voltage* test in which the transformer is itself operated at a voltage and frequency sufficiently in excess of normal to generate in the windings a voltage of the specified magnitude.

These two tests differ radically. The former applies the same voltage to all windings under test whereas in the latter a test between turns and a graduated test between windings and core is obtained. B.S. 171: 1936 distinguishes between transformers having windings with graded insulation and with fully-insulated windings; also between the types of system on which they are to be used and their tendency to be affected by lightning. Depending upon the above and on the number of phases and the system earthing, the induced-voltage test voltages are $1 + 1.4V$, $1 + 1.6V$, $1 + 1.73V$, $1 + 2V$, $1 + 2.8V$, and $1 + 3.46V$ kilovolts, where V = rated service voltage in kV.

11. **Impulse Tests.** Voltage tests at normal frequency are not a reliable guide to the behaviour of a transformer when subjected to lightning and switching transients, for in these cases the electric stress distribution over the insulation is determined by effective capacitances whose influence at low rates-of-change is negligible. Surge phenomena have been discussed in Chapter V, § 13, with their influence on surge-voltage distribution.

Methods of *impulse testing*[*] have been developed to determine the ability of a transformer to withstand the effects of high unidirectional voltages resembling the lightning surge of Fig. 72. It is common for purchasers to specify impulse tests before a transformer is put into service. Because of its expense and complexity, which make it

---

[*] Hagenguth and Meador, "Impulse Testing of Power Transformers" *Trans. A.I.E.E.*, 71 (III), p. 697 (1952); Taylor, *Power System Transients* (Newnes).

unsuitable for routine application, the impulse-testing procedure is applied only as a "type" test. The impulse test consists in the application of a limited number of unidirectional surge voltages to one or two phases of the h.v. and l.v. windings. Typical standard test voltages, applied to the line terminals of star-connected transformers with solidly-earthed neutrals, are—

| Nominal voltage, kV.: | 3·3 | 6·6 | 11 | 33 | 66 | 110 | 132 | 275 |
| Peak test voltage, kV.: | 45 | 60 | 75 | 170 | 250 | 450 | 550 | 1 050 |

The wave of test voltage, derived from a surge-generator (such as that shown diagrammatically in Fig. 84), takes normally the forms

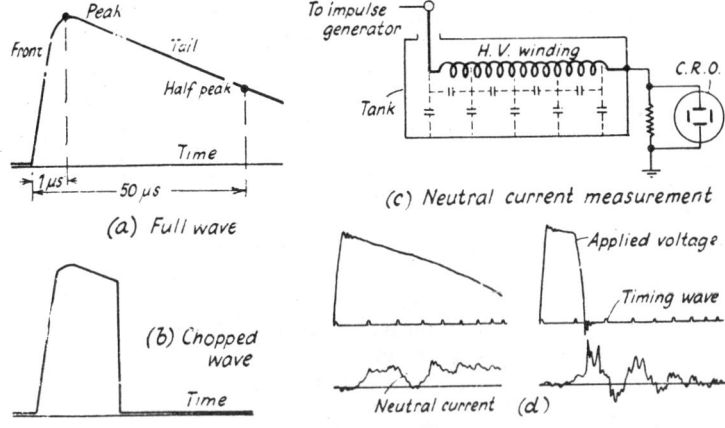

(a) Full wave

(b) Chopped wave

(c) Neutral current measurement

(d)

FIG. 95. IMPULSE TESTING

shown in Fig. 95: (a) is a "full wave," and (b) a "chopped wave" obtained by paralleling the surge generator by a rod-type spark-gap, which has a delayed breakdown. The rise time of the wavefront is arranged to be about 1 μs, and the subsequent decay to half peak value for a full wave is typically about 50 μs. An impulse test schedule might consist of: (1) a full-wave test, followed by (2) two chopped-wave tests, followed by (3) one full-wave test. For each test simultaneous oscillographic records are made of the applied voltage and the neutral current.

The neutral current may be measured by the voltage across a resistor connecting the transformer tank to earth, Fig. 95 (c). If an initial test is made at a voltage substantially below the expected impulse strength of a transformer, then any failure that occurs during the normal test routine will change the shape of the current oscillogram. With modern techniques, and considerable experience, failures can be detected with near certainty even when there are no other signs (such as noise or smoke). The position of such a failure in the winding is much more difficult to find, however. Fig. 95 (d) shows a pair of typical oscillograph records.

CHAPTER VII

TRANSFORMERS: DESIGN

1. **Frames.** The design of a transformer depends obviously on its rating, voltage, number of phases and frequency: less obviously, but no less actually, on its type, service conditions, and the relative costs of copper, iron, insulation, labour, machinery and organization. The traditional practice of the manufacturer and the experience of the designer are further factors which may exert a quite considerable influence on the product, while the requirements of stock frame-sizes may introduce restrictions on a free design. Attention must be given to flexibility, cheapness, lightness, facility for repairs and alterations, and a minimum number of stock parts. The mechanical design is at least as important as the electrical design. In the purely mechanical structure of frames, the use of rolled steel of I-, channel- and angle- section is popular on account of the fact that these sections are standardized and easily obtained.

As is usual in modern electrical manufacture, a "line" of transformers with a definite number of *frame sizes* is developed for standardization. Thus in the three-phase core type (which is more readily adaptable for the purpose) the dimensions W, D and d are fixed for a given frame, while L is varied between limits to give a range of outputs from the one frame (see Fig. 96). A typical series of oil-insulated self-cooled transformers might be arranged as follows, with five frame sizes giving a range of 10–1 200 kVA.—

Frame number:	I	II	III	IV	V
Output limits, kVA.:	10–40	40–100	100–300	300–500	500–1 200

A design might be roughed out for a mean output in each case. Some overlap of rating between consecutive frame-sizes is desirable, the choice near the frame limits being influenced by the voltage and the specification.

The design process is an experienced manipulation of several significant variables, which are interrelated in complex ways. The process can be set down in a form suitable for programming a digital computer. Such a machine can not only make the necessary arithmetical calculations, but can also take logical decisions on the results, and make modifications to fit imposed conditions. A computer can reduce design time, facilitate the incorporation of newly-developed materials, optimize dimensions, embody experience, and free the designer from routine labour. Considerable progress has been made in designing by computer. Some of the factors that must be considered are described in this Chapter.

For large transformers, outside the range of standard frames, preliminary designs are prepared on the basis of previous experience, with due regard to the economical use of materials: thus dimensions may have to be modified in order to avoid wastage of large quantities of sheet steel, or to conform to the transport loading gauge.

In service, any transformer is subject to electrical, thermal and mechanical stresses. As the size of a design is increased, the accommodation of electrical stresses (by the insulation) becomes easier, since for mechanical reasons there will be little difference in the thickness of the insulation for a given voltage, so that the insulation becomes a smaller part of the available space with increase of size. It is quite otherwise, however, with thermal stresses, as the greater masses of material provide longer paths for heat flow, as well as losses increasing as the third power of the linear dimensions. Much more difficulty is experienced in cooling large units: hence the elaboration of cooling ducts, tanks, radiators, etc.

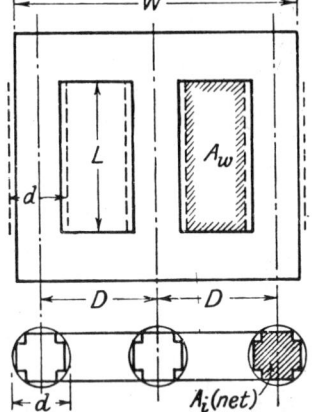

Fig. 96. Main Dimensions of Three-phase Core-type Transformer

2. **Thermal Rating.** The rating of a transformer has an almost exclusively thermal basis, the limitation being the maximum temperature of working for which the insulation will have a reasonable economic life. The critical constituent of most transformer insulations is cellulose, which deteriorates mechanically or physically at a rate determined by temperature, moisture content, electric stress, and oil purity. The most important factor is temperature. Insulating material in oil maintained continuously at a temperature not exceeding 75° C. might be serviceable for half a century, but the low limit would be uneconomical from the rating viewpoint. If the insulation were in oil continuously at 110° C. the life might be less than five years, which would obviously be expensive in replacement cost. The life/temperature relation has received some attention, but the long-term research involved has not as yet produced more than an indication of expected results.

The standard continuous maximum rating (c.m.r.) of a power transformer (B.S. 171: 1936) is based on the rises indicated in the Table on page 129, above standard reference ambient temperatures of 25° C. for water cooling, and 40° C. peak (with 35° C. as a twenty-four-hour average) for air cooling. Thus the *average* winding temperature is limited to about 100° C. (with hot-spot temperatures possibly 30–40° C. higher), and an allowance between 10° and 20° C. is made for the temperature-gradient between windings and oil.

A transformer satisfying the specification, switched on at full c.m.r. when cold (i.e. at ambient temperature) will eventually reach the temperature-rises quoted. The temperature-rise is at first rapid, then slower, and the final steady temperature is approached only after a considerable time. Fig. 146 is typical, and the analysis accompanying it shows that the temperature-rise/time curve is exponential. The heating time-constant depends on the mass, heat capacity and cooling effectiveness of the transformer: its value may lie between one and five hours. Natural cooling and high-voltage insulation produce longer time-constants.

It is clear that a transformer whose temperature-rise is below the specified limit is capable of overload for a restricted period. In any given case the temperature at a given moment may be less than the specified limiting values because: (a) the ambient temperature is lower than 35/40° C.; (b) the transformer has been off-circuit, or has been running on no load with only excitation losses, or has been running on less than rated load, or has been operating on its c.m.r. but has not yet reached the limits of temperature-rise.

Taking first the ambient air temperature or (in the case of water cooling) the inlet water temperature: if these are lower than standard, a margin for the temperature-rise is available. Practically all transformers are capable of delivering 110 per cent of full load continuously when the ambient temperature of the air is 15° C. (or inlet water 10° C.) below standard. Conversely they may have to be rated down for continuous operation at ambient temperatures higher than normal.

Whenever the transformer has a temperature-rise less than its specified limit it may on thermal grounds be subjected to overload. The time for which such overload can be sustained depends obviously on the initial temperature, and less obviously on the time-constant. The greater the mass of a transformer, the greater will be its thermal capacity, and the smaller its temperature-rise for short-time overloads. It would be possible to construct a transformer of small mass to comply with the standard rating in respect of temperature-rise of windings and oil by supplying ample tank cooling-surface and adequate oil ducts in the windings; but its performance on short-time overloads would be unsatisfactory. However, it is not correct to compare the merits of transformers in this respect solely by a comparison of their weights without a knowledge of their cooling: the ability of a transformer to withstand overloads depends on the efficiency of the cooling system as well as on the mass of material available for heat storage. But these two factors determine the time-constant. Their relative importance is a matter of the duration of the overload. The cooling settles the sustained-overload temperature-rise exclusively: the heat-capacity determines the temperature-rise caused by an overload lasting only a few minutes.

Finally, the overload that can be imposed depends on the ratio (full-load I^2R loss/core loss), because the core loss is substantially independent of the load kVA., while the I^2R loss is proportional to its square. At 200 per cent load, a transformer with a 1/1 normal loss ratio (full-load I^2R/core loss) has losses proportional as 4/1, consequently the loss is $(4 + 1)/(1 + 1) = 2\frac{1}{2}$ times full-load value. If the full-load loss-ratio were 3/1, the proportion on 200 per cent load would be 12/1, and would amount to 13/4 times full-load value. The overload losses in the two cases are as $2\frac{1}{2}$: $3\frac{1}{4}$, so that the greater the loss ratio the less capable is the transformer of sustaining overloads.

A list of overload-duration times based on all these considerations may be found in B.S. 171 : 1936.

The conclusion is that it is rational to load a transformer to a level dependent upon its temperature rather than upon the figure marked on its rating plate. A guide is sometimes given by the fitting of a *hot-spot indicator*. This device comprises a dial-type thermometer with its bulb immersed in the region where the transformer oil is hottest. A small heating coil connected to the secondary of a current-transformer serves to circulate round the bulb a current proportional to the load current, and is so designed that it increases the bulb temperature by an amount equal to the greatest winding-to-oil temperature gradient. The indicator therefore registers an approximation to the hot-spot temperature.

MOMENTARY LOAD LIMITATIONS. Transformers in service must be capable of short-circuit at normal line voltage without injury. The duration imposed (B.S. 171: 1936) is 2 sec. for a transformer with 4 per cent impedance, 3 for 5 per cent, 4 for 6 per cent, and 5 sec. for 7 per cent impedance and over. Transformers with an impedance of less than 4 per cent are called upon to withstand the effects of twenty-five times full-load current for 2 sec. The calculated copper temperature must not exceed 250° C. starting from an initial value of 90° C. for water cooling and 105° C. for air cooling. It is assumed that all the heat developed in I^2R is stored in the copper without dissipation. The *rise* is calculated from

$$\theta = at\left[\frac{2T_1 + at}{2T_1} + \frac{620\,K_d}{2T_1 + at}\right] °\text{ C.,}$$

where t = time in sec.; $T_1 = \theta_1 + 234\cdot5°$; θ_1 = initial temperature in ° C.; K_d = eddy-loss ratio at 75° C. (see page 231), and a = $0\cdot0025 \times$ loss in W. per kg. at θ_1, or in terms of the current density δ in A. per mm.^{2}, $a = 1\cdot9\delta^2 T_1 \cdot 10^{-5}$.

EXAMPLE. A 1 000 kVA. ON-type transformer has a ratio of full-load I^2R loss to core loss of $2\frac{1}{2}$: 1. On rated full load (c.m.r.) its winding resistance indicates an average temperature-rise of 24° C. after 1 hr. and 38·5° C. after 2 hr. (*a*) Estimate the final temperature-rise and the heating time-constant. (*b*) What continuous overload can be sustained when the ambient temperature does not exceed 25° C? (c) How long can 100 per cent overload be sustained for 25° C. ambient temperature with the transformer loaded (i) from cold, (ii) after continuous running on no load?

(a) Using eq. (98) and the relevant discussion, page 245,

$$24 = \theta_m(1 - \varepsilon^{-1/\tau}) = \theta_m(1 - k);$$
$$38{\cdot}5 = \theta_m(1 - \varepsilon^{-2/\tau}) = \theta_m(1 - k^2),$$

where $k = \varepsilon^{-1/\tau}$. Hence

$$\frac{38{\cdot}5}{24} = \frac{1 - k^2}{1 - k} = 1 + k = 1{\cdot}6, \text{ so that } k = 0{\cdot}6.$$

$$\varepsilon^{1/\tau} = 1/0{\cdot}6 = 1{\cdot}67, \text{ giving } \tau = 1{\cdot}96 \simeq 2 \text{ hr.}$$

The final steady conductor temperature-rise will be

$$\theta_m = 24/(1 - 0{\cdot}6) = 24/0{\cdot}4 = 60° \text{ C.}$$

(b) Allowing for a 100° C. average copper temperature the permissible rise is $100 - 25 = 75°$ C. The losses can be increased in proportion to the increase of permissible rise, i.e. by $75/60 = 1{\cdot}25$ times. On rated load the losses are proportional to $1 + 2\frac{1}{2} = 3\frac{1}{2}$. On a load x times increased, the losses will be proportional to $1 + x^2 . 2\frac{1}{2}$. Consequently the admissible overload is such that $(1 + 2\frac{1}{2}x^2)/3\frac{1}{2} = 1{\cdot}25$, i.e. $x = \sqrt{1{\cdot}35} = 1{\cdot}16$ or 116%. (For this condition B.S. 171: 1936 permits 10 per cent overload continuously, and 20 per cent overload for $3\frac{1}{2}$ hr.)

(c) (i) 100 per cent overload, $x = 2$, gives losses in the proportion $1 + x^2 . 2\frac{1}{2} = 11$, and the final steady temperature-rise would be $11/3\frac{1}{2}$ times the normal, i.e. $\theta_m = 188{\cdot}5°$ C. The limit is, however, 75° C., so that

$$75 = 188{\cdot}5(1 - \varepsilon^{-t/\tau}) = 188{\cdot}5(1 - \varepsilon^{-t/2})$$

giving, as the permissible duration, $t = 1{\cdot}01$ hr. This is about twice the nominal figure allowed by B.S. 171: 1936: but in the above calculation hot-spots are ignored.

(ii) Continuous operation with core loss only will give a rise of $1/3\frac{1}{2} \times 60° = 17°$ C. The rise when the load is applied is limited to a further $75-17 = 58°$ C. Proceeding as before

$$58 = 188{\cdot}5(1 - \varepsilon^{-t/\tau}) = 188{\cdot}5(1 - \varepsilon^{-t/2})$$

whence $t = 0{\cdot}73$ hr.

The estimates above are legitimate only in so far as the transformer operates as a homogeneous body. The greater rate of I^2R loss on overload causes changes in the oil circulation and thermal heat-transfer paths, so that the transformer cannot be assumed to be working thermally on overload precisely in the same way as on full load. Nevertheless the calculations do show the actual tendencies.

3. Output. It is sometimes necessary for a designer to work to obtain a specified efficiency. A proper design will only be obtained economically on a basis of the rating of the active materials, which is in turn dependent largely on the losses and the cooling. If the design is evolved without the restriction of obtaining a specified efficiency, the losses can be adjusted to suit the tank, or vice versa, and the efficiency will be high in any case: for it is expensive to have to dissipate an undue amount of heat.

The basis of a design is the choice of net core section A_i and net window area A_w, from an estimate of a suitable flux density B_m Wb./m.², current density δ amperes per m.², and window space-factor k_w. The essential sections are shown diagrammatically in Fig. 96 for core-type transformers. The core section A_i depends on

WINDOW SPACE FACTOR

	3 kV.	10 kV.	30 kV.	100 kV.
100 kVA .	0·28	0·20	0·14	—
800 kVA .	0·37	0·27	0·20	0·15
2 000 kVA .	0·40	0·31	0·23	0·16
10 000 kVA .	0·45	0·37	0·28	0·21

the circumscribing circle diameter d, and on the chosen core shape. The larger the transformer, the more nearly is the core shaped to the circular. It will be seen that the window area A_w has to accommodate all the turns in a single-phase unit, and twice the turns per phase in a three-phase unit. Only a small part of A_w can be filled with copper, however: the rest is occupied by insulation, packing and clearances. The higher the voltage and the smaller the transformer, the smaller will be the active area of the window, $k_w A_w$. Typical values are shown above.

For the higher outputs, the cost per kVA. at high voltages will be less than for small outputs, and it is uneconomical to build small h.v. transformers except for testing.

As with machines, there is a simple numerical relation between the output and the main dimensions and loadings. Let

Φ_m = main flux, Wb.;

B_m = maximum flux-density, Wb./m.²;

δ = current-density, amperes per m.²;

A_i = net core section, m.²;

A_w = net window area, m.²;

L = core-length, m.;

D = distance between core-centres, m.;

d = diameter of circumscribed circle round core, m.;

k_w = space-factor of window;

f = frequency, c/s.;

E_t = e.m.f. per turn, volts;

T_1, T_2 = number of primary, secondary turns;

I_1, I_2 = primary, secondary current, amperes;

V_1, V_2 = primary, secondary phase voltage, volts;

a_1, a_2 = primary, secondary conductor section, m.^{2};

L_i = mean length of magnetic circuit, m.;

L_{mt} = mean length of turn of windings, m.;

G_i = weight of active iron, kg.;

G_c = weight of copper, kg.;

γ_i, γ_c = specific cost of iron, copper, per kg.;

p_i = specific core loss, watts per kg.;

p_c = specific I^2R loss, watts per kg.

The e.m.f. per turn from eq. (6) is

$$E_t = 4\cdot44f\Phi_m$$
$$= 4\cdot44fB_mA_i \text{ volts.}$$

With a current density δ in both primary and secondary windings, the copper area in one window is, for a three-phase core-type transformer,

$$2(a_1T_1 + a_2T_2) = k_wA_w,$$

while
$$a_1T_1 = a_2T_2.$$

The primary current is

$$I_1 = a_1\delta = k_wA_w\delta/4T_1,$$

so that for the ampere-turns of primary (or secondary)

$$IT = I_1T_1 = I_2T_2 = \tfrac{1}{4}k_wA_w\delta.$$

The rating in volt-amperes is

$$S = 3V_1I_1$$
$$= 3(V_1/T_1)I_1T_1 = 3E_t \cdot I_1T_1$$

where V_1 and I_1 are phase values.

Hence

$$S = 3 \cdot 4\cdot44fB_mA_i\tfrac{1}{4}k_wA_w\delta$$
$$= 3\cdot33fA_iA_wB_m\delta k_w \text{ VA.} \qquad . \qquad . \qquad . \qquad (28)$$

In the case of a single-phase core-type transformer the active window area is

$$(a_1T_1 + a_2T_2) = k_wA_w$$

so that the output becomes

$$S = 2 \cdot 22 f A_i A_w B_w \delta k_w \text{ VA.} \qquad (29)$$

a result which is obvious if a single-phase transformer be regarded as a three-phase unit shorn of one limb.

For a single-phase shell type transformer, there is a single core with a window on each side of area $A_w = (a_1 T_1 + a_2 T_2)/k_w$. Following a similar line of reasoning, the expression for the output becomes eq. (29), identical with that for the single-phase core type since, as Fig. 9 shows, there is no essential difference between the two.

SPECIFIC LOADINGS. For 50-c/s. power transformers, flux densities of 1·1–1·4 Wb./m.² are generally employed, the limits being the losses and saturation. Whereas efficient cooling suffices to dissipate the core losses due to high induction densities, there is no simple cure for saturation (at present), which remains as a technical limitation in its production of undesirable harmonics. (See Chapter V, § 12.)

Transformers for urban and rural supply, usually termed *distribution* transformers, should be worked with lower densities on account of the core losses and magnetizing kVAr., which are important when the transformer is in continuous operation on small loads.

The distinction between distribution and power service is important. The implication in the use of these terms is that a power transformer may be cut out of circuit at times of light load, whereas a distribution transformer is continuously energized. The effect on operating costs (and therefore on the design) has already been mentioned. (Chapter V, § 4.)

The current density depends intimately on the power-loss dissipation per unit area of coil-surface in contact with the cooling medium. Vertical ducts in coils are more effective than horizontal ducts on account of their thermo-syphon effect. Thus whereas the vertical surface of a coil may be able to dissipate 1·5 kW. per m.² to the oil in contact with it, a horizontal surface may be unable to deal with more than 500 W. per m.² without overheating. Under these limitations, approximate figures for current density δ are 2·0–3·2 MA. per m.² for tanks with tubes or radiators; 5·4–6·3 MA. per m.² for transformers with forced-cooling. (These figures correspond to 2·0–3·2 and 5·4–6·3 A. per mm.²) In power transformers the load at which the unit operates is generally less than full load, so that the $I^2 R$ loss at full load is made greater than the core loss: a ratio of 2 to 4 is common. Since the $I^2 R$ loss is proportional to the square of the current density, the influence of δ on the design is considerable.

4. Allocation of Losses. If p_i be the specific iron loss and p_c the specific I^2R loss at full load in W. per kg., the total full-load loss is

$$P_i + P_c = (p_iG_i + p_cG_c)10^{-3} \text{ kW.}$$

At any fraction x of full load, the loss becomes $(P_i + x^2P_c)$ since the I^2R loss obviously depends upon the square of the load. If S be the rated output, the efficiency at load xS is

$$\eta_x = xS/(xS + P_i + x^2P_c)$$

which will be a maximum when $(d\eta_x/dx) = 0$, or when

$$(xP + P_i + x^2P_c)S = xS(S + 2xP_c),$$

that is, when $\qquad\qquad P_i = x^2P_c$. $\qquad$. $\qquad$. $\qquad$. (30)

so that the efficiency is a maximum when the (variable) I^2R loss is equal to the (constant) iron loss.

If a transformer were required to work on full load all the time it was in circuit, P_c would naturally be made equal to P_i, whence $x = 1$ and the maximum efficiency would occur at the full working load. In practice such a loading is rarely obtainable, and the load for maximum efficiency is less than full load. This controls the relative weights of material used, for if

$$x^2P_c = P_i$$

then $\qquad\qquad x^2p_cG_c = p_iG_i;$

and $\qquad\qquad G_i/G_c = x^2P_c/P_i$ $\qquad$. $\qquad$. $\qquad$. (31)

is the ratio of weights of active material for the criterion of maximum efficiency at the mean working load.

Quite different criteria might be used as a basis; for example, the *cheapest* transformer in first cost is that in which the aggregate cost of material is a minimum (neglecting any variations in construction cost). If γ_i and γ_c are the costs per unit weight of the prepared sheet steel and insulated conductors, then

$$G_c\gamma_c = G_i\gamma_i$$

or $\qquad\qquad G_i/G_c = \gamma_c/\gamma_i$. $\qquad$. $\qquad$. $\qquad$. (32)

This ratio is not as a general rule very different from that in eq. (31). If $B_m = 1.2$ Wb./m.², with $p_i = 2.5$ W. per kg.; and $\delta = 2.75$ A. per mm.², giving $p_r = 17.8$ W. per kg. at 75° C.; and $x = 0.6$, then from eq. (31)

$$\frac{G_i}{G_c} = \frac{0.36 \cdot 17.8}{2.5} = 2.56.$$

If the cost of prepared copper is 24d. per lb., and of transformer steel plates 9d. per lb.,

$$G_i/G_c = 42/9 = 2.7.$$

The cost of materials is, of course, subject to wide fluctuations, but the whole design, works and test experience of a factory cannot be upset on this account.

A third criterion, the logical one, is to design a transformer on a basis of minimum annual cost (i.e. capital charge on cost of transformer + depreciation + cost of energy losses = a minimum). But since this requires a knowledge of the cost of lost energy, the exact load curve, the load factor, etc., it is possible only with the aid of a computing machine.

5. **Design.** The basis for design, which should have reference to the questions raised in the previous section, can be related comparatively simply to the e.m.f. per turn. Since the flux Φ_m determines the core section and the ampere-turns IT fix the total copper area, the ratio $\Phi_m/IT = r$ will be constant for a transformer of a given type, service and method of construction. The output in kVA. per phase is

$$S = 4 \cdot 44fT\Phi_m I \cdot 10^{-3}$$
$$= 4 \cdot 44f(\Phi_m^2/r)10^{-3}$$

whence $\quad \Phi_m = \sqrt{S} \cdot \sqrt{(r \cdot 10^3/4 \cdot 44f)}$.

Substituting this value of Φ in the expression eq. (9) for the e.m.f. per turn,

$$E_t = 4 \cdot 44f\sqrt{S} \cdot \sqrt{(r \cdot 10^3/4 \cdot 44f)}$$
$$= K\sqrt{S} \cdot \quad . \quad . \quad . \quad . \quad . \quad (33)$$

where $\quad K = \sqrt{(4 \cdot 44fr \cdot 10^3)}$, .

and depends on type, material and labour costs, factory organization, etc.

The ratio r does not change with size in a range of transformers of given type.

If the criterion of load for maximum efficiency is applied, K assumes values of the order given in the table below when the rating is in kVA.—

Type			K
Three-phase, shell	.	.	1·3
„	core, power	.	0·6 — 0·7
„	„ distribution	.	0·45
Single-phase, shell	.	.	1·0 — 1·2
„	core	.	0·75 — 0·85

These figures apply to the mean ratings of the frame-sizes in a line of transformers, so that at the limiting ratings, the value of K will differ somewhat from the mean values.

A workable design can now be built up on a basis of experience. Choosing B_m and δ, the product A_iA_w is known from eq. (28),

E_t from eq. (33), and an appropriate value of K. A_i can be calculated from E_t and the chosen value of B_m. The core is designed to conform to a standard frame, from which d and D (Fig. 96) are obtained. The length L of the window is obtained from A_w and the width between core-circumscribing circles.

Typical core sections are shown in Fig. 23, with their appropriate

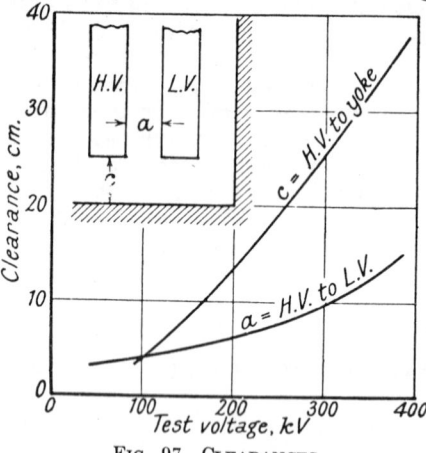

FIG. 97. CLEARANCES

net core areas A_i in terms of the diameter of the circumscribing circle. The factor 0·9 accounts for a 10 per cent loss of gross core area in plate insulation.

6. Coils and Insulation.

The constructive features of the several types of coils have already been outlined (Chapter IV). The insulation of individual turns is rarely called upon to withstand voltages greater than 75 V. under normal conditions of service so that the insulation between successive turns is largely determined by the mechanical characteristics of the insulating material. The end turns of h.v. windings must, of course, be carefully designed and insulated to withstand overvoltages (Chapter VI, § 13). The insulation to earth, between h.v. and l.v. windings, and between coil sections, is determined by the voltage for which the transformer is designed. The clearances, thicknesses, etc., for these purposes are fixed from test combined with experience. Fig. 97 summarizes typical distances and thicknesses for (a) h.v. to l.v. winding, (c) end of h.v. winding to yoke, for oil-immersed three-phase transformers. The clearance between coil sections is of the order of 1 cm. The voltage gradient should be *across* boards, oil ducts and paper layers, in which direction these materials have greatest electric strength.

Tappings, by B.S. 171: 1936, are designed for $\pm 2\frac{1}{2}$ and 5 per cent voltage adjustment.

The reinforcement of end turns depends on the rated voltage; the total number of reinforced end-turns at each end of a winding varies from 3 per cent of the total turns in lower-voltage transformers down to 1·75 per cent for 132 kV. and 0·75 for 220 kV. transformers. The practice of reinforcement, particularly for large h.v. transformers, is now closely integrated with investigations, theoretical and practical on models, into actual surge-voltage distribution.

7. **Reactance.** The estimation of reactance is primarily the estimation of the distribution of the leakage flux and the resulting line linkages with the primary or secondary coils. An accurate solution to this problem is well-nigh impossible—as is, in fact, almost every problem of magnetic field distribution in the neighbourhood of iron masses.

The distribution of the leakage flux depends on the geometrical con-

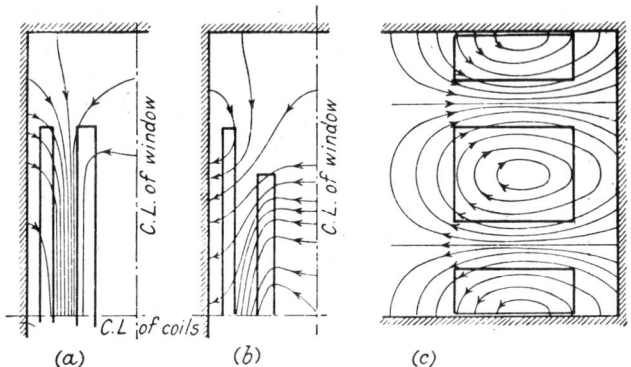

FIG. 98. LEAKAGE FLUX

formation of the coils and of the neighbouring iron masses, and also on the permeability of the latter. The diagrams in Fig. 98* show typical calculated distributions based on simplifying assumptions (such as constant permeability $\mu_r = 10$, two-dimensional symmetry, etc.). In case (a), that of cylindrical core-type coils of *equal length*, it is noteworthy how the leakage field is packed into the space between the windings, and how it runs parallel with the core for nearly the full length of the coils. Where there is an inequality in the coil-lengths, however, the field is very considerably altered, as shown in Fig. 98 (b). For the shell-type transformer with sandwich coils, Fig. 98 (c) shows a typical leakage-field distribution. In cases (a) and (c) the field is sufficiently symmetrical and geometrical to permit of considerable simplification for the sake of a usable approximate expression.

CYLINDRICAL CONCENTRIC COILS, EQUAL LENGTH. For this case the actual leakage field, e.g. Fig. 98 (a), is assumed to consist of a longitudinal flux of uniform and constant value in the interspace between primary and secondary; and a field crossing the conductors, reducing linearly to zero at the outer and inner surfaces. See the "axial component" in Fig. 76 and the AT distribution in Fig. 99 (a). Further, the permeance of the leakage path external to the coil length L_c is assumed to be so large as to require the expenditure of a

* Based on Figs. 96, 97, and 102 of Hague's *Electromagnetic Problems in Electrical Engineering* (Oxford).

negligible m.m.f.; i.e. all the m.m.f. is expended on the length L_c. The effect of the magnetizing current in unbalancing the primary and secondary ampere-turn equality is neglected.

Let (AT) be the ampere-turns of either the l.v. or the h.v. coils on one limb. Then the flux density in the duct of radial width a is $B_a = \mu_0(AT)/L_c$. Taking one-half of the total duct leakage as linking either winding (it makes little difference to the result whether this is strictly true or not), then the duct flux linking each of the T_1 primary turns is approximately $\mu_0(AT)\frac{1}{2}aL_{mt}/L_c$, and the linkages are consequently $\mu_0(AT)T_1\frac{1}{2}aL_{mt}/L_c$. Here L_{mt} is the mean circumferential length of the annular duct, nearly the same as the mean length of a primary turn.

The flux density at the radial distance x from the surface of the primary winding is xB_a/b_1. The flux in an elemental annular ring of width dx and approximate circumference L_{mt} is $L_{mt}(xB_a/b_1)dx$. This flux does not link T_1 turns, but only the outer portion xT_1/b_1: the linkages are therefore $L_{mt}B_aT_1(x/b_1)^2dx$. Summing the total linkages over the radial distance b_1,

$$L_{mt}B_aT_1\int_{x=0}^{x=b_1}\left(\frac{x}{b_1}\right)^2dx = \frac{1}{3}L_{mt}B_aT_1b_1.$$

The total linkages of the primary are

$$L_{mt}B_aT_1\left(\frac{a}{2}+\frac{b_1}{3}\right) = \mu_0(AT)T_1\frac{L_{mt}}{L_c}\left(\frac{a}{2}+\frac{b_1}{3}\right).$$

Similarly the linkages of the secondary are

$$\mu_0(AT)T_2\frac{L_{mt}}{L_c}\left(\frac{a}{2}+\frac{b_2}{3}\right).$$

The reactance of the primary is obtained from the flux linkages per ampere $\times 2\pi f$, or

$$x_1 = 2\pi f\mu_0T_1{}^2\frac{L_{mt}}{L_c}\left(\frac{a}{2}+\frac{b_1}{3}\right)\text{ ohms,}$$

since, per ampere, $(AT)T_1 = T_1{}^2$. For the secondary reactance,

$$x_2 = 2\pi f\mu_0T_2{}^2\frac{L_{mt}}{L_c}\left(\frac{a}{2}+\frac{b_2}{3}\right)\text{ ohms.}$$

From eq. (13), the total reactance in primary terms is

$$X_1 = x_1 + x_2' = x_1 + x_2(T_1/T_2)^2, \text{ which becomes}$$

$$X_1 = 2\pi f\mu_0T_1{}^2\frac{L_{mt}}{L_c}\left(a+\frac{b_1+b_2}{3}\right)\text{ ohms.} \qquad . \qquad . \qquad (34)$$

where L_{mt} may be interpreted as the mean length of turn of primary and secondary together. The per-unit reactance is

$$\varepsilon_x = I_1X_1/V_1,$$

and using the value of X_1 in eq. (34),

$$\varepsilon_x = \frac{2\pi f \mu_0 L_{mt} I_1 T_1{}^2}{L_c V_1}\left(a + \frac{b_1 + b_2}{3}\right)$$

or $\qquad \varepsilon_x = \frac{2\pi f \mu_0 L_{mt}(AT)}{L_c E_t}\left(a + \frac{b_1 + b_2}{3}\right)$ per unit . . (35)

substituting (AT) for the ampere-turns per limb of *either* coil, and $E_t = V_1/T_1$ for the volts per turn.

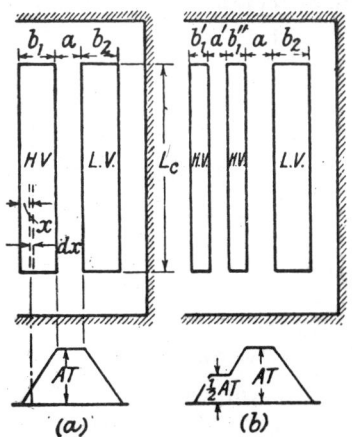

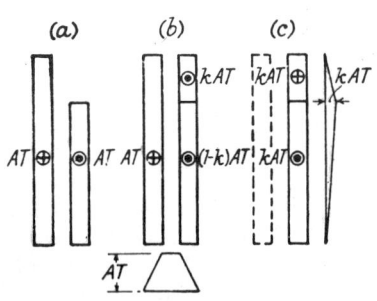

FIG. 99. PERTAINING TO THE CALCULATION OF LEAKAGE REACTANCE OF CYLINDRICAL COILS

FIG. 100. ANALYSIS OF ASYMMETRICAL CYLINDRICAL COILS

In some cases one of the windings (usually the h.v. coil) is split into two approximately equal concentric parts. Fig. 99 (*b*) shows that, with similar assumptions to those made above, the percentage reactance may be written

$$\varepsilon_x = \frac{2\pi f \mu_0 L_{mt}(AT)}{L_c E_t}\left(a + \frac{b_1' + b_{.}'' + b_2}{3} + \frac{a'}{4}\right) \text{per unit (36)}$$

CYLINDRICAL CONCENTRIC COILS, UNEQUAL LENGTHS. Fig. 98 (*b*) indicates that, compared with the case of equal length, considerable difficulty will be found in making suitable simplifying assumptions. Obviously the assumption illustrated in Fig. 99 is quite inadmissible. The leakage field depends on the proportional difference in length, and on where the difference occurs (e.g. at one or both ends, or in the middle, etc.). The case is a very common one, as it may be produced by end-turn reinforcement, tappings, or by the normal small difference of coil length usual in manufacture even in the absence of tappings, etc.

The effect of divergences on the reactance requires to be investigated for each individual case. Fig. 100 shows one method of treating the problem. The coils, each of ampere-turns (AT) are shown in section. The actual arrangement (a) can be considered as equal to the sum of (b), a symmetrical system of longitudinal ampere-turns amenable to the treatment leading to eq. (35), and (c) an asymmetrical transverse system of ampere-turns producing a cross-flux. Having determined the flux distribution due to each of the systems (b) and (c) as in Fig. 76, the reactance is determined from the e.m.f.'s induced in the actual windings due to each of the flux distributions.

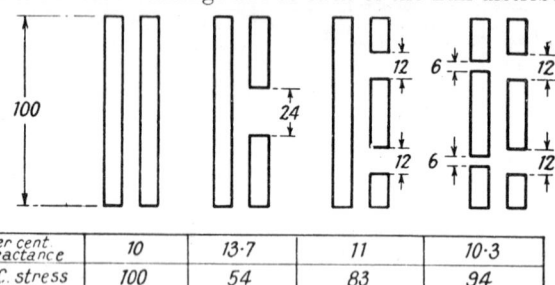

Per cent reactance	10	13·7	11	10·3
S.C. stress	100	54	83	94
Axial stress	0	18	9	5·3

Fig. 101. Arrangements of Asymmetrical Cylindrical Coils

In Fig. 101, a few typical cases are given, with figures of percentage reactance compared with that with symmetrical concentric coils of equal length.

SANDWICH COILS. Fig. 102 shows the simplified case of Fig. 98 (c). It is usual to arrange the winding as in Fig. 29 (b), where the end coils have one-half the turns of the remainder. There will be n h.v. coils, $(n-1)$ l.v. coils of equal ampere-turns, and two l.v. half-coils. Taking two adjacent half-coils as a unit, the reactance of such an arrangement is, by analogy with eq. (34),

$$X_{1n} = 2\pi f\mu_0\left(\frac{T_{1n}}{2}\right)^2 L_{mt}\frac{1}{w}\left(a + \frac{b_1 + b_2}{6}\right) \text{ ohms.}$$

where T_{1n} is the number of turns per *whole* section. For the n sections and $2n$ units, the total reactance is

$$X_1 = 2\pi f\mu_0 2n(T_{1n}/2)^2 L_{mt}\frac{1}{w}\left(a + \frac{b_1 + b_2}{6}\right)$$

$$\simeq \pi f\mu_0 \frac{T_1^2}{n} L_{mt}\frac{1}{w}\left(a + \frac{b_1 + b_2}{6}\right)$$

since $nT_{1n} = T_1$. The effect of subdivision is apparent. The per-unit reactance is

$$\varepsilon_x = \frac{\pi f\mu_0 L_{mt}(AT)}{wnE_t}\left(a + \frac{b_1 + b_2}{6}\right) \text{ per unit} \tag{37}$$

(AT) is the full phase ampere-turns of either primary or secondary.

8. **Mechanical Forces.** The general discussion in Chapter V, § 14, gave for the radial force on a cylindrical coil the expression

$$p_r = \tfrac{1}{2}\mu_0(iT)^2(L_{mt}/L_c) \text{ newtons} \qquad . \qquad . \qquad . \ (38)$$

The stress produced can usually be withstood by the copper provided that the coil is cylindrical. The axial force due to asymmetry was shown to be

$$p_a = \tfrac{1}{2}\mu_0 k(iT)^2 L_{mt}/2(a + b_1 + b_2) \text{ newtons} \qquad . \qquad . \ (39)$$

for an instantaneous current i. For the cases shown in Fig. 101 the resultant axial stress will be lower because the oppositely-directed terms partially neutralize. The two bottom rows of figures give stresses in terms of the radial stress in the symmetrical case.

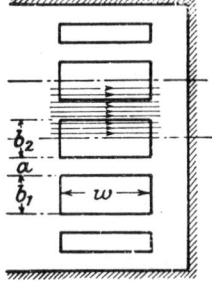

FIG. 102. LEAKAGE OF SANDWICH COILS

9. **Magnetizing Current.** In a single-phase transformer, the ampere-turns $I_0 T_1$ provided by the primary winding on no load (and by the resultant of $I_1 T_1$ and $I_2 T_2$ on load) require to be sufficient to produce the working flux in the complete magnetic circuit. The maximum flux-density B_m in the core is chosen as a compromise between the reduction of the length per turn of the winding and the saturation of the core. A density of $1 \cdot 4$ Wb./m.² is typical of large power transformers: higher densities than this tend to require excessive magnetizing current and introduce undesirable harmonics into this current. A lower density is generally used in the yokes of core-type transformers, as here it is possible to reduce the m.m.f. and core losses without adversely affecting the length of the copper. A 25 per cent increase may be used.

The calculation of the magnetizing current must take account of the small gaps between the core-plates unavoidable with the constructional methods adopted. If l_g is the total equivalent length of such gaps, then the r.m.s. magnetizing current is

$$I_{0r} = (at_c l_c + at_y l_y + 800\ 000 B_m l_y)/(\sqrt{2})T_1$$
$$= at_m L_i/(\sqrt{2})T_1 \qquad . \qquad . \qquad . \qquad . \qquad . \ (40)$$

where at_c, at_y are the ampere-turns per cm. length of core and yoke respectively with the flux-densities employed; l_c, l_y are the lengths of those parts; T_1 is the number of primary turns; L_i is the mean core length and at_m a mean value for the ampere-turns per m. length. The calculation is performed with maximum densities, and on the assumption that the current is sinusoidal (a very rough approximation), the r.m.s. value is obtained by dividing by $\sqrt{2}$.

If the core loss be P_i, the active component of the no load current is

$$I_{0a} = P_i/V_1 \qquad . \qquad . \qquad . \qquad . \qquad . \qquad (41)$$

It is not usual actually to calculate the no-load current or its components. In a transformer of normal design, I_0 will be of the order of 5 per cent of the full-load current, and considerable variation in it may be made without affecting the performance of the transformer on load. The losses, however, are important in determining the efficiency, and the harmonics in the magnetizing current may

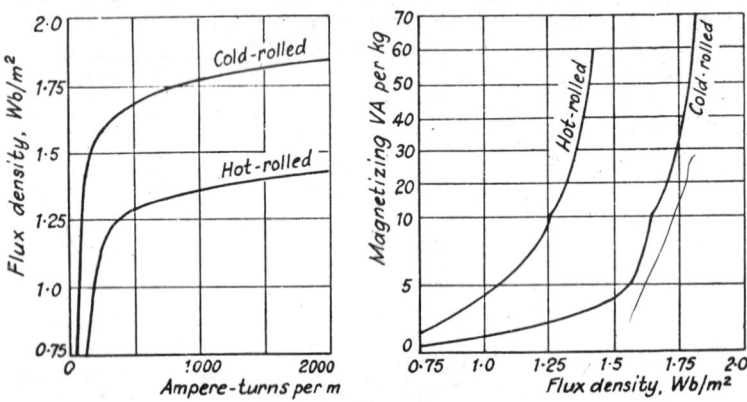

FIG. 103. B/at AND MAGNETIZING VA. CURVES FOR 0·35 MM.
TRANSFORMER STEEL

cause trouble in transmission and distribution transformers on light loads or no-load conditions.

A knowledge of the B_m/at curve for the core material, and of the total weight of iron, permits the estimation of the magnetizing current from curves. Since the primary e.m.f. is

$$E_1 = 4{\cdot}44fT_1B_mA_i \text{ volts,}$$

where A_i is the net core section, then the *magnetizing volt-amperes* will be

$$E_1I_{0r} = 4{\cdot}44fI_{0r}T_1B_mA_i \text{ volt-amperes.}$$

But from eq. (40), the term $I_0 T_1 = at_m I_i/\sqrt{2}$,
whence

$$E_1I_{ir} = 4{\cdot}44fB_mA_i at_mL_i/\sqrt{2} \text{ volt-amperes.}$$

The product A_iL_i is the volume of the core, and 7 500 A_iL_i is its weight in kg. Consequently the magnetizing VA. per kg. is given by

$$4{\cdot}2fB_mat_m \;.\; 10^{-4} \text{ VA. per kg.} \qquad (42)$$

Fig. 103 gives typical curves for transformer steels for 50 c/s., no allowance being made for gaps. The improvement obtained with cold-rolled grain-oriented steel is notable: it permits substantially higher flux densities than the normal hot-rolled steel, so economizing in the amount of core and winding material.

In three-phase core-type transformers each limb is concerned with

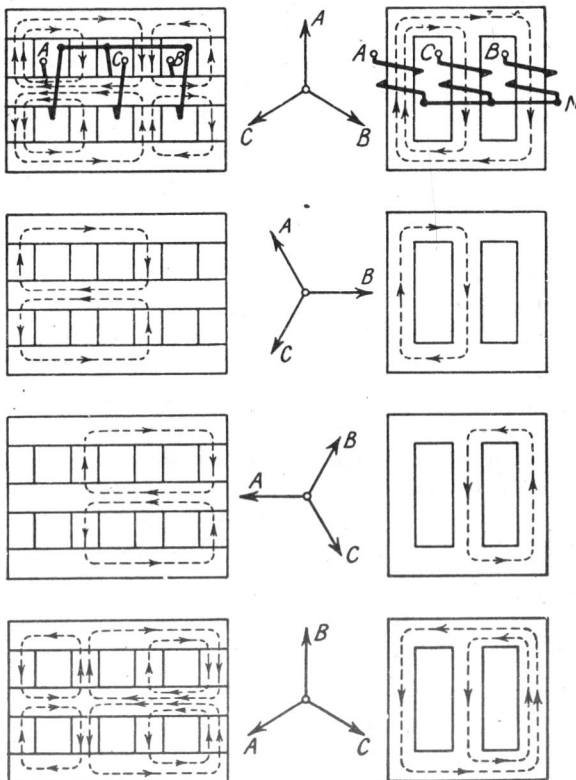

FIG. 104. MAGNETIC CIRCUITS OF THREE-PHASE SHELL- AND CORE-TYPE TRANSFORMERS

one phase, and the magnetic circuit is completed through the other two phases in parallel. The magnetizing current per phase is therefore that required for a core and part of a yoke. Since three-phase transformers have an asymmetrical magnetic circuit, the magnetizing currents differ in the several phases. With the core type the central phase has a complete magnetic circuit of rather lower reluctance than that of either of the outer phases, and in fact a test on such a transformer will show the central phase to have a few per cent less magnetizing current. Generally, the cores are long compared with the yokes, and the yokes are usually of a greater cross-section : the yoke reluctance is thus only a small fraction of the total, so that the unbalance of magnetizing current is small. With very large three-phase core types a five-limbed construction may be used.

The changes of flux are not quite so obvious in the shell type, but may be followed by a consideration of Fig. 104. It is noteworthy that the central phase winding is reversed with respect to the outer two. If this were not done, the effect would be to increase the flux in the common yokes by 50 per cent with a great increase in core loss. For shell type transformers the magnetizing current of the outer phases is of the order of 50 per cent greater than that of the middle phase.

10. **Cooling.** The great majority of transformers are oil-immersed. Self-cooling (ON type) is feasible up to 10 000–15 000 kVA. Thereafter forced cooling (OFB or OFW) is necessary, and may be advantageous at lower ratings, depending on the circumstances. Air-blast transformers (AN or AB) are rare, and have obvious voltage limitations: on the other hand, they occupy less space than transformers in tanks.

As has been shown (Fig. 39), the difference of temperature between the hottest parts of the transformer and the outside ambient air consists principally of (a) the drop at the insulation between the copper and the oil, and (b) the drop through the tank wall and skin of air to the free air. These temperature differences are due to the thermal resistance of the layers through which the heat has to be transmitted. To send p watts per m.² through a thickness t m. of material of thermal resistivity ρ_i (temperature difference in degrees centigrade per watt transmitted through a 1 m. cube between opposite faces), requires a temperature difference

$$\theta = pt\rho_i \ °C.$$

With the thicknesses in normal use, the difference may be written

$$\theta = k_i p \ °C.$$

where p is in watts per m.² and k_i has the values roughly indicated in the following table—

Particulars	k_i
D.C.C. or paper-lapped conductors . .	0·020
Bare conductors	0·017
Edge of laminations	0·017
Tank wall	0·150

It might be expected that bare conductors would cool very much more readily than covered wires, but the skin of oil in immediate contact with the conductors is slowed down by friction, and does not facilitate the conduction of heat away so readily as the circulation of the free oil would suggest. In a similar way, the air in contact with the tank walls tends to rise more slowly on account of friction, and to some extent baffles the transfer of heat to the ambient air.

TANKS. For small transformers, plain-walled tanks large enough to contain the transformer and oil have also sufficient surface to dissipate the losses. The increase of volume with the cube of the linear dimensions, together with the collateral rise of losses, results in an increase of loss to be dissipated per unit area of surface (which rises with the square of the linear dimensions, i.e. less rapidly than the volume). Above a certain size a plain tank becomes inadequate to dissipate the losses and must be augmented in some way. In Britain, tubing is employed up to sizes such that this solution is feasible, above which separate radiators are used. For larger sizes still, radiators are supplemented by forced cooling.

Tank walls, including tubes and radiators, dissipate heat by a combination of radiation and convection.* Experiment shows that natural cooling accounts for about 6 W. by radiation and 6·5 W. by convection per m.² of *plain* surface per °C. difference between tank wall and ambient air or surrounding masses : a total of 12·5 W. Thus with a temperature rise of about 35° C. for the tank walls, a plain tank will dissipate about 0·44 kW. per m².

The addition of tubes can greatly increase the total area of tank surface, but will not give a proportional increase in dissipation. This is because the tubes partially screen the tank wall, preventing some of the radiation that would otherwise take place therefrom. There will in fact, be little difference in the total *radiated* loss. The tubes will, on the other hand, improve the convection in proportion to their surface. Further, the columns of oil in the tubes give rise to more effective heads of pressure tending to circulate the oil, so that the tube surface will now be more useful, area for area, than the wall surface. Experiment shows that an addition of 35 per cent may be made to take into account the syphoning action of the tubes.

Neglecting the top and bottom of the tank, let the cooling surface of the tank be S_t m². This, per ° C. temperature difference between wall and ambient air, will dissipate

$$(6 + 6 \cdot 5)S_t \text{ W. per m.}^2 \text{ per °C.} \quad . \quad . \quad . \quad .(43a)$$

Let the total surface be increased to xS_t by the tube walls. Then the total radiation will be $6S_t$ watts as before; the convection from the walls will be $6 \cdot 5S_t$ watts, and from the tubes will be $S_t(x-1)6 \cdot 5 \cdot 1 \cdot 35$ watts. The total dissipation is therefore

$$S_t[12 \cdot 5 + (x-1)8 \cdot 78] \text{ W.}$$

* A discussion of the physical aspects of cooling is given in [B], Chapter III.

Per m.^{2} of area, the dissipation is this expression divided by xS_t, or

$$8 \cdot 8 + 3 \cdot 7/x \text{ W. per m.}^2 \text{ per } °C. \quad . \quad . \quad . \quad (43b)$$

Thus the use of tubes, although increasing the total dissipating surface, also reduces the possible dissipation per unit area. The limit of natural cooling is more quickly reached, and a forest of tubes tends to defeat its own object. A normal spacing is 5 cm. diameter tubes placed at 7·5 cm. centres, but elliptical tubes of, say, 10 cm. × 2·5 cm. give somewhat greater cooling surface for a smaller volume of oil.

11. Specification. Most transformers are built in accordance with a standard specification. The more important details of B.S. 171: 1936 are given below.*

Limits of Temperature-rise. See p. 129.

Rating is the steady kVA that can be carried continuously without exceeding the specified limits of temperature-rise. It is the continuous maximum rating and no short-time rating is recognized.

Efficiency is to be assessed by the method of loss-summation (see Chapter VI, § 8), and stated for unity power factor.

High-voltage Tests. See Chapter VI, § 10, p. 132.

Tolerances. The tolerance on voltage ratio is 0·5 per cent, or one-tenth of the percentage impedance, whichever is the less. On losses, regulation and impedance there is a tolerance of 10 per cent on the guaranteed values.

Overloads. Transformers rated in accordance with the Specification are not capable of sustaining continuous overloads.

Terminal Markings. See Fig. 46.

12. Example. 300-kVA., 6 600/400–440-V. in 2½% steps, mesh/star, 50-c/s., three-phase, core-type, oil-immersed, self-cooled power transformer, to B.S. 171: 1936. A drawing† is given in Fig. 105 and a view of the complete transformer in Fig. 106.

MAIN DIMENSIONS. The frame size for this output is based on a mean rating of 200 kVA., for which eq. (33) gives

$$E_t = K\sqrt{Q} = 0 \cdot 6\sqrt{200} = 8 \cdot 5 \text{ V.}$$

With a flux-density of 1·35 Wb./m.^{2}

$$A_t = 8 \cdot 5/(4 \cdot 44 \cdot 50 \cdot 1 \cdot 35) = 0 \cdot 0283 \text{ m.}^2$$

A three-stepped core is used, for which (Fig. 23) the net core area

$$A_t = 0 \cdot 6 \, d^2, \text{ whence}$$
$$d = \sqrt{(0 \cdot 0283/0 \cdot 6)} = 0 \cdot 217 \text{ m.} = 21 \cdot 7 \text{ cm.,}$$

---

* B.S. 171: 1936. ELECTRICAL PERFORMANCE OF TRANSFORMERS FOR POWER AND LIGHTING.

† The side view (l.v. side) in Fig. 105 shows connections appropriate for a single-layer l.v. winding. Both ends of each phase are at the top of the limb for the two-layer winding called for in the design following.

rounded off for constructional convenience to $d = 0.21$ m. $= 21$ cm.

The amended net core section is
$$A_i = 0.6 \cdot 0.21^2 = 0.0264 \text{ m.}^2 = 264 \text{ cm.}^2$$

For this frame the core centres are spaced
$$D = 0.34 \text{ m.} = 34 \text{ cm.}$$

The flux is $0.0264 \cdot 1.35 = 0.0356$ Wb., the e.m.f. per turn
$$E_t = 4.44 \cdot 50 \cdot 0.0356 = 7.9 \text{ V.},$$

FIG. 106. VIEW OF 300 KVA. TRANSFORMER
Bruce Peebles)

and the number of secondary turns per phase
$$T_2 = 254/7.9 = 32.2, \text{ say } T_2 = 32 \text{ turns,}$$

since the secondary phase voltage with the maximum flux density is
$440/\sqrt{3} = 254$ V.

The number of primary turns per phase is
$$T_1 = 32 \cdot 6\,600/254 = 830 \text{ turns}$$

for 440 V. output, and an additional 83 turns making 913 turns for
the lower secondary voltage of 400 V.

The window area is obtained from eq. (28). Choosing $\delta = 2.5$ MA.
per m.^{2} (2.5 A. per mm.^{2}), and $k_w = 0.29$ from the table, p. 139,
$$A_w = 300 \cdot 10^3/(3.33 \cdot 50 \cdot 0.0264 \cdot 1.35 \cdot 0.29)$$
$$= 0.07 \text{ m.}^2 = 700 \text{ cm.}^2$$

The window length $L = A_w/(D - d)$
$$= 700/(34 - 21) = \underline{\underline{55 \text{ cm}.}}$$

The main dimensions are thus

$$d = 21 \text{ cm.}; \quad D = 34 \text{ cm.}; \quad L = 55 \text{ cm.};$$
$$W = 2D + 0.9d = 86.9 \text{ cm.}$$

The yoke area is increased by approximately 15 per cent. It is not shaped like the core. The main dimensions are shown in Fig. 105.

MAGNETIC CIRCUIT. The packet sizes for the cores are based on the sizes given in Fig. 23 with $d = 21$ cm.

$$2 \times (0.42 \cdot 0.10) \cdot 21^2 = 2(8.8 \times 2.1) = 37 \text{ cm.}^2$$
$$2 \times (0.70 \cdot 0.14) \cdot 21^2 = 2(14.7 \times 2.9) = 85 \text{ cm.}^2$$
$$1 \times (0.90 \cdot 0.42) \cdot 21^2 = 1(18.9 \times 8.8) = 166 \text{ cm.}^2$$

$$\text{Gross core section} = \underline{\underline{288 \text{ cm.}^2}}$$

The net core section is $0.9 \cdot 288 = 260$ cm.$^2 \simeq A_t$.
The net yoke section is $0.9 \cdot 17.5 \cdot 18.9 = 297$ cm.^{2}

CORE LOSS.

Core volume $= 3 \cdot 260 \cdot 55 = 43\ 000$ cm.^{3};
weight $= 43\ 000 \cdot 7.55 \cdot 10^{-3} = 325$ kg.;
density $= 1.35$ Wb./m.^{2}
specific loss (Fig. 113, hot-rolled plates)
   $p_i = 2.7$ W. per kg.;
loss $\quad = 325 \cdot 2.7 \cdot 10^{-3} = 0.88$ kW.
Yoke volume $= 2 \cdot 298 \cdot 86.9 = 51\ 800$ cm.^{3};
weight $\ = 51\ 800 \cdot 7.55 \cdot 10^{-3} = 392$ kg.;
density $= 1.35 \cdot 260/297 = 1.18$ Wb./m.^{2};
specific loss (Fig. 113),
   $p_i = 1.9$ W. per kg.;
loss $\quad = 392 \cdot 1.9 \cdot 10^{-3} = 0.75$ kW.
Total core loss $= 0.88 + 0.75$ kW. $(+ 7\frac{1}{2}\%) = P_i = \underline{\underline{1.75 \text{ kW}.}}$

MAGNETIZING CURRENT.

Total ampere turns for cores $= 3 \cdot 55 \cdot 10 = 1\ 650$ A.T.
yokes $= 2 \cdot 86.9 \cdot 4 = 690$ A.T.

$$\text{Total} = \underline{\underline{2\ 340 \text{ A.T.}}}$$

Magnetizing current $= 2\ 340/3 \cdot (\sqrt{2}) \cdot 830 = 0.67$ A.

Percentage magnetizing current $= 100 \cdot 0.67/15.15 = \underline{\underline{4.4\%}}$;

or directly from Fig. 103 (*b*), a core density of 1.35 Wb./m.^{2} requires 31 VA. per kg. The corresponding figures for the yoke are 1.18 Wb./m.^{2} and 10 VA. The total is

$$325 \cdot 30 + 392 \cdot 10 = 13\ 670 \text{ VA}.$$

The magnetizing current per phase at 6 600 V. is therefore

$$13\ 670/(3\ .\ 6\ 600) = 0.69\ \text{A}.$$

Allowing for joints, the magnetizing current will be about 5 per cent. LOW-VOLTAGE WINDING. The secondary phase current is

$$I_2 = 100\ .\ 10^3/254 = 394\ \text{A}.$$

With a current-density of $\delta_2 = 2.3$ A. per mm.^{2} (lower than the mean value used in obtaining the main dimensions in order to improve the cooling) the cross-section of conductor is

$$a_2 = 394/2.3 = 171\ \text{mm}.^2$$

A suitable winding is obtained by using flat strip wound in 16 turns down the limb and 16 turns back. A conductor of 171 mm.^{2} would, however, be too big and must be stranded. The cross-over of the strands to reduce eddy-current losses can then be conveniently made at the bottom of the limb. Taking 46 cm. as winding length, space will be required for $16 + 1$ turns, owing to the helical winding. The space per turn is $46/(16 + 1) = 2.7$ cm. A suitable conductor would therefore be four strips 13 mm. $\times$ 3.25 mm. with 0.25 mm. paper between strips axially, and taped over with 0.5 mm. tape wound half lap.

Overall dimensions of conductor, 27 mm. $\times$ 7.75 mm. The two layers of 16 turns are separated by a cylinder of 0.5 mm. pressboard. The overall dimensions of the l.v. winding are as follows—

Axial length,   46 cm.

Radial thickness, $7.75 + 0.5 + 7.75 = 16$ mm. $= 1.6$ cm.

Diameter of circumcircle $\qquad\qquad d = 21$ cm.

L.v. former, 3 mm. thick, inside dia. $= 21.5$ cm.

outside dia. $= 22.1$ cm.

L.v. winding, 1.6 cm. thick, outside dia. $= 25.3$ cm.

mean dia. $= 23.7$ cm.

L.V. I^2R Loss. Resistance per phase at 75° C.:

Cross-section $a_2 = 4\ .\ (13 \times 3.25) = 169$ mm.^{2}

Length of mean turn

$$L_{mt2} = \pi\ .\ 23.7\ .\ 10^{-2} = 0.745\ \text{m}.$$

Resistance $r_2 = 0.021\ .\ 32\ .\ 0.745/169 = 0.00296\ \Omega$.

I^2R loss $I_2{}^2 r_2 = 394^2\ .\ 0.00296 = 460$ W. per phase.

HIGH-VOLTAGE WINDING. The primary phase current is

$$I_1 = 100\ .\ 10^3/6\ 600 = 15.15\ \text{A}.$$

With a current-density of $\delta_1 = 2.8$ A. per mm.^{2}, the conductor section is

$$a_1 = 15.15/2.8 = 5.42 \text{ mm.}^2$$

The area of No. 12 S.W.G. is 5·48 mm.^{2}, and this will be chosen. The diameter is 2·64 mm. bare, insulated to 2·93 mm. diameter with paper wrapping.

The number of turns is 830 with a further 83 turns tapped for the lower output voltage. At the ends of the winding 3 per cent, or 25 turns, are reinforced with additional layers of paper. The end 1 per cent, or 8 turns, are specially reinforced. The winding is sectionalized into 9 coils of 80 turns, two end coils of 54 turns, and two split gunmetal coil-supporting rings forming the end turn at top and bottom: a total of $9 \times 80 + 2 \times 54 + 2 = 830$ turns. An additional central section of 83 turns tapped every 21 turns is included when the output voltage is lowered to 400 V.

The sections of 80 turns are arranged with 10 conductors axially and 8 radially, the turns being held by binding tape at intervals, inserted during the winding. The end coils and the tapped coil have the same overall dimensions, viz.,

Radial width of section, $8 \times 2.93 \simeq 25$ mm.
Axial length of section, $10 \times 2.93 \simeq 31.5$ mm.

These dimensions include slack and tape insertions.

The sections are spaced 6 mm. apart by U-pieces of pressboard 3 mm. thick. The coil depth is made up as follows—

Sections, 12×3.15 = 37·8 cm.
U-pieces, 24×0.3 = 7·2 cm.
End gunmetal rings, 2×1.2 = 2·4 cm.
Slack   = 0·6 cm.
End insulation and bracing = 7·0 cm.

Total = length of core = 55·0 cm.

L.v. winding   outside dia. = 25·3 cm.
Duct $\simeq$ 0·5 cm.
H.v. former, 3 mm. thick, inside dia. = 26·5 cm.
      outside dia. = 27·1 cm.
H.v. winding, spaced from former on pressboard U-pieces and packing, 5 mm. thick
      inside dia. = 28·1 cm.
      outside dia. = 33·1 cm.
      mean dia. = 30·6 cm.
Clearance between h.v. coils on adjacent limbs = $34 - 33.1$ = 0·9 cm.

H.V. I^2R **Loss.** Resistance per phase at 75° C.

$a_1 = 5.42$ mm.^{2}; $L_{mt1} = \pi \cdot 30.6 \cdot 10^{-2} = 0.96$ m.

$r_1 = 0.021 \cdot 830 \cdot 0.96/5.42 = 3.09 \, \Omega.$ for 440 V. secondary voltage.

I^2R loss $I_1{}^2 r_1 = 15.15^2 \cdot 3.09 = 710$ W. per phase.

TOTAL I^2R LOSS,

$$3(460 + 710) = 3510 \text{ W}.$$

Add 7% for connections and stray losses,

$$I^2R \text{ loss } P_c = 3.75 \text{ kW}.$$

EFFICIENCY.

Full-load loss

$$P_i + P_c = 1.75 + 3.75 = 5.5 \text{ kW}.$$

At unity power factor

$$\eta = [1 - (5.5/305.5)]100 = 98.20\%$$

Load xS for maximum efficiency:

$$x^2 = P_i/P_c,$$
$$x = \sqrt{(1.75/3.75)} = 0.684,$$
or $$xS = 205 \text{ kVA}.$$

REGULATION. From eq. (17),

$$\varepsilon_r = P_c/S = 3.75/300 = 0.0125 \text{ p.u.}$$

From eq. (35), using the figures: $f = 50$; $L_{mt} = \frac{1}{2}(L_{mt1} + L_{mt2})$
$= \frac{1}{2}(0.960 + 0.745) = 0.852$ m.; $AT = I_2 T_2 = 394 \cdot 32 = 12\,600$;
$L_c = 0.47$ m. (mean); $E_t = 7.9$ V.; $a = \frac{1}{2}(28.1 - 25.3) = 1.4$ cm.
$= 0.014$ m.; $b_1 = 0.025$ m.; $b_2 = 0.016$ m.:

$$\varepsilon_x = \frac{2\pi 50 \cdot 4\pi \cdot 0.852 \cdot 12\,600}{10^7 \cdot 0.47 \cdot 7.9} \left(0.014 + \frac{0.025 + 0.016}{3} \right)$$
$$= 0.032 \text{ p.u.}$$

Regulation at full load for unity power factor $= 0.0125$ p.u., for zero power factor $= 0.032$ p.u., and for power factor 0.8 lagging

$$\varepsilon = 0.0125 \cdot 0.8 + 0.032 \cdot 0.6 = 0.029 \text{ p.u.}$$

COOLING. The overall dimensions of the transformer are: height over yokes, 90 cm.; width over three limbs, 101.1 cm.; width over one limb, 33.1 cm. Allowing 7 cm. at each end for clearance, the tank length will be, say, 115 cm. A bigger clearance is required at the sides to accommodate the connections and taps: say a total of 20 cm., making the tank width 55 cm. The net height of the transformer is 90 cm.: allowing 5 cm. for base and 25 cm. of oil above, the oil height will be 120 cm. A further 25 cm. is necessary for leads, etc., so that the tank height will be 145 cm.

The area of the tank walls is

$$2(55 + 115)145 \cdot 10^{-4} = 4 \cdot 93 \text{ m}^2.$$

The full-load loss to be dissipated is 5·5 kW. This is much greater than can be expended in the area available, for with 12·5 W. dissipation per m.² per ° C. (eq. (43a)), the temperature-rise of the tank walls would be

$$5\,500/12 \cdot 5 \cdot 4 \cdot 93 = 90° \text{ C.}$$

A suitable temperature rise is 35° C. Using eq. (43b), the total tank + tube area must be increased x times, so that

$$4 \cdot 93x(8 \cdot 8 + 3 \cdot 7/x)35 = 5\,500$$

whence $\qquad\qquad\qquad x = 3 \cdot 25.$

The required area of tube wall is consequently

$$(x - 1)S_t = 2 \cdot 25 \times 4 \cdot 93 = 11 \text{ m.}^2$$

The height below the oil-level limits the tubes to an average length of about 100 cm. Taking 5 cm. diameter tubes, their area is

$$\pi \cdot 5 \cdot 100 \cdot 10^{-4} = 0 \cdot 157 \text{ m.}^2 \text{ each.}$$

The number required is $11/0 \cdot 157 = 70$. The tank conveniently accommodates 80 tubes as shown in Fig. 105, giving some margin on the temperature rise.

WEIGHTS. The weight of active iron is $325 + 392 = 717$ kg. The weight of active copper is

$$3[(74 \cdot 5 \cdot 1 \cdot 69 \cdot 32) + (96 \cdot 0 \cdot 0542 \cdot 913)]8 \cdot 9 \cdot 10^{-3} = 235 \text{ kg.}$$

The weight-ratio is $G_i/G_c = 3 \cdot 05$. Compare eq. (32).

13. **Example.** 125-kVA., 2 000/440-V., 50-c/s., single-phase, shell-type, oil-immersed, self-cooled power transformer.

MAIN DIMENSIONS. From eq. (33)

$$E_t = 1 \cdot 0\sqrt{125} = 11 \cdot 2 \text{ V.}$$

The flux is therefore

$$\Phi_m = 11 \cdot 2/4 \cdot 44 \cdot 50 = 0 \cdot 0505 \text{ Wb.}$$

Suitable densities are $B_m = 1 \cdot 1$ Wb./m.², and $\delta = 2 \cdot 2$ MA. per m.² (2·2 A per mm.²). The window space factor will be about 0·33 for this voltage and output. The net iron section of the core will be

$$A_i = 0 \cdot 0505/1 \cdot 1 = 0 \cdot 046 \text{ m.}^2$$

From eq. (29),

$$A_i A_w = 125 \cdot 10^3/(2 \cdot 22 \cdot 50 \cdot 1 \cdot 1 \cdot 2 \cdot 2 \cdot 10^6 \cdot 0 \cdot 33)$$
$$= 0 \cdot 00141 \text{ m.}^4$$

whence $\qquad A_w = 0 \cdot 00141/0 \cdot 046 = 0 \cdot 0307 \text{ m.}^2$

The dimensions of core and window have now to be fixed. A narrow, long core gives better cooling for the iron, but a greater buried length of conductors. A ratio between 2·3 and 3 : 1 would be suitable, i.e.

$$37 \text{ cm. long} \times 14 \text{ cm. wide} = 518 \text{ cm.}^2 \text{ gross}$$

and $$A_i = 0·9 . 518 = 466 \text{ cm.}^2 \text{ net.}$$

The core is built of 0·35 mm. transformer steel plates.

The window is arranged to obtain suitable coil shapes and sufficient

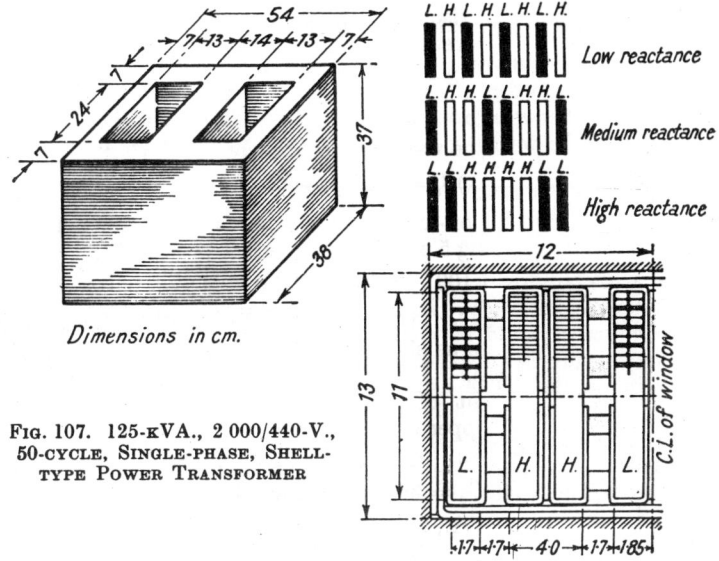

Dimensions in cm.

Low reactance

Medium reactance

High reactance

Fig. 107. 125-KVA., 2 000/440-V., 50-CYCLE, SINGLE-PHASE, SHELL-TYPE POWER TRANSFORMER

subdivision to secure adequately low reactance. A window 13 cm. × 24 cm. = 312 cm.² is suitable. The core dimensions are then as shown in Fig. 108.

WINDINGS. The number of l.v. turns is

$$T_2 = V_2/E_t = 440/11·2 = 39·3, \text{ say } \underline{\underline{40}}.$$

This number is divisible by 4 to allow of the subdivision of the l.v. winding into 4 sections of 10 turns.

The number of h.v. turns is

$$T_1 = 40 . 2 000/440 = 182.$$

This is divided into two coils of 50 turns, and two of 41 turns with reinforced end-turns insulation.

The 8 coils can be arranged in various ways in accordance with the desired reactance. Fig. 107 indicates three possible arrangements.

Details of the conductors are given for the arrangement which gives medium reactance. The currents are

$$I_1 = 125 \cdot 10^3/2\,000 = 62.5 \text{ A};$$
$$I_2 = 125 \cdot 10^3/440 = 284 \text{ A}.$$

The corresponding conductor sections are

$$a_1 = 62.5/2.2 = 28.4 \text{ mm.}^2;$$
$$a_2 = 284/2.2 = 129 \text{ mm.}^2$$

The main insulation lengthwise in the window comprises-

Insulation between end coils and iron	$2 \times 10 =$	20 mm.
Spacers between l.v. and h.v. coils	$4 \times 14 =$	56
Spacers between h.v. sections	$2 \times 6 =$	12
Spacers between l.v. sections	$2 \times 3 =$	6
Slack		6
	Total	100 mm.

The gross window length is 24 cm. The available length for the conductors is $24 - 10 = 14$ cm., i.e. $14/8 = 1.75$ cm. per section. The conductors could be made up as shown in Fig. 107. The h.v. conductor comprises two paper-covered strips each 7 mm. $\times$ 2 mm. (bare) in parallel. The l.v. conductor has 4 strips each 7 mm. $\times$ 4.5 mm. (bare) in parallel, wound together with a 1 mm. strip of fibrous insulation (pressboard) between each set of 4 strips.

LOSSES. The mean length of core from Fig. 107 is 102 cm. The net cross-section is 466 cm.² whence the weight of iron is

$$466 \cdot 102 \cdot 7.5 \cdot 10^{-3} = 356 \text{ kg.}$$

With a density of 1·1 Wb./m.² the specific core loss is $p_i = 1.65$ W. per kg. The core loss is therefore

$$P_i = 356 \cdot 1.65 \cdot 10^{-3} = 0.588 \simeq 0.6 \text{ kW.}$$

The mean length of turn of the winding is 143 cm. The total copper section is

$$a_1 T_1 + a_2 T_2 = 2(7 \times 2)182 + 4(7 \times 4.5)40 = 10\,150 \text{ mm.}^2$$

The copper weight is consequently

$$10\,150 \cdot 10^{-2} \cdot 143 \cdot 8.9 \cdot 10^{-3} = 129 \text{ kg.}$$

The resistivity of copper at 75° C. = 0·021 Ω. per m. and mm.², so that the I^2R loss is

$$P_c = 2.36 \cdot 2.2^2 \cdot 129 \cdot 10^{-3} = 1.47 \simeq 1.5 \text{ kW.}$$

EFFICIENCY. At full load and unity power factor the output is 125 kW. and the losses $1.5 + 0.6 = 2.1$ kW. The efficiency is therefore

$$100\left(1 - \frac{2.1}{127.1}\right) = \underline{98.35\%}.$$

REGULATION. The several quantities in eq. (37) have the numerical values: $b_1 = 0.04$ m.; $b_2 = 0.037$ m.; $a = 0.017$ m.; $w = 0.11$ m.; $L_{mt} = 1.43$ m.; and $n = 2$, as in Fig. 107. Also $E_t = 11.2$ V.; $I_1 T_1 = 182 . 62.5$; $f = 50$. Whence

$$\varepsilon_x = \frac{\pi 50 . 4\pi . 1.43 . 182 . 62.5}{10^7 . 0.11 . 2 . 11.2}\left(0.017 + \frac{0.04 + 0.037}{6}\right)$$

$$= 0.039 \text{ p.u.}$$

for the medium reactance coil arrangement of Fig. 107.

As a comparison, the arrangement marked "low reactance" in Fig. 107, is calculated from the values $b_1 = 0.017$; $b_2 = 0.017$; $a = 0.012$; $n = 4$ giving

$$\varepsilon_x = \frac{\pi 50 . 4\pi . 1.43 . 182 . 62.5}{10^7 . 0.11 . 4 . 11.2}\left(0.012 + \frac{0.017 + 0.017}{6}\right)$$

$$= 0.012 \text{ p.u.}$$

For the arrangement marked "high reactance," the numerical values are: $b_1 = b_2 = 0.11$; $a = 0.02$; $n = 1$. The percentage reactance is

$$\varepsilon_x = 0.133 \text{ p.u.}$$

By arranging the coils unsymmetrically, i.e. l.v. coils one end and h.v. the other, a still greater reactance could be obtained.

TANK. The tank details are worked out as for the example in § 12.

14. Example. 25-kVA., 6 600/440-V., 50 c/s., 3-phase, delta/star, core-type, oil-immersed natural-cooled (ON) transformer. Line current, 2.18/32.9 A. Tappings $\pm 2\frac{1}{2} \pm 5\%$ on h.v. winding. Core dimensions, Fig. 108 (a).

Core

Material	.	.		Stalloy, 0.355 mm.	
E.M.F. per turn	.	E_t		2.12 V.	
Core circumcircle dia.	.	d		127 mm.	

			LIMB	YOKE	
Net iron section	.	A_i	88	108.9	cm.²
Flux density	.	B_m	1.084	0.877	Wb./m.
Flux	.	Φ_m	9.55		mWb.
Weight	.		64	91	kg.
total	.	G_i		155	kg.
specific	.	p_i	1.61	1.03	W. per kg.
totals	.		.103	93.5	W.
total	.	P_i		197	W.

Windings

			L.V.	H.V.	
Current per phase	.	I_{ph}	32.9	1.26	A.
Conductor, bare	.		6.5×2.25	0.81 dia	mm.
ins.	.		7×2.75	1.37 dia	mm.
Insulation	.		Cotton tape	d.c.c. paper	

Conductor section	.	a	14·34	0·510	0·585	mm.*
Current density	.	δ	2·29	2·43	2·15	A. per mm.*
Volts per phase	.	V_{ph}	254	6 600	+ 5%	V.
Turns per phase	.	T_{ph}	120	3 274		
Type of winding	.		Helix	Cross-over		
Coils per limb	.	.	1	4	2	
Turns per coil	.	.	120	641	355	
Turns per layer	.	.	40	34	24	
Layers per coil	.	.	3	19	14	
Winding length	.	.	290	38 × 4 35 ×2		mm.
				Total 252		mm.
Winding depth	.	.	9·5	23·5	23·5	mm.
Insulation between layers			2 × 0·25	4 × 1·5		mm.
			pressboard	paper		
Insulation between coils	.		—	2 × 3 mm. spacers		
Coil diameters, inside	.		133	170		mm.
outside	.		152	217		mm.
Length of mean turn	.	L_m	0·448	0·608		m.
Resistance at 15° C.	.	r	0·0661	65·4		Ω
I^2R loss per limb at 15° C.			71·5	99		W.
Total I^2R loss ⎰15° C. .			225	306		W.
including stray⎱75° C. .	P_c		275	393		W.
Weight of copper	.		21	21·7 + 6·8		kg.
total	.	G_c		50		kg.

Insulation

Between core and l.v. winding: three 0·5 mm. thick wraps of pressboard.
Between l.v. and n.v. windings: 3 mm. thick bakelized-paper cylinder.

Impedance

Reactance, 0·028 p.u.; resistance 0·0267 p.u.; impedance, 0·0386 p.u.

Tank

Temperature-rise of oil	.	.	50° C.
Total losses at 90° C.	.	.	197 + 703 = 900 W.
Dimensions, length × breadth × height	.	.	79 × 32 × 94 cm.
oil level	.	.	74 cm.
Tubes	.	.	none
Weight	.	.	107 kg.
Oil	.	.	33½ gal.

Weight

Total weight of complete transformer, 520 kg. (1 150 lb.)

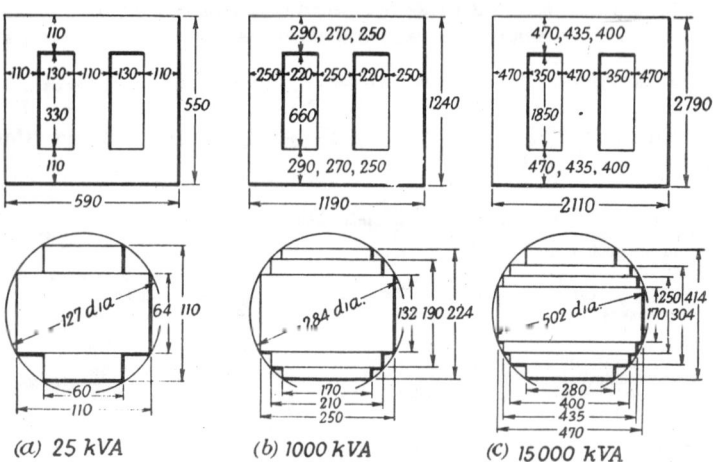

FIG. 108. FRAME SIZES (§§14, 15, 16)
Dimensions in mm. Not to scale

15. Example. 1 000-kVA., 6 600/440-V., 50-c/s., 3-phase, delta/star, core-type, oil-immersed natural-cooled (ON) transformer. Line current, 87·5/1 312 A. Tappings ± 2½% ± 5% on h.v. winding. Core dimensions, Fig. 108 (b).

Core

		Material		
Material		Stalloy, 0·355 mm.		
E.M.F. per turn	E_t	14·1 V.		
Core circumcircle dia.	d	284 mm.		

		LIMB		YOKE	
Net iron section	A_i	458		562	cm.²
Flux density	B_m	1·385		1·132	Wb./m.
Flux	Φ_m	63·6			mWb.
Weight		667		982	kg.
total	G_i		1 650		kg.
Loss, specific	p_i	3·0		1·75	W. per kg.
totals		2 000		1 720	W.
total	P_i		3 720		W.

Windings

		L.V.	H.V.	
Current per phase	I_{ph}	1 312	50·5	A.
Conductor, bare		12 of 14·5 × 4	11·5 × 1·5	mm.
ins.		60 × 13·75	12 × 2	mm.
Insulation		Strips, butt-taped: half-lap tape overall	Half-lap tape	
Conductor section	a	690	17·13	mm.²
Current density	δ	1·9	2·95	A. per mm²
Volts per phase	V_{ph}	254	6 600 + 5%	V.
Turns per phase	T_{ph}	18	492	
Type of winding		Spiral	Disc	
Coils per limb		1	28 + 4 + 4 = 36	
Turns per coil		18	14, 13, 12	
Turns per layer		9	1, 1, 1	
Layers per coil		2	14, 13, 12	
Winding length		604	12, 12, 12 = 558	mm.
Winding depth		28	28	mm.
Insulation between layers		2 × 0·25 mm. pressboard	Paper strip	
Insulation between coils		—	Pressboard spacers	
Coil diameters, inside		296	385	mm.
outside		352	441	mm.
Length of mean turn	L_{mt}	1·016	1·3	m.
Resistance at 15° C.		0·00049	0·655	Ω
I^2R loss per limb at 15° C.		845	1 582	W.
Total I^2R loss }15° C.		3 300	5 700	W.
including stray }75° C.	P_c	3 760	6 650	W.
Weight of copper		350	295	kg.
total	G_c		645	kg.

Insulation

Between core and l.v. winding: three 0·8 mm. thick wraps of pressboard.
Between l.v. and h.v. windings: 3 mm. thick bakelized paper cylinder.

Impedance

Reactance, 0·047 p.u.; resistance, 0·010 p.u.; impedance, 0·048 p.u.

Tank

Temperature-rise of oil	50° C.
Total losses at 90° C.	3 720 + 11 130 = 14 850 W.
Dimensions, length, breadth, height	152·5 × 65 × 183 cm.
oil level	152·5 cm.
Tubes, 5 cm. dia.	2 × 56 + 2 × 27 = 166
Weight	1 650 kg.
Oil	333 gal.

Weight

Total weight of complete transformer, 6 100 kg. (6 tons).

16. Example. 15 000-kVA., 33/6·6-kV., 50-c/s., 3-phase, star/delta, core-type, oil-immersed natural-cooled (ON) transformer. Line current, 263/1 312 A. Tappings ± 2½ ± 5% on h.v. winding. Core dimensions, Fig, 108 (c).

Core

		Material	
Material	. .	Stalloy, 0·355 mm.	
E.M.F. per turn	. E_t	44 V.	
Core circumcircle dia.	. d	502 mm.	

		LIMB	YOKE	
Net iron section	. A	1 505	1 665	cm.²
Flux density	. B_m	1·315	1·19	Wb./m.²
Flux	. Φ_m	198		mWb.
Weight	.	6 160	5 130	kg.
total	. G_i		11 290	kg.
Loss, specific	. p_i	2·57	2·0	W. per kg.
totals	.	15·82	10·26	kW.
total	. P_i	26·08		kW.

Windings

		L.V.	H.V.	
Current per phase	. I_{ph}	758	263	A.
Conductor, bare	.	15 × 4	13 × 3·5 12 × 4	mm.
		5 in parallel	2 in par. 2 in par.	
ins.	.	15·5 × 4·5	13·5 × 4 13·5 × 5·5 mm.	
Insulation	.	0·5 mm.	0·5 mm. 1·5 mm.	
		Cotton tape	paper paper	
Conductor section	. a	297·5	89·7 94·3	mm.²
Current density	. δ	2·54	2·93 2·79	A. per mm.
Volts per phase	. V_{ph}	6 600	19 050 + 5%	V.
Turns per phase	. T_{ph}	150	454	
Type of winding	.	Disc, 5 sections	Disc, 2 sections	
		in parallel	in parallel	
Coils per limb	.	80	80	
arranged	.	5 × 10 : 5 × 6	2 × 28 : 2 × 4 : 2 × 4 : 2 × 2 : 2 × 2	
Turns per coil	.	9, 10	12, 11, 10, 9, 8	
Turns per layer	.	1, 1	1, 1, 1, 1, 1	
Layers per coil	.	9, 10	12, 11, 10, 9, 8	
Winding length	.	1 580	1 580	mm.
Winding depth	.	45	50	mm.
Ins. between layers	.	Paper strip	Paper strips	
Ins. between coils	.	Pressboard spacers	Pressboard spacers	
Coil diameters, inside	.	535	685	mm.
outside	.	625	785	mm.
Length of mean turn	. L_{mt}	1·82	2·31	m.
Resistance at 15° C.	. r	0·0162	0·203	Ω
I^2R loss per limb at 15° C.	.	9·3	13·4	kW.
Total I^2R loss 15° C. }		36·3	50·1	kW.
including stray 75° C. }		41·4	57·6	kW.
Weight of copper	.	2 180	2 330 + 200	kg.
total	. G_c		4 710	kg.

Insulation

Between core and l.v. winding: 5 mm. thick bakelized paper cylinder.
Between l.v. and h.v. windings: 7·5 mm. thick bakelized paper cylinder.

Impedance

Reactance, 0·083 p.u.; resistance, 0·007 p.u.; impedance 0·084 p.u.

Tank

Temperature-rise of oil	50° C.
Total losses at 90° C.	26 + 106 = 132 kW.
Dimensions, length × breadth × height	264 × 112 × 320 cm.
oil level	264 cm.

Radiators: 16 radiators, each with 72 tubes 260 cm. long, mounted on tank, 5 each side and 3 each end.

Oil quantity	2 400 gal.
Conservator	80 cm. dia. × 214 cm.

Weight of tank with radiators, 13 150 kg.

Weight

Total weight of complete transformer, 43 tons.

CHAPTER VIII
ELECTROMAGNETIC MACHINES

1. Energy Conversion. All electromagnetic machines are variations on a common set of fundamental principles, which apply alike to d.c. and a.c. types, to generators and motors, to steady-state and transient conditions.* An electromagnetic machine links an electrical energy system (such as a supply network) to a mechanical one (a prime-mover or mechanical load) by providing a reversible means of energy flow in its common or "mutual" magnetic field. Energy is stored in the mutual field and released as work. A current-carrying conductor in the field is subjected to a mechanical force and, in moving, does work and generates a counter-e.m.f. Thus force-motion is converted to or from voltage-current.

In normal machines, Fig. 5 (a), there are distinct windings. One (the *field* winding) excites the working flux: the other (the *armature* winding) develops rotational e.m.f.'s and mechanical torque.

The balance equation for energy flow in a motor is—

electrical energy input (w_e)

$\qquad =$ mechanical energy output (w_m)

$\qquad\qquad +$ stored magnetic energy (w_f)

For a generator it is only necessary to interchange the electrical and mechanical terms. Conduction and core loss, and non-useful energy stored in leakage fields, all enter into the balance; but if these are separately accounted, the essential energy-conversion process remains. Writing it for convenience in terms of power with $dw_e/dt = p_e$ and $dw_m/dt = p_m$, then

$\qquad$ Motor: $p_e = p_m + dw_f/dt$. $\qquad$ Generator: $p_m = p_e + dw_f/dt$.

The mechanical power term must account for changes in stored kinetic energy, which occur whenever the speed of the machine (and its mechanical attachments) alters.

2. Operating Principles. The physical phenomena concerned in the energy-conversion process are—

(1) The rotational e.m.f. e, generated by the movement of conductors in a magnetic field, eq. (2);

(2) The pulsational or transformer e.m.f. e_p induced by the change of magnetic linkages in a circuit;

(3) The mechanical force set up between a current i and the magnetic field in which it lies, eq. (3);

* Attention is here directed primarily to normal generators and motors, but the same principles apply also to such "machines" as relays, electromagnetic pumps, loudspeakers, and any like system whether in linear, vibratory or rotary motion.

(4) The mechanical force on ferromagnetic material tending to align it with a magnetic field, or to move it into a stronger field.

Principles (1) and (3) were discussed in Chapter II; (2) is the basis of the transformer, and appears also in some rotating machines; (4), though of minor importance here, is worthy of a brief note.

Two-pole Machines. A magnetic field system is by nature bipolar, and the most elementary machine has one N and one S pole. Machines with 4, 6, 8 . . . poles (or, in general, p pole-pairs) can be considered as simple repetitions of 2-pole machines. The treatment that follows is developed in terms of the unit machine.

Reluctance Motors. It is shown in textbooks of electrical technology* that the magnetic pull between ferromagnetic surfaces of area A across which a magnetic field of density B is maintained is given by $f = \frac{1}{2}B^2A/\mu_0$. The expression is derived from the rate of change with gap spacing of the magnetic energy density $\frac{1}{2}B^2/\mu_0$. Such a system, Fig. 109 (a), constitutes an elementary electromagnetic "machine."

The same principle applies to the rotary case (b). Let the rotor have a radius R and axial core-length l, and let the flux be taken as confined within the angle θ to the narrow gap of radial length l_g. The magnetic energy in the two gaps is

$$w_f = 2(\tfrac{1}{2}B^2Al_g/\mu_0) = B^2l \cdot R\theta \cdot l_g/\mu_0.$$

Then the torque will be

$$M = dw_f/d\theta = B^2lRl_g/\mu_0.$$

The armature tends to align itself with the field axis, and so is a simple motor. Suppose $B = 1$ Wb./m.², $l = 2$ cm. $= 0{\cdot}02$ m., $R = 3$ cm., and $l_g = 2$ mm.: then $M \simeq 1$ N.-m. or about $\frac{3}{4}$ lb.-ft. The machine operates by developing a *reluctance torque*. Such a torque can appear in salient-pole synchronous machines.

Machines with Armature Windings. Normal machines do not rely on reluctance torques, but on the principles (1), (2) and (3) above.

Consider a machine running at constant angular velocity ω_r and developing a torque M: the mechanical power is then $p_m = M\omega_r$. The electrical power is $p_e = ei$, where e is the counter-e.m.f. due to the reaction of the mutual field. Then

$$ei = M\omega_r + dw_f/dt \quad . \quad . \quad . \quad (44\,a)$$

at every instant. If the armature conductor a in Fig. 109 (c) is running in a non-time-varying flux of local density B, the e.m.f. is entirely rotational and equal to $e_r = Blu = Bl\omega_rR$, from eq. (2).

---

* E.g., Say, *Electro-technology* (Newnes).

The tangential force on the conductor from eq. (3) is $f = Bli$ and the torque is $M = BliR$. Thus at any instant

$$e_r i = M\omega_r \qquad . \qquad . \qquad . \qquad (44\,b)$$

because $d\omega_r/dt = 0$. This simple equivalence of electrical and mechanical power is useful and important.

Let the armature in (c) be given two conductors a and b: they can be connected in series to form a *turn*. The same current flows through both conductors but in the opposite sense. Provided the

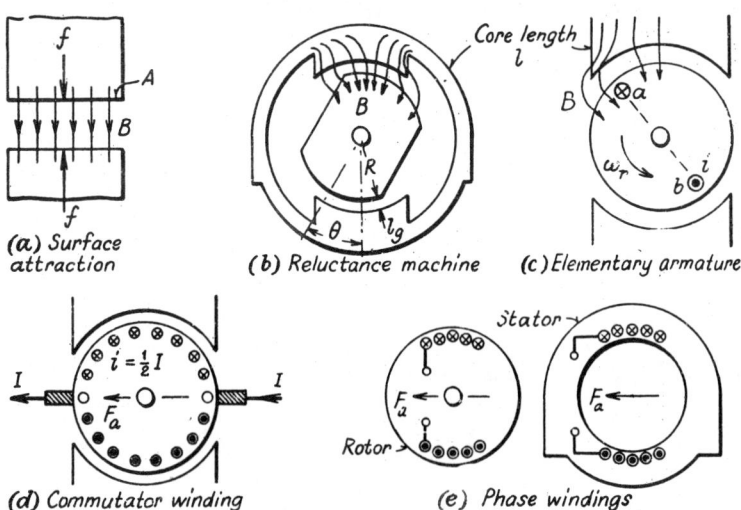

(a) Surface attraction

(b) Reluctance machine

(c) Elementary armature

(d) Commutator winding

(e) Phase windings

FIG. 109. ELEMENTS OF ELECTROMAGNETIC MACHINES

conductors are arranged diametrally (i.e. the turn is of *full pitch*) the torques will always be additive.

A further step is to use more turns in series, forming a *winding* to occupy the available space around the armature surface.

The current in the armature winding will naturally produce its own m.m.f. and flux (called *armature reaction*). The total flux in the machine results from the m.m.f.s of all current-carrying windings, whether on stator or rotor, but the torque arises from that component of the total flux *at right-angles to the m.m.f. axis* of the armature winding.

3. **Windings.** A winding designed for the d.c. excitation of a working field flux is usually concentrated. For an a.c. working flux it will generally be distributed in order to reduce its leakage reactance. Armature windings are of the commutator or phase (tapped) types.

Commutator Winding, Fig. 109 (d). The full-pitch turns form a symmetrically-spread winding closed on itself. Current, led in and out by fixed diametral brushes, divides into the two equal parallel paths so formed. The brushes include between them a constant number of conductors in each path, having always the same current directions and relative positions whatever the armature speed. A particular conductor passes cyclically from one belt of conductors into the other, but its current is reversed by the process of *commutation* as it moves under a brush. The armature m.m.f. therefore coincides always with the brush axis.

Phase Winding, Fig. 109 (e). This has separate external connections. If the winding is on the rotor, its current and m.m.f. rotate

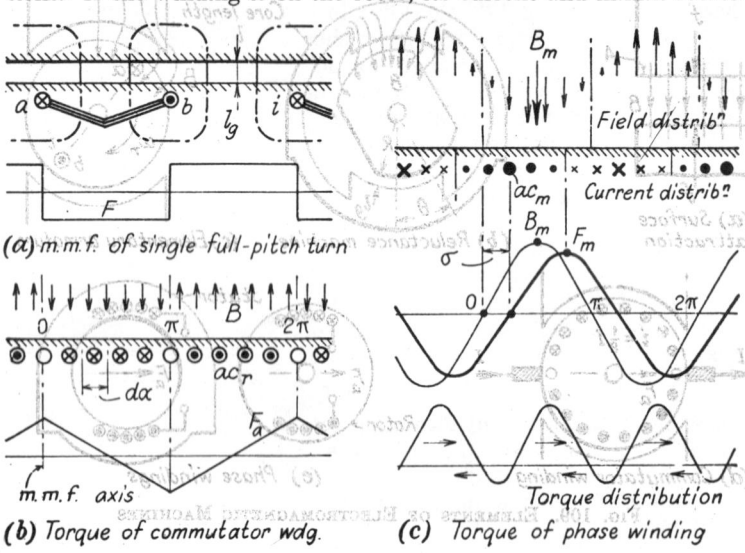

(a) m.m.f. of single full-pitch turn

(b) Torque of commutator wdg.

(c) Torque of phase winding

FIG. 110. M.M.F. AND TORQUE OF ELECTROMAGNETIC MACHINES

with it, and the external connections must be made through sliprings. Two (or three) such windings with 2- (or 3-) phase currents can produce a resultant m.m.f. that *rotates* with respect to the windings.

STATOR AND ROTOR WINDINGS. It is basically immaterial which function, field or armature, is assigned to the stator and which to the rotor, although for practical convenience a commutator winding is almost always used as a rotor.

M.M.F. OF WINDINGS. The magnetic flux in a machine results from the individual m.m.f. contributions of all its windings. A current i in a single full-pitch turn comprising conductors a and b, Fig. 110 (a), produces a total m.m.f. i round any closed path linking the turn—in particular round any path crossing the two air-gaps. The m.m.f. is (in a symmetrical machine) divided equally between the gaps so that the m.m.f. per pole is $F_a = \frac{1}{2}i$. Neglecting core reluctance, this m.m.f. will be expended across the gap-length l_g in producing a flux density $B = \mu_0 H = \mu_0 F_a/l_g = \mu_0 \frac{1}{2}i/l_g$ in each pole-pitch.

A distributed winding will give a distributed m.m.f. Such a case is shown for a phase winding in Fig. 138 (a), and for a commutator winding in Fig. 110 (b).

4. Torque. Unless magnetic asymmetry introduces a reluctance component, a torque is not developed on a winding by its own flux. There must be a second ("field") flux in which the winding lies.

Commutator Winding. Fig. 110 (b) shows a "developed" diagram of a commutator winding arranged for maximum torque production. All the currents in a given direction lie in a field flux density of the same polarity; stated alternatively, the m.m.f. axis of the winding is displaced $\pi/2$ from the field-pole centres.

Let the current strength round the armature surface be ac_r ampere-conductors per radian: that is, ac_r is the number of conductors in one radian of the armature multiplied by the current i carried by each. Further, let the field have a uniform density B. In a small angle $d\alpha$ the current is $ac_r \, . \, d\alpha$, the force on it is $df = Blac_r \, . \, d\alpha$, and the torque is $dM = BlRac_r \, . \, d\alpha$, where R is the effective radius and l the axial length of the armature core. The torque for the whole armature circumferential angle 2π is

$$M = 2\pi BlRac_r.$$

This expression can be put into other terms. A common form uses the flux per pole $\Phi = \pi RlB$, with the input current I and total conductors Z, where $IZ = 4\pi ac_r$, giving $M = IZ\Phi/2\pi$. Alternatively with Φ and the m.m.f. per pole $F_a = \frac{1}{2}\pi ac_r$, then $M = (4/\pi)\Phi F_a$.

These expressions apply directly to a d.c. machine. They also give the mean of the pulsating (but unidirectional) torque of a single-phase commutator machine if Φ and I are r.m.s. values. An additional factor $\cos \phi$ is necessary if the flux and armature current have a time-phase displacement.

Phase Winding. The m.m.f. waves of Figs. 138 and 139 show that the distribution is roughly sinusoidal. An important case relevant to synchronous and induction motors concerns the torque developed by a sine-distributed current in a sine-distributed magnetic flux, Fig. 110 (c).

Let the field flux-density, reckoned from the pole-centre, be $B = B_m \cos \alpha$, and let the armature current distribution in ampere-conductors per radian be $ac_r = ac_m \sin (\alpha - \sigma)$. (These functions represent space-distribution: it is assumed that neither B nor ac alters with time.) In a small angle $d\alpha$ the current is $ac_r \, . \, d\alpha$ and it lies in a flux density B. The torque on the current element is

$$dM = B_m \cos \alpha \, . \, lR \, . \, ac_m \sin (\alpha - \sigma) \, . \, d\alpha.$$

The total torque is twice the integral of this over a pole-pitch (α from 0 to π), giving

$$M = \pi B_m a c_m l R \sin \sigma, \qquad . \quad . \quad . \ (45)$$

neglecting a negative sign showing the torque to act so as to decrease σ. The greater the *torque-angle* σ, the greater is the torque, up to a maximum for $\sigma = \pi/2$ when the condition is comparable to that of Fig. 110 (b).

Because the flux per pole is $\Phi = (2/\pi)B_m l R$, and the m.m.f. is $F_a = a c_m$, an alternative form of eq. (45) is $M = (\pi/2)\Phi F_a \sin \sigma$.

CONDITIONS FOR TORQUE. The maintenance of the torque in a given machine is affected by the form (a.c. or d.c.) of the stator

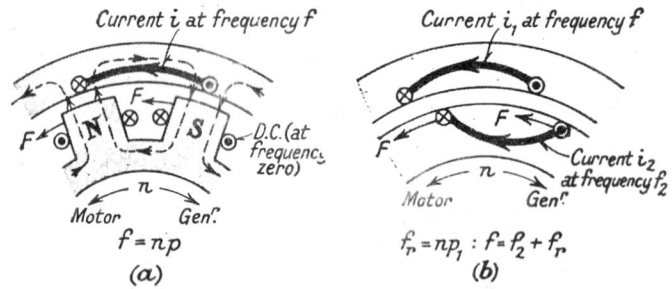

FIG. 111. SYNCHRONOUS AND INDUCTION MACHINES

and rotor currents. Only the polyphase synchronous and induction machines need be considered here.

Synchronous Machine, Fig. 111 (a). The armature (usually on the stator) has phase windings carrying currents of frequency f: the field system is d.c.-excited. If unidirectional torque is to be maintained, the reversals of phase current every half-cycle must synchronize with the movement of the field poles. The rotor can therefore have only one speed, the *synchronous speed*, for which its angular velocity $\omega_r = 2\pi f$, or $n = f$ r.p.s. Thus the rotor must turn through one revolution in one cycle of the phase current in a 2-pole machine. If the machine has p pole-pairs the argument applies to each 2-pole unit and the synchronous speed is $n = f/p$.

Induction Machine, Fig. 111 (b). The stator is like that of the synchronous machine and carries currents of frequency f, but the rotor phase windings are provided with currents of any frequency f_2. In a 2-pole machine running at speed ω_r (or $n = \omega_r/2\pi = f_r$ rev/sec.) the torque can be unidirectional only if $f = f_2 + f_r$. For let the speed be a little less than synchronous; then the rotor gradually *slips* back from the position it would occupy were it running synchronously. But the torque will remain unidirectional if, simultaneously, the rotor current slowly alternates at frequency

$f_2 = f - f_r$. In a machine with p_1 pole-pairs the only change from the conditions in a 2-pole machine is that $n = f_r/p_1$ instead of $n = f_r$.

5. **E.M.F. Production.** The rotational e.m.f. e_r in an armature turn results from the field-flux densities in which its conductors lie at any instant. An additional, pulsational or transformer, e.m.f. e_p appears in the turn if the field flux alternates, providing a means of electrical energy transfer between stator and rotor. The general form of e.m.f., discussed in Chapter XXVI, leads to the expression

$$? = -\frac{d\Phi}{dt} = -\left(\frac{\partial\Phi}{\partial\alpha} \cdot \frac{d\alpha}{dt} + \frac{\partial\Phi}{\partial t}\right) = -\left(\frac{\partial\Phi}{\partial\alpha} \cdot \omega_r + \frac{\partial\Phi}{\partial t}\right) = e_r + e_p,$$

because $\partial\Phi/\partial\alpha = 2BlR$ represents the flux distribution B. In a transformer only e_p appears; in the polyphase synchronous and induction machines the rotating-field concept reduces e to the term e_r, at least under steady-state conditions.

6. **Machine Analysis.** The approach in this book is to develop a physical idea of each machine, and then to formulate its behaviour in terms of e.m.f., current, flux and torque. The method is comprehensive, but particular techniques of limited application are introduced where their use is established practice or otherwise desirable.

EQUIVALENT CIRCUITS. Electric circuit models can be devised to represent currents and voltages in an electrical machine. (This has already been shown for a transformer in Fig. 15.) Such equivalent circuits clarify the construction of complexor ("vector") and locus diagrams; they are almost essential for the treatment of transients. It must be noted, however, that the systems represented will usually be mechanically dynamic, with hidden kinetic energy not explicit in the equivalent circuits. Again, not all the common methods of circuit analysis (such as the reciprocity theorem) are applicable, so some care is necessary in the interpretation.

CHAPTER IX

MAGNETIC CIRCUITS

1. Laws of the Magnetic Circuit.* The usual engineering approach to magnetic circuit design is given here with particular reference to a.c. machines. Let—

Φ = magnetic flux, webers;
A = area of flux path, m.^{2};
l = length of flux path, m.;
$B = \Phi/A$ = flux density, webers per m.^{2};
$H = at$ = m.m.f. per m. (or ampere-turns per m.);
$F = AT$ = m.m.f. (or total ampere-turns);
$\mu = \mu_r\mu_0$ = absolute permeability;
$\mu_0 = 4\pi/10^7$ henry per m. = magnetic space constant;
μ_r = relative permeability;
$S = l/A\mu$ = reluctance, ampere-turns per weber;
$\Lambda = A\mu/l$ = permeance, webers per ampere-turn;
λ = permeance coefficient;
ν = leakage coefficient;
p = specific core loss, watts per kg.;
f = frequency, cycles per sec.

For simplicity, practical magnetic circuits are abitrarily divided into parts, along which the flux density is deemed uniform. For each part

$$\Phi S = BAl/A\mu = (B/\mu)l = Hl = at \cdot l.$$

The total m.m.f. is

$$F = AT = (at_1)l_1 + (at_2)l_2 + \cdots \qquad \cdot \qquad . \ (46)$$

for a series of parts of length $l_1, l_2 \ldots$ along which excitations of $(at_1), (at_2) \ldots$ ampere-turns per m. length are necessary. For air and all non-magnetic materials, $\mu_r = 1$, $B = \mu_0 H$, and

$$H = at = B/\mu_0 \simeq 800\,000 \, B \text{ amp.-turns per m.} \qquad . \ (47)$$

In practice *magnetization curves* relating the flux density to the m.m.f. in ampere-turns per m. are used for the rapid determination of the necessary excitation. Strictly, each specimen has its own magnetization curve, but it is usual to employ a single average curve for each class of material. Fig. 112 gives typical curves.

2. Core Losses. If the flux carried by iron or steel in a magnetic circuit is varied, *hysteresis* and *eddy-current losses* are produced.

---

* See [A], Chapter II; [B], Chapter IV; McGreevy, *The M.K.S. System of Units* (Pitman).

For alternating magnetization these are given in terms of the maximum flux density B_m and r.m.s. density B by—

Hysteresis loss $\quad p_h = k_h f B_m{}^x$

Eddy-current loss $\quad p_e = k_e t^2 f^2 B^2/\rho$ W. per kg.   (48)

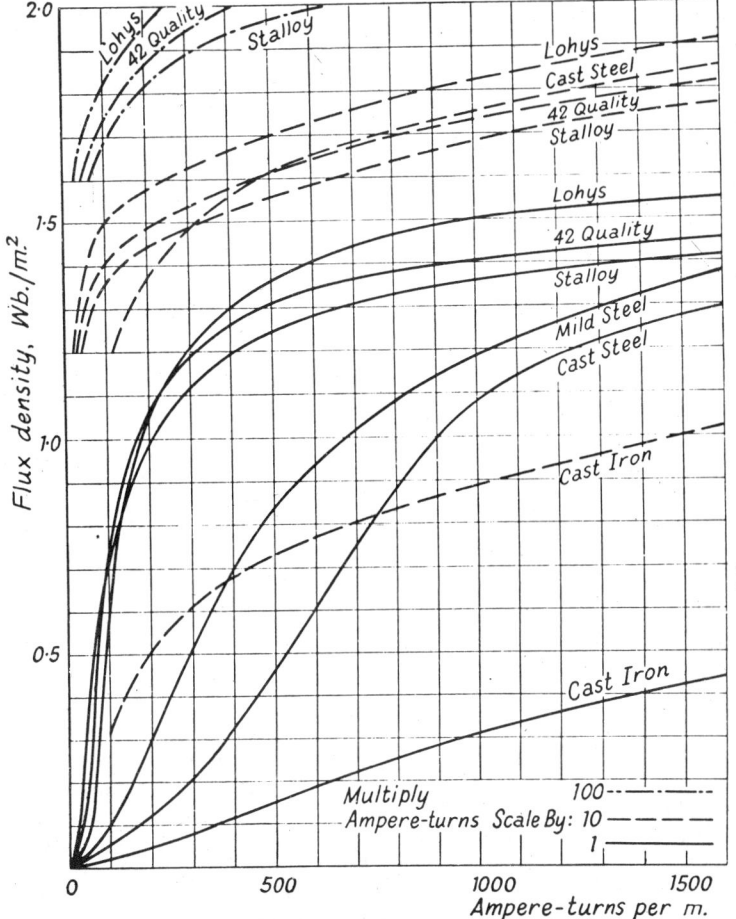

Fig. 112. MAGNETIZATION CURVES

The factor k_e is a numeric; k_h depends on the molecular characteristics of the steel, and upon whether the magnetization is pulsating (i.e. varying in magnitude along a fixed axis) or rotary (i.e. due to a constant flux rotating and changing its direction in the material). The exponent x of the maximum flux density B_m in the hysteresis loss may lie between 0·5 and 2·3: it will generally be between 1·5 and 2 for usual densities.

The thickness t of the laminated material is a factor in the eddy-current loss. Eq. (48) shows that this loss may be reduced by using thinner plates, and a core material of higher resistivity ρ.

Since the expressions of eq. (48) are based on a simplification of the actual conditions, they are unreliable if applied to cases where

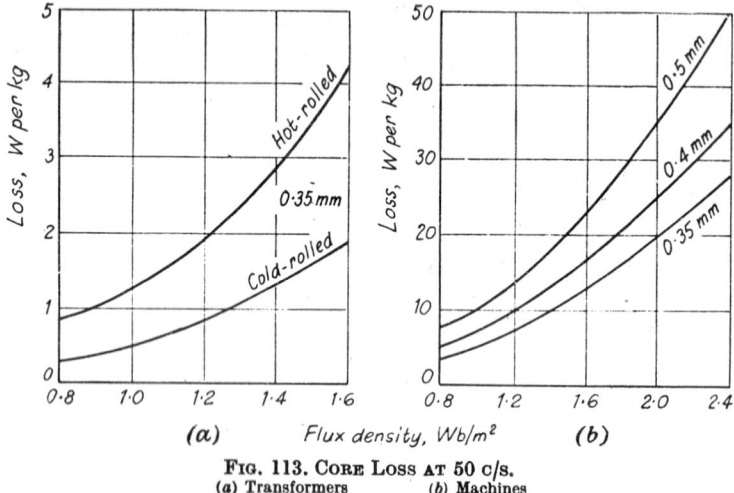

FIG. 113. CORE LOSS AT 50 c/s.
(a) Transformers   (b) Machines

the frequency is abnormal, the laminations thick, or the variation of flux-density not sinusoidal with time.

Core loss is generally estimated from curves based on laboratory tests of prepared specimens. A Table of typical guaranteed loss maxima for the sheet materials whose B/at curves are shown in Fig. 110 is given on page 177, together with values for a special low-loss grade. The loss actually occurring in a built-up core is invariably greater than the figure assessed from test specimens on account of the effects of punching (which markedly hardens the plate edges), short-circuiting between plates due to punching burrs, armature reaction and the difficulty of calculation. To take into account these increases, the designers' curves of iron loss are modified and checked by test figures. Fig. 113 shows curves typical of core losson, in which due allowance has been made.

SURFACE AND PULSATION LOSSES. When the stator and rotor surfaces are very close together, as in an induction motor, additional losses are produced by the rapid changes of local gap reluctance as the stator and rotor teeth pass each other. Further losses may be produced by harmonic fields due to the slotting. Both these losses

are substantially confined to the gap surfaces or the body of the
teeth: they may, however, be surprisingly large, although difficult
to assess. The losses affect a motor by producing a magnetic "drag,"
analogous to friction. Where slot-frequency pulsations are likely to
be large, a-high-quality silicon steel is used.

LOSSES IN WATTS PER KG. AT 50 c/s.

Maximum flux density, Wb./m.^{2}	1·0		1·3	
Plate thickness, mm. . . .	0·35	0·5	0·35	0·5
Lohys	2·90	3·57	4·80	5·88
42 quality	1·93	2·25	3·30	3·81
Stalloy	1·40	1·63	2·35	2·77
Super Stalloy	1·08	—	1·89	—

DESIGN ESTIMATES. In practice the core loss is found to be pro
portional roughly to the square of the density (or even the cube, above
1·2 Wb./m.^{2}).

The simple expressions in eq. (48) are not applicable to machines
in respect of the variation of core loss with frequency, for the highly
complex distribution and variation of density in the several parts
are not susceptible to such direct analysis, which is based on sinusoidal pulsation. Experience suggests that the total core loss in a.c.
machines is roughly proportional to $f^{1·1}$.

3. **Magnetic Materials.** A rough classification for materials used
for the magnetic circuits of electrical machines is in accordance with
whether they carry a pulsating or steady flux. In most cases the
magnetic circuit forms part of the mechanical structure, so that the
mechanical properties are important.

LAMINATED MATERIALS. In all cases of varying flux, the magnetic
material must be laminated. To satisfy this need, magnetic sheet
steels have been developed.

In designing a magnetic circuit, the alternating flux is required
to be produced in the minimum space and with the minimum loss.
The two important magnetic properties are therefore the *permeability*
and the *loss coefficients*.

The permeability is not a directly informative quantity, since
so much depends upon the value of the induction B. The designer
is more interested in the m.m.f. required to produce given values of
B, than in the ratio $\mu = B/H$. Hence the use of "static" B/at curves
such as those in Fig. 112.

The saturation region of a material is frequently important, as
giving an indication of the maximum flux-density that can be
contributed by the material.

The comparatively pure iron sheet used in the early days of electrical engineering construction has given place to alloy steels. The chief alloying constituent is silicon (up to 5 per cent): it increases the permeability at low densities (but decreases it at high densities), reduces the hysteresis loss, and by augmenting the resistivity decreases the eddy-current loss. The greater the silicon content, the better is the steel magnetically from the viewpoint of losses, but the properties are very variable, particularly after punching, shearing, bending, etc. As a rule the exponent x in the hysteresis loss formula (eq. (48)) is $1 \cdot 5 - 1 \cdot 6$ at $B = 1 \cdot 0$ Wb./m.^{2}, rising to $2 \cdot 2$ at $1 \cdot 5$, and falling again to about $0 \cdot 8$ at $2 \cdot 0$ Wb./m.^{2}: these figures are typical of $4 \cdot 5$ per cent silicon steel. The reduction of x at high densities is due to saturation, causing the hysteresis loop *area* to become constant for saturation density and upwards. The resistivity of a $4 \cdot 5$ per cent silicon steel might be of the order 60 $\mu\Omega$.-cm., and less for lower silicon contents down to 12–14 $\mu\Omega$. for comparatively pure iron.

The introduction of silicon has the effect of increasing the tensile strength of alloy steels, but decreases their ductility, making them increasingly difficult to punch or shear. A steel with more than 5 per cent silicon may be too hard and brittle to be readily workable.

Grain-oriented cold-rolled sheet steel in transformers gives important reductions in core loss and magnetizing m.m.f. It has possibilities also in rotating machines, but is difficult to arrange in such a way that the directional property is sufficiently utilized.

Where the core loss is unimportant (as in low-frequency d.c. machines) a low-silicon iron is employed. Similarly with induction motors, where the magnetizing current must be kept low. In large machines, particularly in turbo-alternators, where the size of the machine renders cooling a special problem, a low-loss steel is employed, such as one with 2 per cent silicon. This or a higher grade may be used with totally-enclosed motors to reduce the total losses.

MATERIALS FOR STEADY FLUXES. In this class permeability is important, and the static B/at curve provides a criterion for the choice of material. Mechanical considerations may, however, entail modification of the magnetic properties by the necessity of including alloy materials that increase the mechanical strength but reduce the permeability. Typical is the case of steel castings or forgings for high rotational speeds.

Castings. Low-carbon steels are used for stationary parts (e.g. yokes) not subject to large mechanical stresses. For rotating parts the carbon content is increased, at the expense of the permeability.

The magnetic properties of cast iron are poor, but considerable development has taken place in constructional cast irons.

A 10 per cent nickel, 5 per cent manganese cast iron is nearly non-magnetic ($\mu_r \simeq 1 \cdot 1$), and may be employed when it is desired to avoid shunting a magnetic field.

Forgings. Steel forgings are extensively used for the rotors of turbo-alternators, the rotor being forged and machined from a cast ingot. A high mechanical strength is an absolute essential. Over-hang clamping rings are also forgings, of steel or of non-magnetic chrome-nickel-manganese steel.

PERMISSIBLE FLUX-DENSITIES: Wb./m.^{2}

(1) *Core Plates* at 50 c/s. For teeth: 1·8–2·1. For cores: 0·7–1·1
For transformers: 1·0–1·55, depending on size, service conditions and quality of material.

(2) *Cast Steel.* For poles: 1·9. For yokes, etc.: 1·5.

(3) *Cast Iron.* For yokes, etc.: 0·7.

(4) *Air Gap.* For alternating flux: 0·6. For steady flux: 1·1.

4. Calculation of Magnetic Circuits. Having available the *B/at* curves for the materials employed, the magnetic circuit is divided up into convenient parts in series or parallel, the several flux-densities estimated for a given set of conditions, and the correspond-ing values of *at* found and multiplied by the length of the part. A summation of the total ampere-turns in series gives the total required. Some of the parts forming natural divisions (e.g. the gap or teeth) are rather complex magnetic problems, for which special methods such as those below have been developed.

Let
$L =$ core length, m.;
$L_i =$ net iron length in core, m.;
$l_g =$ gap length, m.;
$y_s =$ slot-pitch, m.;
$w_o =$ slot-opening, m.;
$n_d =$ number of ventilating ducts;
$w_d =$ width of ventilating ducts, m.

AIR GAP. The gap is bounded on either side by iron surfaces, either or both of which may be dentated, provided with duct-openings, or tapered, in such a way as to make it difficult by ordinary simple methods to obtain an accurate estimate of the reluctance of the gap, or the number of ampere-turns to carry a given flux across the gap. The problem is generally solved by use of the analytical results of Carter.* In Fig. 114(*a*), the presence of the slots or ducts increases the gap reluctance by restricting or "contracting" the flux to a degree depending on the width of the opening and the length of the gap. The permeance of the gap over the slot-pitch y_s, per m. axial length of slot, is *less* than if the slots were ignored (Fig. 114 (*b*)) but *greater* than that given by the assumption in (*c*) that the slot portions convey no flux. The effective contracted width of y_s is

$$y_s' = y_s - k_o w_o . \qquad . \qquad . \qquad . \qquad . \qquad . \quad (49)$$

* *J.I.E.E.*, 29, p. 925; 34, p. 47 (1896)

where k_o is a function of the ratio (w_o/l_g), obtained from Carter's curves, Fig. 115.

Similarly, the radial ducts as in Fig. 114 can be accounted by contracting the axial length L of the gap surface to L' where

$$L' = L - k_d n_d w_d \quad . \qquad . \qquad . \qquad . \qquad . \qquad . \quad (50)$$

The coefficient k_d is a function of the ratio (w_d/l_g), from Fig. 115. If, as is frequently the case in practice, the ducts in stator and rotor

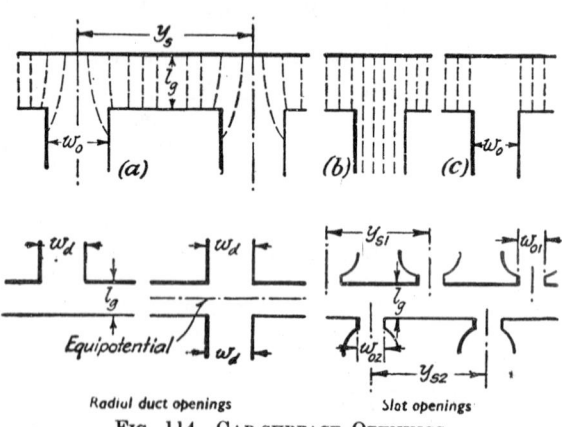

Radial duct openings   Slot openings
FIG. 114.   GAP-SURFACE OPENINGS

coincide, the central plane of the gap may be considered to be an equipotential surface and eq. (50) applied, but with k_d found for the ratio $(w_d/\frac{1}{2}l_g)$: for it would be obviously wrong to contract for each ventilating duct separately.

In certain cases both sides of the gap are slotted. A more convenient way of treating such conditions is to consider the gap length l_g to be *lengthened* to l_g' to account for the increased reluctance due to slot-openings. The following expressions then apply, the Carter function being taken from the curve for "open" or "semi-closed" conditions according to circumstances—

$$l_g' = k_g l_g, \quad . \qquad . \qquad . \qquad . \qquad . \qquad . \qquad . \quad (51)$$

where $k_g = k_{g1} \times k_{g2}$; $k_{g1} = \dfrac{y_{s1}}{y_{s1} - k_{o1} w_{o1}}$; $k_{g2} = \dfrac{y_{s2}}{y_{s2} - k_{o2} w_{o2}}$.

The several quantities are shown in Fig. 114: k_{o1} and k_{o2} are functions respectively of (w_{o1}/l_g) and (w_{o2}/l_g), obtained from Fig. 115.

The ampere-turns for the gap at a salient pole are

$$AT_g = 800\,000 K_g B_g l_g \quad . \qquad . \qquad . \qquad . \quad (52)$$

where l_g is the gap-length at the pole-centre, B_g is the maximum gap density (at the same point), and $K_g = (L/L') \times (y_s/y_s')$ takes account of slot openings and ducts. Here B_g can be taken as the average density $\overline{B}$ multiplied by the ratio (pole-pitch/pole-arc). With long gaps the pole-arc may be considered as extended by, say, $4l_g$, to take account of fringing.

GAP-DENSITY DISTRIBUTION. It is sometimes of advantage to estimate the harmonic content of the gap-density distribution.

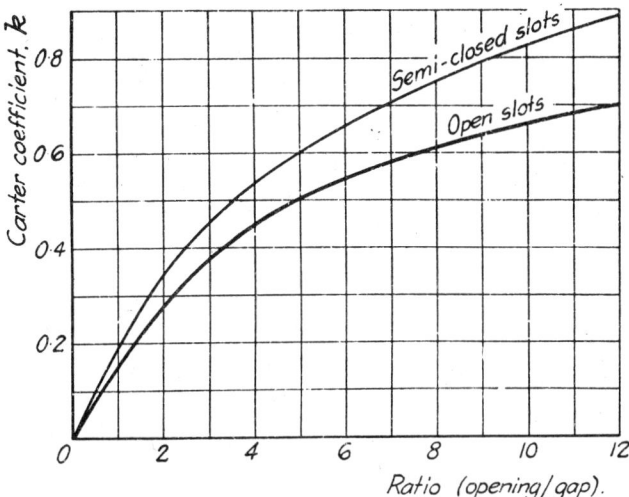

FIG. 115. CARTER COEFFICIENTS FOR AIR GAPS

A method giving such an estimate for odd harmonics up to the seventh (the evens being normally absent) is illustrated in Fig. 116 (a). The outline of the half pole-arc is laid out to a large scale and the stator half-pole-pitch divided into six intervals, with 0 at the interpolar axis and 6 on the pole-centre. From each division-mark is sketched a flux line, leaving and entering the iron surfaces perpendicularly, using as an aid the rules for flux-plotting by curvilinear squares. It is then assumed that the flux-density at the pole-centre, where the gap length is l_g, is 100; and that at other points the density is 100 l_g/l, where l is the length of the flux lines sketched. A plot (b) of the result shows a positive value at position 0. This should be zero, so a straight line is drawn from the ordinate of B at position 0 down to zero at the point where the pole bevel begins. The actual flux-density is taken as the ordinate between the straight line and the original B-plot, on the assumption that the flux so subtracted is a result of the adjacent pole of opposite polarity. The net curve, indicated by the values $b_0(= 0)$, b_1, b_2 . . . b_5, b_6

($= 100$) can now be analysed. The coefficients of the harmonics and fundamental are—

$$B_1 = 0{\cdot}086b_1 + 0{\cdot}167b_2 + 0{\cdot}236b_3 + 0{\cdot}289b_4 + 0{\cdot}323b_5 + 0{\cdot}167b_6;$$
$$B_3 = 0{\cdot}236b_1 + 0{\cdot}333b_2 + 0{\cdot}236b_3 + 0 \qquad\ - 0{\cdot}236b_5 - 0{\cdot}167b_6;$$
$$B_5 = 0{\cdot}323b_1 + 0{\cdot}167b_2 \quad 0{\cdot}236b_3 - 0{\cdot}289b_4 + 0{\cdot}086b_5 + 0{\cdot}167b_6;$$
$$B_7 = 0{\cdot}323b_1 - 0{\cdot}167b_2 - 0{\cdot}236b_3 + 0{\cdot}289b_4 + 0{\cdot}086b_5 - 0{\cdot}167b_6.$$

A closer approximation may be made by dividing the half-pole-pitch into twelve parts. The multipliers required are—

	b_1	b_2	b_3	b_4	b_5	b_6
B_1	0·022	0·043	0·064	0·083	0·102	0·118
B_3	0·064	0·118	0·154	0·167	0·154	0·118
B_5	0·115	0·161	0·154	0·083	− 0·022	− 0·118
B_7	0·132	0·161	0·064	− 0·083	− 0·165	− 0·118

	b_7	b_8	b_9	b_{10}	b_{11}	b_{12}
B_1	0·132	0·144	0·154	0·161	0·165	0·083
B_3	0·064	0	− 0·064	− 0·118	− 0·154	− 0·083
B_5	− 0·165	− 0·144	− 0·064	0·043	0·132	0·083
B_7	0·022	0·144	0·154	0·043	− 0·102	− 0·083

In either case the equation to the flux distribution is

$$B = B_1 \sin\theta + B_3 \sin 3\theta + B_5 \sin 5\theta + B_7 \sin 7\theta.$$

The *average* value is $\qquad \bar{B} = (2/\pi)(B_1 + \tfrac{1}{3}B_3 + \tfrac{1}{5}B_5 + \tfrac{1}{7}B_7)$

and the *r.m.s.* value $\qquad B = \sqrt{[\tfrac{1}{2}(B_1{}^2 + B_3{}^2 + B_5{}^2 + B_7{}^2)]}.$

EXAMPLE. A flux-plot drawn as above, and having the shape shown in Fig. 116 (*b*), yields the following values: $b_1 = 10$; $b_2 = 31$; $b_3 = 68$; $b_4 = 100$; $b_5 = 100$; $b_6 = 100$. Applying the expressions above for the fundamental and harmonics—

$$B_1 = 0{\cdot}86 + 5{\cdot}16 + 16{\cdot}72 + 28{\cdot}9 + 32{\cdot}3 + 16{\cdot}7 = 100{\cdot}6\%;$$
$$B_3 = 2{\cdot}36 + 10{\cdot}32 + 16{\cdot}04 + 0 \quad - 23{\cdot}6 - 16{\cdot}7 = - 11{\cdot}6\%;$$
$$B_5 = 3{\cdot}23 + 5{\cdot}16 - 16{\cdot}04 - 28{\cdot}9 + 8{\cdot}6 + 16{\cdot}7 = - 11{\cdot}2\%;$$
$$B_7 = 3{\cdot}23 - 5{\cdot}16 - 16{\cdot}04 + 28{\cdot}9 + 8{\cdot}6 - 16{\cdot}7 = 2{\cdot}8\%.$$

These components are plotted in Fig. 116 (*c*). The average density is

$$\bar{B} = (2/\pi)\,[100{\cdot}6 - (11{\cdot}6/3) - (11{\cdot}2/5) + (2{\cdot}8/7)] = 60{\cdot}4\%$$

and the r.m.s. value is

$$B = \sqrt{[\tfrac{1}{2}(100{\cdot}6^2 + 11{\cdot}6^2 + 11{\cdot}2^2 + 2{\cdot}8^2)]} = 101{\cdot}9/\sqrt{2} = 72\%.$$

The latter shows that the effect of the harmonics is negligible in the r.m.s. value. Further, the fundamental peak value is nearly the same as the actual maximum density: but this is not always so.

The method described is applicable to other cases of interest, particularly the flux distribution due to quadrature m.m.f. in a salient-pole machine.

TEETH: REAL AND APPARENT FLUX-DENSITY. The flux entering an armature from the air-gap follows paths principally in the iron. Where the tooth-density is high, the magnetomotive force is sufficient to produce an appreciable flux-density in the slots and ventilating ducts. An estimate of tooth-density based on the tooth area alone is consequently erroneous. Calling this apparent density B_t', and the true density B_t, the relation

$$B_t' = B_t + \mu_0 at(K - 1) \qquad (53)$$

enables the true density to be obtained as well as the corresponding m.m.f. per m. expressed in ampere-turns. The second term in eq. (53) refers to the relative amount of flux carried by the non-ferrous part of the magnetic path. The slot-factor K is obtained from

$$K = \frac{\text{combined section of iron and air}}{\text{net section of iron}}$$

$$(53a)$$

K has to be determined for each section taken.*

When reckoning the iron section the space-factor (about 0·9) of the laminated material must be incorporated. In practice K may have values be-

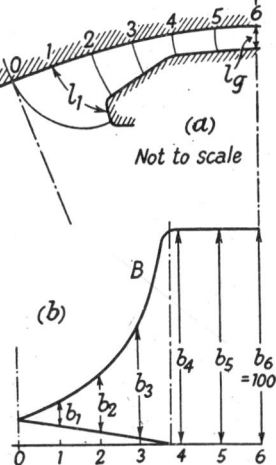

(a)

Not to scale

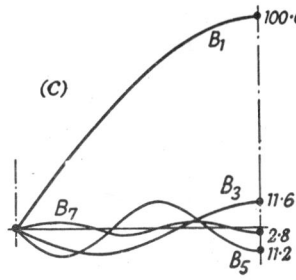

FIG. 116. ANALYSIS OF GAP-DENSITY DISTRIBUTION

tween about 1·5 for machines without ventilating ducts and 4 for the roots of the teeth of turbo rotors.

TAPERED TEETH. The slots punched in the core disks of machines of all but very small sizes are rectangular, so that the teeth are tapered. The high values of saturation in the teeth necessitate a careful estimation of the ampere-turn excitation. Apart from analytical methods, the simplest treatment of the problem is by the graphical or the three-ordinate methods.

(a) *Graphical Method.* From the known value of the flux per

---

* [A], p. 34.

tooth, the density is evaluated for a number of sections along the length of the tooth from tip to root correcting for true density B_t' if necessary by the method above. From a suitable B/at curve, the corresponding values of at are plotted and the mean obtained (e.g. by measuring the area under the curve). This mean value multiplied by the length of the tooth gives the necessary ampere-turn excitation. This method is suitable for turbo-rotor teeth, and is essential for difficult cases where accuracy is important.

(b) *Three-ordinate Method.* For cases of teeth of simple form and with slight taper, Simpson's Rule can be applied to the values of at obtained for three equidistant sections. If these be respectively at_1, at_2, and at_3, the mean value is given by $at = \frac{1}{6}(at_1 + 4at_2 + at_3)$. This is based on the assumption that the curve relating at with tooth-length is parabolic.

(c) *Density at one-third length from narrowest part.* For slightly tapered teeth with low saturation it may be sufficient to find the flux-density $B_{t\frac{1}{3}}$ at one-third of the tooth-length from the narrower end, and to base calculations on the assumption that this density (with its corresponding at value) obtains over the whole length.

(d) *Integration Method.* * An accurate practical method, employing factors obtained by graphical integration, can be applied over the whole range of flux densities and all proportions of teeth and openings.

MAGNETIZING CURRENT. The summation of the ampere-turns for the several series parts of a circuit gives the total excitation to be provided by the exciting winding. The current to be supplied depends upon the number of turns in the winding and upon the way in which it is distributed. The simplest case is that of a concentrated winding completely linked (or nearly so) with the total flux produced, as in the salient-pole machine or the transformer: for then, if AT ampere-turns are needed, a current $I = AT/T_p$ flowing in T_p turns will be required. The more complex cases are deferred until the details of windings have been considered, and only the results will be quoted.

For three-phase uniformly-distributed windings with 60° phase-spread and carrying sinusoidal currents, the m.m.f. curve is very roughly sinusoidal in shape (Fig. 136), and has a peak value F_a related to the r.m.s. magnetizing current per phase by the expression

$$F_a = 1{\cdot}35IT_{ph}K_w/p \text{ ampere-turns per pole} \qquad (54)$$

In the case of induction motors the flux distribution differs from the sinusoidal, and the tooth saturation causes the flux to have a pronounced flattening, Fig. 117. Thus B_m, the peak value, is not deducible from the e.m.f., turns and frequency. For moderate

* Neville, "Reluctance of the Teeth of a Slotted Armature," *Proc. I.E.E.*, 103 (C), p. 338 (1956).

saturation, the flattened flux-density curve can be considered as comprising a fundamental and superimposed third-harmonic flux. The latter will scarcely affect the e.m.f., so that calculations based on the value B_{30}, the density at 30° from the pole-centre (which is common to the actual flattened distribution and to the fundamental component upon which the e.m.f. depends) will give results sufficiently close to the true values for practical calculation. Thus

$$B_{30} = B_{m1} \cos 30°$$
$$= \tfrac{1}{2}\pi(\Phi_1/A)\,(\sqrt{3}/2)$$
$$= 1\cdot36(\Phi_1/A)$$

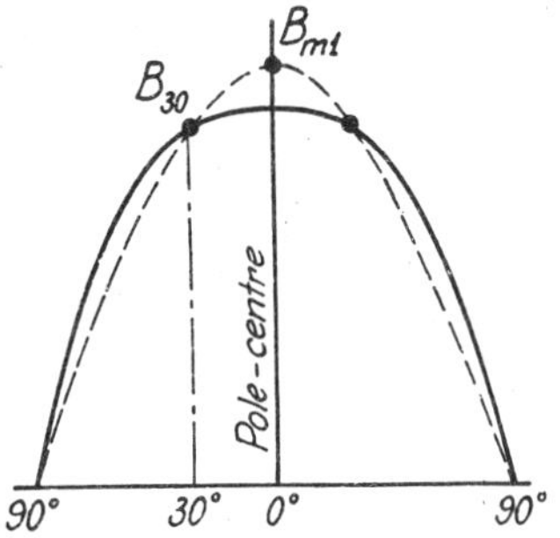

where A is the area through which Φ_1, the fundamental component flux, passes. The r.m.s. magnetizing current in terms of the excitation AT_{30} for B_{30} is

FIG. 117. GAP-FLUX DISTRIBUTION IN INDUCTION MOTOR

$$I = pAT_{30}/1\cdot35K_wT_{ph} \cos 30° = pAT_{30}/1\cdot17K_wT_{ph} \quad (55)$$

Examples in the calculation of magnetic circuits are given in Chapters XVI and XXI on design.

5. **Leakage.*** Leakage flux, in non-useful paths, affects the field excitation in the case of d.c.-excited machines, and the inductive reactance of a.c. windings. Its estimation is always difficult, on account of the complex geometry of the leakage paths, and great accuracy is generally unobtainable. The operating characteristics of induction motors depend intimately on the leakage reactance, however, so that formulae are closely checked against test data.

The flux per ampere flowing in a coil of T_x turns, surrounding a magnetic path having a magnetic reluctance S_x, is

$$\Phi_x = T_x/S_x = T_x\Lambda_x \qquad . \quad . \quad (56)$$

where $\Lambda_x = 1/S_x = \mu A_x/l_x$ is the permeance of the flux path of effective area A_x and length l_x.

The *leakage inductance* is the flux-linkage in weber-turns per ampere. For the coil of T_x turns this is $T_x\Phi_x = T_x{}^2\Lambda_x$ henrys. If the current in the coil pulsates at frequency f, it will have the *reactance*

$$X_x = 2\pi f T_x{}^2\Lambda_x \text{ ohms}. \qquad . \quad . \quad . \quad (57)$$

The difficulty in applying eq. (57) lies in the estimation of Λ_x. It is usual to assume that those parts of the leakage-flux path lying in iron will require a negligible proportion of the coil m.m.f., equivalent to the assumption that the iron has infinite permeability.

* See Carr, *Metro-Vick. Gaz.*, 16, pp. 426–431 (1937); also *J.I.E.E.*, 94 (II), p. 443 (1947).

Attention is then confined to the non-magnetic (or "air") parts of the flux path, and an attempt is made to predict the geometrical pattern of the leakage flux, using the iron surfaces as equipotentials. With various geometrical approximations the air path is then divided into suitable series or parallel sections for which the ratio A_x/l_x (the *permeance coefficient*) can be estimated. The total air-path

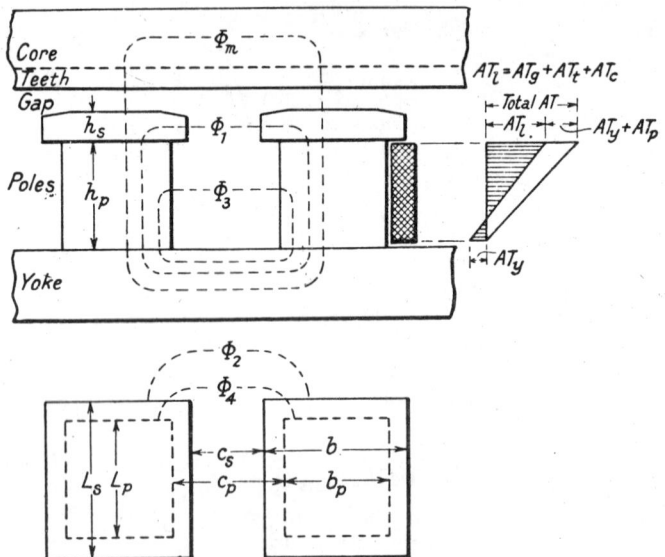

FIG. 118. LEAKAGE OF SALIENT POLES

permeance is now determined, and eqs. (56) or (57) evaluated.*
The permeance is

$$\Lambda_x = \mu_0 A_x/l_x = \mu_0 \lambda_x, \qquad . \qquad . \qquad . \quad (58)$$

and the problem of leakage flux and reactance reduces to the determination of the permeance coefficient λ_x. It will be observed that the absolute permeability μ becomes μ_0 because only non-magnetic flux paths are considered.

The leakage flux per ampere depends on the arrangement and geometry of a winding, not on the use to which the parent machine is put. The *effect* of the leakage on the *operation* does, however, depend on the kind of machine. In synchronous (and also d.c.) machines, the field leakage remains stationary with respect to the field winding and there is no e.m.f. of self-inductance developed by the leakage flux under normal conditions of working. In the induction motor the rotor currents alternate, producing an e.m.f. of

* For analytical solutions of a number of simplified cases, see Hague, *Electromagnetic Problems in Electrical Engineering* (Oxford).

leakage reactance, the equivalent of which appears in the stator winding as part of the total leakage-reactance e.m.f. A résumé of the more important practical cases is given below.

1. LEAKAGE OF SALIENT POLES. This problem is dealt with by approximation. Most low-speed, salient-pole alternators of usual size have many poles on a large diameter, so that the pole-axes are roughly parallel, and a simpler method is made possible.

The leakage flux is split up into the component parts shown in Fig. 118. The ampere-turns AT_l producing leakage flux on no load comprise those for the gap, teeth and core. Approximately, AT_l produces Φ_1 and Φ_2, while $\frac{1}{2}AT_l$ produces Φ_3 and Φ_4. Taking the leakage between facing sides to be in straight lines, and that between parallel faces (Φ_2 and Φ_4) as straight lines + two quarter circles, the estimation is

Leakage between pole-shoes, per pole,

$$\Phi_{sl} = 2\Phi_1 + 4\Phi_2$$
$$= 2 \cdot \mu_0 AT_l \cdot \frac{L_s h_s}{c_s} + 4 \cdot \mu_0 AT_l \frac{h_s}{\pi} \log h \frac{c_s + \frac{1}{2}\pi b}{c_s}$$
$$= \mu_0 AT_l[2L_s h_s/c_s + 2 \cdot 94 h_s \log_{10}(1 + \frac{1}{2}\pi b/c_s)] \qquad . \qquad (59a)$$

Leakage between poles, per pole,

$$\Phi_{pl} = 2\Phi_3 + 4\Phi_4$$
$$= \mu_0 AT_l[L_p h_p/c_p + 1 \cdot 47 h_p \log_{10}(1 + \frac{1}{2}\pi b_p/c_p)] \qquad . \qquad (59b)$$

Φ_{sl} will generally be of the order $20AT_l \cdot 10^{-8}$, while Φ_{pl} approximates in most cases to $80AT_l \cdot 10^{-8}$. Thus the flux at the back of the pole-shoes is the main flux Φ (which crosses the gap) + $20AT_l \cdot 10^{-8}$, while the flux at the root of the pole (next to the yoke) is approximately

$$\Phi + (20 + 80)AT_l \cdot 10^{-8} = \Phi + AT_l \cdot 10^{-6}.$$

2. LEAKAGE OF NON-SALIENT POLES. The non-salient-pole turbo rotor has considerable leakage if the retaining rings clamping the overhang of the winding are of magnetic metal; the amount is dependent upon the saturation flux density of the material, for the induction will reach this amount in all normal cases. A severe disadvantage of the leakage is that it increases the saturation of the rotor core and teeth (particularly in two-pole machines). Large modern machines have a considerable length/diameter ratio, rendering the leakage less important. Further, a relatively non-magnetic manganese steel has been developed for end-ring clamps, greatly reducing the leakage and incidentally the stray load losses.

3. ARMATURE LEAKAGE. The flux threading an armature is determined by the configuration of the parts of the magnetic circuit and of the windings, and by the currents carried by the latter. For

convenience, the whole existing flux is sub-divided into the *main* or *useful* flux which contributes to the energy conversion, and the *leakage* flux which contributes only to reversible energy storage and endows the associated winding with the properties of leakage inductance and reactance. Thus the actual flux, with all its electromagnetic results, is considered as the superposition of working and leakage fluxes, each with its own part to play in the action of the machine. But only in the case of the overhang or end-windings can

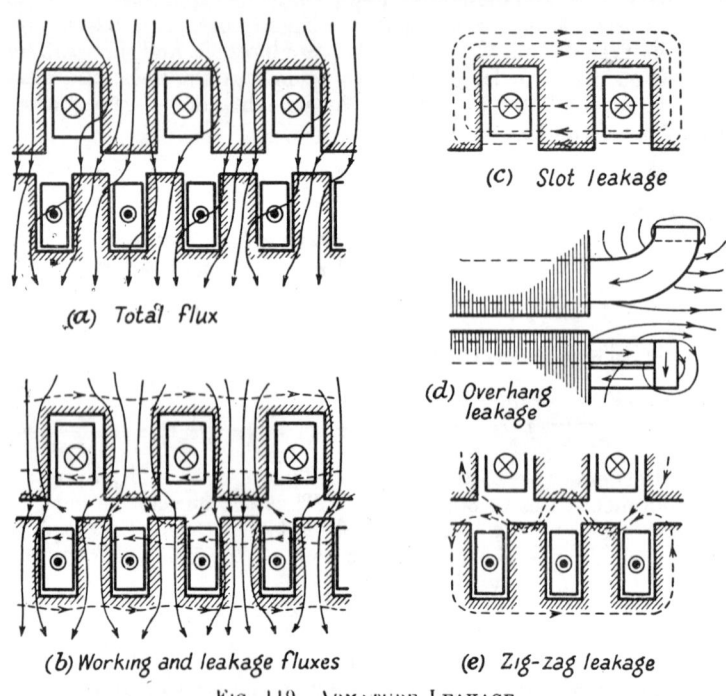

(a) Total flux

(b) Working and leakage fluxes

(c) Slot leakage

(d) Overhang leakage

(e) Zig-zag leakage

Fig. 119.   Armature Leakage

the leakage flux be actually distinguished as a separate entity: within the mass of the armature the leakage and working fluxes are those which, superposed, yield the actual distribution of total flux.

Fig. 119 (a) shows an air gap bounded on each side by a slotted iron mass, the slots carrying conductors with suitable currents. (In the case of the induction motor the currents in the two windings are such as to give an approximate equality of ampere-turns per pole, with opposition of m.m.f.) The lines of total flux are approximately as indicated in (a). In (b) the flux has been resolved into components of working flux and leakage flux which together sum to the total flux in (a) at every point. The working flux is then that which links the windings on *both* sides of the air gap; while lines linking either winding but not both, constitute the leakage flux. Two main physical subdivisions of the leakage flux are: (1) the overhang leakage, Fig. 119 (*d*), and (2) the circumferential gap leakage indicated by the dotted lines in Fig. 119 (*b*), and caused by the "solenoid" action of the two opposing sets of ampere-turns in combination on the gap. A further subdivision of (2) may be made for convenience of calculation, viz.: *slot* and *zig-zag* components.

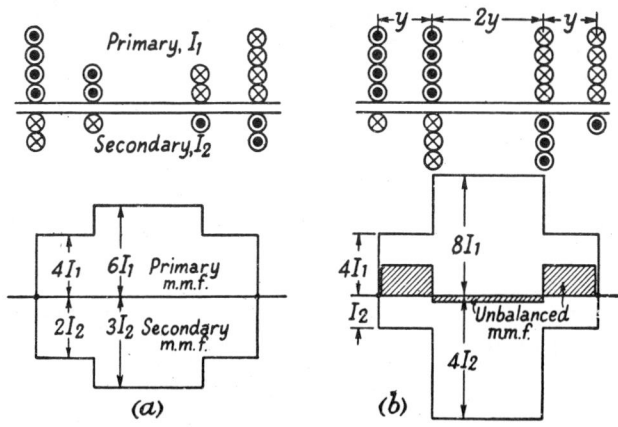

Fig. 120. Differential Leakage

An additional effect, called *differential* or *harmonic* leakage, though of different origin, has an effect comparable to that of a leakage flux.

(1) *Overhang leakage* is a true, separate leakage flux and not an arbitrary component of a complex flux structure. Its value depends on the arrangement of the end-windings and on the proximity thereto of metal—particularly ferromagnetic—masses such as core-stiffeners and end covers. Fig. 119 (*d*).

(2) (*a*) *Slot leakage* crosses the conductors from one tooth to the next. It is in phase with the current for all conductors in a given phase-group, and depends only on the magnitude of the current, neglecting saturation. Fig. 119 (*c*).

(*b*) *Zig-zag leakage* depends on the gap length and on the relative position of the two sets of tooth tips. It tends to "zig-zag" across the gap, utilizing the relatively high permeance of the tooth-tips. Fig. 119 (*e*).

Differential leakage. This is a result of a difference between the conductor distributions or the current distributions on the two sides of the gap. Consider Fig. 120 (*a*), where two windings with identical distributions have a turn-ratio of 2 : 1. It is obvious that if one, the primary, is excited and the other, the secondary, is open-circuited, the e.m.f.-ratio will be 2 : 1. If the secondary has zero slot, overhang and zig-zag leakage, and is short-circuited, it must carry a current I_2 such that its total linkages are zero, and for this $I_2 : I_1 = 2 : 1$. Let the two windings now be distributed as in Fig. 120 (b), with a turn-ratio of 2 : 1 as before, but a dissimilar distribution of conductors. The e.m.f. ratio with the primary excited and the secondary on open circuit will no longer be 2 : 1, because the e.m.f. depends on the distribution. Further, on second-

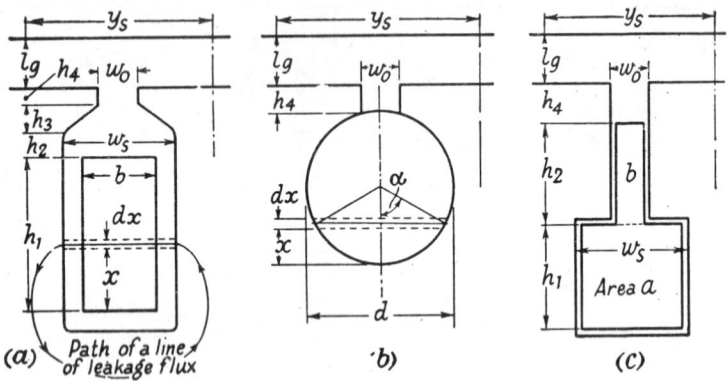

Fig. 121. Slot Leakage

ary short-circuit under the same conditions as before, the secondary total linkages must be zero, i.e.

$$[3(8 \cdot 2y) + 1(8 \cdot 2y + 4 \cdot 2y)]I_1$$
$$- [3(4 \cdot 2y) + 1(4 \cdot 2y + 1 \cdot 2y)]I_2 = 0$$

i.e. $\qquad 72yI_1 - 34yI_2 = 0$, giving $I_2 = (72/34)I_1$,

a ratio differing from 2. The primary linkages, however, are

$$[4(8 \cdot 2y) + 4(8 \cdot 2y + 4 \cdot 2y)]I_1 - [4(4 \cdot 2y) + 4(4 \cdot 2y + 1 \cdot 2y)]I_2$$

i.e. $\qquad 160yI_1 - 72I_2 = (160 - 72^2/34)yI_1 = 7 \cdot 5yI_1.$

Thus while the resultant flux distribution must give zero e.m.f. in the secondary, it does not do so in the primary in spite of the fact that it links both windings. The net primary flux acts as leakage, and is called the *differential* or *harmonic* leakage flux. An alternative approach is by the harmonic resolution of the m.m.f. distributions. The differential leakage is chiefly the result of the higher space harmonics, and for normal windings may be small enough to ignore. In a cage rotor, the currents at every point can balance the primary or stator currents on the opposite gap surface, so that differential leakage vanishes, except for those harmonics whose pole-pitch is comparable with the slot-pitch.

4. Slot Leakage. From Fig. 121(a), assuming that the current in

the slot-conductors is distributed uniformly over their cross-section, the leakage flux produced by the current may be considered to have a path straight across the slot and round the iron at the bottom of the slot,* an assumption that, as usual, is only roughly approximate. The reluctance of the path is assumed concentrated in the slot portion. The permeance coefficient per m. of axial slot-length is the ratio (area/length) of path. For this permeance, a number of parallel parts can be considered, viz. (h_2/w_s) for the part above the conductor; $2h_3/(w_s + w_o)$ for the wedge portion; and (h_4/w_o) for the lip. The slot-height occupied by the conductor has to be treated differently. All the flux *above* the conductor links it as a whole. The flux *within* the height x, however, only links the lower part of the conductor, and contributes to the inductance of that part only. Further, such flux has a lower density than in the part h_2, as only a fraction (x/h_1) of the current is available to provide the m.m.f. for this flux.

At height x, per ampere in the whole conductor, the current linked by a flux passing through the elemental path dx is (x/h_1) ampere, and the flux correspondingly has the value

$$d\Phi_x = \mu_0(x/h_1)\,(dx/w_s)$$

and links only the portion (x/h_1) of the conductor, so that its effective value for inductance is

$$d\Phi_x T_x = \mu_0(x/h_1)^2(dx/w_s).$$

The permeance coefficient of the conductor portion is thus

$$\int_0^{h_1} \frac{x^2}{h_1^2 w_s}\, dx = \frac{h_1}{3w_s}.$$

Adding the several parallel coefficients, the specific slot permeance is obtainable from

$$\lambda_s = \frac{h_1}{3w_s} + \frac{h_2}{w_s} + \frac{2h_3}{w_s + w_o} + \frac{h_4}{w_o} \qquad . \qquad . \qquad . \quad (60)$$

The total slot permeance of a core-length L_s (net) will then be the product $\mu_0 L_s \lambda_s$. Because λ is obtained from ratios, it is unaffected by the units (e.g. mm. or inches) in which h and w are given.

A *round* slot, Fig. 121 (*b*), is treated in the same way. Such a slot may be used for the cage bars of an induction motor rotor, and the conductor may be assumed to fill the circular space. Again considering the flux lines to pass straight across the slot at right angles to the centre-line, the length of an elemental path dx is $d \sin \alpha$, and $x = \frac{1}{2}d(1 - \cos \alpha)$. The ratio of the area of a segment of height x

---

* Many analytically determined flux-distributions similar to this are given by Hague, *Electromagnetic Problems in Electrical Engineering* (Oxford); e.g p. 188.

to the whole circular area is $(2\alpha - \sin 2\alpha)/2\pi$. The specific permeance coefficient of the slot portion is

$$\int_0^d \frac{(2\alpha - \sin^2 2\alpha)^2}{4\pi^2} \cdot \frac{dx}{d \cdot \sin \alpha} = \frac{1}{16\pi^2} \int_0^{2\pi} (2\alpha - \sin 2\alpha)^2 d(2\alpha) = 0.623,$$

which is independent of the diameter. The figure is generally taken as 0.66, so that

$$\lambda_s = 0.66 + \frac{h_4}{w_0} \qquad . \qquad . \qquad . \qquad (61)$$

For a round slot containing no conductor, or a conductor carrying no current, the permeance coefficient is

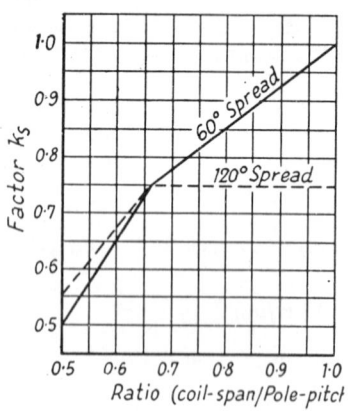

FIG. 122A. SLOT LEAKAGE FACTOR　　　　FIG. 122B. DIFFERENTIAL LEAKAGE FACTOR

$$\int_0^d \frac{dx}{d \cdot \sin \alpha} = \int_0^{2\pi} \frac{\frac{1}{2}d \sin \alpha \cdot d\alpha}{d \cdot \sin \alpha} = \frac{1}{2}\int_0^{2\pi} d\alpha = 1.57.$$

Another slot-shape found in induction-motor rotors is that in Fig. 121 (c), used with skin-effect conductors (see Fig. 189). The current distribution is designedly non-uniform at starting and at low speeds when the rotor frequency approaches that of the stator currents, e.g. 50 c/s. For *running* conditions at normal speeds the frequency is low enough to assume uniform current distribution, and the permeance coefficient is

$$\lambda_s = \frac{1}{(a+b)^2}\left[\frac{h_1}{w_s} \cdot \frac{a^2}{3} + \frac{h_2}{w_0}\left(a^2 + ab + \frac{b^2}{3}\right)\right] + \frac{h_4}{w_0} \qquad . \quad (62a)$$

The total conductor section is $a + b$, the lower part a and the upper part b sharing the current in proportion as their area. For *starting*, the current is confined substantially to b: the permeance estimate is rather complex,* and an empirical expression is

$$\lambda_s = \frac{h_1}{2w_s} + \frac{h_2}{3w_o} + \frac{h_4}{w_o} \qquad (62b)$$

Effect of Chording. Chording a stator or rotor winding affects its winding factor K_w, as shown in Chapter X following. The leakage is also affected because some slots hold conductors whose currents belong to different phases. The effective leakage permeance is reduced by a factor k_s, shown in Fig. 122A.

5. OVERHANG LEAKAGE. Theoretical methods fail completely to deal with the overhang-leakage problem, and only simple empirical formulae are justifiable. Many have been suggested, most designers being able to make an estimate suitable for the machines with which they are familiar. The leakage must be related in some way to the length of the end-connectors and their shapes, the types of winding, the spacing between stator and rotor overhangs, and the proximity of magnetic housings: but to include functions of all these would be impossibly complicated. One form of expression is

$$L_o\lambda_o = k_s Y^2/\pi y_s \qquad . \qquad . \qquad . \qquad (63)$$

where k_s is obtained from Fig. 122A.†

6. ZIG-ZAG LEAKAGE. Fig. 119 (e) shows that zig-zag leakage will depend on the relative widths of teeth, slot-openings and air-gap. One expression (out of many) gives the reactance per phase due to zig-zag leakage in terms of the reactance offered by the main flux. In induction motors this is $x_m \simeq V/I_m$, where V is the applied voltage and I_m the magnetizing current, in each case per phase. The required reactance is then

$$x_z = \frac{5}{6}x_m\left[\frac{1}{g_1^2} + \frac{1}{g_2^2}\right] \text{ ohms per phase} \qquad . \qquad . \qquad (64)$$

where $g_1 = S_1/2p$ and $g_2 = S_2/2p$ are the numbers of slots per pole.

7. DIFFERENTIAL LEAKAGE. An expression for differential, belt, or harmonic leakage is

$$x_h = x_m(k_{h1} + k_{h2}) \qquad . \qquad . \qquad . \qquad (65)$$

where values of the coefficients k_h are obtained from Fig. 122B.† This reactance is ignored for machines with cage windings.

6. **Phase Reactance.** The reactance of a phase of $g'z_s$ turns per pole-pair, with a conductor length of L having specific permeance λ, is, from eqs. (57) and (58).

$$X = 2\pi f T_x^2 \Lambda_x = 2\pi f(g'z_s)^2\mu_0 . 2L(\lambda/g') \text{ ohms,}$$

since the turn length includes two conductors, and the flux has to traverse g' slots, increasing the reluctance. If the turns in all the

* Bruges, *Proc. Royal Soc. Edin.*, 62 (II), p. 175 (1946).

† Alger, "Calculation of the Armature Reactance of Synchronous Machines," *Am.I.E.E., Trans.*, 47, p. 433 (1928).

pole-pairs are in series, the number of turns per pole-pair per phase is T_{ph}/p, and the phase reactance is

$$X = 4\pi f\mu_0 p(T_{ph}/p)^2 L(\lambda/g') = 4\pi f\mu_0(T_{ph}^2/pg')L\lambda$$
$$= 15 \cdot 8 \cdot 10^{-6} f(T_{ph}^2/pg')L\lambda \text{ ohms} . \qquad . \qquad . \qquad . \quad (66)$$

7. Phase Reactance of an Induction Motor.* The total phase reactance of an induction motor referred to the stator is

$$X_1 = x_1 + x_2' = x_{s1} + x_{s2}' + x_o + x_z + x_h \qquad . \quad (67)$$

where x_{s1} = stator slot reactance using the net core-length L_s in eq. (66) with the appropriate permeance from eqs. (60–62);

x_{s2} = rotor equivalent slot reactance calculated as for x_{s1}, then multiplied by $K_{w1}^2 S_1/K_{w2}^2 S_2$;

x_o = overhang reactance from eq. (63);

x_z = zig-zag reactance from eq. (64);

x_h = differential reactance from eq. (65).

For calculations of motor performance from an equivalent circuit, it is quite sufficient to assume that $x_1 = x_2' = \frac{1}{2}X_1$: alternatively the total reactance may be divided in inverse proportions to the numbers of slots, giving $x_1/x_2' = g_2/g_1 = S_2/S_1$. Since the leakage flux—apart from slot-leakage—is a result of the combination of stator and rotor currents, it is not strictly possible to separate it into individual contributions.

8. Phase Reactance of a Synchronous Machine. For a synchronous machine the leakage flux is required to be known to assess regulation. For this purpose a close estimate is not essential. The reactance, however, enters into many problems of much greater complexity, such as short-circuit behaviour and the efficacy of pole-face damping windings. It is not possible to do justice here to these important matters, which have occupied designers for many years.† For the calculation of voltage regulation, the following simplified formulae may be used. The leakage flux is

$$\Phi_l = \Phi_s + \Phi_o \qquad . \qquad . \qquad . \qquad . \quad (68)$$

where Φ_s is the slot leakage flux and Φ_o the overhang leakage flux for full-load current J per conductor. With eq. (56) as a basis, and writing $\mu_0 2L_s\lambda_s$ for Λ_x, the maximum leakage flux associated with the slot portions of a coil of T_c turns is

$$\Phi_x = (\sqrt{2})JT_c \cdot \mu_0 2L_s\lambda_s.$$

---

* A discussion of this subject is given by Carr, *Metro-Vickers Gaz.*, 16, p. 426 (1937); and of leakage reactance in general, *J.I.E.E.*, 94, Part II, p. 443 (1947).

† E.g. Kilgore "Calculation of Synchronous Machine Constants," *Am.I.E.E.*, Trans., 50, p. 1201 (1931); and 54, p. 545 (1935).

For a group of g' coils of the same phase, the turns producing the flux are now $g'T_c$, but the flux has to cross g' slots and its air path is g' times as long. Thus the slot leakage flux is

$$\Phi_s = 2\sqrt{2}\mu_0 JT_c L_s \lambda_s \qquad \cdot \qquad \cdot \qquad \cdot \quad (69)$$

Similarly for the overhang leakage

$$\Phi_o = 2\sqrt{2}\mu_0 JT_c L_o \lambda_o \qquad \cdot \qquad \cdot \qquad \cdot \quad (70)$$

where λ_s and $L_o \lambda_o$ are obtained from eqs. (60–63).

There is no zig-zag leakage in a normal salient-pole synchronous machine because the pole face is an equipotential surface in the quadrature axis, and is not presented at all to direct-axis current components.

CHAPTER X

ARMATURE WINDINGS

1. **Types of Windings for A.C. Machines.** The essential feature of an electrical machine is the electric circuit or *armature winding* in which the working e.m.f. is induced. Where d.c. excitation is required, for the magnetic circuit, a *field winding* is a further requirement.

2. **Armature Windings.** The a.c. armature winding may be defined as one in which, when in motion relative to a heteropolar magnetic field, e.m.f.'s are generated in a number of sections or *phases*. The e.m.f.'s are equal in magnitude and correctly displaced in time-phase relationship. The winding is usually composed of small groups of conductors, in slots spaced round the periphery of the armature, and connected together with the ends brought out to form a phase winding independent of other similar phases.

The a.c. winding differs from the commutator winding of a d.c. machine in that commutation is not required and the question of mechanical balance does not usually arise. The closed connection and the commutator of the d.c. armature winding are necessary so that the e.m.f. between the commutator brushes will be in effect an integration of the flux wave between corresponding points on the armature surface, and will remain a virtually constant e.m.f. although the conductors contributing to it are continually replaced by others. The a.c. winding, however, must produce an alternating e.m.f. corresponding to the actual space distribution of the alternate N- and S-polar flux in the airgap. It thus resolves itself into separate phase groups and could in fact be reduced to a single conductor or turn in each phase.

In a d.c. machine the total flux per pole determines the e.m.f. generated, regardless of its space distribution. But a.c. machines should have a flux distribution as near as possible to a sine wave in order to provide the greatest sinusoidal e.m.f. from a given flux per pole.

CONDUCTORS, TURNS AND COILS. A conductor of length l m. moving laterally at uniform speed u m. per sec. across a magnetic field of density B webers per m.^{2} has an induced e.m.f.

$$e - Blu \text{ volts} \qquad . \qquad . \qquad . \qquad . \qquad (2)$$

If the flux be assumed distributed sinusoidally in space, the e.m.f. increases and decreases with the varying strength B of the flux in which it moves, and reverses when the polarity of the flux reverses: i.e. it is an *alternating* e.m.f. Fig. 123 (a) shows a double pole-pitch of a machine in which the spatial distribution of the gap flux-density over the circumference is sinusoidal. A conductor, l in the position shown has an e.m.f. proportional to the density B in which it is moving. A second conductor, 2, displaced a pole-pitch from 1, moves in a flux density of the same magnitude but opposite polarity. The instantaneous conductor e.m.f.'s e_1 and e_2 are therefore equal but oppositely directed. Let the conductors be joined together at one end to form a single turn: then the e.m.f.'s e_1 and e_2 are additive

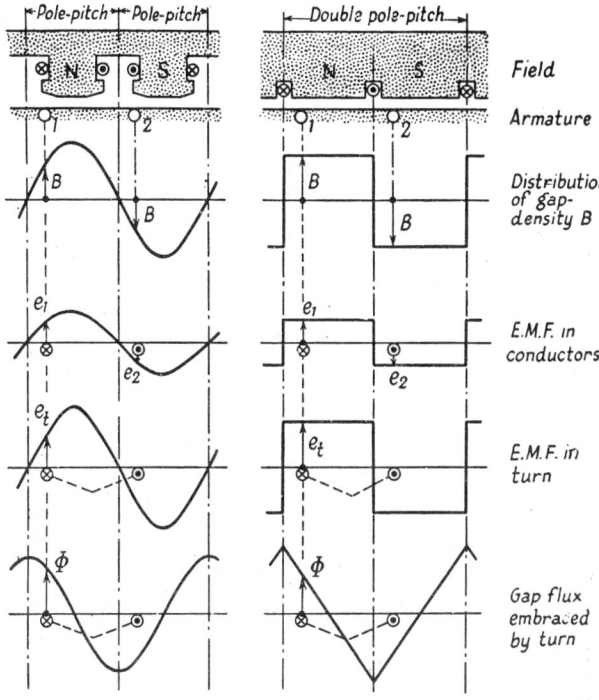

(a) Sinusoidal gap-density (b) Rectangular gap-density

Fig. 123. Rotational E.M.F. in Armature Conductors and Turn

round the turn and sum to the turn-e.m.f. e_t. This is plotted in Fig. 123 (a) in accordance with the position of conductor 1. Alternatively the turn-e.m.f. can be found by applying eq. (1), page 6: $e_t = -(d\Phi/dt)$ volts, on a basis of the rate of change of the linked flux. At the foot of Fig. 123 (a) is plotted the flux Φ embraced by the turn, again in accordance with the position of the turn as defined by conductor 1. For when conductor 1 is at the left-hand magnetic neutral axis the turn embraces the whole of a north-polar flux. Comparison of the curves of e_t and Φ shows that the former is indeed given by the negative rate-of-change of the latter as the turn passes across the poles.

Induced E.M.F.'s. Let the sinusoidally-distributed flux per pole be Φ. The average flux density in the gap will be Φ/YL and the peak density $B_m = (\pi/2)\Phi/YL$. For a speed of rotation of n rev. per sec., the peripheral speed of a conductor will be $u = \pi Dn = 2pYn$ m. per sec. for p pole-pairs and a pole-pitch Y m. The peak e.m.f. of a conductor will then be

$$e_m = B_m Lu = (\pi/2)(\Phi/YL) . L . 2pYn$$
$$= \pi pn\Phi \text{ volts.}$$

Writing $pn = f$, the r.m.s. value will be

$$E = (\pi/\sqrt{2})f\Phi = 2{\cdot}22f\Phi \text{ volts.}$$

For a turn the e.m.f. will be twice this value, and for a coil of T_c concentrated turns the r.m.s. coil-e.m.f. is

$$E_c = 4{\cdot}44fT_c\Phi \text{ volts} \quad . \quad . \quad . \quad . \quad (71)$$

From the linkage viewpoint, it is seen from Fig. 123 (a) that the flux embraced by a turn is $\Phi \cos \omega t$, and the e.m.f. of a coil of T_c turns is

$$e_c = - T_c \frac{d}{dt} (\Phi \cos \omega t) = \omega T_c \Phi \sin \omega t.$$

The peak value is $\omega T_c \Phi = 2\pi fT_c\Phi$ and the r.m.s. value is obtained by dividing by $\sqrt{2}$ to give eq. (71) above, as before.

Since in the example discussed the conductor and turn e.m.f.'s are sine waves, they may be represented by complexors, rotating once for each movement of the conductors across a double pole-pitch.

Fig. 123 (b) shows the e.m.f.'s for a rectangular distribution of the gap density. The discussion is on similar lines, except that eq. (71) does not apply and the e.m.f.'s cannot be represented as complexors unless first analysed into fundamental and harmonic sine wave components.

Connections. In practice the conductors, situated in slots adjacent to the air gap, are universally arranged in series pairs to form turns. A coil may have one or more turns. The connections linking the conductors form the *end-connectors* or, in the mass, the *overhang*. When coil-sides are spaced by exactly a pole-pitch, as in Fig. 123, they are said to be of *full pitch*. The pitch of a coil may be greater or—more usually—less than a pole-pitch, in which case it is described as *short-pitched* or *chorded*.

WINDINGS. The slots in which the coils lie are punched uniformly round the gap surface of the armature in nearly every case. The problem of winding is to arrange and connect the coils in the several slots to obtain the required phase-groupings. In a *single-layer winding* each coil-side occupies a whole slot, whereas in a *double-layer winding* one coil-side lies in the upper position in a first slot while the other occupies the lower position in a second slot spaced a pole-pitch or less from the first. The appearance of a winding may be greatly altered by the use of various shapes of end-connections, but these are only examples of a number of ways of joining in series the conductors themselves, without affecting their electrical grouping.

PHASES. The requirements of an a.c. winding are that it shall produce a symmetrical N-phase system of e.m.f.'s of identical magnitude, wave-shape and frequency, and displaced in time-

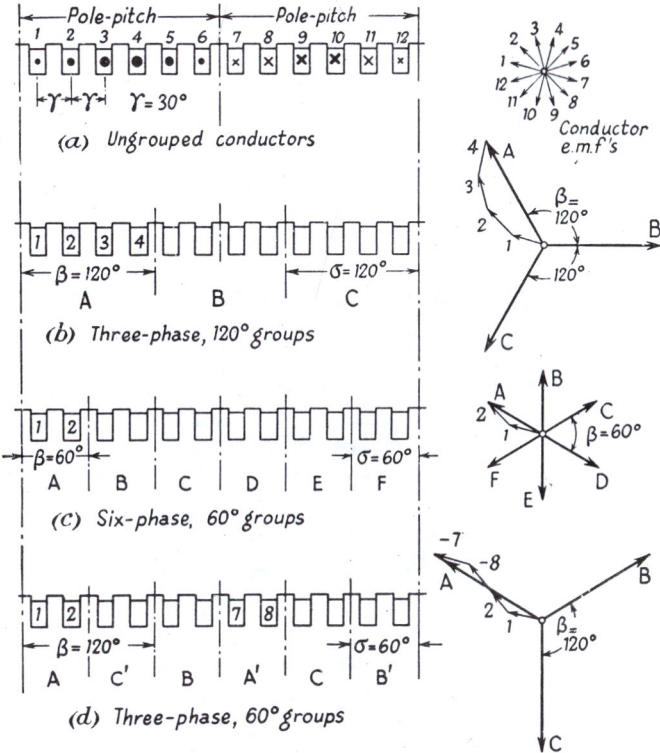

FIG. 124. PHASE GROUPINGS

phase by $\beta = 2\pi/N$ electrical radians. This is secured by having identical windings for all phase-groups, and arranging the groups with an effective space displacement of $2\pi/N$ electrical radians.

The most usual arrangement of phase windings is that in which the spread of each phase is $\sigma = \pi/N$, i.e. $1/N$ of a pole-pitch. The result is a sequence of phase-bands A C B A C B in the double pole-pitch of a three-phase machine.

In Fig. 124 (a) an arrangement of conductors in twelve slots is illustrated in a diagrammatic development of a double pole-pitch.

The corresponding conductor-e.m.f. complexors are shown, it being assumed that the gap flux density is sinusoidally distributed. At (b) a three-phase grouping into phase-bands is made. Each band has a 120° angle of spread, and the summation of conductor e.m.f.'s would yield three phase-e.m.f.'s displaced in time phase by 120°. It will be seen that the resultant phase e.m.f.'s are less than the arithmetic sum of the component e.m.f.'s; this is the effect of spreading the winding. In Fig. 124 (c) is shown a grouping for

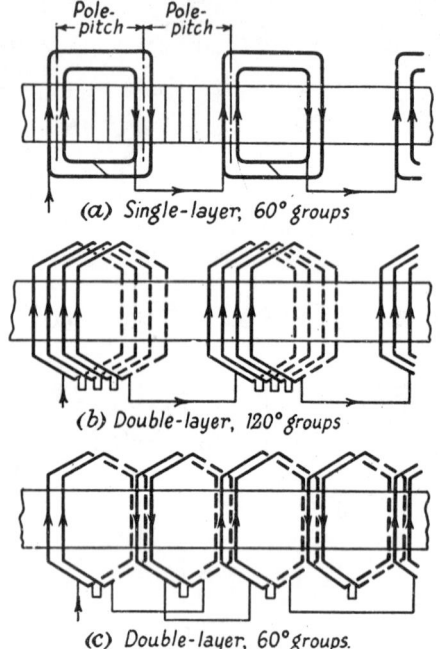

(a) Single-layer, 60° groups

(b) Double-layer, 120° groups

(c) Double-layer, 60° groups.

FIG. 125.   ONE PHASE OF THREE-PHASE WINDING

Six slots per pole, full-pitch (effective) coil-span. The lower sides in the double-layer windings are shown dotted

six phases with 60° spread. Grouping (b) could not be used with a single-layer winding for three phases, nor could (c) be used for six, because the subdivision of the winding leaves no further free conductors a pole pitch apart to form the return conductors. But (c) could be used for *three* phases if D, E and F were employed as the completion of A, B and C. This is shown in (d), where the phase-bands are arranged in this way, and the phase sequence A, B, C obtained by re-lettering. The *phase-bands* follow in the order A, C', B, A', C, B', as already indicated in the preceding paragraph.

CONNECTIONS BETWEEN COIL GROUPS. The overhang of a winding will comprise the end-connections which form the conductors into turns. In addition, the links made at the connection end of a winding comprise (a) coil-to-coil connectors, (b) phase-group connectors, and (c) parallelling connectors. When windings are required for large currents and low e.m.f.'s, the groups will not generally be connected in series round the whole of a multipolar armature, but will be split up into sections each containing the same number of similar coil-groups: the sections are then parallelled by suitable connectors.

Examples of coil and group connectors are given in Fig. 125, for part of one phase of three typical three-phase windings, viz. (a) single-layer, 60° groups; (b) double-layer, 120° groups; and (c) double-layer, 60° groups. A winding of type (c) in a sixteen-pole machine, for example, could have one, two, four, eight, or even sixteen parallel circuits per phase by making provision for the appropriate parallelling connectors. The number of circuits is equal to the number of poles divided by the number of groups per circuit.

NOMENCLATURE. The following symbols are used—
S = total number of slots on a uniformly-slotted armature;
C = number of coils;
e = e.m.f. of a conductor;
e_t = e.m.f. of a turn;
p = number of pole-pairs;
$\gamma = 2\pi p/S$ = slot pitch, electrical radians;*
N = number of phases;
$g = S/2p$ = number of slots per pole;
$g' = S/2pN$ = number of slots per pole per phase;
σ = spread of a phase-group, electrical radians;
$\beta = 2\pi/N$ = angle between e.m.f.'s of successive phases, electrical radians.
S_{ab}, S_{ac} = numbers of slots between given points in phases AB, AC.

3. **Single-layer Windings.** Inspection of Fig. 125 (a) will show that the single-layer winding for more than one phase will require to have an overhang designed to allow the end-connections to be accommodated in separate *tiers* or *planes*. For a three-phase winding the overhang may be arranged in two or three planes. Figs. 126 and 131A.

Single-layer windings fall into two main classes: (i) *unbifurcated* windings in which the coils comprising a pair of phase groups in adjacent pole-pitches are concentric, and (ii) *bifurcated* windings in which each group is split into two sets of concentric coils, each set sharing its return coil-sides with those of another group in the same

---

* A double pole-pitch covers 2π electrical radians. A machine with p pole-pairs has $2\pi p$ electrical radians in its periphery. If there are S equispaced slots on the armature surface, the angular slot-pitch is $\gamma = 2\pi p/S$.

phase. The windings (*a*) and (*b*) in Fig. 126 are unbifurcated (sometimes termed concentric), the former being a continuous and the latter a broken chain arrangement. This affects the number of overhang planes needed to accommodate the end-windings. The winding (*c*) is a bifurcated (or *split concentric*) type, requiring a three-plane overhang.

Single-layer windings still in use have hairpin-shaped coils (Fig. 129 (*a*)) which are pushed through partially-closed slots (Fig. 130), and are then bent at the free ends to be jointed turn-by-turn. They have the advantage that the important slot portions of

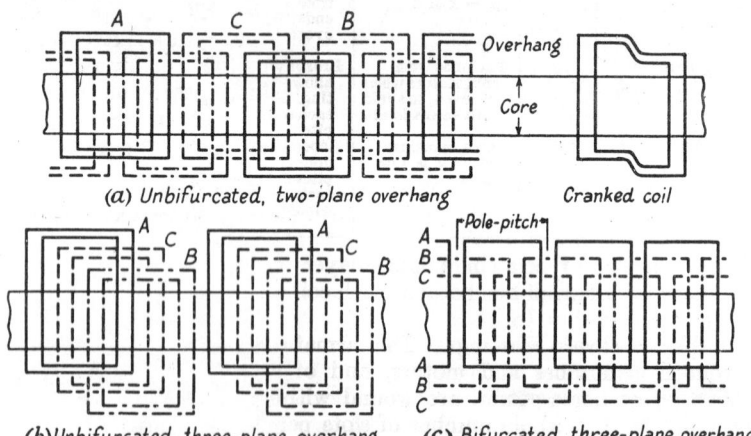

(*a*) *Unbifurcated, two-plane overhang*   *Cranked coil*

(*b*)*Unbifurcated, three-plane overhang*   (C) *Bifurcated, three-plane overhang*

Fig. 126. Single-layer Windings

$g' = 2$ slots per pole per phase, 60° spread

the coil may be carefully insulated before insertion into the armature; and the winding is located in slots which, having a narrow opening, reduce gap excitation, tooth losses and noise. The jointing of individual conductors to form the finished coil is usually done by electrical resistance welding, and each turn must subsequently be suitably insulated over the butt joint. The operation is expensive, which accounts for the decline in the use of single-layer in favour of double-layer windings.

Reference to Fig. 126 and the following Table will make clear the pattern of the commoner forms of single-layer winding. In each case the effective coil span is equivalent to full pitch. A single-layer constant-span coil arrangement is sometimes employed for small induction motors, using circular-section cotton-covered wire for the conductors. This is a *mush* winding. Each coil is first wound on a former making one coil side shorter than the other. In winding, the wires are dropped individually into the slots, first the short side (the wires emerging beneath the long sides of other coils in adjacent slots), then the long sides. Round the armature, long and short sides occupy alternate slots, and the coils are of constant pitch with a whole number of slots per pole per phase.

TYPES OF CONCENTRIC WINDING IN PARTIALLY-CLOSED SLOTS.

Type	No. of O'hang Planes	Use	Advantages	Disadvantages
Unbifurcated— (a) Continuous chain	2	Multipolar machines with $g' = 2$ or 3	All coil groups alike, but inserted from alternate ends: 2-plane overhang	Takes 4 pole-pitches to repeat pattern: special cranked coil needed when p is odd
(b) Broken chain	3	Machines with split stators when p is even, to avoid coils across joints	Broken form allows easy splitting. Repeats each double pole-pitch	Irregular end-windings cause varying overhang leakage and eddy loss in end-covers
Bifurcated	3	Large machines where g' is 4 or more	Repeats each double pole-pitch. Width of each end-winding reduced	Three types of coil groups required

The number of coils in a single-layer winding is one-half the number of slots available, because each coil-side completely occupies one slot: thus $C = \frac{1}{2}S$.

4. Double-layer Windings. The armatures of nearly all synchronous generators and motors, and of most induction motors above a few horse-power, are wound with double-layer windings. Windings with a whole number of slots per pole (g an integer) are called *integral-slot* windings. It is common, however, to find g fractional, and such arrangements (in which the number of slots S is not a multiple of the number of poles $2p$) are termed *fractional-slot* windings. These have a number of advantages, listed in § 5. In all cases the number of coils is equal to the number of slots: i.e. $C = S$.

INTEGRAL-SLOT WINDINGS. In Fig. 127 (*a, b*) some basic arrangements are shown for integral-slot three-phase windings. The letters indicate conductors to correspond with the phases ABC of a three-phase armature. Letters arranged vertically designate conductors in the same slot. The angular distance between successive slots is the angular slot-pitch γ in electrical radians. The coils are former-wound, and all have the same span.

Fig. 127 (*a*) shows a winding with $g = 9$ slots per pole, $g' = 3$ slots per pole per phase, and full-pitch coils. The slot-pitch is $\gamma = 2\pi/18$ radians, or 20°. Each phase-band spreads over 3 slot-pitches so that the phase-spread is $\sigma = 60°$. The conductors b are arranged 120° and conductors c 240° to the right of a, giving $\beta = 2\pi/3 = 120°$ for the angle between successive phase positions and for their corresponding e.m.f.'s. The coils being of full pitch of π radians or 180°, each slot contains coil-sides belonging to one phase only.

In order to reduce the amount of copper in the end-connections and to reduce or suppress certain harmonics in the phase-e.m.f.'s, the coils may be *chorded* or *short-pitched*. Such a winding is shown in Fig. 127 (*b*). It has $g = 9$ and $g' = 3$, also $\beta = 120°$ as before,

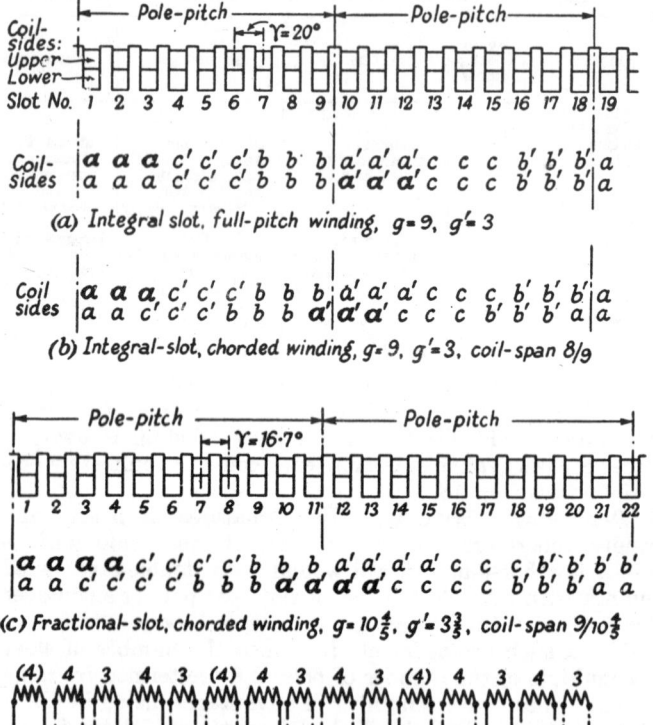

(*a*) Integral slot, full-pitch winding, *g = 9*, *g' = 3*

(*b*) Integral-slot, chorded winding, *g = 9*, *g' = 3*, coil-span 8/9

(*c*) Fractional-slot, chorded winding, *g = 10⅘*, *g' = 3⅗*, coil-span 9/10⅘

(*d*) Layout of a 5-pole unit of winding (c)

FIG. 127. THREE-PHASE DOUBLE-LAYER WINDINGS

but the coil-span is 8 slots instead of 9 as in (*a*), giving a coil span (or coil-pitch) of 8/9 of a pole-pitch, i.e. 160°. Certain slots now hold coil-sides of different phases. Chording reduces the phase-e.m.f., and the coil-span is rarely made less than ⅔ pole-pitch because additional turns become necessary which offset the saving of overhang copper.

FRACTIONAL-SLOT WINDINGS. These have fractional values for the slots per pole g and slots per pole per phase g': the value of g'

is usually an improper fraction such as $\frac{7}{5}$, $\frac{23}{7}$, etc., or in general $S'/2Np'$. In a unit of $2p'$ poles there are S' slots in all, and S'/N slots for each of the N phases with S'/N a whole number. The fraction $\frac{7}{5}$ thus represents 7 slots per phase spread over 5 pole-pitches. In a three-phase machine this means 21 slots for every unit of 5 poles, 42 for 10 poles and so on. All phase windings must be arranged for electrical and magnetic balance, which requires the conductor e.m.f.'s to sum to the same total for each phase, with an exact 120° angle between the vectors of phase e.m.f.

A fractional-slot winding is shown diagrammatically in Fig. 127 (c) for one double pole-pitch. It will be seen that the coils cannot have a full span of 180°, and that the phases are unbalanced. The balance is, of course, secured over the complete unit of poles $(2p' = 5$ in this case), after which the whole layout repeats for the next unit of poles, $2p'$, if any.

The coil-span that can be used has considerable freedom. That in Fig. 127 (c) has a span of 9 slots in $g = 10\frac{1}{5}$ slots per pole: the equivalent $g' = 3\frac{2}{5}$ slots per pole per phase, although this number is not actually found, it representing only the average number in the 5 poles of the unit.

5. **Fractional-slot Windings.** The apparent complication just described has theoretical advantages, reinforced by desirable simplifications in manufacture. These are: (a) the simple but effective reduction of high-frequency slot harmonics, and (b) the use of a total number of armature slots that is not necessarily a multiple of the number of poles. Thus a particular number of slots for which notching gear exists may be used for a range of machines running at different speeds, with consequent saving in equipment and drawings; this is of especial value where a wide range of numbers of poles may be called for, as with synchronous machines.

Fractional slotting is practicable only with double-layer windings. It limits the number of available parallel circuits because phase groups under several poles must be connected in series before a unit is formed and the winding repeats the pattern to give a second unit that can be paralleled with the first.

The simplest cases of fractional-slot windings are those for which g is an integer plus one-half, which employ alternate groups of coils differing by 1. Thus with $g = 7\frac{1}{2}$ slots per pole (or $g' = 2\frac{1}{2}$ slots per pole per phase) in a three-phase winding with 60° spread, alternate phase-groups will contain respectively 2 and 3 slots, and the two phase-groups, under two successive poles, connected in series will, constitute the basic repeatable unit. The number of units is determined by the total number of poles $2p$ in the armature.

As mentioned in § 4, when g' is expressed as the fraction $S'/2Np$ reduced to its lowest terms, S'/N represents the number of slots and coils per phase in the basic unit of $2p'$ poles. Thus for $g' = 2\frac{3}{4} = 11/4$, there are 11 coils per phase distributed among 4 successive poles, and the four groups may contain 3, 3, 3 and 2 coils in series. The dasic unit of $2p'$ poles must naturally be a factor of the total number of poles $2p$, and if a is the number of parallel circuits required, the product $2ap'$ must also be a factor of $2p$.

EXAMPLES

1. A three-phase, 16-pole machine with 108 slots and a 60° phase-spread will have

$$g' = S/2pN = 108/16 \cdot 3 = 9/4 = 2\tfrac{1}{4} \text{ slots per pole per phase.}$$

The groupings in each phase will be 3, 2, 2 and 2 coils in the basic unit of 4 poles. All four 4-pole units may be series connected, or there may be 2 or 4 circuits in parallel.

2. A three-phase, 10-pole machine with 108 slots has

$$g' = 108/10 \cdot 3 = 18/5 = 3\tfrac{3}{5} \text{ slots per pole per phase.}$$

The groupings could be 4, 4, 4, 3 and 3 coils and the basic unit would cover 5 pole-pitches. This permits of one circuit with both units in series, or one pair of parallel circuits only.

REQUIREMENTS OF SYMMETRY. Fractional-slot windings must fulfil the requirements of e.m.f. balance. In each phase the group-sequence must be the same, and the groups must occupy slots so selected as to obtain the required phase-angle $\beta = 2\pi/N$.

Let $S_{ab},\ S_{ac}$ . . . be the number of slot-pitches displacement between a given slot in phase A and similar slots in phases B, C. . . . Then, with an angular slot-pitch $\gamma = 2\pi p/S$ electrical radians (i.e. the total electrical angle in the p pole-pairs divided by the total number of slots)

$$S_{ab}\gamma = 2\pi/N \quad \text{or} \quad 2\pi/N + x\pi$$

gives the position of the required slot in phase B, with x having any positive integral value. For the former expression the "polarity," i.e. the positive direction of the e.m.f., of the group in which the phase B slot lies will be the same as that in phase A. The same is true of the second expression if $x = 0, 2, 4, 6$ . . .; but if $x = 1, 3, 5$ . . . the polarities will be opposite. This must be allowed for in making phase connections. In a similar way

$$S_{ac}\gamma = 4\pi/N \quad \text{or} \quad 4\pi/N + \ldots,$$

and so on for the remainder of the N phases.

As already indicated, a fractional-slot winding may divide itself naturally into basic units of $2p'$ poles containing S' slots each, so that the values $S_{ab},\ S_{ac}$ . . . will be appropriate within each unit. Using for convenience the slots per pole per phase $g' = S/2pN = S'/2p'N = \pi/N\gamma$, the slot-spacings $S_{ab},\ S_{ac}$ can be written

$$S_{ab} = g'(Nx + 2),\ S_{ac} = g'(Nx + 4), \text{ etc.}$$

These always give whole numbers when an appropriate value of x is chosen, except when $2p'$ is a multiple of N. Hence when the number $2p'$, in the improper fraction $g' = S'/2p'N$ reduced to its lowest terms, is a multiple of the number of phases, a balanced winding is not possible.

In three-phase windings ($N = 3$) the grouping of phase B is S'/N slots from that of phase A when $2p' = 2, 5, 8, 11 \ldots$, and the phase C grouping is $2(S'/N)$ slots from phase A. When $2p'$ is $4, 7, 10, 13 \ldots$ these displacements are interchanged.

When the starts (or, depending on the polarity, the finishes) of the various phases have been determined, it remains to fill in a succession of similar groups in each phase from these points. All subsequent slots in each phase will then be correctly located. The sequence of slot groups (e.g. 4, 4, 4, 3, 3 slots in Example 2 above) is determined in such a way as to give the greatest symmetry in the winding. One 5-pole unit of the winding in Example 2 might be tabulated as follows—

Phase A	(+ 4)		− 4		+ 4		− 3		+ 3	
Phase B		+ 3		(− 4)		+ 4		− 4		+ 3
Phase C	− 4		+ 3		− 3		(+ 4)		− 4	

Pole-pitch	1	2	3	4	5

The figures indicate the number of slots containing the *upper* coil-sides in each group with its appropriate polarity, those numbers in brackets being inserted first from the expressions given for S_{ab} and S_{ac}. A layout of the winding is drawn in Fig. 127 (*d*). The diagram of two pole-pitches is shown in Fig. 127 (*c*), the lower coil-sides in the left-hand pole-pitch corresponding to the arrangement of the upper coil-sides of pole-pitch 5 in the Table.

COIL-SPAN. When the grouping of the upper layer of coil-sides has been decided, it follows that the groupings of the corresponding lower coil-sides are determined also, for these have the same group sequence but displaced from the upper groups by the *winding pitch* or *coil-span*. The coil-span must be the same for all coils in the winding, and because g is fractional the span cannot be a full pitch. Thus with $g = 7\frac{1}{2}$ slots per pole the coils might span from slot 1 to slot 9 (8 slot pitches), or 1 to 8 (7 slot pitches): the former is greater and the latter is less than the pole-pitch of $7\frac{1}{2}$ slot-pitches. The span chosen is almost invariably short rather than long. With $g = 7\frac{1}{2}$, a span of 7 slots is 88·9 per cent of full pitch, while 6 slots gives 80 per cent of full pitch. As observed in §§ 4 and 8, chording has technical advantages in the reduction of overhang copper, and of e.m.f. and m.m.f. harmonics.

6. Types of Double-layer Windings. A coil in a double-layer winding represents the entire set of conductors in one slot layer in association with the similar set in the other layer of another slot. The number of coils is therefore the same as the number of slots: $C = S$.

The overhang of a double-layer winding may have a variety of shapes, depending on the method employed to bring about the transition from the upper to the lower slot position. Basically there are only two types, viz. *lap-* and *wave-connected*. The coils may have one turn or several. In the former case single-turn coils require only one conductor in each layer, but this will generally be laminated to facilitate bending in the end-windings and to reduce eddy-current losses. These coils form *bar-windings*, and are commonly

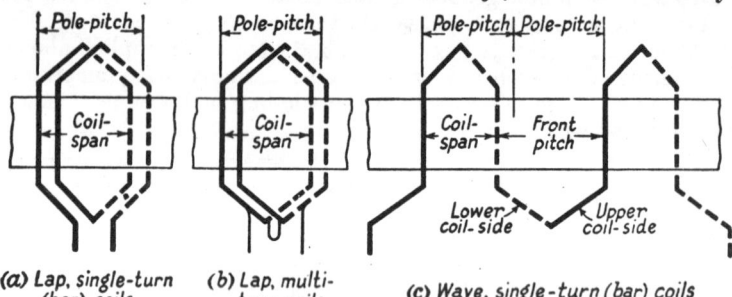

(a) Lap, single-turn   (b) Lap, multi-
   (bar) coils   turn coils   (c) Wave, single-turn (bar) coils

Fig. 128. Double-layer Lap and Wave Coils (Chorded)

found in low-voltage machines with large phase currents. The choice of coil depends on the number of turns per phase required to generate the e.m.f., the current rating, and the number of parallel circuits obtainable.

Fig. 128 shows (a) a pair of single-turn (bar) lap coils, (b) two multi-turn (bar) lap coils, and (c) two bar wave coils. Multi-turn wave coils are not used, because lap coils are more easily connected.

The connection of lap coils in groups is as indicated in Fig. 125 (b) and (c). The wave-connection of bars is not so straightforward, since after one complete tour round the armature (i.e. after series connection of *p* coils) the winding would naturally close on to the start of the first coil. To avoid this, a connection is made to continue the winding into the second coil-side of the first group, and so on until all coil-sides of *alternate* groups in each layer are fully wound. Another independent wave winding is made to utilize the remaining half of the phase. The two parts can then be connected in series or parallel to form the complete phase winding.

When fractional slot wave windings are used, the problem is complicated by the fact that isolated coils are left after the full number of complete turns round the armature has been made, necessitating special connectors.

7. **Choice of Armature Winding.** The following items specify a three-phase winding—

(a) Type of coil: concentric, *lap*, wave.

(b) Overhang: *diamond*, multiplane, involute, mush.

(c) Layers: single, *double*.

(d) Slots: *open*, closed, semi-closed.

(e) Connection: *star* (synchronous machines), *mesh* (induction motors).

(f) Phase-spread: *60°*, 120°.

(g) Slotting: integral, *fractional*.

(h) Coil-span: full-pitch, *chorded*.

(j) Circuits: series, or parallel.

(k) Coils: single-turn, multiturn.

The most usual winding has the features *italicized*. Some of the many types of coil are shown tabulated on page 210.

Multiturn Coil Windings. Machines with small numbers of poles and low values of flux per pole require a relatively large number of turns per phase: this is true also of high-voltage machines. Multiturn coils are needed, and the choice will lie between the double-layer lap winding and the single-layer winding. The former is dropped into open slots in all cases whe. good slot insulation is essential, although in small low-voltage machines the conductors may be fed through a narrowed slot opening into a slot-liner which serves as the main insulation. Single-layer windings are rarely used in large machines except with partially-closed slots and push-through hairpin coils.

The double-layer winding in open slots, compared with the single-layer winding in semi-closed slots, has the following advantages—

(a) ease of manufacture and lower cost of coils, which are all alike;

(b) fractional slotting can be used;

(c) chorded (or short-pitch) spans are obtained.

Against these the single-layer winding has—

(a) higher efficiency and quieter operation on account of the narrower slot-openings;

(b) better space utilization in slots due to elimination of interlayer insulation.

In general it may be said that modern practice favours the double-layer winding except where the slot-openings would be large compared with the length of the air-gap, as in high-voltage induction motors.

Bar Windings. Where single-turn coils are necessary, as with turbo-alternators or multipolar low-voltage machines, the choice lies between double-layer bar lap or wave windings. Both types may have the bars pushed through partially-closed slots and be bent to shape at the other end when the conductor section is moderate, but the heavy bars used in turbo-alternators must be completely formed before being inserted into open slots.

The chorded bar lap winding has the advantage of a shorter overhang at the connection end than the bar wave, and is more suited to cases where several parallel circuits are needed.

Type	Shape	Copper	How wound	Used on	Type of Slot	Recommended max. voltage
Concentric push through		Ribbon; square	Machine	Large A.C. stators	Semi-closed	Any
Concentric drop in slot		Ribbon; square	Mould	Large A.C. stators	Open	Any
Diamond		Round; ribbon; strap	Mould; pulled; former	D.C.armatures A.C. stators	Open	3300
Concentric diamond		Ribbon	Mould	D.C.armatures A.C. stators	Open	600
Short diamond or octagon		Round; strap	Mould; formed	D.C.&A.C. rotors	Open	600
Multipolar or hexagonal		Round; ribbon	Mould	D.C.&A.C. rotors	Open	600
Involute		Strap	Former	D.C.&A.C. armatures	Open	600
Mush		Round	Mould	A.C.rotors A.C.stators	Semi-closed	440

TYPES OF COILS FOR A.C. MACHINES

COIL MANUFACTURE. Fig. 129 (a) gives a view of a single-layer coil for a concentric winding, indicating its general shape before being pushed through a pair of slots. At (b) is shown stages in the production of a multiturn double-layer diamond coil of the *pulled* type. When such a coil is not too heavy it may be wound over two dowels as a flat loop, and the slot portions then gripped in a machine and pulled apart to the required coil span. More massive double-layer coils have to be built up on formers, and the overhang portion may be bent up to make an angle with the slot parts, to form an

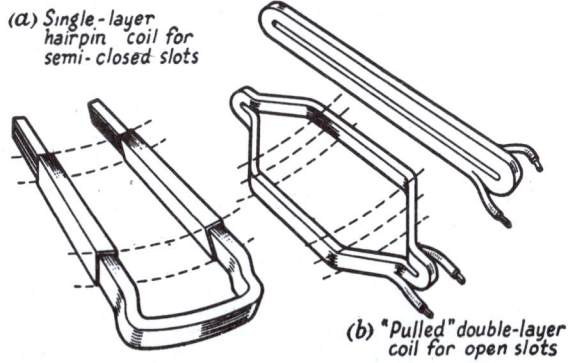

(a) Single-layer hairpin coil for semi-closed slots

(b) "Pulled" double-layer coil for open slots

FIG. 129. EXAMPLES OF SINGLE- AND DOUBLE-LAYER COILS

involute shape. When assembled on the stator the overhang presents a conical surface suitable for bracing.

SLOTS. In very small machines, using round wires of the smaller gauges, the tooth shape is improved if tapered slots are used, Fig. 130. In all but these, parallel-sided slots are employed. Rectangular conductors, of course, demand slots of rectangular shape. The slot-openings serve to avoid excessive slot-leakage, but they cause an effective lengthening of the magnetic path across the air gap, and tunnel slots are occasionally used for small induction motors. Open slots (i.e. with the opening as wide as the slot) must be used where it is desired to complete the coils outside the armature and drop them into the slots.

EXAMPLES. Typical single- and double-layer windings for medium-size machines are illustrated in Fig. 131A and 131B. In the latter the armature is partly wound, and two separate coils are shown.

8. **E.M.F. of Windings.** As discussed in connection with Fig. 123, each conductor in a winding generates an e.m.f. which reproduces, as a variation with time, the space distribution of the flux density in which it moves. The flux in the gap is never exactly distributed in the ideal sinusoidal shape, particularly in salient pole machines and those with very short gaps and high tooth-saturations. The

turbo-alternator, with its comparatively long gap and cylindrical rotor, develops most nearly the sinusoidal distribution.

Now the e.m.f. of a complete phase represents the summation of the e.m.f.'s of a number of phase-groups of displaced conductors, and such a phase is often associated with a second phase in star connection to give a resultant line voltage. The effect of this is always to produce a phase (or line) e.m.f. which approaches much

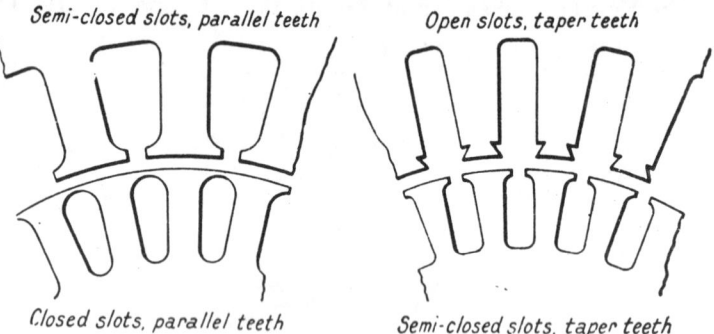

Fig. 130. Types of Slots for A.C. Machines

more nearly to the sinusoidal than do the conductor e.m.f.'s themselves. This effect is shown in Fig. 132 for a salient-pole machine for which, as an example, a pole-arc b of $\frac{2}{3}$rds the pole-pitch Y may produce the gap-density distribution in (b) which approximates to a shortened rectangle. The corresponding e.m.f. e in one conductor, shown at (c), reproduces the gap distribution in its wave-shape. A group of full-pitch coils spread over 60° gives the phase e.m.f. E_{ph1}, the "stepping" due to the summation of conductor e.m.f.'s being ignored (d). This phase e.m.f. subtracted from that of the next phase, E_{ph2}, gives the line e.m.f. E_l in (f). Another method of arriving at the same result is to make a Fourier analysis of the original gap-density curve, as indicated in Fig. 115, page 183, giving

$$B = B_1 \sin \theta + B_3 \sin 3\theta + \ldots + B_n \sin n\theta.$$

A machine can thus be considered to have $2p$ "fundamental" poles, together with families of $6p$, $10p \ldots 2np$ "harmonic" poles, all giving individually sinusoidal flux distributions, as in Fig. 115 (c). Both fundamental and harmonic pole fluxes will generate e.m.f.'s of corresponding frequency in the conductors, but the proportion of harmonics in the phase- or line-e.m.f. wave-form is reduced by grouping.

The magnitude of the fundamental e.m.f. of a phase is affected by—

(a) the fundamental component Φ_1 of the flux per pole Φ;

(*b*) the phase-spread, taking into account the effects of fractional-slotting, if used;
(*c*) the coil-span.
The harmonic e.m.f.'s of a phase are similarly affected by—
  (*a*) the harmonic flux components;
  (*b*) and (*c*) as above;
  (*d*) the phase connection in star or mesh;
  (*e*) the varying gap-reluctance due to armature slotting, which gives rise to "tooth-ripples."

FIG. 151A. SINGLE-LAYER WINDING
Unbifurcated chain winding in semi-closed slots; 2-plane overhang
12 slots per pole, semi-closed
(*British Thomson-Houston*)

Items (*a*) to (*d*) modify the amplitudes of the e.m.f. harmonics derived from the flux-distribution, but item (*e*) introduces additional harmonics independently, which are usually of such order as to be possible causes of interference with telephone circuits. Tooth-ripples are considered separately in § 10.

SINUSOIDAL GAP-FLUX. To introduce the several factors employed to determine the e.m.f. of a phase, it is at first assumed that only a fundamental flux component is present, i.e. that the gap-flux distribution is sinusoidal. In a full-pitch coil of T'_c turns moving through the flux Φ_1 per pole at such a rate as to develop an e.m.f. of frequency f, the r.m.s. value of the e.m.f. is

$$E_c = \sqrt{2}\pi f T_c \Phi_1 = 4\!\cdot\!44 f T_c \Phi_1 \text{ volts} \qquad . \qquad . \qquad . \quad (71)$$

Distribution Factor. A phase comprises a number of coils con-

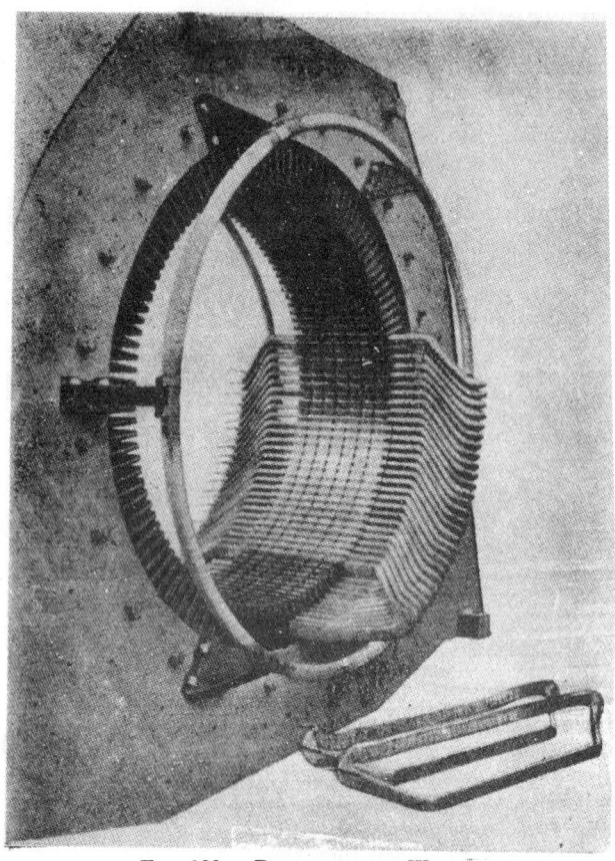

FIG. 131B. DOUBLE-LAYER WINDING
Partly-wound; lap multi-turn coils; open slots
(*British Thomson-Houston*)

nected in series and extending over an arc or spread of σ electrical radians. The e.m.f. of a phase is the complex summation of the constituent individual coil-e.m.f.'s, which will be mutually displaced in angular time-phase by ψ, an angle describing the electrical displacement between *successive* coils (i.e. coils connected in series).

Should electrically successive coils lie in adjacent slots, then $\psi = \gamma$, the electrical slot-pitch angle: this would be typical of a single-layer winding. Electrically successive coils may, however, be displaced by nearly a double pole-pitch as in a double-layer fractional-slot winding, in which case $\psi < \gamma$.

Suppose a phase to comprise m full-pitch coils $(a, b, c \ldots m)$, in which the e.m.f.'s $e_a, e_b \ldots e_m$ are all (naturally) equal, but displaced in time phase by the electrical angle ψ. The phase e.m.f. is the *complex* sum of the m coil e.m.f.'s, Fig. 133A. It is clear that, on account of the displacement angles ψ, the phase e.m.f. is less than the *arithmetic* sum of the coil e.m.f.'s by the factor

$$k_m = \frac{\text{vector sum}}{\text{arithmetic sum}} \text{ of coil e.m.f.'s.}$$

The factor k_m is called the *breadth coefficient* or *distribution factor*. Its value, from Fig. 133A, is

$$\frac{\sum_a^m e}{me} = k_{m1} = \frac{\sin \frac{1}{2}m\psi}{m \sin \frac{1}{2}\psi},$$

and the phase of m full-pitch coils will produce the r.m.s. e.m.f.

$$E_{ph1} = 4 \cdot 44 m T_c k_{m1} f\Phi_1$$
$$= 4 \cdot 44 k_{m1} T_{ph} f\Phi_1 \text{ volts,}$$

writing $T_{ph} = mT_c$ for the total turns in series per phase.

If a winding can be considered as distributed uniformly over a spread

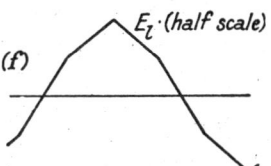

FIG. 132. E.M.F. WAVE-FORM PRODUCED BY RECTANGULAR GAP FIELD

$\sigma = m\psi$ (an approximation to which is afforded by slot skewing or fractional-pitch windings) then the complexor addition reduces to the arc of a circle, Fig. 133B, of which the sum is the chord.

Hence $\qquad k_{m1} = \dfrac{\sin \frac{1}{2}\sigma}{\frac{1}{2}\sigma}$ . . . . (72)

With a phase-spread of 60°, $k_{m1} = 0 \cdot 955$, and for $\sigma = 120°$, $k_{m1} = 0 \cdot 827$. Since the distribution factor directly affects the phase e.m.f., it is evident that the output with a given armature current will be $0 \cdot 955/0 \cdot 827 = 1 \cdot 15$ times as much with the narrow as with the wide spread.

Coil-span Factor. If the coil span is not a full pole-pitch, the conductor e.m.f.'s in the two sides of a coil are not directly additive, but form an addition of vectors whose phase displacement is ε, the angle by which the span departs from its full-pitch value π. The vector sum is reduced by the *coil-span factor*

$$k_{s1} = \cos \tfrac{1}{2}\varepsilon \quad . \quad . \quad . \quad . \quad (73)$$

from its full-pitch value (see Fig. 134 (a) and (b)).

The fundamental e.m.f. of a phase can now be written

$$E_{ph1} = 4 \cdot 44 k_{m1} k_{s1} f T_{ph} \Phi_1 \text{ volts} \quad . \quad . \quad . \quad (74)$$

9. E.M.F. of Windings : General Case. Taking now a more general view of the phase-e.m.f. and its winding factors, the effect of a non-sinusoidal flux-distribution may be discussed.

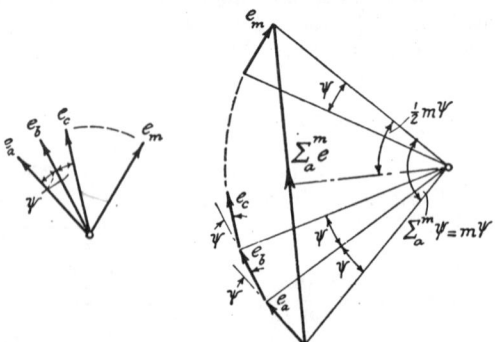

Fig. 133A. Phase E.M.F.

Distribution Factor. When the m coils referred to above form, as is usual, a phase group, then $m = g'$ (the number of slots per pole per phase) and the angle $m\psi = \sigma$ (the phase-spread). The distribution factor becomes

$$k_{m1} = \frac{\sin \tfrac{1}{2}\sigma}{g' \sin \tfrac{1}{2}(\sigma/g')} \quad . \quad . \quad . \quad (75)$$

for the fundamental. Since the "harmonic poies" of the nth space-harmonic have a pitch only $1/n$ of the fundamental pole-pitch, the phase-angles ψ of the harmonic e.m.f.'s will be n times as large: by analogy with eq. (75) the distribution factor for the nth harmonic is

$$k_{mn} = \frac{\sin \tfrac{1}{2}n\sigma}{g' \sin \tfrac{1}{2}(n\sigma/g')} \quad . \quad . \quad . \quad (76)$$

It is necessary to consider the effect of fractional slotting. When g' is the fraction $S'/2Np'$, a layout of a $2p'$-pole unit will be found to contain S'/N coils in S'/N slots per phase evenly distributed over σ electrical radians. The distribution factor is

$$k_{mn} = \frac{\sin \frac{1}{2}n\sigma}{(S'/N)\sin \frac{1}{2}(nN\sigma/S')} \quad \cdot \quad \cdot \quad (77)$$

for the nth harmonic. For example, consider an $N = 3$-phase winding with a spread of $\sigma = 60°$ per phase, with $g' = 2\frac{1}{2}$ slots per pole per phase—

$$g' = \frac{S'}{2Np'} = \frac{S'/N}{2p'} = \frac{5}{2}$$

The unit winding comprises $2p' = 2$ pole-pitches in which there are $S'/N = 5$ coils or slots per phase and $S' = 15$ slots altogether. The effective angular displacement of the coils is $60°/5 = 12°$, and the distribution factor for the fundamental is $k_{m1} = \sin 30°/5 \sin 6° = 0.956$.

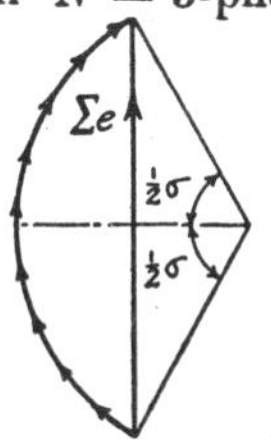

Where $g' > 5$ it is legitimate to assume uniform distribution, giving the effect shown in Fig. 133B, for which

Fig. 133B. Phase E.M.F. of Uniformly-distributed Winding

$$k_{m1} = \frac{\sin \frac{1}{2}\sigma}{\frac{1}{2}\sigma} \text{ and } k_{mn} = \frac{\sin \frac{1}{2}n\sigma}{\frac{1}{2}n\sigma} \quad \cdot \quad \cdot \quad (78)$$

The Table below gives values from eq. (78) for the fundamental and lower-order harmonics, with the more usual phase-spreads. A zero value indicates that the vector arc closes on itself, leaving no resultant.

Values of k_{mn}

No. of phases, N	Phase-spread, σ	k_{m1}	k_{m3}	k_{m5}	k_{m7}	k_{m9}
3 or 6	60°	0·955	0·637	0·191	− 0·136	− 0·212
3 or 1	120°	0·827	0	− 0·165	0·118	0
2	90°	0·900	0·300	− 0·180	− 0·129	0·100
1	180°	0·637	− 0·212	0·127	− 0·091	0·071

The wide spread of 120° eliminates third harmonics, and all others whose order is a multiple of 3—i.e. *triplen* harmonics. The reason for this is connected with the fact that the effective phase-spread to a third harmonic is three times the actual (fundamental) spread: $3 \times 120° = 360°$, and the e.m.f. across the terminals of a winding covering a complete third-harmonic double pole-pitch is zero.

It is of interest to note the distribution factors for the harmonics of a winding with the common spread of $\sigma = 60°$ and an infinitely-distributed winding. These are shown (dots and full-line) in Fig. 135A. The higher harmonic factors are small, and unless very pronounced flux harmonics are present in the gap, the e.m.f.'s to which they give rise will be smaller still. The curve gives values not applicable to integral-slot windings for it can be shown that, with a whole number of slots per pole, i.e. $g =$ integer, the distribution factors for harmonics of orders $n = 6Ag' \pm x$, where A is

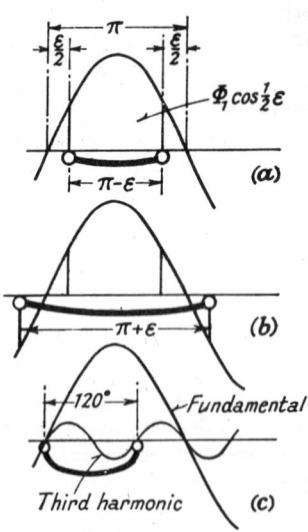

FIG. 134. CHORDED COIL-SPAN

any integer 0, 1; 2, 3 . . . and x is any *odd* number, are the same as those of orders x. Thus with $g' = 3$ slots per pole per phase, and $A = 1$, then any e.m.f.'s of orders 18 ± 1, i.e. 17 or 19, will have distribution factors the same as that of the fundamental. Now in this case the slotting of 18 slots per pole-pair may well give rise to a flux capable of generating 17th and 19th-order harmonics: see § 10. Unless precautions are taken, such as skewing or the substitution of fractional for integral slotting, these harmonics may appear strongly in the phase e.m.f. The crosses linked by the dotted line in Fig. 135A show how the distribution-factors recur for higher harmonics where g is integral. In the case shown, $g = 9$ and $g' = 3$. The lines in Fig. 135A are inserted to link up the points appropriate to each case: they have no other significance.

COIL-SPAN FACTOR. As already discussed, short-pitching or chording a winding reduces its e.m.f., and the appropriate factors are

$$k_{e1} = \cos \tfrac{1}{2}\varepsilon \quad \text{and} \quad k_{en} = \cos \tfrac{1}{2}n\varepsilon \qquad . \qquad . \quad (79)$$

For example, for a coil-span of $\tfrac{2}{3}$rd of a pole-pitch, $\varepsilon = \pi/3 = 60°$, and the factor $k_{e1} = \cos 30° = 0.866$. For the 3rd harmonic $k_{e3} = \cos \tfrac{1}{2} . 3 . 60° = \cos 90° = 0$: thus all 3rd (and triplen) harmonics are eliminated from the coil and phase e.m.f.'s. This is shown diagrammatically in Fig. 134 (c), where it is apparent that whatever 3rd harmonic e.m.f. is generated in one conductor will be balanced out by the other. In general, when a coil is pitched short or long by π/n (or $3\pi/n$, $5\pi/n$ . . . where applicable) no harmonic of order n survives in the coil e.m.f.

The triplen harmonics that may be generated in a three-phase machine are normally eliminated by star-connection of the phases. It is therefore usual to select the coil-span to reduce as much as possible the 5th and 7th harmonics. A pitch of 83·3 per cent ($\varepsilon = \pi/6 = 30°$) is most useful in this respect, as it gives the following factors—

Order of harmonic:	Fund.	3	5	7	9
Coil-span factor, k_{sn}:	0·966	0·707	0·259	0·259	0·707

The 5th and 7th harmonic factors are both small, while the 3rd and 9th harmonic e.m.f.'s will not appear in the line in any case with star connection.

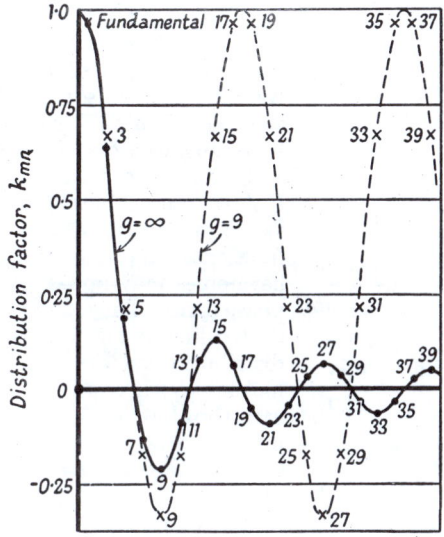

FIG. 135A. DISTRIBUTION FACTOR

Fig. 135B gives a plot of the coil-span factors for harmonics up to the 15th. It will be observed that too short a coil-span may reintroduce higher harmonics. Thus for a span of about 0·72 of a pole-pitch ($\varepsilon \simeq 50°$) both the 13th and 15th harmonic factors are as large as the fundamental factor.

PHASE CONNECTION. Similar remarks apply here to three-phase machines as to three-phase transformers with interlinked magnetic circuits (Chapter V, § 12). The line voltage of a star-connected machine can contain no triplen harmonics unless a star-point circuit is provided for them, which is unusual. Similarly, the mesh connection suppresses triplen harmonic e.m.f.'s by short-circuiting them.

THE E.M.F. EQUATION. The e.m.f. of a phase due to the fundamental component of flux per pole is

$$E_{ph1} = 4{\cdot}44k_{m1}k_{e1}fT_{ph}\Phi_1$$
$$= 4{\cdot}44K_{w1}fT_{ph}\Phi_1 \text{ volts} . \quad . \quad . \quad . \quad (80)$$

where $\Phi_1 = (2/\pi)B_1 YL$, with B_1 the fundamental peak of the spacial distribution of the gap flux density, and $K_{w1} = k_{m1}k_{e1}$ is the *winding factor*. For the nth harmonic

$$E_{phn} = 4{\cdot}44k_{mn}k_{en}nfT_{ph}\Phi_n \text{ volts,}$$

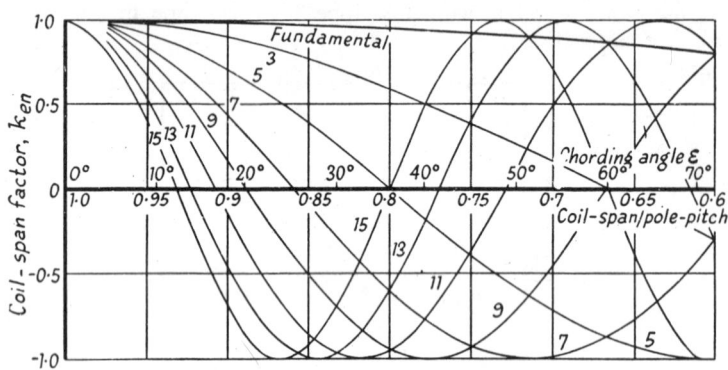

FIG. 135B. COIL-SPAN FACTOR

but a more useful form is

$$E_{phn} = 4{\cdot}44(B_n/B_1)k_{mn}k_{en}fT_{ph}\Phi_1 \text{ volts} \quad . \quad . \quad (81)$$

since $\Phi_n = \dfrac{2}{\pi} B_n \dfrac{Y}{n} L = \dfrac{1}{n} \cdot \dfrac{B_n}{B_1} \cdot \Phi_1$. The nth harmonic and fundamental r.m.s. phase e.m.f.'s are thus related by

$$\frac{E_{phn}}{E_{ph1}} = \frac{B_n k_{mn}k_{en}}{B_1 k_{m1}k_{e1}} = \frac{K_{wn}B_n}{K_{w1}B_1} .$$

The r.m.s. phase e.m.f. will be

$$E_{ph} = \sqrt{(E_{ph1}{}^2 + E_{ph3}{}^2 + \ldots + E_{phn}{}^2)}.$$

For all ordinary cases this is indistinguishable from E_{ph1}. Further, as the example on page 182 indicates, the fundamental flux component Φ_1 differs only slightly from the total flux Φ per pole, so that eq. (80) above can be used for the r.m.s. phase e.m.f. with either flux with little inaccuracy.

The line e.m.f. in star connection is $\sqrt{3} \cdot E_{ph}$, with E_{ph} calculated as above but omitting all triplen harmonic values.

EXAMPLE. A 3 750-kVA., 10 000-V., 3-phase, 50-c/s., 10-pole, low-speed water-turbine alternator has 144 slots containing a two-layer diamond winding with 2×5 conductors per slot. The coil-span is 12 slot-pitches. The flux per pole is 0.116 Wb./m.^{2}, taken as distributed in accordance with the example on page 182. Lay out the winding and calculate the phase e.m.f.

Winding. The slot angle is $\gamma = 2\pi p/S = 10\pi/144 = 0.218$ rad. $= 12\frac{1}{2}°$. The fractional value of slots per pole per phase is

$$g' = S/2Np = 144/3 \cdot 10 = 24/5 = 4\frac{4}{5}.$$

There will consequently be 24 slots per phase, or 72 slots in all, in each of two 5-pole units. The spacing between the starts of A and B will be

$$S_{ab} = g'(3x + 2) = 4\tfrac{4}{5}(3 \cdot 1 + 2) = 24 \text{ slots,}$$

and for C

$$S_{ac} = g'(3x + 4) = 4\tfrac{4}{5}(3 \cdot 2 + 4) = 48 \text{ slots.}$$

To show the arrangement of coil groups, 5, 5, 5, 5 and 4 for each phase, the Table below can be drawn up. The slots are numbered 1, 2, 3 . . . 72 in order, with their angles $0°$, $12\frac{1}{2}°$, $25°$ . . ., values above $180°$ being reduced by $180°$ or integral multiples thereof. Coil-sides for phase A are then allocated to slots with angles from $0°$ up to (but not including) $60°$; phase B from $120°$; and phase C from $60°$. The letters *a*, *b* and *c* in square brackets are those located from the values of $S_{ab} = 24$ and $S_{ac} = 48$ already calculated; i.e. slots 1 for A, 25 for B and 49 for C—

Slot:	1	2	3	4	5	6	7	8	9	10	11	12	13	14	15
Angle:	0	12½	25	37½	50	62½	75	87½	100	112½	125	137½	150	162½	175
Phase:	[a	a	a	a	a	c	c	c	c	c	b	b	b	b	b

Slot:	16	17	18	19	20	21	22	23	24	25	26	27	28	29
Angle:	7½	20	32½	45	57½	70	82½	95	107½	120	132½	145	157½	170
Phase:	a	a	a	a	a	c	c	c	c	[b]	b	b	b	b

Slot:	30	31	32	33	34	35	36	37	38	39	40	41	42	43	44
Angle:	2½	15	27½	40	52½	65	77½	90	102½	115	127½	140	152½	165	177½
Phase:	a	a	a	a	a	c	c	c	c	c	b	b	b	b	b

Slot:	45	46	47	48	49	50	51	52	53	54	55	56	57	58
Angle:	10	22½	35	47½	60	72½	85	97½	110	122½	135	147½	160	172½
Phase:	a	a	a	a	[c]	c	c	c	c	b	b	b	b	b

Slot:	59	60	61	62	63	64	65	66	67	68	69	70	71	72
Angle:	5	17½	30	42½	55	67½	80	92½	105	117½	130	142½	155	167½
Phase:	a	a	a	a	a	c	c	c	c	c	b	b	b	b

The sequence of groups, from each starting point, is seen to be 5, 5, 5, 4, 5. The slots occupied by phase A cover the angle $0° - 57\frac{1}{2}°$ in steps of $2\frac{1}{2}°$, so that the winding is uniformly spread with an angle of $2\frac{1}{2}°$ between coil e.m.f.'s. A winding Table, including polarities, is given below—

Phase A	[+ 5]		− 5		+ 5		− 4		+ 5
Phase B		+ 5		[− 5]		+ 5		− 5		+ 4
Phase C	− 5			+ 4		− 5		[+ 5]		− 5
Pole-pitch	1		2		3		4		5

The corresponding lower coil-sides will lie in slots 12 slot-pitches to the right of their corresponding upper coil-sides.

E.M.F. The total flux per pole is 0·116 Wb./m.^{2} With the analysis shown in Fig. 116 it will be sufficient to accept this figure as the fundamental component Φ_1. The third harmonic flux will be ignored as it will produce no effect in the *line* e.m.f. From eq. (77) for the *distribution factor*,

$$k_{mn} = \frac{\sin \frac{1}{2}(n \cdot 60°)}{24 \sin \frac{1}{2}(n60°/24)} = \frac{\sin \frac{1}{2}(n60°)}{24 \sin \frac{1}{2}(n2\frac{1}{2}°)}.$$

This gives $k_{m1} = 0·956$, $k_{m5} = 0·191$ and $k_{m7} = -0·136$.

The coil span is 12/3 . $4\frac{4}{5} = 4/4\frac{4}{5} = 5/6$ pole-pitch corresponding to 180(5/6) = 150°. The chording angle ε is 30°. From eq. (79)

$$k_{en} = \cos \frac{1}{2}(n30°) = \cos n15°,$$

from which $k_{e1} = 0·966$, $k_{e5} = 0·259$ and $k_{e7} = -0·259$. These figures can also be obtained from Fig. 135B.

The winding factors are then

$$K_{w1} = k_{m1}k_{e1} = 0·956 . 0·966 = 0·925,$$
$$K_{w5} = k_{m5}k_{e5} = 0·191 . 0·259 = 0·0495,$$
$$K_{w7} = k_{m7}k_{e7} = (-0·136) . (-0·259) = 0·0352.$$

The winding arrangement has evidently resulted in a material reduction in harmonic e.m.f. factor. The fundamental e.m.f., from eq. (80), is

$$E_{ph1} = 4·44 . 0·925 . 50 . 240 . 0·116 = 5\ 750\ V.$$

From eq. (81) *et seq.*

$$E_{ph5} = 5\ 750(0·0495/0·925)(11·2/100·6) = 34\ V.$$
$$E_{ph7} = 5\ 750(0·0352/0·925)(2·8/100·6) = 6\ V.$$

These harmonic e.m.f.'s are quite negligible, and the e.m.f. is $E_{ph} = 5\ 750$ V., giving $E_l = 10\ 000$ V. in star.

10. **Tooth Harmonics.** In addition to the e.m.f.'s generated in the windings by those space harmonics of flux which move at the same velocity as the fundamental flux relative to the conductors, certain harmonic voltages of a particularly undesirable order may be produced by the effect of the openings of the slots in which the conductors themselves are located. This effect is aggravated when the slot-openings are large compared with the tooth width and gap length, because these affect the air-gap permeance, to variations of which the harmonics are due. The ripples in the flux wave caused by variation of permeance from point to point in the air-gap are of the form indicated in the oscillogram of Fig. 136 (*a*): but it is important to note that, since the ripples are due to slotting, they do not move with respect to the conductors but glide over the flux-distribution curve, always opposite the slots and teeth which cause them.

Pulsation of Total Gap Permeance. If the main flux wave were assumed to be rectangular (as in a salient-pole machine without fringing) and the ripple flux a pure sine-wave superposed on it, the particular case in which the pole-arc covers an integral-plus-one-half slot-pitches would give a total flux varying between the two extremes shown in Fig. 136 (b) and (c), totalling $\pm$ the flux contained in one loop of the ripple. The frequency of flux pulsation would correspond to the rate at which the slots cross the pole-face, i.e. $2gf$ cycles/sec. Now this stationary pulsation may be regarded as two

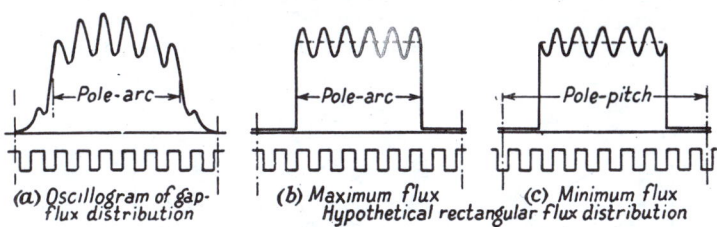

(a) Oscillogram of gap-flux distribution   (b) Maximum flux   (c) Minimum flux
Hypothetical rectangular flux distribution

FIG. 136. TOOTH RIPPLE

waves of fundamental space-distribution rotating at angular velocity $2g\omega$, one forwards and one backwards, where $\omega = 2\pi f$ corresponds to synchronous speed. The component fields will have velocities of $(2g \pm 1)\omega$ relative to the armature winding and will generate therein harmonic e.m.f.'s of frequencies $(2g \pm 1)f$ cycles/sec. But this is only a secondary cause of tooth-ripple.

Rotor Current Pulsation. The prime cause of tooth-ripple is the induction in rotor circuits of currents of frequency $2gf$. These flow in suitably-pitched damper bars in the poles of a salient-pole machine, in the brass or bronze wedges of a turbo rotor, and (to a minor extent) in the main rotor field windings.*

The gap permeance at the stator surface has space undulations, and the variations from the average may be represented approximately by a sine fluctuation $\lambda_1 \sin 2g\theta$. The fundamental rotor m.m.f. acting at the stator bore rotates with respect to the stator slots and may be written, for a given point on the stator, as $F_1 \sin (\theta - \omega t)$. The flux density at the stator surface will have a fundamental corresponding to the average permeance, together with a superposed ripple, due to the fluctuation, given by

$$\lambda_1 F_1 \sin 2g\theta \,.\, \sin (\theta - \omega t)$$

Resolving this into its two oppositely-rotating components gives

$$\tfrac{1}{2}\lambda_1 F_1 \{\cos [(2g - 1)\theta + \omega t] - \cos [(2g + 1)\theta - \omega t]\}$$

With respect to the stator these ripple fluxes vary with position in

* Walker, "Slot Ripples in Alternator E.M.F. Waves," *Proc. I.E.E.* 96 (II), p. 81 (1949).

slot-pitch, and with time at angular frequency ω : thus they cannot contribute any e.m.f. at tooth-ripple frequency.

The same two waves have, however, different velocities relative to the rotor. Any point designated by θ on the stator will require to be designated by $\theta + \omega t$ with reference to the rotor. Making this substitution gives

$$\tfrac{1}{2}\lambda_1 F_1\{\cos\,[(2g - 1)\theta + 2g\omega t] - \cos\,[(2g + 1)\theta + 2g\omega t]\}.$$

As these flux waves have a velocity $2g\omega$ relative to the rotor, they will generate currents of frequency $2gf$ in any closed rotor circuits of suitable pitch. Such currents will superpose an m.m.f. variation

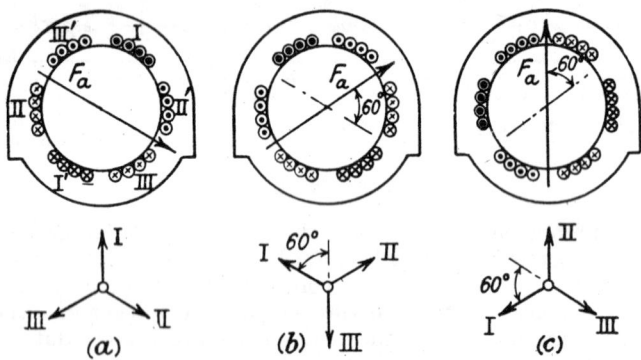

FIG. 137. ROTATING FIELD

of $2gf$ on the resultant pole m.m.f. Resolving these again into forward- and backward-moving components at $2g\omega$ relative to the *rotor*, and therefore $2g\omega \pm \omega = (2g \pm 1)\omega$ relative to the *stator*, then stator e.m.f.'s at frequencies $(2g + 1)f$ and $(2g - 1)f$ will be generated. These constitute the principal tooth ripples.

In the above only the fundamentals of the slot-permeance variation λ_1 and of the rotor m.m.f. F_1 have been considered. Further ripple effect may occur due to the harmonics of each. Quite objectionable ripple may be generated by the second harmonic of the slot permeance.

The suppression of slot ripples is accomplished by : (1) skewing the stator core ; (2) displacing the centre-line of the damper bars in successive pole-shoes ; (3) offsetting the pole-shoes in successive *pairs* of poles : these refer to the mitigation of ripple directly due to slot-permeance fluctuation. To counter the effects of harmonics in the field form, the methods are : (4) shaping the pole-shoes so that the maximum gap length is about twice the minimum, in salient-pole machines ; and (5) the use of composite steel-bronze wedges for the slots of turbo rotors.

11. **Armature Reaction.** The flux actually existing in the air gap

of a machine is due to the m.m.f.'s of all the windings linked by it. On no load with zero armature current the flux is produced by the field ampere-turns alone, but when current circulates in the armature, the *armature reaction*, or effect of the armature m.m.f., is an important second factor in determining the resultant flux. Since the armature winding comprises coils of span approximating to a pole-pitch, its reaction will produce alternate N and S poles.

ROTATING FIELD. A characteristic feature of the three-phase armature is that its self-produced poles *rotate*. In Fig. 137 is shown a number of elementary fixed three-phase two-pole armatures carrying narrow-phase-spread windings, and properly arranged with

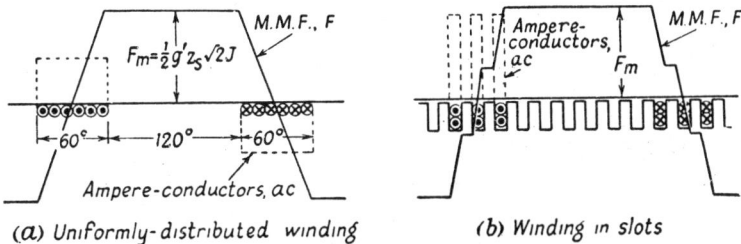

(*a*) Uniformly-distributed winding   (*b*) Winding in slots

FIG. 138   ARMATURE M.M.F. OF ONE PHASE

a phase-pitch of $2\pi/3$ to give phases spaced by 120°. Let the arbitrary positive direction of current in the three phases be upwards (shown by a centre dot on the conductor circles): then considering (*a*), the current distribution will be due to positive maximum current in phase I–I', and one-half negative maxima in II–II' and III–III'. At a subsequent instant (*b*), 60° later in time (one-sixth period), the current in phases I–I' and II–II' will be one-half positive maxima, and the current in phase III–III' will have increased to its negative maximum. The magnetic axis, or direction of resultant m.m.f., is 66° advanced in (*b*) compared with (*a*), corresponding precisely to the advance in time phase. In a similar way, (*c*) shows a further advance of the m.m.f. by 60°, corresponding to a further one-sixth period of time phase. For all three cases, the current distribution is evidently identical, and the magnitude of the m.m.f. will be the same. The resultant m.m.f. thus rotates in synchronism with the change of phase current, completing the passage of one complete double pole-pitch per cycle.

It would be incorrect to assume from Fig. 137 that the rotating m.m.f. maintained a constant value between the 60° intervals, nor even that its rotational speed was uniform therein. Fig. 139, in which the armature is developed, and a time-interval of 30° is taken, shows in fact the two limiting shapes of the m.m.f. distribution. But it is not difficult to show that the *fundamental component* of the m.m.f. reckoned as a Fourier series of harmonics, remains constant and rotates at uniform speed. This is a very important conclusion, and shows that, to a close degree of approximation, the armature m.m.f. can be considered as a constant and uniformly rotating quantity.

MAGNITUDE OF ARMATURE REACTION. The armature reaction of one uniformly-distributed phase is a trapezium of geometrical shape,

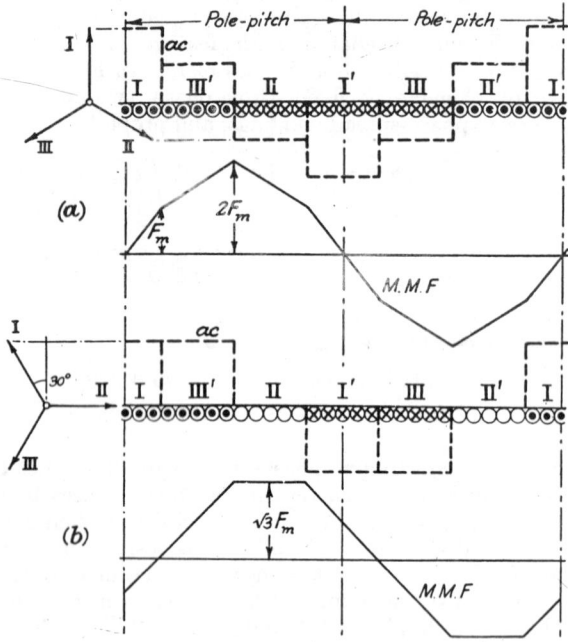

FIG. 139. ARMATURE REACTION OF THREE PHASES

as shown in Fig. 138 (a) for a phase-spread of 60°. Per pole, the amplitude corresponding to the time-maximum current will be F_m, and the corresponding Fourier series is

$$F = \frac{4}{\pi} F_m \cos \omega t \left[\sum_{n=1}^{n=\infty} \frac{1}{n} \left(\frac{\sin \frac{1}{2}n\sigma}{\frac{1}{2}n\sigma} \right) \sin nx \right]$$

when the current varies sinusoidally according to the term $\cos \omega t$, and σ is the phase spread. The term in curved brackets will be recognized from eq. (78) as the distribution factor k_{mn}.

In practical slotted windings the subdivision of the w...ding into sets of slot-conductors results in a "stepping" of the m.m.f. wave, Fig. 138 (b). This does not affect the fundamental and lower-order harmonics, although it will modify the higher harmonics in the Fourier series quoted.

To sum the m.m.f.'s of three such phases, Fig. 139, account must be taken of the time-displacement of their currents and the space-displacement of their axes. For phases I–I', II–II' and III–III', therefore

$$F_{\text{I}} = \frac{4}{\pi} F_m \cos \omega t \left[\sum_{n=1}^{n=\infty} \frac{1}{n} k_{mn} \sin nx \right],$$

$$F_{\text{II}} = \frac{4}{\pi} F_m \cos (\omega t - \tfrac{2}{3}\pi) \left[\sum_{n=1}^{n=\infty} \frac{1}{n} k_{mn} \sin n(x - \tfrac{2}{3}\pi) \right]$$

$$F_{\text{III}} = \frac{4}{\pi} F_m \cos (\omega t - \tfrac{4}{3}\pi) \left[\sum_{n=1}^{n=\infty} \frac{1}{n} k_{mn} \sin n(x - \tfrac{4}{3}\pi) \right]$$

For the resultant m.m.f.,

$$F_a = F_{\text{I}} + F_{\text{II}} + F_{\text{III}}$$
$$= \frac{3}{2} \cdot \frac{4}{\pi} F_m \left[k_{m1} \sin (x - \omega t) + \frac{1}{5} k_{m5} \sin (5x - \omega t) \right.$$
$$\left. - \frac{1}{7} k_{m7} \sin (7x - \omega t) \ldots \right] \qquad . \qquad . \qquad . \quad (82)$$

This result shows that the m.m.f. has a constant fundamental, and harmonics of orders 5, 7, 11, 13, or $6m \pm 1$, where m is any positive integer. All multiples of 3 are absent. Also, the harmonics move with speeds proportional to the reciprocal of their order, either with or against the fundamental: those of order $6m + 1$ move in the same direction as the resultant field, and those of order $6m - 1$ against it.

The pointed curve (a) in Fig. 139 obtains for $\sigma = 60°$ when $\omega t = 0$, and from eq. (82) has the peak value

$$\hat{F}_1 = \frac{18}{\pi^2} F_m \left[1 + \frac{1}{25} + \frac{1}{49} + \ldots \right] = 2F_m ;$$

while the flat-topped curve, (b), obtains when $\omega t = \pi/6$, and has the amplitude

$$\hat{F}_2 = \frac{18}{\pi^2} F_m \cdot \frac{\sqrt{3}}{2} \left[1 + \frac{1}{25} + \frac{1}{49} + \ldots \right] = (\sqrt{3}) F_m.$$

The harmonics are generally (although not always) of negligible importance, so that it is permissible to use the fundamental term alone. If J be the r.m.s. current per conductor, flowing in the z_s conductors of each of g' slots per pole per phase, then (Fig. 138),

$$F_m = \tfrac{1}{2} g' z_s \cdot (\sqrt{2}) J$$

and the fundamental amplitude of the armature reaction is, from eq. (82),

$$F_a = \frac{6}{\pi}k_{m1}F_m = \frac{6}{\pi}k_{m1}\frac{1}{2}g'z_s(\sqrt{2})J = 1\cdot35k_{m1}g'z_sJ$$

ampere-turns per pole. Since $JT_{ph} = Jg'z_sp$, this can be written

$$F_a = 1\cdot35JT_{ph}k_{m1}/p \text{ ampere-turns per pole.}$$

When a short-pitched double-layer winding is used, each layer produces a similar distribution of m.m.f., but these are displaced from

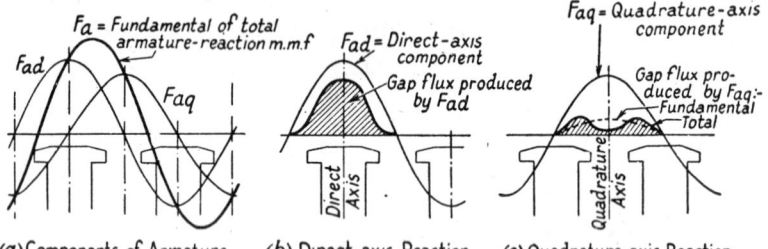

(a) Components of Armature Reaction (b) Direct-axis Reaction (c) Quadrature-axis Reaction

FIG. 140. ARMATURE REACTION IN SALIENT-POLE MACHINE

one another by the chording angle ε. The summation leads to the introduction of the coil-span factor, so that the fundamental term of the armature-reaction in a three-phase winding becomes finally

$$F_a = 1\cdot35JT_{ph}k_{m1}k_{e1}/p$$
$$= 1\cdot35JT_{ph}K_{w1}/p \text{ amp.-turns per pole .} \qquad . \quad (83)$$

FIELD-CIRCUIT EQUIVALENTS OF ARMATURE REACTION. It is necessary in the design of a.c. machines to estimate how many field ampere-turns are equivalent to a given armature reaction. If the field winding be itself a three-phase winding, as in the induction motor, it is only necessary to write

$$1\cdot35J_1T_1K_{w1}/p = 1\cdot35J_2T_2K_{w2}/p \text{ or } J_1T_1K_{w1} = J_2T_2K_{w2},$$

where K_{w1} and K_{w2} are the products of the distribution and coil-span factors of windings 1 and 2 respectively.

When the field winding is of the "single-phase" type carrying a d.c. excitation the problem becomes more complicated, and in any case it will only be possible to obtain the field-circuit equivalent of that component of the sine-distributed armature-reaction which is co-axial with the field pole-centres, i.e. the *direct-axis* armature-reaction.

Direct-axis Reaction. In the non-salient-pole, or cylindrical-rotor, machine (e.g. a turbo-alternator) the field winding is distributed in slots and gives a trapezoidal m.m.f. curve resembling that in Fig. 138 (b), but due to d.c. excitation and therefore of constant amplitude. For this case it suffices to equate the fundamental of the field and armature m.m.f.'s.

For salient-pole machines, Fig. 140, the method of equating m.m.f.'s is not valid because the gap is not uniform in length over the pole-pitch and the field coils are "concentrated" round the poles. The field-circuit equivalent of direct-axis armature reaction can be determined only on the basis of equality between the fundamental *fluxes* which field and armature would separately produce in the airgap. A method of flux-plotting and analysis has already been referred to in Fig. 115, page 183. The results so obtained, with values of the ratio $b/Y = $ (pole-arc/pole-pitch) between 0·65 and 0·75, give approximately

$$F_{ad} = 1 \cdot 07 J T_{ph} k_{m1} k_{e1}/p$$
$$= 1 \cdot 07 J T_{ph} K_{w1}/p \text{ amp.-turns per pole .} \qquad . \quad (84)$$

for the field excitation corresponding to the armature current component J which causes direct-axis armature reaction.

Quadrature-axis Reaction. The component of the armature m.m.f. wave co-axial with the interpolar axis of a salient-pole machine will produce flux in the magnetic circuit that is unaffected by the current in the field winding. There is thus no field-circuit equivalent of *quadrature-axis* armature reaction, F_{aq}. The interpolar gap presents a considerably greater reluctance than does the much shorter gap under the pole-shoes, so that the quadrature-axis flux distribution has a large spatial third-harmonic. The usual star connection of the armature winding prevents any corresponding third-harmonic e.m.f. from appearing at the line terminals. The fundamental component of cross flux, however, although it is small compared with that corresponding to direct-axis m.m.f., can generate an e.m.f. of fundamental frequency in the armature winding, and thus gives quadrature-axis reaction the nature of a pure leakage reactance. The most accurate treatment is by flux-plotting,* but this and other approximations are rather beyond the scope of this book.

PHASE-SEQUENCE COMPONENTS. The whole of the foregoing discussion relates to armature reaction produced by normal, positive-sequence currents. Under fault conditions, or unbalanced loading, or the deliberate application of asymmetrical armature voltages, currents of negative and (in certain cases) zero sequence will flow. Each will produce a corresponding armature reaction.

When negative-sequence currents are present in the armature windings of a three-phase machine, an m.m.f. is developed which rotates at synchronous speed in a direction opposite to that of the pole-system. Double-frequency e.m.f.'s are consequently induced in the field winding by the negative-sequence flux. In a salient-pole machine the flux will not be constant, because of the strongly

---

* See Wiesemann, "Graphical Determination of Magnetic Fields," *J.Am.I.E.E.*, (1927).

variable gap permeance presented to it as it passes in succession the direct and quadrature axes. This is especially true if the machine has laminated pole-faces and no damper windings, for the latter have a compensating effect tending to make the machine equivalent to a cylindrical-rotor construction. The effect is naturally accompanied by considerable pole-face and damper losses.

Zero-sequence currents, being co-phased in time in all three phases, produce no flux in the air-gap.

12. Conductors. The conductors of electrical machines are almost invariably of copper. Their d.c. resistance can be calculated from the figures of the B.S.I. for standard annealed copper, of which the resistance per metre length and square millimetre cross-section is $\rho = 1/58$ Ω. at 20° C. The resistance-temperature coefficient is $1/234\cdot5$ per ° C. at 0° C., so that at 75° C., the resistance of a 1 m. length of copper, 1 mm.^{2} in section, is

$$\frac{1}{58}\cdot\frac{234\cdot5 + 75}{234\cdot5 + 20} = 0\cdot021\ \Omega.$$

If a conductor is worked at a current-density of δ amperes per mm.^{2}, the I^2R loss per m. and mm.^{2} is $\rho\delta^2$, and the loss per kg. of copper is

$$p_c = 1\,000\rho\delta^2/8\cdot9\ \text{W. per kg.} \quad . \qquad . \qquad . \qquad . \quad (85)$$

which has the value $2\cdot36\delta^2$ W. per kg. at 75° C.

EDDY-CURRENT LOSSES. When a conductor of finite section lies in a pulsating magnetic field, it becomes the seat of induced e.m.f.'s, which cause parasitic circulating currents precisely similar to those in steel plates under the same conditions. The additional losses from this cause may easily be very large if suitable precautions are not taken. The obvious step, by analogy with the magnetic circuit, is to laminate or strand the conductors in the direction at right angles to the disturbing field, in order that the flux linkages shall be reasonably small.

The parasitic eddy-currents in an isolated conductor due to its own field are called the *skin effect*. They arise on account of the inductance of the central parts of the conductor exceeding that of the outer parts. The reactance of the centre is therefore greater, and the current flows consequently more readily in the outer layers of the conductor. But any departure from uniform current density increases the I^2R loss over its d.c. value. An alternative explanation is that the greater induced e.m.f. of self-induction in the middle parts of the conductor cause circulating currents which, superimposed on the main current, increase the I^2R loss.

Writing for the *eddy-loss ratio*

$$K_d = \frac{\text{actual } I^2R \text{ loss}}{\text{d.c. } I^2R \text{ loss}} = 1 + \frac{\text{eddy } I^2R \text{ loss}}{\text{d.c. } I^2R \text{ loss}},$$

then the actual loss in a given case is found by calculating the d.c.

I^2R loss (i.e. that loss which would occur if the current were uniformly distributed over the cross-section of the conductor and there
were no stray field effects), and multiplying the amount by K_d.
It can be shown* that the loss factor depends on the geometry of the
conductor, the frequency of pulsation of the current or flux, and the
resistivity of the conductor. The eddy-loss of steel laminations
(eq. (48)) is seen to be obtained from an expression in which these
quantities are represented.

CONDUCTORS IN SLOTS. The usual assumption for the distribution
of slot flux is that of Fig. 121, given in connection with the calculation of leakage reactance (Chapter IX, § 5). The question now under
consideration is, in fact, raised by the unequal inductance produced
by the slot-flux distribution, as already discussed in the context of
eq. (63).

The original solution for the eddy-current losses, due to Field,†
may be applied by means of curves, or by the following formulae.
Loss-ratio for a single solid conductor of depth h cm.—

$$K_d = 1 + \frac{4}{45} (\alpha h)^4 \qquad . \qquad . \qquad . \qquad . \qquad (86)$$

Loss-ratio in the pth layer of a slot containing m layers each of depth
h, with $p = 1$ at the bottom of the slot and $p = m$ at the top:

$$K_{dp} = 1 + \frac{p(p-1)}{3} (\alpha h)^4 \qquad . \qquad . \qquad . \qquad . \qquad (87)$$

Average loss-ratio for m layers:

$$K_{d\,av} = 1 + (\alpha h)^4 (m^2/9) \qquad . \qquad . \qquad . \qquad . \qquad (88)$$

where
$$\alpha = 2\pi \sqrt{\frac{bf}{\rho w_s . 10^9}} \simeq \sqrt{\frac{\text{copper width}}{\text{slot width}}} \qquad . \qquad (89)$$

The ratio (copper width/slot width) is illustrated in Fig. 121(a) by
the ratio b/w_s; ρ is the resistivity of the conductor material in
ohms per cm. cube; and f is the frequency. The approximation
for α in eq. (89) is for $f = 50$ and for copper conductors; and eqs.
(86), (87), and (88) hold only if $\alpha h < 0.7$.

In practice it is unusual to find $\alpha h > 0.7$, as (at 50 cycles) excessive losses would otherwise result. In a slot 8 mm. wide containing
a solid conductor 6 mm. wide and 25 mm. deep, the loss-ratio from
eq. (86) is

$$K_d = 1 + \frac{4}{45} \left[\left(\sqrt{\frac{6}{8}} \right) . 2.5 \right]^4 = 1 + 1.95 = 2.95,$$

that is, the actual loss is nearly three times as great as the calculated

* See [B], p. 106 *et seq.*
† A. B. Field, *Trans. A.I.E.E.*, 24 (1905); also M. B. Field, *J.I.E.E.*, 37,
p. 83 (1906).
See also [A], p. 106; and Taylor. *J.I.E.E.*, 58, p. 279 (1920).

d.c. I^2R loss, and economical design would make such a conductor unthinkable. The value of αh in this case is 2·16.

Where the current to be carried by a conductor necessitates a large cross-section, the liability to excessive eddy loss and heating makes it essential to subdivide the conductor into strips that are sufficiently shallow to keep the loss-factor as near unity as possible. Now the formulae for the loss factors given for laminated conductors are valid only so long as the subdivisions are not connected in parallel in such a way as to permit a path between laminae for eddy currents. Otherwise the utility of subdivision is almost entirely lost.

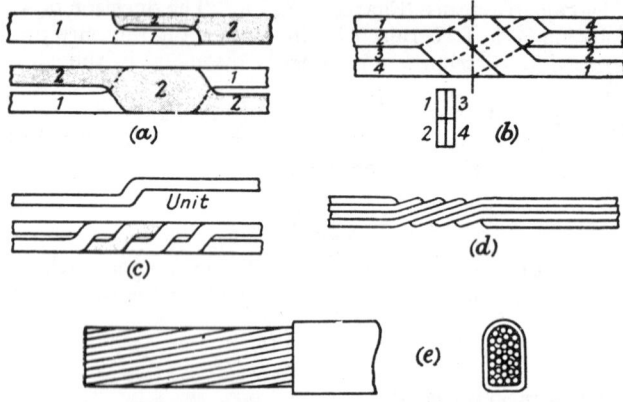

FIG. 141. TWISTED SLOT-CONDUCTORS

Of course, a parallel connection is essential for distributing the conductor current, so that the connection in parallel must be made so as to secure complete balance between the individual subdivisions. This may be obtained by *twisting* or *transposition*.

Twisted slot-conductors have their layers transposed or twisted in the slot so as to obtain symmetrical lengths such that each sub-division occupies all possible layer positions for the same length of slot. In a long-cored turbo-alternator, the twisting may be carried out two or three times in a single slot. The effect is to equalize the eddy e.m.f.'s in all laminae, and to allow the layers to be paralleled at the ends without producing eddy circulating currents between the layers. Considerable ingenuity is shown in the design of the twist, which must be accommodated in the slot space, and generally requires a reduction of the layer cross-section locally. Fig. 141 shows typical twists.* At (a), (b), (c) and (d) the cross-over is obtained by special shaping of the conductors. At (e) the method used is a radical departure; it consists in using lightly insulated round-wire

---

* The question is fully dealt with by Kuyser, "Recent Developments in Turbo Alternators," *J.I.E.E.*, 67, p. 1081 (1929).

compressed-strand conductor assemblies for the slots. The strands are laid up in concentric, helical layers by a special process. A high degree of subdivision is attained, and some approximation to a uniformity of eddy e.m.f.'s in the several strands. The conductor has to be connected at the ends to suitable overhang strips. The loss factor may be of the order 1·5 : it is not easily possible to calculate it.

Transposition in the overhang connectors is a method of making a layer occupy all possible slot positions by twisting the overhang connectors where convenient. A relation between the number of layers and the number of slots per phase is essential. This transposition is obtained automatically in diamond and basket windings, in that the upper conductors of one coil side become the lower conductors of another. This may be sufficient for all but the largest machines, where twisting in slots has to be combined with overhang transposition to keep the eddy loss within bounds.

CONDUCTORS IN OVERHANG. The problem of the estimation of eddy current losses in overhang conductors is very difficult, as the leakage field is not readily estimated either in magnitude or direction. Consider Fig. 142 (a), in which a typical case is illustrated. The actual field distribution may, for analysis, be resolved into approximate fields parallel to and across the conductors, (b) and (c). Since the eddy loss factor is proportional to h^4, the parallel field may frequently be neglected. The length of path of the leakage flux is, in (c), assumed to be $b_1 + h$, and the overhang considered as contained in an imaginary slot $b_1 + h$ wide and $h/2$ deep.

Then
$$\alpha = \sqrt{\frac{nb}{b_1 + h}} \qquad . \qquad . \qquad . \qquad . \qquad (90)$$

at 50 cycles, where n = number of side-by-side strips in a group, b = thickness of strip, and b_1 = overall width of the n strips. According to the type of winding, the various quantities are suitably interpreted. Thus with a basket winding, Fig. 142 (d), the overhang length might be taken in the three parts l_1, l_2 and l_3; with $n = 1$ for l_1, n = two-thirds of the number in the group for l_2, and n = total number in a phase group for l_3. Whence for the overhang

$$K_d = \frac{K_{d1}l_1 + K_{d2}l_2 + K_{d3}l_3}{l_1 + l_2 + l_3} \qquad . \qquad . \qquad . \qquad (91)$$

TRANSFORMERS. Applying the same method as for slot conductors, the eddy loss factor for transformers is found by considering each coil of length L_c as contained in a "slot" of length L. The conductor depth h is replaced by b, and the loss factor is as given by eq. (88), where $\alpha = \sqrt{(L_c/L)}$.

13. **Insulation.** It may be said with some truth that the insulation, or *dielectric circuit*, of electrical machines constitutes their weakest member. Insulating materials are essentially non-metals, and have

a great variety of constituents: they may be organic or inorganic, vegetal or mineral or animal in origin, uniform or heterogeneous in texture, natural or artificial. An ideal insulating material would have—

(*a*) high dielectric strength, particularly at high temperatures;
(*b*) good heat conductivity;
(*c*) permanence, particularly at high temperatures;
(*d*) good mechanical properties, such as ease of working and application, resistance to failure by vibration, abrasion or bending;
(*e*) no attraction for moisture.

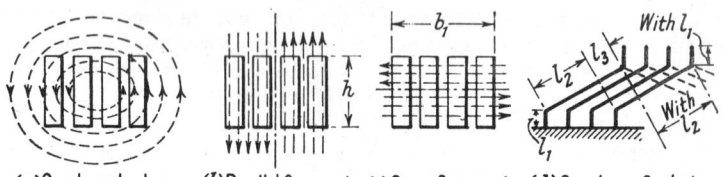

(*a*)Overhang Leakage.   (*b*)Parallel Component Field.   (*c*)Cross Component Field.   (*d*)Overhang Conductors.

Fɪɢ. 142. Overhang Conductors

It is noteworthy that considerations of temperature predominate in this list of criteria. Both copper and iron, the chief remaining materials employed in machine construction, could be worked at temperatures very much higher than is at present customary, without adverse effects; so that the losses could be increased and the amounts of material reduced. The limiting feature is, however, the insulation, which in the present state of the art is liable to deterioration at quite moderate temperatures. Thus the temperature-rise, and therefore the design, is limited almost solely by the restrictions imposed by the insulation.

The British Standards of insulating materials form a convenient classification for electrical machinery and apparatus. The manufacture of machines, etc., is based upon this classification.

Class Y. Cotton, silk, paper, wood, cellulose, fibre, etc., without impregnation or oil-immersion.

Class A. The materials of Class Y impregnated with natural resins, cellulose esters, insulating oils, etc.; also laminated wood, varnished paper, cellulose-acetate film, etc.

Class E. Synthetic-resin enamels, cotton and paper laminates with formaldehyde bonding, etc.

Class B. Mica, glass fibre, asbestos, etc. with suitable bonding substances; built-up mica, glass-fibre and asbestos laminates.

Class F. The materials of Class B with more thermally-resistant bonding materials.

Class H. Glass-fibre and asbestos materials, and built-up mica, with appropriate silicone resins.

Class C. Mica, ceramics, glass, quartz and asbestos without binders or with silicone resins of superior thermal stability.

The classification* is based on the following maximum permitted temperatures:

Insulation class:	Y	A	E	B	F	H	C
Maximum temp., °C.:	90	105	120	130	155	180	> 180

The figures are based on a 20-year working life under average conditions. The life of an insulating material is closely related to the "hot-spot" temperature within the winding it covers.

The cost of the insulation in an electrical machine may be a significant proportion of the total. For large machines the selection of material is particularly exacting, and it may sometimes be found that an expensive material has economic advantage over a cheaper one in higher electric and mechanical strength, superior resistance to "tracking," higher thermal conductivity, or shorter processing times. The behaviour of the material must be closely predictable, because breakdowns can be very costly.

Changes in the insulation of small machines, to exploit newer developments in insulation research and production, can often be achieved quickly because the insulation design is relatively simple. Considerable advances in this field have been made in the past two decades.

MATERIALS. Some notes are given on the available range of machine insulation.

Mica in the virgin or sheet state is difficult to work, and it is used in the form of sheets of splittings with shellac, bitumen or synthetic polyester bonding.

Micafolium is a wrapping composed of mica splittings bonded to paper and air dried. It may be wound on to conductors by hand, then rolled and compressed between heated plates to solidify the material and exclude air.

Fibrous glass is made from material free from alkali metal oxides (soda or potash) that might form a surface coating that would attack the glass silicates. Glass absorbs no moisture volumetrically, but may attract it by capillary action between the fine filaments. Tapes and cloths woven from continuous-filament yarns have a high resistivity, thermal conductivity and tensile strength, and form a good Class B insulation. The space factor is good, but the material is susceptible to abrasive damage. The very moderate elasticity can be improved with varnished tapes by cutting the tape on the bias. Thin glass-silk coverings are available for wires for field coils or mush windings: varnishing is necessary to resist abrasion.

Asbestos is mechanically weak, even when woven with cotton

* B.S. 2757: 1956. "Classification of Insulating Materials for Electrical Machinery and Apparatus on a basis of Thermal Stability in Service."

fibres, and can only occasionally compete with fibre glass. Laminates of asbestos with synthetic resins have good mechanical strength and thermal resistance. Asbestos wire- and strip-coverings have resilience and abrasion resistance, but the space factor is somewhat low.

Cotton fibre tapes woven from acetylated cotton, recently developed, have remarkable resistance to heat "tendering," and are much less hygroscopic than ordinary cotton materials.

Polyamides such as nylon make tapes of high mechanical strength and effect a saving in space by their thinness. Nylon film is one of the few plastic films having adequate resistance to temperature and opposition to tearing.

Synthetic-resin enamels of the vinyl-acetate or nylon types give an excellently smooth finish and have been applied for mush windings, with considerable improvement in winding times and in length of mean turn. Varnishes of the same basic materials give good bonding of windings.

Slot-lining materials have in the past been various mica-composites, but the mica content is easily damaged in forming. With small motors a two-ply varnished cotton cloth bonded to pressboard has been found satisfactory; while three-ply material may serve for heavier windings.

Wood, in the form of synthetic-resin-impregnated compressed laminations, has proved a robust and accurate material for packing blocks, coil supports and spacers. If the electrical properties are not adequate, phenolic paper laminates will be preferred although their cost is greater.

Silicones are semi-inorganic materials with a basic structure of alternate silicon and oxygen atoms. They are remarkably resistant to heat, and as binders in Class H insulation permit of continuous operation at 180°C. Even when disintegrated by excessive temperatures, the residue is the insulator silica. Silicones are water-repellent and anti-corrosive: they have been successfully used in oil-less transformers, traction motors, mill motors and miniature aircraft machines operating over a winding temperature range of 200 to − 40°C. An additional advantage in high-temperature operation is the superior thermal conductivity, improving heat-transfer from conductors and facilitating dissipation.

CHAPTER XI

VENTILATION

1. **Dissipation of Heat.** The losses produced in the core and conductors of electrical machines are converted into heat, which raises the temperature of the several parts. The presence of the ambient air and (in most cases also) a cooling or ventilating system, tends to drop the temperature of the machine. The temperature-rise actually obtaining depends on the relation between the conditions of cooling and the amount of heat produced: the final temperature is reached when the rates of heat production and dissipation are equal.

A hot body dissipates its heat by *radiation, conduction* and *convection.** Most of the heat is removed from electrical machines by a combination of convection and conduction, assisted somewhat by radiation from the outside surface. Briefly, a current of cool air† is passed through the machine to *conduct* the heat away: but the eddying and scrubbing action of *convection* currents is relied upon considerably in order that the air shall take up the heat efficiently. Experiment shows that a rough surface dissipates more readily than a smooth one, and that high air speeds are essential to obtain *turbulence* instead of *stream-line flow.*

The estimation of the cooling of a machine, particularly a large one, is a matter of the first importance, as the rating of the machine depends on the proper ventilation being maintained. Unfortunately, ventilation is far from being an exact science. In fact, it may be considered almost an art, of which the designer is the exponent.

2. **Heat Flow.** In rotating machinery, the heat is dissipated principally by convection to circulating air. For small machines, natural cooling by convection and radiation is relied upon, assisted when the frames are open by the random air-currents set up by the movement of the rotating parts. It is common practice to fit a fan on the rotor to augment the air currents. Baffles may then be used to direct the air more usefully.

For bigger machines, the core masses become too large to be adequately cooled by their outer surfaces, and radial and axial ducts must be used, together with proper means for supplying cooling air to the new surfaces exposed. As one of the most difficult cases, the turbo-alternator with its exceptionally massive core requires a highly elaborate gas-duct cooling system in order that all parts of

---

. * An outline of the physical basis of heat dissipation is given in [B], Chapter III.

† Internal cooling in large high-speed synchronous machines is usually by means of hydrogen: the principles, however, remain unchanged.

the core may be kept to temperatures within the limits imposed by the insulation.

TYPES OF ENCLOSURE. The scheme of ventilation is closely related to the type of motor enclosure. The several types are described as—

Open-pedestal, in which the stator and rotor ends are in free contact with the outside ambient air, the rotor being carried on pedestal bearings mounted on the bedplate;

Open-end-bracket, where the bearings form part of the end-shields which are fixed to the stator housing, a common construction for small and medium-sized machines. The air is in comparatively free contact with the stator and rotor through the openings;

Protected, or end-cover types with guarded openings: the protection may be by *screen* or by *fine-mesh covers*. In the latter case, the machine is regarded as totally enclosed;

Drip-, *Splash-* or *Hose-proof*, a protected machine with the ventilation opening in the end-shields designed to exclude falling water or dirt, or jets of liquid;

Pipe- or *Duct-ventilated*, with end-covers closed except for flanged apertures for connection to pipes along· which the cooling air is drawn;

Totally-enclosed, where the enclosed air has no connection with the ambient air: the machine is not necessarily airtight. Total enclosure may be associated with an internal rotor fan, an external fan, water cooling, or closed-air-circuit ventilation, in which the air is circulated to a refrigerator and returned to the machine;

Weatherproof, or *watertight*, as specified.

Flame-proof, for use in mines.

As a general rule, the more difficult the ventilation the lower must a given machine be rated.

COOLING-AIR CIRCUIT. The arrangement of the path of the cooling air depends on the size of the machine (a larger core requiring more subdivision and more elaborate arrangements for gas distribution) and on its type.

Radial ventilation is commonly employed, because the movement of the rotor induces natural centrifugal movement of the air, augmented as required by rotor fans. Fig. 143A shows (*a*) a typical method suitable for machines below 20 h.p. The end-brackets are shaped to guide air over the overhang and the back of the core, and baffles are fitted to improve the fan efficiency. For larger machines, subdivision of the core is necessary (*b*), and the air paths through the radial ducts are in parallel with those across the overhang convectors. A high rate of heat dissipation is possible in the gap between smooth-bore stators and rotors—unless this gap is very short—on account of high air speeds accompanied by turbulence.

Axial ventilation is suitable for machines of moderate output and high speed. The practically solid rotor construction and restricted spider, necessary to avoid undue centrifugal forces, make it difficult to provide adequate air-paths to radial ventilating ducts. Further, the tendency in design is to increase core lengths and to restrict core

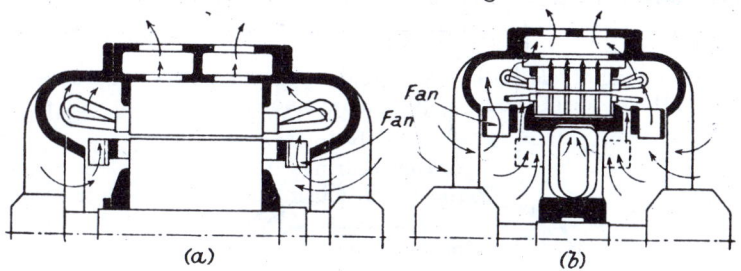

Fig. 143A. Radial Ventilation

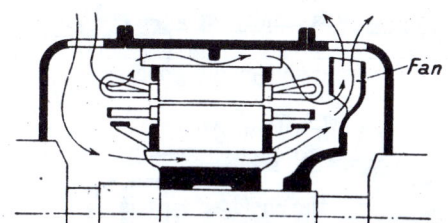

Fig. 143B. Axial Ventilation

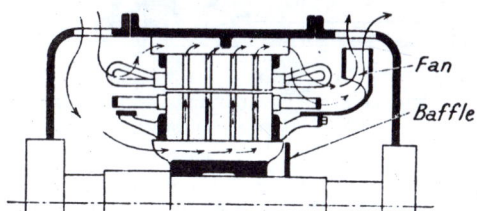

Fig. 143C. Combined Radial and Axial Ventilation

diameters in order to get a cheaper machine. Fig. 143B shows a method of applying axial ventilation to a small machine with plain cores. To increase the cooling surface, holes may be punched in the core plates to form through-ducts where considerable heat-dissipation occurs. This greatly improves the cooling, but requires a larger core diameter for the increased core depth necessary.

Combined radial and axial ventilation, first developed for small turbo-alternators, is employed for large motors. Fig. 143C shows the arrangement of an induction motor for mixed ventilation. The air is drawn in at one end, and encouraged to pass through the ducts by baffling the fan end of the rotor spider. The rotor-mounted fan ejects the air.

Multiple-inlet ventilation. With great lengths of core, there is a tendency to starve the central radial ducts of air. With the multiple-inlet system it is possible to build machines with long cores and obtain effective centre cooling. The stator frame is divided into separate air circuits fed in parallel.

The cooling of *totally-enclosed* machines presents a special problem in that the inside of the machine can have no air-connection with the

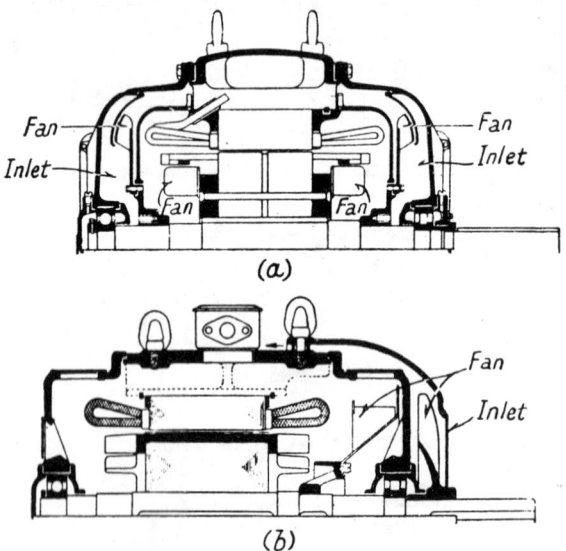

FIG. 144. VENTILATION OF TOTALLY-ENCLOSED MACHINES
(*British-Thomson-Houston*)

outside. Two methods of fan cooling are shown in Fig. 144. At (*a*) a shaft-mounted fan external to the working parts of the machine, blows air over the carcass through a space between the main housing and a thin cover plate. Internal air circulation is produced by an internal fan: this avoids the temperature gradient across the air gap. At (*b*) an internal fan circulates the heat to the carcass. Air is also blown over the outside of the carcass to improve the dissipation.

Forced ventilation is used for the largest machines, which require many tons of cooling air per hour, and permits the cleaning of the air by suitable filters, avoiding the clogging of the ducts which may otherwise occur with such large volumes of air. In an arrangement for a large induction motor, the end covers are airtight, and air from an external fan is fed in from a duct. On high-speed machines rotor-mounted fans may be sufficient. The air is filtered or washed with a water spray, then baffled against flooded scrubbing surfaces to precipitate the dirt. It is then dried by passing over a series of dry scrubbing plates.

A more complete means of securing clean cooling air is the *closed-circuit system*, as employed for turbo-alternators, Fig. 145. The hot air from the machine is passed through a water-cooled refrigerator, then returned to the machine by a centrifugal fan.

QUANTITY OF COOLING MEDIUM.

Let $\quad P =$ loss to be dissipated, kW.;
$\qquad \theta_i =$ inlet temperature of cooling medium $^\circ$C.;
$\qquad \theta =$ temperature-rise of medium, $^\circ$C.;
$\qquad Q =$ volume;
$\qquad H =$ barometric height, mm. of mercury;
$\qquad p =$ pressure.

The specific heat of dry *air* at constant pressure is 0·2375, and 1 kg. of dry air has a volume of $0·775[(\theta_i + 273)/273] . 760/H$ m³. Since 1 kW. $= 240$ cal. per sec., this loss raises the temperature of $240/237·5 = 1·01$ kg. of air by 1° C. per sec. For a loss of P kilowatts,

$$Q = \frac{1·01}{\theta} P . 0·775 \frac{\theta_i + 273}{273} . \frac{760}{H}$$

$$= 0·78 \frac{P}{\theta} . \frac{\theta_i + 273}{273} . \frac{760}{H} \text{ m.}^3 \text{ per sec.} \qquad . \qquad (92)$$

The fan power for Q m.³ per sec. at a pressure p newtons per m.² is pQ watts, and the fan will take

$$P_f = (pQ/\eta . 10^3) \text{ kW.} \qquad . \qquad . \qquad . \qquad . \qquad (93)$$

at an efficiency of η. Typical values for alternator cooling are $p = 1\,000 - 10\,000$ N per m.² $= 10 - 20$ g. per cm.² $= 4 - 8$ in. water gauge, or more in closed machines; $\theta = 20^\circ$ C.; $\theta_i = 25^\circ$ C.; $\eta = 0·2 - 0·4$. With these figures the air volume required is approximately $0·85/P\theta$ m.³ per sec.

In certain cases of very large machines with closed-circuit ventilalation when low windage is needed (e.g. synchronous machines), *hydrogen* may be employed as a cooling medium. Its specific heat at constant pressure is 3·4, and 1 kg. of dry gas occupies $11·3[(\theta_i + 273)/273] . 760/H$ m.³ To absorb P kilowatts, the volume required is

$$Q = 0·8 \frac{P}{\theta} . \frac{\theta_i + 273}{273} . \frac{760}{H} \text{ m.}^3 \text{ per sec.} \qquad . \qquad . \qquad (94)$$

which is nearly the same as for air. But the hydrogen has only one-fourteenth of the density of air, and therefore causes much less windage loss. Further, its heat conductivity being over six times as great, the heat distribution is better. It may be used with a pressure greater than atmospheric (e.g. $H = 1\,000$) to avoid in-leakage of air producing an explosive mixture. As a comparison, taking air

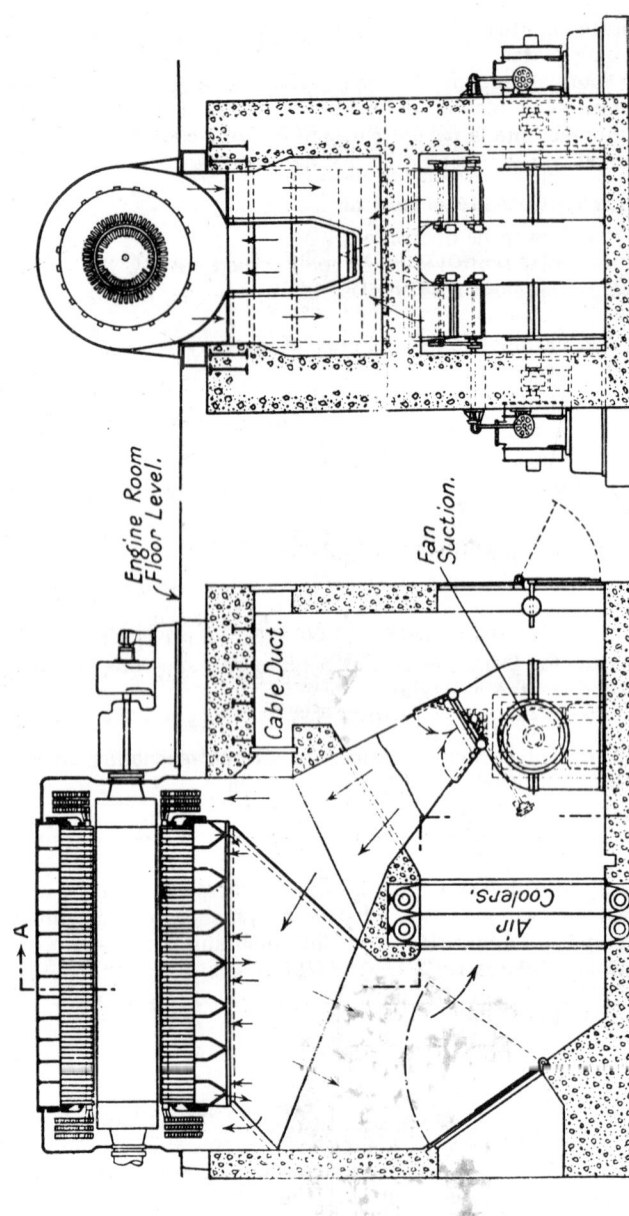

Engine Room Floor Level.

Cable Duct.

Fan Suction.

Air Coolers.

Longitudinal Section thro' Alternator Foundation Block.

Section thro' A.A.

FIG. 145. FORCED VENTILATION

as unity in each case, the *relative* figures for hydrogen are: thermal conductivity, 6·7; density, 0·07; specific heat at constant pressure, 14·3; heat capacity, 0·996; heat conductivity, 1·5.

For cooling the air of closed-circuit machines, or the oil of transformers, a quantity of liquid

$$Q = 0.24P/c_p\theta \text{ litres per sec.} \left.\begin{matrix} \\ \end{matrix}\right\} \qquad (95)$$
$$= 3.17P/c_p\theta \text{ gal. per min.}$$

is required. The specific heat of water is $c_v = 1$ and of oil is $c_v = 0.35 - 0.5$.

FLOW OF HEAT TO COOLING GAS. The circulation of cooling gas has been considered, and if properly designed, the ventilating system will dispose of the heat developed without an excessive bulk of gas being needed. The *internal* flow of heat, from the parts in which it is actually generated to the cooling surfaces from which it is transferred to the gas, is important in determining the hot-spot temperatures and the temperatures to which the insulating material will be subjected. The nature of the machine renders this problem very complex, and, as in many other aspects of design, the designer relies upon empirical constants supported by theoretical analysis and test results. A 50° C. hot-spot temperature-rise might be made up of a rise of 11° in the coolant, 25° between coil-surface and gas and 14° from the hot-spot to the coil-surface.

In its simplest form, the flow of heat reduces to simple conduction from hot parts to cool surfaces, through metal and/or insulation, the latter having by far the greater thermal resistance. For short-time rated machines, a complication is the heat-capacity of the thermal path, but this has no weight after steady conditions are reached, as with continuous ratings.

If simple geometrical shapes are assumed, and symmetrical lines of thermal flow, the temperature difference between two surfaces, required to conduct heat at a given rate from one to the other, is readily estimated. The heat conducted per m.² of path area along a path x m. thick in a material of thermal resistivity ρ_i °C. per W. per m. cube, for a temperature-difference of θ °C. is

$$p_d = \theta/x\rho_i \text{ W. per m.}^2 \qquad (96)$$

Typical values of ρ_i are given below: variations are somewhat wide.

Material	ρ_i	Material	ρ_i
Copper	0·0026	Paper	7·5
Sheet steel— .		Compressed paper . .	8
Along laminae .	0·02	Micanite . . .	8
Through laminae .	0·05 – 0·1	Cotton (dry) . .	17
Mica . . .	3	Air (still) . . .	20
Asbestos . . .	4	Oil	8
Empire cloth . .	4		

The presence of gas pockets will clearly increase the temperature gradient between an insulated conductor and its slot, and has to be rigorously avoided for this as well as other reasons.

In a mass of stampings, the heat flows more readily along than through the laminations. This is fortunate, for the gas picks up heat more readily from edges than from flat surfaces. If a stack of stampings be w m. thick and cooled at both sides, heat will flow from the centre outwards. The heat passing a thickness dx normal to the flow and distant x m. from the centre is $px \times 1 \times 1$ per m.^{2}, and the temperature difference will be

$$d\theta = px\rho_i dx,$$

where p is the heat generated per m.^{3} of material per sec. The temperature-difference between the centre and either side will therefore be

$$\theta = p\rho_i \int_0^{w/2} x \,.\, dx = p\rho_i w^2/8 \quad . \quad . \quad . \quad . \quad (97)$$

In terms of the specific core loss p_i in W. per kg., taking a specific gravity of 7 500, p may be replaced by $7 \cdot 5\, p_i$. For heat flow along and through the laminations, the values from the Table, p. 243, may be used.

3. Exponential Temperature-rise/Time Curves.

It can readily be shown that a homogeneous body, developing heat internally at constant rate, and dissipating heat at a rate proportional to its temperature-rise, will have a maximum rise proportional directly to the heat developed in unit time and inversely to the dissipation per °C. rise per second. From the cold start the final steady temperature-rise is attained exponentially.

Let $p =$ heat developed, J per sec. or W;
$G =$ weight of active parts of machine, kg.;
$c_p =$ specific heat, J per kg. per ° C.;
$S =$ cooling surface, m.^{2};
$\lambda =$ specific heat-dissipation or emissivity, J per sec. per m.^{2} of surface per ° C. difference between surface and ambient cooling medium;
$c = 1/\lambda =$ cooling coefficient;
$\theta =$ temperature-rise, ° C.;
$\theta_m =$ final steady temperature-rise, ° C.
$t =$ time, sec.;
$\tau =$ heating time-constant, sec ;
$\tau' =$ cooling time-constant, sec.

Suppose a machine to have attained after t seconds a temperature-rise θ. In an element of time dt a rise $d\theta$ takes place, while the heat developed is pdt, the heat stored is $Gc_p d\theta$ and the heat dissipated is $S\theta\lambda dt$. The two latter must sum to the total heat developed, i.e.

$$Gc_p d\theta + S\theta\lambda dt = pdt,$$

or
$$\frac{d\theta}{dt} + \theta \frac{S\lambda}{Gc_p} = \frac{p}{Gc_p}.$$

Solving this differential equation, and putting the limits $\theta = 0$ when $t = 0$, and $\theta = \theta_m = p/S\lambda$ when $t = \infty$, gives the expression

$$\theta = \theta_m(1 - \varepsilon^{-t/\tau}) \qquad . \qquad . \qquad . \qquad . \qquad . \qquad (98)$$

where $\qquad \tau = Gc_p/S\lambda$. $\qquad . \qquad . \qquad . \qquad . \qquad (99)$

τ is called the *heating time-constant.*

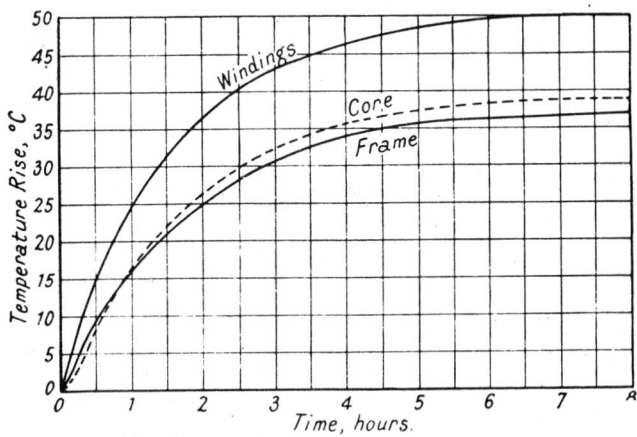

FIG. 146. TEMPERATURE-RISE/TIME CURVES

For short-time ratings the time-constant is important, as it determines the initial rate of temperature-rise at $t = 0$, which is

$$\left(\frac{d\theta}{dt}\right)_{t=0} = \frac{\theta_m}{\tau} \qquad . \qquad . \qquad . \qquad . \qquad . \qquad (100)$$

Interpreted in another way, τ is the time in which the temperature-rise attains 0.632 of its maximum value θ_m, since when $t = \tau$, θ is given by $\theta_m(1 - \varepsilon^{-1}) = 0.632\theta_m$, from eq. (98).

In Fig. 146 are shown typical temperature-rise/time curves taken on a small totally-enclosed induction motor. The exponential curve nearest to the frame-temperature would have a final steady temperature-rise of about 38° C., reaching $0.632\theta_m = 24°$ C. in about 6 000 sec., or $1\frac{3}{4}$ hr., which is the heating time constant. The final rise of the windings is about $52.5°$ C.; 0.632 of this is $33°$ C., reached in about $1\frac{3}{4}$ hr.

It is, of course, an over-simplification to regard a machine as a homogeneous body. It actually comprises several parts, each with a characteristic surface area, mass, heat capacity, thermal conductivity and rate of heat production. The temperature-rise of different

parts—or even of various points in the same part—may be very
uneven, especially on intermittent load or with strongly-cooled
machines.* In design it is necessary to make estimates of hot-spot
temperatures, and to infer the temperature-rise of various parts by
aid of empirical methods.

4. **Cooling Coefficient.** For continuously rated machines, the
curve of temperature-rise is comparatively unimportant. The final
temperature-rise θ_m is the determining factor. On the assumption
that the cooling by convection and conduction and radiation is
proportional to the temperature-rise, then

$$\theta_m = p/S\lambda = cp/S, \quad . \qquad . \qquad . \qquad . \qquad . \qquad . \qquad (101)$$

where $\quad c = 1/\lambda$, the *cooling coefficient*.

The application of eq. (101) can only be made with safety when
ample justification is forthcoming from suitable tests. The form of
the expression for c depends on the speed of the circulating gas, the
speed of the part to be cooled, and the type of surface from which
the heat is to be dissipated.

The following Table gives typical values for c in terms of the
speed u in m. per sec. of the part concerned, S being measured in m.[2]

COOLING COEFFICIENT, c

Part	c	u	Notes
Cylindrical surfaces of stator and rotor	$\dfrac{0.03 \text{ to } 0.05}{1 + 0.1u}$	Relative peripheral speed	Lower figures for forced cooling
Back of stator core	0.025 to 0.04	0	
Rotating field coils	$\dfrac{0.08 \text{ to } 0.12}{1 + 0.1u}$ $\dfrac{0.06 \text{ to } 0.08}{1 + 0.1u}$	Armature peripheral speed	Based on total coil surface Based on exposed coil surface only
Ventilating ducts in cores	$\dfrac{0.08 \text{ to } 0.2}{u}$	Air velocity in ducts	u taken as $\frac{1}{2}$ of peripheral speed of core

5. **Temperature-rise.** The temperature-rise of parts of a machine
may be measured by

(*a*) thermometer, applied to the surface of the part. This gives
the temperature of the *surface* at one point only;

(*b*) embedded temperature detector (thermo-couple or resistance coil), which gives the temperature at one *internal point*;

(*c*) resistance, involving the measurement of the resistance both

* Bates and Tustin, "Temperature Rises in Electrical Machines," *Proc.
I.E.E.*, 103 (A), pp. 471 and 483 (1956).

cold and hot, and estimating the *mean* rise by use of the resistance temperature coefficient: it is available for windings only.

These methods give results that do not refer to the same thing and furnish different bases for estimating hot-spot temperatures. If an embedded temperature detector can be placed, during construction, at a point at which experience shows the temperature to be highest, it will give an indication of the hot-spot temperature. In practice however, the problem of insulation prevents the detectors from being placed too near the conductors.

RATING. A machine is normally rated on a thermal basis of temperature rise. The British Standards Institution (in B.S. 2613: 1955) recognizes three ratings, which are the loads that can be carried by the machine without exceeding the specified limits of temperature-rise set out in the Table below.

LIMITS OF TEMPERATURE-RISE, °C.

for engine-driven a.c. generators; and for cont.-max. and short-time rated machines* of 1 h.p., or 0·75 kW. or kVA., upwards per 1 000 r.p.m.
(B.S. 2613: 1955)

Part	†	Temperature-rise‡		
		A	B	H
Stator windings—				
rated voltage < 1 kV.	T	55	75	
rated voltage 1 — 16·5 kV.:				
< 5 000 h.p. or kVA., core length < 1 m.	R	60	80	
core length > 1 m.	D	60	80	
5 000 h.p. or kVA. upwards .	D	—	80	
Field windings—				
salient-pole, single-layer bare .	R	65	90	
other types	R	60	85	
turbo type .	R	65	90	
Rotor windings—				
connected to commutator or slip-rings	T	55	75	
short-circuited, insulated	T	65	90	
Commutators .	T	55	65	
Slip-rings	T	60	70	
Cores in contact with insulated windings .	T	as for windings		
Short-circuited uninsulated windings; cores not in contact with insulated windings .	Non-injurious value			

* Excluding fractional-h.p. machines; synchronous convertors; traction motors; welding machines; flameproof motors; marine machines.
† Method of measuring temperature: T, by thermometer; R, by resistance; D, by embedded temperature-detector.
‡ A, B, H refer to the B.S. insulation classification.

1. *Continuous Maximum Rating* (c.m.r.), defining the load at which a machine may be operated for an unlimited period. Such a machine is not capable of sustained overload, but can provide considerable momentary overload without injury. The rating is recognized for all* generators and motors of a size exceeding 0·75 kW. or 0·75 kVA. or 1 h.p. per 1 000 r.p.m.

2. *Continuous Rating Permitting Overload* (c.r.p.o.). This is applicable only to d.c. and induction motors of a rating between 1 and 50 h.p. per 1 000 r.p.m., and allows for 25 per cent overload torque for restricted periods, as well as considerably greater 15-sec. excess torques.

3. *Short-time Rating.* This is the output at which a machine can work for 1 hour or for ½ hour without exceeding the specified limits of temperature rise.

The c.m.r. and c.r.p.o. depend on the cooling, but the short-time ratings are largely a matter of thermal capacity. It should be realized that the ratings have only an indirect connection with the service duty of a machine: they form only a conventional method of representing its load capacity.

---

* Except (*a*) generators driven by internal-combustion engines, which have a rating permitting a 10 per cent sustained overload; and (*b*) the classes of machine listed at the foot of the Table, p. 247.

CHAPTER XII

INDUCTION MACHINES: THEORY

1. **Development.** In 1889 Tesla and Ferraris published a description of methods of producing polyphase currents, and the former exhibited a crude type of three-phase motor at the Frankfort Exhibition of 1891. An improved construction, with a distributed stator winding and a cage rotor, was built by Dolivo Dobrowolsky in conjunction with the Maschinenfabrik Oerlikon, and described in 1893.* Dobrowolsky's motors were substantially the same as modern cage-rotor machines. The slip-ring rotor was developed at the turn of the century.

Since then, besides the development and improvement of the cage and slip-ring machines (often referred to as *plain* induction motors), the inherent limitations of the induction motor as regards its speed and its power factor have been the subjects of numerous patents and investigations. The chief variants are the synchronous-induction motor for the improvement of power factor, pole-changing windings and cascade connection for speed control, and the introduction of the commutator for the adjustment of both speed and power factor. These special machines comprise, however, only a small fraction of the total of a.c. motors in use: probably at least 80 per cent of the world's a.c. motors are plain polyphase induction motors, while many of the remainder are single-phase fractional-horse-power motors, of which many thousands are manufactured for small plant.

The general reversibility principle of electromagnetic machines applies to the induction machine. If its rotor is driven at a suitable speed, the machine operates as an *induction generator*. So employed, the machine can serve a useful but restricted role in supply systems.

The theory of the induction machine is first discussed in relation to its action as a motor.

2. **The Action of the Plain Induction Motor.** Although two different types of rotor have been referred to—the *cage* and the *slip-ring*—there is no difference in their operation. The slip-ring rotor carries a normal (usually a two-layer) three-phase winding† comprising coils of span approximating to full pole-pitch. The cage rotor is *electrically* equivalent to the same arrangement. (See § 3.)

The induction motor consists of a fixed core, the *stator*, carrying a

---

* *E.T.Z.*, 1893, p. 185.

† Any number of phases can be used, so long as symmetry is obtained and the m.m.f. rotates; this eliminates the single-phase winding, but permits any other number. Three is usual for slip-ring machines for reasons of economy. Only three rings are necessary.

three-phase winding in slots on its bore. A *rotor*, carrying a cage or slip-ring winding, also in slots, is free to rotate within the stator. The stator winding is connected to the supply, and a uniform rotating magnetic field is produced therein. This induces e.m.f.'s in the rotor, and the currents there produced interact with the rotating field to develop the torque. It would make no difference electrically if the (slip-ring) rotor were connected to the supply instead of the stator.

Consider the elementary arrangement in Fig. 147 (*a*), which shows a two-pole stator and rotor. The windings on the stator (not

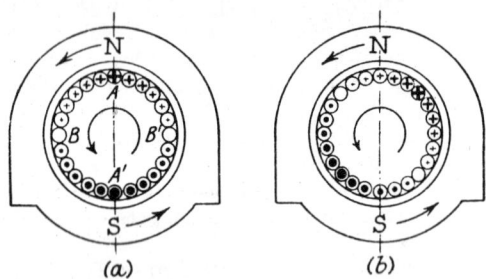

Fɪ 147. Elements of the Induction Motor

shown) are presumed connected to a three-phase supply, and take a current which produces a rotating m.m.f. (Chapter X, § 8), which has a constant and uniformly-moving fundamental component. Acting around the magnetic circuit, the principal reluctance of which resides in the small air-gap between stator and rotor, the stator m.m.f. produces a rotating flux Φ_m. The axis of the flux at a given instant, and the direction of rotation, are shown in Fig. 147 (*a*).

The rotor may be considered to have full-pitch (diametral) turns. The turn *A–A'*, coincident with the instantaneous flux-axis, links no flux, but has the greatest rate of change, and is the seat of an e.m.f. Turn *B–B'* links the whole flux Φ_m, but the rate-of-change of flux linkages in it is zero, and it has therefore no e.m.f. Turns on the rotor occupying intermediate positions have e.m.f.'s depending on the cosine of the angle they make with the instantaneous flux-axis. The e.m.f.'s are indicated on the rotor conductors in Fig. 147 (*a*), the larger dots, etc., indicating the larger e.m.f.'s per turn.

Suppose each turn to be closed on itself, and to have negligible leakage reactance. Then the dots and crosses can represent also the rotor currents, proportional to their respective e.m.f.'s. Now conductor *A* carries a current and lies in a magnetic field produced by the stator. It therefore fulfils exactly the conditions of Fig. 3 (*b*), and a mechanical force tending to move it in the direction of the rotating flux will be generated by electro-magnetic interaction. Conductor *A'* corresponds to Fig. 3 (*c*), so that the turn *A–A'* is acted upon by a couple or torque. Other conductors experience smaller torques, since they carry smaller currents and lie in fields of smaller density: all torques, however, are in the same direction, namely that of the rotation of the stator field.

If the rotor is at rest, the frequency of the induced e.m.f.'s and currents in the rotor is the same as that of the rotating field. The rotor currents therefore produce an m.m.f. which rotates with the stator m.m.f. and is directed in opposition to it. If the rotor is free to rotate, the torque will cause it to move round in the direction of the stator field. The *relative* movement between stator field and rotor conductors is consequently reduced, with a reduction of the rate-of-change of line-linkages of the **rotor coils**. The effect is twofold: (a) the rotor e.m.f. falls, and (b) **the rotor** frequency falls. The torque remains unidirectional, **because the** rotor m.m.f., although now rotating at a slower speed **relative** to the *rotor*, is carried round by the rotor movement so as to give the same speed relative to the stator as before: viz. the speed of the stator m.m.f., or the *synchronous speed*.

If the rotor shaft is not loaded, and the machine has only to rotate itself against its mechanical losses, the rotor speed will rise and approach very closely to the synchronous speed. The speed of rotation, n, of the rotor, is then less than the synchronous speed n_1 by the fraction s, known as the *slip*. The term is descriptive of the manner in which the rotor "slips back" from exact synchronism with the stator field. By definition

$$s = (n_1 - n)/n_1 = p(n_1 - n)/pn_1 = (f - f_r)/f = f_2/f \qquad . \qquad (102)$$

or $n = n_1(1 - s)$. The number of pole-pairs p in the machine is introduced for generality: f is the supply frequency, or the frequency of the stator e.m.f.'s and currents; f_2 is the frequency of the rotor e.m.f.'s and currents; and $f_r = f - f_2$ is the *speed-frequency* or product pn. On no load the slip is generally less than 0·01 or 1 per cent; in large machines with ball or roller bearings it may be much less. Under no-load conditions, then, the rotor currents have a very small frequency—less than one-half-cycle per sec., and the induced e.m.f. is very small, being the fraction s of what it was at stand-still with the same rotating field. The rotor cannot by itself achieve synchronous speed, because if it did, the rotor e.m.f.'s would vanish, and no torque-producing current could flow in the rotor turns.

In Fig. 147 (a), the dots and crosses represent rotor e.m.f.'s, and also currents under the assumption of zero inductance. Actually the rotor windings are associated with appreciable inductance, so that the rotor currents will lag on the e.m.f.'s producing them. Fig. 147 (b) is drawn for the same conditions as (a), except that the rotor windings are presumed to have at standstill a leakage reactance taken for example as equal to their resistance, causing the rotor currents to lag by 45° on the e.m.f.'s. The e.m.f.'s are still as shown in Fig. 147 (*a*), but the currents are as in (*b*). It will be seen that maximum rotor current no longer coincides with maximum gap flux density. Further, there will be a reversed torque on some of the conductors. The effect of the inductance is consequently to reduce the *starting torque* for two reasons: (*a*) the impedance is increased and the current therefore reduced; (*b*) a phase angle is introduced between rotor currents and gap flux. In the case considered tne starting torque has, in fact, been halved. A reduction by $1/\sqrt{2}$ is due to the increase of impedance $\sqrt{(r^2 + x^2)}$ to $(\sqrt{2})r$, and a further reduction by $1/\sqrt{2}$ is produced by the phase-angle of 45° between the flux and the rotor current. Reactance is therefore detrimental at starting.

The reactance becomes less and less as the speed rises and the rotor frequency falls. If x_2 be the rotor reactance per phase at standstill ($s = 1$), it will fall to the value sx_2 for a slip s. At no-load speed, it almost vanishes: the rotor phases are consequently almost non-reactive, and the inductance has no effect. The resistance is now the controlling factor, and the smaller it is, the bigger will be the current at a given small value of slip, thus producing more torque and bringing the speed more nearly to synchronous value.

Thus a high ratio of resistance to reactance is desirable under starting conditions, but the resistance should be low for running near synchronous speed, the normal condition.

Considering an induction motor running on no load at a speed very close to synchronous speed, the e.m.f.'s in the rotor turns will be very small, and so will the currents, giving a torque only sufficient to maintain the rotor in motion. Suppose now that a mechanical load is put on the rotor shaft: the rotor will slow down, and in so doing it will increase the slip, and increase its speed relative to that of the rotating field. The e.m.f.'s of the rotor turns will now increase both in magnitude and frequency, and will produce more current and therefore more torque, as long as the effect of the reactance permits. Conditions of equilibrium are attained when the slip has increased enough to allow a torque to be developed sufficient to meet the mechanical or load torque applied to the shaft. The change of speed, or slip, required is normally small: thus a slip of 0·04 or 4 per cent in a medium-sized motor may be sufficient to allow full-load torque to be developed. The induction motor is therefore a machine of substantially constant speed, and fills the same role as the plain shunt motor among the d.c. machines.

THE GENERALIZED TRANSFORMER. The rotor of the induction motor has no electrically-conducting connection with the stator supply. The power it receives and converts to mechanical output at the shaft is transferred inductively—i.e. by transformer action—from stator to rotor by means of the mutual flux. Depending on the ratio of turns and the slip, the power in the rotor changes in voltage and frequency. For this reason the induction machine has been called the *general transformer*. Since it is possible to have any required number of stator or rotor phases, with voltages adjusted in accordance with the numbers of turns per phase, and frequencies depending on the running speed, all possible variations of voltage, current, frequency, and number of phases can be obtained.

3. Cage and Slip-ring Rotors. Two types of winding are commonly used for plain induction motors, namely (a) the cage, and (b) the slip-ring winding.

CAGE ROTOR. This consists of a series of bar conductors accommodated in the rotor slots, all bars being connected in common to a conducting ring at each end. It may most reasonably be regarded as a series of full-pitch "turns" formed by pairs of conductors or bars a pole-pitch apart, joined together at each end into a closed loop by the end-rings. The latter carry the summation of the conductor currents and affect the resistance of the "turns." Further, in an actual cage, there may not be a whole number of slots per pole so that it may not be possible to find a pair of bars exactly a pole-pitch apart. Nevertheless the conception of full-pitch turns is scarcely affected.

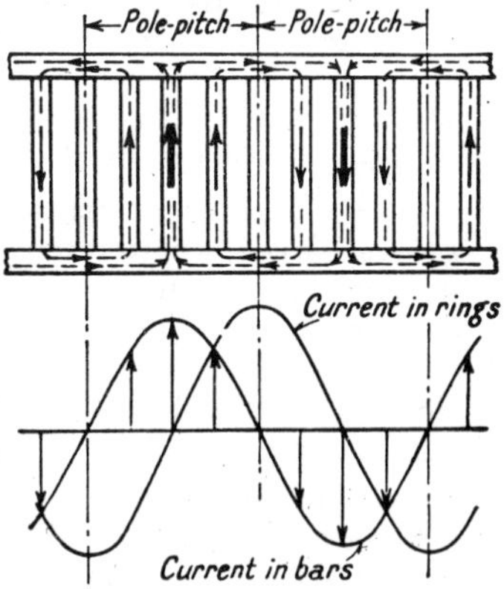

FIG. 148. CURRENTS IN CAGE ROTOR

The distribution of the currents in bars and end-rings for two pole-pitches of a developed cage winding, shown in Fig. 148, follows directly from Fig. 147 on the assumption that the end-ring resistance is negligible compared with the bar resistance. It is seen that the bars carry currents proportional to their instantaneous positions in the gap field, and that the end-ring currents are a maximum at points connected to bars which themselves carry zero current.

The greater the number of bars per pole, the more nearly do their currents produce a sinusoidal current sheet like the current distribution in Fig. 147. It is also clear that the end-ring resistance, if not negligible, will tend to distort the bar current distribution from the sinusoidal. Actually the joints between the bars and the end-rings are difficult to make effective and may be the seat of considerable resistance. The assumption of sinusoidal rotor current distribution is, however, quite justified, for the reasons that (a) only the fundamental sine components contribute to the normal torque, and (b) the harmonics are in most cases small, and contribute only secondary effects.

SLIP-RING ROTOR. A slip-ring or *wound* rotor carries a normal three or six- phase winding, connected in star or mesh and terminated on three slip-rings, which are short-circuited when the motor is running. The winding consists of phases in which all the turns must carry the same current since they are connected in series. The current distribution is that shown in Fig. 139, differing from the sinusoidal. However, the divergence is shown by eq. (82) to be practically insignificant, so that the distribution of the rotor currents is sufficiently described by Fig. 147.

The cage and slip-ring rotors are thus electrically equivalent as long as attention is confined to the fundamental sine-waves of voltage, current, flux, etc. We may proceed on the conclusion that the primary operation of a machine will be the same whichever type of rotor winding it has.

The practical reasons for the choice between cage and slip-ring windings are based in most cases on the following considerations—

(*a*) the cage is permanently closed and its electrical characteristics are fixed, whereas the slip-ring winding permits of the variation of the electrical characteristics by the inclusion of external circuits via the slip-rings; and

(*b*) the cage is adaptable to any number of poles, whereas the slip-ring winding has to be made for one (or possibly two) definite values of p.

Consideration (*a*) is most cogent from the viewpoint of starting the motor against large load torques, where it is advantageous to be able to increase the resistance of the rotor circuit by the inclusion of a rheostat connected across the slip-rings; while (*b*) is of importance in connection with speed control.

MACHINE WITH CONSTANT FLUX

4. Simple Theory of the Induction Machine with Constant Flux. When a voltage V_1 is applied to the stator of an induction machine, voltage drops and e.m.f.'s summing vectorially to V_1 must be developed in the stator. Neglecting drop in resistance, the voltage opposes an equal and opposite e.m.f. developed by the pulsation of a magnetic flux. Only a part of this flux is *mutual*, contributing to the transference of energy from stator to rotor: the remainder is *leakage* flux, which is proportionally much greater than in the transformer, on account of the increase of reluctance in the path of the main flux due to the air gap. The leakage flux is proportional to the stator current, and so increases with the load. The mutual flux must thus decrease with the load, and cannot be assumed constant. Over the load range, however, the variation may not exceed, say, 2–8 per cent, and the assumption of constant mutual flux (or, what is the same thing, the neglect of the stator resistance and leakage reactance) leads to such simple expressions for the characteristics of the induction machine that it will be made for this purpose. Throughout this section, therefore, the *mutual flux is assumed to be constant*. The rotor is considered to have a three-phase winding.

The resultant mutual flux constitutes a rotating field, and it is reasonable to consider the operation of the machine from this viewpoint. It is quite possible, on the other hand, to consider the flux as resolved into the three phase-components which are, of course, individually pulsating. The transformer action of a stator phase on a rotor phase induces in the latter an e.m.f. which in turn produces a current. The interaction of the current on the flux of all three phases produces a torque. All the characteristics of the machine can then be worked out. The method is laborious for three-phase machines, and for this and other reasons, the rotating-field method is employed here.

Let $p =$ number of pole-pairs;
$f =$ supply frequency;
$f_2 =$ frequency of rotor e.m.f.'s and currents;
$f_r =$ rotational frequency $= np$;
$n_1 =$ synchronous speed, r.p.s.;
$n =$ rotor speed, r.p.s.;
$E_1 =$ stator induced e.m.f. per phase;
$E_2 =$ rotor induced e.m.f per phase;
$T_1 =$ stator turns per phase;
$T_2 =$ rotor turns per phase;
$s =$ slip;
$\Phi_m =$ mutual rotating flux, webers.

SLIP. From eq. (102)
$$s = (n_1 - n)/n_1 = f_2/f,$$
and the rotor speed is $n = n_1(1 - s)$ rev./sec.
$$= 60f(1 - s)/p \text{ r.p.m.}$$

On no load the slip is very slight: on full load it is a small percentage. Thus the full-load speed of a 50-c/s., four-pole motor with a 3 per cent slip is $1\,500(1 - 0.03) = 1\,455$ r.p.m. Running as a generator with the same percentage slip (but now negative), the speed would be $1\,500(1 + 0.03) = 1\,545$ r.p.m.

E.M.F.'S. The e.m.f. induced by the rotating field in a stator phase is given by eq. (80),
$$E_1 = 4.44K_{w1}fT_1\Phi_m \text{ volts.}$$

In the rotor phases, the e.m.f. is produced by the same (mutual) flux but at frequency $f_2 = sf$:
$$E_2 = 4.44K_{w2}f_2T_2\Phi_m \text{ volts}$$
$$= 4.44K_{w2}sfT_2\Phi_m \text{ volts.}$$

Analogously with the transformer, it is convenient to replace the actual rotor by an equivalent rotor having the same number of turns per phase, disposed in the same way, as those of the stator. The e.m.f. in an equivalent rotor phase is $E_2' = sE_1$,

so that $\qquad\qquad E_2' = (K_{w1}T_1/K_{w2}T_2)E_2.$

At standstill, the rotating field has the same speed relative to both rotor and stator, so that $E_2' = E_1$. At any slip s, E_2' falls to the

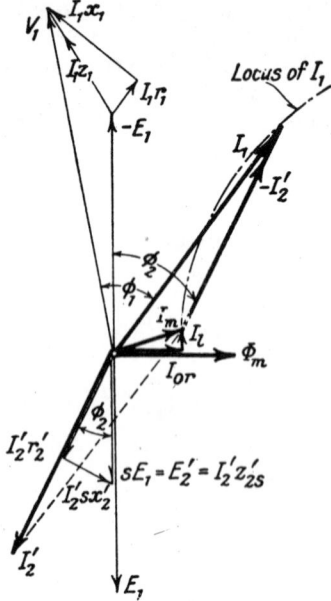

value sE_1, while if the rotor were to attain synchronous speed, the secondary (rotor) e.m.f. would vanish.

CURRENTS. Consider the machine running at a speed corresponding to a slip $+ s$. The rotor phase e.m.f., $E_2' = sE_1$, is applied to the short-circuited rotor phase winding. The windings have each a resistance r_2', and, at standstill, a reactance x_2', corresponding to frequency f. At slip s the rotor frequency is $f_2 = sf$, so that the reactance (which is a function of the frequency) falls to sx_2'. The rotor phase impedance at slip s is therefore

$$z_{2s} = \sqrt{(r_2^2 + s^2x_2^2)} \text{ or}$$
$$z_{2s}' = \sqrt{(r_2'^2 + s^2x_2'^2)}$$

for the actual and equivalent rotors respectively. The rotor current will be

$$I_2' = E_2'/z_{2s}'$$
$$= sE_1/\sqrt{(r_2'^2 + s^2x_2'^2)} \ . \ (103)$$

and it will lag on E_2' by an angle $\phi_2 = \arctan{(sx_2'/r_2')}$.

FIG. 149. COMPLEXOR DIAGRAM OF INDUCTION MOTOR

In Fig. 149, the complexor diagram is drawn. The flux Φ_m is responsible for E_1 in the stator at frequency f and sE_2' in the rotor at frequency f_2. It is permissible to include complexors of different frequencies because the reaction of the rotor on the stator is at fundamental frequency. Whatever the slip or rotor frequency, the current distribution as in Fig. 147 still holds. The rotor e.m.f. produces the current I_2', and is entirely used to circulate this current through the impedance z_{2s}'. The current lags on E_2' by $\phi_2 = \arctan{(sx_2'/r_2')}$, is in phase with the resistance drop $I_2'r_2'$, and is in phase-quadrature with the reactance drop $I_2'sx_2'$.

The stator current must provide a m.m.f. in opposition to that of the rotor for exactly the same reason as in the case of the primary and secondary currents of a transformer: the applied voltage V_1 must be opposed by an e.m.f. E_1 (neglecting resistance and reactance drops in the stator) which is generated by the mutual flux: Φ_m must therefore be maintained. Φ_m is established in the common magnetic circuit by a magnetizing m.m.f., requiring the stator to carry also a magnetizing current I_{0r}, and a core-loss current I_l, totalling I_m. The total stator current therefore consists of two components: (a) — I_2', to oppose the rotor m.m.f.; and (b) I_m, the magnetizing current. These complexors sum to I_1, as shown in Fig. 149. Otherwise described, the stator and rotor m.m.f.'s, proportional respectively to I_1 and I_2', have a resultant m.m.f. represented by the component I_m which maintains the mutual flux.

The applied stator voltage, neglecting stator resistance and leakage reactance, is a vector — E_1, equal and opposite to E_1 the induced e.m.f. If stator drops be included, it suffices to add to — E_1 the voltages $I_1 r_1$ in phase with I_1, and $I_1 x_1$ in leading phase-quadrature with I_1. But it must be observed that, under the assumption of constant flux, E_1 is constant and V_1 must vary with the load current I_1.

POWER. The power input to the stator per phase is, from the diagram in Fig. 149,

$$P_1 = V_1 I_1 \cos \phi_1.$$

Of this, the loss $I_1{}^2 r_1$ is dissipated in the stator windings, and the loss $(- E_1)I_l$ heats the core, due to hysteresis and eddy currents. P_1 may be resolved into the components

$$P_1 = I_1{}^2 r_1 + (- E_1)I_l + (- E_1)(-I_2') \cos \phi_2,$$

since the angle between the vectors — E_1 and — I_2' is that between E_2' and I_2' in the rotor, viz. ϕ_2. Since — E_1 is the voltage component associated with the mutual flux, and — I_2' is the current component equivalent to the rotor current, then $(- E_1)(- I_2') \cos \phi_2$ must be the power delivered by transformer action to the rotor, i.e.

$$P_2 = (- E_1)(- I_2') \cos \phi_2 = E_1 I_2' \cos \phi_2$$
$$= (1/s)E_2' I_2' \cos \phi_2.$$

But $E_2' \cos \phi_2$ is the voltage component $I_2' r_2'$, so that

$$E_2' I_2' \cos \phi_2 = I_2'{}^2 r_2',$$

whence $\qquad P_2 = (I_2'{}^2 r_2')/s,$

or $\qquad I_2'{}^2 r_2' = s P_2$ . . . . . . (104)

Of the power delivered to the rotor, the fraction s is used in the rotor itself and is there lost in heat. The remainder, $(1 - s)P_2$, does not appear in the complexor diagram among the rotor quantities: it is converted into mechanical power and developed at the rotor shaft. The mechanical power is therefore

$$P_m = (1 - s)P_2 \qquad . \qquad . \qquad . \qquad . \qquad (105)$$

The important conclusion is therefore drawn (it is quite general, and not restricted by the assumption of constant flux) that, of the power P_2 delivered to the rotor, the fraction s is lost in I^2R and the remainder $(1 - s)$ appears as mechanical power (including friction and windage), so that

$$P_2 : P_m : I_2{}^2 r_2 = 1 : (1 - s) : s . \qquad . \qquad . \qquad (106)$$

The rotor power will always divide itself in this proportion, so that it is obviously advantageous to run with as small a slip as possible.

TORQUE. The mechanical power developed is $P_m = (1 - s)P_2$ watts, and the motor runs at a speed $n = n_1(1 - s)$ r.p.s., for a fractional slip s. The torque is consequently

$$M = P_m/2\pi n = (1 - s)P_2/2\pi n_1(1 - s)$$
$$= P_2/2\pi n_1 \text{ newton-m.} = P_2 \text{ synchronous W.} \qquad . \ (107)$$

for each of the N phases of the machine. The torque is thus directly proportional to the rotor power input, regardless of speed. The definition of the synchronous watt is that torque which, at the synchronous speed of the machine, would develop a power of 1 W.

The torque of the machine depends on the rotor input power, and therefore on the stator input power, the stator losses being small. The motor input is therefore proportional to the torque: a given torque at a low speed requires the same input as the same torque at a higher speed. At speeds low compared with synchronous speed much power is wasted in rotor I^2R, the mechanical power is small, and the efficiency poor.

Although the torque is produced by a series of conductors, each of which carries a pulsating current and lies in a pulsating magnetic field, the total torque of a polyphase machine has a constant value. This is because the sum of the phase powers in a balanced polyphase system is invariant with time.

GENERATOR ACTION. The discussion applies equally to the induction generator. The rotor is driven at a hyper-synchronous speed so that the slip is now negative. The mechanical power P_m in eq. (106) is equal to the sum of P_2 and $I_2{}^2 r_2$. The rotor e.m.f., Fig. 149, is reversed, so that I_1 now has an active component in *opposition* to V_1, indicative of generator conditions.

TORQUE/SLIP CURVES. For a given main flux and (approximately) stator voltage, the rotor e.m.f, E_2' and current I_2' are settled by the slip, while the phase angle ϕ_2 is a function of r_2' and sx_2' The rotor power P_2 is then also a function of the slip, and so is the torque. The torque/slip relation is evaluated as follows,

From eq. (107),

$$M = P_2 = E_1 I_2' \cos \phi_2.$$

Writing $I_2' = E_2'/z_{2s}' = sE_1/\sqrt{(r_2'^2 + s^2 x_2'^2)}$,

and $\cos \phi_2 = r_2'/z_{2s}' = r_2'/\sqrt{(r_2'^2 + s^2 x_2'^2)}$,

then $$M = \frac{sE_1^2 r_2'}{r_2'^2 + s^2 x_2'^2} \text{ synchronous W. per phase } . \quad (108)$$

This is also a general expression, and valid for actual conditions. If the mutual flux is constant, however, so will E_1 be constant. The torque may be written in terms of the ratio $\alpha = r_2'/x_2' = r_2/x_2$, the quotient of the rotor phase resistance and *standstill* reactance, yielding for an N-phase machine the expression

$$M = \frac{NE_1^2}{x_2'} \cdot \frac{s\alpha}{s^2 + \alpha^2} = K_t \frac{s\alpha}{s^2 + \alpha^2} \text{ syn. W. } . \quad (109)$$

where $$K_t = \frac{NE_1^2}{x_2'} . \qquad . \qquad . \qquad . \qquad . \qquad (110)$$

For constant flux and a given arrangement of rotor winding, K_t is a constant. The torque/slip curve depends on s and α.

Maximum Torque. Eq. (109) for the torque will show a maximum value for any given ratio $\alpha = r_2/x_2$. Writing $(dM/ds) = 0$,

$$(s^2 + \alpha^2) \frac{d(s\alpha)}{ds} - s\alpha \frac{d(s^2 + \alpha^2)}{ds} = 0,$$

giving $\alpha(s^2 + \alpha^2) = 2s^2\alpha$,

so that $\alpha = 0 \text{ or } s^2 = \alpha^2$

are the conditions for $(dM/ds) = 0$. The value $\alpha = 0$, or $r_2 = 0$ gives a *minimum* torque of zero: that is, if the rotor had no resistance it could develop no torque. This is clear from the complexor diagram, Fig. 149, for in such a case I_2 would always lag by 90° on sE_2, and would have no torque component with Φ_m (see Chapter VIII, § 4).

The solution $s^2 = \alpha^2$, or $s = \pm \alpha$, gives the torque a maximum value of

$$M_m = \pm K_t \frac{\alpha^2}{2\alpha^2} = \pm \tfrac{1}{2} K_t \qquad . \qquad . \qquad . \qquad . \quad (111)$$

If $s = \pm \alpha = \pm r_2/x_2$, or $r_2 = \pm sx_2$, then the torque is a maximum: the torque attains its greatest value at a slip such that the resistance r_2 and the actual reactance sx_2 are equal. The solution gives equal positive and negative values of s. The positive value corresponds to some speed between standstill and synchronous speed, $s = 1$ and $s = 0$ respectively. The negative slip refers to a speed exceeding the synchronous, and the torque has a reversed direction. It is consequently a retarding torque, so that the machine must be *driven* to raise its speed to hypersynchronous values, and becomes a generator.

In the constant K_t there is the term $1/x_2'$. The maximum torque depends on the rotor reactance, but given this, it will be independent of r_2'. The speed for maximum torque will, however, depend on the ratio r_2'/x_2', but varying r_2' will merely change the slip at which

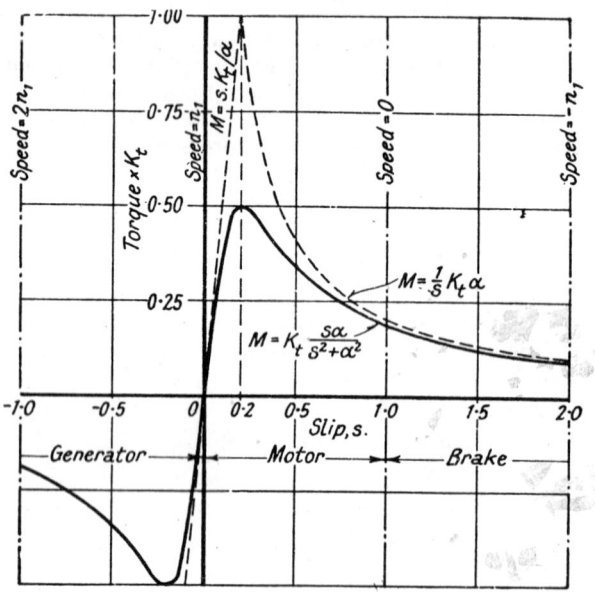

Fig. 150. Torque/Slip Relation

maximum torque occurs, not its magnitude. If $r_2' > x_2'$, the maximum torque will occur at a slip $s > 1$, which corresponds to a *backwards* speed. The forward torque associated with the backwards speed makes the motor act as a brake.

Torque over Speed Range. The normal range of speed of the induction machine operating as a motor is that included in the slip limits of 1 and 0, i.e. standstill and synchronous speed.

At *normal speeds* close to synchronism, s is small (e.g. up to 0.05), and the term s^2 in eq. (109) can be neglected by comparison with α^2 Hence approximately

$$M = K_t s \alpha / \alpha^2 = s \cdot K_t / \alpha \qquad . \qquad . \qquad . (112)$$

For a given machine K_t/α is a constant, so that the torque is directly proportional to s, as indicated by the dotted line marked in Fig. 150. An increase in load torque is developed by a (nearly) proportional increase in slip, giving the machine a speed/torque curve nearly rectilinear, similar to that of the d.c. shunt motor.

At *low speeds* and at starting, s approaches unity. Further, α is usually small in normal motors, a typical value being $r_2/x_2 = 0\cdot2$. The term α^2 is then small compared with s^2 and can be neglected. Thus from eq. (109), approximately,

$$M = K_t s\alpha/s^2 = (1/s) \cdot K_t\alpha \qquad . \ (113)$$

$K_t\alpha$ being a constant for a given motor, the torque is inversely

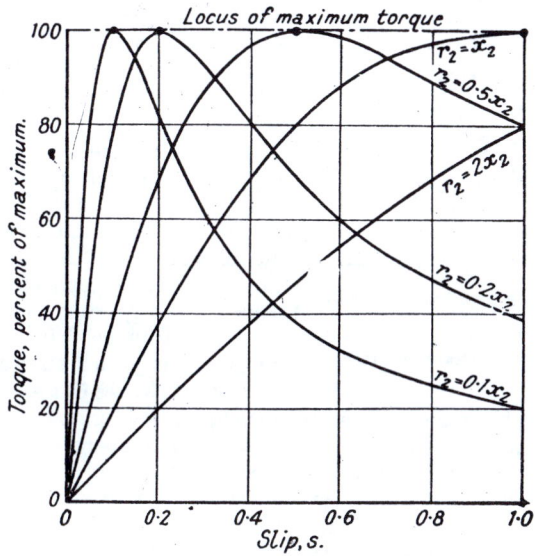

FIG. 151. TORQUE/SLIP CURVES

proportional to slip: i.e. the torque/slip curve is a rectangular hyperbola, as shown in Fig. 150.

Comparing eq. (112) and (113), the two curves will cross when

$$sK_t/\alpha = \alpha K_t/s,$$

i.e. when $s = \alpha$. The value of M corresponding is obviously K_t. The actual maximum torque is $M_m = \frac{1}{2}K_t$, from eq. (111), and this is obtainable from the complete expression for the torque in eq. (109), which can be plotted quite readily.

Fig. 150 gives the approximate straight-line and hyperbolic shapes for the ranges $s \to 0$ and $s \to 1$ respectively, and also the plot of eq. (109), for a case in which $r_2/x_2 = \alpha = 0\cdot2$. The torque is scaled in terms of the constant K_t. The curves are continued for values of s greater than unity (braking), and for negative values (generating), for all of which eq. (109) holds. The most usual range of operation is, of course, with s between 0 and 1.

Effect of Rotor Resistance. Fig. 151 shows curves drawn from eq. (109) for several values of $\alpha = r_2/x_2$, and may be considered to apply to the same machine if x_2 is a constant and r_2 varied. The value of r_2 does not affect the magnitude of the maximum torque: it only controls the slip at which this torque occurs. Thus when $r_2 = 0.2x_2$, the maximum torque occurs at $s = 0.2$ (i.e. 80 per cent of synchronous speed); when $r_2 = x_2$, it occurs at $s = 1$, i.e. at standstill. If $r_2 > x_2$, the maximum torque is not reached in the ordinary motoring range of s, but will occur at some value of s greater than unity, corresponding to backward movement.

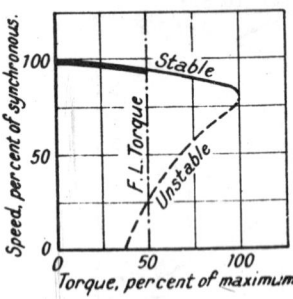

FIG. 152. NORMAL SPEED/
TORQUE CHARACTERISTIC

The normal full-load torque is less than the maximum. It may be one-half or less. In accelerating, therefore, the motor runs through an unstable region in which a rising speed gives rising torque. This may not be important, except where the motor has to start against a considerable load torque. In any case, such a load would probably be driven by a slip-ring motor in which the rotor resistance is variable (see Chapter XIII, § 3).

The unstable region of the torque/slip characteristic is not readily observable in practice, as the motor will not operate stably at any speed within the unstable range, except where special loading devices are used. The most important part of the motor characteristic is shown in Fig. 152, which gives a typical *speed/torque* curve and shows as a thick line the normal load range.

MECHANICAL POWER. To obtain power/slip curves, the torque expression in eq. (109) may be multiplied by $\omega_r = 2\pi n_1(1 - s)$ to give

$$M\omega_r = P_m = K_t \frac{s\alpha(1 - s)}{s^2 + \alpha^2} \text{ watts} \qquad . \qquad . \qquad (114)$$

The shapes of power/slip curves for the output for various values of α are shown in Fig. 153. The curves of mechanical power can be readily obtained by calculating from eq. (114) or by multiplying the ordinates of Fig. 151 by the appropriate values of $(1 - s)$. The mechanical power has a maximum value when $dP_m/ds = 0$. Differentiating $s\alpha(1 - s)/(s^2 + \alpha^2)$ gives

$$(s^2 + \alpha^2)(1 - 2s) - s(1 - s)2s = 0$$

whence

$$s = -\alpha^2 \pm \alpha\sqrt{(1 + \alpha^2)},$$

the positive sign before the root referring to motoring and the

negative to generating conditions. Inserting this value of s in eq.
(114) gives for the value of the maximum mechanical power

$$P_{mm} = \tfrac{1}{2}K_t(\sqrt{(1 + \alpha^2)} - \alpha) = \tfrac{1}{2}K_t(z_2 - r_2)/x_2,$$

where $z_2 = \sqrt{(r_2{}^2 + x_2{}^2)}$, the rotor impedance at $s = 1$. The curves
in Fig. 153 clearly show that for large rotor resistances, the mechani-

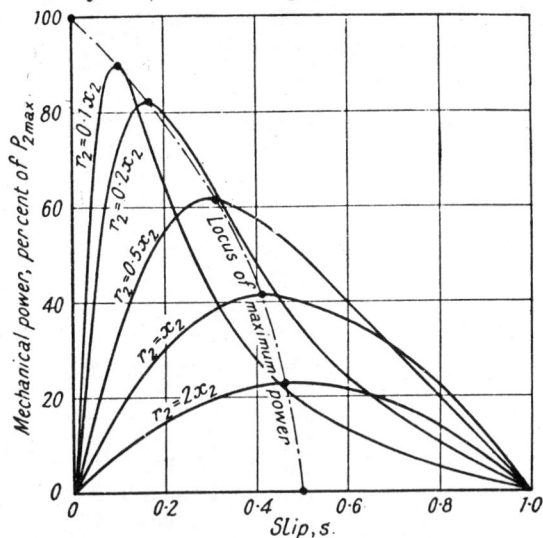

FIG. 153. MECHANICAL POWER/SLIP CURVES

cal power is severely limited. Such large resistances would, in fact,
waste most of the input in I^2R loss.

SUMMARY. Sufficient has now been said for the approximate
characteristic curves of a plain induction motor to be drawn. Eq.
(103) gives the rotor current in terms of the slip and the rotor
resistance and standstill reactance. The stator current is obtained
by adding the magnetizing current complexor. The torque is given
by eq. (109) and the mechanical power by eq. (114), all in terms of the
slip. The same analysis, with slip now negative, serves for generator
action. It must be remembered that stator impedance has been
neglected.

SIMPLE EQUIVALENT CIRCUIT AND CIRCLE DIAGRAM. The rotor
current from eq. (103) is

$$I_2' = \frac{sE_1}{\sqrt{(r_2'^2 + s^2x_2'^2)}} = \frac{E_1}{\sqrt{[(r_2'/s)^2 + x_2'^2]}}.$$

The second expression is the first divided by s. Fig. 154 shows
the equivalent circuit for each The currents in (a) and (b) are the

same, but in (b) the applied voltage E_1 is constant. It is now easy to solve for I_2' for any chosen slip.

Suppose $E_1 = 100$ V., $r_2' = 0.1$ Ω, $x_2' = 0.5$ Ω. For a 5 per cent slip, $s = 0.05$ and $r_2'/s = 0.1/0.05 = 2.0$ Ω, whence

$$I_2' = 100/\sqrt{(2^2 + 0.5^2)} = 48 \text{ A.},$$

at a phase angle $\phi_2 = $ arc tan $(0.5/2) = 14°$ lagging on E_1. Similarly for 20 per cent slip, $s = 0.2$ and $I_2' = 141$ A. at $\phi_2 = 45°$. At standstill $s = 1$, and the current is 196 A. at a lagging phase angle

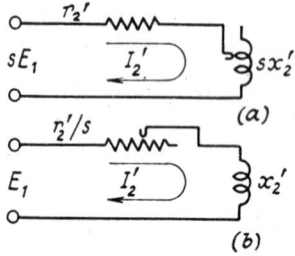

(a)

(b)

Fig. 154. Equivalent Circuit

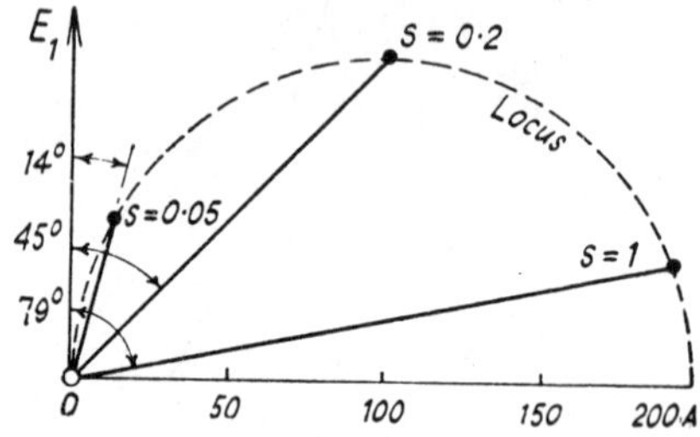

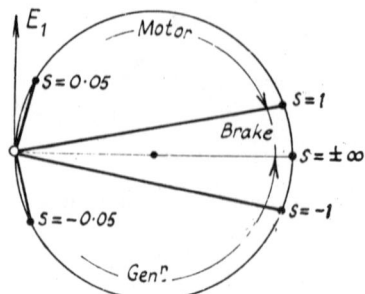

Fig. 155. Circular Current Locus

Fig. 156. Rotor Current Circle

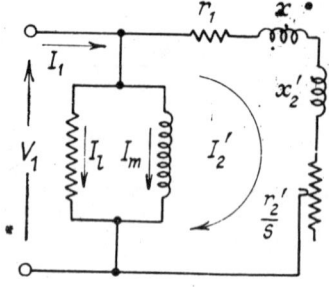

Fig. 157 Simple Equivalent
Circuit

of 79°. If these currents are plotted as complexors, Fig. 155, the locus of their extremities is found to be a circle on the diameter $E_1/x_2' = 100/0.5 = 200$ A.

For synchronous speed, $s = 0$ and the rotor current is zero, while if the machine is driven above synchronous speed the slip is negative and the current has a phase angle exceeding 90° lagging. This implies a reversal of the active current component, resulting in generator action, Fig. 156. The locus of I_2' is thus a complete circle, including regions of motor, generator and brake action.

STATOR CIRCLE DIAGRAM. The effects of the stator can be included in the locus diagram. Working with a constant voltage V_1 at the stator terminals, the stator resistance and leakage reactance with the rotor so that V_1 is impressed across the impedance

$$(r_1 + r_2'/s) + j(x_1 + x_2').$$

As before, this gives a circular locus, the introduction of the stator impedance merely altering the limits of impedance between which the machine operates. The assumed constant flux demands a constant magnetizing current I_m and loss component I_l, and these are represented in the equivalent circuit of Fig. 157 by the shunt resistance and reactance across the stator terminals.

The circle, Fig. 158, is constructed on the diameter $P_m W = AB = V_1/(x_1 + x_2')$, which is the current the rotor would carry at

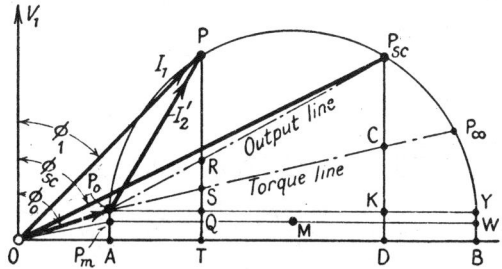

FIG. 158. SIMPLE CIRCLE DIAGRAM

standstill if there were no resistance r_1 or r_2'. $OA = I_{0r}$, the pure magnetizing current, and $AP_m = I_l$ the loss current. $OP_m = I_m$, the no-load current of the stator, and for any rotor current $I_2' = P_m P \simeq P_0 P$, the corresponding stator current is $I_1 = OP$. The vertical axis represents E_1 or V_1. P_{sc} is the extremity of the current vectors at standstill ($s = 1$), and $P_m P_{sc}$ represents the rotor standstill current. OP_{sc} is the total stator "short-circuit" current, the standstill current being so called by analogy with the short-circuited transformer.

NO-LOAD LOSS. When the motor runs on no load, the rotor has a very small current sufficient to develop the friction and windage torque. This current in the rotor, and its stator counterpart, are represented by $P_m P_0$ in Fig. 158. The no load stator current is then $I_0 = OP_0$, of which the active component is AP_0, in phase with V_1, and corresponds to friction, windage and core loss, together with a small stator I^2R loss, usually negligible. The no-load input per phase is

$$P_0 = V_1 I_0 \cos \phi_0 = P_0 A \text{ to scale.}$$

When loaded, the rotor speed falls and with it friction and windage losses. At the same time the rotor frequency increases, introducing more rotor core loss. It can be assumed for simplicity that the sum of the friction, windage and core losses is constant and that

$$P_0 = P_0 A = QT = KD \text{ to scale.}$$

At standstill the friction and windage losses naturally vanish, but the rotor core loss is a maximum.

SHORT-CIRCUIT LOSS. On short circuit, the current in the stator is $I_{sc} = OP_{sc}$. The loss per phase is

$$P_{sc} = V_1 I_{sc} \cos \phi_{sc} = P_{sc}D \text{ to scale,}$$

of which KD represents core loss. The remainder, $P_{sc}K$, must therefore be I^2R loss in the stator and rotor conductors. Divide $P_{sc}K$ at C such that

$$\frac{P_{sc}C}{CK} = \frac{\text{rotor } I^2R}{\text{stator } I^2R}.$$

Join $P_0 P_{sc}$, and $P_0 C$. From any point P, representing a stator current $I_1 = OP$ and rotor current $I_2' = P_m P \simeq P_0 P$, drop a perpendicular.

Then
$$\frac{RQ}{P_{sc}K} = \frac{P_0 Q}{P_0 K} = \frac{P_0 P \cos \angle QP_0 P}{P_0 P_{sc} \cos \angle KP_0 P_{sc}}$$

$$= \frac{P_0 P \cdot (P_0 P / P_0 Y)}{P_0 P_{sc} \cdot (P_0 P_{sc} / P_0 Y)} = \frac{(P_0 P)^2}{(P_0 P_{sc})^2}$$

$$= \frac{\text{total } I^2R \text{ loss for current } P_0 P}{\text{total } I^2R \text{ loss for current } P_0 P_{sc}}.$$

Now since $P_{sc}K$ represents to scale the total I^2R loss for a current $P_0 P_{sc}$, then RQ represents to the same scale the I^2R loss for a current $P_0 P$. Ignoring the difference in magnitude between the stator and equivalent rotor currents, then RQ can be divided at S in the same ratio as $P_{sc}C/CK$ to give RS = rotor I^2R and SQ = stator I^2R to scale, for the working point P.

INPUT AND OUTPUT. For a working point P, the stator current is OP, the angle ϕ_1 is $\angle V_1 OP$, and the input per phase is

$$P_1 = V_1 I_1 \cos \phi_1 = PT \text{ to scale.}$$

QT represents core, friction and windage losses, SQ the stator I^2R and RS the rotor I^2R. The remainder, PR, represents to scale the mechanical output P_m corrected for mechanical losses. Thus the line $P_0 P_{sc}$, on which R lies, may be called the *output line*, and intercepts between it and the circle represent mechanical power output per phase to scale. The *efficiency* is therefore

$$\eta = PR/PT.$$

The slip, from eq. (104), is the ratio

$$s = \frac{I_2'^2 r^{2'}}{P_2} = \frac{\text{rotor } I^2R \text{ loss}}{\text{rotor input}},$$

roughly given by RS/PS. The maximum output power is that corresponding to P_{max} (see Fig. 158) obtained by drawing from the centre of the circle, M, a line perpendicular to the output line P_0P_{sc}. The intercept $P_{max}N$ is then the maximum output to scale.

TORQUE. From eq. (107), the torque is proportional to the rotor input, which is the mechanical output P_m plus friction and windage losses plus rotor I^2R, or approximately P_m plus rotor I^2R, which is PS in Fig. 158. The line P_0C, on which S lies, is therefore called the *torque line*. At starting the torque is $P_{sc}C$. The maximum torque is the maximum intercept between the torque line and the circle, $M_{max}G$, Fig. 168.

If the motor were driven *backwards* at very high speed, the rotor e.m.f. $E_2' = sE_1$ would be large on account of the increase of s. At the same time the rotor reactance sx_2' would increase, and preponderate so greatly over the resistance r_2' that the latter might be considered negligible. The rotor current would then be approximately $I_2' = sE_1/sx_2' = E_1/x_2'$. The input current would be limited by the stator impedance as well as by x_2', so that the stator current (neglecting magnetizing current) would be given by

$$I_1 \simeq I_2' = \frac{V_1}{\sqrt{[r_1{}^2 + (x_1 + x_2{}')^2]}}$$

an expression which can be derived directly from eq. (117) if s is made infinitely large. The stator current corresponding to the conditions described is OP_∞, the current for infinite slip. Since the rotor current is entirely reactive there will be no torque: P_∞ is therefore the intersection of the torque line and the circle, as shown in Fig. 158. That the torque curve tends to zero at high values of slip is suggested by Fig. 150.

The diagram in Fig. 158 is not of the usual shape, on account of the losses and magnetizing current having been exaggerated for clarity. A typical circle diagram for a practical machine is given in Fig. 257. The full-load current is quite small compared with the short-circuit current, the centre of the circle is practically on the horizontal axis, and the working part of the circle (no-load to full-load) is concentrated in the left-hand corner.

MACHINE WITH PRIMARY IMPEDANCE AND VARYING FLUX

5. **Flux Variation.** The characteristics of the induction machine are altered, but not greatly, when a constant applied voltage is assumed rather than a constant flux: at least, not within the usual limits of load. The simple method of Fig. 157, yielding the circle diagram of Fig. 158, is often quite good enough for operation at speeds within, say ± 10 per cent of the synchronous. For wider limits, however, a more "exact" representation is necessary.

The variation of the magnetic flux in the machine is shown in Fig.

159. At (a), the conditions of no-load with negligible rotor current, the majority of the flux produced is mutual. Some flux, however, denoted by Φ_1, does not link the rotor. This is a small proportion, perhaps 2–3 per cent of the total. The total stator flux is $\Phi_m + \Phi_1$, added arithmetically, because, there being no appreciable rotor current, there is only the stator m.m.f. acting on the magnetic

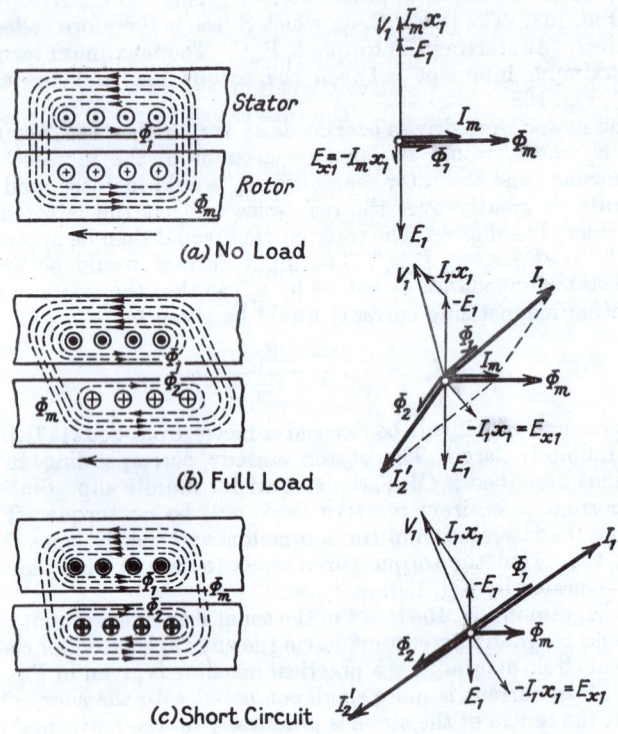

Fig. 159. Magnetic Fluxes in the Induction Motor

circuit. The stator current being the magnetizing current I_m, the fluxes are responsible for the induction of E_1 and $E_{x1} = -I_m x_1$, to the sum of which the applied voltage V_1 is equal and in phase opposition: in other words, the two components of V_1 are $-E_1$ and $I_m x_1$. This neglects stator resistance drop, which is permissible.

Now under any and every condition of load, the stator flux components Φ_m and Φ_1 must sum vectorially to a total which will induce in each stator phase an e.m.f. equal to the applied voltage (neglecting resistance). Thus at full load, Fig. 159 (b), the mutual flux Φ_m has induced an e.m.f. in the rotor, which causes a current to circulate. The ampere-turns of the rotor tend to demagnetize the machine, so that the stator takes an equal and opposite ampere-turn component, which in turn produces a greater stator leakage flux Φ_1. Thus Φ_m must be reduced. The reduction is not equal to the increase of Φ_1, because Φ_1 is produced by the stator current alone, while Φ_m results from the combination of both stator and rotor ampere-turns, and is due to their resultant. Φ_m does, nevertheless, fall appreciably.

At standstill (i.e. on *short circuit*), the rotor e.m.f. and current are large, and so must be the stator current. Much of the stator flux is now leakage, a comparatively small mutual flux (e.g. 50 per cent of no-load value) being left. The case closely resembles that of the transformer on short circuit.

In Fig. 159 (c), we may for simple argument assume that there must be five lines of induction to produce the necessary total e.m.f. equal to the applied voltage V_1. Of these five, only two link the rotor, the remainder being leakage. The same effect at the terminals would be obtained if the stator flux were distributed as at (a), and the additional linkages that the rotor now has are reduced by adding two to its leakage lines. The main flux then becomes the same as at (a), the stator has the same small leakage as on no load, and the rotor has taken over the stator load leakage. Hence by assuming that the stator leakage reactance is transferred almost wholly to the rotor, the main flux may be considered constant. This argument supports the statement already made, and enables the simple formulae of § 4 to be applied with reasonable accuracy after modification.

Fig. 159 does not purport to represent the actual flux distribution, which is much more complex.

6. **Equivalent Circuit Development.** The equivalent circuit of the induction motor is arrived at in a manner analogous to that used for the transformer. In Fig. 160 (a) the stator resistance and reactance are separated from the magnetic circuit common to stator and rotor in which respectively the e.m.f.'s E_1 and sE_2 are induced by the mutual flux. The rotor circuit is closed on to the rotor resistance r_2 (a constant) and reactance sx_2 (proportional to s). Only one phase of stator and rotor is shown, but, suitably interpreted, the performance of the whole motor is directly determinable by multiplying by the number of phases. For the purposes of the equivalent circuit, a cage rotor must be replaced by a slip-ring rotor of like characteristics.

If the magnetizing current and loss current are considered as flowing through the shunt circuits x_m and r_m, where $x_m = E_1/I_{0r}$ and $r_m = E_1/I_l$, then the magnetic circuit of the mutual flux can be considered ideal, and may be represented by a loss-less transformer, Fig. 160 (b).

The final step is to make the substitution shown in Fig. 160 (c), which allows of the direct connection of rotor and stator circuits by changing the rotor impedance from r_2' and sx_2' to r_2'/s and x_2'. From eq. (103)

$$I_2' = \frac{E_2'}{z_{2s}'} = \frac{sE_1}{\sqrt{(r_2'^2 + s^2 x_2'^2)}}$$

The rotor current is that of a circuit containing a constant resistance r_2' and variable reactance sx_2' and across which a variable terminal

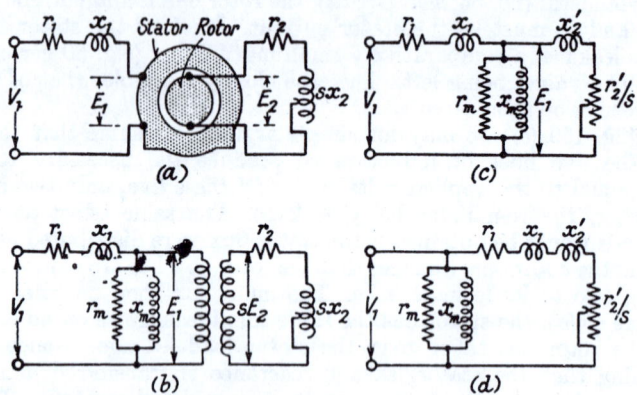

FIG. 160. DEVELOPMENT OF THE EQUIVALENT CIRCUIT

voltage sE_1 at a variable frequency sf is applied: Fig. 160 (b). If the right-hand side of the expression above for I_2' be divided in both numerator and denominator by s, it becomes:

$$I_2' = \frac{E_1}{[\sqrt{(r_2'^2 + s^2 x_2'^2)}]/s} = \frac{E_1}{\sqrt{[(r_2'/s)^2 + (x_2'^2)]}} \tag{115}$$

Eq. (115) represents a circuit containing a constant reactance x_2 and a variable resistance r_2'/s, with an applied voltage E_1 which is not a function of the slip: Fig. 160 (c). The rotor current has the same value in both (b) and (c), but whereas (b) is equivalent to the actual conditions, (c) is a quite distinct circuit having the same current as (b) for the same value of s. Since (c) is easier to deal with than (b) it is generally employed to find the characteristics of the motor.

There is an important difference in the two circuits (b) and (c) of Fig. 160. The rotor power input to (b) is $I_2'^2 r_2'$, while that to (c) is $I_2'^2 r_2'/s$, where I_2' is the same in each case. Thus for (b) the rotor input is that appearing as heat, viz. the I^2R loss: in (c) on the other hand, it is $1/s$ times as much, viz. the whole power P_2 transferred to the rotor including the I^2R loss and the mechanical power P_m. Calculations based on the rotor input in (c) will therefore yield directly the total rotor input and the torque in synchronous watts. If the resistance r_2'/s be divided into the constant r_2' and the variable $r_2'(1 - s)/s$ in series, then the loss in the former represents the true rotor I^2R, while in the latter the power is that actually converted into mechanical power, P_m.

There are numerous defects in the equivalent circuit of Fig. 160 (c). Under short-circuit conditions it unduly reduces the hysteresis and eddy-current losses in cores and teeth, since these losses are actually due to the total and not to the mutual flux, and are therefore not fully represented by the shunt resistance r_m.* However, any attempt to represent actuality in the equivalent circuit is liable to become a laborious academic exercise, and no further complications will be considered. In many cases, indeed, the further simplification of Fig. 160 (d) is used, where the magnetizing and loss currents are made constant by shifting the shunt circuit r_m, x_m to the terminals.

The equivalent rotor resistance and reactance employed in the equivalent circuits are obtained in a way analogous to that used for transformers:

$$r_2' = \frac{K_{w1}^2 T_1^2}{K_{w2}^2 T_2^2} r_2, \text{ and } x_2' = \frac{K_{w1}^2 T_1^2}{K_{w2}^2 T_2^2} x_2 \qquad . \qquad . \qquad (116)$$

if the rotor has a three-phase slip-ring winding. The winding factors $K_w = k_m k_e$ are defined in Chapter X, § 7: see eq. (80). In the case of cage windings the expressions are less obvious. See § 8

DEDUCTIONS FROM THE EQUIVALENT CIRCUIT. The use of the equivalent circuit in Fig. 160 (c) permits the characteristics of the plain induction motor to be more accurately determined, particularly for currents and slips out of the normal load range, where the assumption made in § 4 of constant mutual flux can no longer be maintained.

On no load the rotor current is very small and may be disregarded: the slip s is an almost vanishing quantity, making r_2'/s in Fig. 160 (c) tantamount to an open-circuit. The equivalent circuit reduces therefore to that shown in Fig. 161 (a). The magnetizing impedance preponderates, being so much greater than r_1 and x_1, so that $E_1 \simeq V_1$. The losses are $E_1^2/r_m \simeq V_1^2/r_m$, and consist of core, friction and windage, the two latter actually requiring a small rotor current. On short circuit, Fig. 161 (b), the current in the main circuit is so much greater than the magnetizing current that the latter may for a simple approximation be neglected. The stator current is limited by the combined resistance $R_1 = r_1 + r_2'$ and reactance $X_1 = x_1 + x_2'$. E_1 has fallen to about one-half V_1, the value depending on the ratio of stator and rotor impedance.

The general equivalent circuit, in Fig. 160 (c), will furnish a complete "circuit" explanation of the machine. Using for simplicity the

---

* Hawkins, *J.I.E.E.*, 69, p. 1149 (1931).

complex operational notation, the stator impedance is $z_1 = r_1 + jx_1$, the rotor impedance $z_{2s}' = r_2'/s + jx_2'$.

The admittance of the magnetizing circuit is $y_m = 1/r_m - j(1/x_m)$ $= g_m - jb_m$. The rotor current is

$$I_2' = E_1/z_{2s}',$$

and the magnetizing current

$$I_m = E_1 y_m.$$

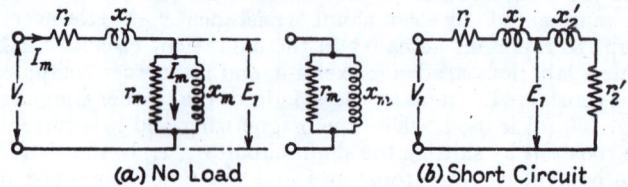

(a) No Load　　　　(b) Short Circuit

FIG. 161. EQUIVALENT CIRCUIT ON NO LOAD AND SHORT CIRCUIT

The stator current is the complex sum of these, or

$$I_1 = E_1(1/z_{2s}' + y_m).$$

The terminal voltage is

$$V_1 = E_1 + I_1 z_1 = E_1[1 + (z_1/z_{2s}') + z_1 y_m].$$

The complex number $z_1 y_m$ is a small fraction, very nearly a small positive scalar value (since it is the product of two complex numbers each with a large phase angle, one positive and one negative). Putting therefore $(1 + z_1 y_m) = c_1$, a plain number slightly greater than unity, then $V_1 = E_1[c_1 + (z_1/z_{2s}')]$ from which[*]

$$E_1 = V_1 \frac{1}{c_1 + (z_1/z_{2s}')} = V_1 \frac{z_{2s}'}{z_1 + c_1 z_{2s}'}$$

$$= V_1 \frac{(r_2'/s) + jx_2'}{(r_1 + jx_1) + c_1[(r_2'/s) + jx_2']}.$$

At synchronous speed, $s = 0$ and $E_1 = V_1/c_1$ so that c_1 is the ratio V_1/E_1 when the rotor is driven at synchronous speed.

The rotor current is

$$I_2' = \frac{E_1}{z_{2s}'} = \frac{V_1}{z_1 + c_1 z_{2s}'} = \frac{V_1}{(r_1 + jx_1) + c_1[(r_2'/s) + jx_2']}, \qquad (117)$$

of which the scalar value is

$$I_2' = \frac{V_1}{\sqrt{\{[r_1 + c_1(r_2'/s)]^2 + (x_1 + c_1 x_2')^2\}}}.$$

[*] An analysis on these lines can be used to represent the induction machine as a transformer with a variable secondary load, so reducing it to the form of Fig. 44. See Morris, "Some Tests of an Exact Practical Theory of the Induction Motor," *Proc. I.E.E.*, 97 (II), p. 767 (1950)

Curves typical of the variations of E_1, I_1 and I_2' with slip are given in Fig. 162, for positive (motor) and negative (generator) values. At synchronous speed, $s = 0$, E_1 is very nearly equal to V_1: it may be within 2 or 3 per cent. It rapidly falls with increase of slip in either direction. The current I_2' is zero at synchronous speed, increases rapidly with small values of slip and thereafter tends to a constant value. The stator current I_1 is the magnetizing current at synchronous speed, but soon reaches values very close to those

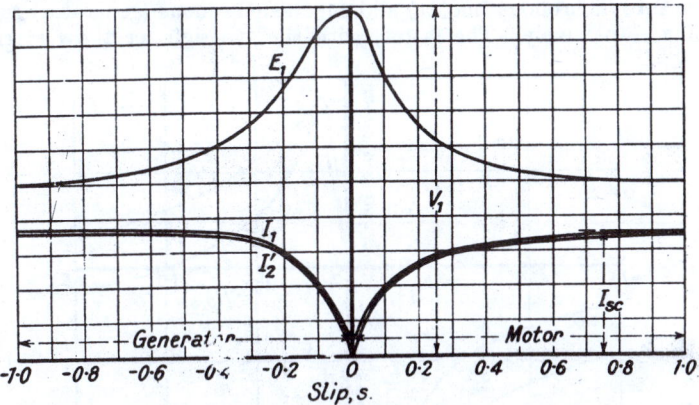

FIG. 162. VARIATION OF E.M.F. AND CURRENTS WITH SLIP

of I_2', since I_m is comparatively small. The mutual flux Φ_m is proportional to E_1.

From eq. (107), the torque in synchronous watts is given directly by the input to the rotor which in the equivalent circuit is

$$M = P_2 = I_2'^2 r_2'/s \text{ per phase.}$$

Putting in the scalar value of I_2' from eq. (117)

$$M = \frac{V_1^2(r_2'/s)}{[r_1 + c_1(r_2'/s)]^2 + (x_1 + c_1 x_2')^2} \qquad . \qquad . \qquad . (118)$$

This is plotted in Fig. 163.

The mechanical output is $(1 - s)M$, while the fraction $sM = sP_2$ is the true I^2R loss in the rotor. The equivalent circuit dissipates the *whole* of P_2 in its resistance r_2'/s, and to get the real I^2R loss in the actual machine the fraction s of P_2 must be taken, the remainder being the mechanical output.

The slip for maximum torque is obtained from eq. (118) above by putting $dM/ds = 0$, yielding

$$s = \pm \frac{c_1 r_2'}{\sqrt{[r_1^2 + (x_1 + c_1 x_2')^2]}} \simeq \pm \frac{r_2}{\sqrt{[r_1^2 + (x_1 + x_2')^2]}} \quad (119)$$

a value which does not greatly differ from $r_2'/(x_1 + x_2')$. Compare the simple expression r_2'/x_2' in § 4. The curves of Fig. 151 hold roughly if the value of α is interpreted as the right-hand side of the expression for s above. The maximum torque is obtained by inserting the critical value of s, eq. (119), into the expression for M in eq. (118), giving

$$M_m = \frac{V_1^2}{2c_1\{\sqrt{[r_1^2 + (x_1 + c_1x_2')^2]} \pm r_1\}} \qquad . \qquad . \quad (120)$$

or subsynchronous (motor) and supersynchronous (generator) running respectively. The latter, with the -ve sign for the r_1 term,

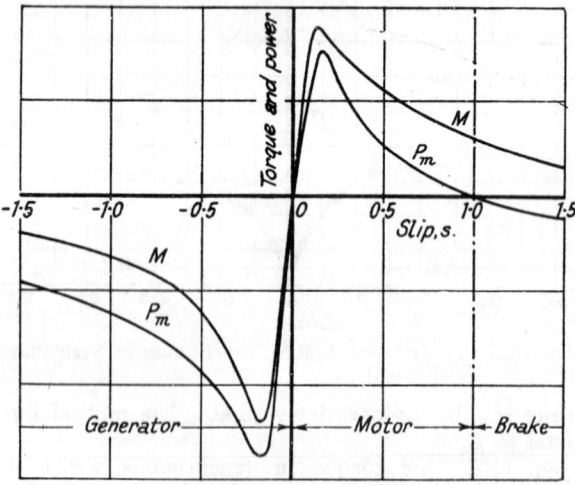

Fig. 163. Variation of Mechanical Power and
Torque with Slip

represents the maximum torque required to drive the machine as a generator. The maximum torque is thus independent of the rotor resistance. For motor, *subsynchronous*, operation it is larger as r_1, x_1 and x_2' are reduced. The rotor resistance does affect the speed at which maximum torque occurs. The lower the rotor resistance, the nearer to synchronous speed does the torque attain a maximum. To get maximum torque at starting, putting $s = 1$ in eq. (110), gives

$$r_2' = \frac{\sqrt{[r_1^2 + (x_1 + c_1x_2')^2]}}{c_1} \simeq \sqrt{[r_1^2 + X_1^2]}, \qquad . \qquad . \quad (120a)$$

where $X_1 = x_1 + x_2'$. For *supersynchronous* or generator operation the maximum torque is seen from eq. (120) to be independent of r_2 as for normal motor conditions, and increases with reduction of both stator and rotor reactance. But an increase of stator resistance now increases the maximum torque. Comparing the two values of maximum torque M_m from eq. (120),

$$\frac{M_m \text{ (supersyn.)}}{M_m \text{ (subsyn.)}} = \frac{\sqrt{r_1{}^2 + X_1{}^2} + r_1}{\sqrt{r_1 + X_1{}^2} - r_1} \simeq \frac{X_1 + r_1}{X_1 - r_1},$$

and the approximation holds so long as r_1 is sufficiently less than X_1 to ignore its presence under the root. Thus if $X_1 = 7r_1$, the ratio of maximum torques becomes $8/6 = 1 \cdot 33$. This is roughly the case in Fig. 163. If the primary resistance is large, the maximum torque running supersynchronously may be very high indeed. (See Chapter XIII, § 11.)

LOSSES AND EFFICIENCY. Neglecting mechanical losses, the efficiency is

$$\eta = \frac{P_m}{P_m + I_1{}^2 r_1 + I_2{}'^2 r_2{}' + I_m{}^2 r_m}$$

where the gross mechanical power is

$$P_m = I_2{}'^2 r_2{}'(1 - s)/s$$

and the core loss is $I_m{}^2 r_m$. Experience shows that the efficiency is a maximum for small values of slip, for which I_m is roughly constant and I_m and $I_2{}'$ are nearly in phase quadrature: consequently

$$I_1{}^2 r_1 \simeq I_m{}^2 r_1 + I_2{}'^2 r_1.$$

The I^2R losses then total to

$$I_1{}^2 r_1 + I_m{}^2 r_m + I_2{}'^2 r_2{}' = I_m{}^2 r_1 + I_2{}'^2(r_1 + r_2{}').$$

Thus for the efficiency

$$\eta = \frac{P_m}{P_m + [I_m{}^2 r_m + I_m{}^2 r_1] + I_2{}'^2(r_1 + r_2{}')}.$$

The part of the denominator in square brackets is nearly a constant. P_m is roughly proportional to $I_2{}'$, so that η is a maximum with respect to I'_2 when the variable losses equal the constant losses, i.e.

$$I_m{}^2 r_m + I_m{}^2 r_1 = I_2{}'^2(r_1 + r_2{}').$$

The left-hand side comprises the no-load loss (omitting friction), so that the efficiency is a maximum when the additional I^2R loss due to the load is equal to the no-load loss P_0: its value is

$$\eta_m = P_m/(P_m + 2P_0).$$

CALCULATION OF PERFORMANCE. The expressions leading to eq. (117) above, derived from the equivalent circuit, may be used to calculate the performance of an induction machine where the several impedance components are known. To obtain full-load figures it is usually necessary to calculate for two or three values of slip in the neighbourhood of the estimated full-load slip, and to extrapolate. An example of the calculations is given below, the full-load slip having been previously assessed.

EXAMPLE. 50-h.p., 420-V., 3-phase, 10-pole, mesh-connected crane motor, totally-enclosed, 1-hour rating. Equivalent circuit, Fig. 160 (c), with: $r_1 = 0.19\Omega$; $x_1 = 1.12\Omega$; $r_2' = 0.29\Omega$; $x_2' = 1.12\Omega$; $r_m = 143\Omega$; $x_m = 16.8\Omega$. The stator and rotor reactances are taken as each equal to one-half of the total reactance $X_1 = 2.24\Omega$.

Magnetizing admittance, $y_m = 0.007 - j0.0595 = 0.06/\underline{-83.3°}\Omega$:

Stator impedance $z_1 = 0.19 + j1.12 = 1.136/\underline{80.4°}\Omega$.

Constant $c_1 = 1 + z_1 y_m = 1 + 1.136/\underline{80.4°} \cdot 0.06/\underline{-83.3°}$

$$= 1.068/\underline{-0.2°} \simeq 1.07.$$

Phase voltage, $V_1 = 420/\underline{0°}$ V. The calculations below are for phase values throughout.

		Full Load	Standstill	
[1] Slip, s	. .	0.026	1.00	
[2] r_2'/s	. . .	11.16	0.29	Ω.
[3] $Z_{2s}' = (r_2'/s) + jx_2'$	.	11.16 + j1.12	0.29 + j1.12	Ω.
		= 11.2/$\underline{5.7°}$	= 1.16/$\underline{75.5°}$	Ω.
[4] $c_1 Z_{2s}'$	. . .	11.91 + j1.19	0.31 + j1.19	Ω.
[5] $Z_1 = r_1 + jx_1$	.	0.19 + j1.12	0.19 + j1.12	Ω.
[6] $Z_1 + c_1 Z_{2s}'$				
$\quad = [4] + [5]$	.	12.10 + j2.31	0.5 + j2.31	Ω.
		= 12.32/$\underline{10.8°}$	= 2.36/$\underline{77.8°}$	Ω.
[7] $Z_{2s}'/(Z_1 + c_1 Z_{2s}')$				
$\quad = [3] \div [6]$	.	0.908/$\underline{-5.1°}$	0.49/$\underline{-2.3°}$	
[8] $E_1 = V_1 \times [7]$	.	381/$\underline{-5.1°}$	206/$\underline{-2.3°}$	V.
[9] $I_2' = E_1/Z_{2s}'$				
$\quad = [8] \div [3]$	.	34.0/$\underline{-10.8°}$	177.5/$\underline{-77.8°}$	A.
[10] Rotor input, $I_2'^2 r_2'/s$	.	12 900	9 150	W.
[11] Rotor losses:				
$\qquad I_2'^2 r_2'$	. .	335	9 150	W.
$\qquad$ Friction, etc.	. .	115	—	W.
$\qquad$ Total	. .	450	—	W.
[12] Rotor output, P_m	.	12 450	0	W.
		= 50/3 h.p.	0	h.p.

		Full Load	Standstill	
[13] Torque, M .	.	. 12 785	9 150	syn. W
		= 100	= 72	%
[14] $y_m Z_{2s}'$	.	. 0·145 − j0·656	0·069 − j0·0095	
[15] $1 + y_m Z_{2s}'$	.	. 1·145 − j0·656	1·069 − j0·0095	
		= 1·32/− 29·8°	= 1·069/− 0·5°	
[16] $I_1 = I_2'(1 + y_m Z_{2s}')$				
	= [9] × [15] .	. 45·0/− 40·6°	190/− 78·3°	A.
[17] $\cos \phi_1$	.	. 0·76	0·20	
[18] Stator input,				
	$P_1 = V_1 I_1 \cos \phi_1$	. 14 350	16 200	W.
[19] Efficiency,				
	$\eta = [12] \div [18]$	. 87	0	%

The performance of the machine as an induction generator can be found in the same way. Let the machine, connected to busbars of rated voltage and frequency, be driven at 616 r.p.m., i.e. at a slip of − 0·026: then the calculation runs as below:

[1]	− 0·026		[10]	Rotor output 13 900 W.
[2]	− 11·16	Ω.	[11]	360 + 115 = 475　W.
[3]	− 11·16 + j1·12		[12]	Rotor input 14 375　W.
	= 11·2/174·3°	Ω.	[13]	14 375　syn. W.
4]	− 11·91 + j1·19	Ω.	[14]	− 0·01 + j0·67
[5]	0·19 + j1·12	Ω.	[15]	0·99 + j0·67
[6]	− 11·72 + j2·31			= 1·20/34·0`
	= 11·95/168·8°	Ω.	[16]	42·4/− 134·8°　A.
[7]	0·94/5·5°		[17]	0·608
[8]	395/5·5°	V.	[18]	Stator output 10 800 W.
[9]	35·3/− 168·8°	A.	[19]	75　%.

7. Current Diagram. The locus of the stator current can be derived from the equivalent circuit by the method of inversion.

INVERSION. If the equivalent circuit of an electrical machine be expressed in the form of a series-parallel network of impedances, then the total circuit current at constant voltage can be obtained by finding the locus of the combined admittance between terminals. For this purpose parallel branches of the network are added as admittances, and series branches as impedances. To find the admittance of a branch from its impedance involves the inversion of the latter since

$$Y = Y \ \underline{/- \theta} = 1/Z = 1/Z \ \underline{/\theta}.$$

The process is simplified when simple geometrical loci have to be inverted, and for present purposes it is sufficient to note :*

(a) The inverse of a straight line AB about a pole O (Fig. 164 (a)) is a circle passing through O with a diameter on the perpendicular OC from O to AB.

(b) The inverse of a circle about a pole on its circumference

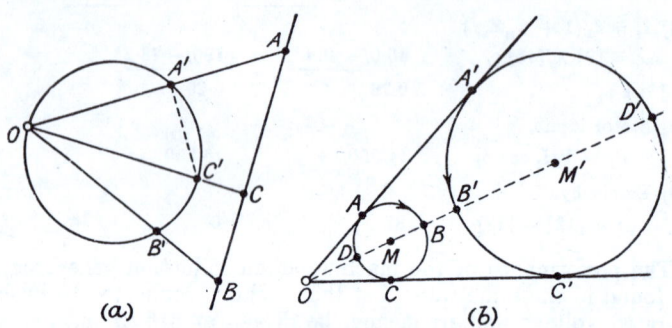

FIG. 164. INVERSION OF STRAIGHT LINE AND CIRCLE

is a straight line perpendicular to the diameter of the circle through O. This is the converse of (a).

(c) The inverse of a circle about any point not on its circumference is another circle with its centre on the line joining the centre of inversion O to the centre M of the circle and included between the same tangents, Fig. 164 (b). The rotation of the points on the two circles is in opposite senses.

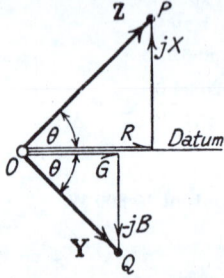

FIG. 165. RELATION OF IMPEDANCE AND ADMITTANCE

These rules are geometrical. For electrical purposes, admittance is obtained from impedance by reflection across the datum, i.e. the axis of resistance or conductance, Fig. 165. In practice only the final reflection need be performed, as the shape is unaltered by the reflection.

The use of inversion to determine the current locus may be shown by a numerical example.

CURRENT LOCUS OF AN INDUCTION MOTOR: EXAMPLE. Find the locus of the stator current of the cage induction motor, one phase of which is represented by the equivalent circuit of Fig. 166 (a). In order to obtain the current diagram in its normal position (as in Fig. 163), the resistance axis is taken vertical, and the reactance to the right. Starting with the rotor circuit (suffix "2") as most

* For proofs, and examples of inversion, see [B], Chapter VIII; also Miles Walker, *Control of the Speed and Power Factor of Induction Motors* (Benn).

remote from the terminals, its variable impedance $z_{2s}' = r_2'/s + jx_2'$ is plotted by drawing a vertical line at distance $x_2' = 0.38\,\Omega$ from the origin $= O_2$, and scaling it in terms of slip. At slip unity, $r_2'/s = r_2' = 0.5\,\Omega$; at slip $s = 0.1$, $r_2'/s = 5\,\Omega$, and so on. The vertical line between $s = 1$ and $s = 0$ (at infinity) therefore represents the impedance locus z_2 of the rotor. In parallel with it is the

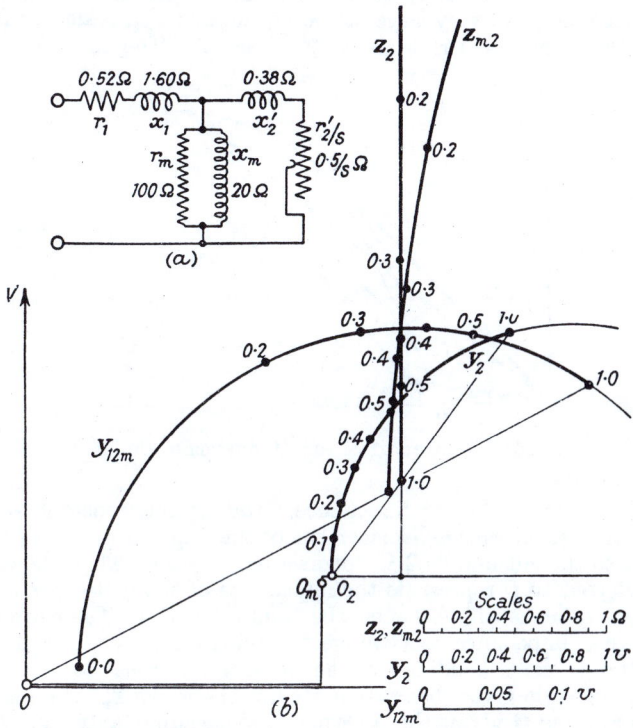

FIG. 166. ADMITTANCE LOCUS OF EQUIVALENT CIRCUIT

magnetizing circuit (suffix "m"). The impedance of the magnetizing and rotor circuits together is found by adding their respective admittances. First find the admittance of the rotor by inverting the rotor impedance locus z_2 about O_2 to obtain y_2, which will be a circle passing through O_2 and with a diameter on the x axis, of length $1/x_2' = 1/0.38 = 2.63\,\mho$ to scale. The scale is chosen arbitrarily to fit a convenient space. By projecting lines from O_2 to points on the z_2 line, the intersections with the y_2 circle indicate corresponding points, so that the admittance circle can be scaled in terms of slip. Adding the magnetizing to the rotor admittance requires merely shifting the origin to O_m by a distance to the left

$1/x_m = 1/20 = 0.05\ \mho$, and downwards by a distance $1/r_m = 1/100 = 0.01\ \mho$, to scale. The circle measured from O_m therefore represents the combined admittances of the rotor and magnetizing circuits, y_2.

It now remains to add the stator impedance. This first requires the inversion of y_2 to find the corresponding impedance: that is, the circle has to be inverted about the point O_m outside it. This leads to another circle of very large diameter, marked z_{m2}, scaled off from y_2 in terms of s in the same way as before. The same scale is

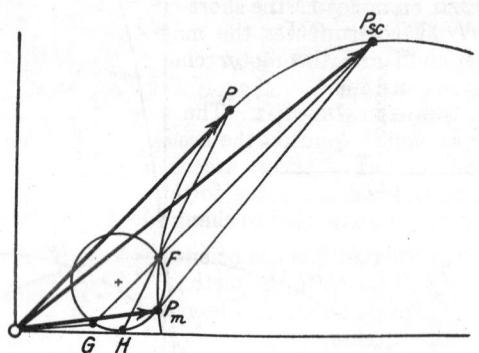

FIG. 167. CONSTRUCTION FOR MAGNETIZING CURRENT

chosen as for the rotor impedance. Adding the constant stator impedance requires the displacement of the origin from O_m to O by $1.60\ \Omega$ to the left and $0.52\ \Omega$ downwards, to scale. Then the circle z_{m2} referred to O represents the impedance of the motor per phase for any value of slip between the limits 0 and 1. The current is (voltage $\div$ impedance) = (voltage $\times$ admittance), so that if the admittance corresponding to the total impedance be found, a change of scale gives the current locus. The curve z_{m2} is therefore inverted about O, giving the circle of combined admittance y_{12m}. Multiplying the admittance scale by the voltage gives the scale of current.

While vectors drawn from O to the locus y_{12m} give the total stator current I_1, the two components I_m and $-I_2'$ are not directly obtainable unless an additional locus for the solution of

$$I_m = E_1 y_m = V_1 y_m \frac{z_{2s}'}{x_1 + c x_{2s}'}$$

be inscribed on the diagram. The result of such an operation leads to an additional circle for the magnetizing current, Fig. 167. This part of the diagram is very small, and it is sufficient to use the following construction*: Draw the short-circuit current OP_{sc} and

* See Carr, *J.I.E.E.*, 66, p. 1174 (1928); and *Electrician*, 103, p. 24 (1929)

the no-load magnetizing current OP_m. Join P_mP_{sc}, and bisect the angle $OP_{sc}P_m$. Project the bisector from P_{sc} to G, where $P_{sc}G = \sqrt{[(OP_{sc}) . (P_mP_{sc})]}$. $P_{sc}G$ cuts the stator current locus circle in F. Then the magnetizing current has a locus given by the circle passing through P_m, G and F. PH is the rotor current for any stator current OP, and is defined by the facts that it passes through F and bears the same ratio to P_mP as GP_{sc} does to P_mP_{sc}. OG, the magnetizing current associated with the mutual flux under short-circuit conditions, is usually about one-half of OP_m, the corresponding current on no load: the short-circuit rotor current is GP_{sc}.

The construction presupposes the magnetizing current and flux to be proportional, and the motor characteristic resistances and reactances to be constant.

PRACTICAL CURRENT DIAGRAM. The current circles obtained in § 4 and by inversion depend on the assumption of constant circuit characteristics. Actually, these are not constant. Apart from variations in ohmic resistance due to temperature-rise, the several characteristics may be expected to change with load, as follows—

r_1: the eddy-current loss coefficient will vary with load current on account of saturation in the teeth;

x_1: will change on account of increasing tooth reluctance with the larger load currents;

r_m: the core loss will change in a complex manner with load;

x_m: will alter with the main flux due to saturation;

r_2': the eddy-current loss coefficient will change with tooth-saturation and with the varying rotor frequency;

x_2': as for x_1.

In general, r_1, x_1, r_2' and x_2' will be less for large than for small currents, with the reactances decreasing probably faster than the resistances. It may therefore be expected that the short-circuit current will be greater and of (probably) higher power factor than as estimated from values of r_1, x_1, r_2' and x_2' that are quite suitable for currents within the load range. It is evident that no single circle will be accurate over the whole range, and the diagram should be drawn with reference to the use to which it is put. If the full-load conditions are required, and maximum torque, the diagram should be drawn for characteristics applicable to the full-load current. It is not generally possible to do more than estimate the changes at very large short-circuit currents.

From the remarks above, it will appear that the current locus of a modern induction motor (in which the saturation is high) tends towards an ellipse with its major axis roughly horizontal.

Although it is quite possible to obtain graphically a considerable range of information from the circle diagram, such information is not readily interpreted, and indeed may be erroneous and misleading. A practical diagram therefore omits such things as slip lines, efficiency scales, loss lines, etc., and concentrates on a diagram which, drawn from test data or design figures, will give the full-load stator and rotor currents, the full-load power factor, and the maximum (pull-out) torque. Such a diagram* is shown in Fig. 168, where—

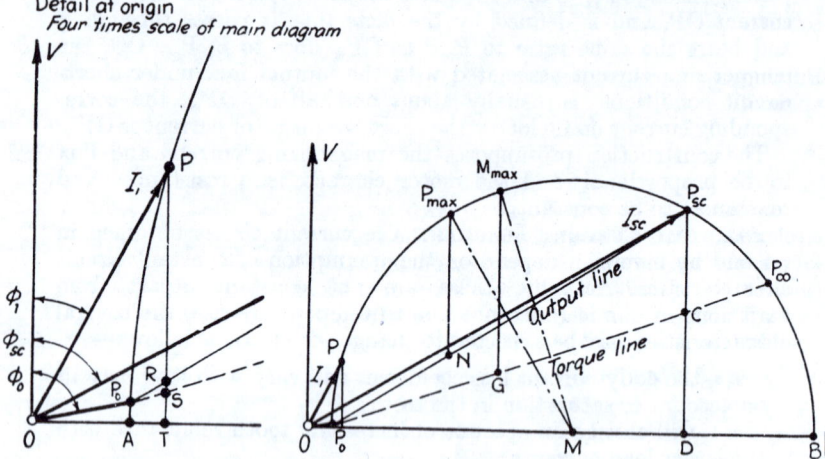

FIG. 168. PRACTICAL CURRENT DIAGRAM

$\overline{OP_0} = I_0 =$ no-load current at normal voltage V_1, at phase-angle ϕ_0;

$\overline{OA} = I_{0r} =$ magnetizing component of no-load current

$\overline{AP_0} = I_{0a} =$ active component of no-load current, for core and no-load mechanical losses;

$\overline{OB} = I_{sci} =$ ideal short-circuit current, neglecting resistance

$\simeq V/(x_1 + x_2') = V/X_1$:

$\overline{AB} =$ diameter of circle, of centre M;

$P_{sc} =$ nominal short-circuit point for normal voltage V_1;

$\overline{OP}_{sc} =$ nominal short-circuit current for normal voltage V_1;

$\phi_{sc} =$ nominal angle of lag of short-circuit current;

$\simeq$ arc tan $[(x_1 + x_2')/(r_1 + r_2')]$;

$\overline{P_{sc}C/CD} = r_2'/r_1 =$ rotor equivalent resistance/stator resistance;

$\overline{P_0P_{sc}} =$ output line;

$P_0C =$ torque line;

$P_{max} =$ maximum output point;

$M_{max} =$ maximum torque point.

* The centre M of the circle is taken as lying on the baseline. For small machines M can be located as in Fig. 158. The ratio P_{sc}C/CD is strictly the ratio of I^2R losses.

From the geometry of the diagram the maximum power output is the intercept $P_{max}N$, which in terms of the radius r of the circle is $r(1 - \cos \phi_{2sc})/\sin \phi_{2sc}$.

But $\qquad r = I_{2sc}'/2 \sin \phi_{2sc}$,

so that $\qquad P_{max}N = \dfrac{I_{2sc}'(1 - \cos \phi_{2sc})}{2 \sin^2 \phi_{2sc}} = \dfrac{I_{2sc}'}{2(1 + \cos \phi_{2sc})}$

Putting $I_{2sc}' \simeq I_{sc} - I_0$, and $\phi_{2sc} \simeq \phi_{sc}$, the stator short-circuit angle of lag, then the maximum output is

$$P_{mm} \simeq V_1 \frac{(I_{sc} - I_0)}{2(1 + \cos \phi_{sc})} \text{ watts per phase }. \qquad . \quad (121)$$

CONSTRUCTION FROM NO-LOAD AND SHORT-CIRCUIT DATA The circle diagram, yielding estimates of the full-load current and power factor, the maximum horse-power output, the pull-out torque, and (indirectly) the full-load slip and efficiency, can be constructed from no-load and short-circuit data derived from design figures or test results.

Design Data. A design will furnish the following quantities—

I_{0r} = magnetizing current per phase, reactive component;

I_{0a} = active component per phase of no-load current corresponding to core loss, windage and friction; the small stator I^2R is usually neglected;

$X_1 = x_1 + x_2'$ = total standstill reactance per phase in primary terms;

$R_1 = r_1 + r_2'$ = total resistance per phase in primary terms;

V_1 = voltage per phase.

Referring to Fig. 168, draw to scale $OA = I_{0r}$ and $AP_0 = I_{0a}$. Calculate $OB = I_{sci} = V_1/X_1$. Bisect AB to find the centre of the circle M which can now be drawn on the diameter AB The point P_{sc} is found by drawing the line OP_{sc} at angle $\phi_{sc} = $ arc tan X_1/R_1 to the vertical axis. Join P_0P_{sc} for the output line. Divide $P_{sc}D$ at C such that $P_{sc}C/CD = r_2'/r_1$. Join P_0C to give the torque line.

If a amperes per unit length (e.g. inch or centimetre) is the scale used for the above, then every unit length represents V_1a volt-amperes per phase, and $3V_1a$ volt-amperes for the machine. Measured vertically upwards, the volt-amperes become watts, so that the power scale for the whole machine is $3V_1a/746$ h.p., or $3V_1a/1\ 000$ kW., per unit length. To find the maximum power P_{mm}, draw the line MP_{max} perpendicular to P_0P_{sc}, and drop the perpendicular $P_{max}N$ to the output line: to scale, this represents the maximum output. The pull-out torque is similarly found by drawing MM_{max} perpendicular to P_0C, and dropping the perpendicular $M_{max}G$ on the torque line. The torque is obtainable in synchronous watts. The full-load current is OP such that the intercept PR on the perpendicular PT to the base line is a length to scale representing the rated output. Having found P, the full-load power factor is obtained by measuring ϕ_1, and the full-load current from OP.

The efficiency at full-load is calculated by measuring the stator and rotor currents to scale, finding the I^2R losses, adding the core, friction and windage losses, to find the input and output. The slip is obtained from eq. (104).

Test Data. From a no-load test at normal voltage, and a short-circuit test at a voltage giving 100–200 per cent of full-load stator

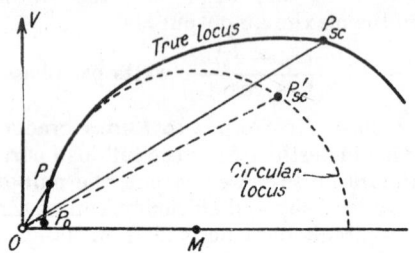

FIG. 169. CURRENT LOCUS FROM COMPLETE TEST

current on short circuit, the complex currents I_0 at ϕ_0 and I_{sc} at ϕ_{sc} are directly obtained, the latter after multiplying the actual measured current by the ratio of normal to test voltage. The points P_0 and P_{sc} are determined by the currents and the circle drawn by joining P_0P_{sc}, bisecting it, drawing at the point of bisection a perpendicular, to cut the base line in M; then constructing a circle with M as centre, having a circumference passing through P_0 and P_{sc}. The remainder of the construction is exactly as above. An example is given in Chapter XIV, § 7.

That the true current locus generally tends to the elliptical is demonstrated by the locus in Fig. 169, drawn from test figures. The motor was a 400-V., three-phase, 50-c/s., six-pole, $1\frac{1}{2}$-h.p. machine with a current small enough to be measured readily at short circuit on normal voltage without detriment. The short-circuit test figures were

Line Voltage V.	Line Current A.	Power W.	Impedance (Calculated)			
			Z	R	X	cos ϕ
61	2·41*	115	44	19·8	39·5	0·45
77	3·01	179	44	19·8	39·5	0·45
92	3·62	259	44	19·8	39·4	0·46
231	10·25	2 152	39	20·5	33·2	0·53
400	19·2	8 128	36	21·9	28·7	0·61

* Full-load current.

In Fig. 169 it is clear that for the purposes of finding full-load values the dotted circle, drawn from the no-load test and the short-circuit test at normal current, is sufficient. For starting torque and current, on the other hand, the circle may give erroneous results.

8. Equivalent Cage Rotor.

The current locus diagram, and the equivalent circuit on which it is based, presuppose a rotor winding having the same number N_2 of phases as the stator N_1, and applies to one of those phases. In a cage rotor the number of phases is not immediately evident, but it will be clear that there must be an N_1-phase slip-ring rotor having the same electrical characteristics, which can be substituted without altering the operation of the motor.

In general, where there is an N_1-phase stator winding with a winding factor K_{w1} and T_1 turns in series per phase, and a corresponding N_2-phase rotor, the rotor phase current will be

$$I_2 = \frac{N_1 K_{w1} T_1}{N_2 K_{w2} T_2} I_2'$$

where I_2' is the primary current in N_1 phases electrically equivalent to the rotor current I_2 in N_2 phases.

A cage can be considered as a polyphase winding in which there are S_2 conductors or bars, connected in pairs a pole-pitch apart to form $S_2/2$ phases of one turn each. More conveniently, since it is not usual to find two bars separated a pole-pitch exactly, and also since it is preferable to have an expression for the current per bar, we may consider each bar to comprise one phase of one-half a turn. Thus the "phase" current I_2 above becomes the current per bar

$$I_b = \frac{N_1 K_{w1} T_1}{(S_2/p) \cdot 1 \cdot \frac{1}{2}p} I_2' = \frac{2N_1 K_{w1} T_1}{S_2} I_2' \quad . \quad . \quad . \quad . \quad (122)$$

since in every double-pole-pitch the bars approximately repeat, so that there are S_2/p bars or phases per double-pole-pitch, and p bars in parallel in each "phase." Using the slots per pole g and conductors per slot z_s, then with $N_1 = 3$, $K_{w1} = 0.955$, $T_1 = \frac{1}{3}g_1 p z_{s1}$, and $S_2 = 2pg_2$,

$$I_b = 0.955(g_1 z_{s1}/g_2)I_2' \quad . \quad . \quad . \quad (123)$$

From Fig. 148 it will be seen that the greatest current in the end-ring is the sum of half the currents in the g_2 bars of a pole-pitch. Assuming a large number of bars and a sinusoidal current distribution round the rotor periphery (Fig. 148), the end-ring current has a peak value equal to $2/\pi$ times the current $\frac{1}{2}g_2 i_{bm}$, i.e. $g_2 i_{bm}/\pi$, so that its r.m.s. value is

$$I_c = I_b g_2/\pi \qquad . \qquad . \qquad . \qquad . \qquad . \qquad . \qquad . \qquad (124)$$

which, with the same conditions as for eq. (123), becomes

$$I_c = 0{\cdot}3 g_1 z_{s1} I_2' \qquad . \qquad . \qquad . \qquad . \qquad . \qquad . \qquad (125)$$

Writing r_b as the resistance of one bar (including two joints with the rings), and r_c as the resistance of one end-ring (considered as a conductor of length π times its mean diameter), the total I^2R loss is

$$S_2 I_b{}^2 r_b + 2 I_c{}^2 r_c = (S_2 r_b + 2 g_2{}^2 r_c/\pi^2) I_b{}^2 \qquad . \qquad . \qquad (126)$$

The equivalent N_1-phase rotor will have a loss $N_1 I_2'{}^2 r_2'$ exactly equal to the above. Hence

$$r_2' = \frac{S_2(r_b + 2 r_c g_2{}^2/\pi^2 S_2) I_b{}^2}{N_1 I_2'{}^2}.$$

Substitute for $I_b{}^2$ from eq. (122), and write $S_2 = 2 p g_2$ so that $2 r_c g_2{}^2/\pi^2 S_2 \simeq 0{\cdot}1 r_c g_2/p$; then

$$r_2' = \frac{4 N_1 K_{w1}{}^2 T_1{}^2}{S_2}\left(r_b + 0{\cdot}1 r_c \frac{g_2}{p}\right) \qquad . \qquad . \qquad (127)$$

For a normal three-phase stator, $N_1 = 3$, and $K_{w1} = 0{\cdot}955$, so that

$$r_2' = 11 \frac{T_1{}^2}{S_2}\left(r_b + 0{\cdot}1 r_c \frac{g_2}{p}\right) \qquad . \qquad . \qquad . \qquad (128)$$

In the same way, the equivalent reactance is

$$x_2' = \frac{4 N_1 K_{w1}{}^2 T_1{}^2}{S_2} x_2 \qquad . \qquad . \qquad . \qquad . \qquad (129)$$

where x_2 is the reactance of a rotor phase. It is usual to calculate only the slot leakage separately; the belt, overhang and zig-zag leakages being obtained in combination with the stator.

CHAPTER XIII

INDUCTION MACHINES : PERFORMANCE AND CONTROL

1. **Performance.** Considering a plain induction motor from the viewpoint of its operating characteristics, its behaviour may be noted from no load to overload. On no load with rotor short-circuited, the machine runs at a speed very close to the synchronous speed at which the rotating field of the stator revolves. The rotor current is very small, because only the loss torque (for windage and friction) has to be developed. The stator takes a corresponding active current which, with the core-loss current, is the active no-load current component. The magnetizing current is much larger than the active current, so that the no-load power factor is low: e.g. 0·1 or (for small motors) 0·25.

If a load torque is applied to the shaft, the rotor tends to slow down. The rotor e.m.f. consequently increases, and with it the rotor current. Since the rotor frequency is very low, the inductive reactance has little effect, and the rotor current is nearly in phase with the rotor e.m.f.: i.e. it is substantially a torque-producing current. (Fig. 110 (c).) The slip increases so as to provide enough rotor current and torque; and, very nearly, the load torque and slip are proportional. Thus both the slip/torque and speed/torque relations are rectilinear. The increase in rotor current necessitates a rise in the corresponding (active) component of stator current, and since the magnetizing current remains practically constant, the power-factor is considerably raised.

On full load the slip may be about 5 per cent in a 2 h.p. motor down to about 1¼ per cent in a 1 000 h.p. motor, and the stator current is about three times the no-load value. The power factor rises to a value between 0·8 and 0·9. The efficiency, which is of course zero at no load, rises to a maximum at or (usually) below full load. This maximum occurs roughly where the I^2R losses in stator and rotor sum to an amount equal to the no-load "constant" losses. At higher loads, the I^2R losses increase more rapidly than the output, so that the efficiency slowly falls.

For moderate overloads the stator current and slip increase roughly in proportion to the load, while the power factor, as the circle diagram (e.g. Fig. 168) indicates, remains sensibly unchanged. For larger overloads both current and slip increase more rapidly, until at maximum load (about 100 per cent overload) the current may be about three times its full-load figure, at a power factor in the neighbourhood of 0·7. At this stage a small increase in overload will cause the motor to stop suddenly. The current will rise to e.g. five times rated full-load value, the torque will be small, and the power

287

factor lower, 0·2– 0·4. The machine will not restart until the load torque has been reduced below that available at starting. The rating of a motor must be reckoned in relation to its pull-out or stalling (maximum) torque, a usual relation of maximum to full-load torque being 2 : 1.

A typical set of characteristic curves over the complete range is given in Fig. 170. Evidently the machine, within its normal range

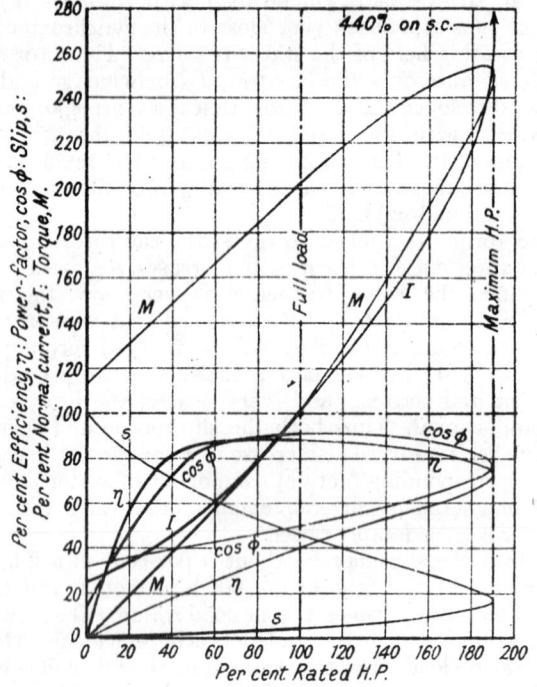

FIG. 170. PERFORMANCE CURVES

of loading, is a constant-speed motor, employing the term "constant" in the same sense as when used with reference to a d.c. shunt motor, the characteristics of which as regards speed and torque are very similar.

The efficiency and power factor obtainable on full load naturally depend on the size of the machine. For large machines both the losses and the magnetizing current are proportionally lower than for small motors, with consequent improvement in efficiency and power factor. An important factor is the speed, which is controlled by the number of poles. On account of the poorer utilization of material and the large number of overhang conductors, a low-speed motor is heavier, less efficient and more reactive than a fast machine. Typical figures are given in the table below.

FULL-LOAD EFFICIENCY AND POWER FACTOR OF 50-c/s. MOTORS

Type of Rotor	Rating h.p.	Efficiency, %			Power Factor, %		
		1 500 r.p.m.	750 r.p.m.	500 r.p.m.	1 500 r.p.m.	750 r.p.m.	500 r.p.m.
Cage	1	72			75		
	2	81	75		82	66	
	5	84	83	81	86	74	70
	10	86	84	82	87	78	71
	20	88	85	83	89	83	75
	50	90	89	87	90	85	82
	100	91	90	89	92	89	84
	1 000	93	93	92	94	93	91
Slip-ring	3	76	74		78	60	
	7	81	80	79	82	67	66
	10	84	83	80	84	74	73
	20	87	85	82	88	80	77
	50	89	88	86	90	83	81
	100	91	89	88	92	87	83
	1 000	93	92	91	94	91	90

TYPES, RATINGS AND APPLICATIONS. A brief summary is given:
Cage Motors. The four main subsections are—

1. Motors with normal starting torque and current, having cages with low resistance and reactance, low full-load slip, good efficiency and power factor, and high pull-out torque.

2. Motors with normal starting torque and low starting current, having larger rotor reactance but the same slip and efficiency as (1). The power factor and pull-out torque are less. These motors may be started by direct switching up to larger sizes than those in (1).

3. Motors with high starting torque and low starting current, using deep-bar or double-cage rotors. A 200 per cent starting torque with moderate current is obtained on full voltage. The efficiency, power factor and pull-out torque are lower than for (1). The motors are useful for starting against load (e.g. reciprocating compressors).

4. Motors with high full-load slip, using a comparatively high-resistance cage, with large starting torque, low starting current and low efficiency. Used for drives with heavy starting but light running duty. If employed for loads with rapidly fluctuating torques (e.g. punch presses) they may be attached to flywheels for load-peak equalization.

Slip-ring Motors. Suitable for heavy, frequent starting and accelerating duty-cycles. A high starting torque is obtained with a low starting current at high power factor. The rotor losses are mainly in the external resistance, easing the problem of rotor cooling.

Multispeed Motors. These are usually *cage* motors with single or double pole-changing windings. They may be classified as—

(a) variable-torque, power output proportional to square of speed (fans, centrifugal pumps);

(b) constant-torque, power output proportional to speed (conveyors, stokers, reciprocating compressors, printing presses);

(c) inverse-torque, power output rating constant (machine tools, lathes, boring mills, drills, planers).

Less frequently multispeed motors are of the *slip-ring* type, as for hoists, conveyors and elevators.

Gear-motors are high-speed motors with integral unit gear construction giving output shaft speeds of 10–500 r.p.m. The motor is generally a 4-pole 1 500 r.p.m. machine. If the service is heavy, the gear must be very robustly designed. Gear-motors may be applied to conveyors, agitators, low-speed fans, blowers and screens.

Special machines include: submersible motors, usually 2-pole machines with long small-diameter vertical rotors, the liquid passing through the rotor itself; motors with solid steel rotors, employed for intermittent starting of other machinery; 2-phase motors for servomechanisms in which one phase is supplied at constant voltage with the other phase provided with a quadrature voltage of varying magnitude and polarity; and selsyns for servo error-detection or control.

MOTOR SPEEDS. The normal running speed of an induction motor is close to the synchronous value settled by the supply frequency and the number of poles. It will clearly be advantageous on grounds of motor weight and cost to utilize the higher available speeds. The effect of speed on operating characteristics in a given case is shown roughly by the figures tabulated below for a range of standard 5 h.p. 50-cycle cage motors.

SERVICE CHARACTERISTICS OF 5 H.P. 50-C/S. CAGE MOTORS

Number of poles . .	2	4	6	8	10	12	16
Synchronous speed, r.p.m. . . .	3 000	1 500	1 000	750	600	500	375
Relative cost, % . .	98	100	125	145	175	200	310
Efficiency, % . .	84	85	84	83	82	81	79
Power factor. % . .	90	86	81	74	73	72	70
Relative weight, % .	85	100	130	160	200	240	440

2. **Starting.** To accelerate a motor from rest and to bring it to normal operating speed requires an input of energy (a) to supply the losses, and (b) to provide the rotating parts with kinetic energy. The losses are considerable: neglecting all friction and core losses, and assuming the motor to accelerate on no load, the heat developed in

I^2R is equal to the stored kinetic energy—an important matter where frequent starting is necessary. When friction is accounted, and when the machine has to start against a load torque, the heat developed in I^2R is considerably greater.

In the design of a motor, consideration has to be given to the conditions under which it will start, how frequently the start is required, and, especially for starting against overload, how long the start will take. Generally a considerable starting torque will have to be developed, and since the regulations of the supply authorities will set a limit to the starting current peaks allowable, some form of startor will generally be needed.

The torque in synchronous watts per phase is
$$M = I_2'^2 r_2'/s \simeq 0 \cdot 8 I_1^2 r_2'/s = k I_1^2/s$$
if we assume $I_2' \simeq 0 \cdot 9 I_1$. For normal full load current I_n the torque is
$$M_n = k I_n^2/s_n$$
where s_n is the full-load slip. At starting with a stator current I_s,
$$M_s = k I_s^2.$$
The ratio of starting to full-load torque is consequently
$$M_s/M_n = (I_s^2/I_n^2)s_n = a^2 s_n \text{ where } a = I_s/I_n \qquad (130)$$
When the motor is *direct-switched* on to a normal-voltage supply, the starting current I_s is the short-circuit current I_{sc} and
$$M_s/M_n = (I_{sc}^2/I_n^2)s_n$$
Thus if $I_{sc} = 6I_n$, $s_n = 5$ per cent, then $a = 6$ and
$$M_s/M_n = 6^2 \cdot 0 \cdot 05 = 1 \cdot 8;$$
the machine developing $1 \cdot 8$ times full-load torque with 6 times full-load current.

Direct-switching is only permitted by supply authorities for small motors below a rating of, say, 10 h.p. It is, however, not uncommon to find cage motors of 100 h.p. or more direct-switched in mining machinery.

It has already been seen (Fig. 153) that the starting torque of the induction motor can be increased by raising the rotor resistance. This is readily feasible with slip-ring motors, but not with cage motors. The inclusion of a high-resistance rotor certainly increases the starting torque of the latter, but decreases the efficiency. A startor may therefore either increase the rotor resistance or decrease the starting current. In the former case the starting torque may be increased and the current simultaneously limited, in the latter the torque must inevitably be reduced.

The methods available for the starting of induction motors are—

A. For slip-ring machines—
   Rotor rheostat.

B. For cage machines—
   Stator rheostat;
   Auto-transformer;
   Star-delta switch;
   Centrifugal devices.

These are detailed in following sections.

3. Starting of Slip-ring Motors. From eq. (120a), maximum torque can be obtained at starting if the referred rotor resistance (including external resistance R) be made approximately equal to the combined reactance, or

$$(r_2' + R') \simeq (x_1 + x_2'),$$

at $s = 1$. It will be clear from eq. (117) that the inclusion of additional rotor resistance will reduce the stator current at standstill and simultaneously raise its power factor. The same is shown by the circle diagram, Fig. 171. The normal short-circuit point is P_{sc1}. An increase of rotor resistance does not affect the diameter of the circle, which depends only on the reactance: the result is a movement of the short-circuit point to a position such as P_{sc2}. The torque is obviously increased, so long as the added resistance is not too much. If the resistance R is adjusted to bring P_{sc2} to M_{max}, then maximum torque is secured at starting. A further increase in R reduces the torque below maximum value, but at the same time reduces the input current and raises the power-factor, so that the current is more efficiently utilized.

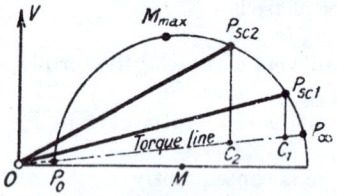

FIG. 171. EFFECT OF ROTOR RESISTANCE ON STARTING TORQUE

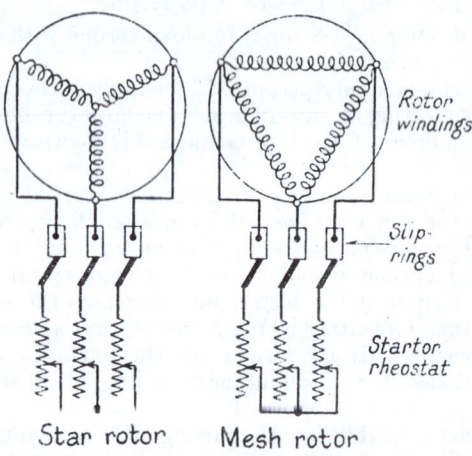

Star rotor   Mesh rotor

FIG. 172. ROTOR STARTER

In this method of starting, the rotor winding is open-circuited, the phase ends being connected to slip-rings, three in number, which are connected to a rheostat, as in Fig. 172. The rheostat may take the form of a stud- or contactor-type arrangement with metallic resistor steps, or carbon or electrolyte for large currents. It may be built to include the main stator switch together with under-voltage and overcurrent releases. A liquid starter permits of gradual, smooth starting.

The starting current is chosen on a basis of the starting torque required and the losses that can be allowed during the starting period. The losses can be reduced by using a larger starting current provided that the motor is operating on a stable torque/slip characteristic.

Where the number of resistance steps is small, the starting rheostats may be mounted on the rotor shaft, and arranged to be

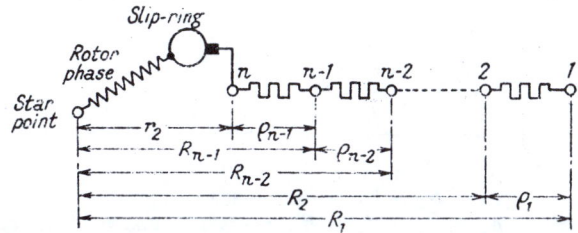

Fig. 173. Pertaining to Calculation of Rotor Startor

cut out by a centrifugal switch. The use of slip-rings and external switchgear is avoided, but the device is complicated if frequent starting is required (and consequently much heat developed), and if more than two resistance steps are needed. A motor fitted with a centrifugal switch will start itself automatically if stalled.

ROTOR RHEOSTAT STARTOR. As is usual in startor calculations, the motor is assumed to start against a constant torque, and the rotor current is presumed to fluctuate between fixed limits I_{2max} and I_{2min}. Let R_1, R_2 . . . be the total resistance per phase in the rotor circuit on the first, second . . . steps, consisting of the rotor resistance r_2 and external section resistances ρ_1, ρ_2, ρ_3 . . . See Fig. 173. At the commencement of each step, the current is I_{2max}, the resistance in circuit is R_1 for slip $s_1 = 1$, R_2 for slip s_2, and so on, so that

$$I_{2max} = \frac{E_2}{\sqrt{[(R_1/s_1)^2 + x_2{}^2]}} = \frac{E_2}{\sqrt{[(R_2/s_2)^2 + x_2{}^2]}} = \cdots$$

where E_2 is the standstill rotor phase e.m.f. on open circuit. Clearly

$$R_1/s_1 = R_2/s_2 = R_3/s_3 = \ldots = R_{n-1}/s_{n-1} = r_2/s_{max} \quad . \ (131)$$

if the slip under normal operating conditions (i.e. rotor short-circuited) is s_{max} when the rotor is carrying I_{2max}. On the first step, R_1 remains in circuit until the motor has started and the slip has fallen from unity to s_2, and at the same time the current from I_{2max} to I_{2min}; vhence

$$I_{2min} = \frac{E_2}{\sqrt{[(R_1/s_2)^2 + x_2{}^2]}} = \frac{E_2}{\sqrt{[(R_2/s_3)^2 + x_2{}^2]}} = \ldots ;$$

so that

$$R_1/s_2 = R_2/s_3 = R_3/s_4 = \ldots = R_{n-1}/s_{max} . \qquad . \quad (132)$$

Combining eq. (131) and (132),

$$s_2/s_1 = s_3/s_2 = s_4/s_3 = \ldots = R_2/R_1 = R_3/R_2 = \ldots = r_2/R_{n-1} = \gamma$$

But $s_1 = 1$, so that the total resistance in circuit on the first step is

$$R_1 = r_2/s_{max} \qquad . \qquad . \qquad . \qquad . \quad (133)$$

while thereafter

$$R_2 = \gamma R_1; \; R_3 = \gamma R_2 = \gamma^2 R_1, \text{ etc.};$$
$$r_2 = \gamma R_{n-1} = \gamma^{n-1} R_1$$
$$= \gamma^{n-1} r_2/s_{max.}$$

Hence $\qquad \gamma = \sqrt[n-1]{s_{max}} \qquad . \qquad . \qquad . \qquad . \qquad . \quad (134)$

The resistance sections are

$$\rho_1 = R_1 - R_2 = R_1(1 - \gamma);$$
$$\rho_2 = R_2 - R_3 = R_1(\gamma - \gamma^2) = \gamma \rho_1;$$
$$\rho_3 = R_3 - R_4 = R_1(\gamma^2 - \gamma^3) = \gamma \rho_2 = \gamma^2 \rho_1, \text{ etc.}$$

If s_{max} is known for the assumed maximum starting current I_{2max}, the $n - 1$ sections can be found. If s_{min} for current I_{2min} is also

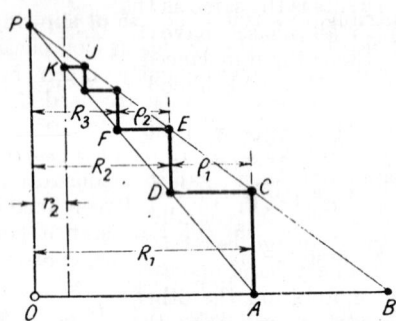

FIG. 174. GRAPHICAL CONSTRUCTION FOR ROTOR STARTOR

known (or can be estimated), a simple graphical construction is possible, Fig. 174. Draw $OA = R_1 = r_2/s_{max}$ to scale, and $OB = r_2/s_{min}$. Raise a perpendicular OP on OA, and join A and B to any point P on it. Draw the perpendicular AC, then

$$AC/OP = AB/OB = 1 - OA/OB = 1 - (s_{min}/s_{max})$$
$$= 1 - s_2/s_1 = 1 - s_2.$$

Draw CD parallel to OB, then $CD = OA . AC/OP = R_1(1 - s_2) = \rho_1$, since $\rho_1 = R_1 - R_2 = R_1(1 - s_2)$ from eq. (131). If the construction is continued by drawing DE, EF, FG . . . , then the horizontal intercepts between the lines AP and BP will represent ρ_1, ρ_2 . . . to scale, and the intercepts between the lines AP and OP will represent R_2, R_3 . . . With given positions of A and B, the construction must be continued until a point such as K falls on the intercept r_2. Then the number of studs n is the same as the number of horizontal lines AB, CD, EF . . .

The number of rheostat sections and contactors (if these are used) may be reduced by cutting out starting resistance one phase at a time, using fewer and larger resistance sections per phase. The resistors are star connected and the first step is represented by the switching-on of the stator. Sections of resistance are then short-circuited in successive phases, the final step cutting out simultaneously the starting resistance remaining in two of the phases. Fig. 175 shows the connections of a startor with 5 steps. Using a modification of the

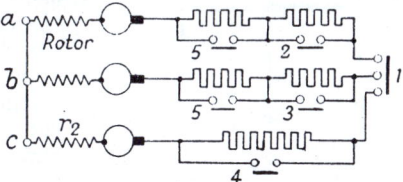

FIG. 175. UNBALANCED ROTOR-RESISTANCE STARTOR

nomenclature in Fig. 173, let the resistance in the rotor phases abc at any setting of the startor be $R_a + r_2$, $R_b + r_z$ and $R_c + r_2$. Then if these total phase resistances bear to one another the relation of a geometrical progression, the torque produced by the three unbalanced phase currents is the same as that produced under balanced conditions in which all phases have the same resistance as the geometric mean. Since the unbalanced currents are changed round at every step, there is little likelihood of trouble with unbalanced heating.

CHOICE OF ROTOR VOLTAGE. Rotor windings of slip-ring machines must be insulated to withstand the full standstill voltage. The choice of this voltage is arbitrary, and controllable by the designer. For small machines there is generally no difficulty in keeping the open-circuit slip-ring voltage below 500 V., which limit should not be exceeded for ordinary hand-operated rotor startors and switch-gear. With large motors, on the other hand, a low rotor voltage entails a large current, complicating the design of slip-rings, brush-gear and startor contacts. Thus for 1 000-h.p. motors, a rotor voltage of 1 000–2 000 V. might be chosen.

4. Starting of Cage Motors. A cage rotor being electrically isolated, no alteration can be made in its characteristics by direct connection, although it is possible to alter them in other ways. Apart from special devices, the starting current of a cage motor can only be reduced by lowering the voltage applied to the stator.

EFFECTS OF REDUCED STATOR VOLTAGE. If the stator voltage of an induction motor be reduced from normal value to the fraction x, the no-load and short-circuit currents will be changed in nearly the same proportion. The main flux which, over the range of normal loads, is roughly constant, is determined by the applied voltage and will reduce substantially in proportion to the reduced voltage. The magnetizing current will similarly fall, so long as the magnetic circuit is not highly saturated. Further, since the core losses are roughly proportional to the square of the flux-density and consequently of the voltage, the active component of the no-load current will be reduced in proportion to the voltage reduction. On

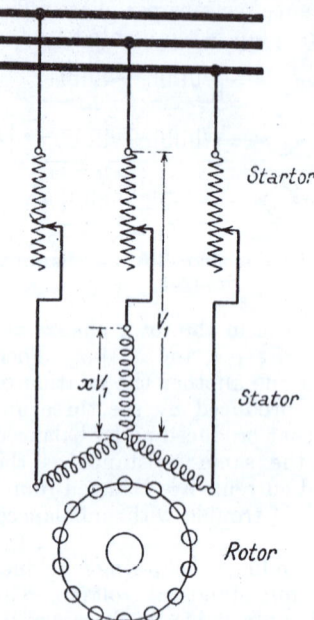

short circuit the current is given by the quotient of applied voltage and short-circuit impedance, and will therefore be to a close approximation a direct linear function of the applied voltage.

The output and torque for a given slip change as the *square* of the applied voltage, as is deducible from eq. (108), Chapter XII, § 4. This is important, for the reduction of the starting current to the fraction x of normal value at full voltage means the simultaneous reduction of the starting torque to x^2 of its normal value: e.g. a reduction of starting voltage to 0·58, reduces starting torque to $0·58^2 = 0·33$ of normal.

The circle diagram of a motor on normal voltage may, from the above considerations, be used for any lower voltage if its current scale be reduced by x and its power and torque scales by x^2.

Fig. 176. Stator Rheostat Starting

METHODS OF STARTING. (a) *Direct Switching.* A cage motor may be switched direct on to a normal-voltage supply, momentarily taking several times full-load current at low power factor. A small torque is produced, of a magnitude depending upon the rotor resistance. The rate of temperature-rise is high, and the motor may be damaged if a start is delayed by (a) an excessive load torque, or (b) insufficient rotor resistance, or (c) a fall in the supply voltage. The direct switching of cage motors is usually the subject of regulations by supply authorities which lay down the greatest rating of motor that may be so started. The limit may be a few horse-power on low-voltage distribution networks. High-voltage motors with low short-circuit currents (e.g. 3 times full-load) may be direct-switched in ratings up to 500 h.p. where mains regulation permits.

For the direct-switching case, if M_{sc} be the torque developed on short circuit at normal voltage with the current I_{sc}, then the starting torque is $M_s = M_{sc}$, and the starting current per phase is $I_s = I_{sc}$. The ratio of starting to full-load torque from eq. (130) is

$$M_s/M_n = (I_{sc}/I_n)^2 s_n \qquad . \qquad . \qquad . \qquad . \quad (135)$$

which, as already shown, results in a comparatively small starting torque with a large starting current.

(*b*) *Stator Resistor or Inductor Startor.* The connections are shown in Fig. 176. The result is merely to reduce the normal voltage per

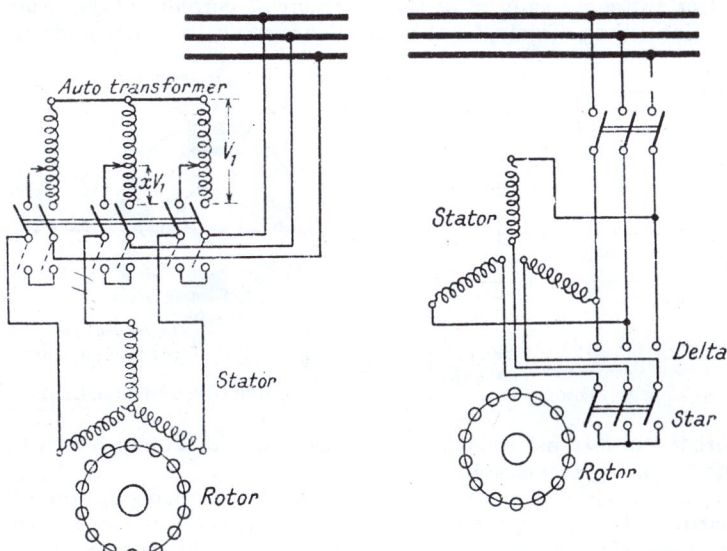

FIG. 177A. AUTO-TRANSFORMER
STARTING

FIG. 177B. STAR-DELTA
STARTING

phase to the fraction x, and to change somewhat the power-factor of the line current. The starting current is $I_s = xI_{sc}$ and the starting torque $M_s = x^2 M_{sc}$. The ratio of starting to full-load torque is

$$M_s/M_n = (I_s/I_n)^2 s_n = x^2 (I_{sc}/I_n)^2 s_n \qquad . \qquad . \quad (136)$$

or $1/x^2$ of that obtainable with direct switching. This method may be used for the smooth starting of small machines, e.g. centrifugal oil purifiers: but the star-delta switch is cheaper, and gives a better torque.

(*c*) *Auto-transformer Startor.* Suppose in Fig. 177A that the auto-transformer is used to reduce the phase voltage to the fraction x of normal value. Then the motor current at starting is $I_s = xI_{sc}$, and the starting torque is $M_s = x^2 M_{sc}$. The ratio

$$M_s/M_n = (I_s/I_n)^2 s_n = x^2 (I_{sc}/I_n)^2 s_n \qquad . \qquad . \quad (137)$$

i.e. exactly as for the previous case with the stator impedor. The advantage of the method lies in the fact that the voltage is reduced by *transformation*, not by dropping the excess in impedance, so that the line current and power input are reduced in comparison with method (*b*).

Neglecting magnetizing current and losses in the auto-transformer, the current on the output side is I_s in the fraction x of the turns, and on the input side will consequently be $xI_s = x^2I_{sc}$ in all the turns of each phase. Thus for a starting torque of x^2 of that obtainable by direct switching, only x^2 of the short-circuit current is taken from the line: alternatively, for the same line current, the starting torque

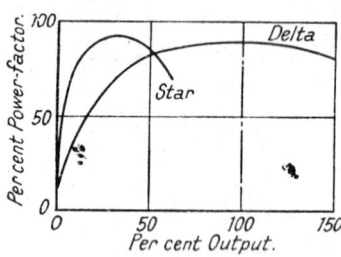

FIG. 178. POWER FACTOR
OF MOTOR IN STAR AND
DELTA

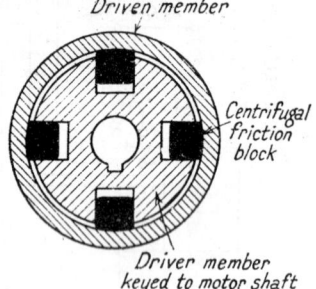

FIG. 179. CENTRIFUGAL CLUTCH

with the auto-transformer is $1/x$ times as great as that obtainable with the stator impedance startor.

The system of connections shown provides voltage steps during starting, the auto-transformer being cut completely out of service after the start has been completed. Since it is only in use for short periods, the current density in the windings may be increased considerably above that normal for power transformers. At the same time, since the starting torque depends on the square of the current, the impedance of the transformer should be kept small. The auto-connection is here of particular advantage.

(*d*) *Star-delta Switch.* For this method (Fig. 177B), a motor must be built to run normally with a mesh-connected stator winding. At starting, the winding is connected temporarily in star. The phase voltage is thus reduced to $1/\sqrt{3} = 0.58$ of normal, and the motor behaves as if the auto-transformer were employed with a ratio $x = 0.58$. The starting current per phase is $I_s = 0.58I_{sc}$, the *line* current is $(0.58)^2I_{sc} = 0.33I_{sc}$, the starting torque is one-third of short-circuit value, and the ratio

$$M_s/M_n = \tfrac{1}{3}(I_{sc}/I_n)^2 s_n .\qquad\qquad . \qquad . \qquad . \quad (138)$$

The method is cheap and effective, so long as the starting torque is not required to exceed about 50 per cent of full load torque. It can therefore be used for machine-tools, pumps, motor generators, etc. The method is unsuitable for motors at voltages exceeding 3 000 V. because of the excessive number of stator turns needed for delta running.

Where induction motors are required to run for considerable periods on small loads, a star-delta switch permits the machine to be star-connected during these periods, with reduction of magnetizing current and increase in efficiency. Fig. 178 shows power-factor curves for a motor worked at normal voltage and at $1/\sqrt{3}$ of normal by means of a star-delta switch.

CENTRIFUGAL CLUTCH. All methods of starting plain cage machines involve a reduction of the starting torque. In order to keep a

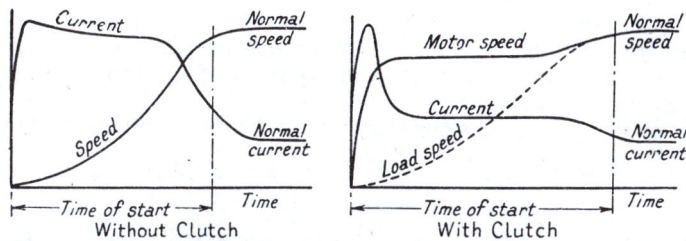

FIG. 179A. STARTING DIRECT WITHOUT AND WITH CENTRIFUGAL CLUTCH

high efficiency and to avoid undue rotor heating, the cage resistance must be restricted, so that the starting torque even with direct switching is inherently low. A commonly-used device, to enable the cage machine to start on very light load, and to take up the load torque when its own torque has increased to values nearer the maximum, is the centrifugal clutch. This is essentially a pulley or coupling with two mechanically separate parts (Fig. 179): the inner one is attached to the motor shaft and is free to revolve at low speeds within the outer hollow part. Mechanical coupling between the inner and outer concentric parts is afforded by weighted blocks which, by centrifugal action, come forcibly into contact with the outer cylinder and lock the two parts together. The blocks may be coated with friction brake material, or may be metal running in oil which is squeezed out by the centrifugal pressure. At starting, the motor commences its rotation without appreciable load, and as it increases its speed, the load is gradually taken up by the decreasing clutch-slip. By spring-loading the blocks, the clutch is prevented from operating until after a pre-determined speed has been reached. The use of a centrifugal clutch will protect the motor against overload, since the machine cannot be called upon to supply a torque greater than that which causes the clutch to slip.

Fig. 179A shows typical current and speed curves to a time base, illustrating the effect of a centrifugal clutch. It will be seen that, although the attainment of normal full speed takes a little longer when the clutch is used, the total amount of heat developed in the motor I^2R losses will be very much reduced.

A similar method of starting employs a hydraulic torque-convertor, or "fluid coupling," which can be used with star-delta or direct switching. Such a drive is controllable, and enables the motor to be rated on running rather than starting torque. There being no metallic or frictional connection between the motor and its load, adjustments and maintenance are greatly reduced.*

EXAMPLE. A small cage induction motor has a short-circuit current of 6 times full-load current, and a full-load slip of 5%. Find the motor and line starting currents in terms of full-load current, and the starting torque in terms of full-load torque, for (a) direct switching; (b) stator resistance startor and (c) auto-transformer startor limiting the starting current to twice full-load value; (d) a star-delta switch. (e) Find the ratio of an auto-transformer to give full-load torque at starting.

Applying the expressions found above for the several methods, the results may be tabulated as below.

	(a) Direct Switching	(b) Stator Resistance	(c) Auto-transformer	(d) Star-delta Switch
Phase-voltage applied at starting, in terms of normal value V_1 . .	1·0	0·33	0·33	0·58
Phase current at starting—				
In terms of short-circuit current I_{sc}	1·0	0·33	0·33	0·58
In terms of normal current I_n	6·0	2·0	2·0	3·48
Line current at starting—				
In terms of normal with mesh connection	6·0	2·0	0·67	2·0
Starting torque—				
In terms of short-circuit torque M_{sc}	1·0	0·11	0·11	0·33
In terms of full-load torque M_n .	1·8	0·2	0·2	0·6

With the conditions here imposed on the rheostat and auto-transformer starting, viz. twice full-load current in the motor at starting, the torque produced is naturally the same in each case but the line current drawn by the auto-transformer is only one-third of that taken with the rheostat startor. If the same *line* current were allowed in these two cases, the figures for auto-transformer starting would be changed to those of the last column.

(e) For full-load torque at starting,

$$M_s/M_n = (I_s/I_n)^2 s_n = 1$$

whence $I_s = I_n/\sqrt{s_n} = 4{\cdot}46 I_n = 0{\cdot}75 I_{sc}$

* Sinclair, "Some Problems in the Transmission of Power by Fluid Couplings," *Proc.I.Mech.E.*, p. 83 (1938).

so that a 75 per cent tapping is required. The line current would be $(0.75)^2 = 0.56$ times short-circuit value, or 4·5 times full-load current.

CURRENT PEAKS DURING STARTING. The initial current flowing when an induction motor is started is the fraction x of short-circuit current corresponding to the fraction x of normal voltage initially applied, together with a transient in each phase in accordance with the discussion in Chapter XVIII. The transients rapidly vanish, but the "steady-state" current does not fall appreciably until the motor has attained over two-thirds of normal speed. It is consequently

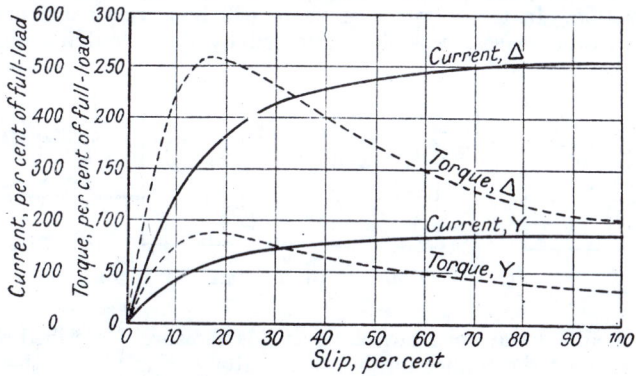

FIG. 180. TORQUE/SLIP AND CURRENT/SLIP CURVES FOR STAR AND DELTA CONNECTIONS

desirable to allow the motor to attain steady speed as nearly as possible, before the change is made to normal running conditions. Thus in Fig. 180, drawn for star-delta starting, the change-over from star to mesh connection should not be made until the speed is within about 10 per cent of synchronous: otherwise there will be a peak *line* current considerably greater than full load—greater in fact than the standstill current with star connection. Centrifugal clutches that operate too early or too suddenly will also produce severe current peaks.

But current surges may occur for another reason even when care is taken to avoid a premature operation of a change-over switch. Any method of starting, such as auto-transformer or star-delta switch, that involves a momentary disconnection of the main supply, may cause transient current surges of a very severe nature. Consider a cage motor running in the star-position ("start") of a star-delta switch connection. In changing over to the "run" or delta connection, the stator winding is disconnected from the supply for an interval usually between 0·1–0·3 sec. The excitation of the rotating magnetic field is withdrawn, but the field does not instantaneously collapse. The short-circuited cage rotor immediately develops induced currents which hold the original field at the position, with reference to the rotor surface, that it held at the moment of interruption. The field, now "frozen" to the rotor, is carried round by it, inducing e.m.f.'s of gradually reducing magnitude in the temporarily open-circuited stator winding. On account of the rotor slip, the stator induced e.m.f.'s will have a frequency less than that of the supply. When the main switch now closes in the delta position, the supply voltages are reconnected to stator windings that have induced e.m.f.'s of comparable size, but with a relative time phase that may be anything between coincidence and opposition. In the former case there will be a very large current surge in all phases, possibly reinforced by the "doubling" effect.

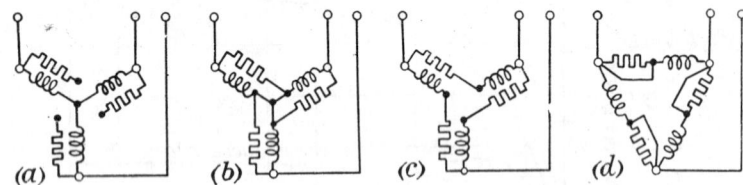

(a)   (b)   (c)   (d)

Fig. 181. Modified Star-delta Switching

A modified star-delta switching avoids the surge described above by retaining the connection between stator and line while the star-to-delta change is made. The sequence is shown in Fig. 181. The motor is switched on in star (a); resistors are parallelled with the phases (b) leaving the motor itself unaffected; the star-point is opened (c) putting the windings in delta with series resistance; and (d) the resistors are short-circuited. The method overcomes surges and mitigates the normal current-rise at change-over.

5. **Dynamics of Starting.** The proper choice of a motor may require a knowledge of its starting duty, especially for high-inertia loads. Inadequate torque may result in excessive run-up time and motor heating.

ACCELERATION TIME. The torque developed by the motor provides for dead-load counter torque (including friction) and for acceleration of the rotating masses:

$$M = M_l + J(d\omega_r/dt)$$

where M = motor torque;[*]
$\quad M_l$ = dead-load counter torque;[*]
$\quad \omega_r$ = angular velocity, rad./sec.;
$\quad J$ = moment of inertia of rotating parts;[*]
$\quad t$ = time, sec.

* A consistent system of units is necessary. With M.K.S. units M is in newton-m. and J in kg.-m.²; with British units, lb.-ft. and slug-ft.², and a conversion factor is required in eq. (141)

The net accelerating torque is $M - M_l = M_a = J(d\omega_r/dt)$, and the running-up time is therefore

$$t = J\int(1/M_a)d\omega \qquad . \qquad . \qquad . \qquad (139a)$$

If M_a is constant between two speeds ω_m and ω_n, the speed interval is accomplished in

$$t = (J/M_a)(\omega_n - \omega_m) \qquad . \qquad . \qquad (139b)$$

The total revolutions of the motor and its load can be found from $(1/2\pi)\int\omega_r . dt$.

Evaluation of the run-up time will generally require step-by-step integration. Fig. 182 shows a graphical method using eq. (139b) in

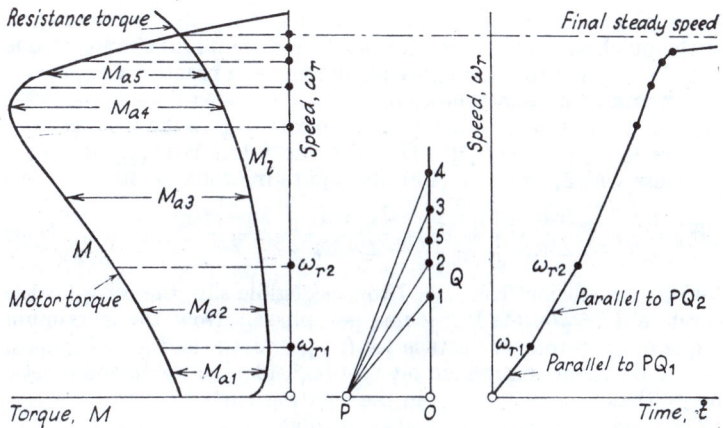

FIG. 182. GRAPHICAL INTEGRATION

the form $\Delta\omega_r/\Delta t = M_a/J$ for the slope of the speed/time curve between selected speed-intervals. A plot of M and M_l to ordinates ω_r gives M_a as the intercept for any speed. Averaging M_a over convenient speed-steps, the slopes of the lines from a pole P to points Q will then be those of the corresponding part of the speed/time curve. The pole P is such that $OP = J$ in the same unit system that $OQ = M_a$.

MOTOR WITH PURE INERTIA LOAD. Suppose that there is no dead-load or friction torque and that the motor torque is taken up entirely with inertia-acceleration. Suppose also that the motor is direct-switched and has constant rotor resistance. Then $M = M_a$ and the conditions can be related to pull-out torque M_m by eqs. (109) and (111), which give

$$M/M_m = 2/[(s/\alpha) + (\alpha/s)].$$

where $\alpha = r_2/x_2$ is the slip for maximum torque. Now $M = M_a = J(d\omega_r/dt)$, so that

$$\frac{M}{M_m} = \frac{J\omega_1}{M_m} \cdot \frac{d}{dt}\left(\frac{\omega_r}{\omega_1}\right) = -\frac{J\omega_1}{M_m} \cdot \frac{ds}{dt}$$

where ω_1 is synchronous angular speed. Combining the two expressions,

$$dt = -\frac{1}{2}\frac{J\omega_1}{M_m}\left(\frac{s}{\alpha} + \frac{\alpha}{s}\right)ds \qquad . \qquad . \qquad . \qquad . \qquad (140)$$

which for acceleration from standstill to slip s gives

$$t = \frac{J\omega_1}{M_m}\left(\frac{1 - s^2}{4\alpha} + \frac{\alpha}{2}\ln\frac{1}{s}\right).$$

If this relation is plotted for various values of α it will be found that, for the quickest start, the motor must produce its maximum torque for a slip $\alpha = 0.4$ to 0.5, a figure likely to give in practice a somewhat uneconomical running condition.

ROTOR HEATING. Eq. (103) can be recast to give the rotor current as $I_2 \simeq I_{2sc}/\sqrt{[1 + (\alpha/s)^2]}$. The rotor heat loss is $\int I_2^2 r_2 \cdot dt$ joules per phase which, from eq. (140) for a pure-inertia start is

$$W_2 = -\frac{I_{2sc}^2 r_2 J\omega_1}{2\alpha M_m}\int s \cdot ds = \frac{I_{2sc}^2 r_2 J\omega_1}{2\alpha M_m}\cdot\frac{1 - s^2}{2} \qquad . \qquad . \qquad (141)$$

During acceleration from $s = 1$ to a negligible slip, the rotor heat is therefore $I_{2sc}^2 r_2 J\omega_1/4\alpha M_m$ joules per phase. Now the maximum torque in synchronous watts is $\frac{1}{2}E_2 I_{2sc}$ per phase, and $\alpha = I_{2sc} r_2/E_2$, so that W_2 reduces quite simply to $\frac{1}{2}J\omega_1^2$, identical with the kinetic energy that has been stored in the rotating parts.

Thus the acceleration of masses involves a rotor-circuit heat loss equal to the kinetic energy stored. This loss is basically unavoidable. The only way of reducing heat loss within the rotor winding is to transfer part of the circuit resistance to an external resistor, a step possible only with slip-ring rotors.

A heat loss comprising approximately $(r_1/r_2')W_2$ will, of course, occur in the stator, and this can be reduced by reduction, where feasible, of the stator resistance.

6. **Cogging, Crawling, and Noise Production in Cage Motors.** With certain relationships between the numbers of poles and of stator and rotor slots in cage motors, peculiar behaviour may be observed when the machine is started. Thus with the number of stator slots S_1 equal to the rotor slots S_2, the machine may refuse to start at all, a phenomenon termed *cogging*. With other ratios S_2/S_1, the motor may exhibit a tendency to run stably at low speed, e.g. one-seventh of normal: this is called *crawling*. In some cases excessive vibration may be set up at certain speeds below normal running speed, accompanied by *noise* resembling a medium-pitched howl. These

phenomena are much less prominent in slip-ring machines, with their higher starting torque.

The essential difference in behaviour of slip-ring and cage rotors in this connection is that the cage rotor will circulate a current under any harmonic e.m.f. that is produced by the gap flux, except that which has a wave-length equal to the pitch of the rotor bars. Slip-ring windings, on the other hand, tend to reduce the effect of most harmonics of pole-pitch different from the coil pitch.

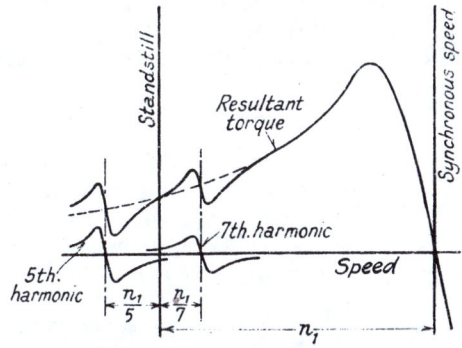

FIG. 183A. HARMONIC "INDUCTION" TORQUES

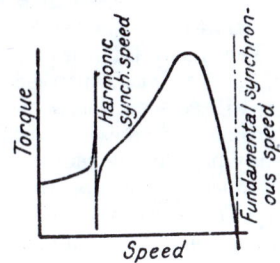

FIG. 183B. HARMONIC "SYNCHRONOUS" TORQUES

Parasitic magnetic fields are attributable to m.m.f. harmonics originating in (a) windings, (b) the slotting, (c) saturation, (d) gap-length irregularity. Minor causes include (e) overhang leakage fields, (f) axial leakage of the main flux. In some cases (g) unbalance of, or (h) harmonics in, the 3-phase supply voltage will produce the trouble. The most important causes are (a), (b), (c) and (d) inherent in the machine. The effects include elastic deformation (e.g. shaft deflection), parasitic torques, vibration and noise.

The causes of harmonics have been briefly noted in Chapter X, § 7. The winding harmonics, introduced into the gap by the m.m.f. wave-form of the three-phase winding carrying sinusoidal currents, are shown in eq. (82) to be of amplitude $1/n^2$ of their order n in terms of the fundamental amplitude. The movement of the harmonics is with or against the direction of the fundamental in accordance with their expression as $n = 6m \pm 1$, where m is any integer: if $n = 6m + 1$, the motion is in the same direction as that of the fundamental, but at $1/n$th of the speed, while if $n = 6m - 1$, the motion is in the reverse direction. The fifth harmonic, of 4 per cent amplitude, rotates against the fundamental at one-fifth synchronous speed and the seventh, of slightly more than 2 per cent amplitude, rotates at one-seventh of synchronous speed in the same direction as the fundamental. It will be clear that these harmonics *alone* will have little effect on the operation of a machine. They may, however, be very greatly augmented by slotting.

Consider, then, that there are present in the gap of a three-phase induction motor fifth and seventh harmonic fields. These may be

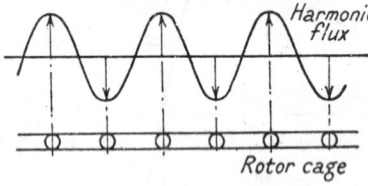

FIG. 184. ROTOR CAGE IN HARMONIC FIELD

deemed produced by sets of additional rotating poles superimposed on the fundamental main poles. They will produce rotor e.m.f.'s and currents, and consequently harmonic torques, having the same general shape as that of the fundamental, but with "synchronous" speeds respectively one-fifth backwards and one-seventh forwards, as shown in Fig. 183A. The resultant torque/speed curve will be a combination of the fundamental, fifth harmonic and seventh harmonic curves, and it will be seen that a marked "saddle" effect is produced. If the harmonic torques are sufficiently pronounced, the seventh harmonic may prevent the motor speed exceeding about one-seventh of normal, since the downward slope of the resultant torque at this speed is a stable running condition over the torque range between the maximum and minimum points. The motor therefore *crawls*.

The rotor m.m.f. has a harmonic content, and with certain ratios of slots and speeds it is possible to get a stator and a rotor harmonic rotating together in the gap, producing a *synchronizing* torque as in a synchronous motor. There will then be a tendency for a sharp synchronizing torque to be established at some subsynchronous speed, Fig. 183B, again producing a tendency to crawl.

HARMONIC INDUCTION TORQUES. In an N-phase winding with g' slots per pole per phase, the e.m.f. distribution factors of harmonics of orders $n = 6Ag' \pm 1$ are the same as those of the fundamental (see Chapter X, § 8): A is any number 0, 1, 2, 3. . . . Also those of orders $6Ag' + 1$ rotate in the same direction as the fundamental, whereas those described by $n = 6Ag' - 1$ rotate in the reverse direction. A motor with four poles and twenty-four slots ($g' = 2$) will thus tend to produce eleventh and thirteenth harmonics most strongly: a four-pole motor with thirty-six slots ($g' = 3$) will encourage seventeenth and nineteenth harmonics. Torque "saddles" will in these instances be observable at $+ \frac{1}{13}$ and $- \frac{1}{11}$, and at $+ \frac{1}{19}$ and $- \frac{1}{17}$ respectively of synchronous speed.

The effect may be augmented by the rotor slotting, S_2. If each half-wave of a harmonic corresponds to a rotor bar, Fig. 184, the harmonic torque will be a maximum. For example, if the number of stator slots $S_1 = 24$, and of rotor slots $S_2 = 44$, the eleventh harmonic will have 44 half-waves in the gap, each corresponding to one rotor bar. Not only will the eleventh harmonic torque be large, but it will vibrate strongly.

It is necessary to avoid values of S_2 exceeding S_1 by 50–60 per cent, as there will otherwise be some tendency towards saddle harmonic torques. Fig. 185 shows typical torque/speed curves taken on a four-pole 50-c/s. motor with various slottings. The eleventh harmonic torque at $- 137$ r.p.m. and the thirteenth at $+ 115$ r.p.m. are strongly marked.

Other typical cases are $S_1 = 36$, $S_2 = 57$ or 63, producing seventeenth and nineteenth harmonic torques; or $S_1 = 48$, $S_2 = 83$, the twenty-fifth harmonic appearing at $1\,500/25 = 60$ r.p.m., in a four-pole 50-c/s. motor.

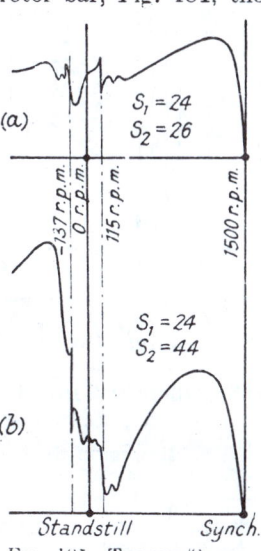

FIG. 185. TORQUE/SPEED CURVES OF FOUR-POLE, 50-CYCLE MOTOR

HARMONIC SYNCHRONOUS TORQUES. In a similar way, the rotor slotting produces its own harmonics, of orders $(S_2/p) \pm 1$, the $+$ referring to rotation with the machine. Consider a four-pole machine with $S_1 = 24$ and $S_2 = 28$ slots. The stator develops a reversed eleventh and a forward thirteenth harmonic: the rotor develops a reversed thirteenth and a forward fifteenth. The thirteenth harmonic is thus produced by both stator and rotor, but with opposite rotation. If n_1 is the synchronous speed of the fundamental, and n the speed of the rotor, then the synchronous speed of the thirteenth harmonic is $+ n_1/13$, or relative to the rotor $- (n_1 - n)/13$. The rotor runs at speed n, and so rotates its own thirteenth harmonic at a speed $- (n_1 - n)/13 + n$ relative to the stator. The stator and rotor thirteenth harmonics fall into step when

$$+ n_1/13 = -(n_1 - n)/13 + n, \quad \text{i.e. } n = \tfrac{1}{7}n_1.$$

Thus the torque discontinuity at one-seventh synchronous speed is here produced not by the seventh harmonic, but by the thirteenth,

Fig. 186 shows a torque/speed curve obtained from a four-pole cage motor with this slotting. The synchronous torque at $n = 1\,500/7 = +214$ r.p.m. is strongly marked: the machine could not fail to crawl at this speed.

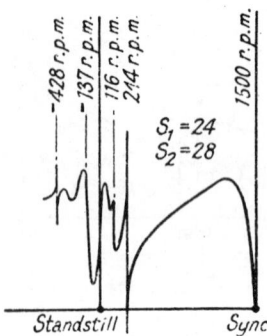

It is noteworthy that with small variations of load, the speed changes slightly if the crawling is due to induction harmonics, but if due to synchronous action, the crawling speed is constant. In Fig. 186 the saddle at $+116$ r.p.m. is due to induction torque by the thirteenth stator harmonic, and that at -137 r.p.m. to the reversed eleventh. There is a small synchronous torque at -428 r.p.m.

Typical synchronous torques produced in four-pole 50-c/s. motors (of synchronous speed 1 500 r.p.m.) are listed in the following Table, for cases in which the slots per pole in the stator differs by 2 from the slots per pole in the stator, or $g_2 = g_1 \pm 2$.

Fig. 186. Torque/Speed Curves of 4-pole, 50-cycle Motor

HARMONIC SYNCHRONOUS TORQUES IN FOUR-POLE 50-c/s. CAGE MOTORS

Stator Slots S_1	Rotor Slots S_2	Harmonics Produced in				Speed of Harmonic Torque	
		Stator		Rotor		$\times n_1$	r.p.m.
24	20	-11	$+13$	-9	$+11$	1/5	-300
24	28	-11	$+13$	-13	$+15$	1/7	$+215$
36	32	-17	$+19$	-15	$+17$	1/8	-187
36	40	-17	$+19$	-19	$+21$	1/10	$+150$
48	44	-23	$+25$	-21	$+23$	1/11	-137

A similar set of figures can be worked out for $g_2 = g_1 \pm 4$, etc.

A special case is that of $S_2 = S_1$, in which precisely the same order harmonics are strongly produced, all rotating at corresponding speeds in both stator and rotor. Each pair of harmonics of a given order thus produce together a synchronizing torque, so that the rotor will remain at rest (*cogging*) unless the fundamental torque is large enough to move the machine from rest, which is unlikely. A similar condition will obtain when S_2 is an integral multiple of S_1.

NOISE. All parasitic fields are natural noise-producers, as they cause vibration. High tooth saturations and zig-zag leakage with unsuitable slot numbers generate unbalanced magnetic pull, which moves round the rotor as the slot-openings fall in and out of register.

The rapidly rotating unbalanced pull encourages vibration and noise. As a result, the rotor shaft may deflect, and even run through a critical speed. Noise is reduced by stiffening the stator and the rotor shaft, and by skewing the rotor slots.

NUMBER OF SLOTS IN CAGE MOTOR. Practice has shown that the most common cause of crawling is the harmonic of order $n = 6g' + 1$ (see page 306). The size of the harmonic—the forward harmonic corresponding to the slotting—is greatest with open slots, or where the lips of semi-closed slots are too thin and saturate easily.

To limit asynchronous harmonic torques, S_2 should not exceed $1.25S_1$. To reduce synchronous torques, S_2 should not equal $6px$ or $6px \pm 2p$, where x is a positive integer. Slot harmonics are mitigated if $S_2 \neq S_1 \pm 2p$, or $S_1 \pm p$ or $\frac{1}{2}S_1 \pm p$. A rule to reduce vibration is $S_2 \pm 6px \pm 1$ or $6px \pm 2p \pm 1$. Even with the choice of an unprohibited number, however, noise and vibration may still be produced unless careful attention is paid to design details.

7. **High-torque Cage Machines.** The great simplicity and robustness of the cage motor are very valuable characteristics of the machine from the standpoint of cost and maintenance. The inherent disadvantage of low starting torque has attracted the attention of numerous investigators, and many modifications have been devised to attempt to obtain good starting without sacrificing the simplicity and cheapness of the machine. The problem is clarified by a comparison of the plain cage and slip-ring motors.

COMPARISON OF CAGE AND SLIP-RING MOTORS. Compared with a slip-ring motor of like rating, the cage motor has the following advantages—

1. A better space-factor in the rotor slots, a short overhang, and consequently less conductor copper* and a smaller I^2R loss.

2. Very small rotor overhang leakage, and a better power factor. The diameter of the circle diagram is increased, and the pull-out torque is therefore greater.

3. A higher efficiency, so long as the cage resistance does not have to be augmented to improve the starting torque.

4. No slip-rings, brush-gear, short-circuiting devices, rotor terminals or starting rheostats. The star-delta startor switch is usually sufficient.

5. A lower cost: a cage motor with star-delta switch may cost 20 per cent less than a slip-ring motor with rheostat startor, of like rating.

6. Better cooling on account of the bare end-rings: also more space for rotor fans.

The disadvantages of the cage motor are almost entirely connected with starting, a small starting torque being produced by a

---

* About one-half of what is required for a slip-ring rotor.

very large starting current, unless the full-load efficiency is sacrificed by conceding more rotor I^2R.

Fig. 187 shows at (a) and (b) the current diagrams for 10-h.p. slip-ring and cage motors respectively. The diameter of the former is 63 per cent that of the latter, the maximum torque is 67 per cent, and the full-load power factor is 0·86 for the slip-ring compared with 0·90 for the cage motor. The full-load efficiencies are respectively 0·86 and 0·87, the cage rotor having been designed for minimum loss and a slip of 3 per cent. The slip-ring motor, with a suitable rotor rheostat, will give a maximum torque of 2·1 times full load for rather less than 3 times full-load current. The cage rotor, on the other hand, would only give 1·1 times full-load torque on

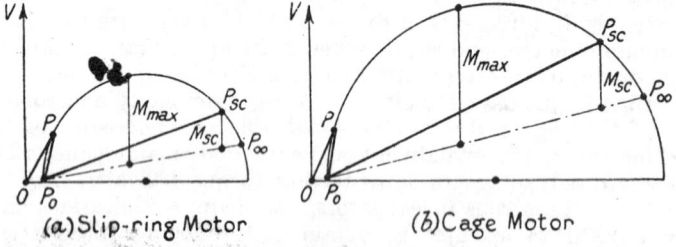

(a) Slip-ring Motor   (b) Cage Motor

Fig. 187. Comparison of Slip-ring and Cage Motors

short-circuit at full voltage, and would then take 6 times normal current. This cage design is not normal: the only considerations are in obtaining the maximum efficiency and output without regard to starting torque and current. The most usual type of standard low-resistance cage motor gives 35–45 per cent of full-load torque with $2\frac{1}{2}$ -$3\frac{1}{2}$ times full-load current, on a start at 60 per cent of normal voltage.

High Torque Rotors. Methods of obtaining increased starting torque with short-circuited rotors may be classified into the following categories—

1. Change of rotor or stator connections.
2. Composite or deep rotor bars, using skin effect.
3. Double-cage construction.

(1) *Special Connections.* These are very numerous. A few typical arrangements are shown diagrammatically in Fig. 188. Method (a) uses pole-changing to lengthen the paths of the rotor currents at starting, thus increasing the resistance. During running, the pole-pitch of the stator is the same as the coil-span of the closed wave connections, and currents circulate along the wave coils. For starting, the number of poles is reduced (see § 8), and the current paths now include the high-resistance end-rings, so that the starting current is reduced and the effective rotor resistance increased. Method (b), due originally to Boucherot, comprises two stators and

a two-part rotor having at each end a low-resistance end-ring, and in
the middle high-resistance connections. One half of the rotor lies
in the field of either stator. At starting, the stator windings are
caused to be a pole-pitch out of phase. The two e.m.f.'s induced in a
rotor bar are therefore in opposition, with the result that the rotor
currents circulate through the high-resistance connections. In
normal running, the two stator fields are brought into phase, and
the machine behaves like an ordinary cage motor. Originally the
phase alteration was obtained by moving one of the stators bodily
round on trunnions. A later form employed star-delta connection
of the stator windings to obtain phase displacement. The disadvantages of these methods, which involve changes in the stator

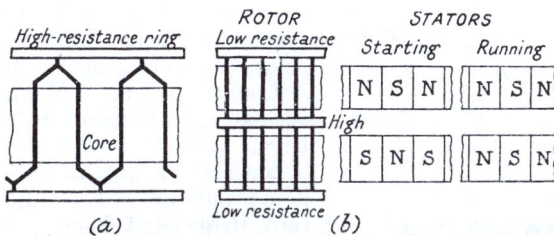

FIG. 188. SPECIAL HIGH-TORQUE ROTOR CONNECTIONS

windings, include complication, low power-factor, and difficulties
with insulation in high-voltage motors.

(2) *Skin-effect Conductors.* The variation of rotor frequency gives
a means of changing the effective rotor resistance, since an inductive
impedance loses almost all of its reactance at the low frequencies
common to induction motor rotors when operating at normal slip.
A number of composite conductors have been developed using this
variable reactance to produce *skin effect* in rotor conductors (bars
or end-rings), resulting in a greater effective rotor resistance at low
speeds. A typical example is Wall's composite conductor, consisting of a central copper rod covered in the slot portion by a seamless
steel sheath. The sheath is copper-plated on the outside, care being
taken that the plating overlaps the ends and makes good contact
with the central rod. Tests show that the resistance to alternating
current is about $3\frac{1}{2}$ times that to direct current, so that a starting
torque at least equal to full-load torque is obtainable. At rest and
at low speeds, the skin effect, assisted by the steel sleeve, tends to
confine the current to the sleeve, so that the uniform distribution is
upset and the effective resistance raised. At normal running speeds
the rotor frequency is so low that the current is carried substantially
by the low-resistance copper rod.

A similar effect is obtained by using iron end-rings, preferably
sectionalized so as to dissipate more readily the heat developed.

A more orthodox construction is obtained with deep solid conductors, where the increase of effective resistance due to eddy current production is encouraged. The development of eddy loss in conductors has already been dealt with in Chapter X, § 11, where

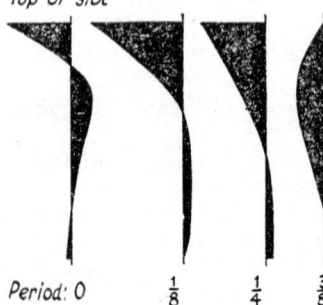

(a) (b) (c)
Flat Taper Sash
bar bar bar

Top of slot

Period: 0 $\frac{1}{8}$ $\frac{1}{4}$ $\frac{3}{8}$

(d)   Distribution of current-density in flat bar at starting

Fig. 189. Skin-effect Conductors

it is shown to be proportional to the fourth power of the conductor depth. The approximate formulae, eqs. (86) and (88), are not applicable to cases such as the present: they hold for cases in which the conductor depth is limited. Nevertheless, they give an indication of the magnitude of the additional I^2R losses that occur with deep conductors. In a typical conductor of this sort, the width is 1·4 mm., tightly fitting the slot, and the depth is 30 mm. From eq. (88), $\alpha = 1$ and $\alpha h = 3$, so that by eq. (86), $K_d = 1 + \frac{4}{45}(3)^4 \simeq 8$. A considerable increase in resistance is thus obtained at starting when the rotor frequency is high.

Fig. 189 shows typical shapes of deep bars and slots. It is necessary to know the effective resistance of conductors of any chosen shape. Expressions have been developed[*] involving hyperbolic functions of αh, resembling in form the well-known formulae for transmission-line propagation. In fact, the basic resemblance to propagation from top to bottom of the slot is shown in Fig. 189 by the current density distributions in a parallel-sided bar at instant 1 (when the total bar current is zero), and instants 2, 3 and 4, respectively $\frac{1}{8}$, $\frac{1}{4}$ and $\frac{3}{8}$ of a period later.

With small motors, there is a limit to the depth of slot usable, as accommodation must be found for adequate cross-section for the core, the shaft, and possibly axial cooling ducts. The increased slot leakage (which is, of course, an essential factor in producing eddy-current loss), affects the full-load power factor adversely, and the efficiency will be lower than that of a plain cage.

***

* Putnam, "The Starting Performance of Synchronous Motors," *Trans. Am. I.E.E.*, Feb. 1927; and Bruges, "Evaluation and Application of Certain Ladder-Type Networks," *Proc. Roy. Soc. Edin.*, 62 (II), p. 175 (1946).

The current being confined chiefly to the upper part of the conductor during starting, this part will become hot; but the thermal conductivity dissipates the heat to the lower portion, especially where a greater section of copper is employed there as in the T-bar of Fig. 189. The T-bar has a mechanical advantage over the parallel-sided deep bar in that the shoulders keep the bar tighter in the slots. Mechanical vibration, together with thermal expansion and contraction due to slip-frequency currents, is very damaging to a slack bar and its insulation. The connection of the rotor slot bars to their end-rings must be carefully designed, because the attachment represents the most likely source of weakness in a cage rotor. Fig. 190 shows a method of connecting deep bars to end-rings.

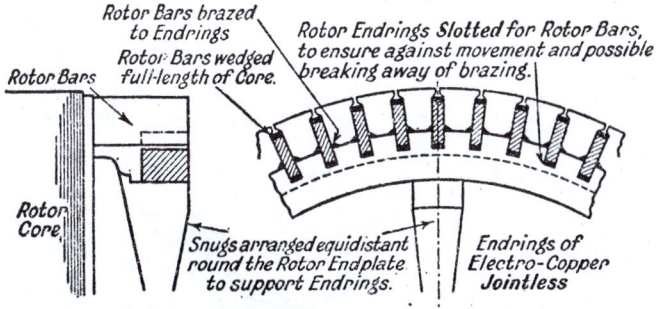

FIG. 190. CONNECTION OF DEEP ROTOR BARS TO END-RING

DOUBLE-CAGE MOTOR. The arrangement of two cages for the production of large starting torques is due to Boucherot. In Fig. 191, cage K_1 is close to the periphery of the rotor, and has a high resistance. The inner cage K_2 has a low resistance, but is set deeply in slots having a considerable leakage flux on account of the long narrow slitted "lip." Thus the outer cage has a high resistance together with the low reactance normal to an ordinary cage winding, while the inner cage has low resistance and large reactance. At starting the leakage reactance of the inner cage K_2 is large enough to cause the rotor current to flow chiefly in the outer cage, the high resistance of which produces considerable I^2R loss, and consequently good starting torque. When the speed is normal, the reactance of both cages is almost negligible, so that the rotor current is carried by the two cages in parallel, giving a low effective resistance. The slits are necessary in the deep slotting to prevent the main flux from missing the inner bars: if the inner bars were completely buried in the rotor iron, the flux would only link the outer cage. The slits thus ensure that the inner cage shall be active.

By a suitable choice of the resistances of the two cages, the number and breadth of the slot openings, and the depth of the inner cage, it is possible to control the shape of the torque/speed curve so as to obtain any desired starting torque, within limits. To a first approximation it is possible to consider the cages as producing torques quite separately as though they were in separate machines. In Fig. 192 the torque slip curves for a typical rotor are given. The total torque is the summation of the two individual torques, and it will

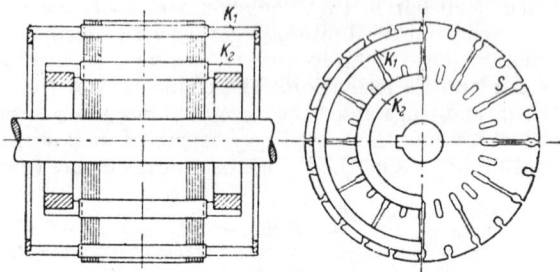

FIG. 191. DOUBLE-CAGE (BOUCHEROT) ROTOR

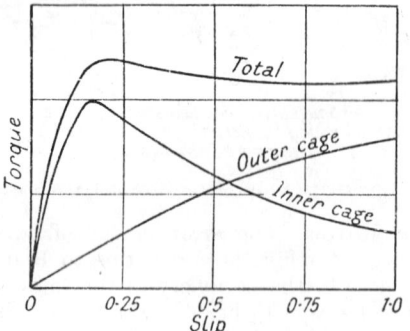

FIG. 192. TORQUE/SLIP CURVE OF DOUBLE-CAGE MOTOR

be clear that by varying the several resistances and reactances, a wide range of operational characteristics is possible.

This flexibility is not obtained without sacrifice of some of the qualities of the plain cage machine. The pull-out torque is much smaller, since one half of the winding (or one cage) has a maximum torque at a speed quite different from that of the other cage. Much heat is developed in the high-resistance cage during starting, since this cage bears the burden of the starting loss. With frequent starts, it may burn out. The additional leakage of the inner cage results in a drop in the power factor at full load. The high resistance of the outer cage increases the loss on full load and lowers the efficiency.*

* Since, however, a normal single-cage machine cannot generally be built for minimum rotor loss, lest the starting torque be too low, the efficiency of the double-cage motor may not be smaller; it may even be greater than that of a normal cage motor.

The use of two cages doubles the possibility of mechanical failure, which occurs most usually at the junction of slot bars with end-rings.

TYPES OF SLOTTING. Three shapes are shown in Fig. 193. In (*a*), the greater inductance of the inner cage is due to the flux crossing the slit connecting the two conductor portions. There will be a proportion of the leakage flux linking *both* cages, i.e. there will be mutual induction between the bars of the inner and outer cages. Thus, in Fig. 193 (*b*) the leakage flux due to a current in the lower

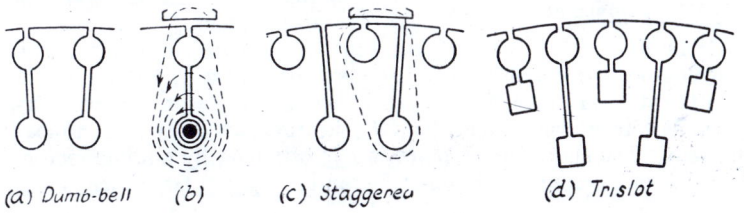

(*a*) Dumb-bell   (*b*)   (*c*) Staggereu   (*d*) Trislot

FIG. 193. DOUBLE-CAGE SLOTTINGS

bar is seen in part to cross the slit and in part to enclose the upper conductor. Similarly, some of the flux due to a current in the upper conductor links also the lower conductor. The amount of mutual induction is approximately calculable.

The slotting of Fig. 193 (*c*) separates the two sets of rotor bars in such a way as to reduce the mutual induction to a comparatively small amount, unless the stator slots are so wide (Fig. 193 (*c*)) as to link magnetically the two cages. This must be avoided by a suitable choice of slot-pitch.

For the "dumb-bell" slotting, Fig. 193 (*a*), the reactance of the outer cage is assessed normally (see Chapter IX, § 5), but the zigzag leakage is reduced, and a smaller value (e.g. one-half) is taken. The reactance of the inner cage is taken as that due to the body of the slot and the connecting slit. In Fig. 193 (*c*), the inner cage can be assumed devoid of zigzag leakage. The stator zigzag may be considered as halved.

The trislot arrangement, Fig. 193 (*d*), permits of an adequate number of inner-cage slots without excessive saturation at the roots of the teeth. The windings in the two inner rows may both be connected at each end to an end-ring, or, better, be of insulated bars connected to form short-circuited short-pitch turns to reduce 7th harmonics.

EQUIVALENT CIRCUIT. Provided both cages completely link the main flux, they may be considered as parallel windings. The equivalent circuit for one phase of a double-cage machine is shown in Fig. 194, in which r_1, x_1, r_m, x_m are respectively the resistance and reactance of the stator and magnetizing impedance; r_2'/s and x_2' are for the outer cage; and r_3'/s and x_3' are for the inner cage. Mutual inductance is neglected. The performance of the motor can

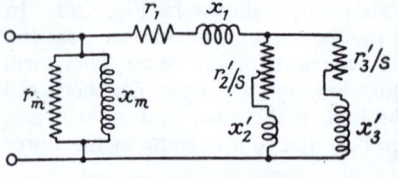

FIG. 194. EQUIVALENT CIRCUIT OF DOUBLE-CAGE MOTOR

be found by drawing two separate circle diagrams for the individual cages, and adding corresponding complexor rotor currents to obtain the effective rotor current, then to the resultant adding the magnetizing current to give the stator current.

If in Fig. 195 P_2 and P_3 correspond to outer- and inner-cage currents at a given slip, their complex sum gives P, a point on the combined rotor current curve, and also on the stator current locus, reckoned from the origin O: OP_0 is approximately the no-load current. The short-circuit point is P_{sc}', obtained by adding vectorially the short-circuit currents P_0P_{2sc} and P_0P_{3sc} for two cages. Since

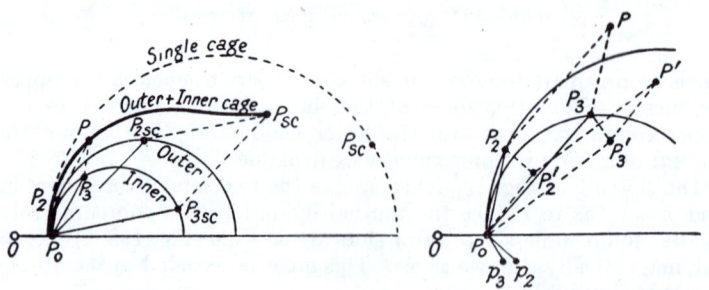

FIG. 195. CURRENT-LOCUS OF DOUBLE-CAGE MOTOR

FIG. 196. CURRENT-LOCUS INCLUDING MUTUAL INDUCTION

the outer cage has the lesser reactance, its circle has the greater diameter, while the short-circuit point P_{2sc} is well up towards the crest of the circle since the resistance is large, and the cage is designed to produce the best starting torque. The circle for the inner cage has a more normal short-circuit point, but its diameter is small on account of its large reactance.

The effective reactance of the double cage rotor is greater than that of the plain cage. While it is readily possible to *increase* the leakage of the inner cage, a compensating reduction of outer-cage reactance is scarcely possible, so that on balance the combination has a reactance greater than normal. The combined rotor current of the double-cage machine must always lie therefore within the circle of the plain motor. This inevitably means that the pull-out torque, the full-load power factor and the efficiency must be lower than the corresponding values for a machine of which the rotor has a plain cage, the stator being identical. The circle of the single-cage machine is shown dotted in Fig. 195.

The torque of the double-cage machine can be obtained from Fig. 195 by drawing the torque lines for each circle separately, and adding the values for corresponding slips.

The diagram in Fig. 195 neglects mutual induction between the cages. With some slottings, particularly that of Fig. 193 (a), the diagram so obtained differs from test results. In Fig. 196, the current P_0P_3 of the lower cage produces a leakage flux in phase with it. By mutual inductance an e.m.f. lagging $90°$ on P_0P_3 is induced in the outer cage, and this generates a current practically in phase with the e.m.f., since the outer cage has a preponderating resistance. The additional current in the outer cage is thus P_0p_2. The current P_0P_2 in the outer cage induces a mutual e.m.f. in the lower and an additional current P_0p_3 lagging on the e.m.f. by an angle depending on the ratio of reactance to resistance (in the diagram it is $45°$). The current of the combined cages is P_0P' instead of P_0P, the value obtained (as in Fig. 195) with neglect of mutual flux. This accounts for the lower starting torque and power factor generally obtained on test compared with the values predicted by the method of Figs. 194 and 195.

CURRENT LOCUS. The equivalent circuit of a double-cage motor, including any mutual inductance there may be, is given in Fig. 197 (a). The suffix 1, as before, refers to the stator; 2 to the outer cage; 3 to the inner cage; 23 to the approximate mutual effect; and m to the magnetizing impedance; all referred to the stator phase. Neglecting the reactance of the outer cage as small, and assuming a constant magnetizing current (i.e. putting x_m and r_m across the stator terminals), simplifies the equivalent circuit to that of Fig. 197 (b). The impedance of the parallel paths 2 and 3, which contain the components variable with slip, is

$$\frac{1}{\dfrac{1}{(r_2'/s)} + \dfrac{1}{(r_3'/s) + jx_3'}} = \frac{1}{s} \cdot \frac{r_2'(r_3' + jsx_3')}{r_2' + r_3' + jsx_3'}$$

$$= \left[\frac{\{r_3'(r_2' + r_3') + s^2x_3'^2\}r_2'/s + jr_2'^2x_3'}{(r_2' + r_3')^2 + s^2x_3'^2} \right].$$

The term in j is the effective reactance, and that without is the effective resistance of the combined rotor circuits.

For values of s close to zero, the effective resistance reduces to

$$\frac{r_2'r_3'}{(r_2' + r_3')} \cdot \frac{1}{s}$$

This is obviously the combined resistance of r_2'/s anu r_3'/s in parallel. For the reactance at $s = 0$, the expression is

$$\frac{r_2'^2 x_3'}{(r_2' + r_3')^2}$$

Thus at low values of slip the motor will behave as if it were a plain cage machine in which the rotor resistance is that given by r_2' and r_3' in parallel, and the rotor reactance is

$$x_{23} + \frac{r_2'^2 x_3'}{(r_2' + r_3')^2}.$$

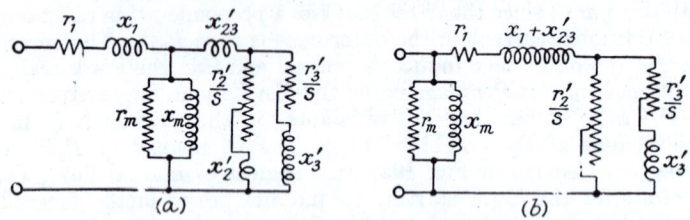

FIG. 197. EQUIVALENT CIRCUITS OF DOUBLE-CAGE MOTOR

At and near standstill, $s \simeq 1$, the resistance will be

$$\frac{r_2'[r_3'(r_2' + r_3') + x_3'^2]}{(r_2' + r_3')^2 + x_3'^2},$$

and the reactance

$$\frac{r_2'^2 x_3'}{(r_2' + r_3')^2 + x_3'^2}.$$

The initial curve at $s = 0$ and at the short-circuit point can be found from the equations above, and inspection will show roughly the type of torque/slip curve to be expected. The current locus may be obtained fully by inversion, a process best illustrated by an example.

EXAMPLE. A 30-h.p., 400-V., mesh-connected, 50-c/s., double-cage motor has an equivalent circuit per phase given by Fig. 198A, in which the cages are in slots like those of Fig. 193 (c). The values of the several impedances are: $r_1 = 0.6 \ \Omega$; $x_1 = 1.9 \ \Omega$; $r_2' = 2.1 \ \Omega$; $x_2' = 1.3 \ \Omega$; $r_3' = 0.4 \ \Omega$; $x_3' = 4.5 \ \Omega$. For the magnetizing impedance, $r_m = 350 \ \Omega$; $r_m = 67 \ \Omega$. If the admittance locus is found, a change of scale converts it to the current locus for a given phase voltage.

Referring to Fig. 198B, co-ordinate axes are drawn about O_{23}, resistances being measured vertically upwards, and reactances horizontally to the right. The locus of the outer cage impedance z_2' is obtained by drawing a horizontal line $1.3 \ \Omega$ to scale, and raising a vertical to $2.1 \ \Omega$ to scale: the point represents the outer cage impedance for $s = 1$. At lower values of s the resistance increases to r_2'/s, so that for $s = 0.75$, $r_2'/s = 2.1/0.75 = 2.8\ \Omega$, etc. In a similar way the locus of the inner cage impedance z_3' is drawn by an intercept $4.5\ \Omega$ to scale horizontally to the right, and a vertical of $0.4\ \Omega$ to scale for $s = 1$. For $s = 0.75$, $r_3'/s = 0.4/0.75 = 0.53\ \Omega$, etc. The rectilinear loci z_2' and z_3' are now inverted about O_{23}, to give the two circular loci y_2' and y_3' with centres M_2 and M_3 respectively. These are marked off in terms of slip by projection

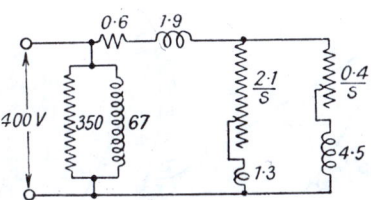

Fig. 198a. Example: Equivalent Circuit

from O_{23} on to corresponding points on z_2 and z_3. The combined admittance of the two cages, $y_2' + y_3'$, is found by adding corresponding complexor values of y_2' and y_3'. In Fig. 198b, y_2' has been added for convenience to y_3', giving the dotted locus y_{23}'. Re-inversion about O_{23} gives the curve marked z_{23}', which is the combined impedance of the two cages. The stator impedance z_1 can now be added. Since it consists of constant resistance and reactance, it can be included by shifting the origin to O_{123}, i.e. by $0.6\ \Omega$ downwards and $1.9\ \Omega$ horizontally to the left, to scale. The impedance locus measured from O_{123} is that of the whole motor excluding the magnetizing impedance. To find the admittance of this part, invert z_{23} about O_{123}, to give y_{123}. The admittance of the magnetizing circuit is now included by shifting the origin to O by a vertical distance $1/r_m$ and a horizontal distance $1/x_m$ to the scale of y_{123}. Then O is the origin for the terminal admittance, and the locus y_{123} becomes the locus of the total stator current by a suitable change of scale, since $I_1 = V_1 y_{123}$, where V_1 is the phase voltage.

It will be seen that at small values of slip the current locus is nearly the arc of a circle of small radius: also that at slips near unity the locus approximates to the arc of a large-diameter circle. This indicates that most of the current is carried by the inner (running) cage at normal speed and by the outer (starting) cage at low speed. If the circles for the two cages are drawn in, an estimate of the torque may be made by drawing the individual torque lines: e.g. through O_{123} a line is drawn giving an intercept with the horizontal through O_{123} equal to the stator I^2R loss per phase for any given stator current. For a slip of 0.1, the stator current by measurement is 65 A., the loss is $65^2 . 0.6 = 2\,540$ W., corresponding to a loss-current of $2\,540/400 = 6.4$ A. Setting this as a small vertical on the axis through O_{123} and vertically beneath the 0.1 slip point, its extremity gives the torque line for the range of $0.0 - 0.1$ slip. For slips exceeding 0.1, the current locus is no longer circular and the torque line therefore not straight. It is quite readily estimated

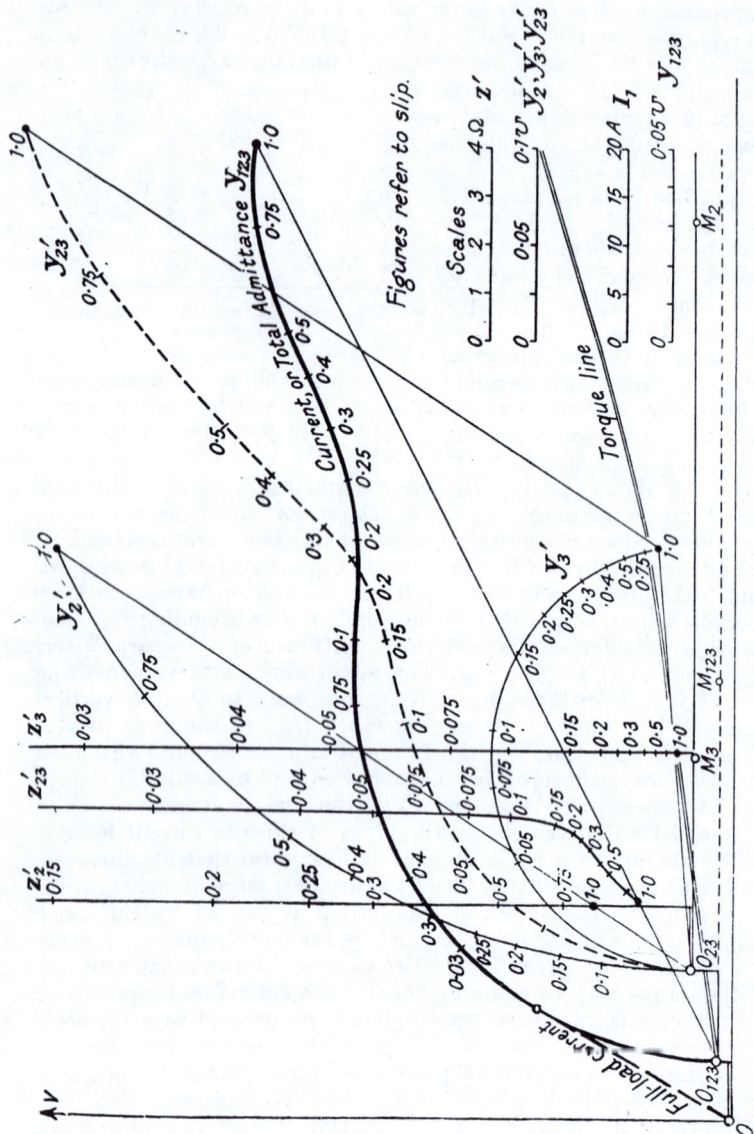

FIG. 198B. EXAMPLE: LOCUS DIAGRAM

by calculating the stator I^2R loss at a few points. Thus at standstill the stator current is 114 A., the loss $114^2 . 0.6 = 7\ 800$ W.; the corresponding loss-current is $7\ 800/400 = 19.5$ A., and this, set off vertically beneath the unity slip point, gives the torque line position. One or two further points similarly calculated enable the torque line to be drawn in.

The torque/slip curve, Fig. 199, shows that a comparatively uniform torque is maintained over the speed range during starting.

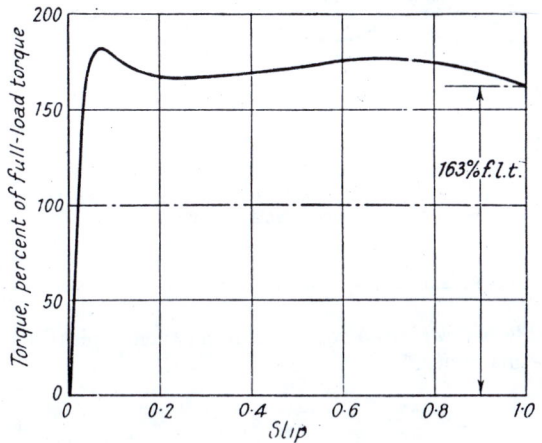

FIG. 199. EXAMPLE: TORQUE/SLIP CURVE

At standstill 163 per cent of full-load torque is obtained with 4·75 times full-load current. With a star-delta switch these figures would be reduced to 54 per cent torque with 1·58 times normal full-load current. The pull-out torque is about 180 per cent of normal.

DESIGN OF DOUBLE-CAGE ROTORS. The allocation of resistance and reactance to the two cages of a double-cage machine is a somewhat intricate matter. It depends largely on the type of torque/slip characteristic required. As already pointed out, the current locus must lie wholly within the circle of the corresponding single-cage machine, the diameter of the circle being fixed by the total short-circuit reactance. The more nearly the inner cage approaches the normal construction, the bigger will be its circular locus, and the more nearly will the pull-out torque approach that of the plain cage motor (Fig. 200 (a)); but at the same time the starting current will increase and the starting torque grow less (Fig. 200 (b)), because of the greater paralleling effect of the two cage resistances at starting. Towards the other extreme is the use of considerable inner-cage reactance and an adjustment of the outer-cage resistance to achieve maximum starting torque (Fig. 201 (*a*)), resulting in a torque/slip curve like that of Fig. 201 (*b*). The pull-out torque is now but little greater than full-load value. This type of characteristic (combined with stator-rheostat control—see the dotted curves of Fig. 201 (*b*)) is applicable to cranes, and gives a rough approximation to the "series" characteristic furnished by a d.c. series motor,

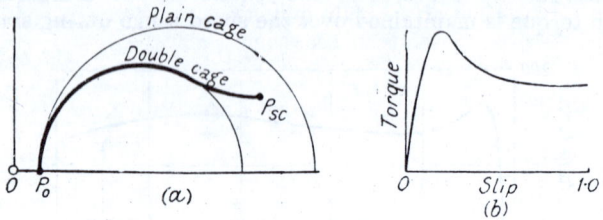

FIG. 200. LOCUS DIAGRAM AND CORRESPONDING
TORQUE/SLIP CURVE

with the added advantage of inherent speed-limitation on light loads.

The Table below gives a comparison between typical slip-ring and cage-rotor machines.*

	Slip-ring Motor	Single-cage Motor	Double-cage Motor: Dumb-bell Slots	Double-cage Motor: Staggered Slots
Efficiency. %				
Full load	88·0	90·0	90·0	90·0
¾ load	88·0	90·0	89·0	90·0
½ load	86·0	89·0	87·0	89·0
Power factor				
Full load	0·85	0·92	0·86	0·88
¾ load	0·79	0·90	0·83	0·87
½ load	0·67	0·82	0·76	0·82
Full load slip, %	4·0	2·8	2·1	2·2
Starting/full-load current				
Direct start	—	6·0	5·6	5·5
Star/delta switch	—	2·0	1·9	1·8
Starting/full-load torque				
Direct start	—	1·3	2·1	2·1
Star/delta switch	—	0·43	0·7	0·7

* A full discussion of cage motors is given by Hoseason, *J.I.E.E.*, 66, p. 410

8. **Speed Control.** The plain induction motor is essentially a constant-speed, shunt-characteristic machine, its action being intimately connected with its synchronous speed. Its nearly constant speed and low power factor constitute its chief disadvantages. Much investigation has been made since the invention of the motor to overcome the restrictions on speed control and power factor without sacrificing the admirable and valuable simplicity and robust construction of the plain induction motor.

Of the several methods in use, some control speed, while others

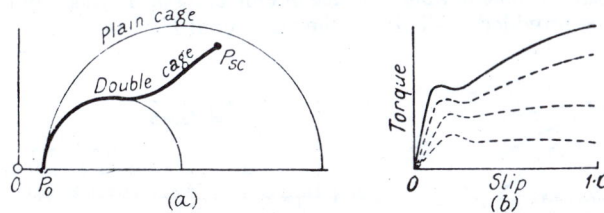

FIG. 201. LOCUS DIAGRAM AND CORRESPONDING
TORQUE/SLIP CURVE

deliberately affect the power factor also. The chief methods of speed control involving no special provision for the compensation of low power factor are—

    (a) Resistance in rotor circuit;
    (b) Pole-changing;
    (c) Cascade connection, and variants of it;
    (d) Change of supply frequency.

The no-load speed of an induction motor closely approaches its synchronous speed

$$n_1 = f/p_1 \text{ r.p.s.}$$

from which it is seen that n_1 may be adjusted by variation of p_1, giving method (b), or of f, giving method (d). Fig. 151 shows that resistance variation in the rotor circuit can be made to provide speed/torque curves falling from no-load speed more or less steeply: this is the essential feature of method (a).

CONTROL BY ROTOR RHEOSTAT. Referring to the simplified theory of the induction motor, eq. (112) from Chapter XII, § 4, shows that near synchronous speed the torque and slip are related by the approximate expression

$$M = sK_t/\alpha = sx_2K_t/r_2 \propto s/r_2.$$

For a given torque, therefore, the slip at which the motor works is proportional to the rotor resistance. Fig. 202 (cf. Fig. 152) shows the torque/speed curves of a slip-ring induction motor in which a rotor rheostat is used to insert additional resistance in the rotor on load. The "shunt" characteristic inherent in the machine is given a greater fall with load according to the amount of added resistance. Since the input to the motor is proportional to torque, it follows that the input for a given torque is the same regardless of the speed, although the lower the speed the lower the output. In brief, a departure from synchronous speed results in a decreased output and an increased rotor-circuit I^2R loss, since the rotor input P_2 appears as mechanical power to the extent of $(1 - s)P_2$, the remainder sP_2 being developed as heat. If speed control is obtained in this way, therefore, a considerable sacrifice of efficiency is entailed, and means must be provided to dissipate the additional I^2R loss.

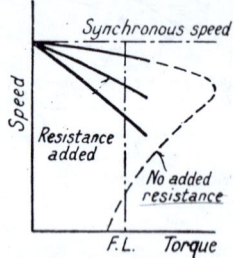

FIG. 202. SPEED CONTROL BY ROTOR RHEOSTAT

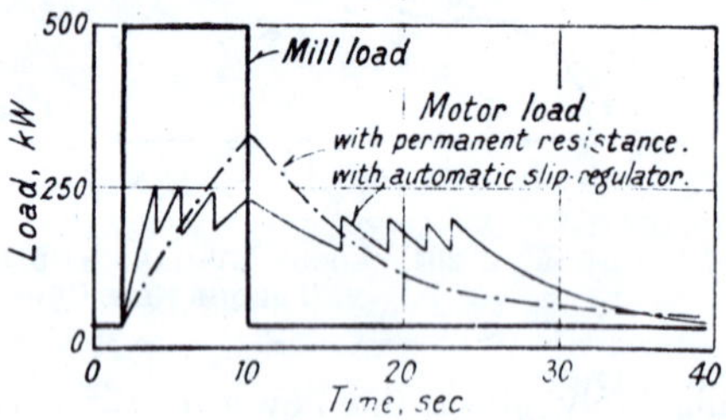

FIG. 203. LOAD EQUALIZATION BY SLIP REGULATOR

EXAMPLE. A 400-V., 50-c/s., 6-pole induction motor gives full-load torque at 970 r.p.m. The input on locked-rotor test at 200 V. is 15 A. and 2·1 kW. What rotor resistance must be added to reduce the speed to 800 r.p.m. for full-load torque?

The data require the assumption of constant gap flux, also that $r_1 = r_2'$ and $x_1 = x_2'$. Then $R_1 = r_1 + r_2'$, $3I_1{}^2R_1 = 2\,100$ W., and $R_1 = 2\,100/3 \cdot 15^2 = 3 \cdot 1\ \Omega$ per phase (equivalent star value); whence $r_2' \simeq 1 \cdot 6\ \Omega$. The impedance is $Z_1 = 200/\sqrt{3} \cdot 15 = 7 \cdot 7\ \Omega$ per phase, giving $X_1 = x_1 + x_2' = 7 \cdot 0\ \Omega$ and $x_2' = 3 \cdot 5\ \Omega$ per phase. Full-load torque, normally at a slip of 0·03, is to be produced at slip 0·2 by increasing the rotor resistance to R'. Applying eq. (108),

$$M_n = E_1{}^2 \frac{0 \cdot 03 \cdot 1 \cdot 6}{1 \cdot 6^2 + 0 \cdot 03^2 \cdot 3 \cdot 5^2} = E_1{}^2 \frac{0 \cdot 2R'}{R'^2 + 0 \cdot 2^2 \cdot 3 \cdot 5^2}$$

syn. W. per ph. This gives $R' = 10 \cdot 7$ or $0 \cdot 05\ \Omega$. The latter value is not possible so that the resistance to be added is $10 \cdot 7 - 1 \cdot 6 = 9 \cdot 1\ \Omega$ per phase. (This is, of course, the equivalent star value referred to the stator.)

In practice, speed-control by rotor rheostat is employed in the following cases: (i) Starting; (ii) Where occasional slight speed drop is required; (iii) For flywheel motor-generator sets, in which sudden loads are supplied from the kinetic energy of the flywheel obtained by dropping its speed, as for rolling mill or winding drives. The speed-drop is controlled by a rotor rheostat in the motor circuit, which gives automatically the necessary speed reduction as soon as a load is applied. When the load is removed, the speed rises, restoring energy in the flywheel. If a contactor arrangement is used for the rotor rheostat, it is possible to secure a degree of equalization in the motor load not possible with a permanently-connected rotor resistance (Fig. 203). An alternative arrangement is a liquid rheostat,

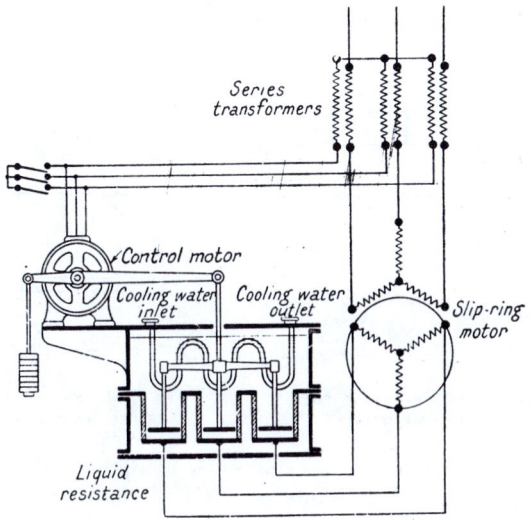

FIG. 204. SLIP REGULATOR

Fig. 204, operated by a small auxiliary motor whose torque depends on the load of the main motor. This is called a *slip-regulator*.

The control motor is shown with simple current control; more elaborate methods use speed and torque adjustment by error-sensitive servomechanisms operating on the slip regulator.

Stator Impedance Control. For small machines a rheostat may be included in the stator supply lines to reduce the speed. More economically, a saturable reactor may be employed. By varying the d.c. excitation the stator applied voltage is varied. Regulation and efficiency are roughly the same as for rotor rheostat control.

POLE-CHANGING. By providing an induction motor with a winding or windings developing alternative numbers of poles, it is given a number of possible synchronous (and therefore also running) speeds. The simplest arrangement is a separate stator winding for each speed desired, rated to suit the particular application, giving a performance on each speed substantially independent of the others. Such an arrangement would be used for lift operation, where the change-over from one winding to another has to be made without losing control of the motor.

Speeds in the ratio 2 : 1 can be produced by single windings on the consequent-pole principle. Different ratios can also be obtained, e.g. 3 : 2, but wide differences generally demand separate windings. The windings are accommodated in the stator slots with the lowest-speed winding as a rule nearest the air gap to secure the minimum reactance necessary for the good design of a low-speed motor. The high-speed winding has usually sufficient margin to permit its location at the bottom of the slots. It is unusual to find a machine

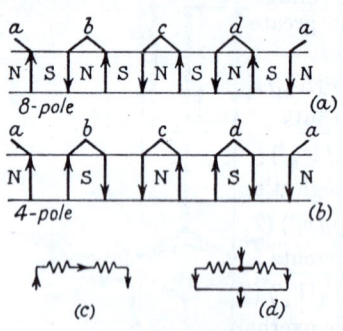

FIG. 205. 8-POLE/4-POLE
CONNECTION

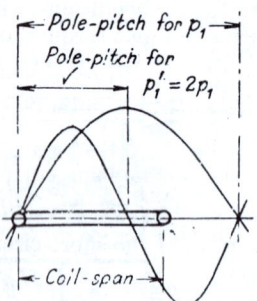

FIG. 206. PERTAINING TO
COIL-SPAN FACTOR

with more than four numbers of poles available, since a greater number involves large amounts of inactive copper and very deep slots.

Let the diagrammatic developed arrangement of Fig. 205 (*a*) represent the coils of one phase of an eight-pole winding. The current directions in the coils give rise to the N S N S . . . polar succession shown. Let coils *b* and *d* be reversed, as in Fig. 205 (*b*). Then the polar succession is N – S – N – S –, i.e. the number of poles is halved. The same process of reversing half the coils is applied to all phases, with the result that the number of poles is halved and the running speed doubled. One method of accomplishing the reversal is to tap the middle of a series of coils (Fig. 205 (*c*) and (*d*)), which clearly reverses the current direction in one half of the coils.

The coil-pitch of a stator winding is naturally fixed. A coil-pitch equal to one-eighth of the stator circumference provides full-pitch coils for an eight-pole machine, $\frac{2}{3}$-pitch for a six-pole and $\frac{1}{2}$-pitch (90°) for a four-pole machine. The coil-span factors for these would be $\cos \frac{1}{2}\varepsilon = 1$, 0·866 and 0·707 respectively, and the winding factors would be these figures multiplied by the appropriate distribution factors. (See Chapter X, § 8.) Clearly it is disadvantageous to work with too short a span, and for a pole-changing winding with speeds in the ratio 2 : 1, a possible arrangement would be to have a coil span of $1\frac{1}{3}$ of a pole pitch for the larger number of poles, which becomes $\frac{2}{3}$ pole-pitch for the smaller number, as shown in Fig. 206: in each case the coil-span factor is 0·866. However, in order to reduce excessive reactance at the lower-speed connection, the spans are often nearer $\frac{1}{3}$ and 1.

In general, let p_1, p_1' be the smaller and larger numbers of pole-pairs. The ratio of the field densities in the gap is

$$B/B' = (T_1 K_{w1}' p_1 / T_1 K_{w1} p_1') \cdot V_1 / V_1' \qquad . \qquad . \ (142)$$

where T_1, K_{w1} and V_1 refer to the turns in series per phase, the winding factor and the phase voltage respectively, while dashes refer to the conditions with the greater number of poles p'. The relative pull-out torques are

$$M_m / M_m' = (B/B')^2 (K_{w1}/K_{w1}')^2 (p_1'/p_1) \qquad . \qquad . \ (143)$$

the relative maximum power outputs

$$P_{mm} / P_{mm}' = (M_m / M_m') (p_1'/p_1) \qquad . \qquad . \qquad . \ (144)$$

and the magnetizing currents (neglecting the effects of saturation)

$$I_{0r} / I_{0r}' = (B/B') (p_1/p_1') (T_1' K_{w1}' / T_1 K_{w1}) \qquad . \qquad . \ (145)$$

The ratio of the short-circuit currents is

$$I_{sc} / I_{sc}' = (V_1/V_1') (T_1'/T_1)^2 \qquad . \qquad . \qquad . \ (146)$$

disregarding any difference in the overhang leakage in the two cases. If the phase voltage is the same for both numbers of poles, then $V_1/V_1' = 1$. The expressions in eqs. (142) to (146) are only approximate: the ratio of magnetizing currents is 10–20 per cent greater and of short-circuit currents 10–15 per cent smaller than as given by the formulae.

The choice of winding arrangement depends on the relative operating characteristics at the two speeds. Two cases may be considered—

1. *Constant Torque Range.* If a constant torque is required at both speeds (i.e. if the same range of torque from no-load to pull-out is to be obtained), the developed power is proportional to the speed. The windings are best utilized if

$$B' \simeq B \quad \text{or} \quad T_1' K_{w1}' V_1 / T_1 K_{w1} V_1' \simeq p_1'/p_1$$

from eq. (142); and the ratio of pull-out torques is then

$$M_m / M_m' = (K_{w1}/K_{w1}')^2 (p_1'/p_1).$$

2. *Constant Power Range.* For constant power on both speeds (torque inversely proportional to speed) the ratio

$$B/B' = p_1/p_1'$$

gives the same overload capacity on each speed and the same current density in the windings.

Other characteristics may be desirable, e.g. in fan drives.

Pole-changing in Ratio 1 : 2. With a typical three-phase *single-layer* winding, the coils have normal span for the smaller pole-pitch, and the coils of each phase are in two groups, Fig. 207 (a). For the smaller number of pole-pairs, both groups are in parallel, while for the larger number they are in series, in each phase. Fig. 207 (b) and (c) show the phase connections, which are in star for each case. For these conditions, $V_1' = V_1$, $K_{w1}' = 0.955 \cdot 1 = 0.955$; $K_{w1} = 0.99 \cdot 0.707 = 0.70$ (since the phase-spread is in effect halved

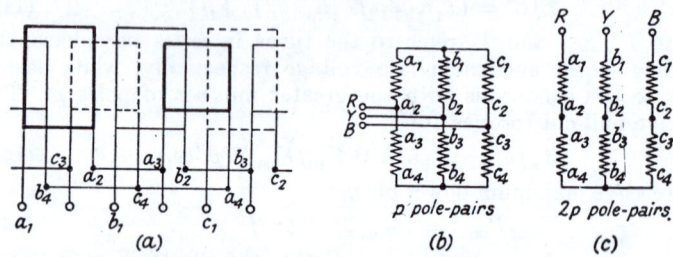

FIG. 207. 1 : 2 POLE-CHANGING WITH SINGLE-LAYER WINDING

for the doubled pole-pitch); and $T_1' = 2T_1$ (since the number of turns in series per phase is halved for the smaller number of pole-pairs); whence

$$\frac{B}{B'} = \frac{2}{1} \cdot \frac{0.955}{0.70} \cdot \frac{1}{2} = 1.36; \qquad \frac{M_m}{M_m'} = 1.36^2 \cdot \left(\frac{0.70}{0.955}\right)^2 \cdot \frac{2}{1} = 2.$$

The ratio of magnetizing currents is

$$\frac{I_{0r}}{I_{0r}'} = 1.36 \cdot \frac{1}{2} \cdot \frac{2 \cdot 0.955}{1 \cdot 0.70} = 1.86,$$

and the ratio of short-circuit currents

$$I_{sc}/I_{sc}' = (2/1)^2 = 4$$

A different characteristic could be obtained by connecting the low-speed winding in mesh instead of star.

Fig. 208 shows a *double-layer* winding arranged for 1 : 2 pole-changing, with a coil-span equal to $\frac{2}{3}$ of the larger and $1\frac{1}{3}$ of the smaller pole-pitch, and giving therefore the same coil-span factor in each case. It could be connected exactly as in Fig. 207 (b) and (c). Since this is a double-layer winding it is possible to have a wide-spread of 120° for the greater number of poles, to reduce third harmonics. With this spread and coil-span, the winding factors are $K_{w1}' = 0.866 \cdot 0.866 = 0.75$; $K_{w1} = 0.955 \cdot 0.866 = 0.827$. If for the larger number of poles the coil groups are in series and the phases in star, and for the smaller number the coil groups are in series and the phases in mesh, then

$$T_1'/T_1 = 1, \quad V_1/V_1' = \sqrt{3}, \quad B/B' = 0\cdot79, \quad M_m/M'_m = 1\cdot5,$$
$$I_{0r}/I_{0r}' = 0\cdot36, \quad I_{sc}/I_{sc}' = 1\cdot73, \quad P_{mm}/P_m{}'_m = 3.$$

This corresponds roughly to "constant torque" conditions. To meet the requirements of "constant power," the greater number of

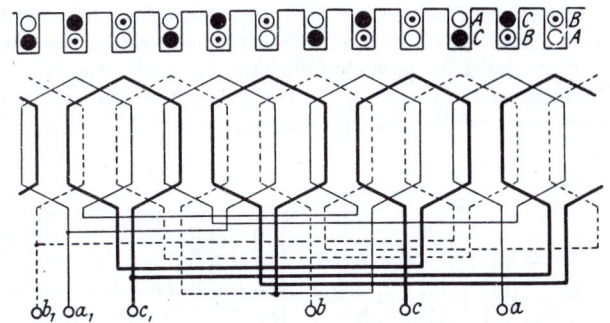

FIG. 208. 1 : 2 POLE-CHANGING WITH DOUBLE-LAYER WINDING

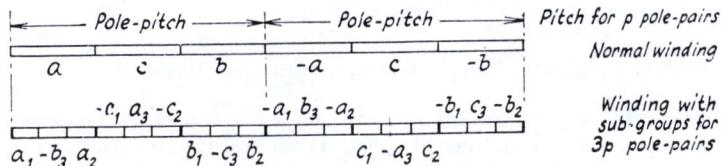

FIG. 209. WINDING FOR 1:3 POLE-CHANGING

poles can be made with series coil groups and mesh connection, the smaller number with parallel groups and star connection. For this

$$T_1'/T_1 = 2, \quad V_1/V_1' = 1/\sqrt{3}, \quad B/B' = 0\cdot52, \quad M_m/M'_m = 0\cdot65,$$
$$I_{0r}/I_{0r}' = 0\cdot47, \quad I_{sc}/I_{sc}' = 2\cdot3, \quad P_{mm}/P_m{}'_m = 1\cdot3.$$

Pole-changing in Ratio 1 : 3. The number of effective poles in a 3-phase winding can be trebled by connecting all phases in series; but the machine is then a single-phase motor without starting torque and is quite unsatisfactory. A method* of securing a reasonable design consists in making three subgroups of each phase winding. Fig. 209 shows at (*a*) two pole-pitches of a normal three-phase winding, and at (*b*) its subdivision into what amounts to nine phases. If subgroups $a_1, -b_3, a_2$ are connected in series they form a normal phase of a machine with 2 poles: the other phases are $b_1, -c_3, b_2,$ and $c_1, -a_3, c_2.$ If now the subgroups $a_1, a_2, a_3; b_1, b_2, b_3;$ and c_1, c_2, c_3 respectively are series connected, they form a three-phase

* Rawcliffe and Jayawant, "The Development of a New 3:1 Pole-Changing Motor," *Proc. I.E.E.*, 103 (A), p. 306 (1956); also Barton, Butler and Sterling, *ibid.*, p. 285.

winding producing 6 poles. In each case the winding has, with respect to its pole-number, a 60° spread. For the same supply voltage the flux density is somewhat low for the 6-pole arrangement, but the density will be increased if one-third of each phase is omitted in the 2-pole connection.

Rotor Windings for Pole-changing. The cage rotor is most convenient for pole-changing motors, since it will assume the same number of poles as the primary field provided that it has a sufficient number of bars. If starting is difficult, a slip-ring rotor becomes necessary, arranged for pole-changing together with the stator.

The resistance of a cage rotor comprises the bar and end-ring resistances, and while the bar-resistance is independent of the

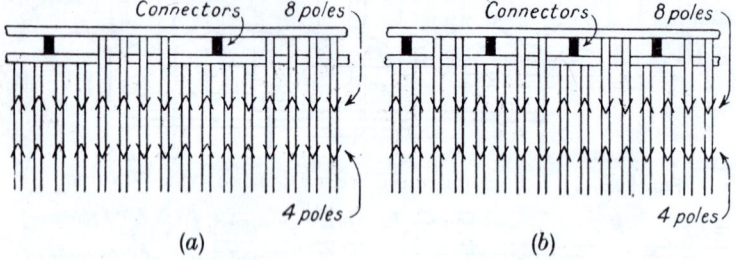

(a)   (b)

FIG. 210. TORQUE CONTROL WITH CAGE ROTOR

number of stator poles, the effective end-ring resistance decreases as the square of the number of poles. If the end-ring resistance is an appreciable fraction of the whole rotor resistance, the torque/slip curve for the machine may be markedly altered by pole-changing. By means of a series-parallel arrangement for the rotor cage, however, it is possible to control the torque/slip curves independently. Fig. 210 (a) shows a possible arrangement for a 1 : 2 speed-change. The bars are divided into four groups, alternate groups being connected to separate rings at either or both ends of the rotor. The rings are interconnected at equi-distant points by connectors. With the eight-pole connection the rotor becomes two cages in parallel and no current is carried by the connectors. With the four-pole stator, each group is in series with the adjacent groups through the connectors, and by suitably adjusting the resistance of these, the effective rotor resistance with the four-pole stator connection can be made as great as, or greater than, that with eight poles. The opposite effect is produced by the arrangement Fig. 210 (b). A typical application is for lift work, where the four-pole connection is used for lifting and the eight-pole for braking. Then the effective rotor resistance can be adjusted to give a large starting torque at rest with the high-speed winding, and a large braking (regenerating) torque near the synchronous speed of the low-speed winding.

CASCADE CONNECTION. If a slip-ring induction motor be run at subsynchronous speed with open-circuited rotor, an e.m.f. can be observed across the rings, of magnitude and frequency dependent upon the slip s. If the motor is to produce a torque at the speed considered, a closed circuit must be provided and the rings may be either—

    (a) short-circuited, in which case the rotor e.m.f. is entirely absorbed in circulating a current in the inherent rotor impedance, and the excess torque must be absorbed by braking;

    (b) connected to external resistances, in which case most of the rotor e.m.f. is expended in circulating the rotor current through them;

    (c) connected to apparatus capable of absorbing the energy that is in (b) wasted in the resistances, in (a) absorbed by the brake.

With (a) the only speed control economically obtainable is the small slip inherent in the normal operating characteristic: with (b) considerable speed-drop is obtainable, but much of the energy given by the stator to the rotor is lost in heat: with (c) this otherwise-wasted energy is utilized—provided that the absorbing apparatus is capable of developing a back e.m.f. of the exact magnitude and frequency demanded by the rotor speed and slip. This is the basis of all "economical" methods of speed control not involving changing the number of stator poles or the supply frequency. The *slip* power or fraction s of the total rotor input P_2, is then used mainly (when s is large) in doing useful work by some suitable means.

The chief difficulty in absorbing the slip power usefully lies in its low voltage and frequency. Either it must be converted to normal voltage and frequency and returned to the supply system, or changed into mechanical power by a second motor capable of working at the slip frequency.

In the cascade arrangement the slip power is supplied to an auxiliary induction motor mechanically coupled to the main or primary motor, Fig. 211 (a). For stability the mechanical coupling (either direct or through belting or gearing) is essential. The supply is connected to the stator of the main motor. The main rotor slip-rings are connected to the auxiliary motor stator, while the auxiliary rotor is a cage winding, or a slip-ring rotor with starting rheostat and short-circuiting device. Alternatively the two rotors may be connected, as in Fig. 211 (b). It is then possible to dispense with slip-rings, although they are generally retained to permit the use of either motor alone. Let p_1, p_2 be the number of pole-pairs of main and auxiliary motor respectively, and f be the supply frequency. The synchronous speed (or speed of the rotating field) of the main motor is $n_1 = f/p_1$ r.p.s. Suppose n be the speed of the set. Then the frequency f_2 in the main rotor and in the auxiliary motor is

$$f_2 = (n_1 - n)p_1.$$

The rotating field of the auxiliary machine has a synchronous speed of

$$n_2 = f_2/p_2 = (n_1 - n)p_1/p_2.$$

On no load the frequency of the short-circuited member of the auxiliary motor (rotor in Fig. 211 (a) and stator in (b)) must be

$$(n_2 - n)p_2 \simeq 0, \text{ whence } n_2 \simeq n:$$

consequently

$$n \simeq n_2 = (n_1 - n)p_1/p_2$$

or

$$n \simeq n_1[p_1/(p_1 + p_2)] \qquad . \qquad . \qquad . \qquad . \quad (147)$$

The no-load speed of the set is thus approximately a speed corresponding to the numbers of poles of the two motors together. This

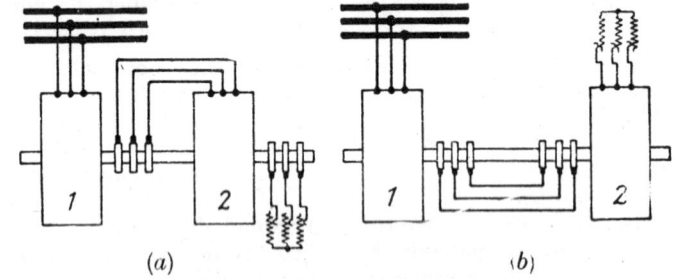

FIG. 211. CASCADE CONNECTION

presupposes that the torques of the two machines are in the same direction : if not, by similar reasoning it can be shown that

$$n \simeq n_1[p_1/(p_1 - p_2)]. \qquad . \qquad . \qquad . \quad (148)$$

which is a possible running speed only if $p_2 \not\prec p_1$.

When the set is started, the supply voltage at frequency f is applied to the stator of motor 1. The same frequency is induced in the rotor of 1 and supplied to motor 2. Both then develop a forward torque. As the shaft speed rises the rotor frequency of motor 1 falls and so therefore does the *synchronous* speed of motor 2. The set comes to a stable speed when the shaft speed and speed of the rotating field of machine 2 are equal. For if the speed rose further, the rotor 2 would be running faster than its rotating field, and would produce a back torque by reversing the rotor currents. This would react on rotor 1 by tending to make it generate as well. The set cannot therefore rise above the speed given by eq. (147).

Running at the synchronous speed of the combination as given by eq. (147) or (148), no current would be induced in the short-circuited secondary of the auxiliary machine. In the rotor of the main motor the magnetizing current for the secondary machine would be induced, while the primary of the main machine would carry the combined magnetizing currents. On load, the speed drops by a few per cent, and normal load currents flow. The electrical power taken from the supply by the first stator, except for I^2R and core losses, is given to the rotor, where it is divided into two parts: one, proportional to n, is converted to mechanical power; the other, proportional to $(n_1 - n)$, is developed as electrical power at slip frequency, and passed to the auxiliary motor which employs it in producing mechanical power and losses. The mechanical outputs of the two machines are therefore approximately in the ratio $n : (n_1 - n)$, or $p_1 : p_2$. The mechanical outputs have the same ratio as the numbers of poles.

With two motors of twelve and eight poles respectively, connected to a 50-c/s. supply, the following no-load speeds can be obtained—

Connection	$n =$	Speed		Output
		r.p.s.	r.p.m.	
Differential cascade .	$f/(p_1 - p_2)$	25·0	1 500	$P_1 - P_2$
Auxiliary motor alone .	f/p_2	12·5	750	P_2
Main motor alone .	f/p_1	8·33	500	P_1
Cumulative cascade .	$f/(p_1 + p_2)$	5·0	300	$P_1 + P_2$

The available outputs in these cases are shown in the last column: P_1 and P_2 refer to the mechanical outputs, for it must not be for-

gotten that the main machine must have an electrical rating to cover the power it supplies to the auxiliary machine.

The current diagram for a cascade set takes the spiral form shown in Fig. 212. F_{sc} is the short-circuit point and P_0 the point for no-load running, in cascade connection. The arc $P_{sc}P_0$ covers the operation of the set from standstill to normal no-load speed. The power factors are low on account of the double magnetizing current. If the

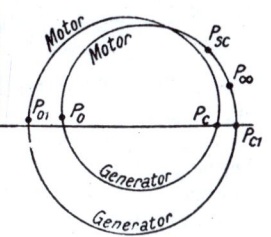

FIG. 212. CURRENT LOCUS OF CASCADE SET

set be driven above its synchronous speed $f/(p_1 + p_2)$, it will begin to generate: the raising of the speed has the effect of reducing the slip, reducing the frequency in rotor 1 and the same in stator 2, lowering the speed of the rotating field in 2 and thus causing a marked negative slip in the second rotor. The rotor e.m.f.'s are reversed so that the second machine gives out energy from its stator terminals, forcing active components of current to flow against the applied stator e.m.f. But the latter is the e.m.f. of the first rotor, so that here, too, the currents are reversed, and with them the directions of the tangential forces on the conductors: thus the set generates.* (See Fig. 213.)

* See § 11, below.

As the speed of the set is raised, the negative slip of the second rotor increases very rapidly, for the speed of its rotating field is decreasing. The reactance of rotor 2 increases and at the same time its stator e.m.f. reduces, resulting in a smaller generating torque. In the first machine the result is to reverse the direction of the rotor currents and this machine again becomes a motor. At the synchronous speed of the main motor, its torque naturally becomes zero:

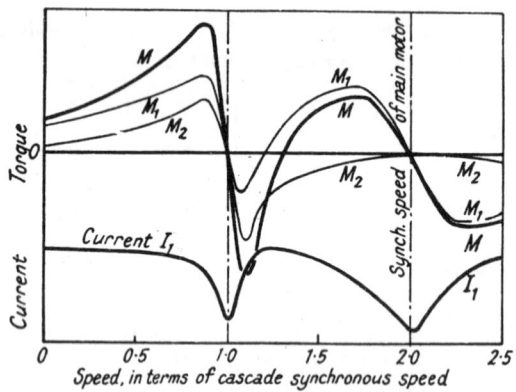

Fig. 213. Torque/Speed and Current/Speed Curves of Cascade Set

so does that of the auxiliary machine, for at this speed its stator field vanishes.

As the speed is raised from the cascade synchronous value $f/(p_1 + p_2)$, the input current locus is the circle from P_0 to P_c below the horizontal axis, and the set again becomes a motor at P_c, but is not stable until it passes over the crest of the circle towards P_{01}, the synchronous speed of the main motor. In this region the second machine acts like an impedance in the rotor circuit of the first. If the set be again driven above n_1, both machines generate, and the current follows the locus P_{01}, P_{c1}. The point P_∞ refers to infinite slip, and the arc $P_{sc}P_\infty$ to slips greater than unity, where the set acts as a brake. The cascaded set has thus two stable running speed ranges, one near P_0 where the synchronous speed is $f/(p_1 + p_2)$ r.p.s., the other near P_{01}, where it is f/p_1. The latter speed cannot be reached by the set with the auxiliary machine short circuited, since there is an intermediate generating range between the two motoring ranges. This range can, however, be reached by using the starting rheostat. It is preferable for this higher speed range to use the main motor alone with its rotor short-circuited and the auxiliary motor running unexcited: in this way a much better power factor and overload capacity are obtained, which is readily understood because of the reduced reactance in the secondary circuit and the consequent enlargement of the diameter of the circle diagram.

A group of characteristic torque, power and current curves to a base of speed is given in Fig. 213, for a cascade set with $p_1 = p_2$. The total torque M is the sum of the primary M_1 and secondary M_2.

Cascade connection suffers from the disadvantages of low power factor, efficiency and pull-out torque. In spite of the complication, only a few economical running speeds are obtained; and the method is very rarely used.

CHANGE OF SUPPLY FREQUENCY. This method of speed control, acting on the obvious principle of changing the synchronous speed of the motor, is only possible in exceptional circumstances, such as on electrically propelled ships. The change of frequency actually involves speed-control of the generator prime-movers, so that the application is restricted to cases where the induction motor is substantially the only load on the generators. By this control, speed changes are very easily made, and in particular reversing is rapid, for the connections to any two of the stator terminals have only to be interchanged. The first steps on the rotor resistance startor could be cut out even before the motor had come to rest.

Portable tools, such as drills, reamers, woodworking appliances and grinders, may be operated with miniature high-speed induction motors. A common rating is 110/220 V. at 180 c/s., or 125/250 V. at 200 c/s., and outputs better than $1\frac{1}{2}$ kg. per h.p. (continuous) or $\frac{3}{4}$ kg. per h.p. (maximum) can be obtained, the weight being that of the active motor material. With 2-pole design, the synchronous speeds are 10 800 and 12 000 r.p.m. The pull-out feature of the induction motor is an advantage in that it prevents excessive overloading, while the normal nearly-constant speed characteristic is useful in maintaining output.

Such machines require a frequency changer, which may be of the induction type, Fig. 214. An auxiliary synchronous or cage-induction motor drives a machine of the slip-ring induction type in a direction opposite to that of the rotating field, so that the rotor frequency is greater than that of the supply frequency. If the changer rotor develops a frequency f_2 with the supply frequency f, the output of the driving motor is the fraction $(f_2 - f)/f$ of the high-frequency output, ignoring losses. Now $f_2 = sf$, so that if the supply frequency is to be multiplied by k (where $k = 4$ for a 200-cycle output from a 50-c/s. supply) then $k = s$ and the rotor of the changer must be turned against the 50-c/s. rotating field with a slip $s = k$, or a *speed* of $s - 1 = k - 1$ times the field speed. Thus to develop a 200-c/s. output, the rotor must be driven at 3 times the speed of the rotating field $n_s = f/p$. The high-frequency supply is taken from the three slip-rings of the frequency-changer rotor in Fig. 214. A fourth, neutral, ring may be provided if required.

The losses comprise those due to I^2R, the primary or stator core loss at frequency f, the rotor core loss at frequency f_2, and the pulsation losses in stator and rotor teeth. The pulsation loss increases considerably with frequency, and it is necessary to design the changer with low flux density and with high-quality steel.

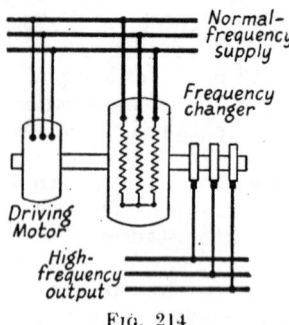

Fig. 214
FREQUENCY-CHANGER

The complexor diagram ot an induction frequency-changer is identical with that of the transformer in Fig. 16, except that $E_2' = sE_1$, and the active component of the no-load current includes friction and windage. The regulation is likely to be large: for convenience o, tool operation it should not exceed 15 per cent. The maintenance of reasonably constant driving speed suggests that the auxiliary motor, if of the induction type, should have a low-resistance cage.*

OTHER METHODS OF SPEED CONTROL. If both stator and rotor are connected to the normal-frequency supply, the rotor will either remain at rest or run at twice synchronous speed, in accordance with the relative directions of the rotating fields produced by the two members. This gives the possibility of a two-speed machine which would, however, require arrangements for accommodating a normal-voltage winding on the rotor, leading currents in to the slip-rings, and accelerating the rotor between the first and second running speeds.

Speed variation is possible by means of a special mechanical construction for the stator, which is carried on trunnions or by some other equivalent method, and is itself rotated by means of an auxiliary motor. If its speed is n' in the same direction as its rotating field (which rotates relative to its windings at speed n_1), then the rotor will revolve at a speed of $n' + n = n' + n_1(1 - s)$, or n' greater than its normal, relative to the bedplate. The auxiliary motor must be rated at the capacity necessary to provide the fraction $n'/(n' + n_1)$ of the output. If the direction n' be reversed, then the rotor speed relative to the baseplate is $n - n.'$ Any speed between $n_1 - n'$ and $n_1 + n'$ is therefore obtainable. The difficulties here are mainly mechanical, the motors being electrically quite normal, unless pole-changing is used to obtain additional speeds.

Based on the above was the Oerlikon motor (Fig. 215) in which an inner rotor is surrounded by an annular member carried on bearings

---

* See Schwartz, "Theory and Application of a Self-Propelled Stator-Fed Frequency Convertor," *Proc. I.E.E.*, 102 (A), p. 56 (1955). This machine is based on a 3-phase commutator motor with both stator and commutator brushes fed from a normal supply.

on the rotor shaft. Surrounding the annular rotor is a fixed statur
with pole-changing windings. The outer surface of the annular rotor

carries a cage winding, and the
inner surface a pole-changing
winding connected to the main
supply through slip-rings. The
stator revolves the annular
rotor at a speed corresponding
to the number of poles in use,
less slip. The inner winding
of the annular rotor acts like
a revolving stator to the main
inside rotor, and up to eighteen
economical speeds can be
obtained by permutations of
the numbers of poles and by
fixing the annular rotor.

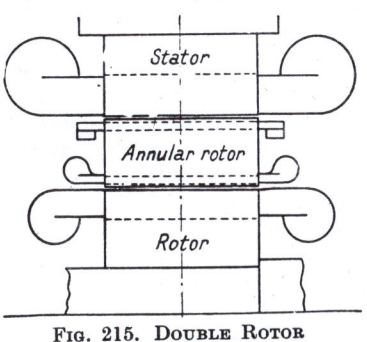

FIG. 215. DOUBLE ROTOR
SPEED-CONTROL

An induction motor may be operated at one-half normal speed by
opening one of its rotor phases, or by connecting to it an asymmetrical
secondary impedance, producing negative-sequence m.m.f.'s.*

A variable speed motor of laboratory size has been developed† on
the shifting-flux principle, using a disc or spherical rotor.

9. Power Factor Control. Induction motors rarely achieve a full-
load power factor much above 0·9, and since for part of the operating
time motors are light-loaded, the average power factor is very much
lower. In cases where motor loads are charged by the supply
authority on a kVA. basis, the problem of power-factor improvement
offers an economic reward for its solution.

The low power factor of a plain induction motor is due to

(a) the reactive lagging magnetizing current necessary to
generate the magnetic flux;

(b) stator and rotor leakage reactance.

At normal operating speeds the former is very much more important than the latter; in fact, to a first approximation the power
factor can be considered to be controlled entirely by the relative
magnitudes of the magnetizing current and the active current
corresponding to the output.

The power factor of a composite (e.g. a factory) load can be compensated as a whole by use of such devices as static or synchronous
condensers. Alternatively all motors, or certain motors, of a group
might have individual methods of phase compensation. Only the
latter will be considered here, under the headings—

(a) static capacitors in shunt with the stator;

* Barton and Doxey, *Proc. I.E.E.*, 102 (A), p. 71 (1955).
† Williams and Laithwaite, "A Brushless Variable-Speed Induction
Motor," *Proc. I.E.E.*, 102 (A), p. 203 (1955).

(b) the synchronous-induction motor;
(c) phase-advancers;
(d) phase-compensated induction motors.

STATIC CAPACITORS. The connection of a set of three static condensers across the terminals of a three-phase induction motor results in the addition of a constant leading component to the current taken. By suitable choice of capacitance, any desired degree of compensation can be obtained, but the power factor varies with the

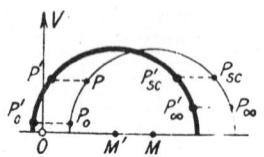

FIG. 216. CURRENT-LOCUS
OF MOTOR WITH SHUNT
CAPACITOR

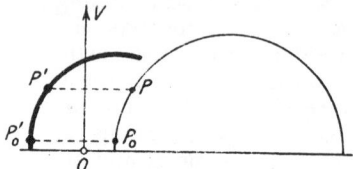

FIG. 217. CURRENT-LOCUS
OF SYNCHRONOUS-INDUCTION
MOTOR

load and is only unity at one particular motor output. The thin circular arc in Fig. 216 represents the current locus of the motor alone. The addition of the capacitor shifts the locus bodily to the left by an amount depending on the capacitance, affecting the power factor at all working loads, but not affecting the motor characteristics in any way. If adequate compensation is provided for full-load conditions, then small loads will be taken with a current at some low leading power factor.

If the induction motor takes P watts and Q lagging vars at full load, corresponding to a power factor $\cos \phi = P/\sqrt{(P^2 + Q^2)}$ $= P/S$, where S is the volt-ampere load, and it is desired to raise the power factor to $\cos \phi'$, then the lagging vars must be reduced from Q to Q' where $Q = P \tan \phi$ and $Q' = P \tan \phi'$, since P is unaltered. The leading vars necessary are

$$Q - Q' = P(\tan \phi - \tan \phi').$$

If the compensation is to be done by means of a capacitor, its rating must be $\omega C V^2 = P(\tan \phi - \tan \phi')$ whence its capacitance is

$$C = P(\tan \phi - \tan \phi')/\omega V^2.$$

The capacitor can be made smaller by raising its voltage, e.g. by means of an auto-transformer, which is quite satisfactory for low-voltage supplies, but does not save for higher voltages since the cost rises sharply with the voltage above about 600 V. The price in any case restricts the method to comparatively small motors.

The switching of a capacitor-compensated motor requires care, to avoid even the momentary possibility of self-excitation as an induction generator, a condition that might result in a sudden and abnormally high stator voltage.

SYNCHRONOUS-INDUCTION MOTOR. This consists of an induction motor which, after it has been started normally, is synchronized by exciting its rotor windings by direct currents introduced at the slip-rings. The motor is described in Chapter XVIII, § 7. For completeness and comparison with other methods, its current locus is given in Fig. 217. The thin line represents the locus as a plain induction motor, the thick line the locus with the machine synchronized. A noteworthy feature is the large leading current taken at small loads.

PHASE ADVANCERS. A *phase-advancer* consists of a device connected into the rotor circuit of the induction motor, but mechanically external to it. Adopting the nomenclature suggested by Professor Miles Walker,* phase advancers may be classified into the following two types—

(i) *Expedor*, that developing a voltage in the rotor which is a function of the secondary current and has some phase relationship to it. It generates or absorbs an e.m.f., which resembles an impedance rise or drop.

(ii) *Susceptor*, that developing a voltage which is a function of the secondary (rotor) open-circuit e.m.f. and has some phase relationship to it. It affects the magnetizing current of the machine, and consequently its magnetic circuit susceptance.

In either case, the basic idea consists in so adding to the rotor induced e.m.f. that the rotor current is advanced in phase, and the consequent reaction on the stator advances the phase of the stator current also. Consider the vector diagram of the plain induction motor in Fig. 149. The rotor current I_2' flows by reason of the production of the e.m.f. E_2', which circulates the current against the impedance z_{2s}'. The rotor current lags on the e.m.f. E_2' by $\phi_2 = $ arc tan (sx_2'/r_2'). The relevant part of Fig. 149 is reproduced in Fig. 218 (a). If, in a slip-ring rotor, the rings are connected to a source of e.m.f. E_c' of the correct slip frequency and of suitable phase relation, the total rotor e.m.f. becomes the vector sum of E_2' and E_c', so that the magnitude and phase of the rotor current may be considerably changed. In Fig. 218 (b) the phase of E_c' has been chosen so as to produce the maximum advance of the phase of the rotor current I_2'. The corresponding component $- I_2'$ of the stator current is similarly affected, and it is seen that the power factor of the input current I_1 can be increased to unity, or made leading.

The problem of phase compensation now reduces to this: how can an e.m.f. of suitable phase relation and exactly the correct

* Rudra and Miles Walker, *J.I.E.E.*, 69, p. 445 (1931).

slip frequency be produced ? The solution lies in the special machine
known as a phase-advancer, which in practice always involves the

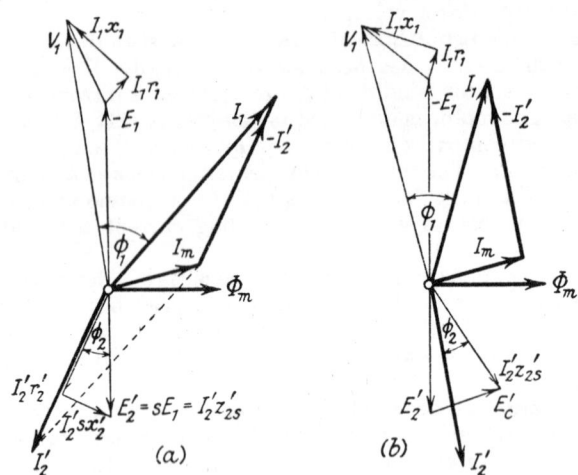

FIG. 218. PERTAINING TO PHASE-ADVANCING

use of a commutator, either as an essential or as a convenient
feature.

The principal types of advancer are—

1. *Expedor*. (*a*) Kapp vibrator;   (*b*) Leblanc or Scherbius
advancer.

2. *Susceptor*. Frequency converter.

In the expedor type the phase-compensating e.m.f. E_c' is proportional to the rotor current, and the advancer behaves as if it

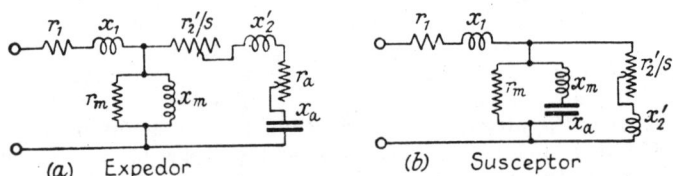

FIG. 219. EQUIVALENT CIRCUITS OF MOTORS WITH PHASE ADVANCERS

were a variable capacitor in the rotor circuit, Fig. 219 (*a*). The
susceptor type, on the other hand, behaves like a capacitor connected in series with the magnetizing impedance, Fig. 219 (*b*).
Neglecting the curvature of the current locus of the induction motor
in the load range, the locus may be expressed as shown by the

vertical line in Fig. 220 (*a*). The locus of the motor with an expedor type advancer is then the sloping line so marked, the compensation being proportional to the rotor current and therefore vanishing at no load. The locus for a motor with a susceptor type advancer is roughly a vertical line parallel to the locus of the plain motor, but shifted to the left by an amount proportional to the degree of compensation. The full current loci actually appear as in Fig. 220 (*b*) and (*c*), from which it will be seen that the susceptor advancer can

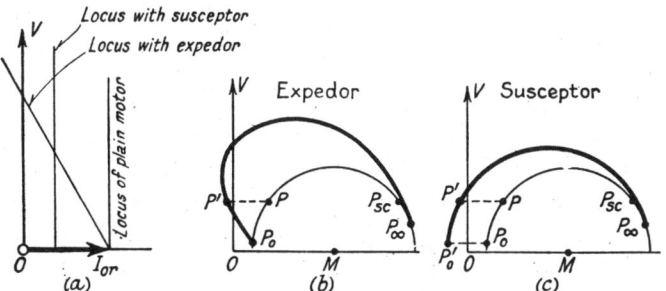

Fig. 220. Locus Diagrams for Motors with Phase Advancers

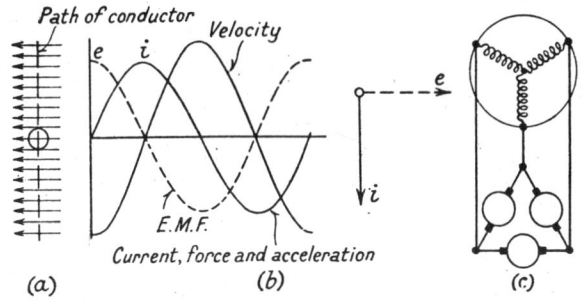

Fig. 221. Pertaining to the Kapp Vibrator

be arranged to give leading power factors, or, by adjustment, unity power factor, at small loads, whereas the expedor type cannot.

Kapp Vibrator. Consider the conductor lying in the uniform magnetic field in Fig. 221 (*a*). Let the conductor be supplied with an alternating current. The force of interaction will similarly alternate, and, if unrestrained, the conductor will move vertically with simple harmonic motion. As it does so, it cuts through the magnetic field and becomes the seat of an induced e.m.f. proportional to its velocity. The e.m.f. will be in phase quadrature to the current, as shown at (*b*), and the voltage applied to the wire to maintain the current against the e.m.f. will lag behind the current by 90°, neglecting the effects of friction, resistance and inductance. Thus the vibrating wire behaves like a capacitor, and is particularly suitable for low frequencies and voltages. The e.m.f. developed corresponds to E'_c in Fig. 218(b).

For practical reasons, the conductor is replaced by coils on an armature, and connection made to them by a commutator and slip-rings. The inclusion of three such devices with separately-excited or permanent fields in the slip-ring connections of an induction motor is the essential feature of the Kapp vibrator, Fig. 221 (c). For starting, the rotor of the main machine is disconnected from the vibrator and a rheostat introduced.

Leblanc or Scherbius Advancer. This comprises in its simplest form a d.c. armature. On the commutator are placed three brush-sets at

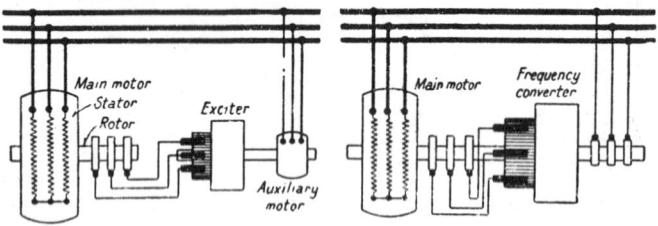

FIG. 222. LEBLANC OR SCHERBIUS.   FIG. 223. FREQUENCY-CONVERTER
ADVANCER   ADVANCER

a spacing of 120 electrical degrees. Let three-phase currents be supplied to the three brush-sets. If the armature is at rest, a rotating field of supply frequency is produced, and a back e.m.f. of the same frequency appears at the brushes. If now the armature be rotated, the relative speed of conductors and field will be changed—reduced if the direction of field and rotation are the same. When the speed is raised to the synchronous speed of the supply frequency, no relative motion exists between the conductors and the field, so that the induced e.m.f. vanishes and the current is maintained merely by a supply voltage equal to the IR drop. When the speed is further raised, and exceeds the synchronous speed, the relative movement of conductors in the field is reversed, and consequently also the e.m.f. which is developed across the brushes. The device therefore acts like a capacitor. It must be remembered that although the magnitude of the e.m.f. changes (at constant supply current it is proportional to the slip), its frequency is always precisely that of the supply on account of the function of the commutator. The device is applied as in Fig. 222. The "supply" frequency is the slip frequency of the main rotor, and the current is the main rotor current. The "synchronous" speed of the device is thus dependent on the slip of the main motor and will be low. The advancer is driven at a higher speed, and injects an e.m.f. proportional to the rotor current and always of the correct frequency, giving phase compensation in the manner described by Fig. 218. The advancer is cut out of circuit at starting until the main motor is running at normal speed.

Frequency-converter. This takes the form of an armature with d.c. type winding, commutator and slip-rings, connected mechanically to the main motor. Three-phase brush gear on the commutator is connected into the main motor rotor circuit, and an e.m.f. at slip-frequency is injected by energizing the advancer armature from the main supply through its slip-rings. The motion of the advancer at the speed of the main motor results in slip-frequency e.m.f.'s being tapped off at the commutator, of phase depending on the brush position. The arrangement is shown diagrammatically in Fig. 223.

PHASE-COMPENSATED INDUCTION MOTORS. The phase advancers briefly described above involve the use of a machine additional to the main induction motor, and although the latter may be a cheaper and more robust machine than a normal motor (e.g. a greater air-gap and increased leakage is quite permissible since the power factor is to be compensated), the use of an external advancer is generally confined to large units.

Successful attempts have been made to incorporate phase-advancing devices in small motors, known as *compensated induction motors*, automatic in action and requiring no special gear. Typical examples are the Torda and Osnos motors, the former under the trade name of "All-Watt," the latter as "No-Lag" and "Kosfi-Leading."

In the Torda motor a Leblanc exciter is incorporated in tunnel slots below the main rotor slots, having a different number of poles from the main winding to avoid interaction with the main flux. The commutator is connected to the d.c. type Leblanc winding, and the three-phase brush sets *e* ∼e connected to the slip-rings of the main rotor winding. The motor then operates as a Leblanc set like that in Fig. 222.

The Osnos motor combines the main motor with a frequency-converter. The primary winding of the induction motor is in this case put on the rotor, and connected to the main supply through slip-rings. An additional rotor winding is placed in slots below the main slots of the rotor, and connected to a commutator. The three brush-sets of the latter are joined to the stator (secondary winding). The main field is used to energize the advancer winding, which produces slip-frequency e.m.f.'s at the commutator brushes because the main field travels past the brushes at slip speed. Neglecting any changes in the magnitude of the main flux with load, the compensating e.m.f. is constant over the load range, so that the machine is over-compensated (i.e. takes a leading current) at small loads.

10. **Control of Speed and Power Factor.** In certain cases, speed and power-factor control are needed at the same time. The matter is generally one of providing a continuously variable speed with shunt

characteristic, resembling the torque/speed relation obtainable from a d.c. shunt motor with field rheostat.

For a restricted speed range, of say 30–40 per cent, an induction motor can be arranged with an auxiliary machine to give a speed range of say 15–20 per cent *above* and *below* synchronism. The auxiliary machine may be similar to the Leblanc exciter, but with the addition of a stator winding; or a frequency-converter; or a three-phase commutator motor.

If a wide speed range is required, such as 3 or 4 : 1, with continuous variation between these limits, it is generally necessary to dispense with the induction motor entirely, and to employ a three-phase commutator motor, from which a shunt characteristic can be obtained with phase compensation over the whole speed range.

For a proper understanding of these arrangements, often termed *variable-speed sets*, it is necessary to inquire into the function of the commutator and d.c. winding used with polyphase currents. Such an investigation is, however, outside the scope of the present book.

11. Induction Generator. The stator current of an induction motor connected to a constant-voltage constant-frequency supply consists of the magnetizing current and a component balancing the ampere-turns of the rotor. The relation is shown in Fig. 149. The rotor current I_2' flows in the short-circuited rotor circuit due to the e.m.f. $E_2' = sE_1$ induced therein by the mutual flux. Just below synchronous speed, E_2' is a small fraction of its standstill value, the rotor frequency is only a fraction of one cycle per second and the rotor current is small and nearly in phase with E_2'. At synchronous speed (supposing the rotor to be driven by external means) the rotor e.m.f. and current vanish. If the rotor speed be raised above synchronous speed, the rotor e.m.f. reappears, but in phase opposition to its subsynchronous position because the rotor conductors are moving *faster* than the stator rotating field, and the line-linkages of the rotor turns change in the reverse sense. The reversal of the rotor e.m.f. reverses also the rotor current, and the stator component $-I_2'$ reverses. The resultant stator current now consists of the magnetizing current as before, and a component in phase opposition to the stator applied voltage. The machine is consequently an *induction generator*.

EXTERNALLY-EXCITED INDUCTION GENERATOR. The induction generator differs from the synchronous generator in certain important respects. Its excitation is by means of the magnetizing current in the stator: this current is drawn from the supply of frequency f to which the machine is connected.

The excitation determines the flux and the frequency, and consequently the synchronous speed above which the machine must run to generate. For proper functioning, the induction generator must therefore be connected to some existing source of supply, to determine its frequency and to provide it with magnetizing voltamperes.

Fig. 224 shows the conditions (a) for subsynchronous and (b) for supersynchronous speeds. The normal subsynchronous conditions are similar to those already shown in Figs. 149 and 156. In the rotor, above synchronous speed, the conditions are electrically similar to those at subsynchronous speeds. As the speed is raised above synchronism, the slip increases negatively, the rotor frequency rises from zero and the rotor e.m.f. similarly increases. The rotor current is given as before by eq. (117), but with a reversed direction. The rotor current locus is consequently a complete circle. The diagram for the stator current is then given by the complete circles in Fig. 224 (a) and (b). Just as for the simple

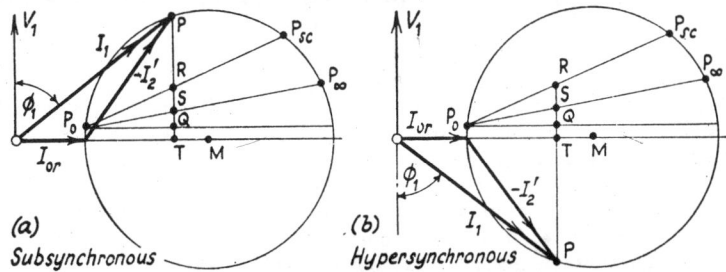

(a) Subsynchronous   (b) Hypersynchronous

FIG. 224. CURRENT-LOCUS OF INDUCTION MACHINE

circle diagram for motoring conditions, Fig. 224 (a) (compare Fig 158), PT represents the input, RS the rotor I^2R, SQ the stator I^2R, and QT the core, friction and windage losses, and PR the output; so RS, SQ, and QT have the same significance when the machine is generating, Fig. 224 (b). The electrical output as a generator is TP: the mechanical input is RP, to scale. The slip for maximum torque is the same as for motor conditions, eq. (119), page 273, taking the negative sign. The maximum generating torque is greater than the corresponding motor value: see eq. (120).

As a generator, the complexor E_1 is the terminal voltage of the machine (neglecting stator impedance drop), and the stator current I_1 is clearly a *leading* current of definite phase-angle ϕ_1. The output is determined by the circular locus, and cannot be arranged to provide a lagging load. This feature emphasizes the necessity for a.c. excitation by synchronous machinery.

The torque/slip characteristic of the machine above synchronism is very similar to that for normal running as a motor at subsynchronous speeds, as indicated in Figs. 162 and 163. Fig. 162 shows the current curves for generating conditions. If the prime mover develops a greater driving torque than the maximum countertorque, the speed rises into an unstable region and the slip increases. At some high value of slip, the generating effect ceases and the machine becomes a brake.

Induction generators need little auxiliary apparatus; are easily run in parallel without hunting, and at any frequency; speed variations of prime movers are comparatively unimportant; the excitation fails on sudden short-circuit, and with it the generator output, giving some self-protection; the rotor can have a robust cage winding. The drawbacks are the need for externally-supplied magnetizing volt-amperes, the short gap, and the moderate efficiency.

The induction generator can utilize waste heat in chemical works, or the power from small-outlying hydro plants. The prime-mover need not be governed, but the generator must then sustain 200-300 per cent overspeed. To minimize transients and heavy motoring currents, the set is brought to synchronous speed by the prime-mover and the stator connections closed to a 3-phase network. Speed may be observed by local- or remote-indicating stroboscope, the latter fed by a permanent-magnet shaft-end generator over pilot wires. Alternatively an automatic switch on the generator shaft may operate at 95 per cent synchronous speed to close the main stator switch.

SELF-EXCITED INDUCTION GENERATOR. The magnetization of an induction generator may be provided by a capacitor bank. No external a.c. supply is then necessary, but the frequency and generated voltage are affected by the speed, load and capacitor rating. If the load is inductive, its magnetic energy circulation must be dealt with by the capacitor, the induction machine being incapable of doing so.

Provided the rotor has sufficient remanent field of suitable polarity, the machine will self-excite. The operating conditions can then be elucidated in terms of the equivalent circuit (a) of Fig. 225.

On no load the capacitor current $I_c = V_1 \omega C = V_1/X_c$ must equal the magnetizing current $I_m = V_1/x_m$, the terminal voltage V_1 reaching the value $V_1 = I_c X_c = I_m X_c$. But V_1 is a function of I_m, so that stable operation requires the line $I_m X_c$ in (b) to cut the voltage/excitation characteristic. Alternatively, if V_1 is plotted as a function of the magnetizing susceptance $1/x_m$ as in (c), the operating point is fixed by the corresponding susceptance $1/X_c = \omega C$ of the capacitor. Clearly, self-excitation is possible only if the capacitor susceptance is at least as great as the minimum (unsaturated) magnetizing susceptance.

On load, the generated power $V_1 I_2' \cos\phi_2$ provides the load power and the loss $V_1^2[(1/R) + (1/r_m)]$. The reactive currents must sum to zero, so that

$$V_1[(1/X) + (1/x_m) + (I_2'/V_1)\sin\phi_2 - (1/X_c)] = 0,$$

which determines the capacitor for a given load and voltage. The load characteristics resemble those of a d.c. shunt generator, as suggested by Fig. 163. The load conditions are shown in Fig.

225 (d). The active component $I_2'\cos\phi_2$ of the stator current provides for I_l in r_m and I_R in the load conductance. The reactive component $I_2'\sin\phi_2$, together with I_x in the load susceptance and I_m for magnetization, are supplied as I_c by the capacitor.

Unwanted self-excitation with dangerous voltages may occur in induction generators working over long transmission lines if synchronous machines at the far end become disconnected; or in motors having power factor correction if the capacitors remain connected to the running machine when, for any reason, the supply to it is cut off.

EXAMPLE. The ideal short-circuit current of a 3-phase induction

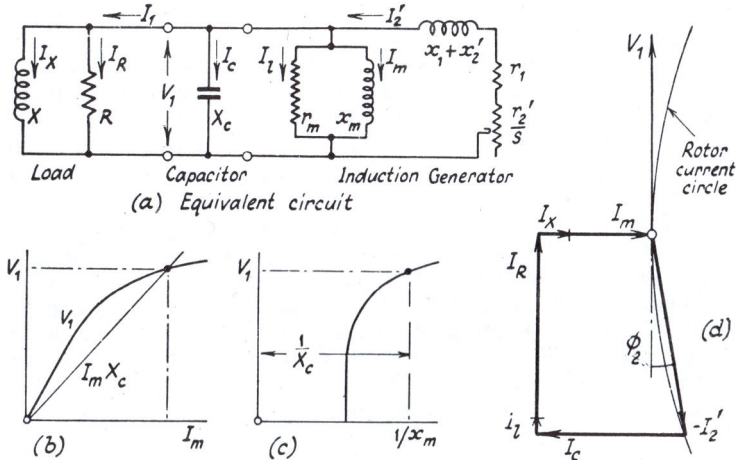

FIG. 225. SELF-EXCITED INDUCTION GENERATOR

machine is 750 A., and the core-loss current is 10 A., at 3 000 V. 50 c/s. The magnetizing current is—

Line voltage, V.:	1 450	2 500	3 000	3 400
Magnetizing current, A.:	20	40	60	90

What star-connected capacitance would enable the machine (i) just to self-excite, (ii) to generate 3 300 V. on no load? (iii) Determine the conditions for a load current of $125 - j20$ A. at 3 000 V.

Working per phase, the magnetizing susceptance is found:

Phase voltage V_1, V.:	840	1 440	1 730	1 960
Magnetizing current I_m, A.:	20	40	60	90
Mag. susceptance I_m/V_1, m℧:	24	28	35	46

(i) The minimum capacitive susceptance for self-excitation is 24 m℧, so that $C = 24 . 10^3/314 = 76\ \mu\mathrm{F}$ per phase at 50 c/s. (ii) Using the method of Fig. 225 (c), interpolation gives 43 m℧ for

3 300 V. line or 1 730 V. phase, whence $C = 137$ μF. (iii) The rotor current circle is drawn, as in Fig. 225 (d), with diameter 750 A. to scale. Then $I_m = 60$ A., $I_l = 10$ A., $I_R = 125$ A., $I_X = 20$ A. In the diagram the active generated current I_2' cos $\phi_2 = 125 + 10 = 135$ A.; and with $I_X + I_m = 20 + 60 = 80$ A., the capacitor current is 104 A., requiring $C = 104 . 10^6/314 . 1\,730 = 190$ μF. The diagrams $(b - d)$ in Fig. 225 are to scale for this example.

12. **Electric Braking.** Where it is necessary to brake a load connected to an induction motor, as in mine hoists and cranes, three electrical methods are possible. These have several features giving some advantage over purely mechanical braking, but in their application the duty-cycle of the motor may be made more severe.

REGENERATIVE BRAKING. This involves the use of the motor as an induction generator, for which it must be over-driven by the load at hypersynchronous speed, i.e. with negative slip. If a rheostat be connected across the slip-rings in the same way as for starting, the braking can be maintained up to the maximum safe limit of motor speed. The amount of rotor external resistance required is usually moderate, and the braking returns useful energy to the supply system. The braking fails when the speed falls to synchronous speed. With pole-changing motors the braking can be continued down to the neighbourhood of the lowest synchronous speed provided by the greatest number of poles.

PLUGGING. This is braking by reversal of the stator connections while the motor is running, causing the reversal of the stator rotating field. The slip is greater than unity and the machine operates purely as a brake. The running conditions are described in a current diagram by the region between P_{sc} and P_∞, Fig. 168. The rotor and stator currents tend to be large. Cage motors of 5 h.p. or less are plugged direct, using the star connection with star-delta started machines: larger motors require stator resistors. Slip-ring machines employ the rotor starting resistance for current limitation.

DYNAMIC BRAKING. The basis of this method is to excite the stator windings with direct current immediately after their disconnection from the three-phase supply. The d.c. excitation may be obtained from a rectifier, and the connection made automatically in the switchgear by the interconnection of the main and brake contactor circuits. With slip-ring motors, resistance can be introduced into the rotor circuit with advantage during the braking operation. The d.c. stator excitation will correspond to an alternating stator current I_1 (see Chapter XVIII, § 8), held constant. The flux produced depends on the ampere-turns which result from the combined action of I_1 and the rotor current I_2, and will generate a secondary e.m.f. which is a function of the flux and the speed n. Finally, I_2 is itself produced by the secondary induced e.m.f. acting through the secondary impedance. The braking effect will be settled by the amount of rotor I^2R loss.

Since the stator d.c. excitation will produce a stationary field, the rotation of the rotor at synchronous speed n_s produces conditions similar to those in a normally-excited induction motor at standstill. Again, when the d.c.-excited motor is at rest, there will be no induced e.m.f., a condition which resembles that under which a normal machine works when its rotor runs at synchronous speed. The operation of a motor may therefore be described by an equivalent circuit whose variables have the factor $S = 1 - s$, where s is the slip. Further, instead of there being a constant stator applied voltage, it is necessary to imagine a constant stator current, Fig. 226 (a).

If leakage reactance be neglected and the rotor impedance is taken as purely a resistance (augmented in practice by external resistors connected to the slip-rings, thus justifying the assumption),

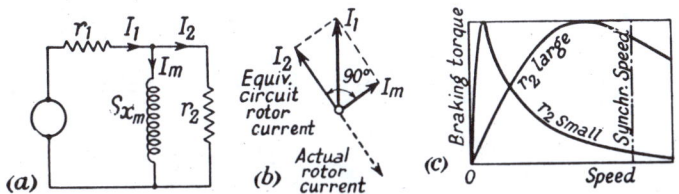

FIG. 226. DYNAMIC BRAKING

the following expressions hold* for the equivalent circuit: the complexor $I_m = I_1 - I_2'$, and from the diagram in Fig. 226 (b), $I_m{}^2 = I_1{}^2 - I_2{}^2$ numerically. I_m produces a flux which in turn generates an e.m.f. in the rotor which can be expressed as

$$E_2 = I_m x_m (n/n_s) = I_m x_m (1 - s) = S I_m x_m,$$

where x_m is the magnetizing reactance chosen to relate magnetizing current and rotor e.m.f. The expression for E_2 and that for I_m lead to the equivalent circuit in Fig. 226 (a), where leakages are neglected. Resolving $S x_m$ and r_2 into their series equivalent gives the resistance term

$$r_{2S} = r_2 (S x_m)^2 / [r_2{}^2 + (S x_m)^2]$$

from which the rotor $I^2 R$ loss per phase for any speed having the fraction S of synchronous speed is $I_1{}^2 r_{2S}$ The braking torque corresponds to $I_1{}^2 r_{2S}/S$ synchronous watts per phase. If the torque/speed curve be analysed as for an induction motor, it is found that the torque is a maximum when $S = r_2/x_m$ and that this maximum depends only on x_m and the stator current. The braking-torque/*speed* curve resembles that of an induction motor torque/*slip* curve, as shown in Fig. 226 (c) for two values of rotor resistance. For dynamic braking, the average torque is of importance, and this can be shown to be greatest when $r_2 = \frac{1}{2} x_m$.

* Based on the treatment by Millar, *Elec. Review*, 133, p. 887 (1943).

One of the chief difficulties in applying the brief analysis given lies in the choice of x_m: it is strongly affected by saturation. The net magnetizing current will be greatest at zero speed, for then there is no rotor current and the stator current becomes in effect all magnetizing: under these conditions the flux and saturation will be greatest, and x_m is bound to be considerably reduced in consequence. At higher values of speed the value of x_m will rise.

It should be noted that the results above are only an approximate indication of the behaviour of the machine because of the neglect of leakage and of the variations in x_m.*

With d.c. excitation equivalent to full-load stator current, the

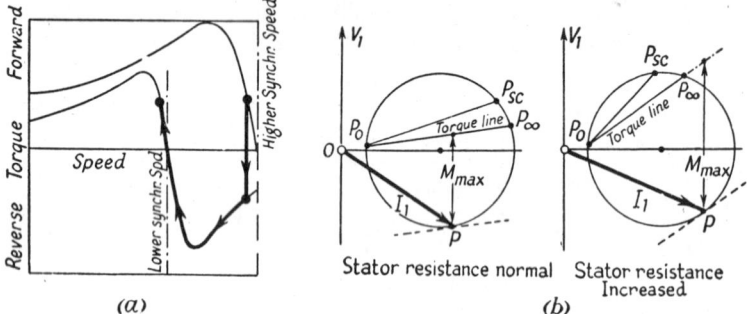

FIG. 227. MOTOR BRAKING CONTROL

peak torque will be about 60 per cent full-load torque for a cage and 100 per cent for a slip-ring machine. Excitation up to $2\frac{1}{2}$ times this value may be used.

The dynamic method gives a smooth stop, less rapid than with plugging, but with less rotor loss and with no tendency to reverse.

OTHER METHODS. A motor may be stopped by applying a friction brake. A variant of this simple method uses a spring-loaded brake with a thrustor. The latter, connected across the rotor slip-rings, develops a push-off thrust inversely proportional to rotor speed. Thus a steady *creep speed* can be obtained (e.g. for mine winders), for a speed rise lowers the rotor frequency and voltage, reduces the thrust, allows the brake to tighten and controls the creep. See also § 9.

BRAKING CONDITIONS. Most induction motors require only occasional starting and stopping, but certain loads such as mills and cranes driven by multispeed pole-changing motors may be retarded by switching from the high-speed to the low-speed connection. In the latter case a negative or braking torque is developed until the machine has dropped to the low-speed synchronous value. The

---

* For a more exact analysis, see Harrison, "The Dynamic Braking of Induction Motors," *Proc. I.E.E.*, 103 (A), p. 121 (1956).

circle diagram in Fig. 224 (*b*), the torque/slip curve in Fig. 163, and the discussion leading to eq. (120) *et seq.*, all indicate that the peak reverse torque may be considerably higher than that for normal motor action. Fig. 227 (*a*) indicates that, for changing down the speed of a two-speed pole-changing motor from fast to slow, the torque will be reversed twice in rapid succession, passing through a severe reverse peak. A duty requiring frequent repetition of these conditions calls for care in design and control. All parts subject to reversing stresses must be liberally dimensioned and slackness avoided, e.g. by using taper shafts for couplings and welded instead of keyed laminations. The stator windings must be well braced and particular attention paid to the mechanical design of cage rotor bars and end-rings. If the motor can be allowed to retard down to its lower synchronous speed before re-connection, much of the stress reversal can be avoided. A simple time-interval is not always satisfactory in that the retardation-time depends on the load. It is to be noted that the inclusion of resistance in the stator connections to mitigate the reverse torque is actually deleterious, for the peak reverse torque is markedly increased thereby. This is shown by eq. (120) and by the circle diagrams in Fig. 227 (*b*).

In crane and mill drives, where the attached load may have considerable inertia, low-speed motors of small diameter may be chosen so that the inertia is not unduly increased by the motor itself. The motors have in consequence inherently high reactance and low pull-out torque, which is offset by the use of Class B insulation and the higher temperature rise of 75° C. The magnetizing current of mill motors tends to be high also because the facility of an easily-repaired stator winding necessitates open slots.

The heating of an induction motor during acceleration is proportional to the stored kinetic energy, and in changing down the energy returned will, if the motor is connected during retardation, be partly expended in the motor resistance. The loss due to reversal by *plugging* is four times that due to acceleration only. Since such losses due to the kinetic energy of the load and the motor rise as the square of the speed, the use of low-speed machines is essential.

13. Operation on Unbalanced Voltage. If the voltages applied to the terminals of a three-phase induction motor are unbalanced, the motor performance will be modified. The most direct approach to the problem is by the analysis of the set of unbalanced voltages into two symmetrical sets of balanced voltages of positive and negative phase-sequence respectively. Each set produces corresponding balanced currents, and the synthesis of the two sets of current vectors represents the actual currents produced in the three stator phases by the original unbalanced voltages. The behaviour of the machine to the positive-sequence voltage is normal. The negative-sequence voltages, however, set up a reverse rotating field so that if the rotor slip is s with respect to the normal sequence, it will be $(2 - s)$ to the negative sequence. The motor behaves as the addition of two separate motors, one running at slip s with a terminal voltage of V_+ per phase, the other with a slip $(2 -- s)$ and a terminal voltage V_-. The power output will be

$$P_m = I_{2+}^2 r_2 (1 - s)/s - I_{2-}^2 r_2 (1 - s)/(2 - s) \text{ W. per phase,}$$

and the torque

$$M = r_2 [I_{2+}^2/s - I_{2-}^2/(2 - s)] \text{ syn. W. per phase,}$$

where the positive- and negative-sequence currents are functions of

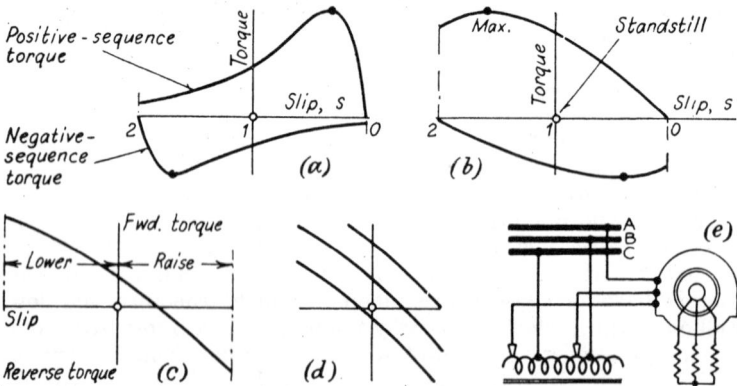

Fig. 228A. Negative-sequence Hoist Control

their sequence voltages, the motor constants r_1, r_2', x_1 and x_2', and the slip s. Thus I_{2+} is obtained from eq. (117) using s and V_+, while for I_{2-} the same equation is used substituting $(2 - s)$ for s and writing V_- for the applied voltage.

MOTOR CONTROL BY NEGATIVE-SEQUENCE TORQUE. Moderate voltage-unbalance affects a machine principally by increasing its I^2R loss, as the negative-sequence counter torque in the forward direction is small: see (a) of Fig. 228A. If with a slip-ring machine the rotor is heavily loaded on resistance, the torque/slip characteristics (b) are considerably modified, and the net torque (c) approaches a linear relation with slip. This method of control is useful in crane and hoist operation to obtain a stable *raising* torque to hold a load when it is being *lowered*. Varying degrees of unbalance give a series of nearly-linear characteristics (d) for close control, thus avoiding lowering on friction brakes.

Unbalanced voltages can be obtained, for example, as shown at (e). The stator winding is connected to line A, and to two points on an auto-transformer connected across lines B and C. A wide range of relative negative-sequence torque, up to single-phasing, can be obtained.

SINGLE-PHASING. This is the result of the disconnection of a phase of a polyphase induction motor when the machine is running. The most usual cause is the blowing of a fuse in one phase; less often, a defective contact in the "run" side of a star-delta or auto-transformer switch. Should single-phasing occur, a fully-loaded three-phase motor will be on the verge of stalling. If it does not

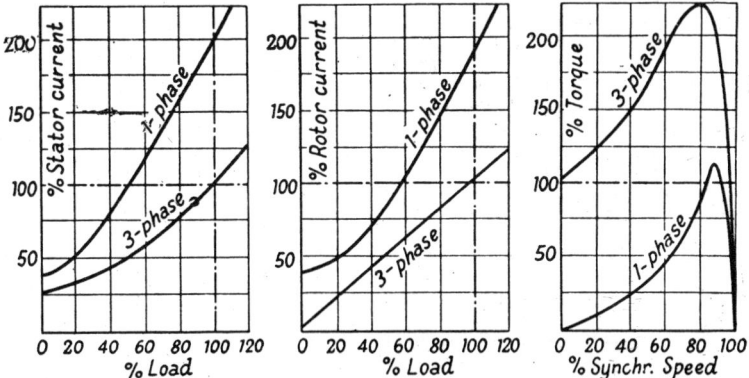

FIG. 228B PERFORMANCE UNDER NORMAL AND "SINGLE-PHASING" CONDITIONS

stall and is carrying more than half load, both stator and rotor will be seriously overheated. The temperature-rise in the stator will be most marked in two phases if star-connected, or in one if mesh-connected. Fig. 228B shows a typical comparison for a machine when running normally and when single-phasing.

Deliberate single-phasing to produce negative-sequence braking is sometimes employed for light duty, such as in the operation of large lathes.

14. Induction Regulators. Induction regulators are used for varying the voltage of an a.c. supply, or for maintaining a constant voltage under varying load conditions. The constructional features closely resemble those of slip-ring induction motors, especially in the three-phase regulator. Since the movable member is not required to rotate like the rotor of a motor, the shaft is generally arranged vertically for convenience: it is brought out to a pinion or gear-wheel engaging with a worm, which provides a means for adjusting the rotor position over (usually) a pole-pitch of arc.

Consider a 3-phase slip-ring induction motor at standstill on open circuit. If the primary (stator) be energized, a magnetic flux is produced which induces e.m.f.'s in the rotor windings. The magnitude of the slip-ring e.m.f.'s depends on the ratio of primary to secondary turns, and their phase with respect to the stator applied voltage is determined by the relative position of the rotor and stator windings. If the e.m.f.'s developed by the rotor secondary are injected into the supply, its voltage may be raised or lowered.

SINGLE-PHASE REGULATORS. Fig. 229 shows diagrammatically the cross-section of a bipolar single-phase induction regulator, and

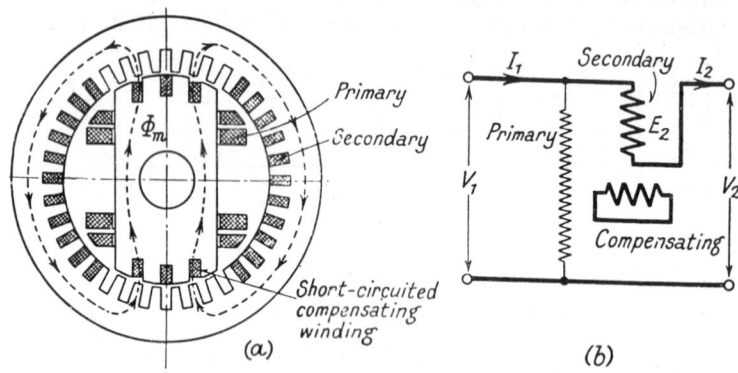

FIG. 229. SINGLE-PHASE INDUCTION REGULATOR

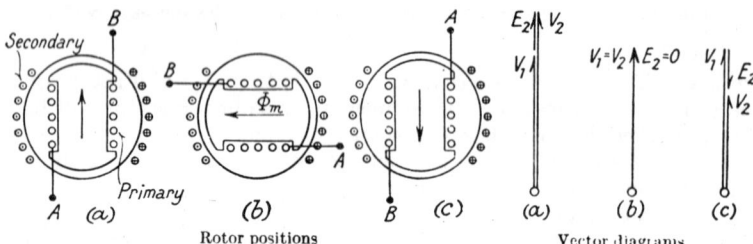

FIG. 230. EFFECT OF ROTOR DISPLACEMENT IN SINGLE-PHASE REGULATOR

its connections. The primary winding is, for convenience, placed on the rotary member and connected across the single-phase supply. The alternating flux Φ_m induces in the secondary, on the stator, an e.m.f. E_2 which varies in accordance with the relative positions of the stator and rotor axes. Fig. 229 (a) corresponds to maximum secondary e.m.f. In Fig. 230 (a) the e.m.f. E_2 provides a maximum increase, or *boost*. A 90° movement of the rotor reduces the line-linkages of the secondary with the primary flux Φ_m to zero—i.e. the mutual inductance vanishes. A further 90° movement again brings a maximum secondary e.m.f., but in a direction such as to reduce or *buck* the resultant supply-circuit voltage. In the complexor diagrams, V_1 refers to the supply voltage, V_2 the output voltage, and E_2 the e.m.f. induced in the secondary by the flux Φ_m. In Fig. 229 (b) I_2 is the load current and I_1 the supply current, which is the complex sum of I_2 and I_0, the magnetizing and core-loss current. The phase of I_2 with respect to V_2 depends of course on the load connected to the output side.

The single-phase regulator is always provided with a short-circuited rotor winding on an axis in quadrature with the primary. If this winding were not provided, the leakage reactance of the stator (secondary) winding would be large when the rotor was in or near the position of zero boost, Fig. 230 (b), so that, for a given

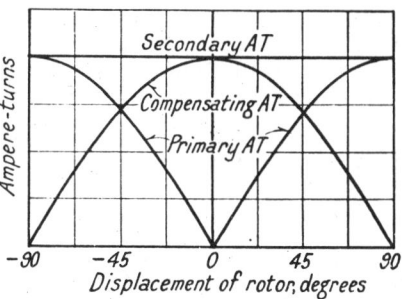

Fig. 231A. Variation of Ampere-turns in Single-phase Regulator

line current, a large voltage would be necessary to overcome the reactance drop. The series secondary winding would operate as a reactor and the core would be saturated. The short-circuited winding

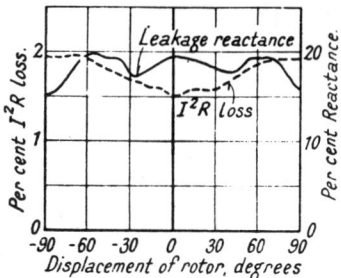

Fig. 231B. Leakage Reactance and I^2R Loss

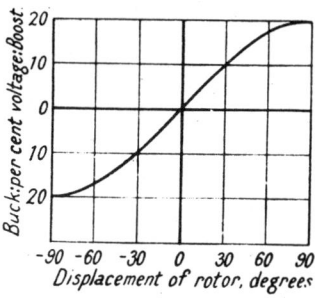

Fig. 231C. Voltage Regulation

has, in this position, maximum mutual inductance with the secondary and the effective reactance thereof is consequently reduced. The compensating winding tends to keep the I^2R losses constant for a given load current and various degrees of boost or buck. The variation of primary and short-circuit ampere-turns with angular displacement of the rotor is shown in Fig. 231A. Since the losses vary as the sum of the squares of the currents, the I^2R loss remains constant for any rotor angle, assuming that the primary and short-circuited windings have the same effective resistance (which is approximately the case in practice). Fig. 231B gives curves typical of leakage reactance and I^2R loss in relation to rotor angle, showing that both are roughly constant.

It is not difficult to appreciate that, with uniformly-distributed windings and a sinusoidal flux, the degree of boost or buck is proportional to the sine of the angle of displacement. Fig. 231c gives a regulation/rotor-angle curve for a single-phase regulator designed for $\pm$ 20 per cent variation.

THREE-PHASE REGULATORS. The electric and magnetic circuits of the three-phase induction regulator are almost exactly similar

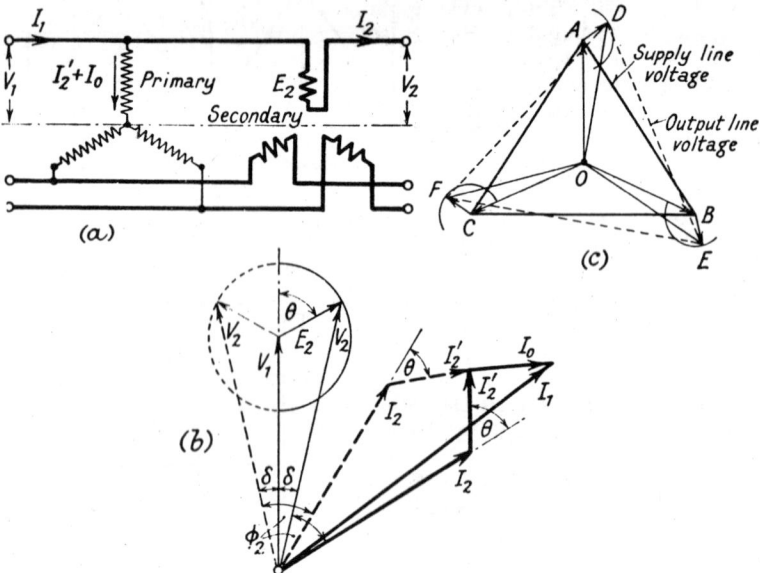

FIG. 232. THREE-PHASE INDUCTION REGULATOR

to those of the slip-ring induction motor. It is not, however, usual to employ slip-rings, since the movement of the rotor is limited: the general practice is to use flexible connections to the rotor which is used as the primary winding. The connections are shown in Fig. 232 (a).

Variation in output voltage is obtained by altering the angular position of the rotor, causing a phase-shift but not a change in magnitude of the secondary induced e.m.f.'s.

If V_1 is the supply voltage, E_2 the e.m.f. induced in the secondary, and V_2 the output voltage, each per phase, the voltage diagram in Fig. 232 (b) consists of the constant V_1 with the addition of E_2 at an angle θ corresponding to the rotor position to give the output voltage V_2. The locus of V_2 is consequently the circle drawn with centre on the extremity of V_1 and of radius E_2. The limiting voltage regulation may be obtained along either half of the locus, the full- or the dotted-line portions, depending on the direction in which the rotor is moved, i.e. either against or with the rotating primary field.

The output current I_2 has a phase displacement ϕ_2 with the voltage V_3, as determined by the load. The corresponding primary current is I_2'. With T_1 primary and T_2 secondary turns per phase,

$$I_2'/I_2 = T_2/T_1 = E_2/V_1 \qquad . \qquad . \qquad . \ (149)$$

neglecting impedance drops. Neglecting losses also, $V_1 I_2'$ cos γ = $E_2 I_2$ cos ψ, where γ is the angle between the primary voltage and the current-component I_2', and ψ is the angle between E_2 and I_2.

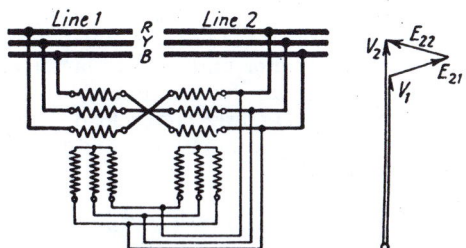

FIG. 233. TWIN THREE-PHASE REGULATOR

The right-hand side of the equation represents the output of the regulator to the line by transformer action from the primary. From eq. (149) it is clear that $\gamma = \psi$, so that the relative phase-positions of I_2 and I_2' are known. The primary has a further component I_0 for magnetizing current and core loss, so that the total primary current I_1 is the vector sum of I_2' and I_0. The current taken from the supply is I_1.

The full-line complexor diagram in Fig. 232 (b) refers to the full-line locus. Comparison with the case of the opposite rotor movement (dotted locus) shows that the *total* supply current is the same in both cases. On the other hand, the *primary* current, the complex sum of I_2' and I_0, is greater for the dotted locus. Hence it is usual to arrange the rotor movement to secure the lesser primary current.

The discussion above refers to one phase of the regulator. The complexor diagram for all three phases is combined in Fig. 232 (c), where ABC, DEF are the input and output voltages respectively.

The supply and output voltages are only in phase at maximum boost and buck. For all other positions there is an angle δ (Fig. 232 (b)) which has a maximum value of arc tan (E_2/V_1). This is unimportant except in certain cases of interconnection between systems where such a phase displacement would give rise to large circulating currents. The displacement can be avoided by using either a twin regulator or a bank of single-phase units.

A twin regulator comprises two coupled three-phase regulators each of half the total required boost kVA. The secondary windings are connected in series and the rotors have opposite phase rotation, so that the resultant output voltage V_2 is always exactly in phase with the supply voltage V_1. The primary windings are connected in parallel in order to share the same voltage and because the primary currents will not be equal, as already shown in connection with Fig. 232 (b). The opposite phase rotation of the two members of a twin regulator makes it advantageous to couple the two mechanically and to move them together, since the induction-motor torques produced are opposed and so neutralize to a considerable extent. In single-unit three-phase regulators the torque has to be taken up by the control devices.

The torque depends on the rotor position, the power factor of the load, and the load current. For unity power-factor on the output side, the torque is a maximum for maximum boost or buck, and zero for $\theta = 90°$: for a zero-power-factor load, it is a maximum at $\theta = 90°$, and zero at the maximum boost and buck positions. In general the torque is given by

$$M = kV_1I_2' \cos (\theta \pm \phi_2).$$

When two three-phase regulators are coupled to form a twin set the effective torque is proportional to $\cos (\theta + \phi_2) - \cos (\theta - \phi_2)$. Thus for a unity-power-factor load the torques always cancel completely. For zero power factor, the neutralization vanishes for $\theta = 90°$, i.e. no boost or buck, and the torque becomes equal to twice that of either regulator alone.

A connection diagram for a twin regulator is shown in Fig. 233. Where the line voltage is too high for direct connection (e.g. > 11 000 V.), it is customary to use transformers to supply the regulator primary and to inject the secondary e.m.f. into the main circuit.

The *rating* of a regulator is the product of its full-load output current, voltage boost, and number of phases. Thus a regulator working on a 6 600 V. three-phase system in which full-load line current is 200 A., to provide a $\pm$ 10 per cent variation, is

$$\sqrt{3} \cdot 6\,600 \cdot \frac{10}{100} \cdot 200 \cdot 10^{-3} = 228 \text{ kVA.}$$

Regulators are productive of noise due to vibration. An improvement is made when, as is the common practice, the regulator is immersed in oil in a tank provided with cooling tubes closely resembling that of a transformer.

15. **Electromagnetic Pumps.** The high conductivity of liquid metals (such as sodium, sodium-potassium alloy and bismuth, employed as coolants in nuclear reactors) permits the use of electromagnetic

pumps. These exploit the electromagnetically-generated pressure within the liquid itself when carrying current in the presence of a magnetic field. Although possibly of lower efficiency than conventional mechanical types, electromagnetic pumps are generally smaller, simpler and more readily connected into the pipework. The obvious method is that based on the "mercury-motor" coulomb-meter formerly used for the metering of domestic d.c. supplies. But current may also be *induced* into the liquid, so that the induction-motor principle is also applicable.

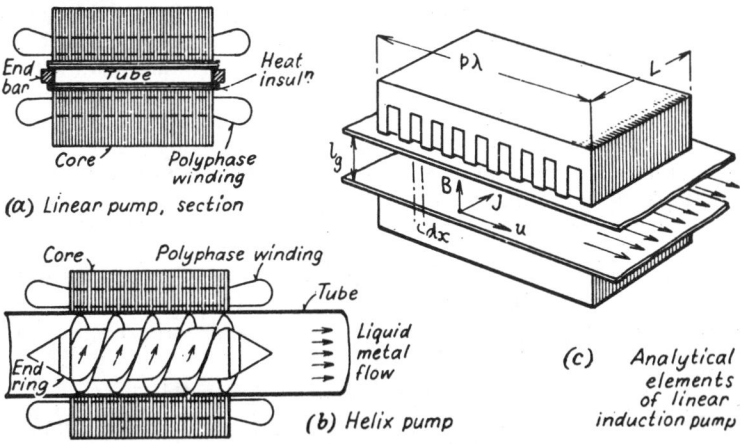

(a) Linear pump, section

(b) Helix pump

(c) Analytical elements of linear induction pump

FIG. 234. INDUCTION PUMPS

INDUCTION PUMPS. A *flat linear* pump, Fig. 234, has a wound stator opposing a second core (which may also be wound) across a gap in which a flat rectangular-section tube carries the liquid metal. Side bars of copper along the narrow edges of the tube walls perform the same function as the end-rings of a normal induction motor, the liquid metal forming the "bars" of the "cage." An *annular* version of the linear pump has a central magnetic core, within an outer core (carrying windings) in the form of separate rings spaced along the tube. The liquid metal occupies the annular gap between inner and outer cores.

The *helical* pump more closely resembles an induction motor, both in appearance and in operation. The rotating field produced by a substantially normal stator winding endows the liquid metal with circumferential motion, and the helical guide develops therefrom an axial component of velocity.*

THEORY OF THE LINEAR PUMP. The operating principles of the ideal linear pump can be based on Fig. 110 (c) and the theory of the

* Blake, "Conduction and Induction Pumps for Liquid Metals," *Proc. I.E.E.*, 104 (A), p. 49 (1957).

cage motor. The system of Fig. 234 (c) shows the elements of a pump with a stator of width L and length $p\lambda$ (corresponding to L and $p \cdot 2Y = \pi D$ in a normal induction motor). The stator has a series of pole-pitches $\frac{1}{2}\lambda$ (corresponding to Y) so that the flux is a travelling wave

$$\Phi_x = \Phi_m \cos(\omega t - 2\pi x/\lambda) = \Phi_m \cos(\omega t - \beta x)$$

at a point x, while the gap flux density is $B_x = B_m \sin(\omega t - \beta x)$ with $B_m = \Phi_m(2\pi/\lambda L)$. The flux wave travels at a synchronous velocity $u_1 = f\lambda$ when the three-phase windings are energized at frequency f. The liquid moves in the same direction at velocity u with a slip $s = (u_1 - u)/u_1$, and the rate of flow is $Q = uLl_g = u_1(1 - s)Ll_g$ m³ per sec. The e.m.f. induced in the liquid in direction L is $e = B_xL(u_1 - u) = B_xLsu_1$: it produces a current-density $J = B_xsu_1/\rho$, where ρ is the resistivity of the liquid. This neglects end-bar resistance and minor leakage reactances.

In an axial elemental length dx the electromagnetic force on the liquid is $dF = B_xJLdx$: i.e.

$$dF = (Lsu_1/\rho)B_m^2 \sin^2(\omega t - \beta x)$$

at any instant. The summation over a wavelength (or pole-pair) gives for the force the average value $F = \frac{1}{2}\lambda Lsu_1B_m^2/\rho$, so that the mechanical output per wavelength is the product pressure × flow-rate

$$P_m = (F/L)Q = [\frac{1}{2}Ll_g\lambda u_1^2 B_m^2/\rho]s(1 - s) = s(1 - s)P_2.$$

The loss in the liquid is the average $J^2\rho$ in each unit of volume, giving per wavelength $I^2R = s^2P_2$. The output is a maximum for $s = 0.5$. The efficiency is $P_m/(P_m + I^2R) = (1 - s)$, which is to be expected in an ideal induction machine.

The stator input is a magnetizing current, together with a component balancing the current induced in the liquid. In a practical pump there are core losses, I^2R losses in the tube walls and a loss of output in hydraulic friction. More important are the end effects. The flux cannot have the postulated form near the axial extremities of the core, and the end slots must carry a winding graded to "match" the transition from the region external to that within the core. If this is not done, considerable I^2R losses occur in the liquid near the ends, with serious reduction in efficiency.

CHAPTER XIV

INDUCTION MOTORS: TESTING

1. **Objects of Tests.** A manufacturer has his motors tested to obtain a brief résumé of the motor characteristics, in order that its guarantees may be satisfied. As most motors are now produced in stock lines, only a few machines may receive anything like a complete test: others will only have inspection for electrical and mechanical soundness. Where new types are developed or changes made in the design, a more detailed investigation is fruitful, for the wise designer bases his judgments on test results.

For the investigation of electromagnetic phenomena and for an insight into the behaviour of a machine, many more or less elaborate tests can be applied. Certain of these are included below.

2. **D.C. Resistance.** The resistance of all available circuits is checked by circulating a suitable direct current and measuring the voltage drop between terminals. The test, if taken with the motor cold (at room temperature) provides a check for the calculated values, or a basis for the estimation of efficiency. (See § 7.)

3. **Voltage-ratio or Open-circuit Test.** This is naturally applicable only to slip-ring motors, and consists in applying normal voltage at normal frequency to the stator winding, and measuring the voltage between the three rings taken two at a time, to confirm the calculated figure, and to check for balance of phases and connections. The power and current input to the stator are noted as a further check, for if low or excessive values are observed, some defect is probable, or the motor is being energized at the wrong voltage or frequency.

The phase voltages V_1 and V_2 in this test will not have the same ratio as the numbers of effective turns per phase, on account of the primary leakage. There may be a difference of a few per cent. An estimate of the turn-ratio may be made if a supply voltage V_2 is applied to the rotor slip-rings and the stator voltage v_1 be measured across its terminals. It will be found that $v_1 < V_1$. The turn-ratio (or ratio of effective turns if the stator and rotor winding factors are not identical) is approximately given by

$$T_1/T_2 = \tfrac{1}{2}(V_1 + v_1)/V_2.$$

4. **Running-up Test.** As has been seen (Chapter XIII, § 6), cage motors are liable to harmonic torques, productive of crawling. A running-up test is made on cage motors to ensure that they are capable of starting against a reasonable load torque without failing to attain normal running speed. The test also reveals noisy running and the presence of loose bars. The load torque applied depends

upon the rating and size of the motor and its method of starting. Thus a continuously-rated motor for star-delta starting might be expected to start with 25–33 per cent of full-load torque at 175 per cent of full-load current in the line.

5. **No-load Test.** One of the most informative tests is the no-load test, which gives the core and pulsation loss, friction and windage loss, magnetizing current and no-load power factor. Further, any mechanical unbalance, noise, faulty connection, etc., are revealed.

The stator connections are made to a supply of normal frequency and variable voltage, and instruments are included to measure the voltage, input power and current. After having been started, the motor is run with its rotor in the normal running condition, i.e. short-circuited, and with the brush-gear raised in the case of slip-ring motors with this equipment.

When the motor has run long enough for its bearings to show distress if faulty, the applied voltage is raised to about 20 per cent over normal, and input power and current observed. The slip is measured. It is not sufficient to measure the *speed*, for the slip is very small and cannot be accurately found from the difference between running and synchronous speeds.* The readings are taken at lower values of voltage down to that at which the current starts again to rise.

Fig. 235 shows curves typical of those drawn from no-load test data. At normal voltage the current I_0 is one-quarter to one-third of normal full-load current. The power factor is low (e.g. 0·15) since the two components I_{0r} and I_{0a} of I_0—for magnetization and losses respectively—differ considerably in magnitude. Usually I_{0r} is 5–10 times as large as I_{0a}, which is the active component for core and mechanical losses. As the voltage is reduced, the power and current fall because the main flux (which is roughly proportional to V_1 at first) reduces simultaneously. The power curve is nearly parabolic at voltages near normal since the core losses are proportional to the square of the flux density and therefore of the voltage. The power factor rises since I_{0r} falls faster than I_{0a}. The latter may actually increase, for the mechanical losses, which are practically unchanged, require a greater current component to satisfy them at the lower voltage.

When the voltage has been reduced to, say, 20 per cent of normal, the magnetizing current is small and so is the core loss. The speed has fallen only by a few per cent, however, and the friction and windage loss is still maintained, calling for a considerable active current component to counterbalance the corresponding small but growing rotor current, which interacts with the reduced flux to keep the machine running. The power factor thus rises, and the slip must be greater to permit the rotor e.m.f. to increase and circulate a

---

* For notes on the measurement of slip, see Parker Smith, *Electrical Engineering Laboratory Manual—Machines*, p. 28 (Oxford).

larger torque current. The power is now almost entirely for the mechanical losses, and if the curve be extrapolated to zero voltage, the intercept represents the mechanical loss. (If there is any doubt about the extrapolation, the P_0 curve should be plotted to base V^2: the relation will be found almost rectilinear, and the extrapolation becomes simply the extension of the line to meet the power axis at $V^2 = 0$.)

If the voltage is reduced still further, the torque is maintained only by increases in slip and current, whence the minimum value in the current curve. Eventually conditions become unstable and the motor stops.

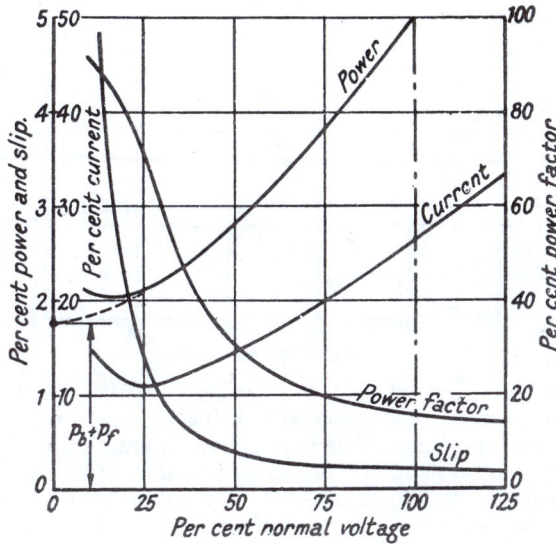

Fig. 235. No-load Test Curves

The power curve actually includes a small I^2R loss. If the no-load current at normal voltage is about one-third of the full-load current, the I^2R due to it will be only one-ninth of normal, and may generally be neglected. If the stator resistance is known, it is, however, quite simple to correct for it.

6. Short-circuit (Locked Rotor) Test. This is analogous to the short-circuit test of a transformer. The rotor is held stationary and short-circuited under its normal running condition. The test consequently reveals no mechanical defects, but is of importance as furnishing the short-circuit current and power factor which, with the no-load current and power factor, enables the current diagram to be drawn. In addition the I^2R losses measured by the test are necessary for the estimation of efficiency by loss-summation.

The stator is supplied with a low voltage of normal frequency, to avoid excessive currents. The position in which the rotor is clamped may affect the current. If so, the variations are noted when the rotor is locked in various positions and a mean position found. Alternatively the rotor may be allowed to rotate very slowly during the progress of the test.

The voltage is raised in steps, with readings of current and power input, until the current reaches not more than twice normal. The readings are taken quickly to avoid overheating.

Fig. 236 gives typical test results, the input power, power factor,

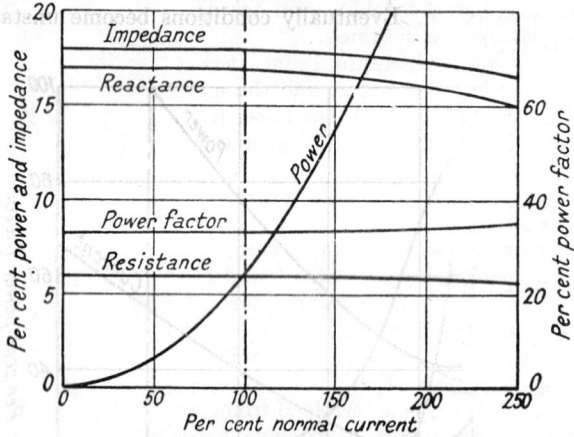

FIG. 236. SHORT-CIRCUIT TEST CURVES

impedance, resistance and reactance being evaluated and plotted to a base of stator current. The power curve is practically a parabola, since the input on short circuit is almost entirely I^2R loss. The main flux is reduced to about one-half of its no-load value on short circuit at normal voltage, and considerably less at the usual short-circuit test voltage: the core loss due to it is reduced still further. Although the leakage is greater, the core loss as a whole will thus be quite small. In particular the pulsation losses naturally vanish, and the mechanical losses are absent.

The impedance is seen to fall with the higher currents. This is due to the reduced leakage reactance consequent upon the saturation of the teeth. For the same reason the parasitic eddy-current losses in the conductors are somewhat reduced, so that the effective resistance may also fall. Thus at the higher current values the current is no longer proportional to the applied voltage, but increases more rapidly at a power factor probably higher than at currents in the region of full-load magnitude.

For deep-bar and double-cage rotor machines, the effective value of x_2', the rotor "standstill" reactance, increases at normal speed above its actual standstill value due to the redistribution of the rotor current in a more uniform fashion. For accurate performance calculations the motor reactance should be determined at low secondary frequency. The locked-rotor test may be repeated at, say, 10 cycles per sec. for a normal-frequency machine. If the reactance is now calculated from the test results, it is multiplied by the ratio (rated/test) frequency, the resulting figure being the effective standstill reactance x_2' corresponding to running conditions.

If a detailed investigation is not required, the data from the tests described in §§ 2, 3, 4, 5, and 6 furnish all that is needed to confirm the soundness of a machine and to allow its full-load operating characteristics to be obtained.

7. Efficiency and Operating Data. It is not convenient to perform a load test on a large motor, and the method of loss-summation is allowed by B.S. 269 : 1927. The losses to be reckoned are—

Fixed loss : (a) core loss; (b) bearing friction; (c) total windage; (d) brush friction.

Direct load loss : I^2R loss (e) in stator winding, (f) in rotor winding, (g) brush loss.

Stray load loss : stray loss (h) in core, (j) in conductors.

Losses (a), (b), (c) and (d) are found by deducting the small no-load I^2R loss from the total measured loss in the no-load test. Losses (e) and (f) for slip-ring motors are calculated from the currents (obtained from the circle diagram) and resistances of the windings. For cage rotors, loss (f) with the small part of (h) and (j) appertaining to the rotor, is determined from the expression—

Rotor loss = (b.h.p. + windage + friction) $\times s/(1 - s)$.

Loss (g) for all brushes together is taken as the slip-ring current $\times$ 1 V. The stray losses, which are not susceptible to exact determination, are covered by deducting 0·5 per cent from the efficiency as calculated on full-load, and *pro rata* for other loads.

The circle diagram is an essential feature of this method, and at the same time it gives the full-load current, power factor and slip, and the maximum output.

EXAMPLE. The following test data refer to a 30-h.p., three-phase, mesh-connected, 500-V., four-pole, 50-cycle, cage induction motor—

No-load Test.

Line Voltage V.	Line Current A.	Wattmeters, kW.	
		1	2
564	9·8	− 2·07	3·69
515	8·6	− 1·55	3·05
480	7·8	− 1·22	2·62
320	5·4	− 0·25	1·40
160	4·5	+ 0·28	0·69
75	7·9	+ 0·36	0·60

Short-circuit Test.

Line Voltage V.	Line Current A.	Wattmeters, kW.	
		1	2
60	18·6	− 0·27	0·83
100	32·0	− 0·75	2·35
140	44·8	− 1·50	4·65

The stator resistance (hot) is 0·68 Ω per phase. Separate the principal losses, and find the full-load current, power-factor; efficiency and slip; the maximum torque and power output.

The wattmeter readings are added algebraically to give the total

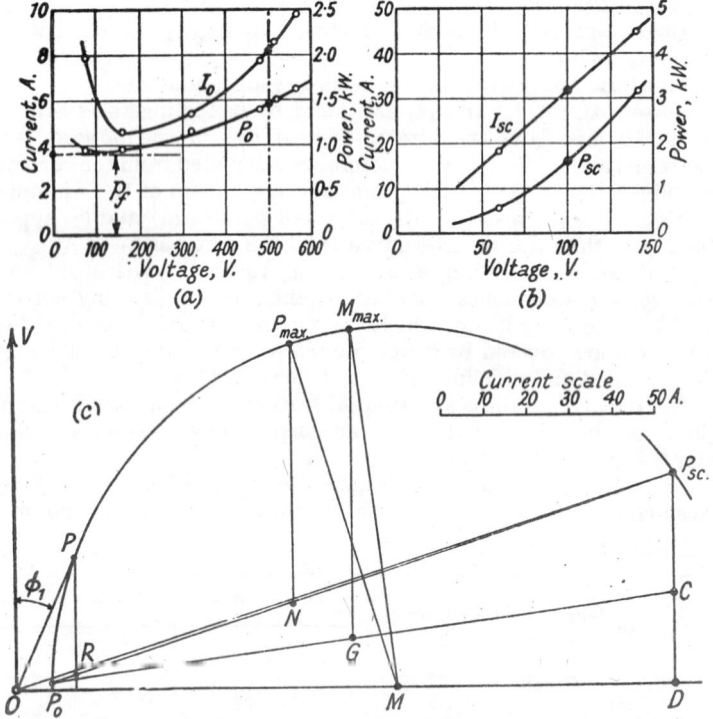

FIG. 237. CIRCLE DIAGRAM FROM TEST RESULTS

power, since they refer to the two-wattmeter method. Plotting the no-load readings, Fig. 237 (a), the no-load normal-voltage input is —
$$V_1 = 500 \text{ V.}; \quad I_0 = 8\cdot2 \text{ A.}; \quad P_0 = 1\cdot45 \text{ kW.};$$
whence $\cos \phi_0 = 1\cdot45 \cdot 10^3 / \sqrt{3} \cdot 500 \cdot 8\cdot2 = 0\cdot202; \quad \phi_0 = 78\cdot4°.$

In Fig. 237 (a), by extrapolation, the friction and windage loss is

$$p_f = 0.9 \text{ kW.}$$

The no-load I^2R loss is $3 \cdot (8.2/\sqrt{3})^2 \cdot 0.68 \cdot 10^{-3} = 0.05$ kW. The core loss* is therefore

$$p_{c1} + p_{h1} + p_v = 1.45 - 0.90 - 0.05 = 0.50 \text{ kW.}$$

Plotting the short-circuit readings, Fig. 237 (b), and taking values for about full-load current (estimated at 32 A.).

$$V_1 = 100 \text{ V.}; \quad I_{sc} = 32 \text{ A.}; \quad P_{sc} = 1.6 \text{ kW.};$$

whence $\cos \phi_{sc} = 1.6 \cdot 10^3/\sqrt{3} \cdot 100 \cdot 32 = 0.29; \quad \phi_{sc} = 73.1°.$

The corresponding values for normal voltage will be

$$V_1 = 500 \text{ V.}; \quad I_{sc} = 160 \text{ A.}; \quad P_{sc} = 40 \text{ kW.}; \quad \cos \phi_{sc} = 0.29.$$

The stator I^2R loss is then $3(160/\sqrt{3})^2 \cdot 0.68 \cdot 10^{-3} = 17.4$ kW., so that the rotor I^2R is $40 - 17.4 = 22.6$ kW.

The circle diagram, Fig. 237 (c), is drawn for *line* values since these are given. The current scale is 10 A. per unit. One unit vertically therefore represents $\sqrt{3} \cdot 500 \cdot 10/746 = 11.6$ h.p. For full load of 30 h.p. a vertical intercept of $30/11.6 = 2.59$ units is necessary between the full-load point P and the output line $P_0 P_{sc}$. This fixes the position of P, and the stator current is $OP = 32$ A. to scale, at a power factor of 0.906. The full-load losses are therefore

	kW.
Friction and windage	0.90
Core	0.50
Stator $I^2R = 32^2 \cdot 0.68 \cdot 10^{-3}$ . . .	0.70
Rotor I^2R (Fig. 240 (b)) $= 1.60 - 0.70$ . .	0.90
Total losses . .	= 3.0 kW.
Output $= 30 \cdot 0.746$.	= 22.4 kW.
Input . . .	= 25.4 kW.

This corresponds to the input from the circle diagram which is

$$\sqrt{3} \cdot 500 \cdot 32 \cdot 0.906 = 25.2 \text{ kW.}$$

The efficiency is

$$\eta = 22.4/25.3 = 88.6\%.$$

The slip is

$$s = 0.9/(0.9 + 0.9 + 22.4) = 0.9/24.2 = 3.72\%.$$

The maximum power is obtained from the vertical intercept between P_{max} and the $P_0 P_{sc}$ line. It measures 5.75 units $\equiv 5.75 \cdot 11.6 = 66.6$ h.p., or 2.22 times full load.

The torque line is obtained by dividing $P_{sc}D$ at C so that $P_{sc}C/CD = 22.6/17.4$, the ratio of the short-circuit rotor/stator I^2R loss. Raising a perpendicular from M to the $P_0 C$ line and producing to cut the circle at M_{max} gives the pull-out torque as 2.5

* See § 11, following.

times full-load torque. The starting torque, $P_{sc}C$ to scale, is approximately equal to full-load torque, but is not a reliable guide.

8. Load Test. In the case of a small motor, an absorption brake or a coupled calibrated d.c. generator may be used to load the machine. The motor is operated on normal voltage and frequency at loads between zero and 50 per cent or 100 per cent overload, readings being taken of voltage, current in all phases, total power and slip. Where guarantees for pull-out torque or starting torque are to be furnished, these may also be investigated.

Fig. 238 gives curves for speed, power factor, efficiency, brake

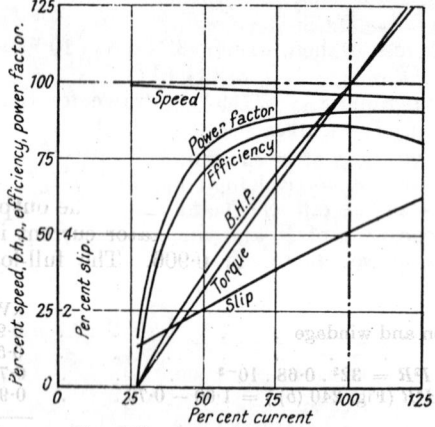

FIG. 238. LOAD TEST CURVES

horse-power, torque and slip, to a base of percentage normal full-load current. The torque/current curve is generally fairly straight, as is the b.h.p. curve, since the speed is nearly constant. As the current increases from no load value, the power factor rises to a maximum near full load. As Fig. 170 shows, the power factor will fall again to short-circuit value if the load is increased and the motor stalls. The efficiency is zero at no load, but rises to a maximum where (roughly) the I^2R losses equal the no-load losses. Thereafter the efficiency falls because the losses increase more rapidly than the output.

9. Temperature-rise Test. Since the temperature-rise limits the output and rating of a motor, its determination is a matter of some importance. In general, a true test of temperature-rise can only be obtained by loading the machine on (a) a direct mechanical load; or (b) a coupled calibrated d.c. or a.c. (induction) generator. Method (a) gives a direct result, but is only possible with small machines, both for reasons of difficulty in absorbing large mechanical powers and of the great cost of energy. Method (b) is more economical if the generator can be loaded on to supply mains: see § 12, below.

Alternative methods are various. One is to drive the machine backwards (i.e. against its rotating field) with the stator voltage reduced to give full load current. The motor works as a brake under conditions corresponding to a point on the circle between P_{sc} and P_∞ in Fig. 168. The mechanical power involved in driving the motor is equivalent only to the losses. The latter are not, however, normal, on account of the subnormal voltage that has to be applied to the stator to limit the current therein to full load value.

The temperature-rise is measured by thermometer and is limited to the values shown in the Table on p. 247.

10. **Insulation Test.** This is normally taken by an ohmmeter (usually with d.c. generator developing 500 V.) between windings and from separate windings to earth or frame. The readings in $M\Omega$ must not be less than 2, or $\frac{1}{500}$ of the terminal voltage, whichever is the greater. Any damage to the insulation or faulty drying or impregnation is shown by a low reading.

The *flash* test is taken after the motor has been found otherwise satisfactory. It consists in applying a specified voltage between the various windings and earth, to disclose weakness of insulation or insufficient clearances. The r.m.s. value of the flash test voltage is given below—

Part	Test Voltage
Stators, for machines of above 1 h.p. but below 3 h.p. per 1 000 r.p.m.	1 000 V. + twice rated voltage between phases
Stators, for machines > 3 h.p. per 1 000 r.p.m.	1 000 V. + twice rated voltage between phases: minimum test voltage 2 000 V.
Slip-ring rotors, non-reversing	1 000 V. + twice open-circuit voltage between rings
Slip-ring rotors, reversing	1 000 V. + four times open-circuit voltage between rings

The tests are applied by means of a high-voltage testing transformer.

11. **Separation of Losses.** Apart from the calculable I^2R loss, the losses occurring in an induction motor include the following—

p_b = brush-friction loss;
p_e = eddy-current loss in iron;
p_f = mechanical losses in windage and bearing friction;
p_h = hysteresis loss;
p_p = pulsation loss.

The suffices 1 and 2 below refer to stator and rotor quantities respectively.

The friction losses p_b and p_f require no comment. The stator loss in hysteresis and eddy currents is $p_{e1} + p_{h1}$ at normal flux.

At standstill the rotor core loss is $p_{e2} + p_{h2}$, when the rotor is open-circuited. At a slip s, the loss becomes $s^2 p_{e2} + s p_{h2}$, since the former is proportional to the square and the latter to the first power of the frequency. The eddy current loss corresponds to a torque proportional to the slip, but the hysteresis loss corresponds to a constant torque, since $s p_{h2}$ is the product of the slip s and a constant power, or the slip-speed $s n_1$ and a constant torque. The rotating field produced by the stator may be conceived as magnetizing the rotor, producing corresponding rotor poles. The hysteretic effect is, however, to retard the rotor poles behind the stator ones, so that a continuous attraction is set up between the stator and induced rotor poles. This *hysteretic torque* urges the machine in the direction of the stator rotating field. The power supply for the loss in rotor hysteresis must naturally be given to the stator in association with the main rotating field : but the loss is employed usefully in producing hysteretic torque which tends to raise the speed.

The pulsation loss is a high-frequency tooth loss in stator and rotor, having the frequency $S_2 n$ in the former and $S_1 n$ in the latter, and produced by variations of gap reluctance as the tooth tips pass each other. The loss depends on the relative slot-pitches in a manner not easily susceptible to calculation.

NO-LOAD TEST WITH ROTOR DRIVEN. Suppose a slip-ring machine to be driven with open-circuited rotor and stator excited from a normal-voltage, normal-frequency supply. As the rotor speed is varied, all the losses will vary with the exception of the stator loss $(p_{e1} + p_{h1})$. Since the friction losses are supplied by the drive, the stator electrical input corrected for $I^2 R'$ loss consists of

$$(p_{e1} + p_{h1}) + (p_{e2} + p_{h2})$$

at standstill, and

$$(p_{e1} + p_{h1}) + s(s p_{e2} + p_{h2})$$

at slip s, for stator and rotor core losses. The rotor losses produce a forward torque, and therefore power, to drive the machine, and this must come also from the stator. From eq. (106) the stator takes in addition

$$(1 - s)(s p_{e2} + p_{h2}).$$

At synchronous speed the torque due to rotor eddy-currents is zero, but that due to rotor hysteresis may have any value between the limits $\pm p_{h2}$, depending on the position of the rotor with respect to the stator rotating poles. The electrical power supplied to the stator therefore changes discontinuously by $2 p_{h2}$ between speeds just below and just above synchronism.

Fig. 239 shows loss/slip curves typical of the conditions described, viz. the motor driven by external means with its rotor open-circuited and the stator energized at normal voltage and frequency. Curve a represents the stator input ; and curve b the constant stator loss $(p_{e1} + p_{h1})$. The intercept between a and b

is the loss given electrically to the rotor $(sp_{e2} + p_{h2})$. The part $s(sp_{e2} + p_{h2})$ which is lost in the rotor is the intercept between b and c: the remainder between c and a is the part developed by the rotor as a forward driving power. The net mechanical power supplied to or by the rotor at various speeds is given by curve d. This is obtained by comparing the mechanical power needed to drive the rotor with the stator first excited then unexcited. The difference indicates what mechanical power the rotor develops from the electrical power it receives from the stator. The intercept between d and the base-line is the mechanical power that must be

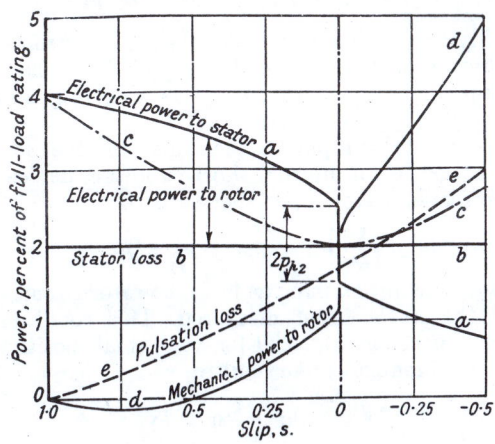

FIG. 239. LOSS CURVES

externally supplied to turn the rotor. Curve e is the sum of d and the intercept between c and a. Since d does not include purely mechanical losses, then e (which is the total mechanical power supplied from drive and stator together) is the pulsation loss, which causes an effect resembling friction.

FRICTION LOSSES. The friction loss can be separated by the method of § 5 (Fig. 235). Alternatively a retardation test with and without the stator excited (for slip-ring machines) will separate the friction from the total no-load loss by comparing the rate of speed drop.*

METHOD OF LOSS SEPARATION. Based on the foregoing and Fig. 239, the following method† separates all losses for slip-ring motors—

1. Measure the power supplied to the stator at normal voltage and

* Parker Smith, *Electrical Engineering Laboratory Manual—Machines*, p. 99 (Oxford).
† Richter, *E.T.Z.*, 42, p. 1 (1921).

frequency with the rotor open-circuited and at rest. The stator input corrected for I^2R is the stator and rotor iron loss

$$P_1 = p_{e1} + p_{h1} + p_{e2} + p_{h2}.$$

This corresponds to the ordinate of curve a, Fig. 239, at standstill.

2. Measure the power supplied to the stator at normal voltage and frequency with the rotor short-circuited and running on no load. The corrected stator input is

$$P_2 = p_{e1} + p_{h1} + p_v + p_f + p_b.$$

The rotor eddy-current loss is very small since it is proportional to s. The rotor hysteresis loss p_{h2} is also supplied by the stator, but practically the whole of it is returned as a driving torque, partly providing for $(p_f + p_b)$.

3. Measure the stator input for the same condition as 2, but with the brushes raised, if an internal short-circuiting device permits of this. The corrected stator input then falls to

$$P_3 = p_{e1} + p_{h1} + p_v + p_f.$$

4. With the machine running as in 2, the rotor circuit is suddenly opened and the stator input measured. This condition is represented by the curves in Fig. 239 for very small positive values of slip. The stator input falls (curve a) to

$$P_4 = p_{e1} + p_{h1} + p_{h2}.$$

The pulsation loss, curve e, and the friction are supplied by the kinetic energy of the rotor, which consequently slows down. The stator power input readings are taken as quickly as possible on this account.

5. Apply to the rotor a voltage such that the stator voltage is $\frac{1}{2}(V_1 + v_1)$ on open circuit (see § 3), and the mutual flux is normal. Measure the rotor input when running on no load with the stator short-circuited; this will correspond exactly to the conditions of 2, except that the functions of stator and rotor are inverted. The corrected rotor power input is

$$P_5 = p_{e2} + p_{h2} + p_f + p_r + p_b.$$

6. With the machine running as in 5, the stator is suddenly open-circuited and the rotor power input falls to

$$P_6 = p_{e2} + p_{h2} + p_{h1}.$$

This corresponds to test 4.

7. By the method of § 5, Fig. 235, the mechanical losses are evaluated.

$$P_7 = p_f + p_b.$$

Having the results of these seven tests available, and corrected for I^2R loss, the separate items are obtained as follows—

$$p_b = P_2 - P_3; \qquad p_{h1} = P_6 + \tfrac{1}{2}[P_2 - P_5 - P_1];$$
$$p_f = P_7 - p_b; \qquad p_{h2} = P_4 + \tfrac{1}{2}[P_5 - P_1 - P_2];$$
$$p_{s1} = P_1 - P_6; \qquad p_v = \tfrac{1}{2}[P_5 + P_2 - P_1] - P_7.$$
$$p_{s2} = P_1 - P_4;$$

The method is subject to certain difficulties, but with care yields fairly accurate results.

12. Back-to-back Test. When two identical induction motors are available an informative input-output test may be made by coupling the machines mechanically together and supplying one at normal voltage and frequency, the other at normal voltage and a slightly lower frequency so that it operates as an induction generator. A repeat with the power flow reversed, i.e. with the second machine running as a motor on a frequency slightly higher than normal and the first machine generating, the position of all instruments being unchanged, gives a check which eliminates meter errors. Some of the losses can be separated by means of the input and output readings and calculated values of stator I^2R loss. The fundamental core loss and the friction and windage losses may be assumed unchanged by the small frequency adjustments.

13. Equivalent Circuit. The no-load and short-circuit tests, §§ 5 and 6, give values from which the equivalent-circuit components r_1, x_1, r_2', x_2', r_m and x_m may be calculated, provided that at least one of the resistances can be measured (e.g. with d.c.) and that the reasonable assumption $x_1 = x_2'$ is made. More expeditiously, if x_0 is determined from the magnetizing current, and x_{sc} from the short-circuit current, the reactance values for constructing the equivalent circuit, Fig. 160 (c), are given by: $x_m = \sqrt{[x_0(x_0 - x_{sc})]}$, $x_1 = x_0 - x_m$, and $x_2' = x_1$. The circle diagram drawn from test should check with calculation based on the equivalent circuit. Such a calculation is given on page 276.

CHAPTER XV

INDUCTION MOTORS: CONSTRUCTION

1. General Arrangement. The three-phase induction motor differs constructionally from the d.c. machine in certain essentials: its short air-gap, absence of a commutator, speed limitation, simple windings (particularly the cage), and laminated stator. It is not possible as in the d.c. machine to use the frame as part of the magnetic circuit. The stator core must be carried in a shell or housing which provides a means for protecting the stator, carrying the end-covers, bearings, and terminal box.

Fig. 240A shows the component parts of a small cage motor. The housing carries the stator core and windings—in this case a mush winding. The end covers receive the ball or roller bearings with their clamping plates. The rotor is of cast copper or aluminium, the end-rings having projections to improve cooling by fanning action. Fig. 240B shows the assembled motor.

2. Frames. The frame of a motor may be cast or fabricated. The former method is still common for small and miniature motors, but medium-sized and (in particular) large machines are almost exclusively fabricated. Rolled steel plates cost about three times as much as cast iron, but the cost of patterns, moulding, casting and cleaning is considerable; setting this against the cost of electrodes and jigs, and allowing for the sale of scrap, the difference in manufacturing costs is not great. It may still be cheaper to use cast iron, however, for repetition work.

The chief advantage of fabricated construction is in its application to new designs and modifications, which can be made without reference to existing patterns. The principal constructional difficulty is the avoidance of distortion when the parts are welded together.

Essentially the frame is a short cylinder with end plates and axial ribs on the inner surface. In larger sizes the housing is made in box form, Figs. 241A and 242. When the ribs have been machined and the ends turned where necessary for the end covers, the frame is ready to receive the stator core, Fig. 248. In Fig. 241B, alternative designs are shown for fabricated construction.* At (a), the length of the core requires an additional centre plate for stiffness. The main frames are cut from one or two plates like Fig. 241A (b), and the wrapper has vent holes cut from it. At (b), the stator stampings are built around an expanding mandrel and compressed tightly between two end-plates which have had holes cut in them. Through the holes are pushed rectangular steel ribs, which are then welded to the

* Carr, *J.I.E.E.*, 78, p. 383 (1936).

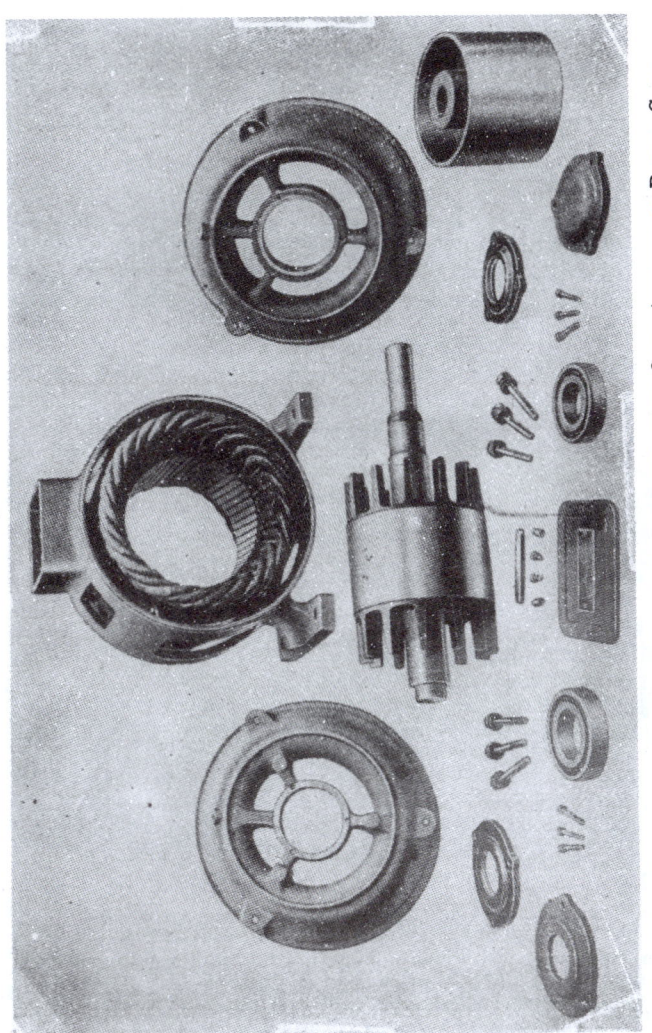

Fig. 240a. Component Parts of Small Cage Rotor with Cast Aluminium Rotor Cage

(General Electric Co.)

end-plates, stampings and to two steel cylindrical stiffeners. Feet are welded on as required. The rotor construction (c) comprises a shaft with, say, six arms and a cross-stiffener welded to it. A large-

FIG. 241B. COMPLETED CAGE MOTOR
(General Electric Co.)

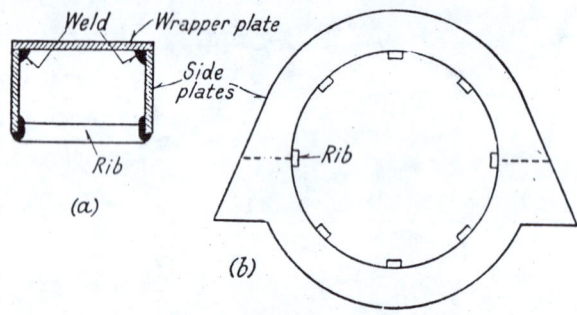

FIG. 241A. FABRICATION OF LARGE STATOR FRAMES

diameter rotor (d) is built up from a cast or forged steel hub, spider plates, cross-stiffener plates, and (at the outer periphery) ribs which are machined to receive the rotor stampings.

3. Cores. The stator and rotor cores are built up of thin sheets of special core steel, insulated one from the other by means of paper, japan, varnish or sprayed china clay. The gap surfaces of the plates have suitable slots punched out, either open, semi-closed, or completely closed. For small machines, both stator and rotor punchings

(Figs. 243A and 243B) are complete rings. The waste of material and difficulty of handling make it necessary to employ sectional plates (Fig. 244), suitably keyed to the housing. Fig. 245 shows a

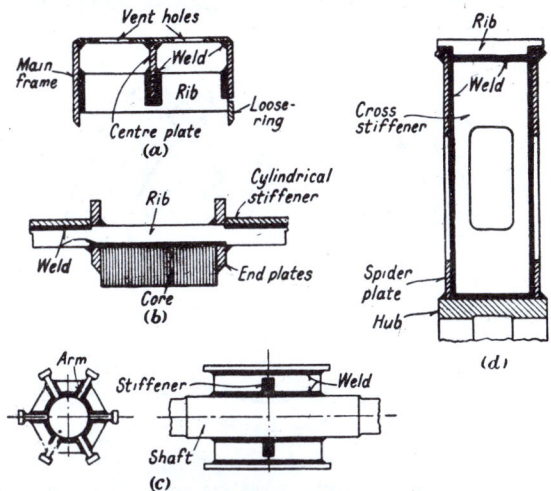

Fig. 241B. Methods of Fabrication

FIG. 242. FABRICATED FRAME WITH STATOR CORE OF 100 H.P. MOTOR
(*Bruce Peebles*)

large rotor being assembled on its spider. In the assembly, spacer plates are used to provide the radial ventilating ducts. At each end of the core an end-plate is fitted to stiffen the structure and to clamp it with a pressure of 30 lb. per sq. in. or more. Fig. 248 shows diagramatically a suitable method, and Fig. 246 a more elaborate method with sector clamping plates on a large fabricated motor.

In small motors the rotor core is mounted directly on the shaft,

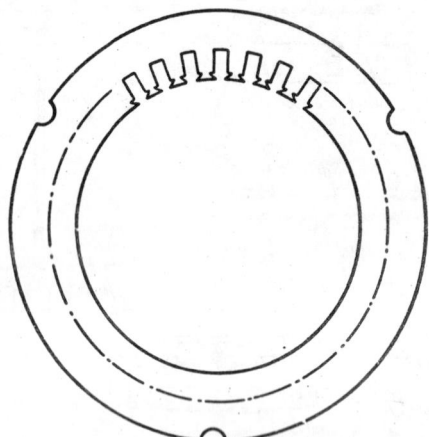

FIG. 243A. RING STAMPING WITH OPEN SLOTS FOR STATOR

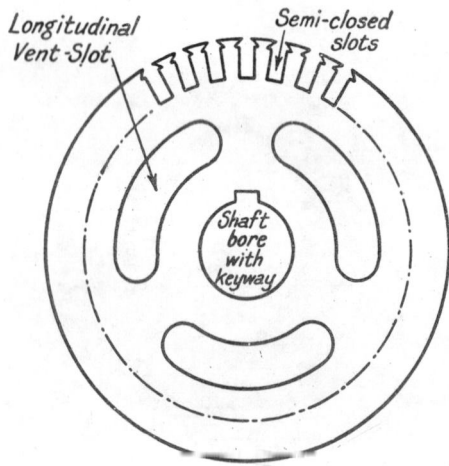

FIG. 243B. RING STAMPING WITH SEMI-CLOSED SLOTS FOR ROTOR

to which it is keyed, and clamped between end plates secured between a shoulder on the shaft and a shrink-ring. Frequently the core is skewed to reduce magnetic pulsations and noise. For large motors the rotor is built up on a fabricated spider, as Fig. 247.

Both rotor and stator core-surfaces have to be ground to obtain an accurate air gap.

4. **Windings.** For large motors or those for high voltage the

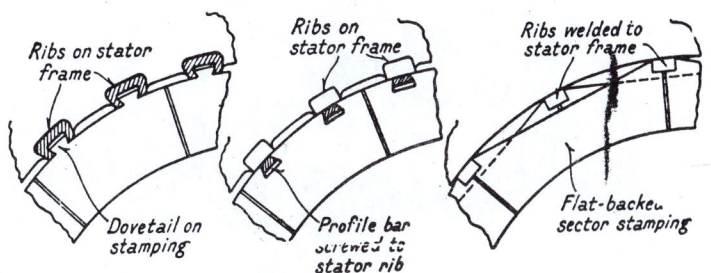

Ribs on stator frame

Ribs on stator frame

Ribs welded to stator frame

Dovetail on stamping

Profile bar screwed to stator rib

Flat-backed sector stamping

FIG. 244. SECTOR STAMPINGS FOR STATOR, AND METHODS OF FIXING

FIG. 245. ASSEMBLING SEGMENTAL CORE STAMPINGS ON A LARGE ROTOR SPIDER
(British Thomson-Houston)

stator phases may be formed by single-layer concentric coils. But even here, as for most medium-sized machines, the double-layer lattice winding is common. Fig. 250 shows a stator partly wound with a double-layer winding. The machine is for a fairly high

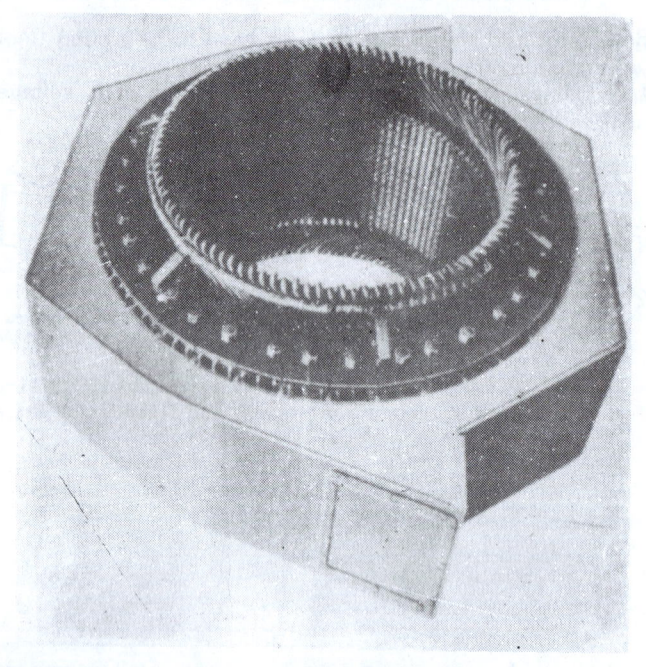

FIG. 246. COMPLETED STATOR SHOWING SECTOR CORE-CLAMPING PLATES
AND TWO-LAYER WINDING
(British Thomson-Houston)

FIG. 247. FABRICATED ROTOR SPIDER FOR 2 500 H.P., 133 R.P.M
(Bruce Peebles)

voltage so that the conductors are small. The former-wound coils are made and introduced into the slots, and the overhang afterwards taped. The small motor in Fig. 249 has a mush winding, still commonly used for motors of a few horse-power working at ordinary

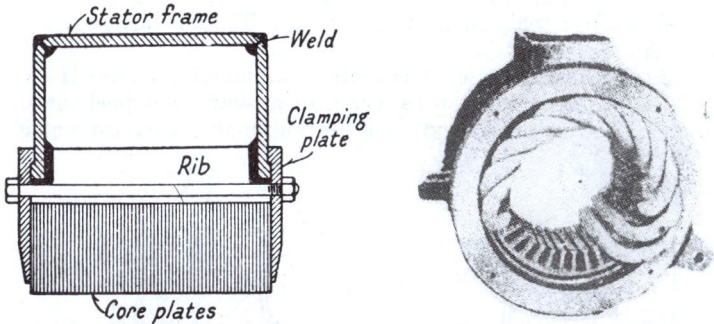

Fig. 248. Core Clamping

Fig. 249. Partly-wound Stator for ½ h.p., 440 V. Motor, 1 000 r.p.m
(Bruce Peebles)

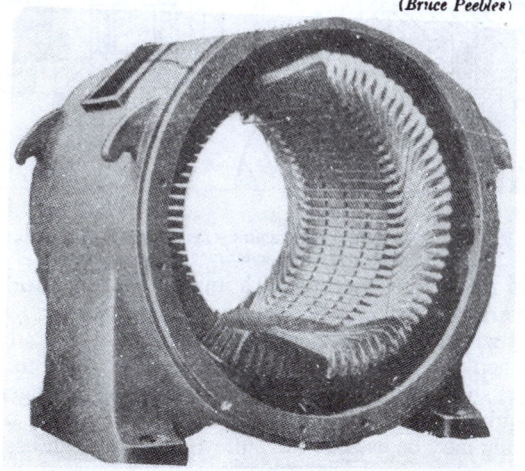

Fig. 250. Partly-wound Stator
(General Electric Co.)

supply voltages. The stator in Fig. 246 has double-layer former-wound coils laid in open slots, so that the coils can be finished before being inserted into the slots. For a lower voltage, bar windings in semi-closed slots might be preferred.

In slip-ring rotors of large size, a bar winding can be used because the choice of voltage is usually free. For smaller machines wire-wound rotors can be used, with coil arrangements similar to those of the stator. Cage rotors in small and miniature motors can be cast in one piece in copper or aluminium, with the end rings moulded to form simple fan blades (Fig. 240). Large rotors have copper or brass bars driven through the slots, the ends being electrically welded or silver soldered to the end rings. The fans are separately mounted.

5. **Shafts, Bearings, etc.** The shaft of an induction motor is short and stiff, in order to keep as small an air-gap as is mechanically possible. The use of ball and roller bearings makes accurate centring

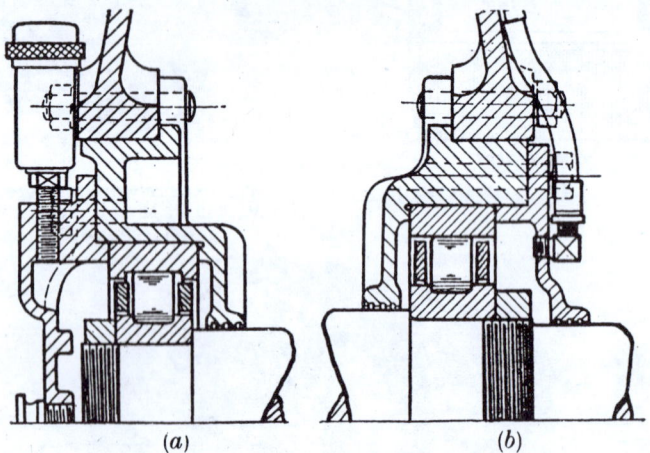

(a)   (b)

FIG. 251. ROLLER BEARINGS ALLOWING AXIAL PLAY
The bearing in (a) has an outer race with limiting lips

of the rotor rather simpler than with journal bearings, and the reduction in overall length is an advantage. Ball and roller bearings are packed with grease and sealed after manufacture and can generally be relied upon to work for long periods without attention. Fig. 251 (a) shows a double-lipped roller bearing arrangement for a motor requiring axial play. Alternatively for small machines a roller bearing may be used at the pulley end and a ball bearing at the other. Where large and heavy rotors call for journal bearings, the self-aligning spherical-seated type is useful.

Slip-rings are in many cases of brass or phosphor-bronze shrunk on to a cast-iron sleeve with moulded mica insulation. The assembly is pressed on to the rotor shaft and located either between the rotor core and the bearing, or on the shaft extension. In the latter case a slip-ring cover is provided and the shaft has to be bored to take the slip-ring leads through the bearing.

Other components are the terminal boxes and, for slip-ring rotors, brush-lifting and short-circuiting gear. Fig. 252 gives the details of a typical arrangement, showing the cover removed. The movement

of the handle clockwise advances a set of three socket contacts on
to a corresponding set of knife-edge contacts by means of the grooved
running-sleeve shown. At the same time the brushes are lifted by
a bar depressed on to the lever ends of each brush box.

FIG. 252. ROTOR SHORT-CIRCUITING AND BRUSH-LIFTING GEAR
(General Electric Co.)

6. **Induction Generators.** Generator construction closely resembles
that of motors. Even in large sizes a cage rotor is used to give low
loss and reactance for high efficiency and power factor. The design
may have to take account of runaway speeds up to three times
normal: the diameter is then less and the axial length more than
for a comparable motor. The rotor bars, of taper or sash-bar section,
are strongly wedged, and special attention given to end-ring connections. The stator may have heaters to keep the insulation dry when
the machine is out of service.

CHAPTER XVI

INDUCTION MOTORS: DESIGN

1. Standard Frames. Apart from a few special machines (such as multi-speed motors) the manufacture of small and medium-sized induction motors is as far as possible concentrated into a range of standard frames to cover a wide load rating. The general practice is to limit the number of rated outputs to a few round figures, e.g.:

$$1, 2, 3, 5, 7\tfrac{1}{2}, 10, 12\tfrac{1}{2}, 15, 20, 25, 30, 40, 50, 60, 75, 100 \text{ h.p.}$$

Intermediate ranges are obtainable. The various possible (synchronous) speeds for 50-c/s. motors are—

$$3\,000, 1\,500, 1\,000, 750, 600, 500, 428 \ldots \text{r.p.m.}$$

All these can be provided with a limited number of frames. A *frame* is the mechanical structure required to house a stator of given

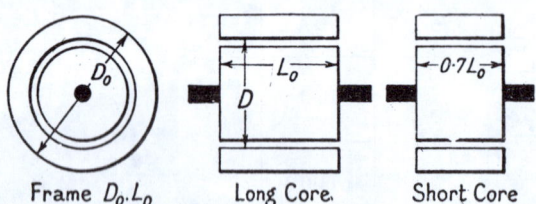

Frame $D_0.L_0$ Long Core.   Short Core

FIG. 253. ILLUSTRATING FRAME SIZE

diameter, together with its bearings, end-covers, terminal blocks, etc. Thus the essential dimensions appropriate to a frame are D, and L_0, Fig. 253. A variation of rating is possible by using alternative core lengths less than L_0; e.g. $0.7L_0$. Further, some variation in stator bore D is permitted by the frame. Thus a moderately limited number of frames suffices to satisfy the requirements of design for an infinitude of possible combinations of rating, voltage and speed. The following list* is typical of a frame series covering a range of 50-c/s. industrial motors between 1 h.p. and 100 h.p.

The two or three diameters given for each frame size are determined by the necessary depth of stator core behind the slots. The greater the number of poles, the less is the flux per pole and the

---

* Based on Fig. 60, Hopkins, *Induction Motor Practice* (Pitman). This book gives a very complete discussion of commercial design. B.S. 2083: 1956 gives standard *external* dimensions for 3-ph., totally-enclosed, fan-cooled, 50-c/s., industrial cage motors with Class A insulation, and continuous maximum ratings between ¼ and 25 h.p. at the four highest speeds.

Frame Designation		Outside Diameter of Stator Stampings, cm. D_o	Gross Core-length, cm. L_u	No. of Poles $2p$	Stator Bore cm. D	D^2L cm.³
A	1	21·0	5·0	2 4	11·0 13·0	600 845
	2		7·0	2 4	11·0 13·0	850 1 180
B	1	24·0	6·5	2 4 6	13·0 15·0 16·5	1 100 1 460 1 760
	2		9·0	2 4 6	13·0 15·0 16·5	1 520 2 020 2 450
C	1	29·0	9·0	2 4 6	15·0 18·5 20·0	2 020 3 080 3 600
	2		12·0	2 4 6	15·0 18·5 20·0	2 700 4 100 4 800
D	1	34·5	10·5	2 4 6	18·5 22·5 24·5	3 600 5 300 6 300
	2		14·0	2 4 6	18·5 22·5 24·5	4 800 7 100 8 400
E	1	41·5	12·0	4 6	27·0 29·0	8 750 10 000
	2		15·0	4 6	27·0 29·0	11 000 12 600
F	1	46·5	12·0	4 6	31·5 34·0	11 900 13 800
	2		16·0	4 6	31·5 34·0	15 900 18 500
G	1	53·0	13·0	4, 6 8	36·0 40·0	16 800 20 800
	2		18·0	4, 6 8	36·0 40·0	23 300 29 000
H	1	59·0	16·0	4, 6 8	40·0 45·0	25 600 32 000
	2		20·0	4, 6 8	40·0 45·0	32 000 40 000

smaller need the core-depth be. Thus for multi-polar motors a larger diameter can be employed and the fall of output of a frame due to fall in speed to some extent offset. Assuming output to be directly proportional to D^2Ln, the outputs obtainable from Frame B are in the proportion—

Short Core.	Small diameter .	.	.	.	.	1·87
	Medium diameter	.	.	.	.	1·23
	Large diameter .	.	.	.	.	1·0
Long Core.	Small diameter .	.	.	.	.	2·6
	Medium diameter	.	.	.	.	1·7‥
	Large diameter .	.	.	.	.	1·38

In designing the ranges it may be possible to economize in core material by making the larger bore of one frame equal to the smaller bore of the next-sized frame.

Motors of ratings in excess of 100 h.p. may be part of a standard range, but tend to be designed more individually, particularly when patterns are rendered unnecessary by fabrication.

The rating of a frame depends, of course, not only upon its size and the speed, but also upon the specific magnetic and electric loadings, the type of enclosure, whether a short-time or continuous loading is required, and upon the frequency of starting and stopping. Theoretically, any increase in the efficiency (and perhaps also the power factor) is of financial advantage to the user of a motor, and the capitalized value of the annual saving on the cost of power is available for the additional cost of a motor designed more liberally to secure the increase. Actually it is scarcely possible to sell *small* motors on anything other than first cost; which may not be unsound, since energy charges have a tendency to become altered unexpectedly. With large (and, particularly, special) motors, the question is more closely investigated since more money is at stake, and the values of efficiency and power factor to be obtained by a motor may be the subjects of clauses in the specification. In such a case, a more fundamental method of design may be used than is customary for small stock machines. The delimited speeds at which induction motors must work by reason of the frequency impose restrictions on the design of these machines. Thus if a 20 h.p. d.c. motor were required for a speed of 600 r.p.m., a four-pole design would be chosen. An induction motor to work at the same speed on a 50-c/s. supply, would, however, be obliged to have ten poles, and would by comparison be uneconomical. The disadvantages are greatest for motors with speeds low for their output.

Frames G and H in the list on p. 385 are provided with radial ventilating ducts in the cores. Long cores tend to overheat unless subdivided and cooled in this way. Since the gap is very small, the rotor and stator should have ducts in the same plane to avoid air throttling. In some cases this leads to noisy running in high-speed motors. The number of ducts used depends on the cooling arrangements: one may be allowed for each 5–6 cm. of core length.

2. **Windings.** A considerable variety of slottings can be used for induction motors, as indicated by Fig. 130. In small motors where round wires of small gauge are used, the taper-slot, parallel-tooth arrangement is useful as giving the maximum area of slot for a given tooth density. The small stator bore gives considerable taper to such slots and helps to keep the depth of the stator stamping in reasonable proportion. In larger machines the taper is very much less: also strip conductors are preferred, which demand parallel-sided slots. For slip-ring rotors bar windings are common, so that it is usual to find parallel slots even for small sizes, in spite of the difficulty of securing adequate tooth width at the root, the narrowest part. The tunnel slot is rarely used for stators since pull-through windings are necessary. The same applies to rotors, but here the free choice of voltage permits fewer, more robust conductors to be used. Large high-voltage motors may have open stator slots to secure the advantage of coils made and completely insulated outside the machine.

Types of windings have been described in Chapter X. The number of slots per pole per phase should normally be not less than 2 to avoid excessive leakage reactance. Strip or bar windings with wave connections and four to eight conductors per slot are common for larger stators, and for most rotors with, however, fewer conductors.

Low leakage reactance implies a large diameter in the current locus circle, a higher frame rating, and more reserve torque. Contrasting four slots with two slots per pole per phase for a given motor, the m.m.f. per leakage path round the slot would be doubled for the larger slots and the magnetic lines would link twice as many conductors, giving four times the reactance per slot and twice the phase reactance. Thus it is better to use as many slots as is economically possible.

For a given slot area, the wide shallow slot gives greater leakage reluctance, but of course restricts the tooth area for the main flux.

Insulation for industrial voltages (e.g. not exceeding 440 V.) is provided by Class A material. The slot insulation comprises a liner of micanite and elephantide extending beyond the core on both sides, and lapping over the coil within the slots, the edges being pressed down by the wedge of fibre strip.

As an example of insulation suitable for use up to 500 V., the slot insulation may be made of flexible micanite and leatheroid each 0·25 mm. (10 mils) thick, and projecting 1 cm. beyond the core at each end. After the insulated wires are introduced through the slot-opening and hammered down, a strip of flexible micanite is inserted, and the slot insulation trimmed and lapped over within the slot, the *cell* thus formed being retained by a fibre wedge driven in from one end. The ends of the coils are taped and impregnated.

In estimating the available slot area, account must be taken of the slight lack of exact register between successive stampings, which reduces the width of slot.

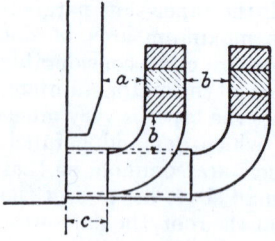

FIG. 254. OVERHANG CLEARANCES

The methods employed for stator-winding insulation will depend primarily on the size of the machine and its working voltage. Small-section round wires are simply enamelled, generally with a synthetic-resin enamel of the polyvinyl-acetate type, with an additional cotton covering for voltages up to 650 V.: in the form of a mush winding the conductors can be fed into semi-closed slots. Some modern enamels give finishes so smooth that slot-filling becomes relatively easy. After winding, the whole stator winding may be impregnated with a thermo-setting varnish and baked.

For large machines diamond coils (or concentric coils) are pre-formed, mica-wrapped, sealed and impregnated before insertion into the slots. Glass and micanite are employed for higher voltages with class B insulation. Progress is being made with Class H insulating materials, for which substantially higher temperature-rises are permitted, with consequent advance in rating.

The clearances at the overhang, dimensions a, b and c, Fig. 254, are approximately $a = 10 + 3V$; $b = 10 + 10 + 2V$; $c = 10 + 6V$ mm., where V is the line voltage in kV.

3. Design. The process of design is to obtain the dimensions and electrical particulars of a machine to satisfy a given specification covering horse-power, speed, temperature-rise and conditions of service. For standard motors the designer's task is considerably lightened by routine and computer methods of using data giving approximate values of quantities that would otherwise have to be calculated. Let D = stator bore, m.; L = stator core-length, m.; n = speed, r.p.s.; S = electrical rating of machine, volt-amperes.

RATING AND DIMENSIONS. The volt-ampere rating of a 3-phase machine in terms of its phase e.m.f. and current is $S = 3E_{ph}I_{ph}$. The e.m.f. is given by eq. (80), and considering only fundamental-frequency components then

$$S = 3 \cdot 4{\cdot}44K_wfT_{ph}I_{ph}\Phi_m \text{ volt-amperes.}$$

For convenience in design, the following specific loadings are defined:

(1) *Specific magnetic loading*, the average magnetic flux density over the whole surface of the air-gap,

$$B = 2p\Phi_m/\pi DL \text{ webers per m}^2$$

(2) *Specific electric loading*, the number of r.m.s. ampere-conductors per unit length of gap-surface circumference, or the r.m.s. peripheral current density,

$$ac = 3 . 2T_{ph}I_{ph}/\pi D \text{ amp.-cond. per m.}$$

$\bar{B}$ is limited by saturation and losses in the teeth, and by the excitation necessary to overcome the gap reluctance. The electric loading ac affects I^2R loss and armature reaction. The losses must be dissipated by ventilation, so that the designer has several restrictions on his choice of specific loadings, which are affected by the size, type, construction, speed, ventilation and insulation of the machine. With given specific loadings, and with $f = pn$,

$$S = 4\cdot44K_w . pn(\pi DL\bar{B}/2p)\tfrac{1}{2}\pi Dac$$
$$= 1\cdot11K_w\pi^2\bar{B}acD^2Ln \text{ volt-amperes} \qquad . (150)$$

Thus the rating is proportional to the average gap density $\bar{B}$, the armature surface current density ac, the volume of the rotating member (represented by the D^2L product, Fig. 255), and the speed of rotation n.

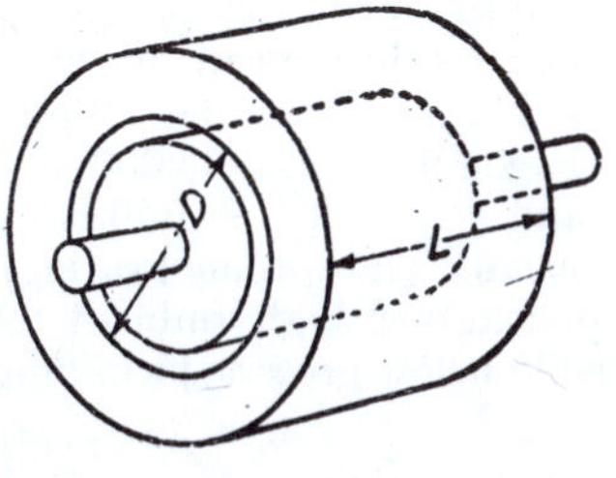

This rating is in fact the product $M\omega_r$ of torque and angular velocity, discussed in Chapter VIII. For the present case eq. (45) applies, and taking $\sigma = 90°$ for the optimum torque conditions, then

FIG. 255. D^2L PRODUCT

$$M\omega_r = p . \pi B_m ac_m LD\pi n \text{ watts}$$

for p pole-pairs. But $ac_m = F_a$, the armature ampere-conductors per pole (or ampere-turns per pole-pair), which from eq. (83) is

$$F_a = (3\sqrt{2}/\pi)K_wT_{ph}I_{ph}/p = (3/\sqrt{2}\pi)K_w\pi Dac/p.$$

Also $\bar{B} = (2/\pi)B_m$, so that

$$M\omega_r = 1\cdot11K_w\pi^2\bar{B}acD^2Ln \text{ watts}$$

which is the same as eq. (150) for unity power factor.

OUTPUT COEFFICIENT. This is defined as

$$G = S/D^2Ln = 1\cdot11K_w\pi^2\bar{B}ac . 10^{-3} = 11K_w\bar{B}ac . 10^{-3} \quad (151)$$

with S for convenience in kVA. By assigning suitable values (obtained from design experience) to $\bar{B}$ and ac, the D^2L product for a given rating S kVA. at speed n is obtained, S referring here to the electrical input. In terms of the mechanical output P_m in kW. or (H.P.) in horse-power,

$$S = P_m/\eta \cos\phi = (H.P.)0\cdot746/\eta \cos\phi \quad \text{kVA.} \qquad (152)$$

where η and $\cos\phi$ are estimated full-load values.

SPECIFIC LOADINGS. $\bar{B}$ directly influences the core-loss and magnetizing current, and thus has an important effect on the power factor: ac determines to a large extent the I^2R loss. Further, on account of the peculiarities of the induction motor, the flux determines the pull-out torque. If a motor were required to give a pull-out torque very much larger than its full-load torque, the requirement could be satisfied by the use of wide teeth and small, narrow slots, giving a large working flux, relatively small leakage, and a short-circuit current many times full-load current. The continuous rating of such a motor would be small for its frame size because the conductor section would be so restricted. The use of deeper, wider slots would increase the rating at the expense of pull-out torque. Further enlargement of the slots would so reduce the tooth area that the gap flux density would be low. The number of stator turns per phase would have to be increased, which would result in a high reactance. The motor would have a low percentage magnetizing current and a large thermal rating, but very little reserve of torque.

For normal 50-c/s. machines, $\bar{B}$ lies between 0·3 and 0·6 Wb./m.², rising to about 0·65 for crane, rolling-mill and similar drives. In the stator teeth a mean density of 1·3–1·7, with a maximum of 2·2–2·4 Wb./m.², may be employed. The electric loading ac varies between 5 000 and 45 000, and the current density may lie between 3 and 8 A./mm.² A pull-out torque of about twice full-load value is usual. The specific loadings have to be reduced for special service applications, high ambient temperatures, total enclosure, etc. The Table below gives typical figures—

SPECIFIC LOADINGS

for 50-c/s. ventilated motors with peripheral speeds between 15 and 40 m./sec. and Class A insulation. For Class B insulation increase loadings by 7½ per cent and current density by 16 per cent.

D m.	$\dfrac{L}{D}$ max.	Slip-ring			Cage		
		$\bar{B}$ Wb./m.²	ac amp.-cond./m.	δ A./mm.²	$\bar{B}$ Wb./m.²	ac amp.-cond./m.	δ A./mm.²
0·1	0·8	0·3	6 000	3·8	0·3	11 000	4·0
0·15	0·75	0·35	10 000	3·6	0·35	15 000	3·8
0·2	0·7	0·4	13 000	3·4	0·4	18 000	3·6
0·3	0·65	0·43	17 500	3·3	0·43	22 500	3·5
0·4	0·62	0·45	21 500	3·2	0·45	26 000	3·5
0·5	0·6	0·46	25 000	3·2	0·46	29 000	3·5
0·75	0·5	0·47	30 000	3·2	0·47	33 000	3·5
1·0	0·42	0·48	32 500	3·2	0·48	35 000	3·5
1·5	0·33	0·5	34 000	3·2			
2·0	0·3	0·51	35 000	3·2			
3·0	0·3	0·53	37 000	3·2			

MAIN DIMENSIONS. The D^2L product obtained from eq. (151) has to be split up into its components D and L. For normal speeds a roughly square pole of pole-pitch $Y = \pi D/2p$ equal to the core-length L gives good electrical design, particularly as regards leakage reactance. This proportionality cannot, however, always be used, as the speed and frequency control the number of poles, so that a square pole may result in a diameter excessive for mechanical reasons. The tendency is to use a restricted diameter and greater core-length in order to decrease the proportion of inactive copper in the overhang, and to obtain a generally cheaper motor.

The diameter having been chosen, the core-length is obtained from the D^2L product by division.

The *gap-length* is determined by the magnetizing current (the avoidance of too low a power factor), mechanical considerations (bearings, shaft deflection, unbalanced magnetic pull, etc.), and other secondary matters such as pulsation losses and cooling from the gap surfaces. The gap is generally made as small as is mechanically possible without undue refinement in respect of bearings and tolerances. In small motors it may be of the order of 0·25 mm. A typical expression for estimating the gap length is

$$l_g = 0 \cdot 2 + 2\sqrt{(DL)} \text{ mm.} \qquad . \qquad . \qquad . \qquad (153)$$

where D and L are in m. The gap can be made a mere clearance in very small motors, and will in general be smaller where ball and roller bearings are used. In large motors the gap-length may be 2 or 3 mm. An expression for journal-bearing machines is

$$l_g = 1 \cdot 6\sqrt{D} - 0 \cdot 25 \text{ mm.} \qquad . \qquad . \qquad . \qquad (153a)$$

where again D is in m.

PERFORMANCE. Use of the following more or less approximate expressions enables the salient operating features of a designed motor to be forecast.

The maximum power factor is

$$(\cos \phi)_{max} = (I_{sci} - I_{0r})/(I_{sci} + I_{0r})$$

where I_{sci} is the ideal short-circuit current omitting resistance, i.e. $V_1/(x_1 + x_2')$. This expression is obtained from the geometry of the circle diagram.

The maximum power from eq. (121) is

$$P_{mm} = N_1 V_1 \frac{I_{sc} - I_0}{2(1 + \cos \phi_{sc})} \text{ watts.}$$

The maximum power is approximately a measure of the maximum torque.

The starting torque from eq. (130) is

$$M_s = M_n (I_{sc}/I_n)^2 s_n.$$

Other information may be derived from the circle diagram drawn on a basis of the no-load and short-circuit currents.

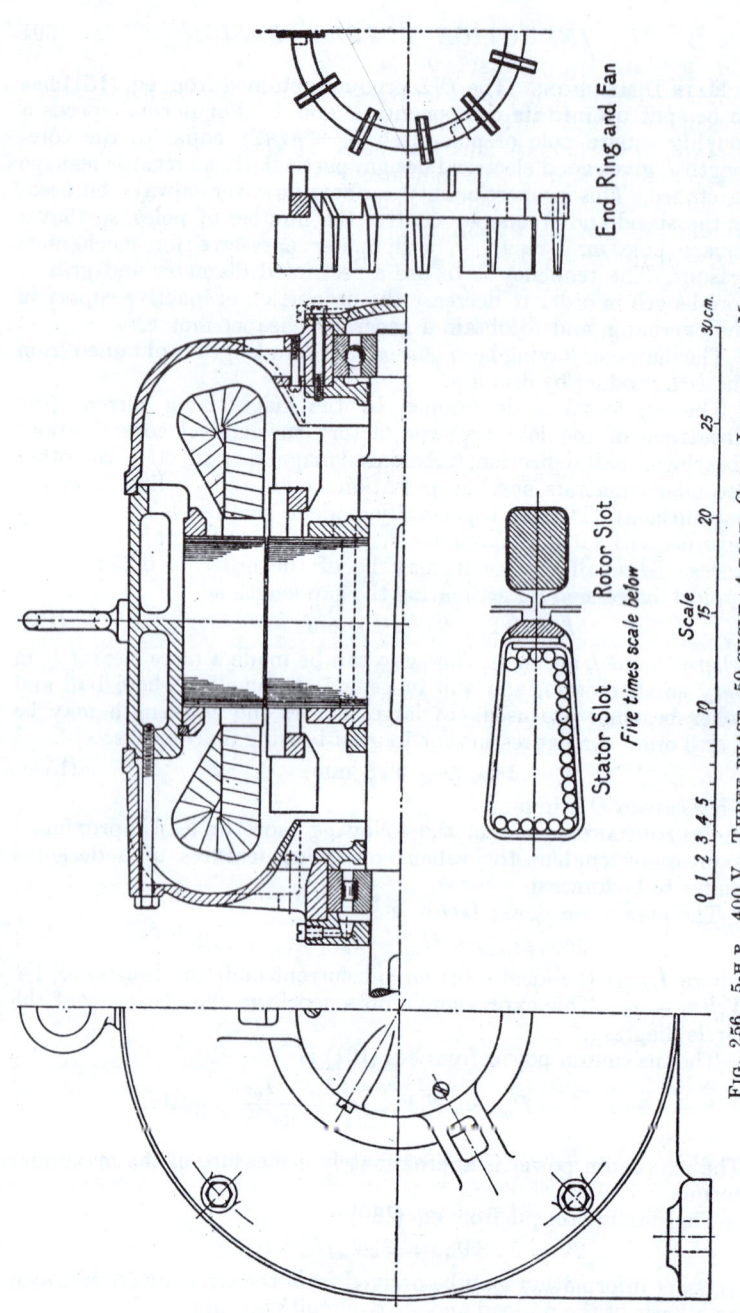

End Ring and Fan

Stator Slot

Rotor Slot

Five times scale below

Scale

0 1 2 3 4 5   10   15   20   25   30cm.

Fig. 256. 5-h.p., 400 v., Three-phase, 50-cycle, 4-pole, Cage Induction Motor

4. **Example**. 5-h.p., 400-V., 3-phase, 50-c/s., cage induction motor; 1 500 r.p.m.; to be arranged for star-delta starting. Specification, B.S. 2613: 1955. Drawing, Fig. 256.

MAIN DIMENSIONS. This four-pole motor is intended for stock at a competitive price. The efficiency and power factor are therefore to some extent sacrificed to light weight. Choosing a full-pitch winding with $K_w \simeq 0.955$,

$\overline{B} = 0.46$ Wb./m.^{2} and $ac = 22\,000$ amp-cond. per m., the output coefficient, eq. (151), is

$$G = 11 . 0.955 . 0.46 . 22\,000 . 10^{-3} = 106.$$

Taking $\eta = 0.83$ and $\cos \phi = 0.84$, the kVA. rating is

$$S = 5 . 0.746/0.83 . 0.84 = 5.35 \text{ kVA.}$$

The main dimensions

$$D^2L = 5.35/106 . 25 = 0.00202 \text{ m.}^3 = 2\,020 \text{ cm.}^3$$

A suitable frame is B2 (see p. 385), where for a four-pole machine a D^2L value of 2 020 cm.^{3} is obtainable, and

$$D = 0.15 \text{ m.} = 15 \text{ cm. ; } L = 0.09 \text{ m.} = 9 \text{ cm. ;}$$
$$Y = 0.118 \text{ m.} = 11.8 \text{ cm.}$$

These dimensions give a pole that does not depart seriously from a square. The six-pole motor in this frame has a diameter of 16·5 cm., a pole-pitch of 8·6 cm., and the same length 9 cm. Again the pole is roughly square, but the pole-pitch is the shorter dimension.

The core is too short to require a radial duct. The net core length is $L_s = 9$ cm., the iron length $L_i = 0.9 . 9 = 8.1$ cm. From eq. (153) the gap length is

$$l_g = 0.2 + 2\sqrt{(0.15 . 0.09)} = 0.432, \text{ say } 0.45 \text{ mm.}$$

STATOR WINDING. The no-load e.m.f. per phase (mesh connection) is $E_1 \simeq 400$ V. The no-load flux is

$$\Phi_m = 0.46 . 0.118 . 0.09 = 4.9 \text{ mWb.}$$

The number of turns per phase is

$$T_1 = \frac{400}{4.44 . 0.955 . 50 . 4.9 . 10^{-3}} = 385.$$

The slot-pitch for mechanical reasons must not be too small—say not less than 1 cm. The number of slots per pole per phase should be at least three to avoid excessive reactance. With $S_1 = 36$ stator slots the slot-pitch is $y_{s1} = \pi . 15/36 = 1.31$ cm. There are thus $q_1' = 3$ slots per pole per phase, and with $z_1 = 64$ conductors per slot,

$$T_1 = 2 . 3 . 64 = 384 \text{ turns per phase,}$$

which is close enough to the estimated figure to disregard the change in Φ_m. The voltage and current are low enough to necessitate a mush winding.

The full-load phase current is estimated as

$$I_1 = 5 \cdot 0{\cdot}746/3 \cdot 400 \cdot 0{\cdot}83 \cdot 0{\cdot}84 = 4{\cdot}5 \text{ A.}$$

Taking a current density $\delta_1 = 4$ A. per mm.^{2}, the conductor area is $4{\cdot}5/4 = 1{\cdot}12$ mm.^{2}: S.W.G. No. 18 has an area $1{\cdot}17$ mm.^{2}, diameter $1{\cdot}22$ mm., insulated with fine d.c.c. to $1{\cdot}40$ mm.

The stator slot is shaped to obtain parallel-sided teeth $0{\cdot}6$ cm. wide. Allowing $0{\cdot}5$ mm. thick slot liner and 1 mm. slack and roughness in the slot width, the slot-depth totals 24 mm., including a slot-opening 3 mm. wide with a 1 mm. lip. The *mean* tooth density (corresponding to the mean gap density) is

$$\overline{B}_t = 4{\cdot}9 \cdot 10^{-3}/3 \cdot 3 \cdot 8{\cdot}1 \cdot 0{\cdot}6 \cdot 10^{-4} = 1{\cdot}12 \text{ Wb./m.}^2$$

The *maximum* tooth-density will be about $1{\cdot}5$ times this (i.e. rather less than $\pi/2$ times on account of saturation) or about $1{\cdot}68$ Wb./m.^{2}

The current-density used is not high, but the use of the next S.W.G. (No. 19) involves a considerable increase in the I^2R loss, although it makes winding into the slots much easier.

Stator Resistance. Length of mean conductor

$$L_{mc} \simeq L_s + 1{\cdot}15Y + 12 \simeq 35 \text{ cm.}$$

Length per phase, $384 \cdot 2 \cdot 35 \cdot 10^{-2} = 269$ m. Resistance at $75°$ C.,

$$r_1 = 0{\cdot}021 \cdot 269/1{\cdot}17 = \underline{\underline{4{\cdot}85 \; \Omega.}}$$

I^2R loss, all phases, $3 \cdot 4{\cdot}5^2 \cdot 4{\cdot}85 = \underline{\underline{300 \text{ W.}}}$

Stator Core. The depth of the core at the back of the slots with an outside diameter of $D_o = 24$ cm. is

$$\tfrac{1}{2}(24 - 15 - 2 \cdot 2{\cdot}4) = 2{\cdot}1 \text{ cm.}$$

The maximum core density is

$$B_c = 4{\cdot}9 \cdot 10^{-3}/2 \cdot 2{\cdot}1 \cdot 8{\cdot}1 \cdot 10^{-4} = 1{\cdot}44 \text{ Wb./m.}^2$$

This density is inherent in the choice of $D_0 = 24$ cm. for the Frame Size B2.

ROTOR CAGE. Choosing $S_2 = 30$ slots, then

$$y_{s2} = \pi \cdot 14{\cdot}92/30 = 1{\cdot}56 \text{ cm.}; \quad g_2 = 30/4 = 7{\cdot}5.$$

$$I_2' \simeq I_1 \cos\phi = 4{\cdot}5 \cdot 0{\cdot}84 = 3{\cdot}8 \text{ A.}$$

From eqs. (123) and (125),

$$I_b = 0{\cdot}955 \cdot 9 \cdot 64 \cdot 3{\cdot}8/7{\cdot}5 = 277 \text{ A.,}$$

and

$$I_e = 0{\cdot}3 \cdot 9 \cdot 64 \cdot 3{\cdot}8 = 655 \text{ A.}$$

Experience shows that a current density of $\delta_2 \simeq 6$ A. per mm.^{2} gives a reasonable starting torque. The area per bar is $a_2 = 277/6 = 46$ mm.^{2} The conductor width must be chosen with due regard

to the mechanical strength of the rotor teeth and their flux density.
Slots 10·5 mm. deep × 6·5 cm. wide with lightly insulated conductors 8·75 mm. × 5·75 mm. with rounded corners and an area of 46 mm.^{2} may be used. The slots are skewed one slot pitch. The length per bar is 12 cm.

The resistance per bar at 75° C. is

$$r_b = 0·021 . 0·12/46 = 0·055 . 10^{-3} \; \Omega.$$

The loss in the bars is

$$S_2 I_b{}^2 r_b = 30 . 277^2 . 0·055 . 10^{-3} = 130 \; W.$$

The rings are cast with fan blades for cooling. The area is approximately 120 mm.^{2} and the mean diameter 11·7 cm. The resistance is

$$r_c = 0·021 . \pi . 11·7 . 10^{-2}/120 = 0·064 . 10^{-3} \; \Omega,$$

and the loss is

$$2 I_c{}^2 r_c = 2 . 655^2 . 0·064 . 10^{-3} = 55 \; W.$$

The total cage $I^2 R = 130 + 55 = 185$ W.

The equivalent resistance from eq. (128) is

$$r_2' = 11 \frac{384^2}{30} \left(0·055 + \frac{7·5}{\pi^2} . \frac{0·064}{2} \right) 10^{-3} = 4·3 \; \Omega ;$$

or more directly, since $3 I_2'^2 r_2' = 185$ W.,

$$r_2' = 185/3 . 3·8^2 = 4·3 \; \Omega.$$

No-load Current. For the reactive component I_{0r}, the magnetizing current, the no-load magnetizing ampere-turns are required. For the gap, using eq. (51)

$$k_{g1} = \frac{13·1}{13·1 - 0·73 . 3} = 1·20 ; \quad k_{g2} = \frac{15·6}{15·6 - 0·45 . 1·0} = 1·03 ;$$

$$l_g' = 1·20 . 1·03 . 0·45 = 0·555 \; mm.$$

Effective gap-area $A_g = YL = 11·8 . 9 = 106$ cm.$^2 = 0·0106$ m.^{2} The estimation of the magnetizing current is based on the procedure described in Chapter IX, § 4.

MAGNETIZATION

$$\Phi_m = 4·9 \; mWb.$$

Part	Area A . m.^{2}	Length l m.	$B = \Phi_m/2A$	$B_{30} = 1·36 \, \Phi_m/A$	at	$AT = at . l$
Stator core .	0·001 7	0·06	1·440	—	1 300	78
,, teeth .	0·004 36	0·024	—	1·525	2 600	62
Gap . .	0·010 6	0·555 . 10^{-3}	—	0·630	501 000	278
Rotor teeth ($\frac{1}{3}$)	0·004 65	0·010 5	—	1·410	1 000	10
,, core .	0·002 4	0·03	1·020	—	200	6

Total, AT_{30}	434

From eq. (55) $I_{0r} = 2 \cdot 434/1 \cdot 17 \cdot 0 \cdot 955 \cdot 384 \simeq 2 \cdot 0$ A. per phase,
$$= 3 \cdot 5 \text{ A. per line.}$$

The core losses are confined to the stator teeth and core. Allowance for the pulsation losses is included in the curves, Fig. 112 (b). The core is built of 0·5 mm. steel plates, No. 42 grade.

Stator teeth: weight $= 3 \cdot 15$ kg.

The specific loss is $p_i = 25$ W. per kg. for $B_{tm} \simeq 1 \cdot 68$ Wb./m.^{2}

Loss in teeth $= 25 \cdot 3 \cdot 15 = 79$ W.

Stator core: weight $= 8 \cdot 75$ kg.

Specific loss $p_i = 19$ W. per kg. for $B_c = 1 \cdot 44$ Wb./m.^{2}

Loss in core $= 19 \cdot 8 \cdot 75 = 166$ W.

Total iron loss $= 166 + 79 \simeq 250$ W.

The friction and windage are difficult to estimate, but will be small since ball and roller bearings are used. Assuming 1 per cent of the input, say 60 W., the no-load loss becomes

$$P_0 = 245 + 60 = 305 \text{ W.}$$

The active current is

$$I_{0a} = 305/3 \cdot 400 = 0 \cdot 25 \text{ A. per phase,}$$

and the no-load current is

$$I_0 = \sqrt{(2 \cdot 0^2 + 0 \cdot 25^2)} \simeq 2 \cdot 0 \text{ A. per phase} = 3 \cdot 5 \text{ A. per line.}$$

The no-load power factor is

$$\cos \phi_0 = 0 \cdot 25/2 = 0 \cdot 125.$$

SHORT-CIRCUIT CURRENT. Following the method outlined in eq. (67), and using the dimensions previously calculated or obtained from Fig. 256—

Slot Leakage. For the stator specific slot-permeance coefficient

$$\lambda_{s1} = \frac{18}{3 \cdot 9} + \frac{2}{8} + \frac{2 \cdot 2}{8 + 3} + \frac{1}{3} = 1 \cdot 61.$$

For the rotor,

$$\lambda_{s2} = \frac{8 \cdot 75}{3 \cdot 6 \cdot 5} + \frac{2 \cdot 0 \cdot 75}{6 \cdot 5 + 1} + \frac{1}{1} = 1 \cdot 65;$$

referred to stator terms this is

$$\lambda_{s2} \cdot K_{w1}^2 S_1 / K_{w2}^2 S_2 = 1 \cdot 65 \cdot 0 \cdot 955^2 \cdot 36/1 \cdot 30 = 1 \cdot 81.$$

The total slot reactance is, from eq. (66)

$$x_s = 15 \cdot 8 \cdot 50 \cdot (384^2/2 \cdot 3) \cdot 0 \cdot 09(1 \cdot 61 + 1 \cdot 81)10^{-6}$$
$$= 19 \cdot 4 \cdot 0 \cdot 09 \cdot 3 \cdot 42 = 6 \cdot 0 \ \Omega.$$

Overhang Leakage. The reactance is obtained by applying eq. (66) to the total permeance from eq. (63)—

$$L_o \lambda_o = 1 \cdot 0 \cdot 118^2/\pi \cdot 0 \cdot 0131 = 0 \cdot 34,$$

and $$x_o = 19 \cdot 4 \cdot 0 \cdot 34 = 6 \cdot 6 \ \Omega.$$

Zig-zag Leakage. The magnetizing current is $2 \cdot 0$ A., the voltage 400 V. (each per phase), and the magnetizing reactance

$$x_m = 400/2 \cdot 0 = 200 \ \Omega.$$

From eq. (64),

$$x_z = 200 \cdot (5/6)[(1/9^2) + (1/7 \cdot 5^2)] = 5 \cdot 0 \ \Omega.$$

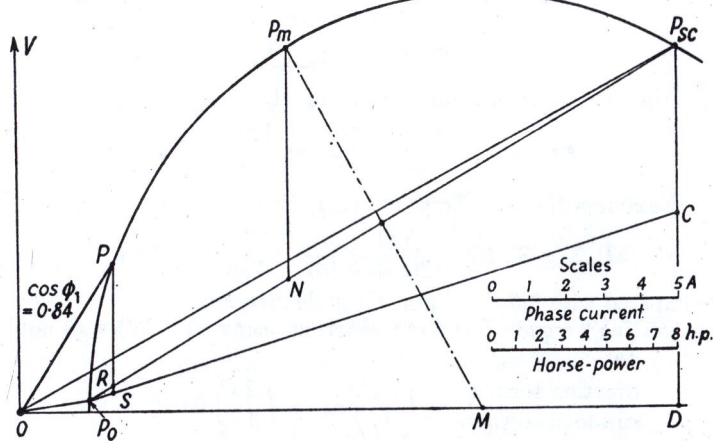

FIG. 257. CIRCLE DIAGRAM FOR 5 H.P., 400 V., CAGE MOTOR

There is no differential leakage because the machine has a cage rotor. The total phase reactance is

$$X_1 = x_s + x_o + x_z = 6 \cdot 0 + 6 \cdot 6 + 5 \cdot 0 = 17 \cdot 6 \ \Omega.$$

The resistance is

$$R_1 = r_1 + r_2' = 4 \cdot 85 + 4 \cdot 3 = 9 \cdot 15 \ \Omega.$$

Hence

$$Z_1 = \sqrt{(9 \cdot 15^2 + 17 \cdot 6^2)} \simeq 20 \ \Omega; \ \cos \phi_{sc} = 9 \cdot 15/20 = 0 \cdot 457.$$

The short-circuit current is

$$I_{sc} = 400/20 = 20 \cdot 0 \ \text{A. per phase, or } 34 \cdot 6 \ \text{A. per line.}$$

CIRCLE DIAGRAM. The circle diagram per phase, Fig. 257, is drawn from the no-load and short-circuit currents and power factors. For full load,

$$I_1 = 4 \cdot 5 \ \text{A.}; \quad I_2' = 3 \cdot 6 \ \text{A.}; \quad \cos \phi = 0 \cdot 84.$$

PERFORMANCE. The estimated currents and power factor having been confirmed by the circle diagram, the results therefrom can be used to calculate specific items of performance.

Efficiency. The full-load losses are—

						w
Stator I^2R: $3 . 4\cdot5^2 . 4\cdot85$	.	.	.	.	.	295
Rotor I^2R: $3 . 3\cdot6^2 . 4\cdot30$	.	.	.	.	.	165
Core and teeth	.	.	.	.	.	250
Friction and windage	.	.	.	.	.	60
						———
		Total loss	.	.	.	770
		Output, $5 . 746$	.	.	.	3 730
						———
		Input	.	.	.	4 500

Efficiency $= 100 (3\ 730/4\ 500) = 83\%$

less $\frac{1}{2}\%$, $\underline{\eta = 82\frac{1}{2}\%}$

Slip. The full-load slip is, from eq. (104),

$$s = \frac{165}{3\ 730 + 165 + 60} = \frac{165}{3\ 955} = \underline{4\cdot2\%}.$$

Maximum Output. From eq. (121) this is approximately

$$P_{mm} = 3 . 400 . \frac{(20 - 2)}{2(1 + 0\cdot457)} . \frac{1}{746} = 9\cdot9 \text{ h.p.}$$

compared with 9·5 h.p. from the circle diagram.

Starting Torque. For direct starting, using eq. (136) with equivalent rotor currents,

$$\frac{\text{Starting torque}}{\text{Full-load torque}} = \left(\frac{I_{sc}}{I_2}\right)^2 s_n = \left(\frac{20}{3\cdot6}\right)^2 0\cdot042 = 1\cdot3$$

For star-delta starting the torque would be one-third of this, or 43 per cent of full-load torque. The values obtained from the circle diagram are the same.

COOLING. In such a small machine the cooling is good, since the rotor end-rings are bare and there are fans at each end. The overhang of the stator contributes largely to the dissipation. From the drawing, Fig. 256, the cooling surface across which air is forcibly moved is approximately a cylinder formed by the inner faces of overhang and core, and the annular faces formed by the coil ends. The back of the stator core also dissipates. The inner area is estimated as

$$\pi . 0\cdot15 . 0\cdot25 + 2 . \pi . 0\cdot2 . 0\cdot04 = 0\cdot167 \text{ m.}^2$$

On account of the relative movement of stator and rotor, this area is $(1 + 0\cdot1u)$ times as effective as the back of the stator. The peripheral speed u is

$$\pi . 0\cdot15 . 1\ 500/60 = 11\cdot75 \text{ m. per sec.}$$

The effective area of the inner surfaces is thus

$$0\cdot167(1 + 1\cdot175) = 0\cdot364 \text{ m.}^2$$

The back of the stator has an area

$$\pi \,.\, 0{\cdot}24 \,.\, 0{\cdot}9 = 0{\cdot}068 \text{ m.}^2$$

The total effective area is $0{\cdot}364 + 0{\cdot}068 = 0{\cdot}432$ m.^{2}. The stator loss is $288 + 245 = 533$ W. Taking an average value of the cooling coefficient c from p. 246 as $c = 0{\cdot}03$, the temperature-rise is

$$\theta_m = 0{\cdot}03 \,.\, 533/0{\cdot}432 = 37^\circ \text{ C.},$$

from eq. (101). The rotor loss is 167 W., and it has a considerable surface of bare copper at the ends as well as the outer and inner

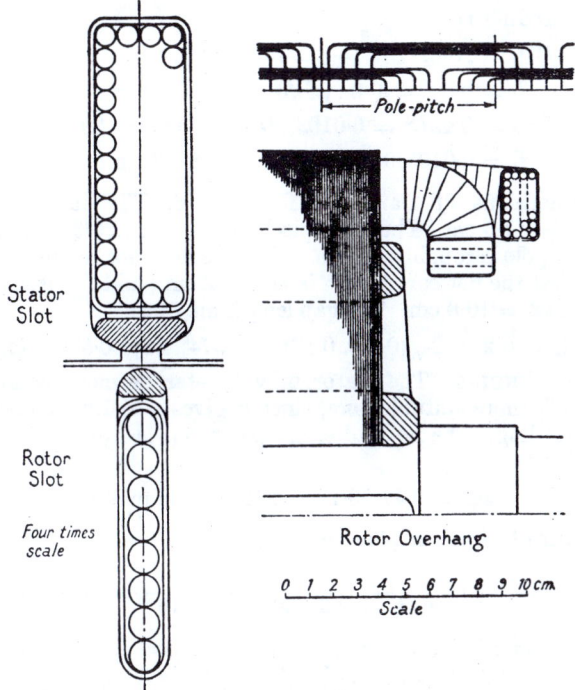

Stator
Slot

Rotor
Slot

*Four times
scale*

Pole-pitch

Rotor Overhang

0 1 2 3 4 5 6 7 8 9 10 cm.
Scale

Fig. 258. Details of 20 h.p., 420 V., Three-phase, 50-cycle, 6-pole
Slip-ring Induction Motor

core surfaces. Its temperature-rise is about 35° C. These figures are comfortably within the specified limits (p. 247).

5. **Example.** 20-h.p., 420-V., 3-phase, 50-c/s., slip-ring induction motor; 1 000 r.p.m. Specification, B.S. 2613: 1955. Drawing, Fig. 258.

Main Dimensions. The synchronous speed is 1 000 r.p.m., so that $2p = 6$ poles.

With $\bar{B} = 0.47$ Wb./m.^{2} and $ac = 24\,500$ amp.-cond. per m., the output coefficient is

$$G = 11 \cdot 0.955 \cdot 0.47 \cdot 24\,500 \cdot 10^{-3} \simeq 120.$$

This is a larger machine than that in § 4, but it has six poles and a tendency to higher reactance, so that the mean density is but little greater. The electric loading is increased, but not greatly, since the cooling is not so good. Taking $\eta = 0.86$ and $\cos \phi = 0.84$, the kVA. rating is

$$S = 20 \cdot 0.746/0.86 \cdot 0.84 = 20.6 \text{ kVA.}$$

The D^2L product is

$$D^2L = 20.6/120 \cdot 16.67 = 0.0103 \text{ m.}^3$$

With $\quad L = 0.8Y = 0.8 \cdot \pi D/6 = 0.42D$, then

$$D^2L = 0.42D^3 = 0.0103, \quad D = 0.29 \text{ m.} = 29 \text{ cm.,}$$

and $\quad L = 0.0103/0.29^3 \simeq 0.122 \text{ m.} \simeq 12 \text{ cm.}$

These dimensions fit frame El for $2p = 6$. The outside stator diameter is 41.5 cm. The pole-pitch is $Y = \pi \cdot 29/6 = 15.2$ cm., so that the pole is roughly square. The core does not require a radial duct, so that the net core length is $L_s = 12$ cm., and the iron length $L_i = 0.9 \cdot 12 = 10.8$ cm. The gap length may be

$$l_g = 0.2 + 2\sqrt{(0.29 \cdot 0.12)} = 0.574, \text{ say } 0.575 \text{ mm.}$$

STATOR WINDING. The stator may be star or mesh connected. The latter is more suitable here, since it gives a section suitable for a mush winding. The no-load e.m.f. per phase is $E_1 \simeq 420$ V. The no-load flux is

$$\Phi_m = 0.47 \cdot 0.152 \cdot 0.12 = 8.56 \text{ mWb.}$$

The number of turns per phase is

$$T_1 = \frac{420}{4.44 \cdot 0.955 \cdot 50 \cdot 8.56 \cdot 10^{-3}} = 232.$$

The number of stator slots per pole per phase is rarely less than 2, and may be at least 3. With $g_1' = 3$, $S_1 = 3 \cdot 3 \cdot 6 = 54$ stator slots of slot-pitch $y_{s1} = \pi \cdot 29/54 = 1.69$ cm., a suitable figure. The T_1 turns of a phase are housed in 18 slots, giving $2 \cdot 232/18 = 26$ conductors per slot, and actually

$$T_1 = 26 \cdot 3 \cdot 3 = 234 \text{ turns,}$$

and the no-load flux falls to

$$\Phi_m = 8.5 \text{ mWb.}$$

The full-load current per phase is

$$I_1 = 20 \cdot 0.746/3 \cdot 420 \cdot 0.86 \cdot 0.84 = 16.4 \text{ A.}$$

With a current density $\delta_1 = 4$ A. per mm.^{2} the required area $16 \cdot 4/4 = 4 \cdot 1$ mm.^{2}, obtainable by connecting two No. 16 S.W.G. wir in parallel, giving $a_1 = 2 \cdot 2 \cdot 07 = 4 \cdot 14$ mm.^{2} area. The diamet of each wire is $1 \cdot 63$ mm. bare, insulated with fine d.c.c. to $1 \cdot 86$ mm.

If a stator slot $10 \cdot 5$ mm. wide is taken, the 26 conductors (52 wires) per slot can be arranged as follows—

Slot Width

		mm.
4 wires, 4 × 1·86 .	. . . . .	7·44
Slot lining, 2 × 0·5	. . . .	1·0
Roughness and slack	. . . .	2·06
Total	. . . .	10·5 mm.

Slot Depth

		mm.
13 layers, 13 × 1·86	. . . .	24·2
Slot lining, 3 × 0·5	. . . .	1·5
Wedge	. . . . .	3·5
Lip .	. . . . .	1·0
Slack .	. . . . .	3·8
Total	. . .	34·0 mm.

The slots are parallel-sided with 4 mm. openings (Fig. 258). The slot pitch at one-third depth from the narrower end is $1 \cdot 82$ cm. The tooth width at this point is $1 \cdot 82 - 1 \cdot 05 = 0 \cdot 77$ cm., and the mean tooth density is

$$\bar{B}_{tt} = 8 \cdot 5 \cdot 10^{-3}/3 \cdot 3 \cdot 0 \cdot 108 \cdot 0 \cdot 77 \cdot 10^{-2} = 1 \cdot 13 \text{ Wb./m.}^2$$

The maximum tooth density will be about $1 \cdot 5 \cdot 1 \cdot 13 = 1 \cdot 7$ Wb./m.^{2}

Stator Resistance. The estimated length of a mean conductor is

$$L_{mc} = 0 \cdot 43 \text{ m.}$$

The length per phase is $2 \cdot 234 \cdot 0 \cdot 43 = 200$ m. The resistance at 75° C. is

$$r_1 = 0 \cdot 021 \cdot 200/2 \cdot 2 \cdot 07 = 1 \cdot 01 \, \Omega.$$

The I^2R loss on full load for all phases is

$$3 \cdot 16 \cdot 4^2 \cdot 1 \cdot 01 = 820 \text{ W.}$$

Stator Core. With the frame dimension $D_o = 41 \cdot 5$ cm., the depth of the stator core at the back of the slots is

$$\tfrac{1}{2}(41 \cdot 5 - 29 - 2 \cdot 3 \cdot 4) = 2 \cdot 85 \text{ cm.}$$

The maximum core density is

$$B_{c1} = 8 \cdot 5 \cdot 10^{-3}/2 \cdot 0 \cdot 0285 \quad 0 \cdot 108 = 1 \cdot 38 \text{ Wb./m.}^2$$

ROTOR WINDING. The motor is not quite large enough to make a bar rotor winding advantageous. A mush winding requires a slot opening to insert the wires. There is some advantage in using closed slots as this somewhat reduces the magnetizing current. A concentric winding is therefore chosen with about eight conductors per slot arranged one above the other. This arrangement entails narrow slots, so that there will be more rotor than stator slots. With $g_2' = 4$, then $S_2 = 4 . 3 . 6 = 72$ slots, and $T_2 = 96$. The slots are closed by a bridge about 0·5 mm. thick, and the ends are semicircular (Fig. 258). The slot-pitch is $y_{s2} = \pi 28·9/72 = 1·26$ cm.

The equivalent rotor current is approximately

$$I_2' \simeq I_1 \cos \phi_1 = 16·4 . 0·84 = 13·8 \text{ A.}$$

With 72 slots and 8 conductors per slot in the rotor, and 54 slots with 26 conductors per slot in the stator, the actual rotor current is

$$I_2 = 13·8(54 . 26/72 . 8) = 33·6 \text{ A.}$$

With a current density of $\delta_2 = 4·9$ A. per mm.², the necessary area is $33·6/4·9 = 6·86$ mm.² Wires of No. 11 S.W.G. have an area of $a_2 = 6·82$ mm.², and a diameter of 2·95 mm. bare, and these will be taken. D.c.c. insulation gives an overall diameter of 3·20 mm.

Using slot insulation (leatheroid and flexible micanite) as in the stator, the slot dimensions will be made up as follows—

Slot Width

		mm.
Conductor	.	3·20
Slot insulation, 2 × 0·5 . . .	.	1·0
Roughness and slack	.	0·8
Total . . .	.	5·0 mm.

Slot Depth

		mm.
8 conductors, 8 × 3·2	.	25·6
Slot insulation, 2 × 0·5 . . .	.	1·0
Filling wedge	.	3·0
Roughness and slack	.	1·4
Total . .	.	31·0 mm.

The slot-pitch at one-third depth from the root is 1·078 cm., so that the tooth width here is $1·078 - 0·5 = 0·578$ cm. The mean tooth density is

$$\bar{B}_{t\frac{1}{3}} = 8·5 . 10^{-3}/4 . 3 . 0·578 . 10^{-2} . 0·108 = 1·13 \text{ Wb./m.}^2$$

The maximum will be about 1·7 Wb./m.²

Rotor Coils. A 2-plane overhang with concentric coils is adopted, as shown in Fig. 258.

Rotor Resistance. There are $24 \times 8 = 192$ conductors in series per phase. The estimated length of mean conductor is 33 cm. The length per phase is $192 . 33 . 10^{-2} = 63.5$ m. The resistance per phase is

$$r_2 = 0.021 . 63.5/6.82 = 0.195 \ \Omega.$$

The full-load I^2R loss, all phases, is

$$3I_2{}^2 r_2 = 3 . 33.6^2 . 0.195 = \underline{665 \text{ W.}}$$

Rotor Core. Taking the same order of core density as in the stator, the depth below the slots must be about 2·85 cm. The inside diameter will be $29 - 2(0.05 + 0.05 + 3.1 + 2.85) = 16.9$, say 17 cm. The actual core depth is 2·8 cm. and the core density is

$$B_{c2} = 8.5 . 10^{-3}/2 . 2.8 . 10^{-2} . 0.108 = 1.4 \text{ Wb./m.}^2$$

No-load Current. The gap coefficient is

$$k_{g1} = \frac{16.9}{16.9 - 0.75 . 4} = 1.215; \quad k_{g2} = \frac{12.6}{12.6 - 0.34 . 1} = 1.028,$$

taking the saturated bridge of the rotor slots as equivalent to a 1 mm. opening.

$$l_g' = 0.575 . 1.215 . 1.028 = 0.72 \text{ mm.}$$

The effective gap area is $A_g = YL = 15.2 . 12 = 182.5$ cm.$^2 = 0.0182$ m.^{2} The following table is drawn up for 0·5 mm. Lohys plates.

MAGNETIZATION

$\Phi_m = 8.5$ mWb.

Part	Area A m.^{2}	Length l m.	$B = \Phi_m/2A$	$B_{30} = 1.36 \ \Phi_m/A$	at	$AT = at . l$
Stator core	0·0031	0·07	1·38	—	550	40
„ teeth ($\frac{1}{3}$)	0·0075	0·034	—	1·54	1 500	50
Gap	0·0182	0·72 . 10^{-3}	—	0·634	505 000	362
Rotor teeth ($\frac{1}{3}$)	0·0075	0·031	—	1·54	1 500	47
„ core	0·0031	0·030	1·36	—	500	15
					Total, AT_{30}	515

Magnetizing current

$$I_{0r} = 3 . 515/1.17 . 0.955 . 234 = 6.0 \text{ A. per phase,}$$
$$= 10.4 \text{ A. per line.}$$

Core Losses. Fig. 112 (*b*). Stator teeth:

weight = 12·5 kg.;

specific loss for $B_{t1} = 1.7$ Wb./m.^{2} (max.), 26 W. per kg.

loss = 26 . 12·5 = 325 W.

Stator core:

   weight = 28 kg.;

   specific loss at $B_c = 1.38$ Wb./m.^{2} (max), 18 W. per kg.;

   loss = 18 . 28 = 505 W.

Total iron loss = 325 + 505 = 830 W.

Friction and windage, 1% input, say 200 W.

No-load loss $P_0 = 830 + 200 = 1\,030$ W.

Active component of no-load current,

$$I_{0a} = 1\,030/3 \, . \, 420 = 0.82 \text{ A. per phase.}$$

The no-load current is

$$I_0 = \sqrt{(0.82^2 + 6.0^2)} = 6.05 \text{ A. per phase,}$$
$$= 10.5 \text{ A. per line,}$$

at a power factor $\cos \phi_0 = 0.82/6.05 = 0.135$.

SHORT-CIRCUIT CURRENT. Applying the formulae from Chapter IX, §7:

$$\lambda_{s1} = \frac{24}{3 \, . \, 10.5} + \frac{5}{10.5} + \frac{2 \, . \, 3}{10.5 + 4} + \frac{1}{4} = 1.9;$$

$$\lambda_{s2} = \frac{25}{3 \, . \, 5} + \frac{1}{5} + \frac{2 \, . \, 3}{5 + 2} + 5\frac{0.5}{5} = 3.2.$$

For the wedge portion, the effective slot-opening is taken as 2 mm,
For the lip permeance, the bridge is considered as 5 mm. wide.
0.5 mm. deep, and with a relative permeability of 5.

$$\lambda_{s2}(K_{w1}^2 S_1/K_{w2}^2 S_2) = 3.2(54/72) = 2.4.$$
$$x_s = 15.8 \, . \, 50(234^2/3 \, . \, 3)0.12 \, (1.9 + 2.4)10^{-6}$$
$$= 4.8 \, . \, 0.12 \, . \, 4.3 = 2.5 \, \Omega.$$
$$L_0\lambda_0 = 1 \, . \, 0.152^2/\pi \, . \, 0.0169 = 0.435;$$
$$x_0 = 4.8 \, . \, 0.435 = 2.1 \, \Omega.$$
$$x_m = V/I_m = 420/6 = 70 \, \Omega;$$
$$x_z = \frac{5}{6}70\left[\frac{1}{9^2} + \frac{1}{12^2}\right] = 1.1 \, \Omega;$$
$$x_\lambda = 70(21.5 + 21.5)10^{-4} = 0.3 \, \Omega.$$
$$X_1 = x_s + x_0 + x_z + x_\lambda = 2.5 + 2.1 + 1.1 + 0.3 = 6.0 \, \Omega.$$
$$R_1 = r_1 + (T_1/T_2)^2 r_2 = 1.01 + 0.195(234/96)^2 = 2.2 \, \Omega.$$
$$Z_1 = \sqrt{(R_1^2 + X_1^2)} = 6.4 \, \Omega.$$
$$\cos \phi_{sc} = 2.2/6.4 = 0.34.$$

The short-circuit current is

$$I_{sc} = 420/6.4 = 65.5 \text{ A per phase} = 114 \text{ A. per line.}$$

This is about four times full-load current.

CIRCLE DIAGRAM. The circle diagram drawn for phase currents, Fig. 259, gives for full load—

$$I_1 = 16\cdot35 \text{ A.,} \quad I_2' = 13\cdot25 \text{ A.,} \quad \cos\phi_1 = 0\cdot835.$$

The maximum power is $P_{mm} = 36\cdot5$ h.p. and the maximum torque is 200 per cent of full-load torque.

PERFORMANCE. *Efficiency.* The full-load losses are

	W.
Stator I^2R: $3 \cdot 16\cdot35^2 \cdot 1\cdot01$	810
Rotor I^2R: $3 \cdot 13\cdot25^2 \cdot 1\cdot16$	610
Core and teeth	830
Friction and windage	200
Total loss . . .	2 450
Output: 20 . 746 . .	14 920
Input	17 370

$$\text{Efficiency} = 100 \ (14\ 920/17\ 370) = 86\cdot0\%$$
$$\text{less } \tfrac{1}{2}\%, \quad \eta = 85\cdot5\%$$

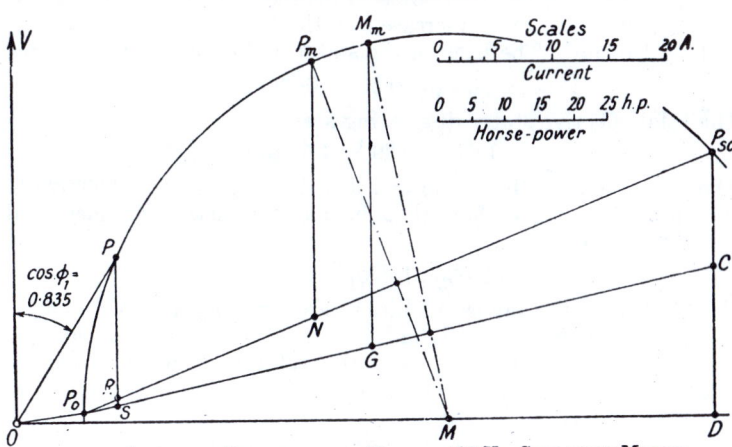

FIG. 259. CIRCLE DIAGRAM FOR 20 H.P., 420 V., SLIP-RING MOTOR

Slip At full load,

$$s = \frac{610}{14\ 920 + 610 + 200} = \frac{610}{15\ 730} = 3\cdot87\%.$$

Maximum Output.

$$P_{mm} = 3 \cdot 420 \ \frac{65\cdot5 - 6\cdot0}{2(1 + 0\cdot34)} \cdot \frac{1}{746} = 37\cdot5 \text{ h.p.}$$

by calculation, compared with 36·5 h.p. from the circle diagram.

STARTER. The starting torque with the rotor short-circuited is of no interest, since the motor has a resistance starter. From Chapter XII, § 3, using a 4-step starter with $I_{max} = $ full-load current, the slip

s_{max} corresponding to full-load rotor current with slip-rings short-circuited is 0·0387. The rotor resistance per phase is 0·195 Ω, say 0·2 Ω to include the rings. Then the resistance in circuit on the first step is

$$R_1 = r_2/s_{max} = 0·2/0·0387 = 5·2 \text{ Ω.}$$

The ratio $I_{min}/I_{max} = \gamma = \sqrt[4]{s_{max}} = \sqrt[4]{0·0387} = 0·44.$

Then

$$\rho_1 = R_1(1 - \gamma) = 5·2 . 0·56 = 2·9 \text{ Ω.}$$
$$\rho_2 = \gamma\rho_1 = 0·44 . 2·9 = 1·3 \text{ Ω.}$$
$$\rho_3 = \gamma\rho_2 = 0·44 . 1·3 = 0·55 \text{ Ω.}$$
$$\rho_4 = \gamma\rho_3 = 0·44 . 0·55 = 0·25 \text{ Ω.}$$
$$r_2 = 0·2 \text{ Ω.}$$
$$R_1 = \overline{5·2 \text{ Ω.}}$$

COOLING. Proceeding as in the previous example, the inner area of the stator bore and end connections is

$$\pi . 0·29 . 0·30 + 2\pi . 0·35 . 0·06 = 0·404 \text{ m.}^2$$

The peripheral speed is $\pi . 0·29 . 1\,000/60 = 15·1$ m. per sec., so that the effective area is increased by the factor $(1 + 0·1u) = 2·51$, making 1·01 m.². The back of the stator and overhang has an area

$$\pi . 0·415 . 0·30 = 0·39 \text{ m.}^2$$

The total effective stator dissipating area is

$$1·01 + 0·39 = 1·40 \text{ m.}^2$$

The stator loss is $810 + 830 = 1\,640$ W. Taking c as 0·033 since the motor is bigger than that in § 4, the maximum measured temperature-rise is

$$\theta_m = 0·033 . 1\,640/1·40 = 38·7°\text{C}$$

For the rotor, the outer cylindrical area of core and overhang is about $\pi . 0·29 . 0·25 = 0·227$ m.³; the end area of the overhang is $2\pi . 0·25 . 0·05 = 0·078$ m.³; and the total is 0·305 m.² This is increased to $0·305 . 2·51 = 0·767$ m.² on account of the rotation. The rotor loss is 610 W., so that

$$\theta_m = 0·033 . 610/0·767 = 26°\text{C.,}$$

disregarding the inner cooling channels in the rotor core.

6. **Example** 150-h.p., 2 000-V., 3-phase, 50-c/s., slip-ring induction motor; 600 r.p.m. Specification, B.S. 2613: 1955.

An outline design is given.

MAIN DIMENSIONS. With $\bar{B} = 0·475$ Wb./m.²

$$ac = 27\,000 \text{ amp.-cond. per m.,}$$
$$\eta = 0·91 \text{ and } \cos \phi_1 = 0·86 :$$
$$G = 11 . 0·955 . 0·475 . 27\,000 . 10^{-3} = 135,$$
$$S = 150 . 0·746/0·91 . 0·861 = 143 \text{ kVA.,}$$
$$D^2L = 143/135 . 10 = 0·106 \text{ m.}^3 = 1·06 . 10^5 \text{ cm.}$$

With $2p = 10$, the diameter/length ratio is too great unless $L > Y$. If $L = 1 \cdot 25 Y = 1 \cdot 25\pi D/10 = 0 \cdot 39D$, then $D^3 = 1 \cdot 06 \cdot 10^5/0 \cdot 39$, $D = 65$ cm., $L = 25$ cm., $Y = 20 \cdot 4$ cm.

Three radial ventilating ducts are used, each 1 cm. wide, so that $L_s = 25 - 3 = 22$ cm., $L_i = 0 \cdot 9 \cdot 22 = 19 \cdot 8$ cm. The gap length

$$l_g = 0 \cdot 2 + 2\sqrt{(0 \cdot 65 \cdot 0 \cdot 25)} = 0 \cdot 2 + 0 \cdot 81 = 0 \cdot 95 \text{ mm., say.}$$

The Lohys steel core-plates used are $0 \cdot 4$ mm. thick.

STATOR WINDING. Star connection is found suitable. On no load $E_1 \simeq 2\,000/\sqrt{3} = 1\,150$ V. From $\overline{B}$, Y and L, the no-load flux is

$$\Phi_m = 0 \cdot 475 \cdot 0 \cdot 204 \cdot 0 \cdot 25 = 242 \text{ mWb.,}$$

whence

$$T_1 = \frac{1\,150}{4 \cdot 44 \cdot 0 \cdot 955 \cdot 50 \cdot 242 \cdot 10^{-3}} = 224.$$

Using a concentric winding in 3 slots per pole per phase, $S_1 = 90$, $g_1' = 3$, and with $z_1 = 15$ conductors per slot,

$$T_1 = 5 \cdot 3 \cdot 15 = 225 \text{ turns per phase.}$$

The full-load current is

$$I_1 = 150 \cdot 0 \cdot 746/3 \cdot 1\,150 \cdot 0 \cdot 91 \cdot 0 \cdot 86 = 42 \text{ A.}$$

Taking $\delta_1 = 4$ A. per mm.^{2}, $a_1 = 42/4 = 10 \cdot 5$ mm.^{2} Suitable conductors 6 mm. $\times$ 1·75 mm. bare, 6·5 mm. $\times$ 2·25 mm. insulated d.c.c., will fit into a slot 10 mm. wide and 42 mm. deep, with a 3 mm. opening.

The stator conductors have a mean length of 70 cm. The resistance at 75° C. is

$$r_1 = 0 \cdot 021 \cdot 2 \cdot 70 \cdot 225 \cdot 10^{-2}/10 \cdot 5 = 0 \cdot 63 \ \Omega.$$

The total stator I^2R loss is

$$3I_1^2 r_1 = 3 \cdot 42^2 \cdot 0 \cdot 63 \cdot 10^{-3} = 3 \cdot 35 \text{ kW.}$$

ROTOR WINDING. It is possible to obtain a two-layer winding for the rotor since there is a free choice of voltage. With $g_2' = 4$ slots per pole per phase, $S_2 = 120$. If there are 2 conductors per slot, the rotor phase voltage on open-circuit is

$$1\,150 \cdot 40/225 = 204 \text{ V. approximately}$$

(actually a little less than this). Such a winding would be quite suitable provided that the rotor teeth were not too highly saturated or too weak mechanically.

The equivalent rotor current is

$$I_2' \simeq 42 \cdot 0 \cdot 86 = 36 \text{ A.}$$

The actual current is

$$I_2 = 36 \cdot 225/40 = 200 \text{ A. approx.}$$

With $\delta_2 = 4 \cdot 5$ A. per mm.^{2}, $a_2 = 200/4 \cdot 5 = 44 \cdot 5$ mm.^{2} Since the working frequency in the rotor is only one or two cycles per sec., the conductors may be solid, say 4 mm. $\times$ 12 mm. bare in slots 7 mm. wide and 35 mm. deep, with 3 mm. opening. The slot-pitch is $y_{s2} = \pi \cdot 64 \cdot 8/120 = 1 \cdot 695$ cm., so that the rotor teeth will not be too narrow.

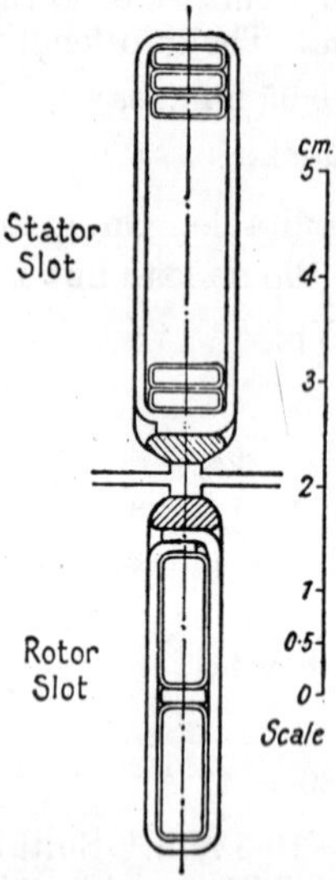

The mean length of conductor is 59 cm., and the resistance hot is

$$r_2 = 0 \cdot 021 \cdot 2 \cdot 40 \cdot 59 \cdot 10^{-2}/48 = 0 \cdot 0206 \; \Omega.$$

The I^2R loss is

$$3I_2{}^2 r_2 = 3 \cdot 200^2 \cdot 0 \cdot 0206 \cdot 10^{-3} = 2 \cdot 47 \text{ kW}.$$

NO-LOAD CURRENT.

$$k_{g1} = \frac{22 \cdot 7}{22 \cdot 7 - 0 \cdot 46 \cdot 3} = 1 \cdot 063 \,;$$

$$k_{g2} = \frac{16 \cdot 95}{16 \cdot 95 - 0 \cdot 46 \cdot 3} = 1 \cdot 087 \,;$$

$$l_g{}' = 1 \cdot 063 \cdot 1 \cdot 087 \cdot 0 \cdot 95 = 1 \cdot 1 \text{ mm}.$$

Effective axial length of gap

$$L' = 25 - 0 \cdot 68 \cdot 3 \cdot 1 \simeq 23 \cdot 0 \text{ cm}.$$
$$A_g = YL' = 20 \cdot 4 \cdot 23 \cdot 0 = 470 \text{ cm.}^2$$
$$= 0 \cdot 047 \text{ m.}^2$$

FIG. 260. SLOT DE-TAILS OF 150 H.P., 2 000 V., SLIP-RING MOTOR

MAGNETIZATION

$$\Phi_m = 242 \text{ mWb}.$$

Part	Area A m.^{2}	Length l m.	$B =$ $\Phi_m/2A$	$B_{30} =$ $1 \cdot 36 \, \Phi_m/A$	at	$AT =$ $at \cdot l$
Stator core .	0·0105	0·09	1·15	—	250	22
,, teeth ($\frac{1}{3}$)	0·0243	0·042	—	1·35	480	20
Gap . .	0·0470	1·1 . 10^{-3}	—	0·70¹	558 000	615
Rotor teeth ($\frac{1}{3}$)	0·0207	0·035	—	1·59	2 200	77
,, core .	0·0107	0·06	1·13	—	200	18

Total, AT_{30} 749

$$I_{0r} = 5 \cdot 749/1 \cdot 17 \cdot 0 \cdot 955 \cdot 225 = 14 \cdot 9 \text{ A. per phase and line.}$$

Core Losses. For the stator teeth, 1 300 W.; for the stator core, 2 180 W.; total, 3·5 kW.

Friction and Windage, 1% kVA. $= 1.40$ kW.

Total no-load loss—

	kW.
Iron	3·5
Mechanical	1·4
Stator $I^2R: 3 . 15^2 . 0.63 . 10^{-3}$	0·4
	5·3 kW.

$$I_{0a} = 5.3 . 10^3/(\sqrt{3}) . 2\,000 = 1.53 \text{ A.}$$

No-load current

$$I_0 = \sqrt{(14.9^2 + 1.53^2)} = 15.0 \text{ A.}$$

$$\cos \phi_0 = 1.53/15.0 \simeq 0.1.$$

SHORT-CIRCUIT CURRENT. With slot data from Fig. 260, the reactances are—

$$\lambda_{s1} = 2.2; \ \lambda_{s2} = 2.5; \ \lambda_{s2}(S_1/S_2) = 1.9;$$
$$x_s = 15.8 . 50(225^2/5 . 3)0.22(2.2 + 1.9)10^{-6}$$
$$= 2.67 . 0.22 . 4.1 = 2.41 \ \Omega.$$
$$L_0\lambda_0 = 1 . 0.204^2/\pi 0.0226 = 0.59;$$
$$x_0 = 2.67 . 0.59 = 1.64 \ \Omega.$$
$$x_m = 2\,000/\sqrt{3} . 14.9 = 77.5 \ \Omega;$$
$$x_z = \frac{5}{6}77.5\left(\frac{1}{9^2} + \frac{1}{12^2}\right) = 1.25 \ \Omega.$$
$$x_h = 77.5(21.5 + 21.5)10^{-4} = 0.33 \ \Omega.$$

The total reactance is thus

$$X_1 = 2.41 + 1.64 + 1.25 + 0.33 = 5.63 \ \Omega.$$

The resistances are

$$r_1 = 0.70 \ \Omega; \ r_2 = 0.0206 \ \Omega; \ r_2' = 0.0206(225/40)^2 = 0.65 \ \Omega.$$

The total resistance is

$$R_1 = 0.70 + 0.65 = 1.35 \ \Omega$$

and the short-circuit impedance

$$Z_1 = \sqrt{(1.35^2 + 5.63^2)} = 5.79 \ \Omega.$$

$$\cos \phi_{sc} = 1.35/5.79 = 0.234.$$

The short-circuit current per phase is

$$I_{sc} = 2\,000/\sqrt{3} . 5.79 = 200 \text{ A.}$$

CIRCLE DIAGRAM. From the diagram, for full-load—

$$I_1 = 42 \text{ A.}; \ I_2' = 35 \text{ A.}; \ \cos \phi_1 = 0.86.$$

PERFORMANCE. *Efficiency*.

		kW.
Stator I^2R: 3 . 42^2 . 0·63	. . .	3·35
Rotor I^2R: 3 . 35^2 . 0·65	. . .	2·40
Core and teeth .	. . .	3·50
Friction and windage .	. . .	1·40
	Total loss . . .	11·05
	Output, 150 . 0·746 . . .	112·0
	Input	123·0 kW.

$$\text{Efficiency} = 100 \ (112/123) = 91\cdot2\%;$$

$$\text{less } \tfrac{1}{2}\%, \quad \eta = 90\cdot7 \ \%.$$

Slip.
$$s = \frac{2\cdot40}{112\cdot0 + 1\cdot40 + 2\cdot40} = 2\cdot1\%.$$

Maximum Output.

$$P_{mm} = 3 . 1 \ 150 \ \frac{200 - 15}{2(1 + 0\cdot234)} \cdot \frac{1}{746} = 360 \text{ h.p.}$$

Maximum Torque. From the diagram, the maximum torque is 2·4 times full-load torque.

7. Examples. Outline designs are given below for two slip-ring machines differing from the normal industrial range. *A* is a motor to drive a centrifugal pump for cooling water: it is provided with Class A insulation and brush-lifting gear. The rating is for a 30° C. rise with ambient air at 55° C. having 100 per cent humidity. The specific loadings are consequently small. Motor *B* is a flame-proof mine-haulage machine, rated for a 50° C. rise, but the severe restriction on ventilation makes it necessary to design for very small I^2R losses, especially in the rotor. The main dimensions of each machine are given in Fig. 261.

Rating		*A*	*B*
Full-load output, h.p.	P_m	360	100
Line voltage, V.	V_1	3 000	3 000
Frequency, c.p.s.	f	50	50 .
Number of phases. . . .	N	3	3
Efficiency, %	η	93	93·8
Power factor, %	$\cos \phi_1$	84	86
Number of poles	$2p$	10	8
Syn. speed, r.p.m./r.p.s.	n_s	600/10	750/12·5
Full-load speed, r.p.m.		590	743
Input, kVA.. . . .	S	347	93
Full-load line current, A.	I_1	66·7	17·8
Loading			
Spec. mag. loading, Wb./m.^{2} . . .	$\overline{B}$	0·388	0·225
Spec. electric loading	ac	30 900	14 900
Output coefficient	G	127	342
D^2L product, m.^{3}	D^2L	0·272	0·218

Main Dimensions

		A	B
Stator-bore, cm.	D	82·5	66
Gross core length, cm.	L	40	50
Periphery. cm.	πD	259	207
Pole-pitch, cm.	Y	25·9	25·9
Ducts, cm.	$n_d w_d$	5 × 1	nil
Net core length, cm.	L_s	35	50
Iron length, cm.	L_i	31·5	45
Gap length, mm.	l_g	1·3	1·2

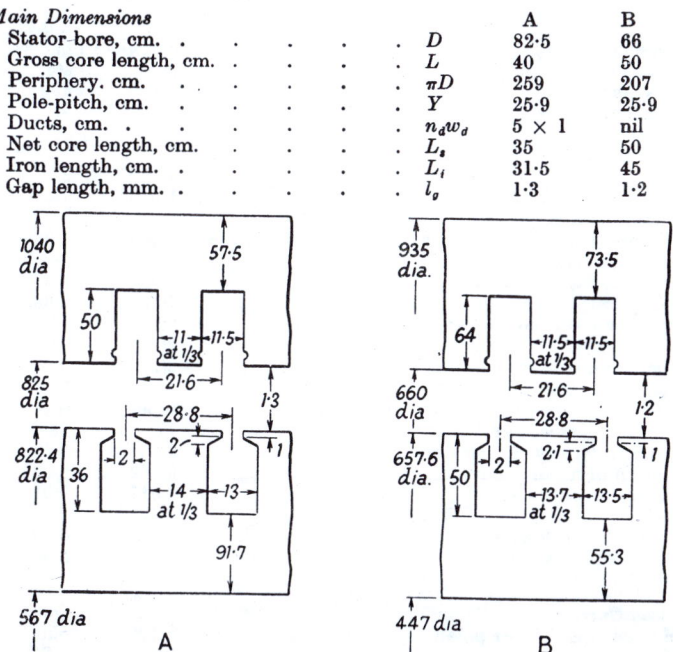

Fig. 261. Slot and Core Details
Dimensions in millimetres

Stator

		2-layer	2-layer
Type of winding		diamond	diamond
Connection		star	star
No-load phase e.m.f., V.	E_1	1 730	1 730
No-load flux, milliwebers	Φ_m	40	29·1
Turns per phase	T_1	200	286
Number of slots	S_1	120	96
Slots per pole	g_1	12	12
Slots per pole per phase	g_1'	4	4
Coil-span, slots		1–13	1–12
Distribution factor	k_{m1}	0·958	0·958
Coil-span factor	k_{c1}	1·0	0·991
Winding factor	K_{w1}	0·958	0·950
Slot-pitch, cm.	y_{s1}	2·16	2·16
Conductors per slot	z_{s1}	2 × 5	2 × 9
Conductor size, mm.		6·5 × 3	6·5 × 2·2
covering		d.c.c.	d.c.c.
area, mm.²	a_1	19·0	14·0
Current density, A. per mm.²	δ_1	3·5	1·27
Length of mean conductor, m.	L_{mc}	0·963	1·032
Phase resistance at 20° C., Ω.		0·35	0·726
75° C., Ω.	r_1	0·426	0·886
Total f.l. I^2R loss, kW.	$3I_1^2 r_1$	5·7	0·84
Gross copper weight, lb.	G_{c1}	472	533

Rotor

		2-layer bar wave star	2-layer bar wave star
Type of winding			
Connection		star	star
Open-circuit s.-r. voltage, V. . .		650	490
Turns per phase	T_2	45	48
Number of slots	S_2	90	72
Slots per pole	g_2	9	9
Slots per pole per phase . .	g_2'	3	3
Coil-span, slots		1–10	1–10
No. of parallel circuits . . .		2	1
Distribution factor . . .	k_{m2}	0·960	0·960
Coil-span factor . . .	k_{c2}	1·0	1·0
Winding factor	K_{w2}	0·960	0·960
Slot-pitch, cm.	y_{s2}	2·88	2·88
Phase current, equiv., A. . .	I_2'	56	15·3
actual, A. . . .	I_2	250	91
Conductors per slot . . .	z_{s2}	2(2 × 3)	2 × 2
Conductor size, mm. . . .		13 × 3	20 × 5
covering		tape	tape
area, mm.² . . .	a_2	77	98·5
Current density, A. per mm.² . .	δ_2	3·32	2·42
Length of mean conductor, m. . .	L_{mc}	0·862	0·963
Phase resistance at 20° C., Ω . .		0·0174	0·0162
75° C., Ω . .	r_2	0·0212	0·0197
Total f.l. I^2R loss, kW. . . .	$3I_2{}^2r_2$	4·0	0·49
Gross copper weight, lb. . .	G_{c2}	385	588

No-load Current

Magnetizing AT per pole . . .		1 100	500
Magnetizing current, A. . . .	I_{0r}	24·6	6·3
Magnetizing reactance, Ω . . .	x_m	70·5	275
Core loss, kW.	P_i	5·7	2·2
Friction loss, kW. . . .		2·9	0·9
No-load loss, kW. . .	P_0	8·6	3·1
Active cpt. of no-load current, A. .	I_{0a}	1·65	0·6
No-load current, A. . . .	I_0	24·6	6·3
No-load power factor . . .	$\cos \phi_0$	0·067	0·095

Short-circuit Current

Total slot permeance coefficient . .	λ_s	4·09	4·87
Slot reactance, Ω	x_s	2·26	9·85
Overhang permeance . . .	$L_o\lambda_o$	0·99	0·93
Overhang reactance, Ω . . .	x_o	1·56	3·75
Zig-zag reactance, Ω . . .	x_z	1·13	4·40
Differential reactance, Ω . . .	x_h	0·30	0·90
Total reactance, Ω . . .	X_1	5·25	18·9
Total resistance, Ω . . .	R_1	0·85	1·57
Short-circuit impedance, Ω . .	Z_1	5·3	19·0
Phase short-circuit current, A. . .	I_{ss}	327	91
Short-circuit power factor . .	$\cos \phi_{sc}$	0·16	0·083

Performance

Full-load losses, kW—

Stator I^2R	$3I_1{}^2r_1$	5·7	0·84
Rotor I^2R	$3I_2{}^2r_2$	4·0	0·49
Brush		0·0	0·14
Core	P_i	5·7	2·20

Performance (contd.)

						2-layer	2-layer
Friction and windage	.	.	.			2·9	0·90
Stray	.	.	.	.	.	0·9	0·33
Total	.	.	.	.	.	19·2	4·9
Output	.	.	.	.	. P_m	268·8	74·6
Input	.	.	.	.	. P_1	288·0	79·5
Efficiency, %	.	.	.	.	. η	93·0	93·8
Full-load rotor input, kW.		.	.	.		276·0	76·3
Slip, %	.	.	.	.	. s_1	1·45	0·083
Full-load speed, r.p.m.	.	.	.	.		591	744
Maximum torque, %	.	.	.	. M_m		270	287

Stator Slot for Machine B

Width, mm.:

			Depth, mm.:		
Copper	6·5		Copper	2·2	
Cotton	0·5		Cotton	0·5	
Tape	0·4			2·7	
Wrap	3·0		× 9	24·3	
Slot liner	0·5		Tape	0·4	
Slack	0·6		Wrap	3·0	
Slot width	11·5			27·7	
			× 2	55·4	
			Slot liner	0·75	
			Layer spacer	2·4	
			Lip and wedge	4·25	
			Slack	1·2	
			Slot depth	64·0	

CHAPTER XVII

SYNCHRONOUS MACHINES: THEORY

1. Types of Synchronous Machine. The alternator, synchronous motor and synchronous (rotary) converter belong to the classification of synchronous machines.

While transformers and induction motors are built by the thousand in small sizes, synchronous machines are generally constructed in large units. The concentration of generating plants in large power stations has resulted in the development of generators capable of producing upwards of 250 000 kVA., or even more. Again, the synchronous motor (for industrial purposes) is rarely used in small sizes on account of the superior characteristics of the induction motor.

Two general types of construction are employed: the *salient-pole* and the *cylindrical* or *non-salient-pole* machines. The salient-pole construction is used for generators and motors of all ranges of output and up to all but the higher speeds. Medium- and large-sized generators for the highest speeds are of the cylindrical-rotor type. The use of a rotating field system is almost universal, since it permits of stationary armatures on which the windings are more readily braced and insulated for high voltages, while slip-rings carrying large currents at high potential differences are avoided. The direct-current excitation for the field system, of course, requires a pair of slip-rings, but the power involved is small.

The speed of all synchronous machines, as also induction motors, is intimately connected with the frequency. For 50-c/s. machines, the higher available speeds are—

No. of pole-pairs, p .	1	2	3	4	5	6	8	10	15	20
Speed, r.p.s., n .	50	25	16·7	12·5	10	8·3	6·25	5	3·3	2·5
r.p.m., $60n$.	3 000	1 500	1 000	750	600	500	375	300	200	150

The highest possible speed for 50 c/s. is 3 000 r.p.m., which is suitable for steam-turbine drives. *Turbo-generators* may be built up to 200 MVA. with only two poles and running at 3 000 r.p.m. for 50-c/s. supply. The four-pole construction, with a synchronous speed of 1 500 r.p.m., is now obsolescent. The other prime movers available for generator drives are water turbines and low-speed internal-combustion engines. With *water-wheel generators* driven by hydraulic turbines the speeds may be of the order of 500 r.p.m. or less, and the number of poles necessary is twelve upwards. Internal-combustion *engine-driven generators*, particularly of large output, have rather low speeds, e.g. 200 r.p.m., requiring thirty poles.

The prime-mover and its speed have a profound influence on the construction and appearance of the machine. The principal features are shown in Chapter XX. In all large machines centrifugal force is important. In turbo-alternators the very high speeds limit the two-pole rotor diameters to about 110 cm., giving a peripheral speed at 3 000 r.p.m. of nearly 175 m. per sec. The salient-pole construction is quite impracticable. The turbo rotor may be a solid steel forging with slots milled out to receive the field windings. Depending as it does on the D^2Ln product, the output is obtained by increasing the length L of the active part of the rotor (and stator) to several metres: this introduces mechanical difficulties concerned with balancing and with the behaviour of a rotating cylinder between bearings a considerable distance apart and having a comparatively low critical speed.

In low-speed machines where large outputs are concerned, it becomes difficult to accommodate the poles on the periphery of the rotor. The pole-core section has usually to be long and narrow, and special means have to be adopted to anchor the poles to the yoke and spider.

Very small machines may follow d.c. practice: in fact a normal d.c. closed commutator winding may be adapted for polyphase generation. The treatment following assumes the armature to be stationary and the poles to rotate throughout.

2. **The Action of the Synchronous Machine.** Consider Fig. 262, which shows the development of a fixed stator carrying windings, and a rotor capable of rotation within it. For simplicity and to bring out the primary features, the resistance and leakage reactance of the stator windings are neglected, and the flux in the magnetic circuit is deemed to be proportional to the resultant ampere-turn excitations, i.e. there is no saturation. Thus the e.m.f. induced is the same as the terminal voltage, and the phase-angle between current and e.m.f. is also the external phase-angle.

At (a) a stator coil is shown in the position that it occupies when maximum e.m.f. is induced in it. The coil links no resultant flux, but is in the position of greatest rate of change. Let the current be in phase with the voltage: then the coil position shown is also that for maximum current. The current in the coil has no effect on the total flux per pole, but causes a strengthening on one side and a weakening on the other side of the pole-shoe. The two conductors therefore find themselves in the circumstances illustrated in Fig. 3, and a torque is produced by the interaction of the main flux Φ_m with the current in the conductors. The torque is seen to be opposed to the direction of motion of the rotor,* so that the electrical power, the product EI, is produced by virtue of the supply of a corresponding mechanical power. The *distortion* of the main flux distribution

---

* The force on the conductors is such as to urge them to the left and by reaction to urge the rotor to the right. The rotation is against this reaction.

is evidence of the conversion of energy from (in this case) mechanical to electrical form, and the machine is a *generator*. As the rotor is moved by external mechanical means, the alternate cutting of N and S polar fluxes by the stator conductors results in alternating e.m.f.'s of frequency equal to the number of pole-pairs passed per

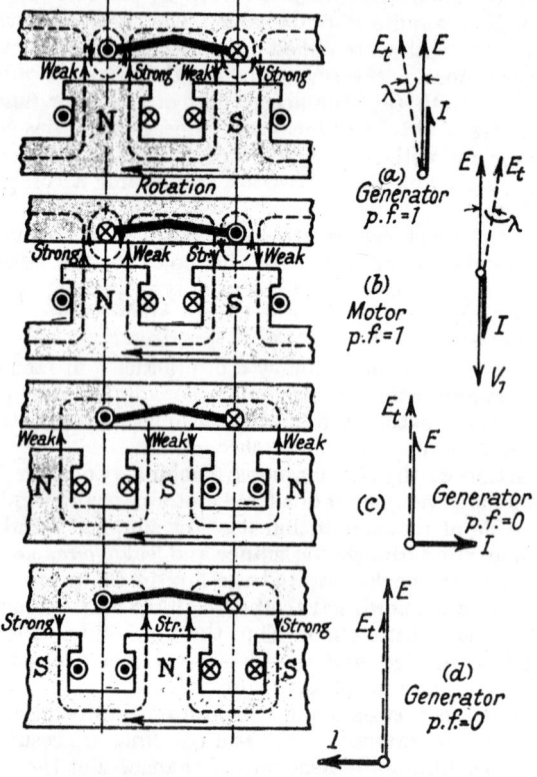

FIG. 262. ACTION OF THE SYNCHRONOUS MACHINE

second. The current alternates with the e.m.f. and thus maintains a unidirectional torque.

The statement that the conditions at (*a*) represent co-phasal e.m.f. and current is not quite true. The strengthening of the resultant flux on the right of the poles and the equivalent weakening on the left in effect shift the main flux axis against the direction of rotation, so that the actual e.m.f. E is a small angle λ behind the position E_t that it would have if the flux were undistorted. The armature currents have evidently shifted the flux axis by *cross-magnetization*.

At (*b*) a motor is shown with operation at unity power factor. The motor is distinguished by the fact that an externally-applied voltage V_1 is needed to circulate in it a current in opposition to the induced e.m.f. *E*. The coil is shown in the position of maximum induced e.m.f. and current, but the current is oppositely directed to that in (*a*). Again the m.m.f. of the coil does not affect the total flux in the common magnetic circuit, but distorts the distribution in such a way as to produce a torque in the same direction as the motion. The machine is a motor by virtue of the electrical input *VI* causing a torque in the direction of motion, i.e. the essentials of mechanical power. The flux distortion causes a shift of the flux axis across the poles, so that the actual e.m.f. *E* is the angle λ ahead of the position E_t that it would have with undistorted flux.

Let the machine working as a generator be connected to a purely inductive load so that the current *I* lags behind the e.m.f. *E* by a quarter-period. Since the coil-position in (*a*) or (*b*) represents that for maximum e.m.f., the poles will have moved through half a pole-pitch before the current in the coil has reached a maximum (Fig. 262 (*c*)). The ampere-turns of the stator coil are now in direct opposition to those on the pole, and work immediately on the magnetic circuit, reducing the total flux and e.m.f. Since the stator and rotor ampere-turns are symmetrically disposed, there is no flux-distortion, no torque, and no mechanical power. This circumstance is in accordance with the fact that there is also no electrical power, *E* and *I* being in phase quadrature. The vector E_t represents the e.m.f. with no demagnetizing armature current, emphasizing the reduction in e.m.f. due to the reduced flux.

At (*d*) the current is presumed to have a phase angle of 90° leading with respect to the e.m.f. By the same argument as in (*c*), the armature current is seen to assist the rotor m.m.f. directly, increasing the total flux and e.m.f.

If the current has a power factor intermediate between unity and zero, a combination of *cross-* and *direct-magnetization* is produced on the magnetic circuit by the stator current. The cross-magnetization is distorting and torque-producing as in (*a*) ; the direct magnetization decreases (for lagging currents) or increases (for leading currents) the ampere-turns acting on the magnetic circuit as in (*c*) and (*d*), affecting the main flux and the e.m.f. in consequence. For a motor the torque is reversed on account of the current reversal, and the direct-magnetizing effect is assisting the field ampere-turns for lagging currents. The action of the stator ampere-turns as described is termed *armature-reaction*: the effect has a far-reaching influence on the behaviour of the synchronous machine, particularly as regards the power factor at which it operates and the amount of field excitation that it needs.

The simple working of the synchronous machine can be summarized as follows.

SYNCHRONOUS GENERATOR. In brief, a synchronous machine driven as a generator produces e.m.f.'s in its armature windings at a frequency $f = np$. The e.m.f.'s when applied to normal circuits produce currents of the same frequency. According to the power factor, more or less field distortion is produced, generating a mechanical torque and demanding an input of mechanical energy to satisfy the electrical output. The torque is unidirectional, for the stator currents change direction in the same time as they are displaced from one magnetic polarity to the next. The torque of individual

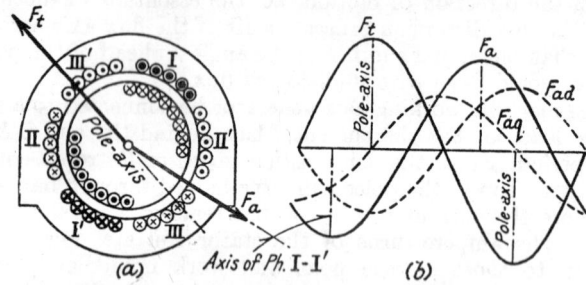

FIG. 263. ARMATURE-REACTION COMPONENTS
The rotation of the rotor is anti-clockwise

phases is pulsating (e.g. see Fig. 110 (c) for precisely the same conditions), but the torque of a three-phase machine is constant for balanced loads.

The division of the fundamental armature reaction into cross-magnetizing and direct-magnetizing components is more convincing with the cylindrical rotor machine, since the uniform gap permits sinusoidal m.m.f.'s to produce approximately sinusoidal fluxes. In Fig. 263 (a) such a machine with two poles is diagrammatically represented. The currents in the three-phase armature winding produce a reaction like that of Fig. 139, having a sinusoidally-distributed fundamental component and an axis coincident, for the instant considered, with that of phase I-I'. The rotor windings, energized by direct current, give also an approximately sinusoidal rotor m.m.f. distribution. The machine is shown in operation as a generator with a lagging current. The relation of the armature reaction m.m.f. F_a to the field m.m.f. F_t is shown at (b). The F_a sine wave is resolved into the components F_{aq} corresponding to the cross-component and F_{ad} corresponding to the direct-component —in this case demagnetization in accordance with Fig. 262 (c). F_{ad} acts in direct opposition to F_t, and reduces the effective m.m.f. acting round the normal magnetic circuit. F_{aq} shifts the axis of the resultant m.m.f. (and flux) backward against the direction of rotation of the field system.

SYNCHRONOUS MOTOR. If a synchronous machine, running at synchronous speed, excited, and connected to a suitable supply, be loaded as a motor, it takes an alternating current which interacts with the main flux to produce a driving torque. The torque remains unidirectional only if the rotor moves one pole-pitch per half-cycle: i.e. synchronous speed is the only possible running speed for a synchronous motor. In a balanced three-phase machine, the fundamental armature reaction is a steady m.m.f. which revolves synchronously with the rotor, its constant cross-component producing a constant torque by interaction with the main flux, while its direct-component affects the amount of the main flux.

An elementary way of regarding a synchronous motor is illustrated in Fig. 264. The stator, like that of the induction motor in Fig. 147, produces a magnetic field rotating at synchronous speed. The poles on the (salient-pole) rotor, excited by direct current in their field windings, undergo magnetic attraction by the stator poles, and are dragged round thereby. On no load the pole axes are practically coincident. When a retarding torque is applied to the shaft, the rotor tends to fall behind. In doing so the attraction of the stator on the rotor becomes tangential to an extent sufficient to develop a counter torque. The

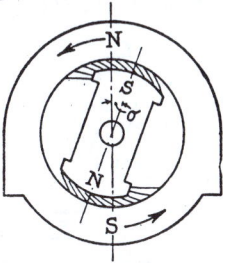

FIG. 264. ELEMENTARY SYNCHRONOUS MOTOR

maximum possible load is that which retards the rotor so that the tangential attraction is a maximum. If the load be increased above this amount, the rotor is too much retarded and the attraction falls, the rotor then pulling out of step.

The magnetic field in Fig. 264 is diagrammatic. Actually the lines of induction must enter or leave the stator and rotor surfaces nearly normally, on account of the high permeability of these members. In a salient-pole machine the torque is developed chiefly on the sides of the poles: in a non-salient-pole machine on the sides of the teeth.*

3. **Synchronous Reactance.** The operation of the synchronous machine can be reduced to comparatively simple expression by the convenient concept of *synchronous reactance*. The resultant linkage of flux with any phase of the armature of a synchronous machine is due, as has been seen, to the combined action of the field and armature currents. For a simple treatment it is convenient to separate the resultant flux into components: (a) the *field flux* due to the field current alone; and (b) the *armature flux* due to the armature current alone. This separation does not affect qualitative matters, but its quantitative validity rests on the assumption that the magnetic circuit has a constant permeability. In brief, the simplifying assumptions are—

(1) All parts of the magnetic circuit have constant permeability,

* See a full discussion of this problem by Howe, *Electrician*, 114, p. 763 (1935).

so that the field and armature fluxes can be treated separately as proportional to their respective currents, and their effects superposed.

(2) The air gap is uniform, so that the armature flux is not affected by its position relative to the poles.

(3) The field flux is sinusoidally distributed in the air gap.

(4) The armature winding is uniformly distributed and carries balanced sinusoidal currents: the harmonics are neglected so that the armature flux is directly proportional to the fundamental component of the armature reaction—i.e. is sinusoidally distributed and rotates at synchronous speed with constant magnitude.

Assumption (1) is roughly fulfilled when the machine works at

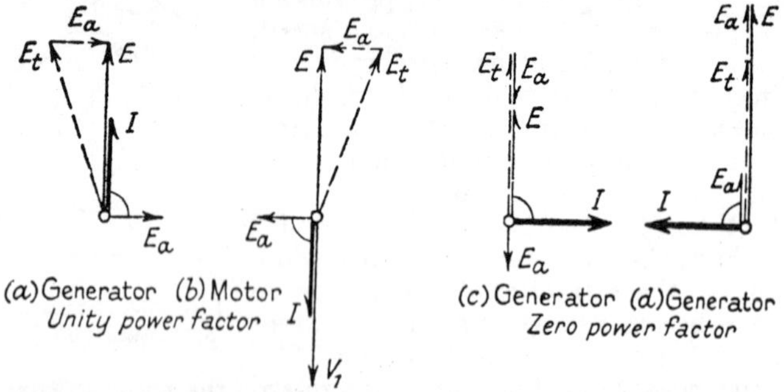

(a) Generator (b) Motor
Unity power factor

(c) Generator (d) Generator
Zero power factor

FIG. 265. PERTAINING TO SYNCHRONOUS REACTANCE

low saturation; (2) and (3) are obviously inaccurate with salient-pole machines; (4) is commonly made and introduces negligible error in most cases. The operation of "ideal" synchronous machines can be indicated qualitatively when the simplifying assumptions above are made.

Referring to Fig. 262, the diagrams for the several conditions contain two e.m.f. complexors, E_t and E. The latter is the e.m.f. actually existing, while the former is that which would be induced under no-load conditions, i.e. with no stator current or reaction E_t is thus the e.m.f. corresponding to the flux produced by the field winding only, while E is that actually produced by the resultant flux developed by the stator and rotor ampere-turns in combination. The actual e.m.f. E can be considered as E_t plus a further fictitious e m f proportional to the stator current

Fig. 265 is drawn in this manner to correspond with Fig. 262, the e.m.f. E_a being introduced such that the complexor $\boldsymbol{E} = \boldsymbol{E_t} + \boldsymbol{E_a}$. It will be seen that E_a is always in phase-quadrature with the current and (under the four assumptions set out above) proportional to it. It therefore resembles the e.m.f. induced in an inductive reactance, so that the effect of armature reaction is exactly as if the stator windings had a reactance $x_a = E_a/I$.

Leakage reactance results from real self-flux linking the armature slot and overhang conductors, producing a corresponding volt-drop Ix; further, it operates during transients. Synchronous reactance is actually an ampere-turn balance between the armature current and part of the field current. Under sudden load change the new balance establishes itself slowly, for it is associated with a mainly iron path. Thus leakage produces true reactance, while the so-called armature-reaction reactance is fictive.

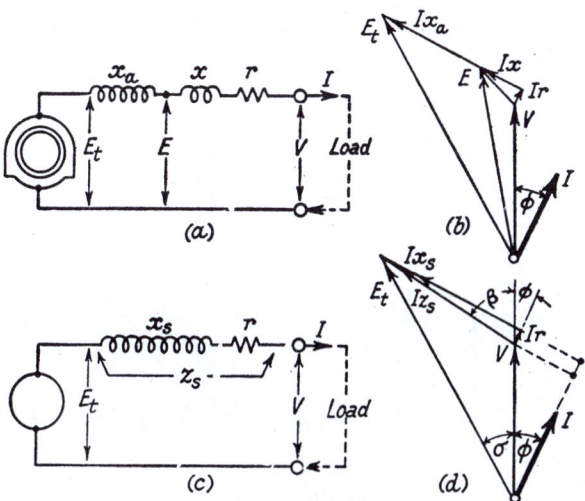

FIG. 266. EQUIVALENT CIRCUITS

If stator resistance and leakage reactance are included, the synchronous machine may be depicted by the equivalent circuit of Fig. 266 (a), in which the resistance r, the true (leakage) reactance x, and the fictitious reactance x_a for armature reaction are separated from the "ideal" machine in which the e.m.f. E_t is induced by the flux produced by the field ampere-turns *alone*. A typical complexor diagram now appears as in Fig. 266 (b), drawn for a generator supplying a lagging current. The e.m.f. developed is considered to be E_t. In x_a a drop Ix_a occurs to take account of armature reaction. The net e.m.f. is E. The terminal voltage is less than E by the further complexor drops in resistance and leakage reactance, Ir and Ix. Obviously V can be obtained directly from E_t and I if r, x and x_a be combined to form an impedance z_s, the components of which are r, and $(x + x_a) = x_s$. The reactance x_s is termed the *synchronous reactance*, and z_s is the *synchronous impedance*. The corresponding equivalent circuit and complexor diagram are shown in Fig. 266.

The ohmic value of the synchronous impedance is readily found. On open circuit, an isolated generator produces no current, while the voltage at its terminals with a given field excitation is $V = E_t$: Fig. 267 (a). On short circuit, the whole of the e.m.f. E_t is consumed in circulating the short-circuit current I_{sc} round the local circuit of impedance z_s, Fig. 267 (b): consequently

$$z_s = E_t/I_{sc} \quad . \qquad . \qquad . \qquad . \qquad . \qquad . \quad (154)$$

where E_t is the open-circuit terminal voltage and I_{sc} is the short-circuit current, taken with the *same* field excitation. Fig. 267 (c)

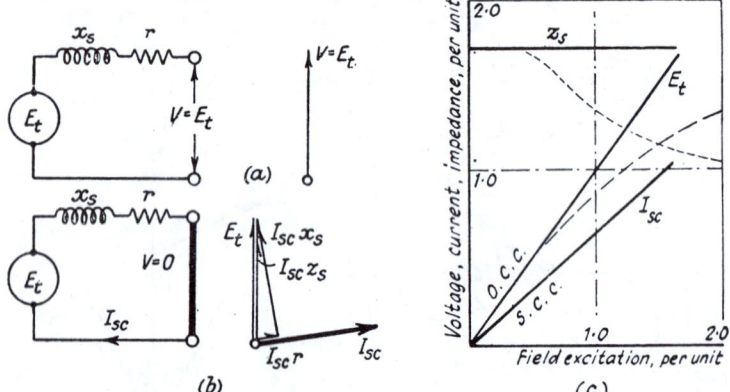

FIG. 267. OPEN- AND SHORT-CIRCUIT CONDITIONS
FOR UNSATURATED MACHINE

shows ideal open- and short-circuit characteristics, from which the curve of z_s is obtained. It is seen that if there is no saturation (i.e. if the open-circuit characteristic relating E_t to the field excitation F_t is rectilinear as shown by the full line), the synchronous impedance z_s remains constant. Actually, however, the onset of saturation at working values of e.m.f. makes z_s very variable: the change in flux due to a given armature-reaction m.m.f. reduces as the saturation becomes more intense. The open- and short-circuit characteristics, and means of estimating suitable values of synchronous reactance, are more fully dealt with in § 11 below.

In large machines, the leakage reactance is of the order 0·10 per unit and the resistance 0·01 per unit. This means that when the machine is generating full-load current, the voltage drop in the resistance is 0·01 per unit of the terminal rated voltage, etc. The reactance x_a is in general very much larger, and the synchronous reactance x_s may be as much as 1·75 per unit. These values of z_s are typical of large modern turbo alternators.

THEORY OF CYLINDRICAL-ROTOR MACHINES

4. Operation of Synchronous Generator with Constant Synchronous Reactance. The synchronous generator, under the assumption of constant synchronous reactance, may be considered as representable by an equivalent circuit comprising an ideal winding in which an e.m.f. E_t proportional to the field excitation is developed, the winding being connected to the terminals of the machine through a resistance r and reactance $(x + x_a) = x_s$, all

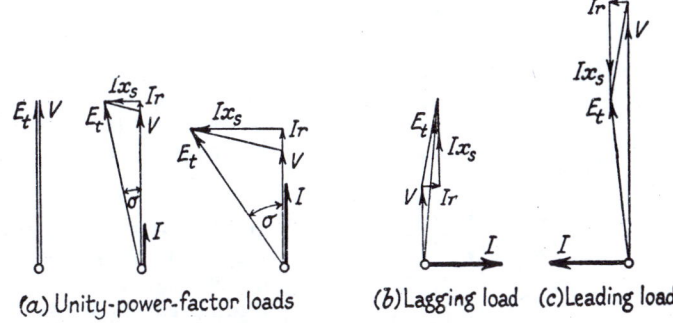

(a) Unity-power-factor loads   (b) Lagging load   (c) Leading load

FIG. 268. VARIATION OF VOLTAGE WITH LOAD AT
CONSTANT EXCITATION

per phase. This is shown in Fig. 266 (c): the principal characteristics of the synchronous generator will be obtained qualitatively from this circuit.

GENERATOR LOAD CHARACTERISTICS. Consider a synchronous generator driven at constant speed and with constant excitation. On open circuit the terminal voltage V is the same as the open-circuit e.m.f. E_t. Suppose a unity-power-factor load be connected to the machine. The flow of load current produces a voltage drop Iz_s in the synchronous impedance, and the terminal voltage V is reduced. Fig. 268 (a) shows the complexor diagram for three values of load. It will be seen that the angle σ between E_t and V increases with load, indicating a shift of the flux across the pole faces due to cross-magnetization. The terminal voltage is obtained from the complex summation

$$V + Iz_s = E_t \quad \text{or} \quad V = E_t - Iz_s.\tag{155}$$

Algebraically this can be written

$$V = \sqrt{(E_t{}^2 - I^2 x_s{}^2)} - Ir$$

for non-reactive loads. Since normally r is small compared with x_s,

$$V^2 + I^2 x_s{}^2 \simeq E_t{}^2 = \text{constant},$$

so that the V/I curve, Fig. 269, is nearly an ellipse with semi-axes E_t and I_{sc}. The current I_{sc} is that which flows when the load resistance is reduced to zero. The voltage V falls to zero also and the machine is on short-circuit with $V = 0$ and $I = I_{sc} = E_t/z_s \simeq E_t/x_s$.

For a lagging load of zero power-factor, the diagram is given in Fig. 268 (b). The voltage is given as before by eq. (155) above, and since the resistance in normal machines is small compared with the synchronous reactance, the voltage is given approximately by

$$V \simeq E_t - Ix_s$$

which is the straight line marked for cos $\phi = 0$ lagging in Fig. 269.

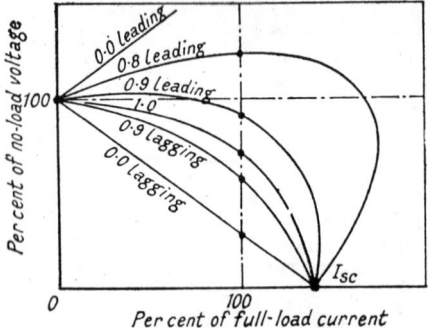

FIG. 269. GENERATOR LOAD
CHARACTERISTICS

The figures refer to power factor

A leading load of zero power factor (Fig. 268 (c)) will have the voltage

$$V \simeq E_t + Ix_s$$

another straight line for which, by reason of the direct magnetizing effect of leading currents, the voltage increases with load.

Intermediate load power factors produce voltage/current characteristics resembling those in Fig. 269. The voltage-drop with load (i.e. the *regulation*) is clearly dependent upon the power factor of the load.

The short-circuit current I_{sc} at which the load terminal voltage falls to zero may be about 150 per cent (1·5 per unit) of normal current in large modern machines.

GENERATOR VOLTAGE-REGULATION. The *voltage-regulation* of a synchronous generator is the voltage rise at the terminals when a given load is thrown off, the excitation and speed remaining constant. The voltage-rise is clearly the numerical difference between E_t and V, where V is the terminal voltage for a given load and E_t is the open-circuit voltage for the same field excitation. Expressed as a fraction, the regulation is

$$\varepsilon = (E_t - V)/V \text{ per unit} \qquad . \qquad . \qquad . \qquad . \qquad (156)$$

Comparing the voltages on full load (1·0 per unit normal current) in Fig. 269, it will be seen that much depends on the power factor of the load. For unity and lagging power factors there is always a voltage drop with increase of load, but for a certain leading power factor the full-load regulation is zero, i.e. the terminal voltage is the same for both full- and no-load conditions. At lower leading power factors the voltage rises with increase of load, and the regulation is negative. From Fig. 266 (*d*), the regulation for a load current I at power factor $\cos \phi$ is obtained from the equality

$$E_t^2 = (V \cos \phi + Ir)^2 + (V \sin \phi + Ix_s)^2 \qquad . \qquad . \qquad . \ (157)$$

from which the regulation is calculated, when both E_t and V are known or found, using eq. (156).

GENERATOR EXCITATION FOR CONSTANT VOLTAGE. Since the e.m.f. E_t is proportional to the excitation when the synchronous reactance is constant, eq. (157) above can be applied directly to obtain the excitation necessary to maintain constant output voltage for all loads. All unity- and lagging-power-factor loads will require an increase of excitation with increase of load current, as a corollary of Fig. 269. Low-leading-power-factor loads, on the other hand, will require the excitation to be reduced on account of the direct magnetizing effect of the zero-power-factor component. Fig. 270 shows typical e.m.f./current curves for a constant output voltage. (They may be derived also from the general load diagram, § 5.) The ordinates of Fig. 270 are marked in percentage of no-load field excitation, to which the e.m.f. E_t exactly corresponds when saturation is neglected.

GENERATOR INPUT AND OUTPUT. For any load conditions represented

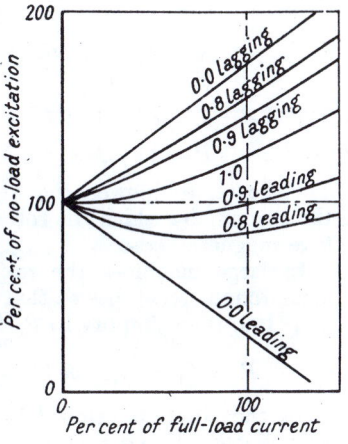

FIG. 270. GENERATOR EXCITATION FOR CONSTANT VOLTAGE
The figures refer to power factor

e.g. by Fig. 266 (*d*), the *output* per phase is $P = VI \cos \phi$. The electrical power *converted* from mechanical power input is per phase

$$P_1 = E_t I \cos (\phi + \sigma).$$

Resolving E_t along I,

$$P_1 = E_t I \cos (\phi + \sigma) = VI \cos \phi + Ir \cdot I = VI \cos \phi + I^2 r.$$

The electrical input is thus the output plus the $I^2 R$ loss, as might be expected. The prime mover must naturally supply also the friction, windage and core losses, which do not appear in the complexor diagram.

For a given load current I at external phase-angle ϕ to V, the magnitude and phase of E_t are determined by z_s. The impedance angle θ is arc tan (x_s/r), and using Fig. 271—

$$I = (E_t - V)/z_s = (E_t/\sigma - V/0)/z_s/\theta$$
$$= (E_t/z_s)/\underline{\sigma - \theta} - (V/z_s)/\underline{-\theta}$$

when referred to the datum direction $V = \overline{V/0}$ Converting to the rectangular form:

$$I = (E_t/z_s)[\cos(\theta - \sigma) - j \sin(\theta - \sigma)] - (V/z_s)[\cos\theta - j \sin\theta]$$
$$= \left[\frac{E_t}{z_s}\cos(\theta - \sigma) - \frac{V}{z_s}\cos\theta\right] + j\left[\frac{E_t}{z_s}\sin(\theta - \sigma) - \frac{V}{z_s}\sin\theta\right].$$

These components represent $I\cos\phi$ and $I\sin\phi$. The power converted internally is the sum of the corresponding components of the current with $E_t \cos\sigma$ and $E_t \sin\sigma$, to give $P_1 = E_t I \cos(\phi + \sigma)$:

$$P_1 = E_t \cos\sigma [(E_t/z_s)\cos(\theta - \sigma) - (V/z_s)\cos\theta]$$
$$+ E_t \sin\sigma [(E_t/z_s)\sin(\theta - \sigma) - (V/z_s)\sin\theta]$$
$$= E_t[(E_t/z_s)\cos\theta] - E_t[(V/z_s)\cos(\theta + \sigma)]$$
$$= (E_t/z_s)[E_t \cos\theta - V \cos(\theta + \sigma)] \text{ per phase} \qquad . \text{ (158)}$$

The output power is $VI\cos\phi$, which is given similarly by

$$P = (V/z_s)[E_t \cos(\theta - \sigma) - V \cos\theta] \text{ per phase} \qquad . \text{ (159)}$$

The greater the power, the larger must σ be: this angle represents the amount by which the rotor magnetic axis is ahead of the armature magnetic ax·s.

In large machines the resistance is small compared with the synchronous reactance so that $\theta = \arctan(x_s/r) \simeq 90°$. Eqs. (158) and (159) then simplify to $P_1 = P$, where

$$P = P_1 = E_t I \cos(\phi + \sigma) \simeq (E_t/x_s) \ V \sin\sigma \quad . \qquad . \text{ (160)}$$

Thus the power developed by a synchronous machine with given values of E_t, V and z_s is proportional to $\sin\sigma$; or, for small angles,

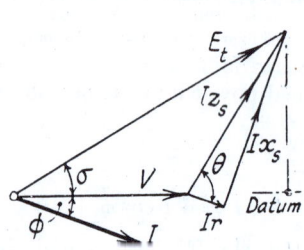

to σ, and the displacement angle σ representing the change in relative position between the rotor and resultant pole-axes is proportional to the load power. The term *load-*, *power-* or *torque-angle* may be applied to σ.

An obvious deduction from eq. (160) is that the greater the field excitation (corresponding to E_t), the greater is the output per unit angle σ: that is, the more stable will be the operation.

FIG. 271. POWER CONDITIONS

PARALLEL OPERATION OF TWO GENERATORS. When two synchronous generators are connected in parallel, they have an inherent tendency to remain in step, on account of the changes produced in their armature currents by a divergence of phase. Consider identical machines 1 and 2, Fig. 272 (*a*), in parallel and working on to the same load. With respect to the load, their e.m.f.'s are normally in phase: with respect to the local circuit formed by the two armature windings, however, their e.m.f.'s are in phase-opposition, (*b*).

Suppose there to be no external load. If machine 1 for some reason accelerates, its e.m.f. will draw ahead of that of machine 2. The resulting phase-difference 2δ causes the e.m.f.'s to lose phase-opposition in the local circuit so that there is in effect a local e.m.f. E_s which will circulate a current I_s in the local circuit of the two armatures. The current I_s flows in the synchronous impedance of

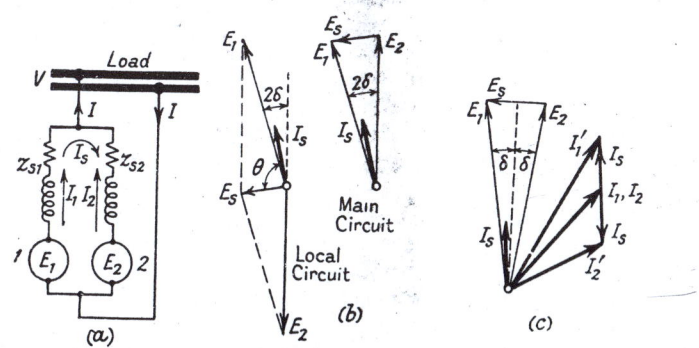

FIG. 272. PARALLEL OPERATION

the two machines together, so that it lags $\theta = \text{arc tan } (x_s/r) \simeq 90°$ on E_s on account of the preponderance of reactance in z_s. I_s therefore flows out of machine 1 nearly in phase with the e.m.f., and enters 2 in opposition to the e.m.f. Consequently machine 1 produces a power $P_s \simeq E_1 I_s$ as a generator, and supplies it (I^2R losses excepted) to 2 as a synchronous motor. The *synchronizing power* P_s tends to retard the faster machine 1 and accelerate the slower, 2, pulling the two back into step. Within the limits of maximum power, therefore, it is not possible to destroy the synchronous running of two synchronous generators in parallel, for a divergence of their angular positions results in the production of *synchronizing power*, which loads the forward machine and accelerates the backward machine to return the two to synchronous running.

The development of synchronizing power depends on the fact that the armature impedance is preponderatingly reactive. If it were not, the machines could not operate stably in parallel: for the circulating current I_s would be almost in phase-quadrature with the generated e.m.f.'s, and would not contribute any power to slow the faster or speed up the slower machine.

When both machines are equally loaded on to an external circuit the synchronizing power is developed in the same way as on no load, the effect being to reduce the load of the slower machine at the same time as that of the faster machine is increased. The conditions are shown in Fig. 272 (c), where I_1, I_2 are the equal load currents of the two machines before the occurrence of phase displacement, and I_1', I_2' are the currents as changed by the circulation of the synchronizing current I_s.

The argument above has been applied to identical machines. Actually, it is not essential for them to be identical, nor to have equal excitations nor power supplies. In general, the machines will

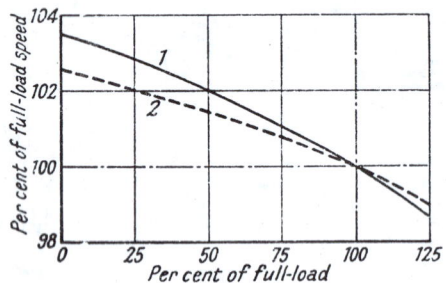

FIG. 273. GOVERNOR CHARACTERISTICS

have different synchronous impedances z_{s1}, z_{s2}; different e.m.f.'s E_1 and E_2; and different *speed regulations*. The governors of prime movers are usually arranged so that a reduction of the speed of the prime mover is necessary for the increase of the power developed. Unless the governor speed/load characteristics are identical, the machines can never share the total load in accordance with their ratings. The governor characteristics take the form shown in Fig. 273. If the two are not the same, the load will be shared in accordance with the relative load values at the running speed, for synchronous machines must necessarily run at identical speeds.

Considering two machines with identical speed/load characteristics in parallel, Fig. 272 (a), the common terminal voltage is

$$V = E_1 - I_1 z_{s1} = E_2 - I_2 z_{s2};$$

whence $E_1 - E_2 = I_1 z_{s1} - I_2 z_{s2},$

$$E_2 = I_2 z_{s2} + IZ = I_2(z_{s2} + Z) + I_1 Z;$$
$$E_1 = I_1 z_{s1} + IZ = I_1(z_{s1} + Z) + I_2 Z;$$

since the total current $I = I_1 + I_2$, and $V = IZ$ where Z is the load impedance. The above expressions give

$$I_1 = \frac{(E_1 - E_2)Z + E_1 z_{s2}}{Z(z_{s1} + z_{s2}) + z_{s1} z_{s2}}; \quad I_2 = \frac{(E_2 - E_1)Z + E_2 z_{s1}}{Z(z_{s1} + z_{s2}) + z_{s1} z_{s2}}.$$

and $I = \dfrac{E_1 z_{s2} + E_2 z_{s1}}{Z(z_{s1} + z_{s2}) + z_{s1} z_{s2}}.$ . . . . . (161)

The terminal voltage is

$$V = IZ = \frac{E_1 z_{s2} + E_2 z_{s1}}{z_{s1} + z_{s2} + (z_{s1} z_{s2}/Z)} \qquad (162)$$

It is also obvious that

$$I_1 = (E_1 - V)/z_{s1} \text{ and } I_2 = (E_2 - V)/z_{s2}.$$

The circulating current under no-load conditions is

$$I_s = \frac{E_1 - E_2}{z_{s1} + z_{s2}}$$

which is the complex difference of the e.m.f.'s divided by the total impedance of the local circuit $(z_{s1} + z_{s2})$ when there is no external load (i.e. $Z = \infty$). If there be an external load of impedance Z, the synchronizing current is affected thereby since the terminal voltage and the voltage drops in z_{s1} and z_{s2} are altered.

EXAMPLE. Two parallel-running synchronous generators have e.m.f.'s of 1 000 V. per phase. The phase synchronous impedances are $z_{s1} = (0.1 + j2.0)\,\Omega$, and $z_{s2} = (0.2 + j3.2)\,\Omega$. They supply a load of impedance $(2 + j1)\,\Omega$ per phase. Find their terminal voltage, load currents, power outputs, and no-load circulating current for a phase divergence of $10°$ (electrical).

Take $E_1 = (1\,000 + j0)$ V., then $E_2 = 1\,000\,(\cos 10° - j \sin 10°)$

$$= (985 - j174) \text{ V.}$$

then $I_1 = \dfrac{(15 + j174)\,(2 + j1) + 1\,000(0.2 + j3.2)}{(2 + j1)\,(0.3 + j5.2) + (0.1 + j2.0)\,(0.2 + j3.2)}$

$$= 160 - j159 = 224/\underline{-44.9°} \text{ A.}$$

from eq. (161). Similarly

$$I_2 = 46 - j96 = 106/\underline{-64.3°} \text{ A.}$$

The total output current is

$$I = I_1 + I_2 = 206 - j255 = 327/\underline{-51.1°} \text{ A.}$$

The terminal voltage is

$$V = (206 - j255)\,(2 + j1) = 667 - j304 = 730/\underline{-24.5°} \text{ V.}$$

The output of the first machine is

$$P_1 = 730 \cdot 224 \cdot 10^{-3} \cdot \cos(44.9° - 24.5°) = 153 \text{ kW. per phase}$$

at a power factor of 0.937 lagging.

The output of the second is

$$P_2 = 730 \cdot 106 \cdot 10^{-3} \cos(64.3° - 24.5°) = 60 \text{ kW. per phase}$$

at a power factor of 0.768 lagging.

The load takes $P_1 + P_2 = 153 + 60 = 213$ kW. More directly

$P = 730 . 327 . 10^{-3} \cos (51 \cdot 1^\circ - 24 \cdot 5^\circ) = 213$ kW. per phase.

The divergence of 10° causes the first machine to produce

$1\,000 . 24 \cdot 7 . 10^{-3} . \cos 22 \cdot 4^\circ = 22 \cdot 8$ kW. per phase.

The second machine receives

$1\,000 . 24 \cdot 7 . 10^{-3} . \cos 12 \cdot 4^\circ = 24 \cdot 1$ kW. per phase

by synchronizing power circulation. The latter is not the difference between P_1 and P_2: $P_1 > P_2$ even for zero divergence on account of the differing impedances. See p. 84 for an analogous case.

The no-load circulating current is

$$I_s = \frac{E_1 - E_2}{z_{s1} + z_{s2}} = \frac{15 + j174}{0 \cdot 3 + j5 \cdot 2} = 34 - j1 \text{ A.}$$

This is substantially a power current. The leading machine has an output of 34 kW. per phase, while the lagging machine takes in slightly less.

Parallel-Generator Theorem. A solution for any number of generators (or transformers) in parallel is obtainable by application of the theorem of Kouwenhoven and Pullen.[*] Referring to Fig. 272 (a), let the load be I amperes at V volts, such that $V/I = Z$. Then

$$V = (I_1 + I_2)Z = \left(\frac{E_1 - V}{z_{s1}} + \frac{E_2 - V}{z_{s2}} \right) . Z$$

$$= \left(\frac{E_1}{z_{s1}} + \frac{E_2}{z_{s2}} \right) Z - V \left(\frac{1}{z_{s1}} + \frac{1}{z_{s2}} \right) Z,$$

whence $$V \left(\frac{1}{Z} + \frac{1}{z_{s1}} + \frac{1}{z_{s2}} \right) = \frac{E_1}{z_{s1}} + \frac{E_2}{z_{s2}}$$

The bracket on the left-hand side contains the sum of the reciprocals of the impedance of the generators and the load, and therefore represents their combination $1/Z_0$ (from the usual rule for the parallel combination of impedances). The terms E_1/z_{s1} and E_2/z_{s2} are the short-circuit currents of the generators taken separately, summing to I_{sc}. Consequently

$$V/Z_0 = I_{sc}, \text{ or } V = I_{sc}Z_0 \qquad . \qquad . \qquad . \text{ (163)}$$

The theorem holds for any number of generators.

EXAMPLE. Taking the problem on page 429, and applying the parallel-generator theorem—

$$\frac{1}{Z_0} = \frac{1}{2 + j1} + \frac{1}{0 \cdot 1 + j2} + \frac{1}{0 \cdot 2 + j3 \cdot 2} = \frac{1}{0 \cdot 365 + j0 \cdot 830}$$

$$= \frac{1}{0 \cdot 905 \underline{/66 \cdot 3^\circ}} \ \Omega.$$

[*] *Elect. Eng.*, **52**, p. 76 (1933).

$$I_{sc} = (1\ 000 + j0)/(0\cdot1 + j2) + (985 - j174)/(0\cdot2 + j3\cdot2)$$
$$= (25 - j498) + (- 35 - j310) = - 10 - j808$$
$$= 808/\!\!-\ 90\cdot7°\ \text{A}.$$

$$V = I_{sc}Z_0 = 808/\!\!-\ 90\cdot7° \cdot 0\cdot905/66\cdot3°$$
$$= 667 - j303 = 730/\!\!-\ 24\cdot4°\ \text{V}.,$$

which compares with the value obtained in the previous example.
For the currents—

$$I_1 = (E_1 - V)/z_{s1} = (1\ 000 - 667 + j303)/(0\cdot1 + j2)$$
$$= 160 - j159 = 224/\!\!-\ 44\cdot9°\ \text{A}.\ ;$$

$$I_2 = (E_2 - V)/z_{s2} \triangleq (985 - j174 - 667 + j303)/(0\cdot2 + j3\cdot2)$$
$$= 46 - j96 = 106/\!\!-\ 64\cdot3°\ \text{A}.,$$

as before.

5. **Synchronous Machine on Infinite Bus-bars.** Up to this point
we have considered only a single generator working on an isolated
load, or two such machines operating in parallel. On account of the
great increase in size of interconnected transmission and distribution
systems in the last decades, and the concentration of generating
plant into a few large stations, the plant capacity connected to a
system may total several hundred thousand kilovolt-amperes. The
behaviour of one single machine connected to a large network is not
likely, therefore, to disturb the voltage and frequency provided the
rating of the machine is only a fraction of the total connected generating plant. In the limit, we may conceive a network at all points of
which the generating plant maintains an invariable voltage and
frequency, i.e. has zero impedance and infinite rotational inertia.
A machine connected to such a network is said to be operating on
infinite bus-bars.

The characteristics of a synchronous generator on infinite bus-bars
are quite different from those when it operates on its own local load.
In the latter case, a change in the excitation changes the terminal
voltage, while the power factor is determined by the load. When
working on infinite bus-bars, on the other hand, no alteration of the
excitation can change the terminal voltage, which is fixed by the
network: the power factor, however, is affected. In both cases the
power developed by a generator (or received by a motor) depends
solely upon the mechanical power provided (or load applied to it).

Practically all synchronous motors and generators in normal
industrial use on large supply systems can be considered as connected to infinite bus-bars, the former because they are relatively
small, the latter on account of the usual arrangements for holding
the voltage automatically constant.

GENERAL LOAD DIAGRAM. Consider a synchronous machine
connected to constant-voltage, constant-frequency bus-bars of

phase voltage V, Fig. 274 (a). Let the machine run on no load with mechanical and core losses only supplied. If the e.m.f. E_t be adjusted to equality with V, no current will flow into or out of the armature on account of the exact balance between the e.m.f. and the bus-bar voltage If the machine be under-excited, E_t will tend to be less than V, so that a leading current I_r will flow which will add to the field ampere-turns by direct armature reaction. Under the assumption of constant synchronous impedance, this is taken into account by $I_r z_s$ as the difference between E_t and V. The current I_r must be completely reactive because no

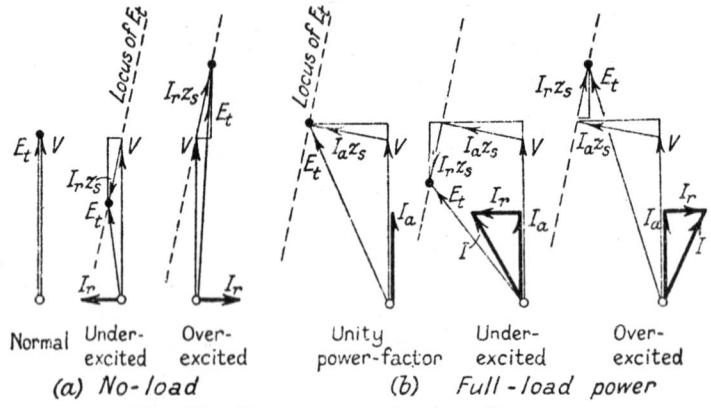

	Normal	Under-excited	Over-excited	Unity power-factor	Under-excited	Over-excited

(a) *No-load* (b) *Full-load power*

FIG. 274. GENERATOR ON INFINITE BUS-BARS

electrical power is being supplied to or by the machine. If now the excitation be increased, E_t will tend to be greater than V. A current will therefore be circulated in the armature—this time a lagging demagnetizing current, which will reduce the net excitation so that the machine again produces at its terminals a voltage equal to the constant bus-bar voltage. The synchronous impedance drop $I_r z_s$ is, as before, the difference between E_t and V, and its phase demands a zero-power-factor lagging current.

Suppose the machine to be supplied with full-load mechanical power. Then as a generator it must produce the equivalent in electrical power: i.e. the output current must have an active component I_a corresponding to full-load electrical power. For an output at exactly unity power factor, the excitation must be adjusted so that the voltage triangle E_t, V, $I_a z_s$, satisfies the conditions required, Fig. 274 (b). If the excitation be reduced, a magnetizing reactive component is supplied in addition, i.e. a leading current I_r, which assists the field winding to produce the necessary flux. If the machine is over excited, a lagging reactive demagnetizing current component is supplied, in addition to the constant power component.

In Fig. 274 (*b*) the Iz drop has been added in components corresponding to the current components I_a and I_r. For all three diagrams of (*b*), I_a and $I_a z_s$ are constant, since the electrical power supplied is constant. Only the component $I_r z_s$ (and therefore I_r) varies with the excitation. Thus the excitation controls not the power, but the *power factor* of the current supplied by the generator to the infinite bus-bars.

The extremities of the E_t complexor (indicated by dots) are seen to lie on the straight line shown dashed. Since all three diagrams refer to full-load power, the dotted line becomes the locus of E_t and of

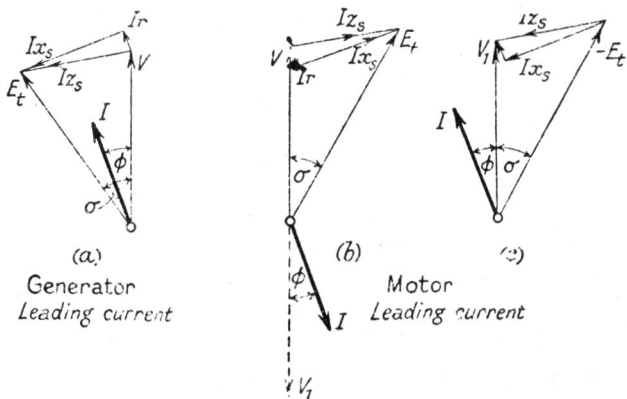

(a)
Generator
Leading current   (b)   (c)
Motor
Leading current

Fig. 275.   Generator and Motor on Infinite Bus-bars

the excitation to scale for constant power output. This is the basis of the *electrical load diagram*, Fig. 276.

A generator working on infinite bus-bars will become a motor if its excitation is maintained and the prime mover replaced by a mechanical load. The change is shown vectorially in Fig. 275 (*a*) and (*b*). V is the *output* voltage of the machine, furnished by the e.m.f. generated. For the motor, the current is in phase-opposition to V, since it is forced into the machine against the output voltage. For convenience, the *supply* voltage V_1 (equal and opposite to V) may be used when the motor is considered, and the diagram then becomes that of Fig. 275 (*c*). The retarded angle σ of E_t or $-E_t$ is descriptive of the fact that when the shaft of the machine is loaded, it falls slightly back relative to the stator rotating field in order to develop the torque.

The power-angle σ, Fig. 275, plays an important part in the operation of the synchronous machine. Changes in load or excitation change its magnitude; when a machine alters from generator to motor action, σ reverses; and when σ is caused to increase excessively, the machine becomes unstable.

ELECTRICAL LOAD DIAGRAM. This is shown in Fig. 276. The com-plexor V represents the constant voltage of the infinite bus-bars. At the extremity of V is drawn an axis showing the direction of the $I_a z_s$ drops—i.e the voltage drops for unity-power-factor output currents. This axis must be drawn at the angle $\theta = $ arc tan (x_s/r) to V. To scale along the axis is a distance corresponding to, say, full load at unity power factor. At this point a line is drawn at right angles to the axis. It is the locus of the E_t values for constant power, or

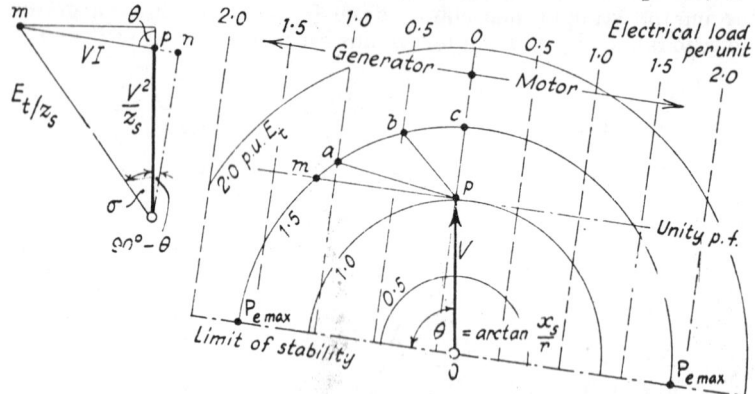

FIG. 276. ELECTRICAL LOAD DIAGRAM

constant-electrical-power line. Other parallel lines are drawn for other loads, one through the extremity of V itself corresponding to zero power output, others on the right-hand side of V corresponding to negative power output, i.e. input to the machine as a motor.

The diagram solves eq (159). Consider the full-load unity-power-factor case in Fig. 274 (b), and multiply each complex voltage by the constant (V/z_s). This gives the inset in Fig. 276, from which $VI = P = mp = mn - np$. Now $mn = (E_t V/z_s) \sin (90° - \theta + \sigma)$ and $np = (V^2/z_s) \cos \theta$, so that P is given directly by eq. (159).

If the excitation be fixed, the extremity of the e.m.f. vector E_t will have a circular locus as indicated by the circular arcs struck with O as centre. Taking $1·0$ per unit E_t as that for which $E_t = V$ on no load and no current, the per-unit excitation for any other loading condition can be found from the diagram. Thus with $1·5$ per unit excitation, the machine will work on full-load power as a generator with a power factor of cos 8° lagging; on half-full-load power with a power factor of cos 42° lagging; and on zero power output with a power-factor of zero lagging, as shown by the lines *pa*, *pb* and *pc*. The variation of the power output (controlled by the input from the prime mover in the case of a generator and by the load applied to the shaft for a motor) with constant excitation is thus accompanied by changes in the load power factor.

If the generator be provided with greater mechanical power with say, 150 per cent (or 1·5 per unit) excitation, then the output power increases with rising power factor from lagging values until, with an output (for this case) of 1·2 per unit power (see Fig. 276), the power factor becomes unity. Thereafter the power increases with a reducing power factor—now leading. Finally the excitation will not include any more constant-power lines, for the circle of its locus becomes tangential to these. If more power is supplied by the prime mover, the generator will be forced to *rise* out of step, and synchronous running will be lost. The maximum power that can be generated is indicated by intercepts on the *limit of stability*. The typical point $P_{e\,max}$ on the left of the load diagram is f^r an excitation of 1·5 per unit.

Similarly, if a motor is mechanically overloaded it will *fall* out of step, because of its limited electrical power intake. The point $P_{e\,max}$ in the motor region again corresponds to 1·5 per unit excitation, and all such points again lie on the limiting-stability line. This maximum power input includes I^2R loss, and the remainder—the mechanical power output—in fact becomes itself limited before maximum electrical input can be attained.

MECHANICAL LOAD DIAGRAM. The mechanical load, or electro-magnetically-converted power P_1 of eq. (158), is for a generator the net mechanical input. For a motor it is the gross mechanical output including core friction and windage loss. A diagram resembling that of Fig. 276 could be devised* by resolving the current along E_t to give $P_1 = E_t I \cos(\sigma + \phi)$. But as the terminal voltage V is taken to be constant, a new circle with another centre is needed for each value of E_t selected. The following method obtains the mechanical loading from the difference I^2r between P and P_1.

The input to a motor is $P = V_1 I \cos\phi$. The electro-magnetic or converted or developed power, which includes the losses due to rotation, is $P_1 = V_1 I \cos\phi - I^2r$. From the latter,

$$I^2 - V_1 I \cos\phi/r + P_1/r = 0,$$

giving $$I = \frac{V_1 \cos\phi}{2r} \pm \sqrt{\left[\left(\frac{V_1 \cos\phi}{2r}\right)^2 - \frac{P_1}{r}\right]} \qquad (164)$$

For each power factor $\cos\phi$, and given voltage V_1 and electro-magnetic power P_1, there are two values of current, one leading and one lagging. The complexor diagrams, Fig. 277 (*a*, *c*), show that there will be two corresponding values of excitation E_t, one large and one small, associated respectively with leading and lagging reactive current components $I_r = I \sin\phi$. At the same time the increased I^2R loss for power factors less than unity requires the active component $I_a = I \cos\phi$ to be larger. The locus of I then forms an

* Rissik, "Synchronous Machine Circle Diagrams," *Electrician*, 125, pp. 111, '91, 338 (1940).

O-curve, while the plot of the current magnitude to a base of excitation E_t gives a *V-curve*, Fig. 277 (*d*).

The *O*-curves are circular arcs, because eq. (164) represents the equation to a circle. Writing

$$(I \cos \phi)^2 + (I \sin \phi)^2 - (V_1/r)(I \cos \phi) + P_1/r = 0,$$

it is seen that I must lie on a circle centred at a point distant $V_1/2r$

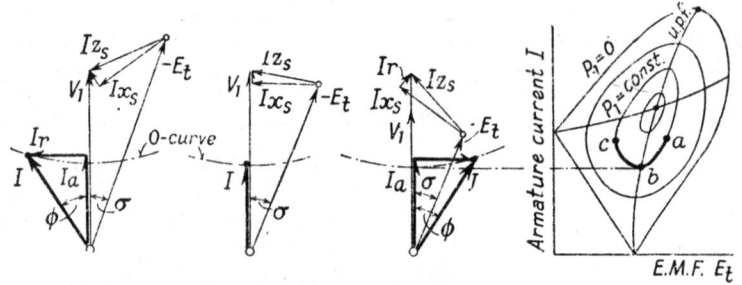

(*a*) *Leading current*   (*b*) *Unity p.f.*   (*c*) *Lagging current*   (*d*) *V- curves*

FIG. 277. SYNCHRONOUS MOTOR WITH CONSTANT OUTPUT AND VARIABLE EXCITATION

from the origin on the axis of $I \cos \phi$, the radius of the circle being $\sqrt{[(V_1^2/4r^2) - (P_1/r)]}$.

The construction of the mechanical load diagram is given in Fig 278. Let OM $= V_1/2r$ to scale: draw

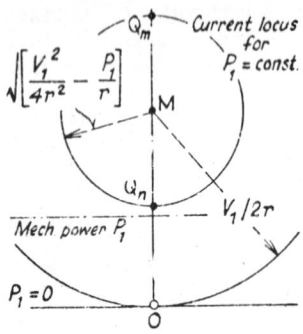

FIG. 278. PERTAINING TO O-CURVES

with M as centre a circle of radius OM. This circle, from eq. (164), corresponds to $P_1 = 0$, a condition for which the circle radius is $V_1/2r$. The circle thus represents the current locus for zero mechanical power. Any smaller circle on centre M represents the current locus for some constant mechanical power output P_1.

For unity power factor

$$I = (V_1/2r) \pm \sqrt{[(V_1/2r)^2 - (P_1/r)]}.$$

Again there are in general two values of current for each power output P_1, the smaller OQ_n in the working range, the greater OQ_m above the limit of stability. If $P_1/r = V_1^2/4r^2$, there is a single value of current $I = V_1/2r$ corresponding to the maximum power $P_{1m} = V_1^2/4r$. The power circle has shrunk to zero radius and becomes in fact the point M. The efficiency is 50 per cent, the I^2R loss being equal to the mechanical output. Such a condition is well outside the normal working range, not only because of heating but also because the stability is critical. The case corresponds to the requirement of the *maximum-power-transfer* theorem, commonly employed to determine maximum-power-output conditions in telecommunication circuits.

The completed mechanical load diagram is shown in Fig. 279 (a), with the addition of OR $= V/z_s$ drawn at angle arc cos (r/z_s) to OM. Circles drawn with R as centre represent constant values of E_t/z_s, or E_t, or the field excitation.

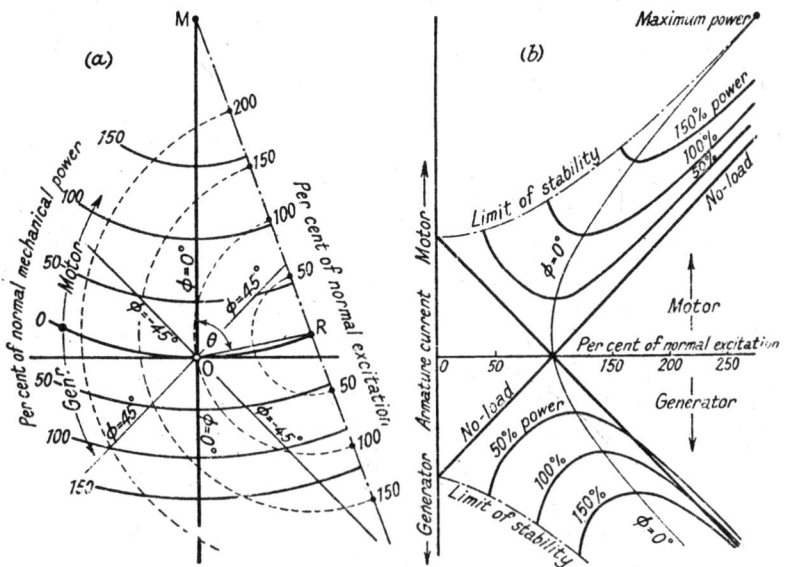

FIG. 279. LOAD DIAGRAM: O-CURVES AND V-CURVES

O-CURVES AND V-CURVES. The current loci in Fig. 279 (a) are continued below the base line for generator operation. The horizontal lines of constant mechanical power are now constant input (from the prime mover) and a departure from unity-power-factor working, giving increased currents, increases the I^2R loss and lowers the available electrical output. The whole system of lines depends, of course, on constant bus-bar voltage. The circular current loci are called the *O-curves* for constant mechanical power. Any point P on the diagram, fixed by the percentage excitation and load, gives by the line OP the current to scale in magnitude and phase.

Directly from the *O*-curves, Fig. 279 (a), can be derived the *V*-curves, relating armature current and excitation for various constant mechanical loads. These are shown in Fig. 279 (b). The variation of power factor with excitation is again shown, and the limit of stability (beyond which the mechanical power supplied to the machine as a generator or taken from it as a motor forces it out of step with the

infinite bus-bars) is seen to depend on the excitation. The stronger
the excitation, the more stable is the machine.

6. **Power Angle : | Generator and Motor Action.** Consider
Fig. 275 (*a*), which shows the complexor diagram of a synchronous
generator from the synchronous-impedance viewpoint. The e.m.f.
E_t is fixed by the excitation, and this in turn settles the power
factor, i.e. cos ϕ. The active component of current is determined
by the supply of mechanical power from the prime mover. For
generator action *VI* cos ϕ may be taken as positive : a negative

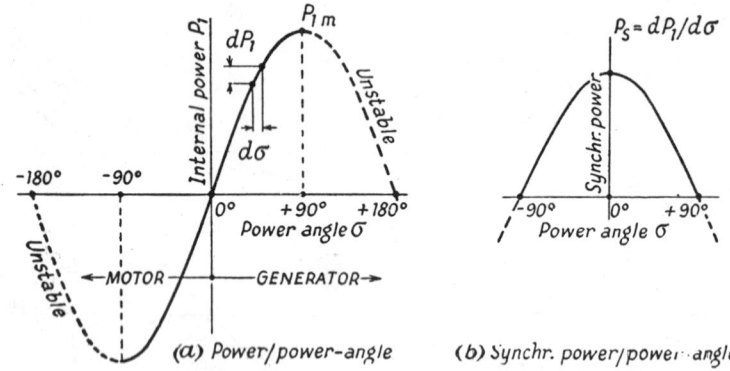

FIG. 280. POWER AND POWER-ANGLE RELATIONS FOR CONSTANT
EXCITATION
The effect of resistance is ignored

value then indicates electrical power intake resulting in motor
action. The angle σ is positive for a generator under normal conditions of load, and becomes negative for a motor, Fig. 275 (*b*).
For a given excitation and terminal voltage, the electrical power
VI cos ϕ cannot alter without σ changing simultaneously.

Eqs. (158) and (159) have been established for the electromagnetic power P_1 (i.e. that concerned with direct conversion
between the mechanical and electrical forms); and the terminal
power P, which differs from P_1 only in the term I^2r—

$$P_1 = (E_t/z_s)[E_t \cos \theta - V \cos (\theta + \sigma)],$$

$$P = (V/z_s)[E_t \cos (\theta - \sigma) - V \cos \theta],$$

where $\theta = $ arc tan (x_s/r). When r can be neglected, both expressions
reduce to that of eq. (160)—

$$P = P_1 = (VE_t/z_s) \sin \sigma = (VE_t/x_s) \sin \sigma.$$

This shows that no change in power can occur with a given excitation unless a corresponding alteration takes place in the power
angle σ. The power/power-angle relation is shown in Fig. 280 (*a*),

being a simple sinusoid. Positive and negative values of σ refer respectively to generator and motor operation. There are clear maxima for each regime, corresponding to the limit of stability for the given excitation, at angles σ approaching 90°.

If an unloaded machine be gradually loaded, the power angle increases at first almost in proportion. As the load approaches the limit, however, a given power increment demands a greater increase in σ: this is significant in the maintenance of stable running and in the development of synchronizing power.

SYNCHRONIZING POWER. Suppose a generator has a mechanical drive and excitation so that the operating conditions correspond to some value of σ (Fig. 281 (a)). Let for some cause the angle σ become $\sigma \pm \delta$. Since the voltage V can be considered as held rigidly constant, the additional e.m.f. introduced by the divergence is $E_s = 2E_t \sin \frac{1}{2}\delta$, which will produce an additional current $I_s = E_s/z_s$, as in Fig. 281 (a). Before the divergence δ, the total internal power from eq. (158) was

$$P_1 = (E_t/z_s)[E_t \cos \theta - V \cos (\theta + \sigma)];$$

it is now

$$P' = (E_t/z_s)[E_t \cos \theta - V \cos (\theta + \sigma \pm \delta)]$$

The difference, the increased power developed when δ leads and the accelerating power when δ lags, is

$$P_s = P' - P_1 = (E_t V/z_s)[\cos (\theta + \sigma) - \cos (\theta + \sigma \pm \delta)]$$

$$= (E_t V/z_s)[\sin (\theta + \sigma) \sin \delta \pm 2 \cos (\theta + \sigma) \sin^2 \tfrac{1}{2}\delta]$$

per phase. If δ is considered to be small, the terms in $\sin^2 \frac{1}{2}\delta$ may be neglected, and the *synchronizing power* P_s becomes

$$P_s \simeq (E_t V/z_s) \sin (\theta + \sigma) . \sin \delta \qquad . \qquad . \quad (165)$$

The synchronizing power developed by a given divergence δ depends on the load (which controls σ). Since in large generators the synchronous impedance is almost entirely reactive, $\theta \simeq 90°$; so that $\sin (\theta + \sigma) \simeq \cos \sigma$, and

$$P_s \simeq (E_t V/z_s) \cos \sigma . \sin \delta \qquad . \qquad (165a)$$

A case of particular interest is synchronizing an unloaded machine on to constant-voltage bus-bars. Then for proper operation $\sigma = 0$ and E_t should be coincident with V. If it is not, Fig. 281 (b), the divergence δ results in the development of the synchronizing power $P_s = (E_t V/z_s) \sin \theta \sin \delta$ per phase. When δ is small enough,

$P_s \simeq (E_t V/z_s) \sin \theta \cdot \delta$. Replacing $\sin \theta$ by unity since $\sin \theta$ $= x_s/z_s \simeq 1$, and E_t/z_s by I_{sc}, then

$$P_s \simeq VI_{sc}\delta \text{ per phase}$$

is the synchronizing power. The short-circuit current I_{sc} is that obtained on a short-circuit test with the same field excitation as is necessary to produce E_t on open circuit.

The term $\cos \sigma$ in eq. (165) is important. As pointed out in the preceding subsection, a given divergence δ produces a smaller synchronizing power when the machine is loaded than when it is on no load. Theoretically the synchronizing power vanishes when $\sigma \simeq 90°$; that is, when the output is a maximum, for if an attempt is made to increase the output further, σ increases and the power actually falls. The simple cosine relation between P_s and σ could be obtained by solving $dP_1/d\sigma$ from eq. (158), giving the *slope* of the power/power-angle curve, Fig. 280 (b).

The synchronizing power supplied by a machine tending to run fast, or to a machine tending to run slow, is associated with a synchronizing torque M_s, where $M_s \omega_r = P_s$, so that

$$M_s = P_s \text{ syn. W.} = P_s/2\pi n \text{ N-m.} \qquad . \qquad . \quad (166)$$

The causes of divergence are referred to later. (See Chapter XVIII.)

EXAMPLE. A 2 000 kVA., 3-phase, 8-pole star-connected synchronous generator runs on 6 000-V. 50-c/s. infinite bus-bars. Find the synchronizing power and torque per mechanical degree of displacement (a) for no load with excitation adjusted to give 6 000 V. on open-circuit; (b) for full load at a power factor of 0·8 lagging. Resistance 0·01 p.u.; synchronous reactance 1·20 p.u. These cases correspond to Fig. 281 (b) and (a).

Since $p = 4$, 1° mechanical $\equiv$ 4° electrical.

(a) From eq. (165) with $E_t = V$:

$$P_s = (V^2/z_s) \sin \theta \cdot \sin \delta.$$

Writing $V^2/z_s = V \cdot V/z_s = VI_{sc}$, then since $z_s \simeq x_s$ is 1·20 p.u., $VI_{sc} = S/1·2$, where S is the full-load rating of 2 000 kVA.; $\theta = \text{arc} \tan 120$, $\sin \theta = 1·0$; $\delta = 4°$, $\sin \delta = 0·07$; whence

$$P_s = (2\ 000/1·2) \cdot 1·0 \cdot 0·07 = 116 \text{ kW}.$$

(b) For full load at power factor 0·8 lagging,

$$E_t = V[1 + (0·01 + j1·20)(0·8 - j0·6)] = V \cdot 1·97/29°.$$

Thus $\sigma = 29°$, $E_t = 1·97V$. From eq. (165 a),

$$P_s = (2\ 000/1·2) \cdot 1·97 \cdot 0·875 \cdot 0·07 = 200 \text{ kW}.$$

The corresponding synchronizing torques are

(a) $M_s = 116 \cdot 10^3/2\pi \cdot 12 \cdot 5 = 1\ 470$ N-m $= 1\ 080$ lb.-ft.

(b) $M_s = 200 \cdot 10^3/2\pi \cdot 12 \cdot 5 = 2\ 550$ N-m $= 1\ 880$ lb.-ft.

Note that the synchronizing torque is greater in (b) than in (a) because of the increased value of E_t.

7. Synchronous Motor. The production of a driving torque in a synchronous motor is due, as discussed in connection with Fig. 262 (b), to the quadrature armature m.m.f. developed by an active armature current component. A constant torque is maintained only if the field system and the armature m.m.f. rotate together and have zero *relative* speed. As the load on the motor changes, there is an internal adjustment in phase relations resulting in an alteration of the

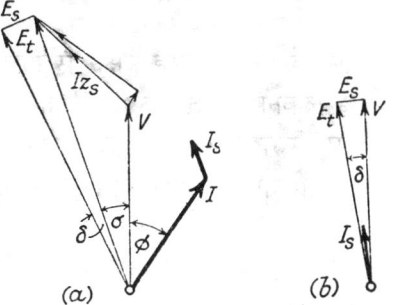

FIG. 281. SYNCHRONIZING POWER

power angle and a momentary acceleration or retardation. The motor then settles down again to the synchronous speed determined by the supply frequency and the number of poles, possibly after a transient period of "swinging."

The operating characteristics of the machine are expressed by the V- and O-curves, Figs. 277 and 279. The relations for current and power can be manipulated to give a number of additional geometrical relations, but saturation makes the effects quantitatively misleading. However, with certain reservations regarding the value and constancy of the synchronous impedance, a load chart for a large machine can be constructed (Fig. 296) to give an overall assessment of the performance.

A brief statement follows on two important characteristics, namely the effect of (a) change in excitation, and (b) change in load.

(a) *Load constant, excitation varied.* A constant electromagnetic load P_1 implies a constant driving torque (including mechanical losses) at the inherently constant synchronous speed. If resistance were neglected the active component I_a of the input current would also be constant. When the I^2R losses are included in the input P, the active current component has to increase slightly as the power factor departs from unity in consequence of the necessary rise in total input current I. As already discussed, the locus of I is a large-diameter O-curve.

Any change in excitation can, apart from I^2R loss, affect only the reactive current component I_r, which must, by direct-axis reaction, make up any deficiency in field ampere-turns for small excitation, and oppose the excess when the field is overexcited.

Fig. 277 shows the power-factor conditions so determined, the armature reaction effects, both of cross- and of direct-axis magnetization, being included as the major part of the synchronous reactance voltage Ix_s. At (a) is shown the case of overexcitation. The current must be such as to satisfy the complex equation $V_1 = -E_t + I(r + jx_s)$ and to have an active component I_a to satisfy the load requirements and the small I^2R loss. The leading reactive component I_r is taken by the armature to provide direct-axis demagnetization. The angle σ is negative (that is, $-E_t$ lags V_1) indicating motor action. If the excitation is reduced, so is I_r, and for a particular value of E_t it vanishes, as shown in (b). The input is now at unity power factor. (c) Further reduction of excitation results in the re-introduction of the reactive current, this time lagging to give a direct-axis magnetization. The active current I_a remains *nearly* constant throughout, representing the constant output P_1 (including rotational losses), together with the small but variable I^2R. The relation between I and E_t is the V-curve shown in Fig. 277 (d), the points a, b, c corresponding to the three cases considered.

There will be a maximum and a minimum value of E_t for a given power output. Outside of these limits the power cannot be maintained. From eq. (158)

$$P_1 = (E_t/z_s)[E_t \cos \theta - V \cos (\theta + \sigma)],$$

where $\theta = $ arc tan (x_s/r). The power is a maximum for a given excitation when no further increase in power angle σ will produce any further output: i.e. when $dP_1/d\sigma = 0$. Applying this condition, $\tan \sigma = \tan \theta = x_s/r$, and the maximum power becomes

$$P_{1m} = (E_t/z_s)[V_1 - E_t r/z_s] \qquad . \qquad . \qquad . \quad (167)$$

To obtain a given output P_1, therefore, the excitation must be such as to satisfy the equation above, whence

$$E_t = (z_s/2r)[V \pm \sqrt{(V^2 - 4P_1 r)}].$$

The two values of excitation which satisfy this expression for a given output power P_1 are the excitation limits. For zero power the lower limit is zero excitation: the working flux is produced by direct-axis reaction, the current having a large value at a lagging power factor of nearly zero. A motor running under these conditions would be at the limit of stability.

The change in load angle σ with excitation within the range illustrated by the complexor diagrams in Fig. 277 is not very marked, but from eq. (160), ignoring I^2R loss,

$$P_1 = P = V_1 I \cos \phi = (V_1 E_t/x_s) \sin \sigma.$$

and it will be seen that

$$I \cos \phi = I_a = (E_t/x_s) \sin \sigma \simeq \text{constant.}$$

Thus as E_t falls σ must increase.

Summarizing: with constant load, the change of exciting current in a synchronous motor (a) changes the power factor, which may be made leading or lagging at will; (b) affects the power angle, which increases as the excitation is reduced so that for low excitations the motor is less stable Further (c) there are maximum and minimum excitations for a given power output: if the excitation is reduced below the minimum the motor will fall out of step, while

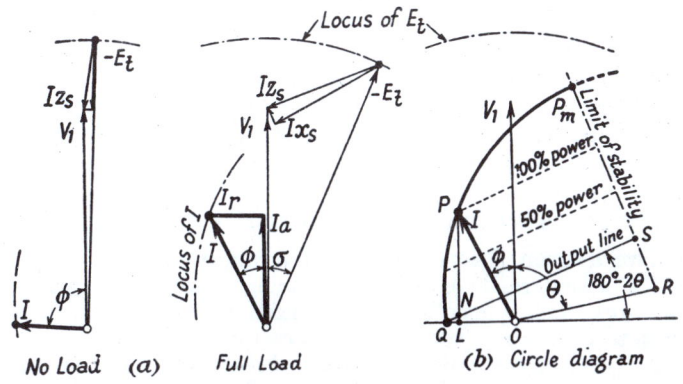

Fig. 282. Synchronous Motor with Constant Excitation
AND Variable Output

the maximum could scarcely be reached in practice in any normal machine.

(b) *Excitation constant, load varied.* Let the motor be strongly excited and running light. The active component I_a of input current will be small, as it corresponds only to I^2R losses. There will be a large leading reactive component for the reasons already discussed in (a) above. As load is applied to the shaft, the development of the necessary torque demands an active, cross-magnetizing, torque-producing current component. The rotor position retards relative to its no-load state to introduce a suitable load-angle σ. The excitation being constant, the vector E_t remains of constant length. These changes are illustrated in Fig. 282 (a). E_t follows the chain-dotted locus struck from the origin, while the Iz_s drop is the intercept between the same locus and the extremity of V_1. Since the current I is proportional to Iz_s and bears to it the constant phase relation $\theta = \text{arc tan} (x_s/r)$, the current locus must also be circular. As the load increases the current comes into phase with the voltage, and subsequently lags: its variation corresponds to one of the dotted circles of constant excitation in Fig. 279 (a).

The *circle-diagram* of the synchronous motor is the current locus described. It is shown in Fig. 282 (b), as the circular arc struck from centre R on a line OR making the angle θ to V_1, where OR $= V_1/z_s$. The radius of the current circle is E_t/z_s. The input power of the motor per phase is $P = V_1 I \cos \phi$, which is proportional to PL for a given current OP $= I$ at phase-angle ϕ to V_1. The electromagnetic power $P_1 = V_1 I \cos \phi - I^2 r$ is the intercept PN between the current circle and a line QN making an angle $\beta = 180° - 2\theta$ to the abscissa axis.[*] Q is located approximately by the intersection of the current circle with the abscissa axis. Lines parallel to the *output line* QN represent various loads. Maximum load occurs at a point P$_m$ on the diameter of the circle perpendicular to QN, and of value P$_m$S to scale.

If the machine be under-excited, the diameter of the current circle shrinks, and the currents may be lagging for all loads. The maximum power will obviously be much reduced: i.e. the machine will be less stable against overloads.

THEORY OF SALIENT-POLE MACHINES

8. **Two-reaction Theory.** In the foregoing discussion, §§ 4–7, the m.m.f.'s of the armature and field windings have been summed as complexors, and it has been emphasized that this procedure is legitimate only when both m.m.f.'s act upon the same magnetic circuit and when saturation effects are absent. The e.m.f. produced can then be obtained by use of the no-load e.m.f. combined with a new e.m.f. proportional to the current and related to it by a constant fictitious reactance co-efficient x_a. Leakage reactance x can, in spite of its different origin, be included with x_a as the synchronous reactance $x_s = x + x_a$. Finally, adding the relatively small armature resistance, the whole comprises the synchronous impedance z_s.

It is shown in Chapter X, § 11, that the direct- and quadrature-axis armature-reaction components in a salient-pole machine cannot be considered as acting on the same magnetic circuit. In fact, while the direct-axis component F_{ad} operates over a magnetic circuit identical with that of the field system and produces a comparable effect thereto (Fig. 140, page 228), the quadrature-axis component F_{aq} is applied across the interpolar space, producing per ampere an altogether smaller effect and, in addition, a flux distribution markedly different from that of F_{ad} or the field m.m.f. It is therefore not surprising that the application of cylindrical-rotor theory to salient-pole machines gives results not conforming to observed behaviour, even allowing for the neglect of saturation.

[*] Wall, "Theory of the Three-phase Synchronous Machine," *J.I.E.E.*, 52. p. 280 (1914).

The two-reaction theory* considers the result of the cross- and direct-reaction components separately, and if saturation is neglected, accounts for their different effects by assigning to each an appropriate value of armature-reaction "reactance," respectively x_{aq} and x_{ad}. The resistance and true leakage reactance may be treated separately, or may be added to the reaction coefficients on the assumption (approximately justified) that they are the same

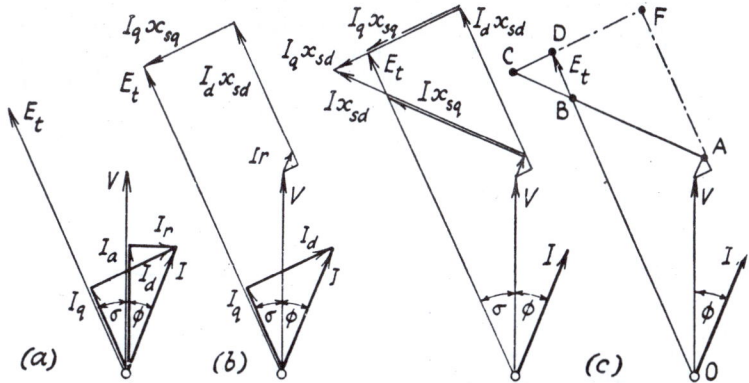

FIG. 283. GENERATOR: TWO-REACTION DIAGRAM FOR LAGGING CURRENT

for any component of the main current. Thus combined reactance values become

$$x_{sd} = x + x_{ad} \text{ and } x_{sq} = x + x_{aq}$$

for the direct- and cross-reaction axes respectively.

The cross- or quadrature-axis reactance in a salient-pole machine is smaller than the direct-axis reactance, because a given current component in that axis gives rise to a smaller flux, the reluctance of the magnetic path including the interpolar spaces.

It is essential to distinguish clearly between the quadrature- and direct-axis components I_q and I_d of the current I in the armature, and the reactive and active components I_r and I_a. Both pairs are represented by complexors in phase quadrature, but the former are related to the excitation e.m.f. E_t while the latter are referred to the terminal voltage V. When σ represents the angle between E_t and V, and ϕ is the terminal output phase angle, then

$$I_q = I \cos (\sigma + \phi), \ I_d = I \sin (\sigma + \phi), \ I = \sqrt{(I_d^2 + I_q^2)};$$

and $I_a = I \cos \phi,$ $\qquad I_r = I \sin \phi,$ $\qquad I = \sqrt{(I_a^2 + I_r^2)}.$

These are shown in the diagram, Fig. 283 (a).

* Blondel, *L'Éclairage Électrique,* 1895.

DETERMINATION OF DIRECT- AND QUADRATURE-AXIS SYNCHRONOUS REACTANCE. The unsaturated values of x_{sd} and x_{sq} for a three-phase synchronous machine may be found by applying low values of balanced voltage to its armature, and driving its rotor mechanically at a speed differing slightly from normal synchronous speed, the field circuit being opened. The rotating armature m.m.f. axis gradually changes, on account of the "slip," between coincidence with the pole-axis and interpolar axis successively. The reluctance of the magnetic circuit varies cyclically between an upper and a lower limit, and the armature current consequently changes in the reverse sense. The ratio of applied voltage to armature current gives the synchronous reactances, using the minimum ratio for the quadrature-axis reactance x_{sq}, and the maximum for x_{sd}. The latter is the same value as would be obtained from normal no-load and short-circuit tests. Typical values for the axis reactances are given in the Table below—

DIRECT- AND QUADRATURE-AXIS SYNCHRONOUS REACTANCE

Type of Machine	x_{sd} p.u.	x_{sq} p.u.
Synchronous motor, high-speed . .	0·8	0·65
low-speed . .	1·1	0·8
Synchronous condenser . . .·	1·6	1·0
Synchronous generator, low-speed .	1·0	0·65

9. Synchronous Generator. A complexor diagram for a salient-pole synchronous generator on a lagging load, utilizing known values of direct- and quadrature-axis synchronous reactance, is shown in Fig. 283 (b). The two reactance voltages are $I_d x_{sd}$ and $I_q x_{sq}$, each in quadrature with its respective current component. The resistance voltage drop Ir is added in phase with I, but could have been drawn as the components $I_d r$ and $I_q r$. The components I_d and I_q are not known until σ is known, but σ depends on $I_d x_{sd}$ and $I_q x_{sq}$. A graphical construction to resolve the difficulty is shown in Fig. 283 (c): it is assumed that V, I, ϕ, x_{sd} and x_{sq} are known. Draw V and add Ir. Draw $Ix_{sd} = $ AC and $Ix_{sq} = $ AB at the extremity A of Ir, and at right-angles to I. Draw OB and extend it to D. Drop the perpendicular CD on OD. Then OD $= E_t$: for if the perpendicular AF is dropped on OD, the intercept FD $=$ AC sin $(\sigma + \phi) = Ix_{sd}$ sin $(\sigma + \phi) = I_d x_{sd}$; and

$$\text{AF} = \text{AB} \cos (\sigma + \phi) = Ix_{sq} \cos (\sigma + \phi) = I_q x_{sq}.$$

The electromagnetic power of a generator, with resistance neglected to simplify the expressions, is $P_1 = VI \cos \phi$. From Fig. 283 (a), $I \cos \phi = I_q \cos \sigma + I_d \sin \sigma$, whence

$$P_1 = V(I_q \cos \sigma + I_d \sin \sigma).$$

Further, $V \cos \sigma = E_t - I_d x_{sd}$ and $V \sin \sigma = I_q x_{sq}$, so that

$$P_1 = V\left[\frac{V \sin \sigma}{x_{sq}} \cos \sigma + \frac{E_t - V \cos \sigma}{x_{sd}} \sin \sigma\right]$$

$$= \frac{VE_t}{x_{sd}} \sin \sigma + V^2\left(\frac{\sin \sigma . \cos \sigma}{x_{sq}} - \frac{\sin \sigma . \cos \sigma}{x_{sd}}\right)$$

$$= \frac{VE_t}{x_{sd}} \sin \sigma + \frac{V^2(x_{sd} - x_{sq})}{2x_{sd}x_{sq}} \sin 2\sigma \quad . \qquad . \qquad . \qquad . \qquad (168)$$

neglecting the resistance r, compared with $P_1 = (VE_t/x_s) \sin \sigma$ for a cylindrical-rotor machine, eq. (160). A term in 2σ, i.e. double the power angle, is introduced into the power/power-angle relation,

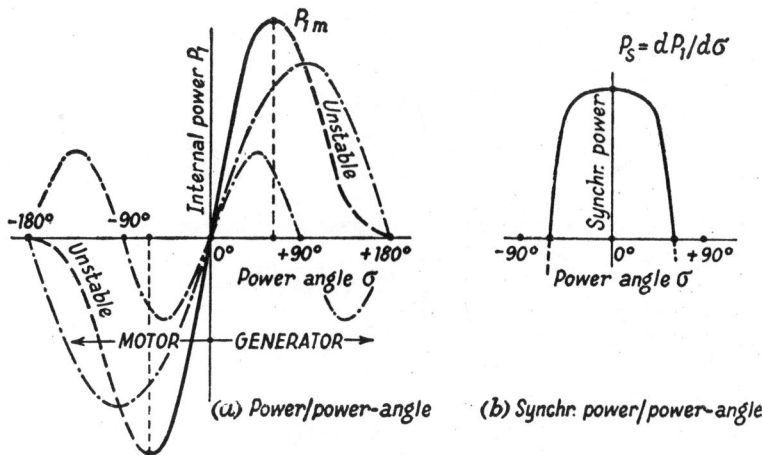

FIG. 284. POWER AND POWER-ANGLE RELATIONS FOR CONSTANT
EXCITATION
The effect of resistance is ignored

shown in Fig. 284. For maximum power the power-angle is now less than 90° (compare Fig. 280): the rise of load with power/angle, $dP_1/d\sigma$, is more rapid, making the machine "stiffer" and improving stability. The reduction in load angle is evident in Fig. 283 (c), where σ is the angle between V and OD: with a cylindrical rotor machine the angle would be that between V and OC. Again, Fig. 283 (b) brings out the dependence of the power-angle on the cross-reaction effect $I_q x_{sq}$ on which the torque depends; while the excitation required for given working conditions—that is, the length of E_t—is largely settled by the direct-reaction $I_d x_{sd}$. For a leading current the direct-reaction vector $I_d x_{sd}$ reduces or reverses so that E_t is smaller. So long as the power output is the same, however, the cross-reaction is not much affected and σ remains nearly the same. Thus the power factor of the generator is clearly controlled by the excitation when the machine works on infinite bus-bars.

The double-angle term in eq. (168) shows that it is possible to generate even if the excitation is removed, i.e. $E_t = 0$. The power

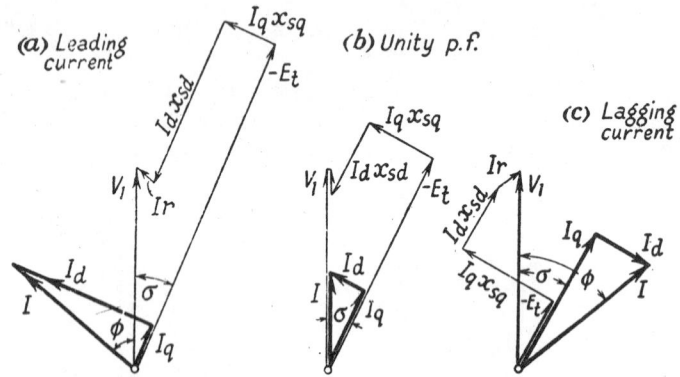

Fig. 285. Motor: Two-reaction Diagrams

is, of course, very limited and the machine would have transient instability. It has been called the *reluctance power*, being due to the saliency of the rotor.

The Table on page 446 shows that in most cases $x_{sq} \simeq 0\cdot6x_{sd}$. With this relation, eq. (168) becomes

$$P_1 = (VE_t/x_{sd}) \sin \sigma + \tfrac{1}{3}(V^2/x_{sd}) \sin 2\sigma.$$

For small angles only, $\sin \sigma \simeq \sigma$ and $\sin 2\sigma \simeq 2\sigma$, and with an excitation such that $E_t = 2V$, the power is $P_1 = (V^2/x_{sd})(2 + \tfrac{2}{3})\sigma$, showing the considerable contribution made by the pole saliency. For $E_t = 0$, $P_1 = \tfrac{2}{3}(V^2/x_{sd})\sigma$.

The synchronizing power, from eq. (168) is

$$P_s = \frac{dP_1}{d\sigma} = \frac{VE_t}{x_{sd}} \cos \sigma + \frac{V^2(x_{sd} - x_{sq})}{x_{sd}x_{sq}} \cos 2\sigma \qquad (169)$$

a relation shown in Fig. 284 (b). For $x_{sq} = 0\cdot6x_{sd}$ this is

$$P_s = (VE_t/x_{sd}) \cos \sigma + \tfrac{2}{3}(V^2/x_{sd}) \cos 2\sigma.$$

Again, the reluctance power contributes to stability.

10. **Synchronous Motor.** Fig. 285 shows two-reaction complexor diagrams for the synchronous motor on the same lines as for the generator in § 9, with excitations giving leading, unity and lagging power factors, at constant output. The diagrams may be compared with those of Fig. 277 for the non-salient-pole machine. The direct magnetizing or demagnetizing effect of the component I_d is clearly marked in the position of $I_d x_{sd}$. The general shape of the V-curves is still the same as shown in Fig. 277 (d), but there are a few second-order effects, particularly in the production of finite reluctance power with zero excitation.

The values of x_{sd} and x_{sq} differ considerably, and because the direct- and cross-magnetizing armature currents I_d and I_q change with alteration of the excitation and power factor, the overall effective synchronous reactance (made up of varying proportions of x_{sd} and x_{sq}) depends on ϕ and σ. The "circle diagram," Fig. 282 (b), loses its circularity and becomes a limaçon, the re-entrancy being evident at low excitations. (See Chapter XVIII, § 1.)

Using the applied voltage V_1 in place of the output voltage V, and employing positive values of σ, eqs. (168) and (169) apply to the motor when I^2R loss is ignored. The motor thus produces a small amount of mechanical power when unexcited. The reluctance power production is discussed in Chapter VIII, § 2, Fig. 109 (b).

SATURATION EFFECTS

11. **Open-circuit and Short-circuit Characteristics.** The *open-circuit characteristic* or o.c.c. relates the terminal voltage on open circuit at normal speed to the amount of field excitation. Since under these conditions the terminal voltage measured is the induced e.m.f., which depends on the total flux Φ_m linking the armature, the o.c.c. is a measure of the saturation in the magnetic circuit. At low values of field excitation the o.c.c. is rectilinear: the iron parts of the magnetic circuit have only a small magnetic loading and are highly permeable, so that the chief reluctance in the circuit is that of the gap. For a density of, say, 0·8 Wb./m.², a magnetic circuit of laminations requires only about 1 AT. per cm. length, while an air gap requires 6 400 AT. per cm. for this density. The part of the excitation devoted to the gap is defined by the *air-line*. Fig. 286 shows an o.c.c. typical of a large modern turbo-generator: it should be compared with the "ideal" case, Fig. 267 (c).

For very high flux densities the iron parts of the magnetic circuit suffer a considerable decline in permeability, and the o.c.c. for the upper range of field excitations has a comparatively small slope. The air-line and the comparatively flat region are linked by a curve showing a more or less definite "knee" (the apparent position of which, however, is largely dependent on the relative scales of the two axes). At normal voltage, corresponding to 1·0 per unit field excitation, it is seen from Fig. 286 that about 20 per cent of the exciting ampere-turns are expended on the iron parts of the circuit and the remainder on the gap, a ratio of 1 : 4. At 125 per cent of normal voltage the ampere-turns for the gap have risen (proportionately) to 100 per cent, but the iron now demands about 70 per cent, a ratio of 1 : 1·4. At higher voltages and saturations, the excitation expended on the iron paths exceeds that on the gap.

The o.c.c. is calculated in the design office by assuming a series of gap fluxes and finding the necessary excitations. Since the slope

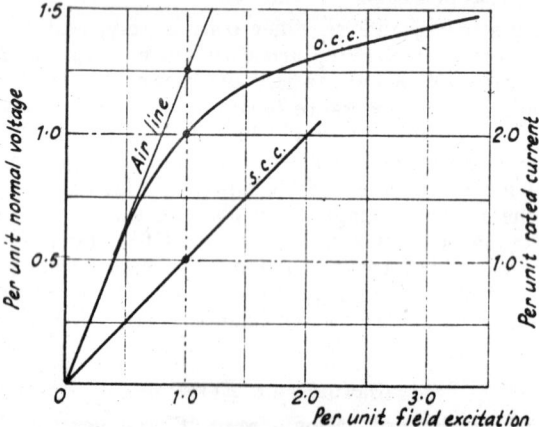

FIG. 286. OPEN- AND SHORT-CIRCUIT CHARACTERISTICS

of the air-line can be easily calculated and the general shape of the curve is known, the determination of a few points is all that is necessary. On test, the machine is driven at or near normal speed and the terminal voltages are measured for a series of field current values.

The *short-circuit characteristic* or s.c.c. relates principally the armature reaction to the field excitation. If there were no leakage reactance nor resistance, the short-circuit current in the armature would furnish the a.c. equivalent of the d.c. rotor excitation. The field excitation has to exceed this equivalent value by the amount necessary to produce the small flux that generates the e.m.f. for circulating the short-circuit current through the resistance and leakage reactance. So long as this flux is small enough the s.c.c. remains a rectilinear relation between armature and field currents. At high values of short-circuit current, however, the field current is high and the consequent leakage produces saturation in the poles, after which a given increase in the exciting current produces less increase in the stator current. In most machines the curvature of the s.c.c. is only slight, and starts well above full-load rated current: thus a single point, together with the origin, suffices for drawing this characteristic.

For full-load current, the short-circuit field excitation must be $F_{sc} = F_s + F_a$, where F_s is an excitation to magnetize the machine sufficiently to generate an e.m.f. equal to the internal drop Iz, and F_a will approximate roughly to $1 \cdot 35 J T_{ph} K_{w1}/p$ ampere-turns per pole, from eq. (83).

The internal power factor on short circuit is practically zero since, apart from $I^2 R$ loss and incidental stray losses, the machine can develop no power. From another viewpoint, the stator and rotor m.m.f.'s are in direct opposition, which implies a zero lagging power factor.

12. Complexor Diagram. In §§ 4–10 the synchronous machine was discussed on a basis of an assumed constant synchronous impedance, in order to bring out most simply its operating characteristics. The fact that no normal machine has a synchronous impedance that is even approximately constant renders it necessary to investigate more closely what actually happens within the machine. The construction of a more exact complexor diagram introduces incidentally a number of points that will be explained as the construction of the diagram proceeds.

Referring back to Fig. 263 (a), a generator is shown by three-phase narrow-spread stator wind-ings and a cylindrical rotor, each member producing an assumed sinusoidally-distri-buted m.m.f. along its axis. The axis of the rotor is ob-vious: that of the stator is coincident with the axis of the phase that carries the maximum current. Strictly speaking, the assumption of sinusoidal m.m.f.'s, currents, e.m.f.'s and fluxes is not justi-fied, but in the practical treatment of the problem the harmonics have no appreci-able effect on the r.m.s. value of the e.m.f., and their sec-ondary effects can be taken account of separately.

Consider Fig. 287A for a non-salient-pole machine. On no load, the flux Φ_m is produced

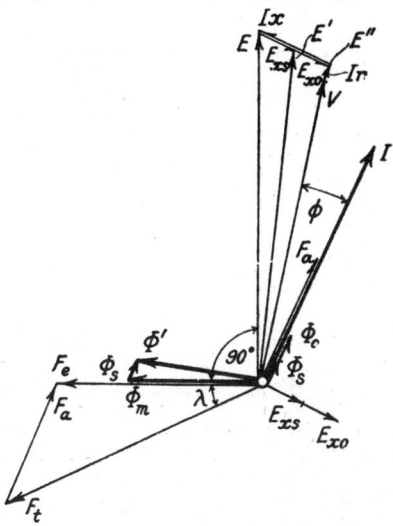

FIG. 287A. COMPLEXOR DIAGRAM OF NON-SALIENT-POLE GENERATOR

by the ampere-turns or m.m.f. F_e on the rotor poles, inducing the e.m.f. E in the armature phase considered. In conformity with the conventions enunciated in Chapter II, Fig. 6, the e.m.f. E, being due to the rate of change of flux-linkages in the phase windings, is represented as lagging by 90° on the flux Φ_m.

When, on load, a current I is produced by the machine, the three phases together produce a reaction m.m.f. F_a in phase with I. This m.m.f. has a demagnetizing component since the current lags, and the pole excitation will have to be increased to F_t and the pole-axis advanced by an angle λ so that the net excitation is again F_e.

The passage of a load current through the armature windings produces in the slots a leakage flux Φ_s in phase with itself. This may be considered to change the total flux from Φ_m to Φ'. The two fluxes, Φ_m and Φ_s, do not have a separate existence in the teeth: the effect of Φ_s is to shift and reduce Φ_m (since it is not in phase with Φ_m). The e.m.f. is now actually E', 90° behind Φ'. It can be deemed as derived from E by the addition of an e.m.f. E_{xs} of self-induction due to the pulsation of the slot-flux Φ_s separately.

A further effect of the load current is to produce a leakage flux Φ_o in the overhang space. This flux has substantially a separate existence, and may be taken into account by adding to the e.m.f. E' the e.m.f. E_{xo} of self-induction produced in the stator winding by the pulsation of Φ_o. The actual e.m.f. left in the machine is thus E''. The terminal voltage V is the same e.m.f. less the Ir drop in the windings, Ir being a complexor in phase with I.

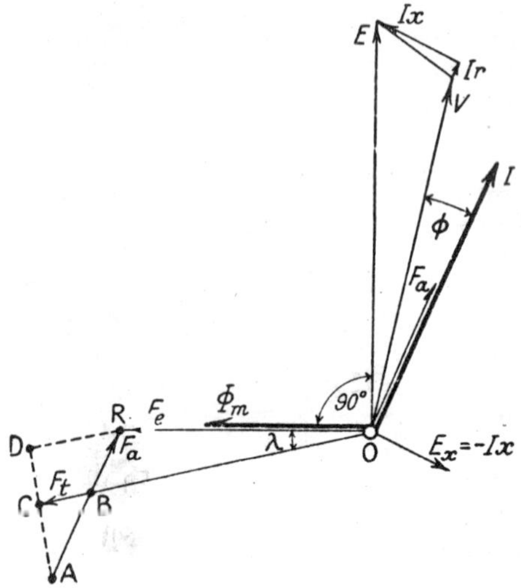

It is usual to lump $E_{xs} + E_{xo}$ together and to consider them opposed by Ix, where x is the *leakage reactance* of the stator windings, for slots and overhang together. In design, however it is necessary to take account of both leakage fluxes. The overhang leakage Φ_o is used to estimate the true component of the reactance: $\Phi_s + \Phi_o$ to calculate the effective leakage reactance; Φ_m for the flux crossing the gap from rotor to stator; and Φ' for the flux in the stator core. The reason for this is that the flux-densities existing in the several parts of the magnetic circuit depend on the actual resultant fluxes therein, and

Fig. 287B Complexor Diagram of Salient-pole Generator

not on a scheme of more or less arbitrary components coexistent in the same place.

The diagram on the same lines as Fig. 287A, but now for motor operation, is drawn by reversing I and the vectors F_a, E_x, Φ_s, Φ_o and Ir which are directly dependent on it. It is usual (as has been mentioned already) to consider the applied voltage V_1 rather than the terminal voltage developed by the machine, and this is introduced by putting V_1 diametrically opposed and equal to V.

In a salient-pole machine, the strongly "polarized" field system resists the displacement of the flux-axis from the field axis: put another way, the cross-reaction (which is responsible for the shift) has to work on a path of high reluctance including the interpolar gaps, instead of one having substantially uniform gap-length and reluctance as in the non-salient pole machine. The effect of the cross-component of armature reaction, F_{aq}, is therefore to a considerable extent reduced. Comparing the two effects in Fig. 140, it is seen that the reluctance of the path on which F_{aq} acts includes the interpolar spaces, which will greatly reduce the cross flux that F_{aq} generates. Although not strictly susceptible to graphical treatment, it is possible to take some account of the effect in the diagram by reducing F_{aq} by an empirical factor obtained experimentally. Referring to Fig. 287B, complexors Φ_m, V, E, I, Ir and Ix are inserted as before. The details of Φ_s, Φ_o, etc., are omitted and their effects taken account of in the leakage reactance drop Ix. The m.m.f. F_s, which is the resultant ampere-turns to produce Φ_m which generates the armature reaction m.m.f. $F_a = $ RA added parallel to I as already shown in Fig. 287A. F_a is divided at B such that RB $= K_r F_a$, where K_r is the fraction expressing the relative permeances of the paths of the cross and direct components respectively. Draw OB and project to C: drop the perpendicular AC. Then OC $= F_t$: for with the external phase-angle ϕ, the armature reaction is divided into a direct component RD $= F_{ad}$, which is added completely to F_s, and a cross component DA $= F_{aq}$, which is added only to the

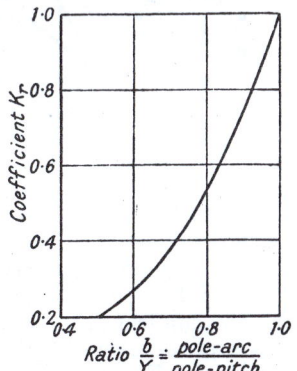

FIG. 288. CROSS-REACTION COEFFICIENT

amount DC $= K_r F_{aq}$. The angle λ by which the pole-axis leads on the flux axis is thus reduced considerably.

The values of K_r depend on the construction of the machine—particularly on the ratio of the pole-arc b to the pole-pitch Y. A curve showing the approximate empirical relation is given in Fig. 288.

13. **Assessment of Reactance.** Regulation, excitation and stability problems can be solved by the methods of §§ 4–10 only if the reactances are precisely known, but these depend on the overall flux distribution and are affected by saturation.

SHORT-CIRCUIT RATIO. Eq. (80) shows that the armature e.m.f. depends on the product $\Phi_m T_{ph}$. A large flux requires a bigger core area (which with turbo-type machines means a longer core, the diameter being limited by peripheral speed) but less armature copper. By contrast, with more turns a smaller core is needed but

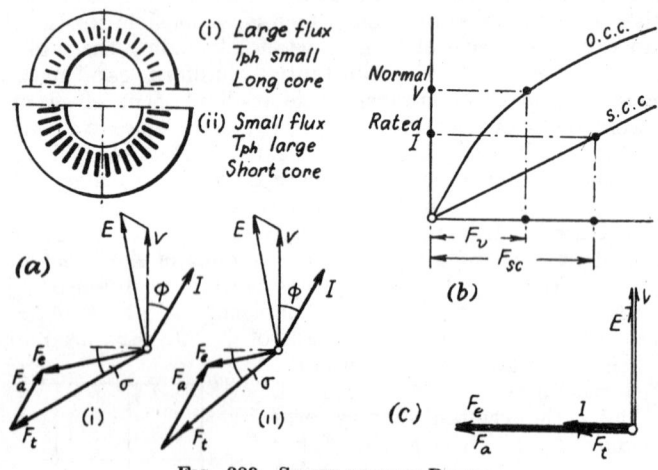

(i) *Large flux*
T_{ph} small
Long core

(ii) *Small flux*
T_{ph} large
Short core

FIG. 289. SHORT-CIRCUIT RATIO

with greater armature conductor section and deeper slots, Fig. 289 (*a*).

Now a fundamental thermal limit is the ampere-turn rating F_t of the field system, part of which has to oppose the armature m.m.f. F_a, the remainder F_e developing the resultant gap flux on which the e.m.f. E depends. With fewer armature turns F_a is reduced, so that the gap length is increased to absorb F_e. The opposite is true for a larger number of armature turns. Comparison of the complexor diagrams in Fig. 289 (*a*) shows that the load angle σ for the former case is less than for the latter, making the machine "stiffer" and more stable.

The relative flux/turns proportionality is embodied in the short-circuit ratio, Fig. 289 (*b*)—

$$r_{sc} = \frac{F_v}{F_{sc}} = \frac{\text{p.u. excitation for normal voltage on o.c.}}{\text{p.u. excitation for rated armature current on s.c.}}$$

Modern turbo-generators normally have r_{sc} between 0·5 and 0·6, but this must be raised to 1·0–1·5 if the loading is likely to be capacitive (as with long transmission lines or extensive h.v. cable networks). The latter range is also common for low-speed generators. On leading zero-power-factor load the armature m.m.f. F_a is direct-magnetizing, Fig. 289 (c), causing self-excitation. To retain voltage control the machine must need positive rotor excitation, so that the gap must be lengthened to increase F_e, and the stator turns reduced to decrease F_a.

The short-circuit ratio is the reciprocal of the synchronous reactance, if x_{sa} is defined in per-unit value for normal voltage and rated current. But x_{sd} for a given load is affected by the saturation conditions that then exist, while r_{sc} is specific and univalued for a given machine.

SYNCHRONOUS REACTANCE. Ignoring resistance, the per-unit synchronous reactance from eq. (154) is

$$x_{sd} = \frac{\text{p.u. voltage on open circuit}}{\text{p.u. armature current on short circuit}}$$

for a given field excitation. For normal (1·0 p.u.) voltage then $x_{sd} = 1/r_{sc}$. Thus if normal open-circuit voltage is given by a field excitation that produces 0·57 p.u. full-load current on short circuit, then $r_{sc} = 0·57$ and $x_{sd} = 1/0·57 = 1·75$ p.u., as shown in Fig. 290 (a). For any other conditions, however, x_{sd} will have a different value.

To avoid ambiguity, the unsaturated synchronous reactance, related to the air line, is defined:

$$x_{sdu} = \frac{\text{p.u. voltage on air-gap o.c.c.}}{\text{p.u. armature current on s.c.c.}}$$

for any excitation. Thus x_{sdu} is constant, and rather greater than $1/r_{sc}$: see Fig. 289(b). It is the value normally quoted for modern machines.

ADJUSTED SYNCHRONOUS REACTANCE. The choice of x_{sd} for a given loading condition requires judgment. Certain graphical estimates have been suggested. Two such are illustrated in Fig. 290. In (b), let $OF = F_e$ be the field excitation on the o.c.c. for an internal e.m.f. $E = V + I(r + jx)$ for a given load. Then CF is the open-circuit e.m.f. E, while AF is the value E_u that would be developed were there no saturation. Join OC. Let $OG = F_a$ be the armature m.m.f.: then $BG = E_{au}$ is the air-line armature-reaction e.m.f. corresponding. The saturated value is taken as $EG = E_a$, and the appropriate synchronous reactance is

$$x_{sd} = x + E_a/I = x + x_a.$$

In (c), the o.c.c. is in per-unit terms, with 1·0 p.u. excitation corresponding to 1·0 p.u. air-line voltage. The e.m.f. E is calculated

for normal voltage and current and for chosen power factor as in (b)
above. From it the o.c.c. excitation OF $= F_e$, and air-line excitation OH $= F_{eu}$, are found. Then OF/OH $= F_e/F_{eu} = k_1$, a saturation factor. A tangent to the o.c.c. is drawn through C and produced
to cut the ordinate axis in T, dividing EO in the ratio TO/ET $= k_2$.
Then the effective synchronous reactance is

$$x_{sd} = x + (x_{sdu} - x)/\sqrt{[k_1(1 + k_2)]}.$$

POTIER REACTANCE. The Potier complexor diagram is that of
Fig. 287. The *Potier reactance*, which gives the voltage drop Ix, is

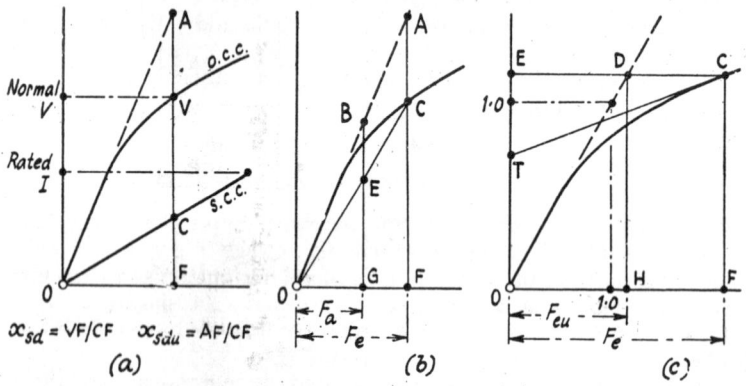

FIG. 290. SYNCHRONOUS REACTANCE

the leakage reactance obtained in a particular way from a test on
the machine at full-load current of power factor zero lagging. Such
a test, requiring but little power, furnishes the excitation for short
circuit and for normal rated voltage, in each case with full-load
z.p.f. current.

Consider an unsaturated machine with negligible resistance. Its
o.c.c. is the straight line in Fig. 291 (a). Point S on the abscissa axis
is such that OS $= F_{sc}$ is the field excitation producing full-load
current on short circuit (which is naturally at a phase angle of 90°
lag). F_{sc} must be sufficient to counter the direct-demagnetizing
armature reaction F_a, also to circulate the current through the
leakage reactance, which requires an induced e.m.f. AX $= Ix$ and
a corresponding field excitation OX $= F_x$. Then $F_{sc} = F_a + F_x$.
Now for a given z.p.f. current both F_a and F_x are constant, and will
apply equally when the total excitation is increased to F_t: for F_t
reduced by F_{sc} gives the remainder F_v producing the air-line terminal
voltage V. The triangle O'A'S' is identical with OAS. Thus AS and
A'S' are equal and parallel, and the dotted line SS' becomes the
voltage/excitation characteristic for full-load current at z.p.f. Then
A'X' $=$ AX is again the Ix drop in leakage reactance, giving the

internal e.m.f. *E*. On open circuit the terminal voltage for an excitation OF $= F_t$ would be TF $= E_t$, so that TS$'$ $= Ix_s$, the synchronous-reactance voltage drop on z.p.f. load. These points are illustrated by the complexor diagram in Fig. 291 (*a*).

Triangle AXS in (*b*) is the *Potier triangle*. Derived from the o.c.c. and the short-circuit point for full-load current, it is assumed to apply equally to a normal saturated case, shown in (*c*). Both the o.c.c. and the z.p.f. characteristics are now curved, but the triangle OAS must everywhere fit between the two characteristics. It follows that the z.p.f. curve should be the o.c.c. shifted bodily by

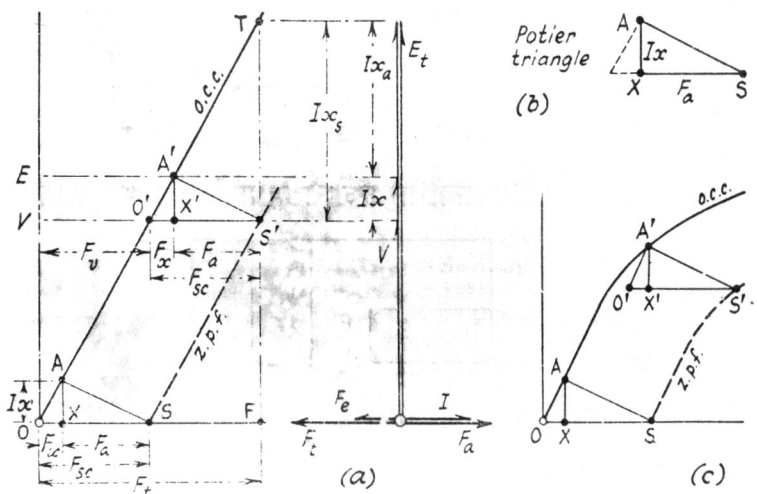

Fig. 291. Potier Reactance

the distance AS $=$ A$'$S$'$. However, points A$'$ and S$'$ do not quite correspond to A and S, for although the gap flux is substantially the same, the greater total field excitation increases the field leakage and causes more pole saturation, so that the o.c.c. and z.p.f. curves have slightly different slopes. As a result A$'$X$'$ is found to be a little greater than AX, and the value of *x* obtained from it (known as the *Potier reactance*) exceeds the value based on pure armature leakage flux.

The Potier triangle can be determined from the o.c.c., the field excitation for full-load current on short circuit, and the leakage reactance *x*. If *x* is not known, a z.p.f. characteristic is needed. Two points suffice: S for short circuit, and S$'$ for normal voltage. Then from S$'$ the horizontal line S$'$O$'$ $=$ SO is drawn, and O$'$A$'$ parallel to OA to cut the o.c.c.

14. Voltage Regulation. The *regulation* of a synchronous generator is the rise in terminal voltage of an isolated machine when full load at given power factor is removed from the machine, the field excitation and speed remaining constant. Thus if V be normal rated voltage, and the terminal voltage rises to E_t when full load is thrown off, then the *per-unit regulation*, eq. (156), is

$$\varepsilon = (E_t - V)/V \text{ per unit.}$$

The inherent voltage regulation of a generator depends on the resistance and leakage reactance, and more particularly on the armature reaction. The values obtained depend also on the power factor of the load, since this determines the direct-magnetizing component of the armature reaction. A demand for close (i.e. small) regulation makes a machine uneconomical to build, since it generally involves a long gap and much field copper. In most cases close inherent regulation is not necessary, since automatic voltage control is commonly used (see Chapter XVIII, § 3), and this avoids the fluctuation of voltage with load.

The regulation is an important characteristic and its value must be known in connection with protection, voltage control and parallel running. Apart from a direct test, the regulation can be estimated by methods such as the following—

1. *From design data—*

   (a) Ampere-turn method;
   (b) Synchronous-impedance method;
   (c) Armature-reaction method.

2. *From test data—*

   (a) Zero-power-factor saturation-curve method;
   (b) Methods based on (1) above, the necessary quantities being measured.

The regulations for a cylindrical-rotor and for a salient-pole machine, each having the same value of direct-axis synchronous reactance x_{sd}, are not likely to differ appreciably. It can be seen from Fig. 283 (c) that for a cylindrical-rotor machine the e.m.f. E_t, corresponding to full-load field excitation on the o.c.c., is given by the length OC, while for a salient-pole machine it is OD: the difference is insignificant. Thus where a simple method suffices it can be applied to either case. The saliency affects chiefly the load angle.

No simple method of predicting regulation is precise, since to take into account every change occurring in the machine between full load and no load would involve the assessment of the exact magnetic state of all parts of the magnetic circuit under both conditions—a very difficult and laborious task. Fortunately the indication given by a careful application of a simple method is generally adequate.

SIMPLE AMPERE-TURN METHOD. This requires a knowledge of the open-circuit characteristic (§ 11), obtained from the design or by direct test. Referring to Fig. 287A, this method identifies E with V on open circuit and builds up the F_t, F_a, F_e triangle as shown in Fig. 292A. Setting off the m.m.f. F_e of field excitation required to produce normal voltage V on open circuit (from the o.c.c.), add at an angle (90° + ϕ) to it the m.m.f. F_a, the field excitation to produce rated current (from calculation). The angle ϕ is the phase angle of the assumed load at the terminals. The resultant F_t is then the total field excitation necessary to produce normal voltage and full-load current at power factor cos ϕ. The construction is as easily made if F_t, F_a and ϕ are known. E_t is the e.m.f. (from the o.c.c.) produced by F_t on open circuit.

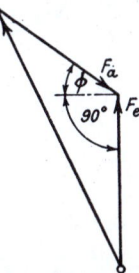

This method gives a regulation that is smaller than the actual value for normal machines.

For a tested machine F_a is taken as the short-circuit field excitation direct from the s.c.c.

SYNCHRONOUS-IMPEDANCE METHOD. The synchronous reactance $x_s = \sqrt{(z_s^2 - r^2)}$, where r is the effective resistance (i.e. inclusive of eddy-loss effects). The complexor diagram of Fig. 266 (d) is constructed so that

FIG. 292A
REGULATION BY
SIMPLE AMPERE-
TURN METHOD

$$\sin \sigma = \sin (\theta - \phi) \,.\, Iz_s/E_t;$$
and $$V = E_t\,(\theta - \phi - \sigma)/\sin(\theta - \phi);$$
where $$\theta = \text{arc tan } (x_s/r), \text{ and } (\theta - \phi) = \beta.$$

There will usually be some doubt as to the choice of z_s: with test results the value at the largest short-circuit current may be taken, although even so the saturation conditions will be lower than for normal load. Alternatively a graphical estimate may be made (Fig. 290). Generally regulations rather greater than those actually occurring are given by this method.

ARMATURE REACTION METHOD. This is a direct application of the diagrams in Fig. 287. For simplicity in drawing, it is convenient to conventionalize the diagram somewhat, by drawing F_e along E instead of in phase-quadrature with it; and consequently F_a in phase-quadrature with I. Fig. 292B illustrates the method of construction for non-salient- and salient-pole machines respectively. The leakage reactance x must be known, as must of course the o.c.c.

Drawing the terminal voltage V and current I at the required phase relation ϕ, the Ir and Ix drops are added to obtain E. Along E is drawn F_e from the o.c.c. The calculated armature reaction F_a is added perpendicular to I, i.e. at an angle (90 + ψ) to F_e: ψ is the internal phase angle between E and I. The complexor F_t

gives E_t from the o.c.c. For salient-pole machinés the F_a complexor is divided as described in § 12 and Fig. 287B. In either case, the regulation is $\varepsilon = (E_t - V)/V$ per unit.

If the o.c.c., s.c.·., and the leakage reactance are known, the total short-circuit field m.m.f. F_{s_2} for (e.g.) full-load current is sub-divided into components F_z and F_a, the former for supplying the Iz drop and the latter the true armature reaction, as follows. Calculate the e.m.f. required to overcome the drop in resistance and

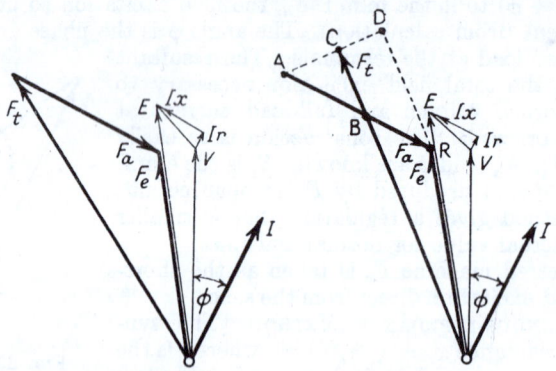

(a) Non-salient-pole generator   (b) Salient-pole generator

FIG. 292B. REGULATION BY ARMATURE-REACTION METHOD

leakage reactance. Then F_z is the field excitation to produce this e.m.f. on open circuit, from the o.c.c. If F_{sc} be the excitation to produce the current on short circuit from the s.c.c., then $F_a = F_{sc} - F_z$. Alternatively, the Potier method can be used.

ZERO-POWER-FACTOR SATURATION-CURVE METHOD. This was dis-cussed in § 13. It employs the results of a z.p.f. full-load current test characteristic to evaluate the Potier leakage reactance x and the armature-reaction m.m.f. F_a'. These quantities are then used to find the regulation by the preceding method.

EXAMPLE. Find the regulation for full load at power factor (a) 0·8 lagging and (b) 0·8 leading for a 500-kVA., 3 300-V., 3-phase, star-connected, 50-c/s., salient-pole synchronous machine. Resis-tance, 0·02 p.u.; leakage reactance, 0·08 p.u.; ratio (pole-arc/pole-pitch), 0 07, reaction coefficient (Fig. 288), $K_r = 0·34$; o.c.c. and s.c.c., Fig. 286, where 1·0 p.u. voltage is 3 300 V. line, and 1·0 p.u. excitation is 5 000 AT.; full-load current on short circuit is circu-lated by a field excitation of 5 000 AT.

The short-circuit ratio is $r_{sc} = F_v/F_{sc} = 1·0$. The unsaturated synchronous reactance is $x_{sdu} = 1·25$, from the air line on the o.c.c.

(1) Simple AT Method. Taking $F_e = F_v = 1·0$ p.u., and $F_a = F_{sc} = 1·0$ p.u., the diagram is drawn as in Fig. 292A for the lagging

power factor with $\phi = $ arc cos 0·8, and with ϕ reversed for the leading power factor. The results are

(a) $F_t = 1\cdot78$ p.u., whence $E_t = 1\cdot26$ p.u., and

$$\varepsilon = 1\cdot26 - 1\cdot0 = 0\cdot26 \text{ p.u.} = \underline{\underline{26\%}}$$

(b) $F_t = 0\cdot89$, $E_t = 0\cdot94$, $\varepsilon = -0\cdot06 = -\underline{\underline{6\%}}$

(2) *Synchronous-impedance Method.* At full-load current the synchronous impedance is $z_s \simeq x_s = E_t/I_{sc} = 1\cdot0/1\cdot0 = 1\cdot0$ p.u. The diagrams are constructed as in Fig. 266 (d). Then for (a) $E_t = 1\cdot80$, and for (b) $E_t = 0\cdot90$ p.u., giving

$$(a) \quad \varepsilon = 1\cdot80 -- 1\cdot0 = 0\cdot8 \text{ p.u.} = \underline{\underline{80\%}};$$

$$(b) \quad \varepsilon = 0\cdot9 - 1\cdot0 = -0\cdot1 \text{ p.u.} = -\underline{\underline{10\%}}$$

(3) *Adjusted Synchronous-impedance Method.* The construction of Fig. 290 (c) is used. For full-load current at power factor 0·8,

$$E = 1\cdot0 + (0\cdot8 \mp j0\cdot6)(0\cdot02 + j0\cdot08) = 1\cdot06 \text{ and } 0\cdot97.$$

(a) $OF = F_e = 1\cdot40$ p.u.; $OH = F_{eu} = 1\cdot06$ p.u.; $F_e/F_{eu} = k_1 = 1\cdot32$; $TO/ET = 1\cdot51$; $x_{sdu} = 1\cdot25$; then

$$x_{sd} = 0\cdot10 + (1\cdot25 - 0\cdot10)/\sqrt{[1\cdot32(1 + 1\cdot51)]} = 0\cdot72.$$

With this value and Method (2), $E_t = 1\cdot55$ p.u., and

$$\varepsilon = 0\cdot55 \text{ p.u.} = \underline{\underline{55\%}}$$

(b) $E = 0\cdot97$; $OF = F_e = 1\cdot18$; $OH = F_{eu} = 0\cdot97$; $k_1 = 1\cdot18/0\cdot97 = 1\cdot21$; $TO/ET = 0\cdot76$; whence

$$x_{sd} = 0\cdot10 + (1\cdot25 - 0\cdot10)/\sqrt{[1\cdot21(1 + 0\cdot76)]} = 0\cdot87.$$

Then $E_t = 0\cdot85$ and $\varepsilon = -\underline{\underline{15\%}}$.

(4) *Armature-reaction Method.* The full-load impedance drop is $\sqrt{(0\cdot02^2 + 0\cdot08^2)} \simeq 0\cdot08$ p.u. The construction is given in Fig. 292B, for a salient-pole machine.

(a) $E = 1\cdot06$ p.u., and $F_e = 1\cdot18$ p.u. from the o.c.c.; $I_z = 0\cdot08$ p.u., requiring an o.c.c. excitation of $0\cdot07$ p.u., so that $F_a = F_{sc} - F_z = 1\cdot0 - 0\cdot07 = 0\cdot93$ p.u.; $RB = 0\cdot34\,RA$, and F_t is found to be $1\cdot87$ p.u., giving $E_t = 1\cdot28$ p.u. from the o.c.c. The regulation is $\varepsilon = \underline{\underline{28\%}}$.

(b) $E = 0\cdot97$, $F_e = 0\cdot98$, $F_a = 0\cdot93$, $F_t = 0\cdot88$, $E_t = 0\cdot94$ p.u.; hence $\varepsilon = -\underline{\underline{6\%}}$.

Summarizing the results—

Percentage regulation, ε					(a)	(b)
Ampere-turn method	.	.	.	.	26	− 6
Synchronous-impedance method		.	.		80	− 10
Adjusted synchronous-impedance method					55	− 15
Armature reaction method		.	.	.	28	− 6

15. V-curves. The variations of armature current with field current on constant power form curves of V shape, as has already been shown in Fig. 277. Although V-curves are obtained for both generator and motor working, their importance is confined principally to motors and synchronous capacitors.

The constant loads for which V-curves are drawn may refer to constant electrical power input, or mechanical power output. Dealing exclusively with the latter case, then it has been shown (§ 5, Fig. 278) that the current loci are the arcs of concentric circles —the O-curves. The constant mechanical power outputs are represented by straight lines denoting constant active current components, and include constant losses such as friction, windage and core loss. The corresponding electrical power inputs must include, in addition, the I^2R losses which increase with fall of power factor: whence the circular loci of input current to obtain the rectilinear loci of mechanical output.

When constant synchronous reactance is not assumed (that is, in all *practical* cases), the V-curves are found by drawing the O-curves for the several constant powers required, finding the currents corresponding to a range of power factors, then applying the construction detailed in Fig. 292B either for cylindrical-rotor or for salient-pole machines (modified for motor operation as necessary), to obtain the field-excitations F_t. A plot between the current and the total field excitation then gives the V-curves.

EXAMPLE. The machine in the previous example is rated as a synchronous motor at 500 h.p. Plot V-curves for 500 h.p., 250 h.p. and no load. Resistance 0·02 p.u. = 0·435 Ω per phase; reactance 0·08 p.u. = 1·74 Ω per phase. Core loss, 8 kW.; friction and windage loss, 7 kW. assumed constant.

The O-curves may be drawn, but the I^2R loss is so small a part of the total that the curvature is immaterial for full- and half-load.

Full Load. The output is 500 h.p. = 373 kW. Adding the 15 kW. of (assumed) constant losses, the total mechanical power developed plus core loss is 373 + 15 = 388 kW. At unity power factor the corresponding current is 388 . $10^3/(\sqrt{3})$. 3 300 = 68 A., and the I^2R loss is 3 . 68^2 . 0·435 . 10^{-3} = 6 kW. The full-load input at unity power-factor is therefore 388 + 6 = 394 kW., and the corresponding current is 69 A. If the power factor were 0·5, the mechanical power of 388 kW. would correspond to 776 kVA., the current to 136 A. and the I^2R loss to 24 kW., the input to 388 + 24 = 412 kW., and the active current component to 412 . $10^3/(\sqrt{3})$. 3 300 = 72 A. The total current at 0·5 power factor is therefore 72/0·5 = 144 A. (The actual I^2R is 27 kW.) If the increased I^2R loss due to low power factor had been neglected, this current would have been 138 A. The difference is small, because the O-curve has a very large

radius and approximates to a straight line between the limits of
0·5 power factor leading and lagging.

Half Load. In a similar manner, the half-load input at unity
power factor is $\frac{1}{2}$. 373 + 15 + 1·5 = 203 kW., and the corresponding current is 35·5 A. At a power factor of 0·5, the I^2R loss

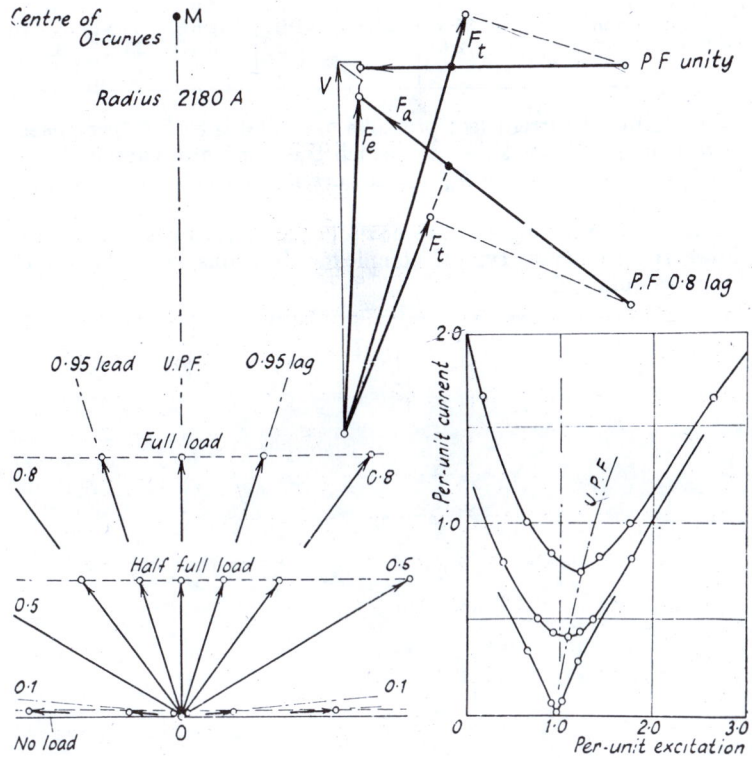

FIG. 293. O-CURVES AND V-CURVES

rises by 4·5 kW. to 6 kW., and the active current component to
36·3: the total current is then 36·3/0·5 = 72·6 A.

No Load. The power input consists of losses only, so that the
I^2R loss is important, particularly at low power factors. The
current can be found from the no-load O-curve; or, more directly,
for an assumed current I, the I^2R loss is $P_c = 3I^2$. 0·435 . 10^{-3} kW.,
and the total loss is $(P_c + 15)$ kW. The corresponding active current component is $I_a = (P_c + 15)10^3/(\sqrt{3})$. 3 300. The power
factor is I_a/I.

From the calculations above, the following Table of currents at given power factors is obtained—

Power Factor	1·0	0·95	0·8	0·5	0·1	0·05
Full-load current, A. .	69·0	73·0	87·0	144·0	. . .	. . .
Half-load current, A. .	35·5	37·5	45·0	72·6	. . .	. . .
No-load current, A. .	2·6	2·7	3·3	5·3	28	85

The full-load current (see previous example) is 87·5 A. per phase corresponding to 500 kVA., for which the armature reaction F_a is 4 700 AT. The values of F_a for the currents in the Table above are in proportion.

The complexor diagrams are now drawn for each case. Fig. 293 shows the O-curves, typical complexor diagrams, and the three V-curves obtained.

CHAPTER XVIII

SYNCHRONOUS MACHINES: OPERATION AND CONTROL

1. Performance of Alternators. The performance of an alternator is a function of the load put on it (both as regards kilovolt-amperes and power factor), and upon the conditions of working—that is, whether working on an infinite bus-bar system or isolated. The load curves are substantially as already given in Figs. 269 and 270. The assumptions there made of constant synchronous reactance do not affect the general shape of the characteristic curves. Fig. 294

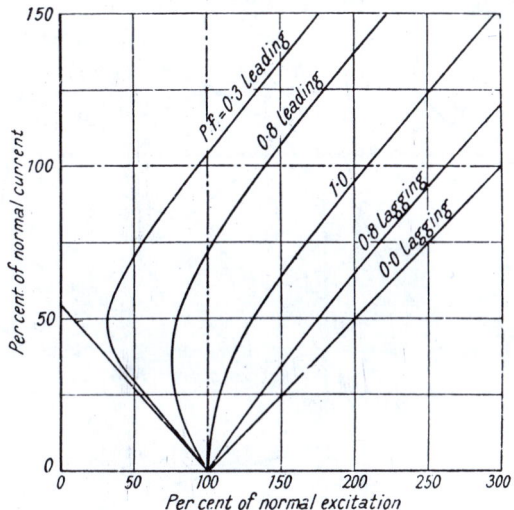

FIG. 294. CURRENT/EXCITATION CURVES FOR NORMAL
VOLTAGE AT VARIOUS POWER FACTORS

(compare Fig. 270) gives the characteristic curves of a typical turbo-alternator with a 175 per cent synchronous impedance; the curves show the percentage of normal current to a base of excitation for various power factor loads at constant normal terminal voltage.

OPERATING CHARTS FOR LARGE GENERATORS. When selecting a large generator the main factor is the rated MVA and power factor, chosen to suit the expected share of the network load. In design, the greatest allowable stator and rotor currents must also be considered, as influencing mechanical stresses and temperature-rise. Other factors include operation at leading power factors, and the general problems of stability. The limiting parameters in the operation are brought out by means of an *operating chart.**

* Szwandler, *J.I.E.E.*, 91, Part II, p. 185 (1944); and *Elec. Review*, 135, p. 513 (1944).

To avoid undue complexity the effects of saturation and of resistance are neglected, and an unsaturated value of synchronous reactance selected. The machine is assumed to be connected to infinite bus-bars. The basis of the chart is the construction shown in Fig. 276, with $\theta = $ arc tan $(x_s/r) = 90°$ and the right-hand (motor) side of the diagram omitted.

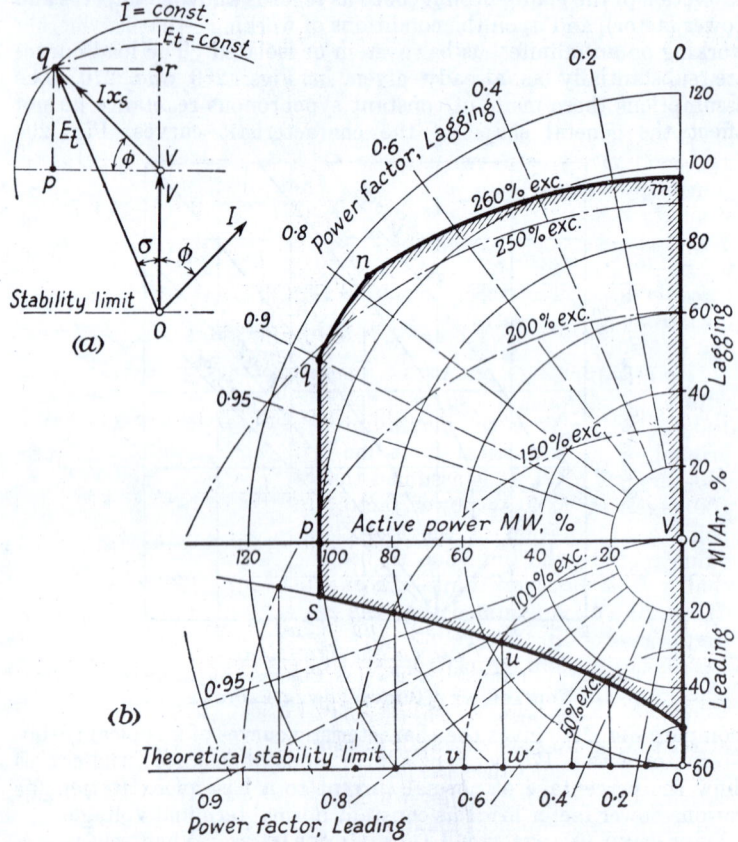

FIG. 295. OPERATING CHART FOR LARGE TURBO-ALTERNATOR

Turbo-generator Chart. The basis is given in Fig. 295 (a). For a given terminal voltage V and current I at phase angle ϕ (i.e. for a given MVA output) the e.m.f. E_t is obtained by adding Ix_s to V at 90° to I. For constant current and MVA, Ix_s is constant and its locus is a circle with centre on the extremity of the voltage vector OV. Constant excitation implies that the locus of E_t is a circle with centre O. The angle between E_t and V is the power-angle: ϕ is the angle between Ix_s and a horizontal line through the extremity of vector OV. Thus Vq is proportional to the MVA, qp to the MVAr, and Vp to the MW, all to a single scale. The latter is found from the synchronous reactance: for with zero excitation $E_t = 0$, $Ix_s = V$, $I = V/x_s$ and is purely reactive (leading), corresponding to *VI* vars and *VI* . 10^{-6} megavars per phase. For $\sigma = 90°$ the static limit of stability is reached (see Fig. 280), so that the horizontal through O represents this limit.

The chart, Fig. 295 (*b*), is drawn for a synchronous reactance of 167 per cent. For zero excitation the current is $100/167 = 60$ per cent of full-load value, representing therefore 60 per cent of full-load MVA in the form of leading MVAr. This fixes all the MVA scales. The vertical OV is drawn and scaled to 60 per cent MVAr, and continued beyond V for lagging MVAr. From V a horizontal line, similarly scaled, gives power in MW. Circles (full-line) drawn with V as centre and similarly scaled as regards radius give stator current in per cent of full load value. Dotted radii from V indicate power factor values. Circles of percentage excitation are drawn in chain-dotted lines from centre O: 100 per cent excitation corresponds to the fixed terminal voltage OV.

It now remains to put in the working area such that any point within it lies inside the assigned limits of operation. Taking 100 per cent MW as the maximum allowable *power* (settled by the prime-mover), a vertical limit-line *spq* is inserted through *p* at 100 per cent MW. It is assumed that the machine is rated to give 100 per cent MW at power factor 0·9 lagging: this determines point *q*. Limitation of the stator current to the corresponding value requires the limit-line to become the circular arc *qn* about centre V. At *n* a new limit—the exciting current—has to be introduced, and it is assumed that the rotor current must not exceed that corresponding to an e.m.f. $E_t = 260$ per cent of *V*. The circular arc *nm* is therefore struck from centre O at this value of excitation. The upper limit *pqnm* is this completed. The line *qp* cannot be continued downwards to the theoretical stability limit because the latter represents a condition of instability where the smallest increment of load will cause the machine to fall out of step. A more satisfactory limit is that for which the loading is, say, 10 per cent less than the theoretical maximum for a given excitation. Consider point *v* on the theoretical stability-limit line for 100 per cent excitation. Reduce O*v* to O*w* by *vw* = 10 per cent of the rated MW. Then *wu*, cutting the 100 per cent excitation circle at *u*, fixes a point for which there is 10 per cent MW in hand as a safeguard against falling out of synchronism. The completed working area, shown with a shaded outline, is *mnqpsut*. A working point placed within this area at once defines the MVA, MW, MVAr, current, power factor and excitation. The load angle σ is found if required by measurement.

Salient-pole Generator Chart. Consider the two-reaction complexor diagram of a generator in Fig. 283 (c), with the Ir drop omitted. For a given terminal voltage V, current I and phase-angle ϕ, the field excitation corresponds to DO $= E_t$. If a circle of diameter $V[(x_{sd}/x_{s_q}) - 1]$ is constructed as in Fig. 296 (a), then the intercept CO′ on the line CO′Q will be equal and parallel to DO in Fig. 283 (c). Dividing all lengths in (a) by x_{sd} gives diagram (b), in which the lengths become: I, given in magnitude and phase; $V/x_{sd} = I_{sc}$, the

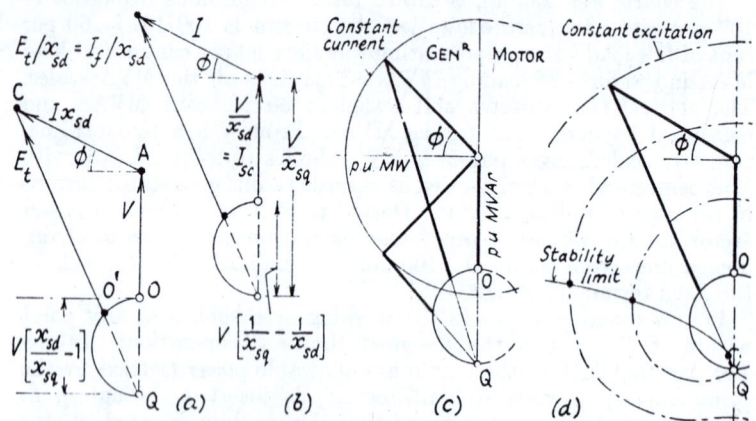

FIG. 296. SALIENT-POLE GENERATOR CHART

current under steady short circuit; and $E_t/x_{sd} = I_f/x_{s_a}$ corresponding to the excitation in per-unit terms, provided that the direct-axis synchronous reactance is assumed to be constant (i.e. to be the unsaturated value). The additional complexor $I\,[(1/x_{sd}) - (x_{s_q}/x_{sd}{}^2)]$ represents the contribution made by the saliency to the internal e.m.f. and field current. The diagram in (c) is now a simple relation* between I, ϕ and I_f.

The conditions obtaining for constant armature current are shown in (c). A circle of radius equal to full-load current (to scale) will decide the scale of MW. and MVAr. In (c) the excitations corresponding to full-load current at a leading and at a lagging power factor are drawn: it may be observed that at low leading power factors the excitation will have to be reversed on account of the effect of direct magnetizing armature reaction.

For constant excitation, the appropriate locus is found by drawing rays from Q, as shown in (d), and marking on each a constant length, such as O′C, *external* to the small saliency circle. Loci of constant excitation are not circles but limaçons, the shape becoming more

---

* Walker, "Operating Characteristics of Salient-Pole Machines," *Proc. I.E.E.*, 100 (II), p. 13 (1953).

evident for small excitations. Each locus will have a point corresponding to a maximum power, the curve through these points constituting the limit of stability.

It is now possible to construct the generator load chart, Fig. 296 (e). It is substantially the same as for the turbo-alternator of

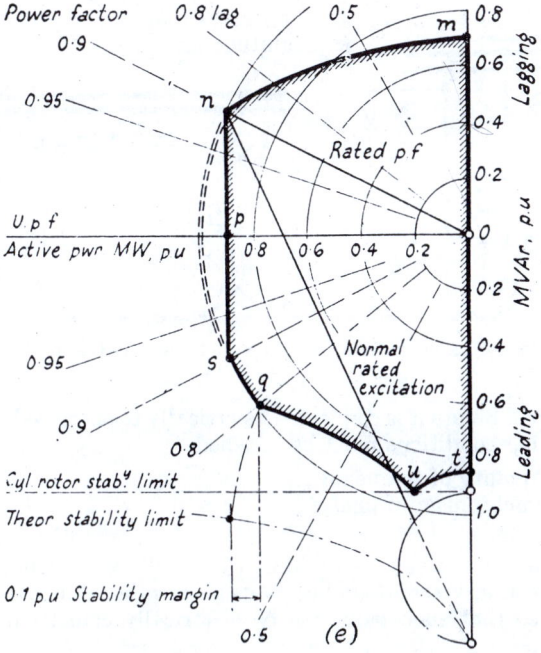

FIG. 296 (CONTD.). SALIENT-POLE GENERATOR CHART

Fig. 295 except in the regions of low excitation. The machine is presumed to have a full-load rating at a power factor of 0·9. Then *nn* represents the excitation limit imposed by rotor heating; *nps* is the prime-mover mechanical limit; *sq* is imposed by stator heating; *qu* is the practical stability limit; and *ut* is a restriction imposed by the provision that there shall always be a positive field excitation.

Charts for Large Motors and Synchronous Capacitors. Fig. 296 (*c*) shows that motor action only requires the circle to be completed to the right of the vertical axis of *V*, to give an operating chart having features in every way comparable with those of a generator.

2. **Starting, Synchronizing and Control of Alternators.** According to the type of prime mover, the preparations for the running-up of a synchronous generator may include warming up the steam-pipes

and turbine, raising oil pressure in the bearings to replace the film of oil squeezed out when the machine is at standstill, etc. The amount of energy stored in the rotating masses of a large unit may be very great, so that the attainment of running speed may take some time.

A machine requires to be *synchronized* if it is to run in parallel

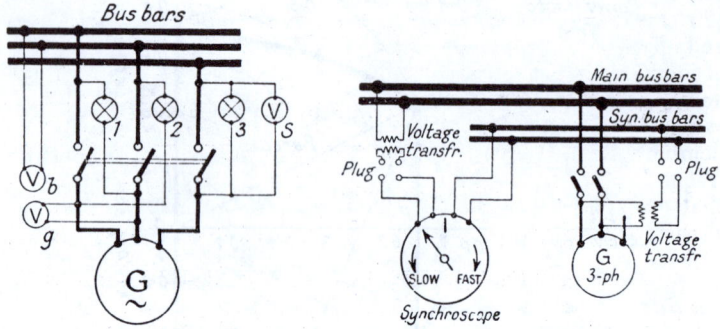

FIG. 297. SYNCHRONIZING AND CONNECTIONS OF SYNCHROSCOPE

with others. Before it is connected electrically to energized bus-bars, the following conditions must be satisfied—

(a) Equality of frequency;
(b) Synchronism of phases;
(c) Equality of voltage.

With these requirements fulfilled, there will be no voltage difference between any corresponding pairs of terminals of machine and bus-bars, so that such pairs can be electrically connected without disturbance.

SYNCHRONIZING. The essential features of synchronizing equipment are shown in their simplest form in Fig. 297. The incoming machine G has lamps connected across the open three-pole switch, Nos. 1 and 2 being cross-connected while No. 3 is directly across the switch. Voltmeter V_b shows the bus-bar and V_g the generator voltage: V_s gives the voltage across one pole of the switch. When the generator has a frequency slightly different from that of the bus-bars, the three lamps slowly brighten and darken in cyclic succession, in a direction depending upon whether the incoming machine is fast or slow. The main switch can be closed safely when the lamps remain steady, and $V_s = 0$.

For high-voltage machines voltage-transformers are used, the lamps are generally augmented by some form of synchroscope, and the incoming machine is synchronized on *synchronizing bus-bars* in the first place, to avoid disturbance due to faulty synchronizing. Automatic devices may be used whereby the main switch mechanism cannot be moved until the conditions for synchronizing are correct. Fig. 297 gives a diagram of essential connections of a rotary synchronizer, in which only one phase is used, and synchronization is indicated by parallel connection to auxiliary synchronizing bus-bars. The synchroscope comprises a spindle with three iron vanes spaced 120° apart and set in a magnetic field produced jointly by the voltages of the incoming machine and the bus-bars. When these are of identical frequency, the spindle remains stationary in a position determined by the relative phase-displacement of the two voltages. When the voltages are coincident in frequency and phase, the spindle pointer stands on a vertical marking on the dial. Any difference in frequency causes the spindle to rotate at a speed corresponding to that difference. The direction of rotation is determined according as the incoming machine is slow or fast. The dial of the synchroscope is marked to correspond.

CONTROL. After synchronizing, an alternator is loaded by adjusting the governor setting to admit more steam (in the case of a turbo-alternator) or water or fuel. This tends to advance the rotor and a synchronizing or load current flows from the machine to the bus bars, corresponding to an exact balance of driving and resisting torques, keeping the machine in step. The greater the rate of admission of steam, etc., to the prime mover, the greater the load assumed by the generator. The governor setting is altered by remote control from the switchboard, so as to give any desired rate of steam admission. Alteration of the excitation does not alter the power output, but changes the amount of kilovar load provided.

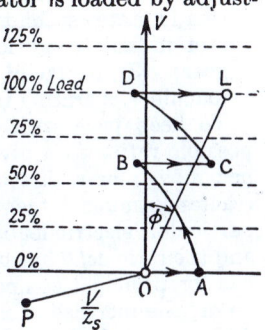

FIG. 298. PERTAINING TO LOADING OF GENERATOR

In a generating station, the loads are shifted from one machine to another by manipulating the governor settings and the kilovars are controlled by the excitation. An increased excitation to one machine causes it to take an increased share of the *lagging* kilovars. At the same time there is a tendency to raise the bus-bar voltage since the other machines are relieved of some of their lagging current components.

Consider the diagram in Fig. 298. The point O represents the working point of a machine just synchronized with its e.m.f. equal to the bus-bar voltage. (The diagram is part of the general load diagram, Fig. 279, for generators.) The excitation is increased so that the machine provides a zero-power-factor lagging current, and the working point moves to A. Steam is now admitted by adjustment of the governor, the excitation remaining constant: the working point swings through the arc AB drawn with P as centre.

(Locus of current with constant excitation.) At B, the excitation and steam supply combine to cause the machine to produce (in this case) about 60 per cent of full-load power with a small leading phase-angle. Increasing the excitation again carries the power factor through unity to a lagging value at C, without altering the power generated. A further admission of steam brings the working point to full-load power at D, again with a leading power factor. Adjustment of the excitation moves the working point to L, which corresponds to full-load power at a power factor cos ϕ as desired. By the process described or by variations of it, the load and power factor at which a generator works (on infinite bus-bars) is controllable by the station operator. The power is observed on a watt-meter, the power factor by means of a power factor meter, or indirectly by an ammeter.

3. **Voltage Control.** The aims of an excitation system are:

(1) to control voltage so that operation is possible nearer to the steady-state stability limit; (2) to maintain voltage under system-fault conditions to ensure rapid operation of protective gear; (3) to facilitate sharing of the reactive load between machines operating in parallel.

To keep pace with turbine design and to obtain the greatest possible rating in a given bulk of material, turbo-generator design has moved in a direction that gives unavoidably a synchronous reactance much larger than that of the low-speed alternator. In particular, electric loadings have been increased by deepening slots, and the ratio ac/B has been raised. Since the synchronous reactance has for principal component the armature reaction which depends on ac, the increase of x_s is the obvious result. Typical figures are: 0·15 per unit leakage reactance and 1·6 per unit armature reaction, giving 1·75 per unit synchronous reactance at a saturation corresponding to normal voltage on no load. Typical o.c.c., s.c.c. and synchronous impedance curves have been given in Fig. 267.

With reactance figures of this order, the field excitation at full load, 0·8 power factor lagging, must be more than $2\frac{1}{2}$ times that for no load at the same voltage. The current range of the exciter must therefore be 2·5 : 1, the corresponding voltage range the same; or, allowing for a mean temperature rise of 90° C. on the rotor copper of the alternator, and 15 per cent voltage in hand, the exciter must have a voltage range of 4 : 1 over which it must possess a stable voltage characteristic. A tendency to instability is produced when, on light loads, the armature reaction has a greater effect on the field system of the exciter, the field being weak. It is generally sufficient to have an air-gap long enough to necessitate a strong field and to enable the wide voltage range to be achieved without excessive saturation.

Normally, the complete excitation circuit chain comprises (*a*) an auxiliary exciter running saturated (or alternatively a battery) so as to provide a substantially constant and stable voltage for the field circuit of the main-exciter; in this circuit the alternator voltage control is arranged; (*b*) the armature of the main exciter, connected to (*c*) the alternator field. Instead of the arrangement (*a*), the main exciter may be self-excited. But this is inherently less stable.

An indication of the rapidity of response of an exciter is given by the numerical value of the average rate of rise from nominal slip-ring voltage in the first half-second following the opening of the exciter armature circuit, expressed in terms of nominal slip-ring voltage. For steam turbo-alternators a response of 0·5 is adequate: e.g. a mean rate of rise of 100 V./sec. for a nominal 200 V. The rate of rise required for waterwheel generators may be considerably greater, in consequence of the overspeed to which these machines may be subjected.

When a sudden change of the alternator load occurs, demanding a change in its excitation to give the same voltage under the new conditions, the automatic regulating equipment must first operate on the exciter field; the exciter voltage changes, and then the alternator field current. The rapidity of response to a load change thus depends on a chain of operations in each of which there is a tendency to delay. Fluxes cannot build up or decay in highly-inductive field systems without an appreciable lapse of time. In the alternator itself the main flux has an appreciable time-constant, and the change of flux with change of load is slow enough to allow the exciting current to be adjusted by the exciter.

It is noteworthy that changes in the alternator load current are reflected in its field current, in so far as the load changes involve changes in the direct-magnetizing component of the armature reaction. Thus if a low-power-factor lagging load be thrown off or diminished, the demagnetizing effect is reduced, tending to cause a flux increase. This is delayed, however, by eddy-currents in solid parts of the magnetic circuit and in the field winding, and by the field inductance, but the slow increase of flux induces a back e.m.f. in the field winding so that, even with constant exciter voltage, the field current suffers a sudden decrease, compensating the decrease in stator *de*-magnetizing ampere-turns. The time-constant of a turbo-alternator may be 5 or 6 sec., varying with the load.

The response of a voltage-controlled alternator's terminal voltage is thus a function of the responses of the voltage-regulator (which is usually very rapid), the exciter, and the main alternator flux. Fig. 299 shows typical curves for a machine working on 40 per cent load at power factor 0·8, and undergoing at the instant $t = 0$ a sudden increase of load to 80 per cent. The initial drop of about 4 per cent is due to the change in the leakage reactance and resistance drops. Thereafter the voltage falls slowly as the main flux changes to the value determined by the new armature reaction. After rather more than 1 sec., the rising exciter current begins to restore the alternator voltage. A shunt-connected exciter gives the slowest restoration, separate excitation is more rapid, while if super-separate excitation is employed (i.e. the application of double normal exciter voltage) the delay in the restoration of the alternator voltage is little more than 2 sec. This is called "field forcing."

With alternators driven by water turbines the problem is complicated by the fact that the inertia may be lower than that of turbine sets, so that the speed is liable to change rapidly when large

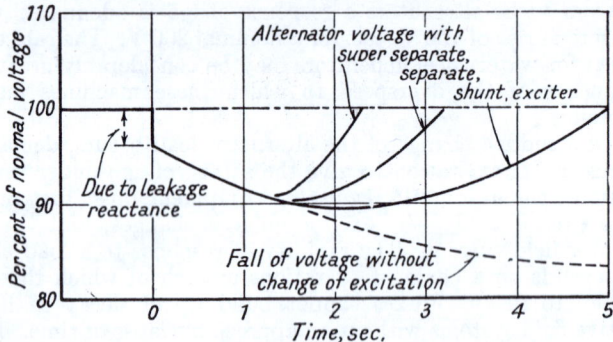

FIG. 299. VOLTAGE/TIME CURVES OF SUDDEN INCREASE OF LOAD FROM 40 PER CENT TO 80 PER CENT AT POWER FACTOR 0·8 LAGGING

alterations of load occur. Further, the working fluid (water) has a very much greater inertia than steam, so that it takes much longer to govern the inlet water to suit the new load. If load is thrown off, this involves upwards of 100 per cent increase in speed. A greater rapidity of exciter response is called for here.

The sudden dropping of lagging load will cause overvoltages in any synchronous generator. Where overspeed is also a result, the voltages developed may be such as to endanger the insulation. Fig. 300 (a) gives the conditions produced by suddenly unloading a 20 000 kVA. waterwheel generator working on a power factor of 0·75 lagging, with a normal speed of 100 r.p.m. The load is thrown off at zero on the time base. The gate, governing the water supply to the turbine, begins to close after 0·5 sec. Meanwhile the speed rises, reaching about 35 per cent overspeed in 5 sec. The voltage regulator reduces the field current initially, but the overspeed and the changing rotor flux prevent it from keeping the voltage down. The generated voltage rises to 175 per cent of normal. The curve marked "Transformer voltage" refers to the voltage on the l.v. side of the transformer to which the generator is directly connected. The rise of this voltage is less than that of the generator on account of the large magnetizing current drawn by it when oversaturated. Fig. 300 (b) shows curves for the same conditions as before, except that an over-voltage, over-frequency relay is employed to open the field circuit. There is still considerable voltage rise.

AUTOMATIC VOLTAGE-REGULATORS. To secure rapid action, modern regulators employ the *overshoot* principle. If the bus-bar voltage falls, the regulator changes the excitation to secure more than a compensating rise. Before the voltage has risen appreciably above normal, however, the regulator acts to reduce it again. If a

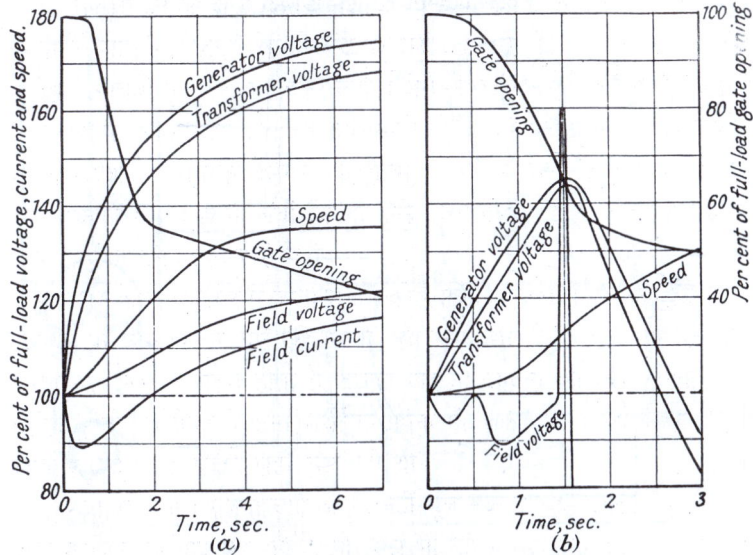

FIG. 300. EFFECT ON WATER-WHEEL GENERATOR OF THROWING-OFF FULL LOAD AT POWER FACTOR 0·8 LAGGING

sudden voltage change takes place, the regulator is thus prepared to restore normal voltage more quickly.

Generator regulators may be (1) *Direct-acting*, in which direct adjustment of resistance is made in the control-field circuit; (2) *Pulse,* in which the control field is fed with current pulses of variable magnitude or duration to give the required mean value; and (3) *Quiescent*, a combination of (2) with a motor-operated field rheostat.

The general principle of regulation is based on the use of a voltage-sensitive element acting against a stable reference force to change the effective resistance in the field of the exciter. The motion is damped, but spring-loaded to provide overshoot, i.e. a degree of over-regulation. The voltage-sensitive element may be a force-producing electromagnet relay, an electrical detector circuit or a balanced relay with floating contact. All regulators operate on the error between a reference setting and the actual voltage generated. Classes (1) and (2) above impose continuous control. Class (3)

operates only when the generated voltage is more than about 1 per
cent in error, so that they have a 2 per cent dead-band insensitivity.

Direct-acting Regulators. In the *rolling-sector* regulator, Fig. 301A,
the exciter field rheostat is connected across a series of contacts
over which sectors *s* roll when driven by a Ferrari disc motor,
energized from the voltage to be regulated, against a spiral spring *f*.
The movement is damped by the disc *o* rotated by quadrant *p*
between the poles of permanent magnets *m*.

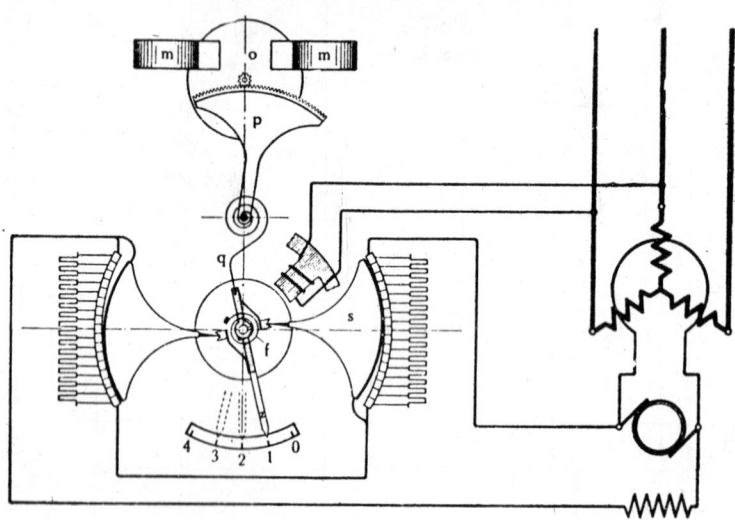

FIG. 301A. DIRECT-ACTING VOLTAGE-REGULATOR

The *carbon-pile* regulator is a variant, using a voltage-operated
coil, supplied through a rectifier, to operate damped and spring-
loaded leverage to tilt the contacts between stacks of carbon plates
forming the exciter field rheostat.

Pulse Regulator. An example of this class is shown in Fig. 301B.
The cam is rotated at about 8 r.p.s. by a single-phase induction
motor fed from the voltage transformer. It causes contact *a* to
vibrate. Spring-supported contact *b* engages with *a* except so far
as its upward movement is limited by the control arm, which is
located by a voltage-sensitive electromagnet and damped by a
spring-loaded dashpot. Contacts *c*, across the exciter field rheostat,
are closed by a spring except when coil *d* is energized. In normal
operation contacts *a* and *b* make and break, and the contacts *c*
break and make, so cutting the field rheostat in and out. The average
field current depends on the relative make and break periods. A
drop in output voltage causes the control arm to fall, and contacts
a and *b* are held open for a longer interval.

Quiescent Regulator. This is basically a pulse-type regulator, but with relay control of a motor-driven rheostat control. It suits base-load generating stations for which the voltage changes are slow.

Compounding. Voltage regulation may be made responsive to load current as well as to voltage. In the circuit of Fig. 301b, the

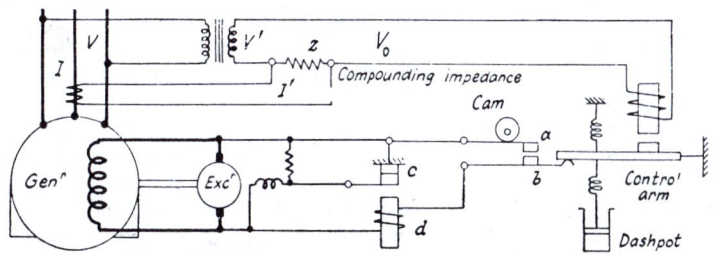

FIG. 301B. PULSE-TYPE VOLTAGE-REGULATOR

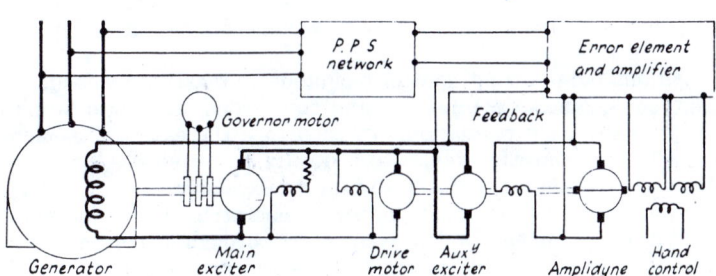

FIG. 301C. SERVO-TYPE VOLTAGE-REGULATOR

voltage V_0 comprises a component V' proportional to the generator output voltage V, and the drop $I'z$ proportional to its current I. Then $V_0 = V' + I'z$ is proportional to $V + IZ$, where Z is some suitable load-circuit impedance. If z is given in effect a phase angle of 90°, the voltage regulator operating from the input voltage V_0 will permit load-sharing of reactive volt-amperes between machines. If the angle is made 180°, a rising characteristic for line-drop compensation can be obtained.

Servo-type Regulator. Fig. 301c shows the main features of a regulator using electronic and rotating amplifiers. The main exciter, on the alternator shaft, has a saturated magnetic circuit and in consequence a substantially constant output voltage. Variation of the main exciting current is obtained by an auxiliary series buck/boost exciter, the field of which is fed from an amplidyne. The auxiliary exciter and the amplidyne are driven by a d.c. motor fed from the main exciter.

The alternator output voltage is taken through a voltage-transformer to a voltage-sensitive bridge, one arm of which contains a saturated diode. Any deviation in the bridge output voltage produces a corresponding change in the output of an electronic amplifier having the amplidyne control field as its load. The amplidyne output then alters the auxiliary exciter field, so varying the main exciting current and regulating the alternator output voltage in the appropriate direction. The several time-lags in the control circuit require feedback stabilization, taken from the auxiliary exciter field and armature, and the amplidyne field.

Other features of the equipment are: a supply for the synchronous governor motor, taken from slip-rings on the main exciter (which in this respect operates as an inverted synchronous converter); an additional battery-fed main-exciter field winding for field-forcing when starting the set, so as to give the governor motor adequate drive; a motor-operated main exciter-field rheostat for initial adjustments or abnormal conditions; field-suppression switches; standby hand control for the amplidyne field; and a positive-sequence network to ensure proper voltage regulation under fault conditions.

Various light-current control techniques, involving transductors, transistors, magnetic amplifiers and thyratrons, have been applied to alternator field regulation. In each case the essential elements are a voltage-sensitive device to control a power source, overshoot for rapid action and feedback to prevent excessive rise.

Excitation Systems. All modern generators have individual excitation, by machines driven direct or through gearing from the alternator shaft, or by separately motor-driven exciters. In the latter case the excitation may be lost if the busbar voltage is suddenly lost due to fault, and to mitigate the effect flywheel storage may be provided, or duplicate supplies.

Machines of very large output may have to be provided with separate motor-driven sets, as their excitation demand may reach 1 000 kW.

The term "voltage regulation" is scarcely apt under modern conditions, for no one machine can much affect its busbar voltage on a large supply network. More suitable terms would be "excitation regulation," or even "power factor regulation."

4. Dynamics of Synchronous Machines. A mechanical rotary system possessing inertia J and a torque M_0 per radian tending to restore its position when given an angular displacement, has a natural frequency $f_0 = (1/2\pi)\sqrt{(M_0/J)}$, and a corresponding natural period $t_0 = 2\pi\sqrt{(J/M_0)}$, neglecting any damping effect.

A synchronous machine working in parallel with other machines forms such a system. Under steady-state conditions it rotates at an angular velocity $\omega_1 = 2\pi f/p$ corresponding to the number p of its pole-pairs and the frequency f of the electrical network; further, it has a constant load angle σ between the rotating pole-axis of its stator and rotor. But if this relative position is disturbed, a synchronizing power flows in the machine to develop a synchronizing restoring torque. Fig. 264 suggests the "elastic" nature of the magnetic link between stator and rotor, from which the oscillatory tendency following a displacement can be inferred.

The oscillation of synchronous machines has considerable practical importance. It affects the *transient stability* of a machine on load; the oscillations consequent upon pulling into step, i.e. *synchronizing* a machine on to live busbars; and the effect on a machine of cyclic fluctuations of mechanical torque, as with a generator driven by an internal-combustion engine, producing *forced oscillations*.

STABILITY. Reference has already been made to the *steady-state stability* limit of a synchronous machine (Chapter XVII, § 6) as the greatest possible power under given conditions of operation and excitation as the load is gradually increased. The *transient stability* limit is the maximum power that the machine can carry and remain synchronized when subjected to a transient condition such as a sudden change in load, a network fault, or a switching operation. The general stability of a machine thus concerns not only normal operating equilibrium, but also the ability to regain it after a disturbance.

The power/power-angle diagrams in Fig. 302 are derived as described in connection with Fig. 280. The greatest power, from eq. (167), ignoring the stator resistance r, is $P_{1m} = VE_t/x_s$. For any load angle the power is $P_1 = P_{1m} \sin \sigma$. A change in load causes an alteration in load angle σ. Suppose the machine to be working on a load P_a with angle σ_a, and the load is suddenly increased to P_b with the equilibrium angle σ_b. The acceleration of the rotor from σ_a to σ_b occupies a time-interval during which it gains an increment of kinetic energy. As a result its speed of rotation rises above the normal synchronous, it passes through the new equilibrium angle σ_b and reaches a more advanced position σ_c which may perhaps lie beyond the critical angle $\sigma = 90°$. In this region a retarding torque is developed on account of the excess of output over mechanical input and a retardation ensues. Oscillation continues until damping has dissipated the oscillation energy. The transient stability limit is reached with that value of P_b which makes the first swing of the rotor terminate at an angle $\sigma_c = 180° - \sigma_b$ for which the power deficit $P_{1m} \sin \sigma_c - P_b$ becomes zero; for in this case there is no tendency for the rotor retardation to continue and swing it back towards the stable equilibrium position. The criterion of transient stability is that the vertically-shaded areas in Fig. 302 must not be greater than the horizontally-shaded areas. For the former represents the kinetic energy gained during the accelerating period from σ_a to σ_b, while the latter is the energy released during retardation between σ_b and σ_c. This is the reason for the placing of the stability limit in Fig. 295.

Under fault conditions the bus-bar voltage V may fall, so that P_{1m} is also reduced. A machine operating at a stable equilibrium may become suddenly overloaded if the voltage falls because the constant input demands a greater power angle for the lower voltage: the input may actually exceed the new value of P_{1m} so that loss of

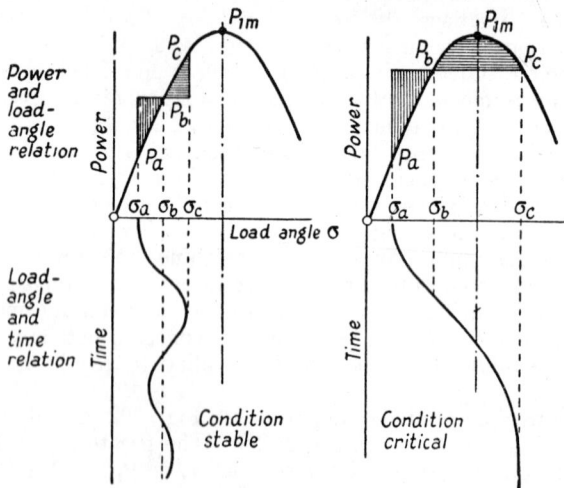

FIG. 302. STABILITY: LOAD-ANGLE/TIME CURVES

synchronism is inevitable unless the fault is cleared rapidly enough to limit over-swing.

DYNAMICAL ANALYSIS. If the equilibrium state of a machine corresponds to a power P_b at load angle σ_b, in Fig. 302, then any other load—say P_a at σ_a—will result in a difference power ΔP and a corresponding torque $M \simeq \Delta P/\omega_1$ acting to accelerate or retard the machine's inertia J. This torque does work* by shifting the rotor relatively by $\sigma_b - \sigma_a$; that is, $w = \int M \cdot d\sigma$. The integral gives the shaded areas in Fig. 302, and represents the basis of the *equal-area criterion* of stability,

It is difficult to analyse a case where the angular swing is large, and in practice such calculations are made by integraph or computer In the following the analysis is confined to small angular deviations for which simplifying assumptions can be made.

Consider the machine to be unloaded. Any divergence σ will call

---

* A change of σ implies a very small change of speed ω_1, but the effect of this is secondary, and for small changes of load angle can be ignored.

into being a synchronizing power $P_s \simeq (E_t V/x_s)\sigma$, corresponding to a torque $(E_t V/x_s \omega_1)\sigma = M_0\sigma$, where M_0 is the torque per radian. If now a sudden constant disturbing torque M_d is applied to the shaft, it will cause angular acceleration $d^2\sigma/dt^2$ of the inertia J, work against a damping torque $D \cdot d\sigma/dt$ assumed proportional to angular velocity, and oppose the synchronizing torque $M_0\sigma$ developed by the displacement. Hence

$$M_d = J(d^2\sigma/dt^2) + D(d\sigma/dt) + M_0\sigma.$$

This second-order differential equation solves to give the angle σ for any time t after the application of the disturbing torque:

$$\sigma = (M_d/M_0)[1 - \exp(-\alpha t)\{\cos \omega_{0d} t + (\alpha/\omega_{0d}) \sin \omega_{0d} t\}], \quad (170)$$

where $\alpha = D/2J$ is the decay factor of the oscillations of natural damped frequency $\omega_{0d} = \sqrt{[(M_0/J) - (D^2/4J^2)]}$. With $D = D_c = 2\sqrt{(M_0 J)}$ the damping is just sufficient to prevent oscillation: this condition is unlikely to be realized. With $D = 0$, that is, no damping, the load-angle becomes a sustained oscillation

$$\sigma = (M_d/M_0)[1 - \cos \omega_0 t]$$

at the natural undamped angular frequency $\omega_0 = \sqrt{(M_0/J)}$.

So long as $M_0\sigma$ properly represents the synchronizing torque. eq. (170) will hold with σ_1, an initial load angle, added to the right-hand size. M_0 may be adjusted for σ_1 by eq. (165).

CAUSES OF DIVERGENCE. Under steady load, the driving and load torques balance for a steady load angle σ. Any sudden change of condition will set up a transient oscillation superposed on the steady mean speed. The incidence of a disturbing force is unimportant unless (a) the machine is running in parallel with other synchronous machinery, and (b) the frequency of the disturbing force is close to the natural undamped frequency f_0. In this connection it must be remembered that a non-sinusoidal disturbing force, having a fundamental frequency quite different from f_0, may nevertheless contain harmonics, one of which has a frequency close to or identical with the natural frequency of the rotating parts of the machine.

The causes of disturbing forces are

(a) Cyclic irregularity in the driving torque of the prime mover, such as occurs with internal-combustion engines.

· (b) Cyclic irregularity in the load torque on a motor, such as with an air-compressor.

(c) Cyclic irregularity on other parallel-running machines, giving the same effect as (a) or (b).

(d) Governor hunting due to lag between the inlet of fuel and the rise in speed due to it (e.g. the fuel admitted by the governed valves in a four-stroke engine is not ignited until after two strokes).

(*e*) Transient effects, such as sudden changes of load, misfire or pre-ignition of the prime mover, synchronizing, etc.

However caused, the unbalanced phase-displacement δ releases an excess torque to act on a mechanical oscillatory system, producing the possibility of phase-swinging or *hunting.*

The possible conditions are as follows.

Suppose a sudden change of load or other transient condition to occur, for example, a sudden reduction of load on a generator. The angle σ is too great by δ, and the driving torque is excessive. The rotating parts accelerate, increasing δ, which establishes a synchronizing torque. This grows as δ increases, until a maximum forward swing is reached, after which the strong synchronizing torque retards the rotor. The latter slips back behind the new steady value of σ, producing a divergence δ in the opposite direction. This creates a forward synchronizing torque which grows with δ and eventually accelerates the rotor again. The rotor thus takes up a to-and-fro swing which normally dies out, the final position being a steady value of σ appropriate to the new reduced load.

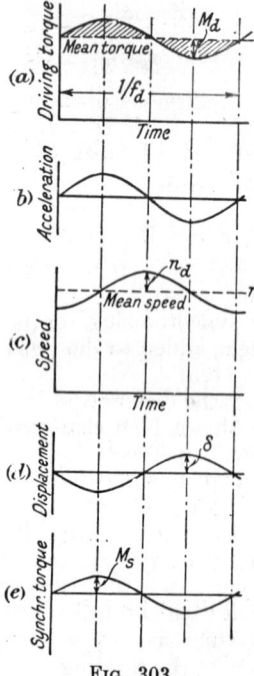

FIG. 303

DISPLACEMENT CURVES OF GENERATOR WITH CYCLIC FLUCTUATION OF TORQUE

If the disturbing force is cyclic, it is manifested as producing a forced vibration, increased (see below) by the synchronizing torque. If the damping is small and the disturbance has a frequency approaching f_0, the amplitude of the rotor oscillation may grow until the machine swings out of step. This is usually the more dangerous case.

CYCLIC DISTURBING TORQUE. Let, for simplicity, a rhythmically-varying driving or load torque be represented by a constant mean torque with a superimposed sinusoidal variation of amplitude M_d and frequency f_d. Neglecting damping, the torque M_d (Fig. 303 (*a*)) produces an acceleration in phase with itself (*b*), since it is absorbed in accelerating or retarding the rotating mass about its steady mean speed. The cyclic acceleration contributes a cyclic speed change n_a about the mean speed n, as shown by Fig. 303 (*c*). This leads further to an angular divergence δ about the steady lead σ of the rotor on the stator pole axis, (*d*). Since the generator is presumed to be running in parallel with other machines, the divergence δ calls into play a proportional synchronizing torque of amplitude M_s, (e). This is seen to be coincident with the disturbing torque M_d, so that the sum $M_s + M_d$ is acting on the rotating masses. Although the synchronizing torque always acts against the displacement δ, it has the effect of increasing this displacement when the disturbing torque is itself cyclic.

The equation of motion will again be

$$M_d = J(d^2\sigma/dt^2) + D(d\sigma/dt) + M_0\sigma,$$

with the disturbing torque now represented by the harmonic function $M_d \sin \omega_d t$ at the cyclic disturbance frequency f_d. Interest lies only in the steady-state solution which, by analogy with an a.c.

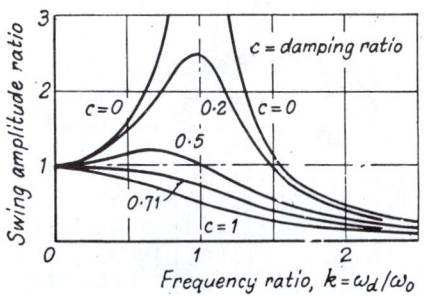

FIG. 304. PHASE-SWING AMPLITUDES

circuit of series parameters RLC, is obtained by writing $j\omega_d$ for d/dt:

$$M_d = (-\omega_d^2 J + j\omega_d D + M_0)\sigma,$$

giving directly the complexor solution for σ as

$$\sigma = M_d/[(M_0 - \omega_d^2 J) + j\omega_d D].$$

Introducing the undamped natural frequency of oscillation $\omega_0 = \sqrt{(M_0/J)}$, the decay factor $\alpha = D/2J$, and the damping ratio $c = D/D_c = \alpha/\omega_0$, then the peak amplitude of swing corresponding to the peak disturbing torque M_d is

$$\sigma = (M_d/M_0)/[(1 - k^2) + j2ck] \tag{171}$$

where $k = \omega_d/\omega_0 = f_d/f_0$ is the ratio of the disturbing frequency to the natural undamped oscillation frequency of the system. Clearly, if the damping is small and f_d approaches f_0, the divergence may grow large enough to swing the machine out of step. Fig. 304 shows, for various fractions c of the critical damping, how the oscillation amplitude will change with the ratio $k = f_d/f_0$.

The cyclic speed-irregularity caused by hunting is usually less than 0.5 per cent. The limit of divergence is a few degrees, the value depending on the type of load and number of poles, for with many poles a small mechanical divergence will represent p times as much electrically.

DAMPING AND FLYWHEEL EFFECT. The swinging of the rotor results in a to-and-fro movement of the gap-flux across the stator superimposed on the steady rotary sweep. This low-frequency movement produces eddy-currents in the pole-faces, armature core and conductors, dissipating some of the kinetic energy of the swing

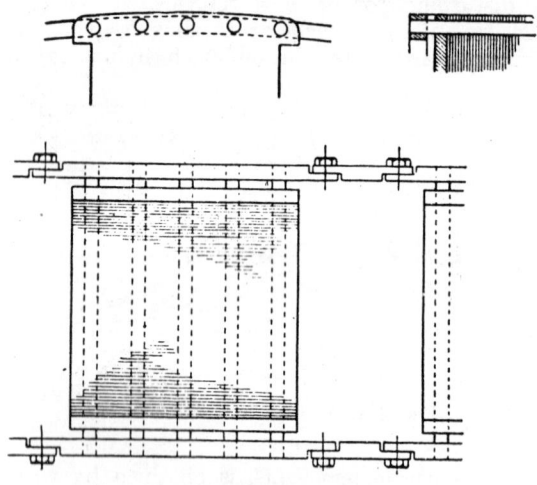

FIG. 305. DAMPER WINDINGS

and reducing its amplitude. For improving parallel running, it is usual to fit additional *damper windings* in low-speed machines. These comprise bare copper bars embedded in slots or holes in the pole-shoes, and brazed or welded at the ends to bars or rings like those of a cage induction motor, Fig. 305. The bars become the seat of e.m.f.'s induced by the flux-swing, and carry currents that by dissipating I^2R loss reduce the energy in the oscillation. For greatest effectiveness they should be of low resistance and should have end-connections formed into continuous rings right round the machine. It is possible to have the resistance too low, in which case the I^2R loss falls.

The eddy currents and their damping torque effect are proportional to the eddy e.m.f., which depends on the relative velocity $d\sigma/dt$. It is thus legitimate to take the damping torque in eqs. (170) and (171) as $D(d\sigma/dt)$.

The damping alone cannot be relied upon if f_d is of the same order as f_0. If $f_d > f_0$, the phase-swinging can be reduced by adding to the rotating mass in some way, e.g. by means of a flywheel: this reduces

f_0 and so removes it farther from f_d. On the other hand, if $f_d < f_0$ (in exceptional cases) improved running is obtained with a reduction in the rotating mass. As a general rule it is necessary to increase the flywheel effect.

It must not be forgotten that a disturbing torque may have a large variety of harmonics, and that one of them, even if very small, may cause resonance and excessive phase-swing. Lower harmonics are as a rule more dangerous. In oil engines, for example, there is likely to be an unnoticed harmonic of frequency corresponding to the camshaft speed (i.e. *half* engine speed), which may be troublesome.

5. **Short Circuit.** The possibility of sudden short circuit of a generator when running fully excited has a profound influence on its mechanical design, particularly as regards the bracing of the stator end-connections against the mechanical forces brought into being by short-circuit currents. The forces depend on the square of the current value, so that there is the likelihood of pressures up to several tons being developed between adjacent stator conductors. In the slots the surrounding steel is sufficient to support the conductors, but in the overhang an elaborate system of bracing clamps must be provided.

The precise train of events within a machine subjected to a sudden short circuit depends on a number of factors, including—

(a) at what instant in the cycle the short circuit is applied;

(b) the load and excitation of the machine immediately before;

(c) the extent of the short-circuit: whether one, two or three phases are involved and whether the short-circuit fault occurs close to the machine terminals or more remotely in the network;

(d) constructional features of the machine, especially those affecting leakage and damping.

The evaluation of the sudden-short-circuit current for given conditions is an elaborate and somewhat empirical process, depending on quantities (such as self- and mutual-inductances and effective resistances) that are themselves variable and difficult to assess.

The time immediately following a short circuit may usefully be divided into three periods: (1) a very short period, covering one or two cycles, during which the conditions are largely dependent upon the flux linking stator and rotor windings at the moment of short circuit; (2) a subsequent and longer interval of transient decay of short-circuit current amplitude affected by damping and by the rise of armature reaction; and (3) a final period during which *steady* short-circuit conditions obtain. The generator will normally be open-circuited before period (3) is reached, except on test.

INITIAL CONDITIONS. An insight into the physical behaviour of a generator immediately following a short circuit can be greatly facilitated by making use of the *theorem of constant linkages.** For any closed circuit with resistance r and inductance L and no source

* Doherty and Shirley, *Trans. A.I.E.E.*, 37, p. 1209 (1918).

of e.m.f. the equation $ri + d(Li)/dt = 0$ can be written. If the resistance is negligible, then $d(Li)/dt = 0$; that is, the linkages Li must remain constant. In a generator the effective inductance of the windings of both stator and rotor may be considered large compared with the resistance, and at any rate for the first two or three cycles the effect of the latter may be neglected. The rotor winding is closed through its exciter, and the stator is closed by the short

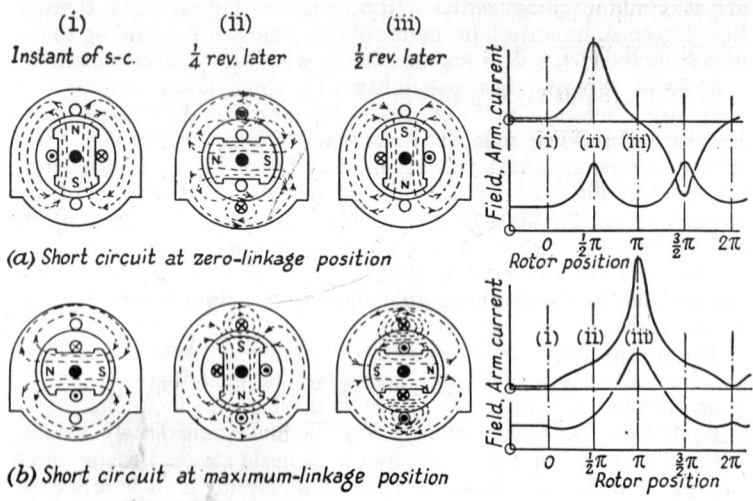

(i) Instant of s.-c.   (ii) ¼ rev. later   (iii) ½ rev. later

(a) Short circuit at zero-linkage position

(b) Short circuit at maximum-linkage position

FIG. 306. ELEMENTARY SINGLE-PHASE SHORT-CIRCUIT.
ANTICLOCKWISE ROTATION

circuit; after which the fluxes linking either winding must remain constant regardless of rotation.

Consider an elementary *single-phase* machine, Fig. 306 (*a*), comprising a bipole rotor field system and a concentrated stator winding. Let the stator be short-circuited at the instant (i) when there are no stator linkages. As the rotor rotates to a new position (ii) at right-angles to the first, it tends to establish full normal linkages in the stator winding: this the stator must oppose by a current in the direction shown, forcing the rotor field into a leakage path. Simultaneously the rotor current must increase to maintain its own flux constant despite the now increased reluctance of its magnetic circuit. At position (iii) the stator current is again zero and the rotor current sinks to normal. A similar argument for subsequent instants gives the current/position relation shown.

Suppose now that the position of the rotor at the instant of stator short circuit is that at (i) in Fig. 306 (*b*). The stator has maximum linkage. Rotation of the rotor into position (ii) removes its linkages from the stator, which consequently assumes a current to produce

.its own, in leakage paths. In position (iii) the rotor attempts to produce full reversed flux in the stator, and is counteracted by the stator current rising to a high value to maintain unchanged its original linkages. The peak current is determined by the reluctance of the leakage paths, i.e. by the leakage reactance. The rotor in position (iii) has its flux forced into a leakage path mainly inside the stator bore, so that the rotor current rises too. The current changes are shown in Fig. 306 (*b*). In actuality, the rotor current variations take place partly in the rotor winding, partly in other closed paths available for current flow, such as the rotor body (if solid), wedges and end-covers, or damper windings.

If the stator winding is distributed, the effect is to round off the severity of the peak currents, without making any essential difference to the waveform. An actual oscillogram, Fig. 309 (*b*), for a case intermediate between those in Fig. 306, supports the analysis based on the constant-linkage theorem.

The short-circuit current is seen to be solely a function of the relative position of stator and rotor, and immediately after short circuit the currents can be assessed in terms of rotor angle.

An analysis of short-circuit torque may be made, for, neglecting saturation, the energy stored in the stator and rotor circuits is $W = \frac{1}{2}\phi_a i_a + \frac{1}{2}\phi_f i_f$, where ϕ_a and ϕ_f are the constant linkages of armature and field respectively. Since all energy changes must come from mechanical power input, the torque is simply the variation of the energy W with the position-angle θ: thus at any given rotor position the electromagnetic torque will be

$$M = \frac{dW}{d\theta} = \frac{1}{2}\left(\phi_a\frac{di_a}{d\theta} + \phi_f\frac{di_f}{d\theta}\right)$$

which is proportional to the slope of the rotor and stator current waveforms. The torque will fluctuate strongly, and will cyclically reverse.

A *three-phase* short-circuit may be dealt with by the theorem on the same lines as for the single-phase machine (or one phase of a three-phase machine) above. At the moment of short-circuit the flux linking the stator from the rotor is caught and "frozen" to the stator, giving a stationary replica of the main-pole flux. For this purpose each phase will in general carry a d.c. component. The rotor maintains its own poles, and accordingly there are two systems of similar poles, one rotating with the rotor and the other fixed with the stator. Each time the rotor poles are in the position they occupied at the moment of short circuit the rotor (field) current is normal and all the stator phase currents zero (if the machine were previously unloaded). One-half cycle later the stator and rotor poles are again axially coincident, but are opposite in sense: yet constant linkages must be maintained in each, and current peaks result. Further, the stationary field held by the stator induces a normal-frequency e.m.f. in the rotor. The resultant a.c. component in the effective rotor field circuit develops a fundamental-frequency flux, which due to rotation produces in the stator windings a *double-frequency* or *second-harmonic* current. Thus the general form of the transient phase current is a wave comprising fundamental, second-harmonic and d.c. components. The rotor current has a large fundamental-frequency fluctuation, part of which will flow in eddy-current paths. Fig. 307 shows curves of stator and rotor currents

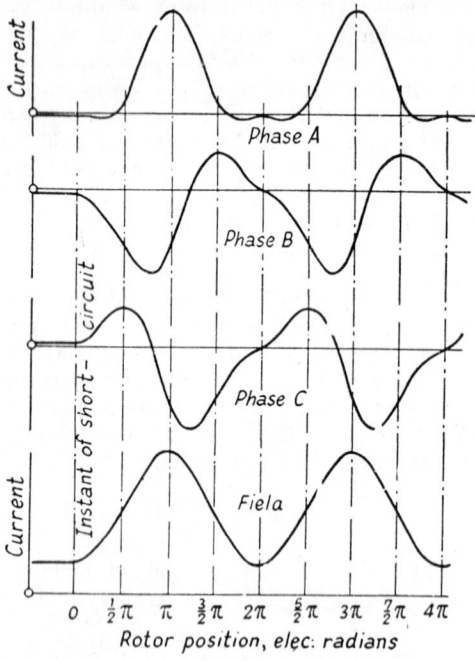

Fig. 307. Elementary Three-phase Short Circuit with Maximum Linkages in Phase A

calculated for an assumed leakage coefficient for a three-phase short circuit at an instant giving complete asymmetry in phase A. Compare Fig. 309 (*a*).

It will be observed that the stator currents on short circuit are similar to those that flow in an inductive circuit when an alternating voltage is suddenly applied, except for the distortion of wave-form from the pure sine. It is customary to consider the currents from this simpler viewpoint, which makes the introduction of resistance and damping effects much more straightforward. In what follows, therefore, it is assumed that the outstanding phenomena can be dealt with on the basis of a leakage reactance.

ARMATURE CURRENT DURING THREE-PHASE SHORT CIRCUIT. Consider a generator without resistance, excited and running on open circuit. If its terminals are suddenly short-circuited, the e.m.f. is provided with a closed path comprising the stator windings, any will circulate a short-circuit current through them. The case mad be considered analogous to that of a sinusoidal e.m.f. suddenly applied to an inductive circuit. The reactance to be taken into account initially is not the synchronous reactance, but a true leakage reactance x. As has been explained (Chapter XVII, § 3), the effect of the armature reaction on *steady* short circuit is to cause a reduction of the flux on account of the opposition of the armature ampere-turns that it represents. The flux links the windings of the stator and rotor and it is not possible to make sudden changes in it on account of the inductance of the windings and the eddy-currents induced in various parts of the magnetic circuit tending to hold it against rapid change. At the moment of sudden short circuit therefore, armature reaction is inoperative in reducing the main flux, and the fictitious reactance x_a representing it must therefore be omitted from consideration: only a true leakage reactance is operative in limiting the short-circuit current. Increased I^2R, core and stray losses constitute a secondary limitation.

The armature reaction begins at once to take effect, but some seconds may elapse before the machine operates under conditions resembling *steady* short-circuit as already described (Chapter XVII, § 11). The short-circuit current commences, therefore, at a large value, while after the elapse of a few seconds it has settled down to the steady-short-circuit value in which the armature reaction exerts its full effect. This presumes that the speed of the machine and its excitation have been maintained, which in practice is naturally not always the case.

The initial short-circuit current is therefore a function of the e.m.f. and the leakage reactance. It is also considerably affected by the instant in the cycle at which short circuit occurs.

It has been shown (Fig. 1 (c)) that a sine voltage $e = e_m \sin \omega t$, suddenly applied to an inductive circuit of inductance L and resistance r, produces in the circuit a current

$$i = (e_m/z) \sin (\omega t - \phi) - (e_m/z) \sin (\omega t_1 - \phi)\varepsilon^{-(r/L)(t - t_1)}$$

where $z = \sqrt{(r^2 + \omega^2 L^2)} = \sqrt{(r^2 + x^2)}$; $\phi = \arctan (x/r)$; and $t_1 =$ instant at which the voltage is applied.

For the case under consideration, r may be considered small compared with L, so that, approximately

$$i = - (e_m/x) \cos \omega t + (e_m/x) \text{ c s } \omega t_1 . \varepsilon^{-(r/L)(t - t_1)}$$

As discussed in Chapter I, § 3, the initial amplitude of the current is affected by the instant at which short-circuit occurs. For short-circuit at an instant near a voltage zero the doubling effect is prominent, and there will inevitably be some asymmetry in at least two phases of a three-phase machine.

The expressions above apply only to the initial conditions of short circuit. In fact, the voltage *e* actually falls with time on account of the growth of armature reaction. Fig. 308 shows the decay of the current peaks produced by the gradual rise of the reaction. The rapid attainment of symmetry and the slower shrinking to steady short-circuit conditions each depend on resistance/inductance ratios, the exact values varying considerably due to satura-

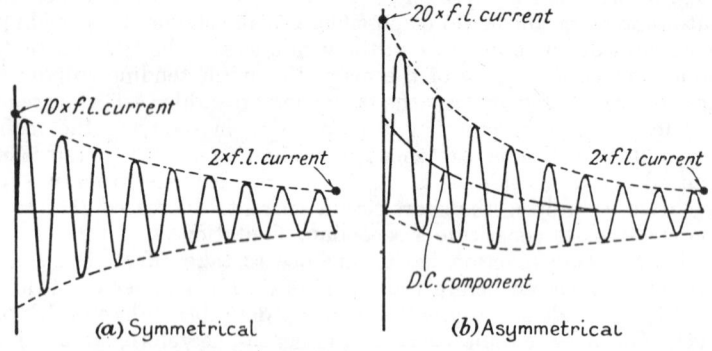

FIG. 308. SYMMETRICAL AND ASYMMETRICAL SHORT-CIRCUIT CURRENTS
IN SYNCHRONOUS GENERATOR

tion, pole position, etc., and the type of construction. Thus in a solid rotor (as with a turbo-generator) the eddy-currents set up considerably reduce the effective rotor winding resistance by shunting. For salient-pole machines the transient may die out within five cycles.

In brief, the short-circuit current is characterized as follows—

(*a*) The initial a.c. component is determined by the leakage.

(*b*) The initial d.c. component is determined by the instant in the cycle at which the short circuit occurs.

(*c*) The total initial current is in general asymmetrical. If the short circuit occurs at zero e.m.f. it is completely asymmetrical, and reaches a peak value nearly double that of the a.c. component.

(*d*) The d.c. component attenuates rapidly and vanishes within a few cycles.

(*e*) The a.c. component attenuates more slowly to a value determined by armature reaction as well as by leakage reactance. It becomes the normal steady-short-circuit current.

In the above, the effect of resistance is disregarded except as contributing to the attenuation. Fig. 308 illustrates the various features for a case in which the leakage reactance is 0·1 per unit and the synchronous reactance is 0·5 per unit: these figure gives initial-short-circuit currents of ten times full-load current when symmetrical, twenty times when unsymmetrical, or any intermediate value; and a steady-short-circuit current of about twice full-load current.

Fig. 309 shows a series of typical oscillographic records of single- and three-phase short circuits on various synchronous generators.

FIELD CURRENT DURING SHORT CIRCUIT. The armature and field windings are linked with the common magnetic circuit of the machine. Any change in the condition of the armature current must therefore influence the field current by mutual induction.

Considering a three-phase short circuit, there will be in the armature currents (a) a suddenly increased alternating-current, and (b) a

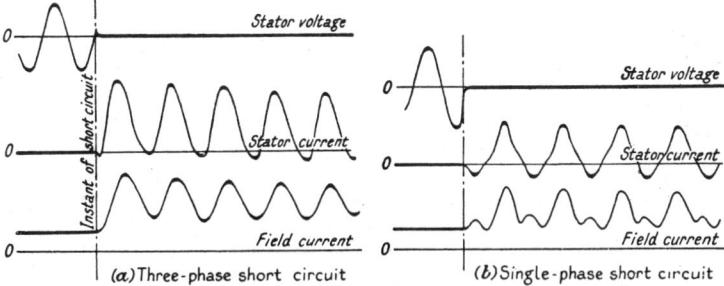

FIG. 309. OSCILLOGRAMS OF STATOR VOLTAGE, STATOR CURRENT, AND FIELD CURRENT DURING SHORT-CIRCUIT

transient direct-current component. The a.c. components combine to form an increased demagnetizing armature reaction, moving synchronously with the rotor poles and in direct opposition to the field ampere-turns. The total flux linking the field winding cannot suddenly change, so that an approximately equal reaction is produced in the field circuit in which the direct current grows to a value counterbalancing the increased armature reaction. The production of this additional field current is due to the slight fall in linked flux, which induces in the field circuit (closed through the exciter armature) a forward e.m.f. which circulates the additional current If there were no damping, eddy-currents, resistance, nor leakage, the total flux in both armature and field systems would be maintained constant. Actually the result is greatly to increase the leakage of both members. The field winding maintains a substantially constant flux, but a smaller proportion of it links the stator.

Thus the effect of the a.c. component of the stator short-circuit currents is to produce an increase in the field current, which is maintained by the falling field flux until the conditions of steady permanent short-circuit are attained, by which time it has sunk back to a value determined only by the excitation.

The d.c. components in the stator windings produce *stationary poles*. As the field system rotates, it passes stationary poles of successively opposite polarity. Thus a North pole on the rotor is confronted with a d.c. excited South pole on the stator, and a half-cycle later a North pole. The field flux has now superimposed on it a new flux pulsating with respect to the field windings at normal machine frequency, which induces in the field coils an e.m.f. circulating a field current component of fundamental frequency.

In Fig. 309 will be observed oscillograms of field current, in which the increase of field current above normal and the fundamental-frequency component are clearly seen.

SINGLE-PHASE SHORT CIRCUIT. Considering a short circuit to take place between one terminal and the neutral, the short-circuit current will flow in this phase only, and the other two phases may be ignored. The stator current being a single-phase one, there will be a pulsating armature m.m.f. The a.c. component of the short-circuit current may be imagined as resolved into two equal uniform m.m.f.'s rotating in opposite directions, one with and the other against the rotor. The former will give rise to a steady demagnetizing m.m.f., while the other will produce a reaction on the field that pulsates at twice the machine frequency. The d.c. component of the short-circuit current will, as in the three-phase case, produce stationary magnetic poles, reacting as before on the field circuit at the machine frequency. The rotor current will therefore exhibit—

(a) A general increase to balance the forward rotating component of the a.c. reaction:

(b) A double-frequency current due to the backward rotating component of the a.c. reaction; and

(c) A normal-frequency current produced by the reaction of the d.c. transient in the stator current, which will vanish if the phase short-circuit current is symmetrical.

Fig. 309 (b) exhibits these reactions in the field current.

SHORT-CIRCUIT REACTANCE. On a three-phase dead short circuit, the a.c. components of the phase currents—that is, the currents left after extracting the d.c. transients—occupy identical envelope shapes like the symmetrical case in Fig. 308. The envelope is substantially exponential in shape with a time-constant τ' of one or two seconds. At the beginning of the short circuit, however, there is a much more rapid decay of the current envelope with time, the corresponding time-constant τ'' being of the order 0·2 sec. or less. A typical envelope shape is shown in Fig. 310. Extrapolation of the main part of the envelope back to zero time, the instant of short circuit, gives the current peak e_m/x' that would obtain but for the abnormally-rapid initial decrement. The actual current peak at zero time, given by e_m/x'', is greater.

The value of τ', called the *transient time-constant*, is determined by the rate of growth of armature reaction: x', the *transient reactance*, is dependent on the leakage flux. At the instant of short circuit the full leakage path is not available, the leakage flux per stator ampere is less than subsequently, so that the reactance is less. For this condition x'', the *subtransient reactance* obtains.

The reason for the increase of effective leakage during the first fraction of a second after short circuit is that the leakage path comprises two main regions—(1) that comprising the overhang, top of the slots and air-gap; and (2) the path across the gap, into the

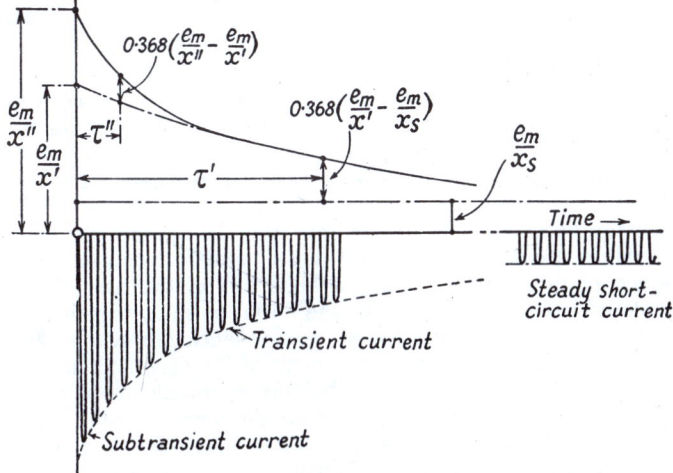

Fig. 310. Analysis of Symmetrical Short-circuit Current

rotor surface and back to the stator. Solid metal in the rotor initially opposes the production of leakage flux in region (2) by the induction of eddy currents, but the eddy losses are such that the delay is only brief. Full leakage per ampere is established in about $\frac{1}{2}$ sec. Thus $x'' < x'$: typically x'' is 0·12 per unit and x' is 0·18 per unit.

Prediction of the short-circuit current waveform, Fig. 310, requires an estimate of the initial and subsequent leakage path permeances, the armature reaction, and the power factor of the short circuit. A three-phase *terminal* short circuit will give a very low power factor, for the leakage reactance is usually very much larger than the resistance. All the reactances—the subtransient, transient and synchronous—then refer to the direct-axis values. If the short circuit occurs at a point remote from the generator terminals into the connected network, the power factor may be higher. Cross-axis conditions now appear, and further reactance values for the cross axis are required. A full statement of the general problem on a reactance basis then requires a knowledge of x_d'', x_d', x_{sd}, and x_q'', x_q', x_{sq}, together with appropriate effective resistances to estimate τ'' and τ'. American literature is particularly rich in analyses on such a basis. Again, if the short circuit is unbalanced, i.e. does not involve the three phases equally, the usual method of approach is by the use of positive-, negative- and zero-sequence symmetrical components. For each component it is possible to employ subtransient, transient and synchronous reactance coefficients.

The development of the subject* has followed two distinct lines: (a) Mathematical analysis based on a classic paper by R. H. Park, who developed the operational equations of an "ideal synchronous machine"; (b) Practical methods due to Crary, Doherty and Nickle, and others, in which the analysis is based on a combination of results for a variety of simplified conditions. By use of both these methods, and by the development of equivalent circuits, the analysis can be reduced to the consideration of three direct-axis circuits and two quadrature-axis circuits. The matter is discussed in Chapter XXVII.

A large number of reactances may thus be needed. In any case the values are at best only estimates, and a much more direct (though empirical) approach can be found in the use of a single calculated leakage reactance for the initial current, together with a standard decrement curve based on experience. In nearly all cases the initial short-circuit current is of chief interest, and its value is substantially independent of the balance or unbalance of the short-circuit fault. The final steady short-circuit current is approximately the same for a two-phase as for a three-phase short circuit, while that for a single-phase short-circuit is about 50 per cent higher.

A series of per-unit figures for direct-axis reactances is given in the Table opposite. The negative-sequence reactance is a rough average between x_d'' and x_q''. The zero-sequence reactance is small, and is strongly affected by the coil-pitch and phase-spread: for if the windings were uniformly distributed and the same current circulated through all phases, there would be zero resultant flux except in the overhang and slots.

FIELD SUPPRESSION. When a short circuit occurs on a machine in operation, it is desirable not only that the machine be disconnected from the bus-bars, but also that its generated voltage be eliminated as quickly as possible, to avoid the aggravation of possible internal faults. The field must be interrupted by an automatic field switch, which may act by opening the main field circuit, opening the exciter field circuit, reversing the exciter field, or short-circuiting the main field and exciter.

If a field circuit be disconnected from its exciter and discharged

* Adkins, "Transient Theory of Synchronous Generators Connected to Power Systems," *Proc. I.E.E.*, 98 (II), p. 510 (1951).

through a discharge-resistance equal to the field resistance, the e.m.f. induced in the field cannot exceed twice the normal voltage; but the field current may persist for a considerable time—e.g. of the order of a minute in a large turbo-alternator. This is too long,

DIRECT-AXIS PER-UNIT REACTANCES OF SYNCHRONOUS MACHINES

	Turbo-generator		Capacitor	Waterwheel generator	
	2-pole	4-pole		With dampers	Without dampers
Synchronous x_{sd}	2·0 1·4–2·5	1·45 1·35–1·65	1·25 0·75–1·8	0·9 0·5–1·5	0·9 0·5–1·5
Transient x_d'	0·19 0·11–0·25	0·27 0·24–0·31	0·21 0·12–0·27	0·23 0·14–0·32	0·23 0·14–0·32
Subtransient x_d''	0·13 0·08–0·18	0·19 0·15–0·23	0·13 0·09–0·15	0·16 0·1–0·27	0·18 0·16–0·3
Negative-sequence x_2	0·16 0·09–0·23	0·28 0·24–0·31	0·12	0·16 0·1–0·27	0·23 0 12–0·37
Zero-sequence x_0	0·08 0·02–0·15	0·28 0·22–0·31	0·03	0·08 0·06–0·1	0·08 0·06–0·1
Inertia constant H	4·7 2·6–6	4·8 4·3–5·7	1·2 0·7–1·8	3·4 2–5	3·4 2–5
Open-circuit time constant τ_{do}	9·5 3·5–16	5·5			

since it allows the generated voltage to persist, and feed energy into an internal fault.

A much more rapid method is to open-circuit the field winding. The eddy-current paths in the rotor magnetic circuit constitute a discharge resistance, which is low enough in a solid-rotor turbo machine to limit the induced field e.m.f. to about five times normal. The field current now takes only a fraction of a second to vanish.

Exciter reversal has initially little advantage over short-circuiting the rotor slip-rings, but reduces the field current rather more rapidly in later stages. If the exciter is itself separately excited from a constant voltage source, the field current can be rapidly reduced to zero. It is then desirable to open the main field circuit as soon as the stator fault current has dropped to zero, to prevent building up a reversed main field.

SHORT-CIRCUIT FORCES. A generator with a sub-transient reactance of 0·1 p.u. may have, on short circuit with full asymmetry, an initial current of 20 times full-load peak, and mechanical forces 400 times normal. With very large generators (e.g. 200 MW.) the forces can be formidable, especially in the overhang end-windings of the stator: the slot conductors are well embedded and repulsion is readily sustained by slot wedges. The overhang structure must be adequately braced and packed without unduly impairing the ventilation.

The resultant force on a coil is due to its own electromagnetically-produced forces, combined with those transferred to it mechanically from its neighbours through the packing blocks and binding. The total moment on a coil about its points of emergence from the stator core (where distortion will particularly affect the insulation) is therefore considerably greater if the end-connections are long. The forces on part of a coil will be those of attraction due to its neighbours in the same phase, and of alternate attraction and repulsion by coils of the other two phases; variants of these forces arise because of conductors in the other layer of plane of the winding. The interplay of the developed forces varies with the kind of short circuit (e.g. symmetrical three-phase, or phase-to-phase, or phase-to-earth, the two latter giving rise to the more severe conditions).* The rhythmic and rapid fluctuation of large forces can produce severe mechanical strains on the end windings and cause insulation failure.

6. **Surge Voltages.** The impact of a high-voltage surge on a synchronous generator has an effect comparable with that on a transformer (Chapter V, § 13). A stator winding is, however, considerably more complex than a transformer winding. It is an elaborate arrangement of coils, insulated with high-permittivity material and embedded in iron for about one-half their total length. With the usual two-layer winding there will be at least two conductors, conductively separate but coupled both capacitively through their insulation and inductively by their mutual slot flux.

Consider a bar-type stator winding constructed with single-turn coils so that each slot has two conductors. Ignoring the discontinuity introduced by the end-windings. the contents of one slot may be approximately represented by the equivalent circuit of Fig. 311 (a). The self-inductances L of the bars have mutual coupling M, and there are mutual capacitances K between conductors, C to earth, and c between coils in adjacent slots due to their proximity in the overhang. These parameters are expressed per unit length of conductor. From the differential equations relating voltages and currents in this network† it is found that the voltage conditions are

---

* Young and Tompsett, "Short-circuit Forces on Turbo-alternator End-windings," *Proc. I.E.E.*, 102 (A), p. 101 (1955).

† Robinson, *Proc. I.E.E.*, 100 (II), p. 453 (1953); 101 (II), p. 335 (1954); 103 (A), pp. 341 and 355 (1956). See also Taylor, *Power System Transients* (Newnes).

satisfied by two sets of travelling waves, Fig. 311 (b). The first set comprises a pair of equal voltages v_1 accompanied by currents i_1 associated with the "transmission-line" system formed by a conductor and the earthed slot-wall. The second set consists of a pair of equal but oppositely-polarized voltages v_2 with currents i_2. The respective currents and voltages are related by the surge impedances presented to them—

$$Z_{01} = v_1/i_1 = \sqrt{[(L + M)/C]} \text{ and}$$
$$Z_{02} = v_2/i_2 = \sqrt{[(L - M)/(C + 2K)]}$$

The velocities of propagation also differ. They are

$$u_1 = 1/\sqrt{[(L + M)C]} \text{ and } u_2 = 1/\sqrt{[(L - M)(C + 2K)]}.$$

There is an upper limit to the propagation of travelling waves, however, due to the series capacitance c which resonates with

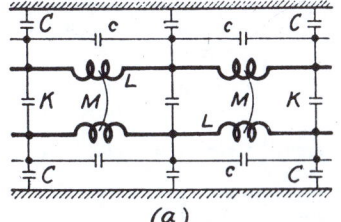
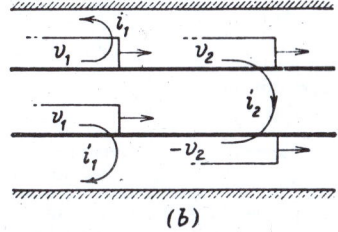

$$(a) \qquad\qquad\qquad (b)$$

FIG. 311. SURGE DISTRIBUTION IN BAR-TYPE STATOR WINDING

$(L + M)$ to form a rejector network. If, therefore, a step-function surge is analysed into a constant term plus a frequency spectrum, only those components below the frequency of rejector resonance can pass through the winding as travelling waves. (The effect of this limitation is substantially to slope the wavefront to a finite initial steepness, and to make the steepness progressively less as the waves advance into the winding by slowing down the higher-frequency components.)[*] The upper-frequency components in the spectrum, which are rejected, form an additional standing voltage component distributed substantially in accordance with the capacitances of the equivalent network: this leads to a high voltage-gradient in the turns of the stator winding near the line terminal.

Where the winding has two or more turns per coil, high transient voltages occur between the comparatively weak insulation of the adjacent turns, intensified by the comparatively large ratio C/K. With concentric-conductor h.v. stator windings, reflection effects occur at the junctions of the "bull," "inner" and "outer" layers due to the abrupt change in surge impedance.

* Taylor, *Power System Transients* (Newnes), Chapter VI.

With considerable simplification a generator may be represented, from the surge viewpoint, as a cable of the appropriate surge impedance and propagation velocity. The surge impedances vary from about 600 Ω, down to 50 Ω for large turbo-generators and water-wheel machines, and the velocity is of the order of 20 m. per μs.

7. Generator Protection. A generator may require to be protected against the consequences of a variety of possible faults, including: overspeed; over-voltage; over-current; unbalanced load; insulation failure; bearing failure; prime mover failure; excessive coolant temperature; system surges; and fire. The Table below summarizes the protective means usually adopted.

METHODS OF GENERATOR PROTECTION

Fault	Turbine and Engine-driven Machines	Water-wheel Machines
Overspeed	Mechanical overspeed device on prime mover (not necessary for diesel engines)	Mechanical overspeed device on alternator shaft
Generator over-voltage	—	Overvoltage relay
Overcurrent	Overcurrent relay or differential protection	Overcurrent relay
Unbalanced load	Phase-unbalance network and relay, or alarm	Alarm
Stator insulation failure	Differential protection	Differential protection
Rotor insulation failure	Alarm from earth-fault detector	Alarm from earth-fault detector
Loss of oil pressure	Automatic transfer to emergency oil supply	Contacts on governor oil system. Standby pump for bearings
Loss of vacuum or steam pressure	Atmospheric relief valve. Vacuum load suppression	—
Failure of prime mover	Reverse-power relay	—
Overheated bearing	—	Electric contact thermometer
Excess ventilating air temperature	Alarm for air/water heat-exchanger	Alarm for air/water heat exchanger
Surges	Supply-network protection sufficient	Supply-network protection sufficient

Equipment for the measurement and recording of shaft distortion and rotor expansion has been developed. Large steam turbines have rotors weighing several tons running at speeds of 3 000 r.p.m., and steam inlet temperatures in the region of 550° C. (1 000° F.). These rotors must be accurately balanced and remain geometrically true. A distortion that caused the mass-centre to deviate only 0·1 mm. from the axis of rotation would, at 3 000 r.p.m., produce a centrifugal force equal to the weight of the rotor. Uneven heating or temporary misalignment may cause displacement; and the length of a turbo-alternator set may increase by an inch or more between the cold and running-temperature condition.

The methods of detection of rotor eccentricity derive from telecommunication techniques, in which either the amplitude or the frequency of a basic carrier signal is modulated by pick-up from inductors located close to the rotor shaft.* The carrier signal may be derived from audio-frequency sources, such as small permanent-magnet generators or valve oscillators. The modulated signal is amplified and fed to indicators and graphic recorders.

8. **Performance of Synchronous Motors.** The load or torque angle σ is one of the most important characteristics of the synchronous motor. From eq. (160) the relation between power developed and load angle is, for constant synchronous reactance, a sine function. With non-salient-pole synchronous motors this is roughly true, but with salient-pole machines the strongly polarized field system causes maximum torque to develop at an angle rather less than 90°, Fig. 284 and eq. (168).

The operating features of the synchronous motor are its constant speed, the possibility of power factor control by variation of the field excitation, and its low starting torque. Fig. 312 shows typical load-current/field-current curves. The synchronous motor can be satisfactorily built from the highest possible speed (e.g. 3 000 r.p.m. on 50 c/s.) to the lowest speeds likely to be required in practice. Induction motors, however, suffer a considerable handicap in performance at speeds below about 500 r.p.m. The poor performance of low-speed induction motors may be of sufficient importance to make synchronous motors preferable. Besides this, the advantages of a better power factor may have economic importance. The chief reason why the use of synchronous motors has been retarded lies in their low-starting torque. Almost all modern developments have been directed towards the elimination of this undesirable feature.

The Table on page 501 compares induction and synchronous motors on a number of points of design and control.

The comparison may be further extended to the power factor and kVA. rating. The high power-factor on load of the synchronous machine reduces its kVA. rating considerably for a given output, and

---

* Antrich, Gardiner and Hilton, *Proc. I.E.E.*, 102 (A), p. 121 (1955); also Ashworth, Hall and Gray, *ibid.*, p. 131.

the leading current on lighter outputs may be of use in phase-advancing parallel inductive loads.

STARTING. The synchronous motor can develop *synchronous* torque only at synchronous speed. For starting it is therefore necessary to employ other methods of developing torque: the d.c. winding on the rotor, as such, is useless. The methods in use are—

(a) Pony induction motor;

(b) Clutch and brake gear;

(c) Starting as induction motor.

(a) *Pony Motor.* Early machines were started by a small direct-connected induction motor, either a machine with a very low rotor resistance and the same number of poles as the synchronous machine;

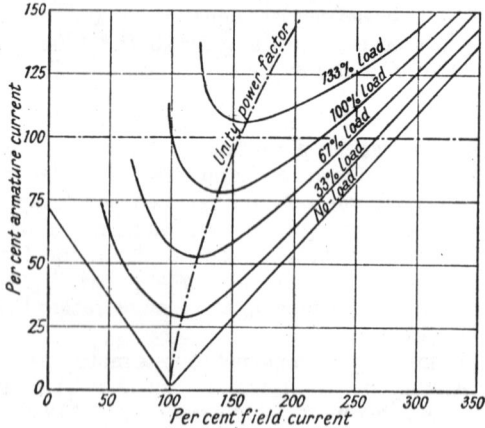

FIG. 312. ARMATURE-CURRENT/FIELD-CURRENT (V-) CURVES

or a motor with a pole-pair fewer and a rotor rheostat; or with a solid unwound rotor. The method involves running the machine up to a speed close to synchronous speed, and synchronizing by hand. Automatic synchronizing is effected by connecting the stator of the synchronous motor first on to low-voltage auto-transformer tappings, then through reactors on to full voltage, lastly short-circuiting the reactors: alternatively, the starting motor may be connected in series with the main machine. The method is used in general only for large machines such as synchronous condensers. The starting torque is limited to that which will accelerate the mass of the machine.

(b) *Clutch.* Where starting torques greater than full-load torque are required, the motor may be run up on no load and synchronized, the load being then taken up by means of a suitable clutch. A modification of this is a construction in which both stator and rotor are free to rotate. The rotor is coupled to the load, and when the

COMPARISON OF INDUCTION AND SYNCHRONOUS MOTORS

	Induction		Salient-pole Synchronous
	Cage	Slip-ring	
Stator Winding	Form wound insulated coils placed in slots	Same.	Same, except that for equal rating there are usually fewer coils, thus providing air space between ends of coils, and improved insulation.
Rotor Winding	Cage winding carries starting and running current.	Form wound insulated coils placed in slots, carry starting current at high voltage and running current at low voltage.	(1) Cage winding carries starting current only. (2) Definite pole, low voltage, direct current exciting winding, easy to insulate, carries relatively small constant running current, regardless of load.
Slip-rings	None.	Three, that carry starting current at high voltage and running current at low voltage.	Two, that carry direct current excitation at low voltage.
Air Gap	Small.	Small.	Three to five times as great as that in induction motors.
Automatic Control	Three switches: starting, magnetizing, and running; an auto-transformer; three control relays and two overload relays.	Three or more switches: one primary and two or more to cut out secondary resistance; four or more operating relays and two overload relays: resistance grids.	One or more switches for starting and running duty; a field contactor; two operating relays and two overload relays. But direct-on-line starting is becoming usual.

stator is energized, the load holds the rotor at rest while the stator runs up to full speed. After synchronizing, a brake is applied to the stator so as to slow it down, simultaneously accelerating the rotor and load. When the stator has been brought to rest, the rotor with its coupled load is running normally, at synchronous speed. Considerable starting torque can be developed in this way.

(c) *Starting as Induction Motor.* This method is accomplished by supplying the stator with three-phase currents at normal or sub-normal voltage, and utilizing the induction torque developed in pole-shoes or damper-windings on the rotor. As a logical development, the machine may be designed as a *synchronous-induction motor*, with a three-phase rotor winding on a non-salient pole core resembling that of a normal induction motor. The two methods are considered in more detail below.

9. Induction-starting a Synchronous Motor. Starting a synchronous motor by utilizing an induction torque requires the rotor to have either solid pole-shoes, or laminated shoes carrying a cage type of damper winding. The three-phase supply at full or reduced voltage is switched on to the stator, and the rotating field interacts with currents induced in the pole-shoes or cage to provide a torque sufficient to raise the speed of the rotor close to the synchronous. Up to this point the rotor field winding is closed through a resistor to limit the alternating voltage developed therein: it is now connected to the exciter or other d.c. supply, and the rotor pulls into step on account of the slow pulsations of synchronizing torque produced as the rotor poles slip past the stator poles.

STARTING ON FULL VOLTAGE. If full normal voltage is applied to the stator at starting, the switchgear required is simple and cheap, for only a main switch and the usual field control gear need be provided. The starting current may, however, be from 3 to 7 times full-load current. Such a current may not always be permissible, in which case a three-phase damper winding, brought out to slip-rings, may be utilized with a short-rated liquid or grid pattern of starter resistor. Machines of several hundred horse-power can be direct-switched to h.v. supplies without undue reduction of the line voltage.

STARTING ON REDUCED VOLTAGE. If the lower torque available is still sufficient, reduced voltage may be applied initially, with consequent reduction of starting current. Methods of reducing the voltage comprise—

1. Series reactors.
2. Auto-transformer.
3. Star-delta switching.
4. Transformer tappings.

1. *Series reactors.* Inductive reactors are connected between the line and the stator of the synchronous motor for starting, and are short-circuited after the machine has been synchronized.

2. *Auto-transformer.* This method may be applied in a manner like that for an induction motor (page 297), but the interruption of the circuit during the starting sequence may produce severe switching transients in the supply system and the possibility of loss of synchronism when passing from "start" to "normal running" connections. These disadvantages are overcome in the *Korndorfer* method, in which the motor is given about half normal voltage in

the starting position from three star-connected auto-transformers. When up to speed, the star-point is opened, leaving the auto-transformers operating as chokes of comparatively low inductance (because they are highly saturated). The motor voltage consequently rises to a value nearer to normal without being interrupted. As a last step the auto-transformers are by-passed.

3. *Star-delta switching.* The same considerations apply as in induction motors. The disadvantage is the momentary interruption of the supply. See page 298.

4. *Transformer tappings.* This method is used for very large

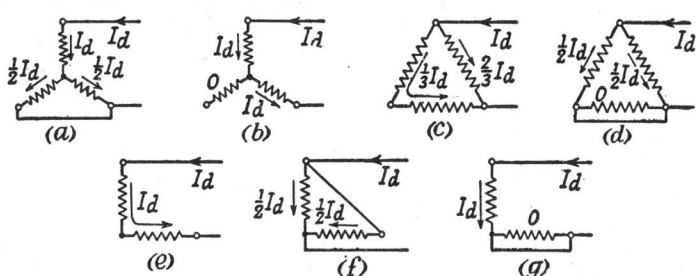

Fig. 313. Connections for Secondary Excitation

synchronous motors or capacitors, where the size justifies the greater cost of starting gear.

10. **The Synchronous-induction Motor.** The idea of synchronizing a slip-ring induction motor by direct current excitation is not new. Danielson suggested it as a method of power factor improvement in 1901.*

Consider a normal slip-ring induction motor with a three-phase rotor. Let the latter be connected as in Fig. 313 (a) to a d.c. source, such as an exciter, which furnishes a current I_d. The current distribution is shown in the diagram: one phase carries I_d, the other two in parallel conduct $\frac{1}{2}I_d$ each. Reference to Fig. 139 will show that this will give ampere-conductor distributions exactly resembling case (a) of that diagram. Thus the d.c. excitation gives rise to alternate N and S poles on the rotor just as do the three-phase alternating currents carried by the rotor when the machine works as an induction motor. The essential difference is that the direct-current excitation is fixed, so that the pole axes are also fixed relative to the rotor winding, and do not shift as when the rotor carries alternating currents.

It is thus possible to use normal slip-ring induction-motor windings to produce fixed magnetic poles, permitting the machine to be used as a synchronous motor. The process of starting and synchronizing is shown in Fig. 313A. The machine is started on a rheostat

* Danielson, *Elektrotechnische Zeitschrift* (1901).

connected to the slip-rings. When this has been fully cut out and the motor is running with a small slip, the connections are changed over to those of Fig. 313 (a), with the exciter in series with the rotor windings. Alternatively the exciter may be permanently in circuit.

The connection of the machine at starting as a slip-ring induction motor renders possible the development of any torque up to about twice full-load value (or higher if specially designed for), so that the synchronous-induction motor is essentially one which fills the demand for a constant-speed motor with large starting torque, low starting current, and power factor correction.

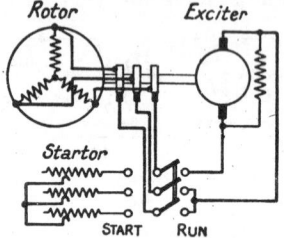

Fig. 313A
STARTING AND RUNNING
CONNECTIONS FOR SYNCHRO-
NOUS-INDUCTION MOTOR

Although in Fig. 313A the d.c. excitation is applied to the rotor, it may be more convenient to "invert" the machine, putting the primary or a.c. winding on the rotor and supplying the d.c. excitation to the secondary winding on the stator. In the normal induction motor the primary (stator) winding is responsible for carrying the magnetizing as well as the load current and is consequently the more heavily-loaded member: in the synchronous-induction machine the secondary carries the magnetizing current and it may usefully be placed on the stator to secure the greater available winding space therein. Other advantages of inversion are: (1) simplification of control gear; (2) better facilities for secondary insulation, permitting of higher voltages and lower d.c. excitations. The latter does not apply to so-called "high-voltage" motors, for it is difficult to provide adequate insulation on a rotor subjected to a line voltage of several thousand volts.

To cover all cases, the expressions *primary* and *secondary* are used below in place of *stator* and *rotor* respectively.

SECONDARY WINDING. The designer's object in deciding on a secondary (d.c.) winding is to obtain the greatest m.m.f. with moderate d.c. excitation power, distribute the loss evenly over the winding, obtain a substantially sinusoidal m.m.f. distribution, provide for damping against phase-swing, and to give satisfactory starting as an induction motor.

The m.m.f. required depends almost entirely on the gap length. A long gap gives a "stiffer" machine with larger overload capacity and less variation of power factor with load: but it is more wasteful of excitation power and gives impaired starting. A short gap gives the opposite effects, and to secure high overload capacity and to limit variations of power factor some method of automatic increase of excitation with load may be employed.

Even distribution of *secondary heating* depends on the type of connection, of which there are many. With the arrangement in Fig. 313A, one secondary phase carries the full d.c. excitation current, while the other two, connected in parallel, each carry one half. If all phases are of the same resistance the rates of heat development are as 4 : 1 : 1. A more or less considerable unbalance of heating is inevitable with *normal* three-phase windings. It is, however, quite possible to employ two-phase or (for excitation only) single-phase secondaries in which even heating is obtainable.

A *sine-distributed m.m.f.* is necessary to avoid harmonic production and noise. Primary chording may be employed to reduce the effect of prominent harmonics.

A synchronous-induction motor requires *damping* against the tendency to hunt on pulsating loads. The low secondary resistance is conducive to good damping, but because the most effective damping winding must lie in the position of maximum flux and the most effective exciting winding is situated in regions of low flux, it is not possible to secure both good damping and effective m.m.f. production in the same winding. Further, as the secondary moves relatively to the primary field owing to hunting oscillations, low-frequency e.m.f.'s are induced in all secondary phases. If the connection is such as to encourage the flow of damping currents, the hunting tendency will be well suppressed; if not, the damping effect is impaired or lost.

Short-time *induction operation* requires the secondary connection to retain a connection appropriate for such working. Some connections, although good for synchronous operation, fail in this respect.

A high standstill *secondary e.m.f.* raises questions of insulation, and may be avoided by the use of a low number of turns per phase. Unfortunately this conflicts with the desirable reduction of the exciting current.

EQUIVALENT SECONDARY CURRENT. In Fig. 313 (*a*) the secondary connection gives currents in the ratio $1 : \frac{1}{2} : \frac{1}{2}$, corresponding to the instants in the cycle of an a.c.-excited winding when the current is a positive maximum in one phase and one-half negative maximum in each of the other two. Consequently the d.c. excitation I_d gives a magnetizing current corresponding to an a.c. peak of $i_{am} = I_d$, or r.m.s. $I_a = I_d/\sqrt{2}$. Thus for an m.m.f. appropriate to an a.c. of I_a, the d.c. excitation supplied to the secondary must be $I_d = \sqrt{2}I_a$ $= 1.41\, I_a$. A similar distribution applies to (*c*), but here the current in the heaviest-loaded phase is only $\frac{2}{3}I_d$, so that $I_d = \sqrt{2}(3/2)I_a$ $= (3/\sqrt{2})I_a = 2.12 I_a$. Case (*b*) has a distribution corresponding to an instant $\frac{1}{6}$th period later, and for it $I_d = (\sqrt{3}/\sqrt{2})I_a = 1.23 I_a$. With the values given the fundamental flux produced by I_d is the same as that due to a given I_a.

The unequal heating of the secondary winding may be avoided, or mitigated, by a variety of special arrangements. In large machines two separate windings, one for d.c. excitation and an additional 2- or 3-phase starting winding, may be provided. The Table below summarizes the comparative qualities of some of the connections that

SECONDARY CONNECTIONS FOR SYNCHRONOUS-INDUCTION MOTORS

Case	(1) I_d	(2) R	(3) V_d	(4) P_d	(5) Heat Distribn.	(6) Start	(7) Damping	(8) Indn. Action	(9) Use
(a)	1·41	1·5	2·12	3·0	2 : 0·5 : 0·5	Normal	Good	Good	Small machines
(b)	1·23	2	2·46	3·0	1·5 : 1·5 : 0	Poor	Poor	1-ph.	Steady drives
(c)	2·12	0·67	1·41	3·0	2 : 0·5 : 0·5	Normal	Poor	1-ph.	Seldom
(d)	2·45	0·5	1·22	3·0	1·5 : 1·5 : 0	Normal	Excell.	Good	Modified
(e)	1·06	3	3·18	3·38	1·69 : 1·69	Normal	Poor	1-ph.	Seldom
(f)	2·12	0·75	1·59	3·38	1·69 : 1·69	Normal	Good	Good	Wide
(g)	1·5	1·5	2·25	3·38	3·38 : 0	Normal	Excell.	Normal	Patented

(1) Relative d.c. excitations for a given m.m.f.
(2) Secondary input resistance in terms of resistance per phase.
(3) Relative exciter voltage, (1) × (2).
(4) Relative exciter power, (1) × (3).
(5) Distribution of (4) between the secondary phases.
(6) Starting qualities. (7) Damping qualities.
(8) Action as induction motor when overloaded.

have been devised.* All the three-phase connections require the same exciter power for a given m.m.f., turns per phase and phase resistance, and all the two-phase arrangements need 12½ per cent more. The two-phase windings (e) and (f) are the only ones giving uniform heating. Connections (a), (d), (f) and (g) give effective damping—particularly (d) and (g) in which one phase is short-circuited on itself. The two-phase windings produce rather larger harmonics.

SALIENT-POLE SYNCHRONOUS-INDUCTION MOTORS. The disadvantage inherent in the motor described above is that if the secondary winding is designed to give a sufficiently low induced e.m.f. at standstill, the exciting current has to be very large and the voltage low. The excitation losses are likely to be large on account of the restricted space available for copper in a distributed winding. The *salient-pole synchronous-induction* motor combines the high inherent efficiency and low excitation loss of the salient-pole synchronous motor with the starting characteristics of an induction motor. The d.c. exciting winding is of normal construction. An independent starting winding connected to slip-rings occupies slots in the pole-shoes. Each winding can be designed almost independently to

* Rawcliffe, *J.I.E.E.*, 87, p. 284 (1940).

fulfil its allotted function, with consequent improvement in synchronous-running efficiency and exciter conditions.

CURRENT DIAGRAM. When operating synchronously, the locus of the stator current for constant rotor excitation is a circle like that in Fig. 282. The line $OR = V/z_s$ at angle $\theta = $ arc tan (x_s/r) is fixed when the synchronous impedance is known. The impedance of an induction motor run at synchronous speed is practically identical with that of the magnetizing circuit x_m and r_m in the equivalent circuit of Fig. 160 (d), so that the no-load current vector OP_0 in the

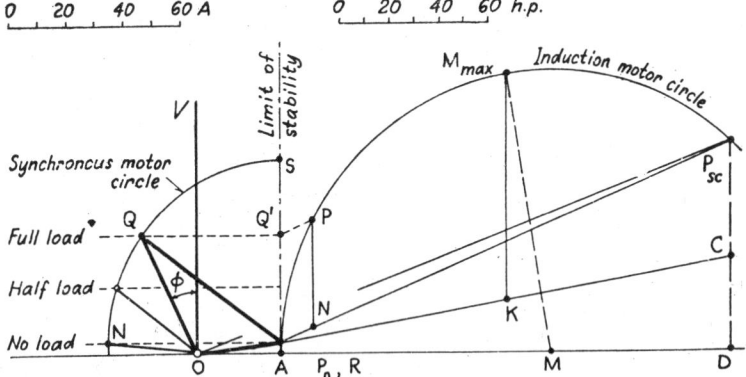

FIG. 314. CURRENT DIAGRAM OF SYNCHRONOUS-INDUCTION MOTOR

circle diagram of the induction motor fixes the point P_0 corresponding to R in Fig. 282, as the centre of the current circles for synchronous operation. The current diagram for the synchronous-induction motor has therefore the form shown in Fig. 314. The current OP_0 is I_0, the no-load current as an induction motor. OP is the full-load current for the same condition, i.e. PN to scale represents full-load output. The limit of stability as a synchronous motor is P_0S, which meets the voltage axis produced at a point $V/2r$ from O: since this distance for normal machines is very great, it is sufficient to draw P_0S parallel to the voltage axis. The rotor excitation is measured from P_0, and the synchronous-motor circle drawn with P_0 as centre has a radius determined by the connection of the rotor as given on page 503, and by the ratio of stator to rotor turns per phase. The constant-output-power lines are parts of O-curves, but are usually so nearly straight that they may be so drawn without appreciable error. The no-load line is drawn horizontally through P_0, and the full-load line is parallel to it at a distance equal to PN.* The stator currents are measured from O. In Fig. 314 the excitation is such as to give leading power factors up to about 125 per cent of full load.

* Strictly, these horizontal lines should be parts of large-diameter O-curves.

The maximum load as a synchronous motor is P_0S to scale, approximately 150 per cent full load.*

A well-designed induction motor makes a poor synchronous motor when its rotor is excited with direct currents. The no-load current ON in Fig. 314A is nearly equal to the full-load current OQ, and the power factor and efficiency are low except near the limit of stability. Low-speed induction motors, on the other hand, have large magnetizing currents, and make much better machines when synchronized. The reason lies in the value of the synchronous impedance, i.e. the magnetizing-circuit impedance. This is very large in a good induction motor in order to secure the smallest possible magnetizing current. But a large synchronous impedance in a synchronous motor entails an excessive change of power factor with load.

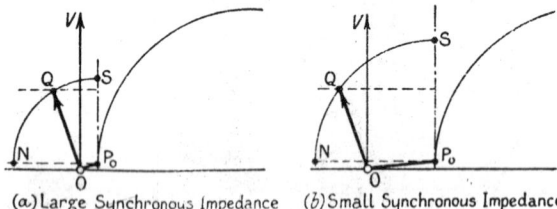

(a) Large Synchronous Impedance   (b) Small Synchronous Impedance

Fig. 314A.   Effect of Synchronous Impedance

Comparative figures of synchronous impedance are 0·2–2·0 per unit in synchronous motors, and 3 per unit in induction motors.

Fig. 314A compares the current diagrams of a synchronous-induction motor with (a) a large, (b) a small synchronous impedance. Clearly, to secure adequate overload capacity P_0S, and to restrict excessive leading kilovars on light loads, a small synchronous impedance is to be preferred. This is obtained by a longer air gap, say 100 per cent greater than is usual in a plain induction motor, and a stronger excitation.

EXAMPLE. A 40-h.p., 500-V., 50-c/s., 3-phase, star-connected, synchronous-induction motor has a short-circuit current of 200 A. at a power factor of 0·35, and a no-load current as an induction motor of 30 A. at a power factor of 0·15. Find the starting, stalling and pull-out torques, and the rotor exciting current, for a leading power factor of 0·9 when running synchronized. Find also the current and power factor at one-half load and no load. Ratio stator/rotor turns per phase, 1/1·3. Ratio stator/rotor resistance per phase, 1/1·2. Rotor connection, Fig. 313 (a).

The test currents given are line values, and therefore serve also as phase values. The induction-motor circle diagram is drawn as described in Chapter XII, § 7. See Fig. 314. To obtain the synchronous-motor circle, draw a line OQ making an angle arc cos

* The diagram assumes all losses constant and equal to P_0A to scale.

$0.9 = 25.9°$ to the vertical through the origin. Draw a horizontal line through P_0, and lines above it at distances equivalent to 40 h.p. and 20 h.p. The 40 h.p. line cuts the line OQ in Q. Then P_0Q is the radius of the synchronous-motor circle.

The full-load torque is P_0Q' to scale; the pull-out torque is P_0S; the stalling torque is $M_{max}K$ and the starting torque is $P_{sc}C$ to scale. Starting torque $= 1.11$ f.l.t.; stalling torque $= 2.17$ f.l.t.; pull-out torque $= 1.74$ f.l.t. The starting torque is actually increased by the use of the startor rheostat up to 2.17 f.l.t. or less, as required.

The rotor exciting current is $P_0Q = 60$ A. equivalent alternating current. For connection (a), $I_d = 60/0.71 = 84.5$ A. This is the exciter current for unity turn-ratio. Since the rotor has 1.3 times as many turns as the stator, the actual rotor current is $84.5/1.3 = 65$ A. From the diagram: full-load current is 43 A. at p.f. 0.9 leading; half full load, 35 A. at 0.62; no load, 30 A. at 0.15 leading. Somewhat closer results are obtained if the full-load stator I^2R loss be included in the no-load loss: also the rotor I^2R and exciter losses if the exciter is on the main shaft.

PERFORMANCE. The performance characteristics of the synchronous-induction motor are readily obtainable from the current diagram. Three different values of torque must be given consideration in choosing or designing a motor. These are: (a) the starting torque, representing the ability of the motor to start the load; (b) the pull-in torque, representing the ability of the motor to maintain operation during the change-over from induction to synchronous running; and (c) the pull-out torque, on which depends the maintenance of synchronous running at peak load. The two former are characteristics of the machine as an induction motor and are closely related. The pull-out torque is a characteristic of the motor operating synchronously. It is not closely related to the other two, but does to some extent affect the starting current. There is naturally a close interrelation between the power factor on full load and the pull-out torque: thus a motor running at 0.9 leading power factor on full load might have a pull-out torque 150 per cent of full load value, while if the machine were designed to work at unity power-factor on full load, the overload torque and exciter rating would be considerably less. Fig. 315 shows a number of characteristic curves for typical synchronous-induction motors.

Where a synchronous-induction motor may be called upon to sustain heavy and sudden overloads, it may be permanently over-excited. This, however, means that the power factor is low over the normal load range, the current is large and the I^2R loss is considerable. Exciters compounded by means of current-transformers and metal rectifiers, or some form of power-factor regulator, may be used so that the power factor of the machine is kept at or near unity over all load ranges.

The synchronous-induction motor is not built for outputs much below 30 h.p., on account of the cost of the exciter: power-factor correction is not particularly advantageous in small units. The application of the motor is to cases where a constant speed is desirable or suitable, such as compressors, pumps, fans, blowers, etc. It has the following advantages compared with an ordinary salient-pole machine: it will start and synchronize against loads up to nearly

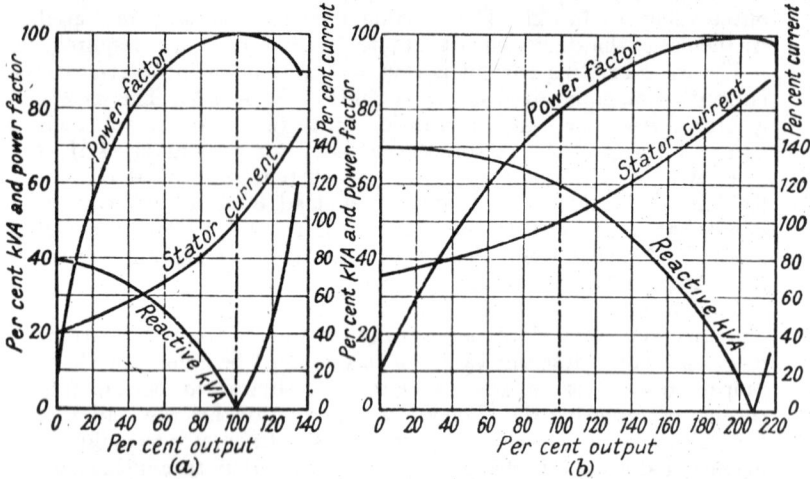

Fig. 315. Performance Curves of Synchronous-induction Motors
(a) Full load at unity power factor; (b) Full load at 0·8 power factor leading

twice full-load torque; its air gap is smaller so that the exciter may be a smaller unit; no separate damper winding is needed; it is somewhat cheaper.

If the load applied exceeds the synchronous pull-out torque, the machine falls out of step and runs as an induction motor with fluctuations of torque and slip due to the d.c. excitation. When the load torque has fallen sufficiently, the motor will automatically re-synchronize.

PULLING INTO STEP. A synchronous-induction motor running at a small slip will pull into step when provided with d.c. excitation.

During transition the sum of the synchronizing and induction torques combines to accelerate the rotor (and its connected load) against the resisting torque. The synchronizing torque with constant d.c. excitation varies roughly sinusoidally with the electrical angle σ between the positions of the instantaneous rotor and stator poles, giving the approximate relation

$$M_s = M_{sm} \sin \sigma.$$

The induction-motor torque is a function of the slip, $M_i = M's$ very nearly if s is small. The acceleration of the rotating masses from the speed corresponding to slip s up to synchronous speed is

$J(d^2\sigma/dt^2)/p$. The load torque is M_l. Thus during the process of pulling into step $M_{sm} \sin \sigma + M's = -(J/p)(d^2\sigma/dt^2) + M_l$ or, with s as the rate of change of σ with respect to synchronous speed,

$$M_{sm} \sin \sigma - M_l + kM'(d\sigma/dt) + (J/p)(d^2\sigma/dt^2) = 0$$

The solution of this equation* is rather complex, even when sinusoidal variations and other simplifications are made.

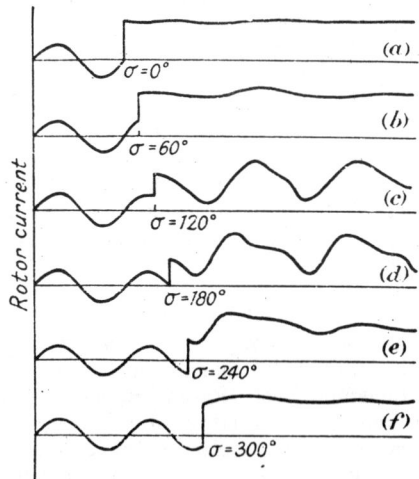

FIG. 316. ROTOR CURRENTS DURING SYNCHRONIZING

When the d.c. excitation is switched on, the pulsating synchronous torque is superimposed on the steady induction torque. The motor may pull into step almost immediately: it may continue to run as an induction motor but with speed and current fluctuations: it may revolve several times and then pull into synchronism. The required condition is to have the motor synchronize on its first swing through the steady-state operating position. The inertia of the rotor and its connected load is a very important factor. The angle σ obtaining at the instant of switching in the d.c. excitation also affects the pulling into step. If the angle is such that the machine generates (i.e. if σ is an angle corresponding to a forward position of the rotor) the speed decreases because initially the generated energy is abstracted from the kinetic energy of the rotating masses. The rotor swings into the motoring region, upon which an additional motoring torque is produced to speed it up again.

The field excitation is an important factor as determining the

* E.g. See Edgerton and Zak, "The Pulling into Step of a Synchronous-Induction Motor," *J.I.E.E.* 68, p. 1205 (1930).

maximum synchronizing torque; the larger the field current, the more certain is the pulling into step.

Fig. 316* shows the rotor current variations in a typical case and describes the effects of switching on the d.c. excitation with various angles σ. (a) With $\sigma = 0$, the rotor current is passing through a zero and increasing; the effect is to add M_s to M_i so that the rotor synchronizes immediately, the rotor being at the instant of switching exactly in the right position for continuing to run synchronously. (d) With $\sigma = 180°$, the d.c.-excited poles oppose the a.c.-excited poles, or M_s opposes M_i. The rotor loses speed until it has slipped back by one pole-pitch, after which M_s and M_i work in conjunction: but the phase-swing produced by the augmented slip is too great to allow the rotor to synchronize, so the torque continues with a pulsating value $M_i + M_{sm} \sin \sigma$ and the machine fails to synchronize. This is shown by the continuance of the rotor a.c. component, which naturally vanishes when the machine runs stably at synchronous speed. With a smaller load or stronger d.c. field, the machine would eventually synchronize. Current curves for intermediate angles are also shown in Fig. 316. It will be seen that pulling into step is most rapid when the d.c. excitation is applied at or just before the position which the rotor will eventually occupy with respect to the stator field when synchronized.

11. Choice of Constant-speed Motors. Except where power factor correction is desired, the ordinary induction motor, with single-cage, double-cage or slip-ring rotor is the cheapest and simplest machine. Where considerations of power factor rule out the plain induction motor, the choice should fall on the salient-pole synchronous motor, if the starting duty falls within its capacity. This motor as ordinarily designed can start against 40–50 per cent of its full torque without excessive currents from the line, and can pull into synchronism against about 15 per cent of full-load torque. When running it has small losses, excellent power-factor characteristics and about 75 per cent reserve of torque. Special construction involving a more elaborate damper winding can be employed to increase starting torque.

Where the starting conditions are more severe, the choice lies between the induction motor with phase-advancer, and the synchronous-induction motor. The differentiation is then the desired power-factor/load characteristic or the torque reserve. The induction motor has a 100 per cent reserve, the synchronous-induction motor about 50 per cent, although the latter figure can be increased by temporary over-excitation. If considerable leading kilovar loads are required for all power outputs, the synchronous-induction motor is to be preferred.

The cost of a phase-advancer (or a.c. exciter) is rather greater

* Based on Cotton, "The Synchronous-Induction Motor," *World Power*, 1, p. 329, and 2, p. 45 (1924). Fig. 14.

than that of a corresponding d.c. exciter, but when the a.c. exciter is used, the designer has much less restriction on specific loadings, and a greater inherent magnetizing current can be allowed.

It might appear that this applies to the synchronous-induction motor as well, but it is here necessary that the rotor currents, which are fixed apart from manual adjustment, should be greater in order to provide a reasonable reserve of torque. If, say, a 50 per cent torque reserve is needed, the rotor loading must be about 40 per cent greater than in the plain induction motor. Thus the rating of a

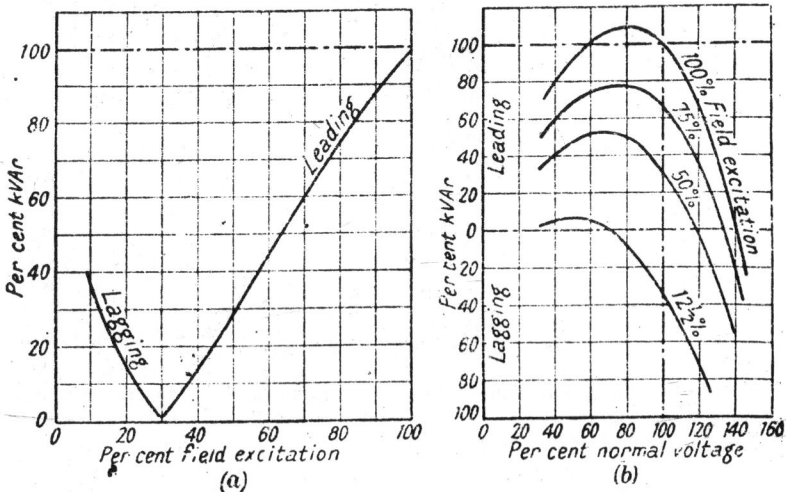

FIG. 317. CHARACTERISTIC CURVES OF SYNCHRONOUS CAPACITOR

given frame may be less as a synchronous-induction than as a plain induction motor.

12. **The Synchronous Capacitor.** A synchronous motor designed to run unloaded and over-excited to serve as a load on the supply system of very low leading power factor, is called a *synchronous capacitor.* These machines may be used to compensate for the phase lag of other parallel loads or in large sizes to control the voltage on transmission lines by the receiving-end power factor.

The design of a synchronous capacitor is characterized by its large synchronous reactance and field windings, since its function is to develop zero-power-factor currents with the least possible expenditure of power in losses. Fig. 317 shows typical characteristic curves. At (a) is the V-curve for constant output voltage. In this case normal excitation, 100 per cent, gives full-load leading kilovars. At an excitation of about 30 per cent the current falls to a minimum value corresponding to the losses. At lower excitations the machine becomes a synchronous inductor. It is usual to limit the lagging kilovars to about one-third of the maximum leading kilovar rating, since the machine may lose synchronism when working on a weak field if disturbed by some circuit fault or surge.

The curves in Fig. 317 (*b*) refer to the performance on variable bus-bar voltage and constant excitation. The excitation is expressed on the same basis as (*a*), i.e. 100 per cent field current gives full load kilovars leading at normal voltage.

The synchronous capacitor is usually built with two or four poles, and as there are no shaft extensions it can conveniently be hydrogen-

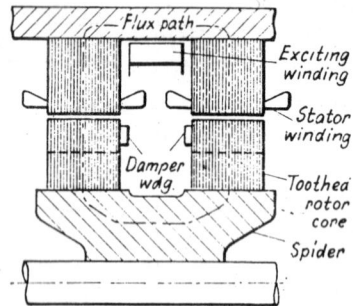

Fig. 318. Homopolar Inductor Alternator

cooled to reduce windage losses. The full-load power factor may be as low at 0·02.

13. The Inductor Alternator. This machine is built to generate a.c supplies of frequencies from 250 c/s. to 200 kc./s. for industrial high-frequency-heating loads. For a given speed and number of rotor "poles" (or teeth), the machine will produce a frequency twice that of a comparable synchronous machine. The absence of rotor windings allows the speed to be raised to the safe limit of rotor centrifugal stress.

The *homopolar* machine has two stator and two rotor cores forming a single magnetic circuit, energized by an annular d.c. exciting winding, Fig. 318.

The *heteropolar* type has a single stator and rotor core. The short shaft makes possible a very short gap, an important advantage in determining the output. The stator and rotor carry varieties of slotting, and the stator periphery is divided into heteropolar zones, Fig. 319A.

Both types have the same basic principle, i.e. the pulsation or variation of the magnetic flux linking the stator coil system, resulting from cyclic changes of gap permeance round the stator bore. The permeance is most effectively varied by use of rectangular rotor teeth and a short gap. Movement of the rotor through one tooth-pitch then produces one complete cycle of gap-permeance variation.

If S_2 is the number of rotor slots (or teeth), and n is the rotor speed in r.p.s., then the frequency generated is $f = nS_2$. This is independent of the stator tooth arrangement.

To utilize fully the variation of flux, the stator coils should have a pitch of one-half the rotor pitch. Fig. 319A (a) and (b) show two arrangements for homopolar machines. One has a substantially conventional winding: the other has coils round individual stator teeth. A type of heteropolar arrangement is shown at (c). Here the *total* working flux remains substantially constant so that no appreciable alternating e.m.f. is induced in the d.c. exciting coils.

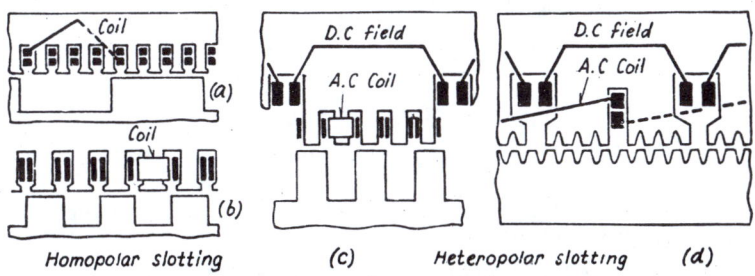

Homopolar slotting   *(c)*   Heteropolar slotting   *(d)*

FIG. 319A. SLOT ARRANGEMENTS FOR INDUCTOR ALTERNATORS

The rotor slot-pitch in a machine of rotor diameter D is $y_2 = \pi D/S_2 = \pi D n/n S_2 = u/f$, so that with the peripheral speed u limited mechanically by centrifugal stress, the slot-pitch must be very small for high frequencies. It then becomes necessary to use stator and rotor teeth of equal pitch, Fig. 319A (d), producing a pulsating flux; with the stator coils housed in larger slots of span equal to any suitable number of the smaller teeth.

THEORY. The gap permeance fluctuates as the stator and rotor teeth move in and out of register. Let it be represented by $\Lambda_0 + \Lambda_1 \cos \theta$, with θ the electrical angle of rotor displacement from the position of maximum permeance. Neglecting iron-circuit reluctance, the no-load flux will be $\Phi_0 = F_t(\Lambda_0 + \Lambda_1 \cos \theta)$ for a d.c. excitation F_t ampere-turns. On load the stator carries a current $i \sin(\theta - \phi - \alpha)$ corresponding to an armature reaction $F_a \sin(\theta - \phi - \alpha)$, where ϕ is the (lagging) power-factor angle of the external load, and α is an additional angle of lag caused by a shift in the angular position of the maximum flux caused by armature reaction. The load flux is then

$$\Phi = [F_t + F_a \sin(\theta - \phi - \alpha)][\Lambda_0 + \Lambda_1 \cos \theta].$$

This product gives a constant term (the average flux), a second-harmonic term (neglected for simplicity) and the term giving the fundamental alternating flux

$$\Phi_m = F_t \Lambda_1 \sin(\theta + 90°) + F_a \Lambda_0 \sin(\theta - \phi - \alpha) = \Phi_t + \Phi_a$$

to which two e.m.f.'s E_t and E_a correspond, giving the output e.m.f. E as their resultant. The complexor diagram in Fig. 319B shows the relation between these quantities. The machine can be represented by the equivalent circuit comprising an e.m.f. source E_t with a

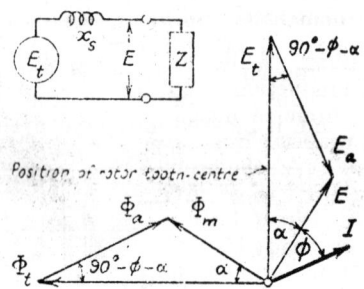

Fig. 319B. Complexor Diagram and Equivalent Circuit

synchronous reactance $x_s = E_a/I$ and a load Z of phase-angle ϕ. The volt-ampere output is

$$EI = E_t^2 Z/[(Z\cos\phi)^2 + (Z\sin\phi + x_s)^2]$$

which is a maximum when $Z = x_s$. The power output is a maximum for the same condition:

$$P_m = (E_t^2/4x_s)[\cos\phi/(1 + \sin\phi)].$$

In this analysis both saturation and losses have been neglected. The core loss is difficult to assess with any accuracy, and stray losses tend to be large.* The second-harmonic fluxes cause additional loss. In order to limit stator I^2R losses the conductors may be made from "litz" subdivided stranding.

* Walker, "Theory of the Inductor Alternator," *J.I.E.E.*, 89 (II), p. 227 (1942); and 94 (II), p. 67 (1946).

CHAPTER XIX

SYNCHRONOUS MACHINES: TESTING

1. **Methods of Testing.** While small and medium-sized synchronous and synchronous-induction motors are made in fair numbers, the class of synchronous machines also includes generators and motors of very large size. Since generators are now made for ratings exceeding 100 000 kVA., the problems of testing are clearly of quite a different order from those involved in small transformers and induction motors. Direct load tests on very large machines are very difficult, since steady load and boiler power are not obtainable, while the cost of performing the tests may well be prohibitive. Indirect tests are of necessity used, the efficiency and other characteristics being calculated from the results.*

2. **Resistance.** The resistance of armature and field windings is taken at known temperature. The measurement is made with direct current, so that the variation of stray loss with load can be observed.

3. **Open-circuit Test.** This is an important test since on it depends calculations of regulation, exciter current and core loss. The machine is driven at constant normal speed with excitation varied over the complete range, yielding the open-circuit characteristic as in Fig. 286. The core loss and friction and windage loss can be measured by driving the machine with a calibrated motor, reading the input at constant speed with various field excitations. A curve such as Fig. 320(a) is obtained when the corrected power is plotted to a base of field excitation. The separation of the power input into its component is made as shown by using the intercept of the power curve on the ordinate axis.

In the case of a turbo-alternator or closed-air-circuit machine it is possible to make measurements of cooling air volume and temperature-rise. By running the alternator on no load without excitation, the measured loss is the windage. If the machine is now excited, the core loss is added and measured, so that the two may be separated by drawing a curve similar to that of Fig. 320(a). The details of the method are given in § 8 below.

4. **Short-circuit Test.** This is to find the short-circuit characteristic and the I^2R loss (with stray loss). The machine is run at or near normal speed with the terminals carefully short-circuited through ammeters, and the field excitation slowly increased from zero. A curve similar to that in Fig. 286 is obtained.

* For further details of testing synchronous machines, see Parker Smith, *Electrical Engineering Laboratory Manual—Machines* (Oxford)

It is not necessary to keep the speed exactly normal because the major part of the field excitation is required to balance the armature ampere-turns, which are dependent on the short-circuit current. The short-circuit current is circulated through the reactance and resistance of the machine by the small induced e.m.f. Both the reactance and the e.m.f. are directly proportional to the speed, so that their ratio, the current, is independent of speed since the resistance is always a small fraction of the impedance. Substantial departures from normal speed make practically no difference therefore to the s.c.c.

If a calibrated driving motor is available, or means for measuring the volume and temperature-rise of the cooling air in the case of a

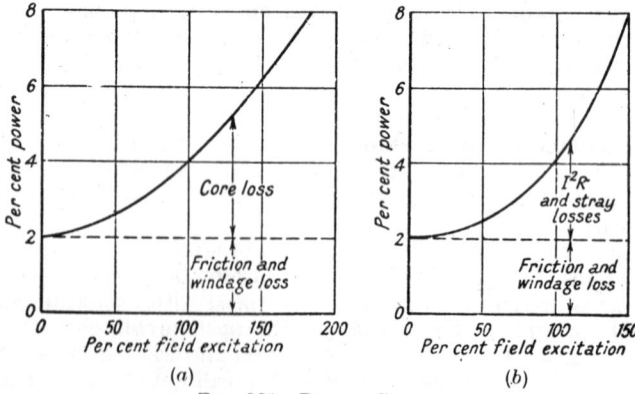

Fig. 320. Power Curves
(a) From open-circuit test. (b) From short-circuit test.

turbo-alternator or similar enclosed machine, the input on short-circuit can be divided into I^2R losses (including stray losses) and friction and windage losses, as shown in Fig. 320(b). The corrected (i.e. net) input is plotted to a base of field excitation. The intercept of the curve on the ordinate of zero excitation gives the mechanical losses which are constant with constant speed. In the short-circuit test, there may be a certain amount of core loss included in the I^2R: this is usually small, and is lumped with the stray loss.

5. **Zero-power-factor Test.** Very large machines, and those that cannot be conveniently loaded for some reason other than one, may be subjected to a full-load test at zero power factor in order to ascertain the temperature-rise. The advantage of this test is that the core loss, I^2R loss and stray loss are normal, and the excitation loss is greater than normal, but is calculable. The power required to drive the machine is, however, only that for the losses. The load may take the form of reactors or synchronous motors with adjusted excitation. It is thus possible to test a very large machine for

temperature-rise under conditions very close to those obtaining on normal full load.

6. Temperature-rise. The specified temperature-rises allowed in synchronous machines are given in Chapter XI, § 5. In large machines it is necessary to have embedded temperature detectors to indicate hot-spot temperatures, read by potentiometer.

7. Efficiency. The estimation of the efficiency of synchronous machines by loss-summation is recognized in B.S. 269: 1927. The losses to be included are—

Exciting Circuit Loss:

        (a) field I^2R;
        (b) rheostat;
        (c) brushes;
        (d) exciter loss.

Fixed Loss:

        (e) core loss;
        (f) bearing friction;
        (g) total windage;
        (h) brush friction.

Direct Load Loss:

        (j) I^2R loss in armature windings.

Stray Load Loss:

        (k) in iron;
        (l) in conductors.

All I^2R losses are computed for a mean winding temperature of 75° C.

Loss (c) is taken for each slip-ring as the slip-ring current multiplied by 1 V., unless metal brushes are used. Losses (a), (b) and (c) together are computed from the exciter voltage and current. Item (d) includes all exciter losses where the exciter is driven from, and is a part of, the complete machine.

The fixed loss is calculated for rated speed and voltage: (e) is the core loss at no load, rated voltage and normal speed. Item (g) includes the windage loss of the machine and its exciter, and any power absorbed in ventilation by internal or external means. Items (e), (f), (g) and (h) are obtained together from a no-load test.

The direct-load loss (j) is calculated from the current and resistance of the windings. The stray losses (k) and (l) are caused by changes in flux distribution due to the load, and their exact values are not readily determined. For the purposes of the specification, they are taken as the amounts indicated by a short-circuit test, i.e. the difference between the total measured mechanical input on short circuit at a given current, and the sum of the friction, windage and I^2R losses. The machine is separately excited for this test.

EXAMPLE. Typical figures for the losses of a turbo-alternator obtained from tests and used to determine the efficiency are given below.

Rating: 23 400 kVA. at power factor 0·8.

Losses: Bearing friction, measured by volume of lubricating oil circulation and temperature-rise; 68 kW.

Windage, from open-circuit test unexcited, measuring the cooling-air volume and temperature-rise; 220 kW.

Core loss, from open-circuit test with excitation to give calculated induced e.m.f. at normal rated load, corrected for windage; 165 kW.

Winding loss from full-load zero-p.f. test, 200 kW.: calculated d.c. I^2R loss in stator, 62 kW.: hence stray loss, $200 - 62 = 138$ kW.

Exciter loss, calculated for normal load at 0·8 power factor, 14 kW.

Rotor I^2R loss, calculated for normal load at 0·8 power factor 96 kW.

Efficiency—

	kW.
Friction loss	68
Windage	220
Core loss	165
Stator I^2R	62
Stray load loss	138
Exciter loss	14
Rotor I^2R	96
Total loss	763 kW.
Output, 23 400 × 0·8	18 750 kW.
Input	19 513 kW.

The efficiency is therefore $100(18\,750/19\,513) = \underline{96 \cdot 09\%}$

The total stator loss on short-circuit test at normal current was found to be 243 kW., giving a stray loss of $243 - 62 = 181$ kW. The stray loss from the full-load zero p.f. test is 138 kW. This is less than on short circuit because in the latter case the iron parts are comparatively unsaturated and greater eddy-current losses are produced therein.

STRAY LOSSES. Stray load losses are due to leakage—

(*a*) Stray fields in solid iron parts near the stator, particularly coil-supports and end-shields in high-speed turbo-alternators.

(*b*) Field distortion by cross magnetization, increasing the iron loss in the teeth.

(*c*) Field pulsation due to harmonics in the armature m.m.f., setting up additional iron losses in the pole faces.

(*d*) Eddy current losses in stator conductors due to part of the main flux being carried through the slots on account of saturation in the teeth.

8. **Loss Measurement.** Electrical, mechanical and calorimetric methods may be used. All auxiliary machines (exciters, starting motors, etc.) are disconnected electrically and, if necessary, mechanically, the excitation being obtained from a separate source.

ELECTRICAL METHODS. A separate (usually d.c.) driving motor can be employed to run the test machine at rated speed. By measuring the input to the motor and deducting its losses, the total loss in the test machine can be determined with good accuracy. Occasionally the exciter has a rating large enough to drive the machine under test. The addition of a few temporary series turns to the exciter field may result in adequate starting torque.

An alternative method is to run the test machine as an unloaded synchronous motor from a variable-voltage rated-frequency source. With care, especially in the measurement of power at very low power factors, the method will give an accuracy approaching that of the separate-drive method. See § 11.

Where a pair of identical machines is available they may be operated in a back-to-back test with rotors mechanically coupled and given a suitable angular displacement.

MECHANICAL METHOD. Dynamometers may be employed, and torsion meters promise a simple and effective method.

CALORIMETRIC METHOD. This method is applicable to machines in which the heat developed by losses can be led away through trunking for measurement of the air volume and temperature-rise.* The volume is obtained from air-velocity measurements by a calibrated anemometer or other means, in a special discharge trunk in which perforated metal screens or "honeycomb" arrangements of sheet metal are used to obtain a more even flow of air. The trunk orifice is divided into a number of equal areas by stretched wires, and the air-velocity and temperature found for each area, from which readings the total air volume and temperature rise is calculated. The relation between the loss and the air volume is that given in eq. (92).

An alternative method dispensing with the measurement of volume is to measure the temperature-rise of the air when a known loss is being dissipated—e.g. the stator core loss, which is measurable separately if the machine is driven by means of a calibrated motor. It is necessary to maintain a constant weight of air through this and subsequent tests in order that the relation between loss and temperature-rise shall be maintained constant.

With large turbo-generators, and other machines of similar construction (e.g. turbo-type synchronous condensers) the method gives reliable results.

9. **High-voltage Tests.** The principal provisions of B.S. 2613: 1955 for synchronous machines concern the application of a sine

* Parker Smith and Barclay, "The Determination of the Efficiency of the Turbo-alternator " *J.I.E.E.*, **57**, p. 293 (1919).

voltage of frequency between 25 and 100 c/s. between the windings and the frame, in accordance with the Table below.

10. Voltage Regulation. When a generator is small enough to be given a direct load test, it is a simple matter to make a direct observation of the voltage regulation. Otherwise, the methods given in Chapter XVII, § 14, are used, appropriate test figures having been obtained.

11. Synchronous Motor. A synchronous motor (or condenser, or generator running as a motor) may be tested as follows without loading. The machine is run on no load on a constant-normal-frequency supply over the maximum voltage range that will permit

HIGH-VOLTAGE TESTS

Part	Test Voltage r.m.s.
Stator windings	1 000 V. + twice rated voltage: minimum 2 000 V.
Generator field windings . .	Ten times excitation voltage: minimum 2 000 V., maximum 3 500 V.
Field windings for synchronous motors and condensers started with the a.c. windings energized—	
(a) Started with field windings short-circuited	1 000 V. + twice rated voltage: minimum 2 000 V.
(b) Started with field windings divided or open-circuited or in series with resistance	1 000 V. + twice r.m.s. induced voltage

the input power to be maintained at unity power factor by adjustment of the field excitation. The input current, voltage and power, and the field current are read. The test is then repeated, but with the field current adjusted so that the machine takes full-load current at a power factor near to zero lagging.

The voltage for unity power factor is plotted to a base of field excitation from the first test. The voltage curve is extrapolated to the origin. The voltage for zero power factor is plotted on the same sheet and extrapolated parallel to the o.c.c. to cut the base line. The two curves can be used to separate the leakage reactance drop from the armature reaction, as in Fig. 291.

The power input in the first test is corrected for armature, field and exciter I^2R losses. The remainder is core, friction and windage. This, plotted to a base of field current, enables the core loss to be separated (Fig. 320(a)) and its value at normal voltage to be read from the curve. The I^2R losses are calculated from d.c. measurements of resistance, corrected for temperature. The exciter current is estimated by the armature-reaction method (Chapter XVII, § 14), and the corresponding exciter losses calculated or separately measured. The losses are then summed as in §7 above.

A.c. bridge methods may be used·for the measurement of losses and power factor of synchronous capacitors under full-load conditions.

12. **Reactance.** The direct- and quadrature-axis synchronous reactances, x_{sd} and x_{sq}, may be found from a "slip test." The test machine is coupled to a second machine or operated on no load,

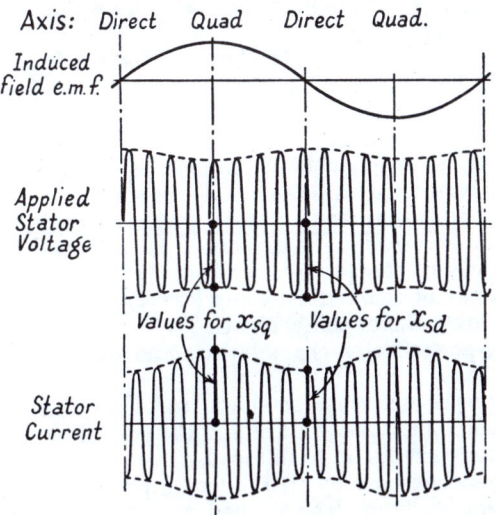

Fig. 321. Measurement of Synchronous Reactances
The slip is exaggerated

with the field circuit open and with a normal positive-sequence voltage applied to the stator. The voltage or the torque is adjusted so that the machine falls from synchronism with a low value of slip. Oscillograms are taken of: (a) e.m.f. induced in the field winding; (b) applied voltage; and (c) armature current. See Fig. 321. Then x_{sd} is the maximum and x_{sq} the minimum ratio of stator voltage to stator current. The method has the advantage of direct measurement.

The stator voltage required may be 20–30 per cent of normal, at normal frequency, and must be capable of close adjustment. The slip must be very small: otherwise the measurements will be in error on account of eddy currents in pole faces or damping windings. The slip tends to pulsate because of fluctuation of torque with relative pole position, with the result that there is a tendency for x_{sd} to be underestimated.

13. Sudden-Short-Circuit Test. This demonstrates the ability of the machine to withstand a sudden short circuit across the stator terminals at rated speed and from normal voltage on open circuit; it also enables some of the reactances and time-constants to be evaluated. The stator is clamped down, then run up to overspeed and the excitation adjusted. The driving motor is de-energized and the set allowed to retard. A multi-track oscillograph, recording phase and field currents, is started, and a remote-controlled short-circuiting circuit-breaker is operated as the speed of the test machine runs through normal value. The oscillograms are evaluated as indicated in Chapter XVIII, § 5.

14. Factory Tests on Large Machines.* Tests at full rated load on large synchronous machines are impracticable. Yet failure to comply with a specification may have serious and costly consequences. The erection of a machine for test may take some weeks; or a machine may be so large that it cannot be erected in the factory and may be tested only after assembly on site.

When testing at the manufacturer's works is feasible, the following characteristic curves and loss measurements are made—

(*a*) Open-circuit characteristic and loss (§ 3).

(*b*) Short-circuit characteristic and loss (§ 4).

(*c*) Zero-power-factor characteristic (§ 5).

(*d*) Unity-power-factor characteristic and loss (§ 11): this is ar alternative to (*a*).

(*e*) Zero-power-factor characteristic and loss: this resembles (*d*) with the minimum stator voltage and the excitation varied to give various stator currents up to full load.

(*f*) Temperature-rises by (i) full-load z.p.f. over-excited run, or (ii) equivalent heat run. The method (ii) comprises a rise measurement on the stator with excitation but no stator current, followed by a test with rated stator current but minimum excitation. The normal rise is then obtained by superposing the individual rises.

(*g*) Sudden-short-circuit test (§ 13).

(*h*) Retardation test for the moment of inertia of the rotating parts.

(*i*) Overspeed test, performed in the case of high-speed machines in a special enclosure. As the windage losses vary as the cube of the speed, a large water-wheel machine may take several thousand horsepower to drive it at the specified overspeed.

(*j*) High-voltage tests (§ 9).

(*k*) Insulation-resistance tests, made before and after (*j*) with a normal 1 000-V. insulation tester.

(*l*) Waveform and interference factor.

(*m*) Gap length, magnetic centrality, balance, vibration, bearing currents.

---

* Ross, "Factory Testing of Large Synchronous Machines." *Elec. Times,* 127, pp. 383, 563 and 705 (1955).

Not all of the tests listed are necessarily made on a given machine. Fig. 322 shows the curves obtained.

TEMPERATURE-RISE. The following assumptions may be made in calculating the temperature-rise from the results of heat runs—

(i) The stator rise is the sum of three individual rises, respectively proportional to windage, core, and I^2R plus stray loss.

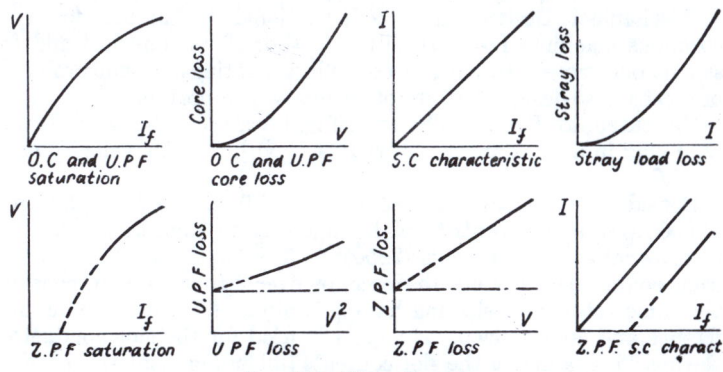

FIG. 322. TEST RESULTS

(ii) The stator core loss is a function of terminal voltage, and is determined by test (a) above.

(iii) The stray loss is the same on short circuit, zero power factor and rated power factor, and is independent of temperature.

(iv) The rotor rise is the sum of two individual rises, respectively proportional to windage and I^2R.

(v) Heat re-distribution after individual heat runs is negligible.

(vi) The gap is a thermal insulator, so that the rises on stator and rotor are independent.

CHAPTER XX

SYNCHRONOUS MACHINES : CONSTRUCTION

1. Mechanical Design. The constructional techniques for synchronous machines fall naturally into those for cylindrical and for salient-pole types. In both cases the larger ratings present problems of mechanical design that are of paramount importance.

CENTRIFUGAL FORCE. The centrifugal force on a body of mass m revolving in a circular path of radius R with angular velocity ω_r is $m\omega_r^2 R$.

Consider a cylindrical rotor of diameter 1·0 m. rotating at 50 r.p.s. (3 000 r.p.m.): the centrifugal force acting on each kilogram mass of material at the surface is $(2\pi 50)^2 . 0·5 \simeq 50\ 000$ N. The force on each pound mass is 4 600 lb. force, or over 2 tons. Again, consider the same rotor with slotting having a pitch of 1/30 of the circumference and 20 cm. deep. If copper is used for the slot conductors the specific gravity of the slot contents (including insulation) can be taken as approximately that of steel (7 800 kg./m.^{3}). In a slice of the rotor 1 m. thick, an element of radius r and thickness dr has, within a rotor slot-pitch, the mass

$$m = 7\ 800(2\pi r/30)dr,$$

and the centrifugal force on the whole slot depth is

$$f_c = \frac{7\ 800 . 50^2 . 8\pi^3}{30} \int_{r\ =\ 0·3}^{0·5} r^2 dr = 5·3 . 10^6 \text{ N}.$$

If the slots are 35 mm. wide, this force is spread over a section $2\pi 0·3/30 - 0·035 = 0·0278$ m.^{2}, and the stress is $190 . 10^6$ N/m.^{2} (or 1 950 kg./cm.^{2} or 12·5 tons/in.^{2}). With an overspeed of 25 per cent this stress is raised by the factor 1·56, to become nearly 20 tons/in.^{2}

HOOP STRESS. Annular rings (such as turbo-generator end-bells or the rotor rims of water-wheel machines) rotating about their axis develop hoop stress, due to centrifugal forces both self-developed and imposed by any inner masses that have to be retained. Consider an annular ring of unit cross-section. The radial centrifugal force due to the ring itself per unit length of circumference is $G\omega_r^2 R$, where G is the specific gravity of the material. If the ring were split across a diameter, the halves would be held together by a force balancing the sum of all the outward radial forces on one-half of the ring, amounting to $G\omega_r^2 R . \pi R(2/\pi) = 2G\omega_r^2 R^2$.

As an example, take a 3 000-r.p.m. turbo rotor with a diameter of 1 m., having retaining rings 35 cm. × 4·5 cm. of mean radius 0·49 m. enclosing copper windings of mass 300 kg. and radius 0·35 m. The self-produced hoop stress is

$$7\,800(2\pi50)^2 \cdot 0·49^2 = 186 \cdot 10^6 \text{ N/m.}^2 = 1\,900 \text{ kg./cm.}^2$$
$$= 27\,000 \text{ lb./in.}^2 = 12 \text{ tons/in.}^2$$

The centrifugal force on half the copper is $150(2\pi50)^2 \cdot 0·35 = 5·2 \cdot 10^6$ N, and this will produce a further hoop stress in the rings of $5·2 \cdot 10^6 \cdot (2/\pi) \cdot 0·49/(0·35 \cdot 0·045) = 100 \cdot 10^6 \text{ N/m.}^2 = 6·5 \text{ tons/in.}^2$ The total hoop stress in the retaining rings is therefore $12 + 6·5 = 18·5$ tons/in.², of which two-thirds is self-produced.

CRITICAL SPEED. A rotor is a structure with mass and elasticity, and so has a natural frequency of vibration. Reducing the problem to its most elementary terms, a shaft possessing only stiffness and running in two bearings without constraint, carries a disc of mass m and no elasticity. If the mass-centre of m is displaced ε from the centre of rotation, then the centrifugal force exerted on the mass will be $m\omega_r^2\varepsilon$. This will cause a deflection δ of the shaft, increasing the centrifugal force to $m\omega_r^2(\delta + \varepsilon)$. The deflection is resisted by the elastic shaft which develops a counter force equal, say to $k\delta$. For equilibrium the two forces are equal, giving

$$\delta = m\omega_r^2\varepsilon/(k - m\omega_r^2).$$

At some value of angular velocity ω_c the denominator becomes zero and the deflection infinite: δ is actually limited by frictional and deflectional losses, but will nevertheless tend to be large at the *critical* or *whirling* speed ω_c. If there is zero eccentricity ε, then δ becomes indeterminate at the critical speed, and a machine can indeed run through its critical speeds without developing noticeable vibration. However, chance disturbances may initiate a momentary deflection so that it is not practicable to operate a machine with a normal operating speed near to the critical.

The simple structure considered does not, of course, hold without modification in practical cases.* There may be more than one shaft span; nodes may be developed on one or more spans; the bearings impose constraint on shaft-flexure; and the structure may not be equally stiff across all diameters (as in turbo rotors where the slotting may not be regular round the whole periphery). The designer aims to make a machine operate as far as possible away from its critical speeds, although it may have to run through two or three critical speeds from standstill to reach normal speed of running.

BALANCING. The static and dynamic balance of a large rotor must clearly be secured to avoid dangerous conditions at critical

---

* Mitchell, "The Critical Speeds of Rigidly-coupled Rotors," *G.E.C. Journ.*, 23, p. 96 (1956).

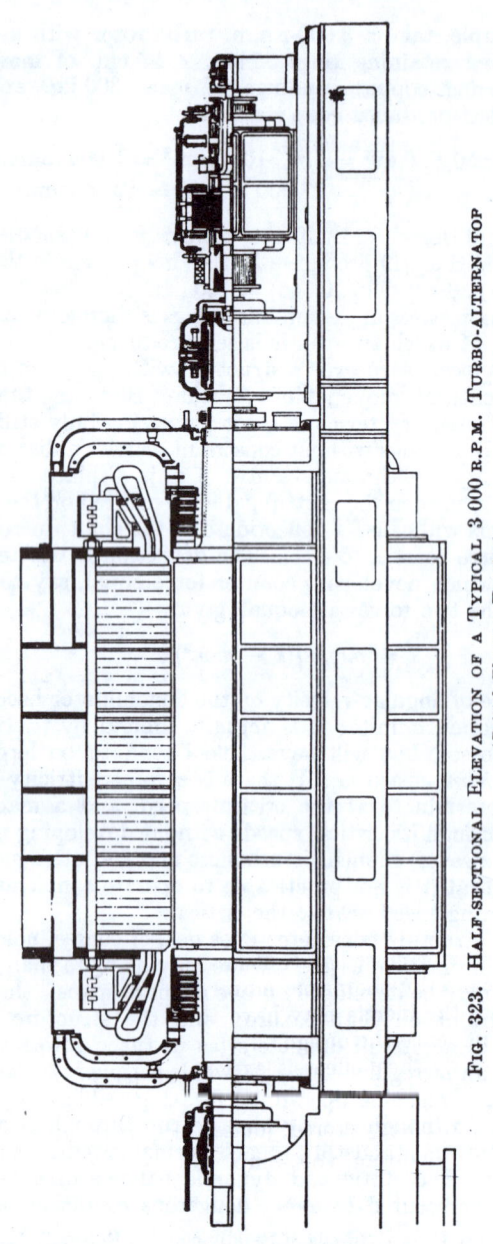

FIG. 323. HALF-SECTIONAL ELEVATION OF A TYPICAL 3 000 R.P.M. TURBO-ALTERNATOR

(British Thomson-Houston)

speeds. Static unbalance is usually adjusted at a critical speed because the deflection of the shaft is most sensitive for this condition. When static unbalance has been corrected the machine is run at full speed to adjust dynamic balance. The method may be by trial and error, or by use of balancing machines which indicate vibration amplitude and phase.

BEARINGS. In turbo-generators the journal rubbing speed may be as much as 65 m./s. and the bearing pressure 15 kg./cm.^{2} (200 lb./in.^{2}) of projected area. With water-wheel generators the thrust bearing is of major importance, for it takes not only the weight of the rotor and runner in a vertical-shaft machine, but also the hydraulic thrust which may more than double the total dead load. A thrust bearing of 2·5 m. diameter may take a total load exceeding 1 000 tons with a rubbing speed of 20 m. per sec. normally, or more than twice this for overspeed.

In a 60-MW. machine the bearing friction may account for a loss of the order of 120 kW. Local bearing pressures are very high. The rotation of the journal drags oil between the bearing surfaces to form a lubricating film. The oil feed-pressure is quite negligible by comparison with the bearing pressures, and serves only to keep the inlets flooded and to circulate cooling oil over the top of the bearing.

Bearings for horizontal-shaft water-wheel machines are normal except that a thrust bearing may be required to withstand un-balanced axial hydraulic thrust. In vertical-shaft machines a thrust bearing is always required. Oil is supplied by motor-driven pumps, and artificially cooled by external circulation, or by cooling coils incorporated in the white metal of guide and the oil pot of thrust bearings.

FLYWHEEL EFFECT. The inertia of the rotating parts of a machine affects its stored kinetic energy when running. A large and heavy machine will therefore take a considerable time to start and stop. In water-wheel generators the effect goes beyond this, for a minimum inertia is necessary because of the sluggishness of hydraulic govern-ing. If sudden load changes are not to cause excessive change of speed, the generator must have not less than a given inertia, as the contribution made by the prime mover is very limited. With low-speed machines the inertia may have to be very great and it may considerably influence the design.

The requisite flywheel-effect is expressed in terms of an inertia constant H, defined as

$$H = \frac{\text{energy stored, joules}}{\text{rating, volt.-amp.}} = \frac{\tfrac{1}{2}J\omega_r{}^2}{S} \text{ sec.}$$

where J is the moment of inertia. In British units with Wk^2 as the rotor inertia in millions of lb.-ft.^{2} and N as the normal speed in r.p.m., then $H = 0.23(Wk^2)N^2/kVA.$ sec. The value of H normally required is between 2 and 9 sec.

FIG. 91. INSERTING COILS INTO 2-POLE MOTOR

(General Electric Co.)

A. Turbo-alternators

The two-pole and four-pole machines differ considerably in construction. At 50 c/s. the former run at 3 000 r.p.m. and the latter at 1 500. The useful range of two-pole machines has been extended to 300 MVA., and in consequence the four-pole construction is obsolete. Fig. 323 shows in outline a typical two-pole construction.

2. Rotors. Rotors are most generally made from solid forgings of alloy steel. The forgings must be homogeneous and flawless. Test pieces are cut from the circumference and the ends to provide information about the mechanical qualities and the microstructure of the material. A chemical analysis of the test pieces is subsequently made. One of the most important examinations is the ultrasonic test, which will discover internal faults such as cracks and fissures. This will usually render the older practice of trepanning along the axis unnecessary.

The rotor forging is planed and milled to form the teeth. About two-thirds of the rotor pole-pitch is slotted, leaving one-third unslotted (or slotted to a lesser depth) for the pole centre. Fig. 324 shows a typical forged rotor.

Windings. The normal rotor winding is of silver-bearing copper. The heat developed in the conductors causes them to expand, while the centrifugal force presses them heavily against the slot wedges, imposing a strong frictional resistance to expansion. Ordinary copper softens when hot, and may be subject to plastic deformation. As a result, when the machine is stopped and the copper cools, it contracts to a shorter length than originally. The phenomenon of *copper-shortening* can be overcome by preheating the rotor before starting up. With new machines the use of silver-bearing copper, having a much higher yield point, mitigates the trouble.

Concentric multi-turn coils, accommodated in a slot number that is a multiple of four (e.g. 20, 24, 28 or 32), are used, the slot-pitch being chosen to avoid undesirable harmonics in the waveform of the gap density. The slots are radial and the coils formed of flat strip with separators between turns. The coils may be preformed. The insulation is usually micanite, but bonded asbestos and glass fabric have both been used.

As much copper (or, in some cases, aluminium alloy) as possible is accommodated in the rotor slots, the depth and width of the slots being limited by the stresses at the roots of the teeth, and by the hoop stresses in the end retaining rings. The allowable current depends on cooling and expansion. Comparatively high temperature-rises are allowed: the hot-spot temperature may reach 140° C.

Cooling. Passage for as much cooling gas as possible is provided. In small machines the main cooling takes place at the outside cylindrical surfaces of the rotor body and retaining rings. It is usual to have a large gap (e.g. 45 mm.) allowing for the flow of large quantities of coolant over the rotor surface. For larger machines provision must be made for cooling the bottom of the slot, the conventional method being that shown in Fig. 325. It is practicable to pass an appreciable volume of gas at high velocity close to the windings, but the temperature-g.. ''ent over the slot-insulation is still a dominant factor. For the la.... .t ratings elaborate ventilating

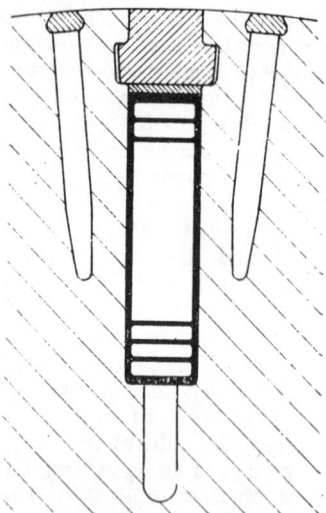

FIG. 325. CONVENTIONALLY-
COOLED ROTOR SLOT AND DUCTS

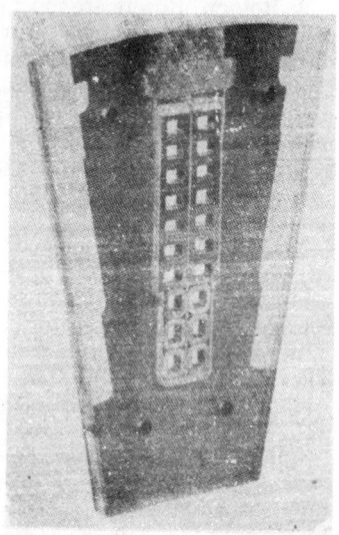

FIG. 326. ARRANGEMENT OF
CONDUCTORS FOR DIRECT COOLING
(*General Electric Co.*)

arrangements are necessary, and for machines of 100 MVA. upwards methods are being developed for direct contact between the rotor coils and the gas coolant. One such method is shown in Fig. 326.* The essential feature is the use of straight rectangular tubes for the slot conductors, ventilated on a cooling circuit separate from that of the stator, the hydrogen gas coolant being circulated through the tubes by a centrifugal impeller mounted on the outboard end of the rotor. The slot tubes are connected at both ends by suitably shaped copper bars forming inlet or outlet ports. The conductors are hard-drawn electrolytic silver-bearing copper, with synthetic-resin-bonded glass cloth laminate insulation.

OVERHANG. The appearance and arrangement of the overhang can be seen in Fig. 327, which also shows the spigot on which the retaining ring is centred. The conductors are sloped conically inward to provide sufficient space for the thickness of the ring, and

* Eaton, "Design of Large Turbo alternators," *G.E.C. Journ.*, 22, p. 155 (1955).

are strongly braced to fit closely within the ring without movement other than that due to unavoidable expansion. Non-magnetic magnesium and magnesium-nickel steels are used for the retaining rings, the non-magnetic property being useful for avoiding excessive magnetic leakage and stray load loss.

FIG. 327. OVERHANG ARRANGEMENT OF 2-POLE ROTOR
(English Electric)

FANS. The rotor may carry centrifugal or axial (propeller) fans, the former being more common. However, restrictions on rotor diameter and the great length between bearings may make separate fans essential for large machines. They may be coupled to a shaft extension or be separately driven.

SLIP-RINGS. Slip-rings are required for conveying the exciting current to and from the rotor winding. Rings of steel, shrunk over micanite, may be placed one at each end of the rotor, or both at one end, inside or outside the main bearing. Fig. 328 shows a typical completed two-pole rotor.

3. **Stators.** Large stator housings comprise a series of annul ı ıgs flame-cut from steel plate, joined by tubes and longitudinal

FIG. ... S. ......D ROTOR
(Parsons)

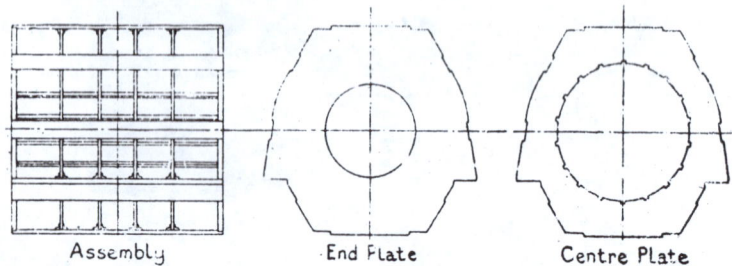

Assembly   End Plate   Centre Plate

FIG. 329. METHOD OF FABRICATING TURBO-GENERATOR STATOR HOUSING

bars, and carrying ribs to take the stator core laminations. Fig. 329 shows a simple stator housing requiring two end plates and four intermediate plates, held apart by tie bars. The core stampings are built up in the frame, the end plates placed in position, and the whole stator clamped together by bolts. The frame is then covered with sheet steel. Fig. 330 shows an assembled stator frame of orthodox design, in which the arrangements for multiple-inlet

FABRICATED STATOR OF 35 700 kVA. GENERATOR
(British ...)

FIG. 331. COMPLETED 60 MW. HYDROGEN-COOLED GENERATOR
(English Electric Co,)

ventilation are seen; and Fig. 331 shows a similar stator complete with a braced basket winding.

CORE. The active part of the stator consists of segmental laminations of low-loss alloy steel. The slots, ventilation holes and dovetails or dovetail keyholes, are punched out in one operation. The stampings are rather complicated on account of the number of holes and slots that have to be produced.

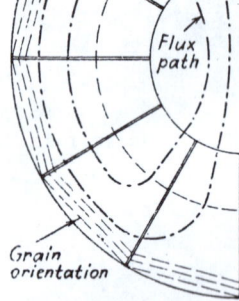

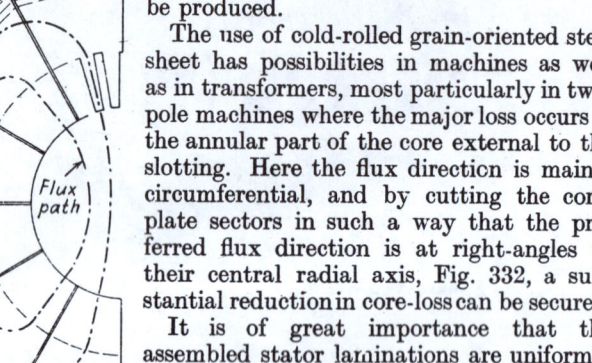

The use of cold-rolled grain-oriented steel sheet has possibilities in machines as well as in transformers, most particularly in two-pole machines where the major loss occurs in the annular part of the core external to the slotting. Here the flux direction is mainly circumferential, and by cutting the core-plate sectors in such a way that the preferred flux direction is at right-angles to their central radial axis, Fig. 332, a substantial reduction in core-loss can be secured.

It is of great importance that the assembled stator laminations are uniformly compressed during and after building, and that the slots are accurately located. The core plates are assembled between end plates with fingers projecting between the slots to support the flanks of the teeth. The end plates are almost invariably of non-magnetic material, for this greatly reduces stray load loss. The end packets of core plates may be stepped to a larger bore for the same reason.

FIG. 332. USE OF GRAIN-ORIENTED STEEL

WINDINGS. The windings of two-pole machines are comparatively straightforward. The number of slots must be a multiple of 3 (or 6 if two parallel circuits are required). Single-layer concentric or two-layer short-pitched windings may be used.

The single-layer concentric winding is readily clamped in the overhang, but causes a higher load loss because the end-connections run parallel to the stator end-plates. Chording is not possible so that flux harmonics have full effect.

The two-layer winding is more common, chorded to about a 5/6 pitch which practically eliminates 5th and 7th harmonics from the open-circuit e.m.f. wave. The end windings are packed, and clamped or tied with glass cord.

It is the invariable practice with two-layer windings to make the coils as half turns and to joint the ends. The conductors must always be transposed to reduce eddy-current losses. The conductors are insulated in many cases with bitumen-bonded micanite, wrapped on as tape, vacuum dried, then impregnated with bitumen under pressure and compressed to size. The process is illustrated in

Fig. 333. Each copper bar A forming part of a conductor is insulated with mica tape, B and C. A set of bars forming one conductor is assembled and pressed, D. The conductor is insulated with layers of mica tape, E; then the conductors are assembled to form a slot bar, F, and pressed to the required dimensions. Fig. 334 shows

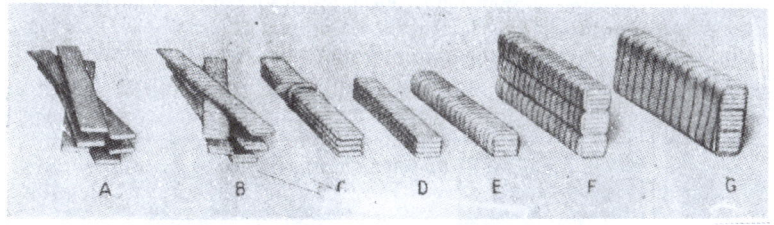

FIG. 333. STAGES IN THE MANUFACTURE OF A LAMINATED CONDUCTOR
(British Thomson-Houston)

FIG. 334. SECTION OF STATOR CONDUCTORS
(Metropolitan-Vickers)

typical conductors. Synthetic resins have now replaced bitumen.

Within the slots, the outer surface of the conductor insulation is at earth potential: in the overhang it will approach more nearly to the potential of the enclosed copper. Surface discharge will take place if the potential gradient at the transition from slot to overhang is excessive, and it is usually necessary to introduce voltage grading by means of a semi-conducting (e.g. graphitic) surface layer, extending a short distance outward from the slot ends.

The slot inductance is increased by setting the windings more deeply into the slots. This has the incidental advantage of spacing the overhang farther away from the rotor end-rings.

4. Ventilation. Forced ventilation and total enclosure are necessary to deal with the large-scale losses and high rating per unit volume. The primary cooling medium is air or hydrogen, which is in turn passed through a water-cooled heat-exchanger.

AIR-COOLING. The arrangement is that of Fig. 145. The water coolers are normally in two sections, so that one can be cleaned while the machine is operating. Fans on the rotor, or separate fans, may be employed, the latter in large machines where bearing-spacing or limitation of the diameter makes integral fans inadequate.

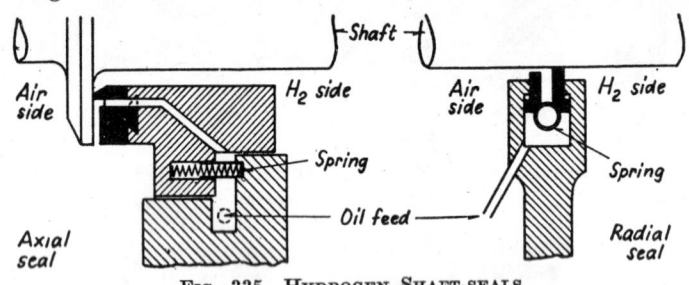

FIG. 335. HYDROGEN SHAFT-SEALS

With integral fans mounted on the rotor, the air is fed to the space surrounding the stator overhang, and pipes and channels convey a proportion towards the centre of the stator core. Therefrom it flows readily inward to the air gap, then axially to the end outlet compartments. With separate fans, however, air can be fed directly to the middle as well as to the ends, as shown in Fig. 145.

HYDROGEN COOLING. Compared with air, hydrogen has 1/14 of the density, reducing windage loss and noise; 14 times the specific heat; $1\frac{1}{2}$ times the heat-transfer, so more readily taking up and giving up heat; 7 times the thermal conductivity, reducing temperature gradients; reduces insulation corona; and will not support combustion so long as the hydrogen/air mixture exceeds 3/1. As a result, hydrogen cooling at 1, 2 and 3 atmospheres absolute can raise the rating of a machine by 15, 30 and 40 per cent respectively.

The stator frame must be gas-tight and explosion-proof. Oil-film gas-seals at the rotor shaft ends are necessary. Two forms are shown in Fig. 335: each must accommodate axial expansion of the rotor shaft and stator frame. Oil is fed to the shaft and the flow is split, part towards the interior (gas) side and part to the air side. The latter mingles with the bearing oil, while the former is collected and degassed.

Fans mounted on the rotor circulate hydrogen through the ventilating ducts and internally-mounted gas-coolers. The gas pressure is maintained above atmospheric by an automatic regulating and reducing valve controlling the supply from normal gas cylinders. When filling or emptying the casing of the machine, an

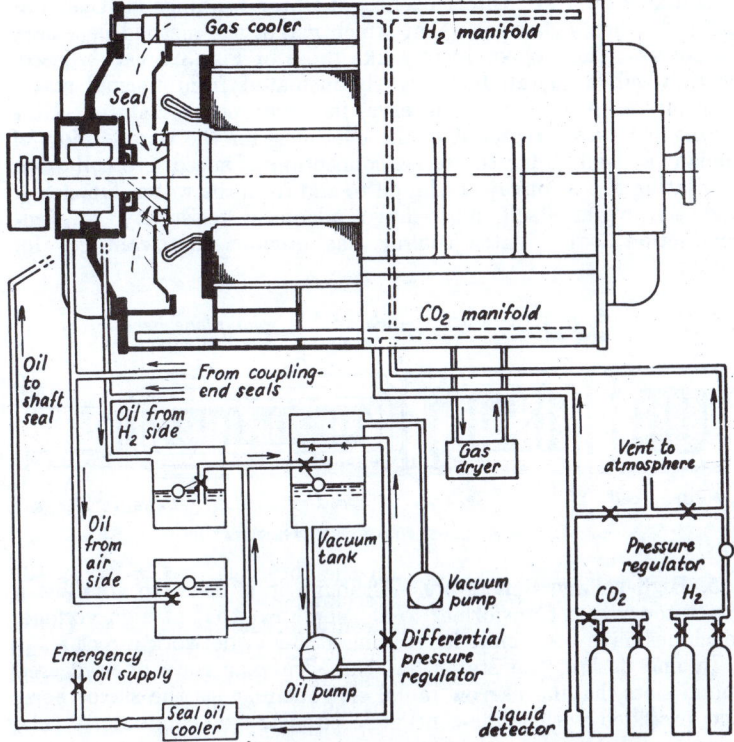

FIG. 336. ARRANGEMENT OF HYDROGEN-COOLING SYSTEM

explosive hydrogen-air mixture must be avoided, so that air is first displaced by carbon dioxide gas before hydrogen is admitted: the process is reversed for emptying. It is usual to provide a drier to take up water vapour entering through seals. The hydrogen purity is monitored by measurement of its thermal conductivity.

Turbo-alternators operating at hydrogen pressures just above atmospheric (so that leaks will be outwards) require about 0.03 m.^{3} per MW. of rating per day. This rises to about 0.1 m.^{3} for hydrogen pressure of 2 atm. abs. The gas consumption of synchronous capacitors which do not need shaft seals, is very much less.

Hydrogen cooling results in substantial increase in MW. rating for a given temperature-rise, and the reduction in windage may add $0.5-1.0$ per cent to the efficiency of a 100-MW. machine. Fig. 336 gives a diagram of the auxiliary equipment required for a hydrogen-cooled machine, and Fig. 331 shows the end of a completed stator showing the end-covers of the twin water-coolers.

DIRECT COOLING. Direct cooling of stator windings is applied at ratings rather higher than that which makes the method necessary for rotors. Tubular conductors like those of Fig. 326 can be used, or thin-walled metal ducts lightly insulated from normal stator conductors. A similar design serves for *water-cooling* a stator. Here arrangements are required in the overhang for the parallel flow of coolant as well as for the series connection of successive coil-sides. Insulating tubes convey the liquid to and from the water "headers," and the water itself must have adequate resistivity to limit conduction loss. Water cooling has obvious disadvantages for rotors.

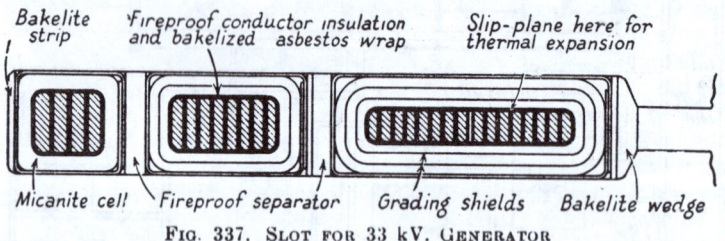

Bakelite strip   Fireproof conductor insulation and bakelized asbestos wrap   Slip-plane here for thermal expansion

Micanite cell   Fireproof separator   Grading shields   Bakelite wedge

FIG. 337. SLOT FOR 33 kV. GENERATOR

5. **High-voltage Generators.** Although it is usual to combine a generator with a transformer to develop an output at high voltage, machines have been built for feeding a 33-kV. network direct.

In one design, the stator has three circular rows of staggered round slots, having narrow radial slot openings to the stator bore, and provided with triple-concentric circular slot conductors. The insulation between conductors, and between the outer conductors and slot walls, is flexible micanite. The electric stress imposed on the insulation is no greater than in a machine built for normal voltage. The innermost (or "bull") conductor in every slot forms that third of the winding connected to the line terminals, while the "outer" conductors are at the star-point end. The conditions for heat dissipation from the central "bull" conductor are rather unfavourable, and a low current-density is necessary. The slot reactance tends to be high.

Fig. 337 shows the slot of a 33-kV. machine of more orthodox design. Three completely separate windings insulated respectively for 11, 22 and 33 kV. (line) are used, the 11-kV. section being at the bottom of the slots. Each conductor is made in the form of a capacitor bushing, with conducting shields buried in the insulation to control radial electric stress. The thicker insulation of the higher-voltage windings requires the copper to be deeper and more extensively laminated. Longitudinal stress control at the ends of the conductor outside the core is facilitated by the grading shields.

B. SALIENT-POLE MACHINES

The machines discussed are generators designed for water-wheel and internal-combustion engine or gas-turbine drives, and salient-pole synchronous motors. Synchronous capacitors may resemble either turbo or salient-pole machines, usually the former.

6. General Construction. Salient-pole generators operate at speeds between 80 and 600 r.p.m., so that their diameters may be very large to accommodate the poles and they may require to be made in sections for transport. Governing and speed regulation, or the transient stability of the associated power network, decides the amount of flywheel effect of the set, the major part of which has to be incorporated in the generator, occasionally by means of a flywheel.

The governing of water turbines is made difficult by the inertia of the moving mass of fluid, and adjustment must be gradual. Sudden changes of load therefore cause rapid change of speed. If the load is lost and the governing fails, the speed may rise to more than twice normal, and generators must be built to run on overspeed for short periods in an emergency. Centrifugal forces therefore dominate the design.

Water-wheel machines may be roughly classified as low-speed (80–200 r.p.m.) and high-speed (300–600 r.p.m.). Both horizontal- and vertical-shaft designs are used, the former with impulse turbines and Pelton wheels, the latter with large reaction and Kaplan turbine drive. High-speed machines have rotors up to 4 m. diameter with cores up to 5 m. long. Low-speed generators have been built with diameters up to 14 m. and bearing loads as much as 1 500 tons.

The inertia constant H is adequate in most large low-speed generators, but with smaller machines it may be necessary to modify the "natural" dimensions of a machine as decided by its rating, voltage, speed and limits of temperature-rise. The designer therefore apportions the D^2L product so as to increase the diameter up to limits imposed by centrifugal force.

To accommodate the many poles of a low-speed machine the diameter is made, as far as is feasible, such that the core length does not exceed three times the pole-pitch, otherwise the pole and field design become uneconomic.

The characteristics of a water-turbine influence the design in the following ways: its low speed necessitates many poles, which must then be salient; the overspeed imposes centrifugal-force limitations on the diameter; natural dimensions may have to be modified to secure adequate flywheel effect; the critical speed must, for safety, be greater than the overspeed; thrust bearings are needed to resist the hydraulic thrust and, in vertical-shaft machines, the weight of the rotor and turbine runner as well; physical size introduces transport difficulties.

Some of the above apply to engine-driven generators, but in

general the design problems are considerably easier than with water-wheel machines.

7. Rotors. The poles are carried on a rotor body. For high-speed machines the body is (a) machined with its shaft from a single forging; or (b) machined separately with stub-shafts then attached; or (c) built up from thick discs shrunk on to a shaft. Medium- and low-speed machines have a larger diameter: method (c) may be

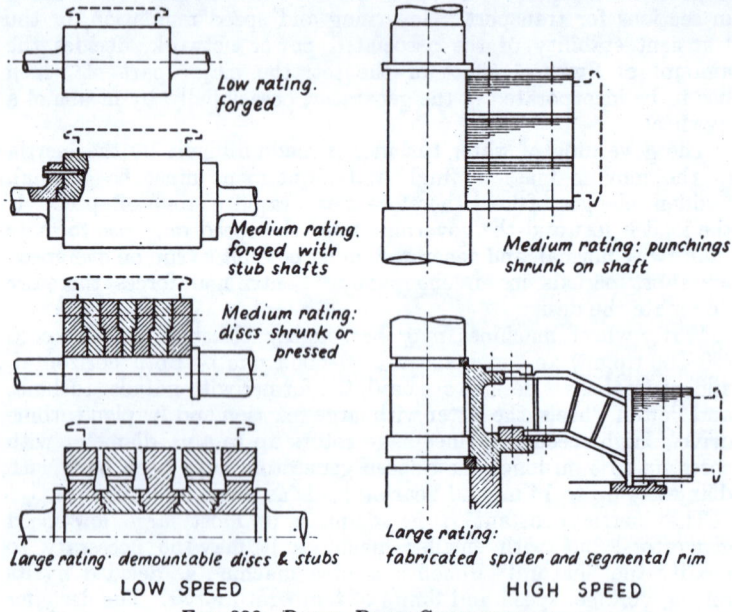

Low rating.
forged

Medium rating.
forged with
stub shafts

Medium rating:
discs shrunk or
pressed

Medium rating: punchings
shrunk on shaft

Large rating: demountable discs & stubs

Large rating:
fabricated spider and segmental rim

LOW SPEED   HIGH SPEED

FIG. 338. ROTOR BODY CONSTRUCTION

used for diameters not exceeding about 4 m.; or (d) the body may be fabricated from a cast-steel spider mounted on the shaft, and carrying a laminar ring of overlapping segmental plates tightly bolted to resist slipping under hoop stress. The rim rests on the spider arms and is driven from them by floating keys, thus relieving the spider of centrifugal force other than that of its own mass. Fig. 338 shows these methods, which in large sizes may require modification for transport in sections.

Up to a peripheral speed of about 80 m./sec. it is possible to use a segmented spider. Above this a solid disc becomes inevitable, and a normal peripheral speed of 110 m./sec. (220 m./sec. on overspeed), using solid steel poles and aluminium field coils, is possible.

Fig. 339 shows a solid-forged medium-speed rotor. Small rotors can be built up of thick punchings, Fig. 340. The poles have open slots for damper windings.

POLES. Laminated poles are used on all medium- and low-speed machines. They comprise thin steel plates clamped between heavy end-plates and secured by studs or rivets. The poles may be bolted

FIG. 339. ROTOR BODY AND SHAFT FOR 3 000-KVA., 6 600-V., 750-R.P.M. WATER-WHEEL ALTERNATOR
(General Electric Co.)

to the rotor body if the peripheral speed does not exceed about 25 m./sec. For higher speeds the poles are dovetailed as in Fig. 340 and, for large machines, as in Fig. 341.

WINDINGS. The field coils are formed from rectangular-section strap, generally of copper and in sizes up to 10 cm. × 1 cm. Coils for short-cored machines are strap wound on edge, with inter-turn asbestos or asbestos-rubber strips, cured and consolidated under a pressure exceeding that due to centrifugal force at runaway speed. The coils of large machines are fabricated from punched straight lengths of copper, Fig. 342 (*a*). They are much easier to make and wind, as a higher width/thickness ratio can be used without the

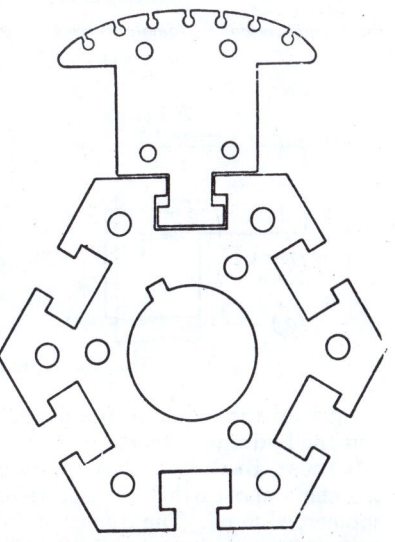

FIG. 340. SPIDER AND POLE PUNCHINGS FOR 50-H.P., 1 000-R.P.M. SYNCHRONOUS MOTOR

thickening at corners that occurs when bending strip-on-edge. The width of strap can be varied to provide cooling fins, and to reduce the width at the pole tips. The choice of pole-section is eased, and silver-bearing copper used with long cores to resist high-temperature creep. The method also facilitates the construction of "canted" field windings which improve resistance to centrifugal forces, Fig. 342 (*b*).

DAMPER WINDINGS. In most cases copper rods in the pole shoe are adequate without interpolar cage connections. The ends are brazed to copper end bars attached to the pole end-plates. Solid poles may render damper windings unnecessary: their use is magnetically permissible because the gap length is at least 2/3 of the stator slot-pitch.

A completed rotor is shown in Fig. 343.

8. **Bearings.** Bearings for small horizontal-shaft machines are conventional. Vertical-shaft generators offer unusual features

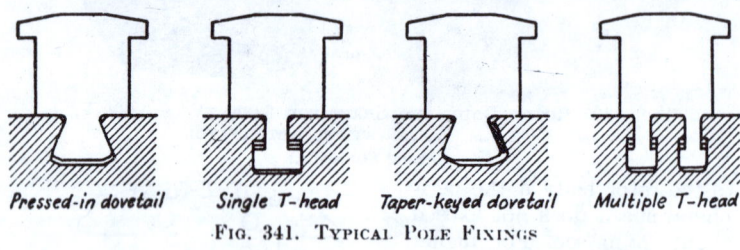

| Pressed-in dovetail | Single T-head | Taper-keyed dovetail | Multiple T-head |

FIG. 341. TYPICAL POLE FIXINGS

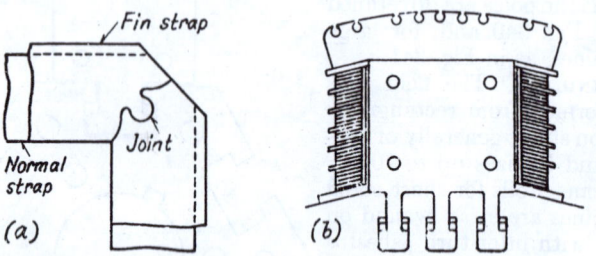

FIG. 342. FIELD WINDINGS

because of the requirements of the weight of the rotor and runner, and the hydraulic thrust.

THRUST BEARING. This comprises a collar rigidly mounted on the shaft and carried on a stationary member. The bearing runs submerged in oil. The runner or collar part is of mild steel or close-grain cast iron, ground and lapped to optical accuracy. The surface of the stationary support member is of white metal on steel pads which form segments of the bearing surface and are free to tilt slightly during running to form a wedge shaped oil film between the rubbing surfaces. Fig. 344 shows the construction of a typical bearing.

The thrust bearing may be above or below the rotor body. If above, it must be removed from the shaft before the rotor can be lifted out, and the stator frame carries the whole rotor load. With the bearing below, the vertical load can be carried on the concrete foundations through a robust thrust-bearing spider.

FIG. 343. ROTOR OF 8-POLE ENGINE-DRIVEN ALTERNATOR
(*British Thomson-Houston*)

FIG. 344. THRUST BEARING FOR 130-MVA. ALTERNATOR
(*English Electric Co.*)

BRAKING AND JACKING. A large machine may take half an hour
to stop if not braked. Vertical-shaft generators are normally fitted
with air- or oil-operated brakes so that the machine may not
run for long periods at low speeds (e.g., 1 r.p.m.) which is highly

detrimental to thrust bearings. The brakes also provide a jacking system to lift the rotor a few mm. to allow the thrust-bearing surfaces to be flooded with oil before run-up.

The mechanical assembly, with the operating and control system is shown in Fig. 345 for a vertical shaft Francis turbine and generator

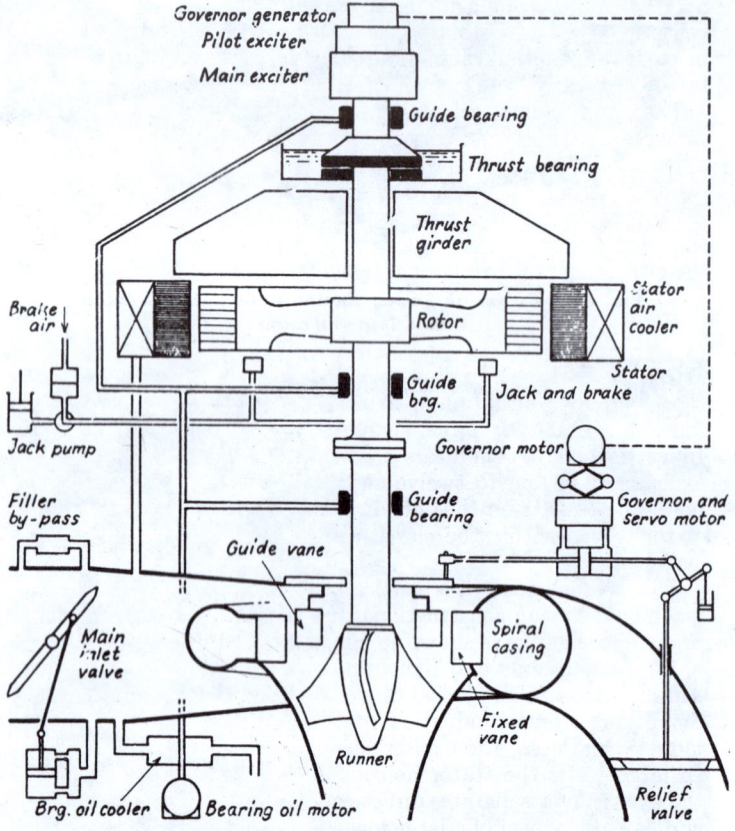

FIG. 345. CONTROL ARRANGEMENTS OF WATER-WHEEL SET

9. **Stators.** Stator frames are fabricated from steel plates and bars, with special provision if the frame carries the thrust bearing. With vertical-axis machines the frame seats on foundation caps, and is located by radial jacks.

CORES. These do not differ essentially from those of turbo machines, except that a very large number of core sectors is necessary in machines of large diameter. If grain-oriented steel is used, the best arrangement is that shown in Fig. 346. The core is supported by dovetail keybars forming part of the stator frame.

WINDINGS. These follow the general pattern described in Chapter X. In large generators the large airgap permits the use of open slots with two-layer diamond coils. The pitch is short for low-speed machines, which have unusually large numbers of coils. Provision may have to be made for removal of coils, or for the assembly of the stator in parts. Insulation offers no unusual features.

The high reactance, narrow slot-pitches and short end-connections of a low-speed machine may permit of simple bracing, such as a ring with packing blocks. In high-speed machines the overhang is longer and the coils less stiff, so that bracing must be more comprehensive.

10. **Ventilation.** Small synchronous machines are cooled in the same way as corresponding induction machines. Closed-circuit ventilation, with integral fans, is normal for large generators. In vertical-shaft generators the heat-exchangers in four to twelve units are bolted on to flats on the outside of the stator frame, within the closed-circuit system.

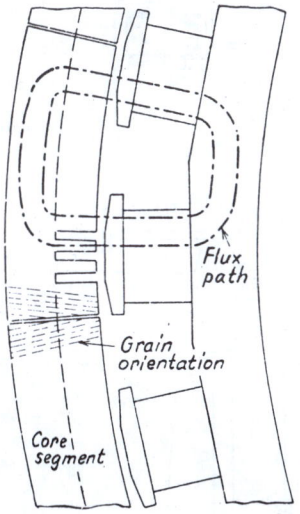

FIG. 346. USE OF GRAIN-ORIENTED STEEL

The arrangement of a vertical-axis machine, showing constructional framework, bearings, stator, rotor, brake and exciters is shown in Fig. 347.

11. **Synchronous and Synchronous-induction Motors.** Figs. 339, 340 and 343 are as typical of synchronous motors as of generators. Small synchronous-induction motors are very similar to standard induction motors, the exciter being overhung from the end shield. In larger sizes the stator housing is supported on a bed-plate and the exciter is a separate, direct-coupled shunt generator. Figs. 348 and 349 are typical of a large low-speed fabricated motor. The rotor of this machine is shown in Fig. 247.

The secondary windings of synchronous induction motors have to be designed on a basis of uniform heat distribution in spite of the unequal loading. This can be done by using more slots per pole per phase for one of the phases as for either of the other two, and employing double-section conductors for this phase. A high open-circuit voltage is generally chosen to keep down the current and cost of the exciter. Large machines may have open-circuit slip-ring voltages of 2 000 V. at standstill, and exciter voltages up to 50 V.

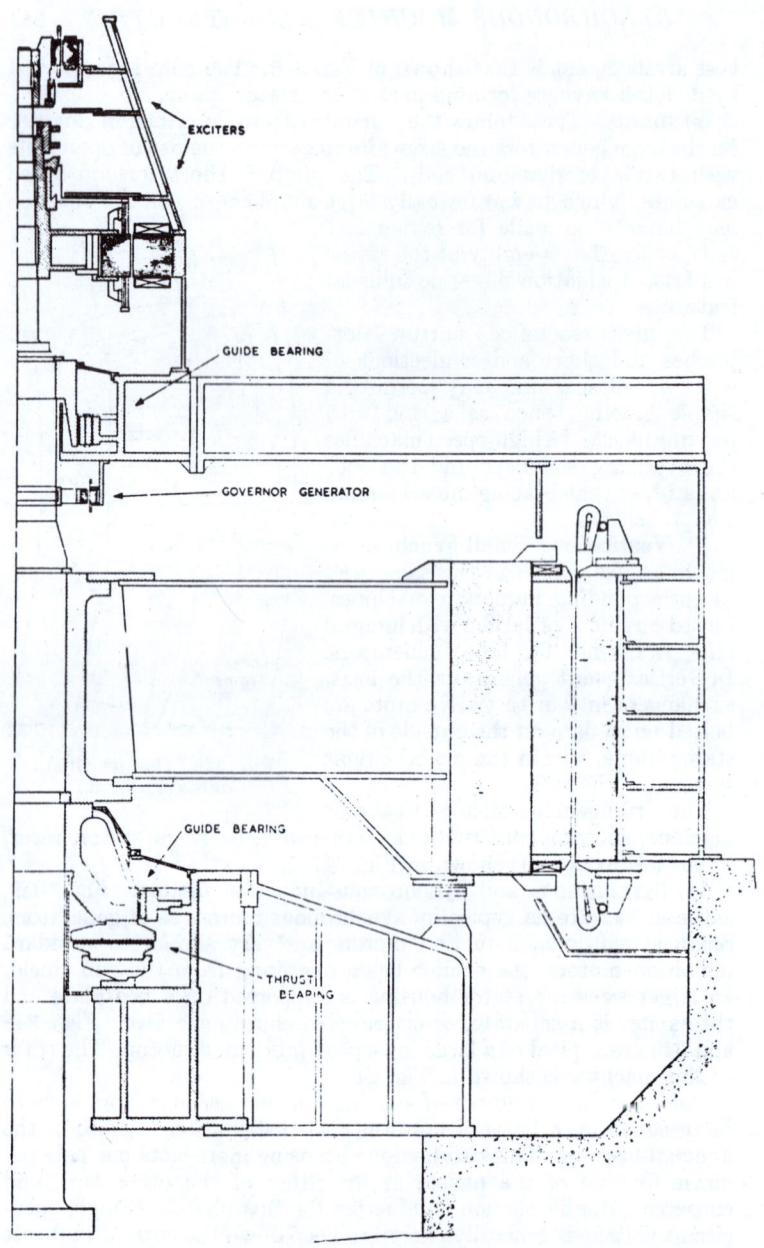

FIG. 347. HALF-SECTION OF MEDIUM-RATING VERTICAL-
SHAFT WATER-WHEEL GENERATOR
(*British Thomson-Houston*)

C. Exciters, etc.

12. **Exciters.** The reliability of an exciter is a first essential, and all modern machines have their own exciters, to avoid the serious consequence of a failure of a common supply. Direct-coupled exciters are preferred, but their output is limited to about 350 kW. at a turbo-alternator shaft speed of 3 000 r.p.m.; however, geared

Fig. 348. Fabricated Stator Frame for 2 500 h.p., 2 300 V., 54-pole Synchronous-Induction Motor

(Bruce Peebles)

drive may be used, although it is not favoured for the larger outputs. Future machines of very large output will require excitation powers of the order of 1 MW., and separate motor-driven units will probably be used.

The conditions under which exciters work are not ideal, as a rule, for d.c. machines. The speed may be rather high for turbo-alternators, or somewhat low for multipolar water-wheel machines. The voltage range is also exceptionally wide, so that the machines are large for their rating.

Main and pilot exciters are normally employed. Vertical machines carry the exciters at the top of the main shaft. If the main exciter is arranged commutator up, the pilot above it with commutator down, and the rotor sliprings between, the whole assembly can be conveniently enclosed with all the brushgear in one place for inspection, and an overall reduction of height.

13. **Auxiliaries.** Generating-station auxiliaries are invariably motor-driven. As a safeguard, each item may be duplicated and the

driving motors connected to either of two separated sources, such as (1) a "unit" transformer supplied from the generator, or (2) a

Fig. 349. Completed 2 500 h.p. Synchronous-Induction Motor and Exciter
(*Bruce Peebles*)

carried out by means of the 'house" transformer where the main busbars can be energized from an external source.

Sets have been built with a small alternator coupled to the shaft of the main machine to supply half of the auxiliary motors, the other half being fed from the "unit" or "house" transformer. Each auxiliary motor can be fed from either source, but the two sources are never paralleled.

As indicated in Fig. 345, the hydraulic governor of a water-turbine-driven machine is driven by a small self-starting synchronous motor fed from a permanent-magnet auxiliary generator carried on the shaft of the main machine.

CHAPTER XXI

SYNCHRONOUS MACHINES: DESIGN

1. **Types.** For the purposes of design, synchronous machines are classified according to speed. To the generator classes—turbo, water-wheel and engine-driven—correspond similar methods of constructing synchronous condensers or motors, while the synchronous-induction motor has constructional features similar to those of induction motors of similar rating and speed.

For machines with six poles and upwards, the salient-pole arrangement is employed, the peripheral speeds of about 30 and 110 m. per sec. in engine-driven and water-wheel types being not too great for bolted-on or dovetailed pole-attachment. For four- and two-pole machines (apart from small sizes) the cylindrical rotor becomes necessary, and the peripheral speed can be raised to about 170 m. per sec., to take advantage of high-speed turbine prime movers.

2. **Specification.** The starting point of a design is naturally the specification to which the machine must be built. The specification fixes the kVA. rating and overload capacity; number of phases; frequency; voltage; speed; efficiency; regulation; temperature-rise; method of ventilation; nature of the load; type and coupling prime-mover; method of excitation; characteristics of plant ʌning in parallel; conditions as to erection, testing and provision of spare parts. All these have a bearing on the design.

3. **Rating and Dimensions.** The output of a synchronous machine is based on the specific magnetic and electric loadings, the speed and the main dimensions. The same approach as in Chapter XVI, § 3, gives the rating by eq. (150) as

$$S = 11 K_w \bar{B} ac D^2 Ln \text{ volt-amperes.}$$

The output coefficient, from eq. (151), with S in kVA. is

$$G = S/D^2 Ln = 11 K_w \bar{B} ac . 10^{-3}. \qquad . \qquad . (174)$$

Since the limiting feature of large machines is in many cases the peripheral speed, it is of interest to express S in terms of the speed $u = \pi Dn$, giving

$$S = 1 \cdot 1 K_w \bar{B} ac L(u^2/n) . 10^{-3} \text{ kVA.} \qquad . \qquad . (175)$$

LOW-SPEED MACHINES. Taking as higher values $B = 0 \cdot 6$ Wb./m.^{2}, $ac = 45\,000$ ampere-conductors per m., $u = 80$ m./s., $K_w = 0 \cdot 95$ and $L = 0 \cdot 5$ m., the output obtainable by eq. (175) is

$$S \simeq 90\,000/n \text{ kVA.}$$

Thus at 375 r.p.m. (16 poles for 50 c/s.), $n = 6 \cdot 25$ r.p.s., and an output of 144 000 kVA. could be obtained. The rotor diameter would be about 4 m. A 40-pole machine with $n = 2.5$ r.p.s. would have a rating of 180 000 kVA. and a diameter of 10 m. Machines of comparable ratings have been built for hydroelectric generation.

TURBO-GENERATORS. Only 2-pole machines need be considered. For 50-c/s. machines, with $\overline{B} = 0.6$ Wb./m.^{2}, $ac = 60\,000$, $u = 170$ m./sec., the diameter is

$$D = u/\pi n = 1.08 \text{ m.} = 42\tfrac{1}{2} \text{ in.,}$$

and the output is

$$S/L \simeq 22\,000 \text{ kVA. per m. of active core-length.}$$

To obtain large outputs it is thus essential to use very long rotors and to raise the electric loading of both stator and rotor by direct cooling. The value of ac can be doubled in this way.

Turbo-generator ratings and characteristics have been standardized (B.S. 2730: 1956), as shown in the Table below. The air-cooled machines have insulation of Class B or better, closed-circuit ventilation and a maximum inlet temperature of 40° C. The higher ratings have hydrogen cooling, and similar provisions regarding insulation and inlet gas temperature.

STANDARD RATINGS OF 50-C/S. 3 000-R.P.M. TURBO-ALTERNATORS

Air-cooled—						
Rating, MW.	10	20	30	30		
MVA.	12·5	25	37·5	33·3		
power factor	0·8	0·8	0·8	0·9		
kV.	6·6/11	6·6/11	11·8	11·6		
Short-circuit ratio	0·55	0·55	0·55	0·8		
Subtransient reactance, p.u.	0·125	0·125	0·125	0·125		
Hydrogen-cooled—						
Rating, MW.	60	60	100	100	120*	120*
MVA.	75	67	125	111	150	133
power factor	0·8	0·9	0·8	0·9	0·8	0·9
kV.	11·8	11·6	To suit machine			
Short-circuit ratio	0·55	0·8	0·55	0·8	0·55	0·8
Subtransient reactance, p.u.	0·125	0·105	0·125	0·105	0·125	0·105

* Hydrogen pressure 2 atm. abs.; remainder 1 atm. abs.

SPECIFIC LOADINGS. Typical conservative loadings for salient-pole machines are given in the Table on the following page. The current density is 3·75 A./mm.^{2}, and the figures refer to machines with Class A insulation. They are therefore subject to some increase with Class B insulation, for which the limits of temperature-rise are about 20° C higher.

D m.	0·2	0·3	0·5	0·75	1·0	1·5	2·0	3·0
L_i/D (max.)	0·7	0·65	0·6	0·5	0·42	0·33	0·3	0·3
$\bar{B}$ Wb./m.²	0·44	0·47	0·53	0·57	0·59	0·60	0·62	0·64
ac A.-C./m.	14 500	19 300	27 500	33 000	38 500	41 000	42 000	44 000

MAIN DIMENSIONS. The product $\bar{B}ac$ determines the output per unit volume, eq. (174). An increase of one of the loadings generally involves a decrease of the other, since a wider tooth necessitates a narrower slot for a given slot pitch. Other considerations apart, the copper width in a slot of fixed depth and pitch should be roughly equal to the tooth width at one-third depth from the narrower end. A high value of ac, however, produces a large armature reaction and demands increased field excitation. By increasing the slot depth, ac can be raised without narrowing the teeth or affecting the stator bore: but more insulation is needed on the slot conductors, the leakage reactance is increased, there are larger core losses in the teeth, a further increase is necessary in the field excitation and the cooling is more difficult.

The subdivision of the $\bar{B}ac$ product is influenced by the loading conditions of the machine, and by the consequent optimum short-circuit ratio. This aspect is discussed on p. 454.

STRAY LOSSES. The losses in a synchronous machine comprise the windage and friction, the stator and rotor core losses (all of which are calculable), and an additional *stray* loss that occurs on load. Although the stray loss is often taken as the difference between the measured total loss on short-circuit and the sum of the calculated losses, the stray loss on load may differ considerably from the figure obtained in such a way. The stray loss is not only difficult to assess, but it may be comparable to the whole of the stator I^2R loss.

Stray loss is caused by alternating stray fluxes that appear, when a machine is loaded, in the pole faces, teeth, core and overhang. They increase eddy-current loss, produce rotational hysteresis and develop eddy loss in the end-plates. They may be augmented by variations in the quality of the core laminations, by segmental core construction and by magnetic stator stiffener plates and rotor end-rings. Non-magnetic materials for the latter may reduce the total core loss by 15 per cent. Further reduction can be achieved by avoiding flux harmonics with a chorded winding, and by stepping out the stator bore in the end packets of core plates. Another important factor is the spacing of the overhang conductors from the end-plates.*

* Richardson, "Stray Losses in Synchronous Electrical Machinery," *J.I.E.E.*, 92 (II), p. 291 (1945).

4. Design. The design of large motors and, particularly, generators, involves a great number of highly specialized problems. It should be emphasized that the mechanical problems to be solved in a large machine are of paramount importance; in fact the electrical design is quite subordinate in this respect. All the aspects of the mechanical design, however, are outside the scope of this book, and only sufficient will be given to indicate the more salient features.

5. Example. 1 250 kVA. at power factor 0·8 lagging, 1 000-kW., 3-phase, 50-c/s., 3 300-V., 300-r.p.m., engine-driven synchronous generator.

SPECIFICATION. Efficiency on full load to be not less than 93·0 per cent.; regulation not to exceed 20 per cent on throwing off rated load; temperature-rise in accordance with B.S. 2613: 1955 for Class B insulation. The machine must operate in parallel with others.

The machine is to be direct-coupled to an engine by means of a flange coupling forged in one piece with the generator shaft. One bearing is to be provided, and an extension of the shaft is to be provided for carrying the exciter armature. A short-circuit ratio of about 1·2 is required.

MAIN DIMENSIONS. Choosing provisionally $\bar{B} = 0.58$ Wb./m.^{2} and $ac = 33\,000$ amp.-cond. per m., and putting $L \simeq 1.1\,Y$, then from eq. (174)

$$G = 11 \cdot 0.955 \cdot 0.58 \cdot 33\,000 \cdot 10^{-3} = 201.$$

For a speed of 300 r.p.m., $2p = 20$ so that $L = 1.1\pi D/20 = 0.173D$. Then $D^2L = 0.173D^3 = S/Gn = 1\,250/201 \cdot 5 = 1.24$ m.^{3}, and $D^3 = 7.2$, $D \simeq 1.9$ m. $= \underline{190\text{ cm.}}$

The core length is $L = 1.24/1.9^2 = 0.335$ m. $= \underline{\underline{33.5\text{ cm.}}}$

Four ventilating ducts each 1 cm. wide will split the stator core into sections about six centimetres wide. The slot length is $L_s = 33.5 - 4 = 29.5$, and the net iron length is $L_i = 0.9 \cdot 29.5 = 26.5$ cm.

The pole-pitch is $Y = \pi \cdot 190/20 = \underline{\underline{29.8\text{ cm.}}}$

The peripheral speed is $u = \pi Dn = \pi \cdot 1.9 \cdot 5 = 29.8$ m. per sec. This is a satisfactory figure, suitable for a simple bolt-on pole construction.

STATOR WINDING. A winding in many slots reduces the inductance and facilitates the conduction of heat from the copper, but requires a greater total quantity of insulation; whereas the concentration of the winding into relatively few slots increases both inductance and the difficulty of dissipating the heat, and is liable to cause more magnetic disturbance in the gap flux.

As a guide, about 900 to 1 500 A. can be accommodated per slot. A choice of $g' = 3$ slots per pole per phase, with about 1 100 ampere-conductors per slot, is suitable.

Winding. The current per phase on full load is

$$I_{ph} = 1\,250 \cdot 10^3/\sqrt{3} \cdot 3\,300 = 219 \text{ A.}$$

The connection is naturally star, which is almost invariably used for generators. With $g' = 3$, the number of stator slots is

$$S_1 = 3 \cdot 3 \cdot 20 = 180.$$

The flux will be

$$\Phi_1 = \bar{B}YL = 0.58 \cdot 0.298 \cdot 0.335 = 0.058 \text{ Wb.}$$

The phase e.m.f. on open-circuit at 100% excitation is $3\,300/\sqrt{3}$ = 1 910 V. Hence the number of turns per phase is

$$T_{ph} = 1\,910/4.44 \cdot 0.955 \cdot 50 \cdot 0.058 = 156.$$

A more convenient number is 150, giving $z_s = 5$ conductors per slot, so that

$$T_{ph} = pg'z_s = 10 \cdot 3 \cdot 5 = 150 \text{ turns.}$$

For this winding a concentric arrangement is suitable, using semi-closed slots, of pitch

$$y_s = \pi \cdot 190/180 = 3.31 \text{ cm.}$$

Slot Dimensions. Semi-closed slots are usual with concentric windings. To find the slot-width, the necessary tooth-width is found, and what remains of the slot pitch is then the slot width.

A no-load density of 1·7 to 1·8 Wb./m.² will be found to be high enough for 50-c/s. machines, otherwise the tooth losses become excessive.

A sketch of the pole arc on the lines indicated in Fig. 115 gives a total flux estimated at 57·6 mWb. and a mean density of 0·875 Wb./m.² over the pole-face.

The pole-arc is $0.66 \cdot 29.8 \simeq 19.5$ cm., and the number of teeth per pole-arc $19.5/3.31 \simeq 6$: these carry almost the whole of the flux. If the tooth density is not to exceed 1·8 Wb./m.², the tooth width must not be less than 2 cm.

This permits a slot-width not exceeding about $3.3 - 2 \simeq 1.3$ cm.

Slot Insulation. The main insulation is of built-up mica.

The conductors themselves are insulated with linen tape, or with mica tape having paper or fabric backing, adding about 0·7 mm. to the depth and thickness of the conductor. Strips of insulation are inserted between conductors, 0·3 mm. micanite being suitable for 3 300 V. After the conductors are laid together they are bound by strong linen tape to facilitate the wrapping process. This binding adds about 0·25 mm. all round the bunch of conductors.

Here the width and depth in the slot taken up other than in the copper might be made up as follows (see Fig. 350)—

Width	mm.	*Depth*	mm
Conductor insulation	0·7	Tooth lip	1·5
Binding tape	0·5	Wedge	4·0
Cell, 2 × 1·75	3·5	Cell, 2 × 1·75	3·5
Clearance and slack	1·2	Binding tape	0·5
		Conductor insulation,	
		5 × 0·7	3·5
		Micanite separators,	
		5 × 0·3	1·5
		Leatheroid strip	0·75
	5·9 mm.		15·25 mm.

Conductors. The assumed current density is 4 A. per mm.² This requires a conductor area of 219/4 = 55 mm.² Taking a slot-width

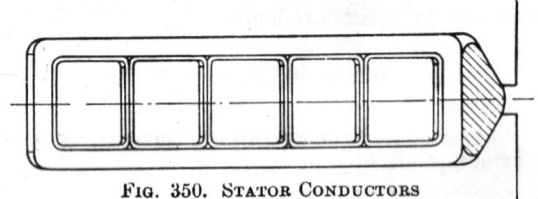

Fig. 350. Stator Conductors

of, say, 1·35 cm., and insulation as above of 5·9 mm. total thickness, the conductor can be 13·5 − 5·9 = 7·8, say 7·5 mm. × 7·5 mm. with rounded corners, giving an area of 54 mm.² The five conductors require a total depth of 37·5 mm., which, with a total insulation, wedge and lip thickness of 15·25 mm. as above, permits a slot depth of 37·5 + 15·25 = 53 mm., say. The slot mouth is made 3 mm. wide.

Overhang. With three slots per pole per phase the overhang can be arranged in two planes. With p even there are no cranked coils.

Resistance. From an estimate, the mean length per conductor is 54 cm. in the overhang and 33·5 cm. in the core, or 87·5 cm. The resistance per phase at 75° C. is $r = 0.021 \cdot 0.875 \cdot 300/54 = 0.102\,\Omega$. Calculating the eddy-current loss-factor, from eq. (89) $(\alpha h)^4 = 0.097$. The average loss-factor from eq. (88) is $K_{dav} = 1 + 0.097 \cdot (25/9) = 1.27$, and the loss in the top layer is $K_{d5} = 1 + 0.097(20/3) = 1.66$. These figures show that the eddy loss will be reasonably small. A more important estimate is the stray load loss. Here 20 per cent will be added to the I^2R loss + eddy-current loss—

I^2R loss: $3.219^2 \cdot 0.102 \cdot 10^{-3}$.	14·7 kW.
Eddy loss: $0.27 \cdot 14.7$	3·0 kW.
Stray loss: $0.2\,(14.7 + 3)$	3·5 kW.
I^2R + stray load loss	21·2 kW.

The IR drop is $219 \times 0.102 \cdot 1.27 = 28.4$ V. The effective resistance is $28.4 \cdot 100/1\,900 = 1.5$ per cent or 0·015 per unit.

Leakage Reactance. From eqs. (60) and (69) the slot leakage Φ_s is found. Using the dimensions in Fig. 350, with $J = 219$ A., $T_c = 5$, $L_s = 0.295$ m.; and

$$\lambda_s = \frac{41.7}{3 \cdot 13.5} + \frac{4}{13.5} + \frac{2 \cdot 2.5}{16.5} + \frac{1.5}{3} = 2.13,$$

Therefore $\Phi_s = 2\sqrt{2} \cdot 4\pi 10^{-7} \cdot 5 \cdot 0.295 \cdot 2.13 = 2.45$ mWb.

The overhang leakage flux, from eqs. (63) and (70), is given by

$$L_o\lambda_o = 1 \cdot 0.298^2/\pi \cdot 0.0331 = 0.856:$$

$$\Phi_o = 2\sqrt{2} \cdot 4\pi 10^{-7} \cdot 219 \cdot 5 \cdot 0.856 = 3.9 \text{ mWb}.$$

The total leakage flux is $\Phi_s + \Phi_o = 6.35$ mWb. Expressing this in terms of the main flux,

$$6.35/57.6 = 0.11 \text{ per unit.}$$

STATOR CORE. For a core density of about 1.1 Wb./m.^{2}, the depth h below the slots must accommodate half the flux: whence $h \simeq 10$ cm. The stator bore is $D = 190$ cm., the depth of two slots is 10.6 cm., so that the outside diameter of the stator core may be

$$D_o = 190 + 10.6 + 20 \simeq 220 \text{ cm.} = 2.2 \text{ m.}$$

AIR GAP. The ratio of the field m.m.f. on open circuit for normal voltage to the armature m.m.f. at full load (i.e., the short-circuit ratio) is 1.2. From eq. (83), the armature m.m.f. is

$$r'_a = 1.35 \cdot 219 \cdot 150 \cdot 0.955/10 = 4\,240 \text{ AT. per pole.}$$

With a no-load tooth density of 1.8 Wb./m.^{2}, about 700 AT. are required by the teeth. The pole core density is estimated at 1.5 Wb./ m.^{2}, requiring 2 300 AT. per m. Its length is estimated at 18 cm., so that 400 AT. will be required for the pole. The stator core and rotor yoke together will take about 100 AT. If the rotor no-load excitation is $1.2 \cdot 4\,240 = 5\,100$ AT. per pole, the amount for the gap is $5\,100 - 1\,200 = 3\,900$ AT. The mean gap density over the pole-arc is 0.875 Wb./m.^{2} approximately, so that the gap length can be

$$l_g = 3\,900/800\,000 \cdot 0.875 = 0.0055 \text{ m.} = 0.55 \text{ cm.}$$

This figure would be quite suitable, as there will be no damping winding and the out-of-balance magnetic pull will not be unduly great.

The gap magnetic coefficients can now be worked out. Using the methods of eqs. (49) and (50),

$$y_s' = 3.31 - 0.12 \cdot 0.3 = 3.27 \text{ cm}$$

There are approximately 5.9 slot pitches in the pole-arc, so that the effective pole arc is

$$b' = 5.9 \cdot 3.27 \simeq 19.5 \text{ cm.,}$$

allowing for fringing: i.e. the small slot-openings make practically no difference. The effective core length is

$$L' = 33.5 - 0.25 \cdot 4 \cdot 1 = 32.5 \text{ cm.}$$

The effective gap area is therefore

$$A_g = b'L' = 19{\cdot}5 \,.\, 32{\cdot}5 = 635 \text{ cm.}^2 = 0{\cdot}0635 \text{ m.}^2$$

The conservative estimate of 620 cm.^{2} for the gap area will compensate the assumption that all the flux passes across a 0·55 cm. gap. Actually the ends are chamfered.

For no-load, the flux of 57·6 mWb. gives a gap density over the pole face of

$$57{\cdot}6 \,.\, 10^{-3}/0{\cdot}062 = 0{\cdot}933 \text{ Wb./m.}^2$$

requiring 800 000 . 0·933 = 745 000 AT. per m., or 4 100 AT. for the gap length of 0·55 cm.

ROTOR. The full-load e.m.f. will be 0·078 p.u. greater than the no-load values, requiring about 4 500 AT. for the gap. The pole will require about 600 AT., the core and yoke about 200 AT., and the teeth 1 200 AT., making in all about 6 500 AT. To this must be added the armature reaction by a graphical construction such as that of Fig. 292B, the armature reaction being 3 360 AT. from eq. (84). The total pole-excitation from this estimate comes to about 9 000 AT., and the rotor must be capable of giving this continuously.

Each pole is built of 1·5 mm. punchings, rivetted between 8 mm. flame-cut end plates, and secured by two bolts screwed into pin bars. No plate-insulation is needed. (Fig. 351.)

The shank width is fixed from the flux and the desired flux-density, say 1·6 to 1·7 Wb./m.^{2} on no load.

The axial length of the pole is taken the same as the gross core length and

$$\text{shank width} = \frac{57{\cdot}6 \,.\, 1{\cdot}094 \,.\, 1{\cdot}15}{10^3 \,.\, 0{\cdot}9 \,.\, 0{\cdot}335 \,.\, 1{\cdot}7} = 0{\cdot}14 \text{ m.}$$

where 1·15 is the estimated leakage co-efficient.

Field Winding. The choice of exciter voltage is open, and can be made with a view to a strip-on-edge field winding. A voltage of 110 V. (i.e. 80 V. with 30 V. reserve) is chosen, and a strip width of 3 cm. is tried.

With narrow pole shanks the strap is easier to bend on edge if the coil ends are made micircul sear, giving a mean length of turn of about 1·25 m. = L_{mt}. At 75° C., with 80 V. on the rotor winding, i.e. 4 V. per pole, the resistance is

$$r = V/I = 4/I = 0{\cdot}021 \,.\, 1{\cdot}25 T/a$$

whence $$a = 0{\cdot}021 \,.\, 1{\cdot}25 \,.\, IT/4 \text{ mm.}^2$$

IT is estimated above as about 9 000 AT., so that

$$a = 0{\cdot}021 \,.\, 1{\cdot}25 \,.\, 9 \ 000/4 = 60 \text{ mm.}^2$$

Take a section 2 mm. × 30 mm., bare strip wound on edge. The winding must have sufficient surface to dissipate the developed heat. From Chapter XI, § 4, a cooling coefficient

$$c = 0.075/(1 + 0.1u) = 0.075/(1 + 2.98) = 0.0188$$

may be taken. Estimating the surface of the coil as L_{mt} times the coil depth of, say, 15 cm., the surface for dissipation is 0·18 m.² per

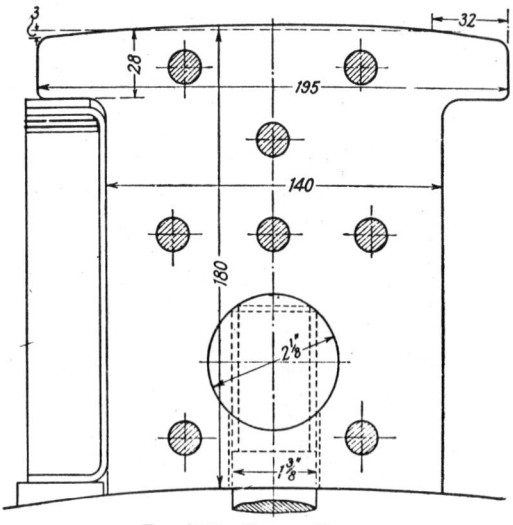

FIG. 351. ROTOR POLE

pole, so that with a temperature-rise of 65° C. a total power dissipation of

$$p = \theta_m S/c = 65 \cdot 0.18/0.0188 = 620 \text{ W.}$$

can be allowed per coil. The voltage is 4 V. so that the current is 620/4 = 155 A. The number of turns can be 9 000/155 = 60, say.

With 60 turns each 2 mm. thick, with interleaved mica 0·3 mm. thick, the total coil depth is 13·8 cm., and 1·2 cm. for packing gives a total depth of 15 cm. The current density with 9 000 AT. is 155/60 = 2·58 A. per mm.²

Poles. The pole dimensions are shown in Fig. 351. The depth of the polar horn is enough to carry the flux emanating from it without undue saturation. A steel washer covers the top of the coil and distributes the forces from the end retaining hooks, which are fixed to the body of the rotor, serving to support the semicircular ends of the coils against centrifugal forces. The width of the polar horn is 19·5 cm., or 0·656 of the pole-pitch.

Per-unit normal voltage			1·0				1·1				1·2			
E.m.f. E, volts Main flux Φ_m, milliwebers			1 910 57·6				2 100 63·5				2 285 69·2			
Part	Area m.²	Length m.	Φ	B	at	AT	Φ	B	at	AT	Φ	B	at	AT
Core . . .	2 × 0·0257	0·17	57·6	1·125	240	40	63·5	1·238	330	60	69·2	1·350	500	90
Teeth (⅓) .	0·0323	0·053	57·6	1·785	8 000	425	63·5	1·965	19 000	1 000	69·2	2·140	45 000	2 400
Gap . .	0·0620	0·0055	57·6	0·933	745 000	4 100	63·5	1·025	815 000	4 470	69·2	1·120	890 000	4 900
AT_1 . .						4 565				5 530				7 390
Pole (min.) .	{ 0·0422	0·18	58·9	1·385	1 200	} 280	65·0	1·520	4 000	} 905	71·3	1·650	8 300	} 2 100
Pole (max.) .			61·7	1·455	2 300		68·4	1·615	7 000		75·8	1·790	17 000	
Yoke . .	2 × 0·030	0·11	61·7	1·020	220	25	68·4	1·130	310	35	75·8	1·255	520	60
Total AT. per pole, F_s .						4 870				6 460				9 550
Field current $I_f = F_d/60$.						81				107·5				159

Rotor Bolts. The copper weight per pole is about 85 lb., and the pole itself weighs about 140 lb. The total weight is 225 lb. and the centrifugal force of 40 000 lb. can be taken by two $1\frac{3}{8}$ in. diameter bolts, in a pin bar of $2\frac{1}{8}$ in. diameter.

OPEN-CIRCUIT CHARACTERISTIC. The flux per pole on no load, $E = 1\ 910$ V., is 57·6 mWb. For any other value of E it will be $0·03E$ mWb.

The pole-leakage, from eqs. (57) and (58), and Fig. 118, is found with the following dimensions: $h_9 = 15$ cm.; $h_4 = 2·8$ cm.; $L_4 = L_9$ $= 33·5$ cm.; $c_s = 9·7$ cm.; $c_p = 12·7$ cm.; $b = 19·5$ cm.: $b_p = 14$ cm.

Leakage between shoes, $\Phi_{sl} = 18AT_l/10^8$.

Leakage between poles, $\Phi_{nl} = 61AT_l/10^8$.

The useful flux is Φ_m; at the top of the pole $\Phi_m + 18 . 10^{-8}AT_l$; at the bottom of the pole $\Phi_m + 79 . 10^{-8}AT_l$; in the yoke $\Phi_m + 79 . 10^{-8}AT_l$. The ampere-turns producing leakage are the sum of those expended on gap, teeth and stator core.

The magnetization curve can now be evaluated. The o.c.c. is drawn in Fig. 352. The armature reaction $F_a = 3\ 360$ AT., and from Fig. 288, $K_r = 0·33$. Using the construction shown in Fig. 352, the full-load field current is 148 A., the corresponding e.m.f. $E_t = 3\ 920$ V. (line) $= 1·19$ p.u. The regulation is therefore $\varepsilon = 0·19$ p.u.

SHORT-CIRCUIT CHARACTERISTIC. This is shown together with the o.c.c. in Fig. 352. For a reactance drop of 0·11 per unit, the necessary field excitation on short circuit with full-load current is 9 A. The armature reaction is 3 360 AT., equivalent to a field current of 56 A. Full-load current on short circuit will be obtained by a field excitation of $56 + 9 = 65$ A.

EFFICIENCY. The I^2R loss in the stator is 14·7 kW. The eddy and stray losses together are 6·5 kW. The weight of the teeth is 390 kg. The density is 1·96 Wb./m.² on full load, and the specific loss (Fig. 112 (b)) for 0·4 mm. plates is 25 W. per kg. The tooth loss is 9·75 kW. The weight of the core is 1 260 kg., and its loss at 11 W. per kg. is 13·9 kW.

The field resistance is 0·525 Ω, say 0·55 Ω to allow for connections. The current is 148 A. and the loss is 12 kW. The brush loss is 300 W. with a drop of 1 V. at each brush. Taking an exciter efficiency of 88 per cent the exciter loss is 1·7 kW. and the total excitation power is $12 + 0·3 + 1·7 = 14$ kW. The friction and windage loss is estimated from experience to be about 1 per cent, say 10 kW.

COOLING. The core loss and part of the buried stator I^2R loss are dissipated from the stator core surface. The buried I^2R loss is $14·7(29·5/87·5) + 3·0 = 8$ kW. The total loss in the stator

		kW.
Stator I^2R		14·7
Eddy + stray loss		6·5
Core loss: teeth		9·7
core		13·9
Rotor I^2R		12·3
Exciter loss		1·7
Windage and friction		10·0
	Total losses	68·8 kW.
	Output	1 000·0 kW.

Efficiency $= 100[1 - (68\cdot8/1\,068\cdot8)] = 93\cdot6\%$

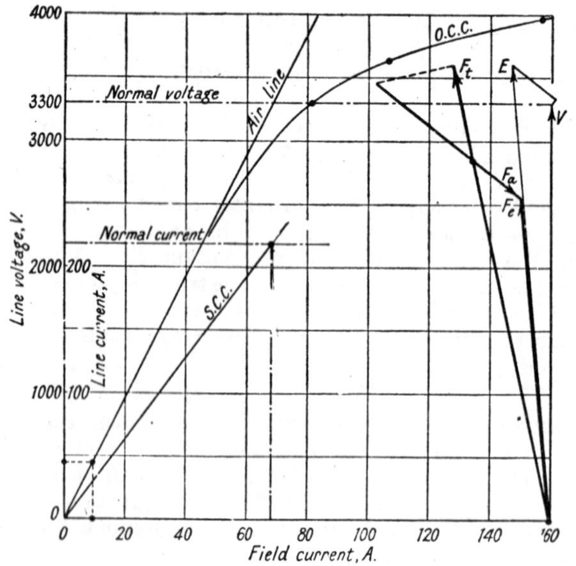

Fig. 352. Open- and Short-circuit Characteristics, and
Construction for Full-load Excitation

body is $8 + 9\cdot7 + 13\cdot9 = 31\cdot6$ kW. The external surface (outside cylindrical area and two end plates) is 3·36 m.² which with 50° C. rise and $c = 0\cdot027$ dissipates 6·2 kW., leaving 25·4 kW. for the gap and duct surfaces. The gap surface is 1·76 m.²; with

$$c = 0\cdot033/(1 + 0\cdot1u) = 0\cdot0079,$$

then $1\cdot76/0\cdot0079 = 220$ W. loss per °C. rise. For the 8 duct surfaces the area is 10·5 m.²; with $c = 0\cdot1/u = 0\cdot0033$, then $10\cdot5/0\cdot0033 = 320$ W. loss per °C. Taking the surface temperature as uniform, then

$$\theta_m = 26\,600/(220 + 320) = 50° \text{C}.$$

for the temperature-rise, which is within the limits.

Design Schedule.

Rating

1 250 kVA., power factor 0·8 lagging, 1 000 kW., three-phase, star-connected, 50-c/s., 3 300 V., 300 r.p.m. Line current 219 A.

Main Dimensions

Number of poles, $2p$	20
Stator bore, D	190 cm.
Stator periphery, πD	596 cm.
Pole-pitch, Y	29·8 cm.
Air-gap, l_g	0·55 cm.
Core-length, L	33·5 cm.
Vent ducts	4 × 1 cm.
Slot length, L_s	29·5 cm.
Net core-length, L_i	26·5 cm.
Plate thickness	0·4 mm.
Peripheral speed of rotor, u	29·8 m. per sec.
Output coefficient, G	20·6

Stator

Number of slots, S	180
Slot-pitch, y_s	3·31 cm.
Size of slot	13·5 mm. × 53 mm.
Type of slot	Semi-closed
Type of winding	Concentric
Slots per pole per phase, g'	3
Turns per coil, T_c	1
Conductors per slot, z_s	5
Turns per phase, T_{ph}	150
Conductor size	7·5 mm. × 7·5 mm.
Current per conductor, J	219 A.
Current density, δ	4·05 A. per mm.²
Length of conductor	0·875 m.
Resistance per phase at 75° C., r	0·102 Ω.
Per-unit resistance	0·015
Per-unit reactance	0·137
Core density, B_c	1·23 Wb./m.²
Tooth density, $B_{t\frac{1}{3}}$	1·95 Wb./m.²

Rotor

Pole size	14 cm. × 33·5 cm.
Shoe size	19·5 cm. × 33·5 cm.
Type of winding	Strip-on-edge
Length of mean turn, L_{mt}	1·2 m.
Conductor size	2 mm. × 30 mm.
Number of layers	1
Turns per layer	60
Turns per pole	60
Full-load exciting current	148 A.
Current density, δ	2·47 A. per mm.²
Resistance at 75° C.	0·55 Ω.
Excitation voltage	100 V.
Yoke size	300 cm.²

Specific Loadings

Specific electric loading, ac	33 100 a.-c. per m.
Specific magnetic loading, $\bar{B}$	0·576 Wb./m.²

6. Example. 25 000 kVA. at power factor 0·8 lagging, 20 000-kW., 3-phase, 50-c/s., 6 600-V., 4-pole, 1 500 r.p.m. turbo-alternator.

This design is given as an example of the method. Modern turbo-alternators are almost invariably two-pole machines.

MAIN DIMENSIONS. Phase voltage, star connection, 3 810 V. Phase current $I_{ph} = 25\ 000 \cdot 10^3/(\sqrt{3}) \cdot 6\ 600 = 2\ 185$ A.

Since the temperature guarantees are stringent, values of $\bar{B}$ and ac are chosen conservatively. Taking $ac = 43\ 000$ amp.-cond. per m., $\bar{B} = 0.525$ Wb./m.^{2}, and $u \divideontimes 105$ m. per sec., then

$$G = 235; \qquad D^2L = 4.26 \text{ m.}^2$$

With a gap length of 2·5 cm., a suitable stator bore would be $D = 1.4$ m., and rotor diameter 1·35 m. Whence

$$L = 2.2 \text{ m.}; \quad Y = 1.1 \text{ m.}; \quad u = 105 \text{ m./sec.}$$

STATOR. The main flux is

$$\Phi_m = 0.525 \cdot 1.1 \cdot 2.2 = 1.27 \text{ Wb.}$$
$$T_{ph} = 3\ 810/4.24 \cdot 50 \cdot 1.27 = 14.15, \text{ say } \underline{14.}$$

There will be one conductor per slot, and 84 stator slots, suitable for a 12-section stator stamping. The slot-pitch is $y_s = \pi \cdot 140/84 = 5.25$ cm: The specific electric loading is actually $ac = 41\ 700$, and the flux is about 1·3 Wb.

Magnetic Circuit. With axial and radial ventilation the stator laminations need only be subdivided into packets of about 10 cm. With 22 ducts each 1 cm. wide, spaced more closely in the middle, the core length is made up from 23 packets totalling 198 cm., and 22 1-cm. ducts, giving $L = 2.2$ m.

Slot length $L_s = 198$ cm.; net iron length $L_i = 0.85 \cdot 198 = 168$ cm. A factor of 0·85 is used on account of the difficulty of tightening the core.

With a core density of 1·05 Wb./m.^{2}, the depth of iron behind the slots must be

$$\tfrac{1}{2} \cdot 1.3/1.05 \cdot 1.68 = 0.37 \text{ m.}$$

About 20 per cent is lost by axial vent ducts, so that the depth must be raised to 0·46 m. The stator core will then have an outside diameter of $1.40 + 0.92 = 2.32$ m.

Ventilation Holes. Assuming an efficiency of 96 per cent, the losses at rated load amount to 800 kW. Take a temperature rise of 30° C. In the stator, with 70 per cent of the total air passing through it at a speed of 25 m. per sec. in the ducts. The air required, from eq. (92), is

$$0.78 \frac{800}{30} \cdot \frac{303}{273} = 23 \text{ m.}^3 \text{ per sec.}$$

The total area of axial holes is therefore $\tfrac{1}{2} \cdot 23 \cdot 0.7/25 = 0.32$ m.^{2}, the air entering the core at both ends.

The stator stamping is shown in Fig. 353.

Slots. To obtain a high leakage reactance, spaces are left above the conductors. The slot is semi-closed to decrease harmonic effects. With 27·5 mm. for the reactance space and wedge, the slot-pitch at that diameter is 5·45 cm. If the tooth density here, the narrowest part, is not to exceed 2 Wb./m.², the minimum tooth width must

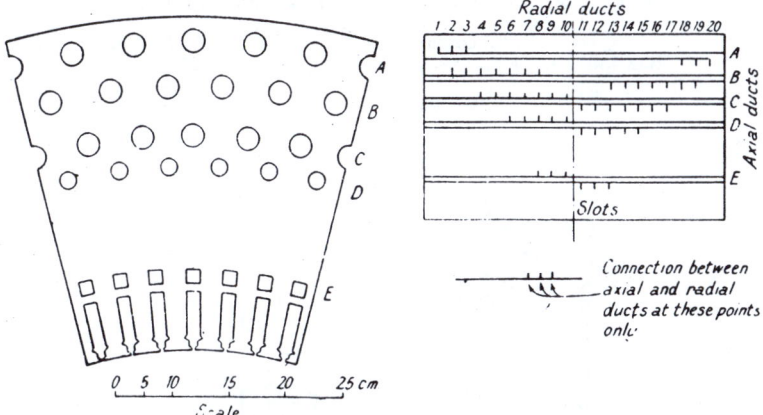

FIG. 353. STATOR CORE PLATE AND DUCT ARRANGEMENT

be about 2·9 cm., so that the maximum slot-width is 2·5 cm. For the main slot insulation, a 3·5 mm. thickness of mica tube (1 200 V. per mm.) will suffice. With these figures, the slot-width is made up as follows—

Slot liner: 2 × 0·25	0·5 mm.
Mica tape: 2 × 0·125	0·25 mm.
Mica tube: 2 × 3·5	7·0 mm.
Clearance	2·25 mm.
Copper width of bar	15·0 mm.
	25·0 mm.

Thus copper bars 15 mm. wide can be used. Taking the current-density as 2·75 A. per mm.², the required copper depth is 2 185/15 . 2·75 = say 55 mm. The eddy-current loss must be checked. The value of $\alpha = \sqrt{(15/25)} = 0.775$. If the loss-ratio in the top layer is not to exceed about 2·5, it is found necessary to have the conductor split into 7 layers of 7·5 mm. each: then the loss in the top layer is $K_{d7} = 2.58$, and the average loss ratio in the slot is $K_{d\ av} = 1.62$.

Each lamination is taped with 0·125 mm. mica tape, and a strip of 0·75 mm. mica is included on one side of the lamination. The total area of copper is 7 . 15 . 7·5 = 787 mm.², say 760 mm.² to allow for rounding.

The slot-depth is made up as follows—

Slot liner	0·25 mm.
Mica tape: 14 × 0·125	1·75 mm.
Mica separators: 7 × 0·75	5·25 mm.
Mica tube: 2 × 3·5	7·0 mm.
Top and bottom packing strips: 2 × 1·0 .	2·0 mm.
Copper: 7 × 7·5	52·50 mm.
Slack	3·75 mm.
Total conductor space . .	72·5 mm
Reactance space and wedge . . .	27·5 mm.
Slot depth . . .	100·00 mm.

The slot is shown in Fig. 354.

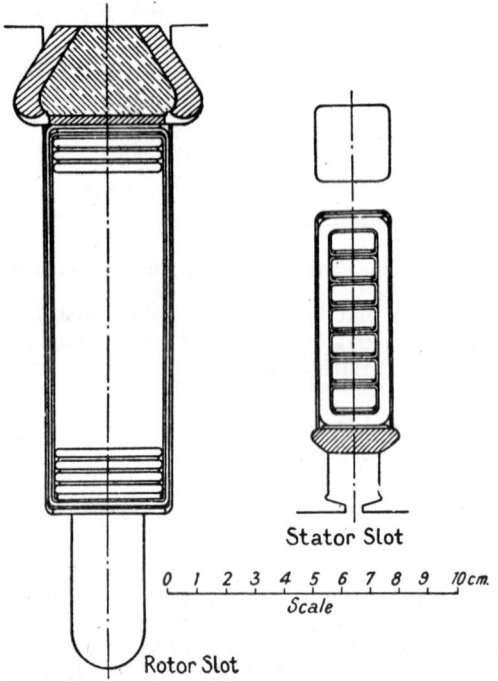

Stator Slot

0 1 2 3 4 5 6 7 8 9 10 cm.
Scale

Rotor Slot

FIG. 354. ROTOR AND STATOR SLOTS

Overhang. With a current density of 4 A. per mm.², the area per lamination is 2 185/4 . 7 = 78 mm.², say, 4 × 20 mm.², giving a total area of 560 mm.² allowing for rounding at the corners.

Resistance. Mean length of slot conductor = 275 cm., there being a projection of 27·5 cm. at each end of the core. The length of a conductor in the overhang is 185 cm. The resistance hot is therefore

$$r = 0·021 . 7 . 4 . [(2·75/760) + (1·85/560)] = 0·00404 \, \Omega.$$

The ohmic I^2R loss is $3 . 2\,185^2 . 0{\cdot}00404 = 58$ kW., which becomes 94 kW. with the eddy loss factor 1·62.

Reactance. Eqs. (60), (63), (69) and (70) give for the slot and overhang leakages

$$\Phi_s = 0{\cdot}048 \quad \text{and} \quad \Phi_o = 0{\cdot}112 \text{ Wb.}$$

The total leakage flux is 0·16 Wb. and the reactance is therefore $0{\cdot}16/1{\cdot}3 = 0{\cdot}123$ p.u.

AIR GAP. The full-load armature reaction is

$$F_a = 1{\cdot}35 . 2\,185 . 14 . 0{\cdot}955/2 = 19\,700 \text{ AT. per pole.}$$

The full rated load will require an excitation about twice the armature reaction, say 40 000 AT.

ROTOR. With a current density of 2·5 A. per mm.^{2}, the copper area per pole will be $2 . 40\,000/2{\cdot}5 = 32\,000$ mm.^{2} This will require an area of about 53 000 mm.^{2} in the slots, with a copper space factor of 0·6. The slotting of the rotor should be of a pitch having no common factor with the stator. The possibilities are—

No. of slot-pitches .	43	47	53	55	59
Wound slots, 70% .	30	33	37	38	41
Coils per pole .	3·75	4·1	4·6	4·7	5·1

The choice lies between 47 and 59 slots: the wider will be taken as giving a mechanically better overhang. Thus 32 slots will be used, having a pitch of 1/47 of the circumference.

A check on mechanical stress at root of tooth gives 8 500 lb. per in.^{2} which is quite reasonable.

Ventilation. This is by channels at the bottom of 24 slots, giving ducts of 12·5 cm.^{2} area. The total gap + duct area at each end is about 4 750 cm.^{2}, to take the required 23·2 m.^{3} of air per sec. at a speed of 25 m. per sec.

Winding. With a standard exciter voltage of 220 V., and with 20 per cent reserve, the rotor voltage will be about 180 V. The length of mean turn is estimated at 6·5 m., and the total copper area per pole has been found to be about 32 000 mm.^{2} There are 8 wound slots per pole. With T_c turns per coil, the conductor area is $32\,000/8T_c$. To obtain 40 000 AT., the number of turns per coil will be

$$T_c = 32\,000 . 180/4 . 0{\cdot}021 . 0{\cdot}65 . 40\,000 . 8 = 32.$$

The nearest convenient number is 36. The conductor area is $32\,000/8 . 36 \simeq 110$ mm.^{2}, or say 38 mm. × 3 mm.

Rotor Slot. The insulation must withstand great mechanical stresses, and the effects of the different coefficients of linear expansion.

To prevent the possibility of the dust affecting the insulation the whole of the slot copper and insulation is enclosed in a 0·6 mm. steel cell. At the bottom of the slot is a 2 mm. spring steel plate to support the winding over the vent ducts. Within the steel cell is first a 2 mm. flexible mica insulation cell and then a 0·5 mm. hard mica cell. The outer flexible mica is to allow the coils to be forced into position and to give freedom of expansion. The turns are separated from each other by a 0·3 mm. pressed mica separator. The whole winding in the slots is then heated under pressure and the wedges placed in position See Fig. 354.

Slot Width

Copper conductor	. . . . . .	.	38·0 mm.
Hard mica: 2 × 0·5	. . . . .	.	1·0 mm.
Flexible mica: 2 × 1	. . . . .	.	2·0 mm.
Steel cell: 2 × 0·6 .	. . . . .	.	1·2 mm.
Slack . .	. . . . .	.	1·8 mm.
	Slot width . . .	.	**44·0 mm.**

Slot Depth

Copper conductors: 36 × 3	. . . .	.	108·0 mm.
Wedge	. . . .	.	30·0 mm.
Copper strip under wedge	. . . .	.	3·0 mm
Hard mica: 3 × 0·5	. . . .	.	1·5 mm.
Flexible mica: 3 × 1·0	. . . .	.	3·0 mm.
Separators: 35 × 0·3	. . . .	.	10·5 mm.
Mica bottom strip .	. . . .	.	1·5 mm.
Spring steel plate .	. . . .	.	2·0 mm.
Steel cell . .	. . . .	.	1·2 mm.
Slack . .	. . . .	.	1·8 mm.
	Slot depth . . .	.	162·5 mm.
	Ventilation duct . .	.	52·5 mm.
	Total . . .	.	**215·0 mm.**

Rotor Overhang. The rotor overhang is as close as possible to the core. The steel cell in the slots does not continue the whole length, but is replaced by a mica cell where the copper leaves the slot. The copper is bent down towards the shaft and then shaped to the overhang.

Alternate turns of each coil are hand taped with 0·35 mm. mica tape from where the copper leaves the mica cells, this operation being carried out before the coils are placed in position, at each end. When the rotor is completely wound the coils are connected together in series ; 1 mm. mica sheets are then taped to each coil of the overhang with mica tape, these sheets being cut to fit the shapes of the overhang. The coils are then taped with half-lap asbestos tape. Between the coils are placed bakelized-asbestos packing pieces. The inner coil is spaced from the polar horn by a bakelized-asbestos moulding which allows free passage of air to the ventilation holes. The slip-rings are mounted at opposite ends of the core, so that in order to bring out the ends of the field winding at the correct positions one coil has one-half turn less. The total length of copper in the rotor winding is estimated as 3 610 m.: its section is 109 mm.[2] and its resistance hot is 0·695 Ω. The copper weight is 3 400 kg.

Retaining Rings. It is necessary to calculate the size of the end rings before the open-circuit characteristic can be determined, because of the large leakage path that this bell constitutes.* The external diameter must be such that it may pass through the bore of the stator.

The limiting feature of the rings is the mechanical stresses set up by the centrifugal force of their own weight and that of the overhang. The ring is shown in Fig. 348. At one end, the thickness is reduced and the bell centred by the core. The other end of

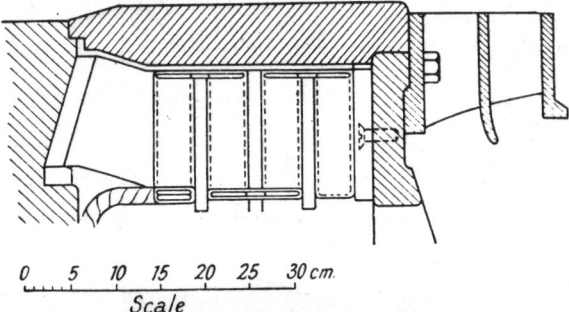

0   5   10   15   20   25   30 cm.
Scale

FIG. 355. END RETAINING RING AND ROTOR OVERHANG

the bell is centred by a bracket mounted on the shaft and kept in position by a lock ring. This bracket also carries the fan. The bracket must be sufficiently rigid to prevent the end bell moving, as the bent coils tend to force the end bell away from the core; it must also permit air to pass into the ventilation ducts on the rotor.

MAGNETIZATION. The saturation curve for each part of the magnetic circuit is first determined on the assumption of no leakage. The curves are then corrected and combined to give a saturation curve for the whole circuit per pole, with the exception of the rotor core. The ampere-turns for the rotor core must be calculated from a knowledge of the field form. These ampere-turns are not effective in producing flux through the rest of the magnetic circuit, so that as an approximation they may be added to the total ampere-turns at the end of the o.c.c. calculations.

Stator and Rotor Teeth. The excitation for suitable densities is found graphically, with allowance for slot flux in each case.

Polar Horn. This is treated as a large tooth.

Stator Core. The core flux is approximately $\frac{1}{2}\overline{B}YL \simeq 0.7\,B_{gm}$, where B_{gm} is the maximum gap density. The core area is 0.623 m.² An approximation to the average path length is two-thirds of the pole pitch at the mean diameter, i.e. 0.54 m.

* Almost all manufacturers now use non-magnetic steel, for which the end-ring leakage is much reduced.

Air Gap. The various gap coefficients are—

Stator slots: opening/gap = $0.6/2.5$; $k_{o1} = 0.05$;

  Slot-pitch = 5.25 cm.; $k_{s1} = 5.25/(5.25 - 0.05 . 0.6) = 1.005$.

Stator ducts: opening/gap = $1.0/2.5$; $k_1 = 0.07$,

  duct-pitch (mean) = 9.6 cm.; $k_{d1} = 9.6/(9.6 - 0.07 . 1.0) = 1.006$.

Rotor slots: opening/gap = $2.0/2.5$; $k_{o2} = 0.15$,

  slot-pitch = 9.04 cm.; $k_{s2} = 9.04/(9.04 - 2.0 . 0.15) = 1.032$.

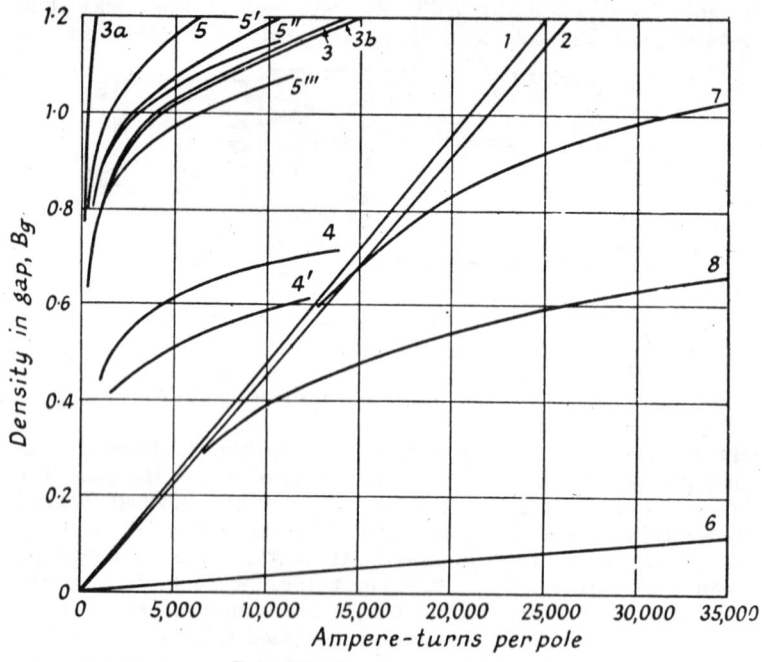

FIG. 356. SATURATION CURVES

Rotor ducts: opening/gap = $2.0/2.5$; $k_2 = 0.14$,

  duct-pitch (mean) = 13.75 cm.; $k_{d2} = 13.75/(13.75 - 2.0 . 0.14)$
    $= 1.02$.

For the unslotted part of the gap,

$$k_g = k_{s1}(k_{d1} + k_{d2} - 1) = 1.031.$$

For the slotted part,

$$k_g = (k_{s1} + k_{s2} - 1)(k_{s1} + k_{s0} - 1) = 1.10$$

Air Lines. Gap ampere-turns = $800\,000 B_g l_g k_g$.

Taking $B_g = 0.8$ Wb./m.², then on the unslotted part of the rotor

$$AT = 800\,000 . 0.8 . 0.025 . 1.031 = 16\,500$$

and on the slotted part

$$AT = 800\,000 . 0.8 . 0.025 . 1.10 = 17\,500.$$

End Ring Leakage. The end rings are assumed to be saturated at

$2\cdot1$ Wb./m.^{2}, and their leakage is estimated to add about 1 per cent to B_g. The saturation curves are corrected by reducing the ordinates by this amount.

Rotor Slot Leakage. This is assumed to run from one polar horn, across 8 slots, to the next polar horn. The permeance coefficient per slot is $\lambda_{s2} = 2\cdot72$. The m.m.f. is that for the polar horn, gap and stator, i.e. $350 + 17\,000 + 800 + 150 = 18\,300$. Across each slot this is $AT_s = 18\,300/4 = 4\,575$. The total slot leakage per pole is $\Phi_{s2} = 2AT_s \cdot \mu_0\lambda_{s2}L_i = 2 \cdot 4\,575 \cdot 4\pi \cdot 10^{-7} \cdot 2\cdot72 \cdot 1\cdot9 = 0\cdot06$ Wb. This is equivalent to an increase of gap density over the polar horn (area $0\cdot975$ m.^{2}) of $0\cdot06$ Wb./m.^{2}

Saturation Curves. These are drawn (Fig. 356) as follows—
(1) Air line, unwound part of rotor;
(2) Air line, wound part of rotor;
(3a) Stator core;
(3b) Stator teeth;
(3) Stator core and teeth;
(4) Wound part of rotor;
(4') (4) Corrected for end-ring leakage;
(5) Unwound part of rotor;
(5') (5) Corrected for end-ring leakage;
(6) Slot leakage;
(5'') (5') Partly corrected for slot leakage;
(5''') Unwound part of rotor corrected for end-ring and slot leakage;
(7) Magnetization curve for unwound part $= (1) + (3) + (5''')$;
(8) Magnetization curve for wound part $= (2) + (4')$.

FIELD FORM AND FLUX PER POLE. The distribution of B_g in the gap is now found for various values of rotor excitation. The ampere turns on the rotor core do not affect the distribution, and will be corrected for later. With any given value of rotor ampere-turns, the flux-density in the gap can be found, and the flux curve drawn. The flux per pole will then be proportional to the area of the flux curve. The calculation is made as follows—

Pole AT.	Tooth 3	Tooth 2	Tooth 1	Polar Horn	Φ	$\dfrac{\bar{B}}{B_{max}}$	B_1
	B_g	B_g	B_g	B_g			
Air Line $1\cdot0$	$1\cdot1$	$2\cdot25$	$3\cdot4$	$4\cdot7$	$0\cdot725$	$0\cdot651$	$0\cdot508$
$1\cdot4$	$0\cdot155$	$0\cdot300$	$0\cdot405$	$0\cdot640$	$0\cdot961$	$0\cdot634$	$0\cdot676$
$1\cdot7$	$0\cdot190$	$0\cdot350$	$0\cdot451$	$0\cdot750$	$1\cdot120$	$0\cdot630$	$0\cdot786$
$2\cdot0$	$0\cdot225$	$0\cdot392$	$0\cdot490$	$0\cdot838$	$1\cdot250$	$0\cdot629$	$0\cdot880$
$2\cdot4$	$0\cdot260$	$0\cdot440$	$0\cdot530$	$0\cdot910$	$1\cdot370$	$0\cdot634$	$0\cdot965$
$2\cdot8$	$0\cdot300$	$0\cdot475$	$0\cdot562$	$0\cdot966$	$1\cdot468$	$0\cdot639$	$1\cdot010$
$3\cdot2$	$0\cdot330$	$0\cdot505$	$0\cdot590$	$1\cdot010$	$1\cdot545$	$0\cdot643$	$1\cdot060$

The last column is the fundamental term in each case.

Field Form Curves. These are shown in Fig. 357. The ordinates are of gap density B_g, the maximum value drawn for the polar horn being 1.01 Wb./m.[2]

Rotor Core Excitation. From the Table, page 571, the various fluxes through the rotor body can be summed. The gap, slot and end-ring leakage flux must be taken into account. The saturation curve is:

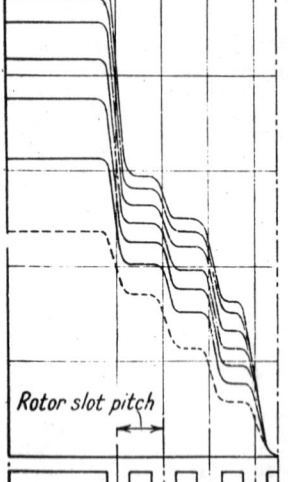

FIG. 357. FIELD-FORM CURVES

Gap density B_{gm}	Rotor core AT
0·5	60
0·75	200
0·9	500
1·0	1 250
1·05	1 750 .

OPEN-CIRCUIT CHARACTERISTIC. For a given maximum gap density over the polar horn, the exciting ampere-turns are obtained from the saturation curves, Fig. 356, with the addition of those for the rotor core. The corresponding e.m.f. is obtained from the fundamental component B_1:

$$E_1 = \sqrt{3} . 4.44 . K_{w1} f T_{ph} \Phi_1$$
$$= 5\,150 \Phi_1$$
$$= 5\,150 . B_1 . 0.636\pi . 1.375 . 2.2/4$$
$$= 77\,600 B_1 \text{ volts}$$

The o.c.c. is then (Fig. 358)—

AT. on Gap, Poles, and Stator	AT. on Rotor Core	Total AT.	B_{gm}	Φ	B_1	E_1	Field Current = $AT/144$
Air Line 10 000	—	10 000	0·470	0·725	0·508	3 940	69·5
14 000	140	14 140	0·640	0·961	0·676	5 250	98·2
17 000	220	17 220	0·750	1·120	0·786	6 090	119·8
20 000	340	20 340	0·838	1·250	0·880	6 820	141·2
24 000	620	24 620	0·910	1·370	0·965	7 470	171·2
28 000	1 000	29 000	0·966	1·468	1·010	7 820	202·0
32 000	1 430	33 430	1·010	1·545	1·060	8 240	232·0

Rotor slot pitch

SHORT-CIRCUIT CHARACTERISTIC. From eq. (82) the armature ampere-turns are $F_a = 1\cdot29 \ . \ 7 \ . \ 1 \ . \ 2\ 185 = 19\ 700$ AT., cor responding to $19\ 700/144 = 137$ A. field current.

The reactance is $0\cdot123$ per unit, or 810 V. (line). The field current to give this voltage on open circuit is $2\ 000/144 = 13\cdot9$ A.

The short-circuit field current for full-load current on short circuit is therefore $137 + 14 = 151$ A., equivalent to 2 180 AT.

REGULATION. Using the method (a) of Fig. 292B, the full-load

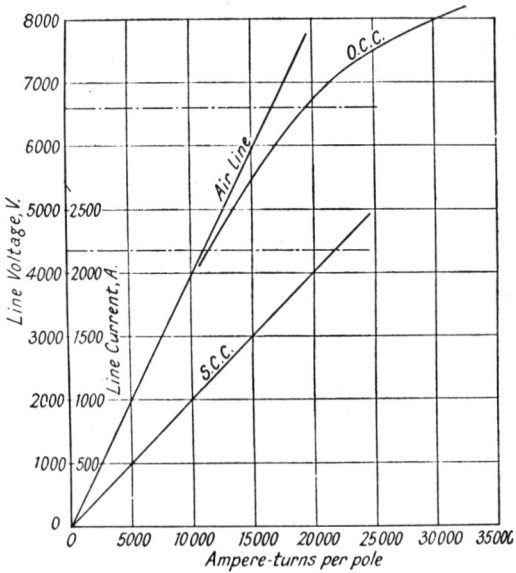

FIG. 358. OPEN- AND SHORT-CIRCUIT CHARACTERISTICS

excitation is 39 000 AT., which gives an e.m.f. of 8 550 V. on open circuit. The regulation is

$$\varepsilon = (8\ 550 - 6\ 600)/6\ 600 = 0\cdot295 \text{ p.u.}$$

LOSSES. Calculated from the core weight and density, the no-load stator *core loss* is 190 kW. Assuming the density to be proportional to the e.m.f., the full-load loss is 235 kW. The stator *teeth* (taken to the bottom of the first row of vent holes) have a loss on no load of 105 kW. For load conditions the problem is not so simple, on account of the distortion of the gap flux due to armature reaction. The cross armature reaction, superimposed on the (gap + stator) field ampere-turns, increases the density on one side of the polar horn and reduces it rather more on the other side. An estimate shows the maximum gap density to be raised to about 1 Wb./m.^{2} on one side of the polar horn. Calculating the corresponding tooth density, the tooth loss from appropriate loss curves is found to be 140 kW., The total no-load iron loss is 295 kW., and on full load 375 kW.

The *stator* I^2R loss is 58 kW. The eddy loss is 36 kW. The flux distortion in the gap causes some flux to pass down the slots. For this a further 6 kW. is allowed, making a total of 100 kW.

The *end-plate* losses are considerable. They are derived from an expression having terms in the size of the overhang, frequency and ampere-turns, and the mean distance of the overhang from the end-plate. The expression gives a total of 25 kW. loss.

The *total stator losses* are—

No load, 295 kW.; Full load, $375 + 100 + 25 = 500$ kW.

The rotor I^2R on no load, field current $= 135$ A., is $135^2 . 0.695 . 10^{-3}$ $= 12.7$ kW: on full load, field current $= 270$ A., the I^2R loss is $270^2 . 0.695 . 10^{-3} = 50$ kW.

The *friction* and *windage* loss is taken as 240 kW., about 1 per cent of the output.

EFFICIENCY. Taking core-density as proportional to e.m.f., and core loss to (density)^2; I^2R and end-plate losses to (load)^2; field current increase proportional to load current; and friction and windage constant—

EFFICIENCY AT POWER FACTOR 0·8

Per-unit Load	0	0·25	0·5	0·75	1·0
Friction and windage . .	240	240	240	240	240
Core loss . .	190	193	201	215	235
Tooth loss . .	105	107	114	125	140
End plate loss .	0	2	6	13	25
Stator I^2R loss .	0	6	25	56	100
Rotor I^2R loss .	13	15	22	34	50
Total losses .	548	563	608	683	790
Output, kW. .	0	5 000	10 000	15 000	20 000
Input, kW. .	548	5 563	10 608	15 683	20 790
Efficiency, % .	0	89·99	94·27	95·64	96·20

COOLING. Fans are mounted at each end of the rotor to draw air into the machine from a duct beneath, force it through the stator ventilating spaces and discharge it from the top of the stator frame. The cooling air for the rotor is set in circulation by the fan effect of the rotor ducts. The power necessary to drive the fan with a pressure of, say, 20 cm. (8 in.) of water and 23·2 m.^{3} per sec. at an efficiency of 0·4 is

$$P_f = 0.1 . 20 . 23.2/0.4 = 117 \text{ kW}.$$

which is part of the windage loss.

Ventilation of the Stator. A large proportion of the air is sent direct to the centre of the stator, as this is usually a hot part. The air distribution (see duct diagram in Fig. 353) is obtained by the use of special spacers, half and whole rings either giving or denying access to the radial ducts.

The full-load loss is 790 kW. Subtract 20 kW. for bearing friction (which is removed by the oil). Assume a temperature-rise of 30° C. The air required is $770 . 10^3/30 . 1\,150 = 22.3$ m.³ per sec. For ease in calculation the inlet temperature is assumed to be 0, so that the numerical temperature at any point is the rise. Before the air enters the air-gap or ducts, the losses to be dissipated are—

End-connectors and overhang . . .	50 kW.
End-plates	25 kW.
Rotor I^2R	50 kW.
Windage and fan, say	235 kW.
	360 kW.

The specific temperature-rise is $30/770 = 0.039$ ° C. per kW., hence the rise of the air before entering the ducts is $360 . 0.039 = 14°$ C.

The cooling from the *gap surface* is based on $c = 0.033/(1 + 0.1u)$ and the stator gap surface (gross) is $\pi DL = 9.7$ m.² The area of the stator vent holes is 3 115 cm.²; of the rotor 300 cm.², and of the gap and slot clearance is 1 340 cm.²; a total of 0.475 m.² The proportion of the air in the gap is $(1\,340 + 300)/4\,755 = 0.345$, and the air volume is therefore $0.345 . 22.3 = 7.6$ m.³ per sec. Assume the loss dissipated by the gap surface to be 75 kW. The rise in temperature of the air will be $22.3 . 0.039 . 75/7.6 = 8.6$ ° C. The maximum gap air temperature will be $14 + 8.6 = 22.6°$ C., and the mean $14 + 4.3 = 18.3°$ C. The machine has been designed for a rise of 40° C., so that the mean temperature-difference between air and gap surface is $40 - 18.3 = 21.7°$ C.

The velocity of the rotor is 106 m. per sec., so that, from eq. (101),
$$p = S\theta_m/c = 9.7 . 21.7 . (1 + 10.6)/0.033 = 74 \text{ kW.}$$
This is near enough to the original estimate to be accepted.

The air entering the *axial ducts* from the stator ends has a temperature of 14° C. above inlet. The volume is $3\,115 . 22.3/4\,755 = 14.6$ m.³ per sec. Assume the loss dissipated in the ducts to be 165 kW. The temperature-rise is therefore 9.8° C. and the mean temperature of the duct air is $14 + 4.9 = 18.9°$ C. The mean difference between duct surface and air is $40 - 18.9 = 21.1°$ C. The total area of the air passages at each end of the machine is 4 755 cm.² The mean velocity in the ducts is $22.3/2 . 0.4755 = 23.5$ m. per sec. $= u_d$. Take $c = 0.25/u_d$, and estimating the cooling surface of all the axial ducts as 83.3 m.², then the heat dissipated from the axial ducts is
$$p = 83.3 . 21.1 . 23.5/0.25 = 165 \text{ kW.,}$$
agreeing with the first estimate.

Considering now the *radial ducts* with $c = 0.1/u_d$. Two currents of air enter the radial ducts, i.e. from the axial ducts and from the gap. From the gap, 7·6 m.[3] per sec. enter at 22·6° C; from the axial ducts, 14·6 m.[3] per sec. at temperature 23·8° C. The mean air temperature is consequently 23·3° C. The temperature-rise of the air leaving the stator is 30° C., the mean temperature in the radial ducts is $\frac{1}{2}(23.3 + 30) = 26.7°$ C., and the mean difference between iron and air is $40 - 26.7 = 13.3°$ C. Estimating the duct surface as 100 m.[2] and the velocity in the ducts as $u_d = 0.1u = 10.6$ m. per sec., the heat dissipated is $p = 100 \cdot 13.3 \cdot 10.6/0.1 = 140$ kW.

The heat dissipated from the *back of the core* is found from the temperature-difference, $40 - 30 = 10°$ C.: the mean air velocity, about 6 m. per sec.; and with a cooling coefficient of $c = 0.075/u_c$, the heat dissipated is $p = \pi \cdot 2.52 \cdot 2.2 \cdot 10.6/0.075 = 14$ kW.

From the *end-plates* of the stator, of area 1·2 m.², taking c as, say, $0.025/u$ where u is about 12 m. per sec., and a rise of $40-14 = 26°$ C., then $p = 26 \cdot 1.2 \cdot 12/0.025 = 15$ kW.

Total Heat Dissipation. The air volume is 22.3 m.³ per sec., the temperature-rise of the air is 30° and of the iron is 40° C. The calculated dissipation is—

					kW.
From end rings, rotor, etc.	.	.	.	.	360
„ air-gap	.	.	.	.	74
„ axial ducts	.	.	.	.	165
„ radial ducts	.	.	.	.	140
„ back of core	.	.	.	.	14
„ end-plates	.	.	.	.	15
	Total	.	.	.	768 kW.

Hot Spots. In the *rotor winding* a calculation based on the dissipation of the I^2R loss through the insulation to the iron of the teeth gives an estimated difference between copper and iron of 25° C. The gap air has a temperature of 18·6° C. and the iron is estimated to have a rise of 35° C., making the hottest part of the rotor slot winding $35 + 25 = 60°$ C. above inlet temperature. In the overhang the ventilation is rather restricted. An estimate based on conduction from the overhang to the slots gives 100° C. rise for the hot spot in the overhang. A similar figure is obtained for the stator.

Mechanical Design. The weight of the complete stator is 66 tons and of the rotor 34 tons. The mechanical design takes into account—

    (a) the stress in the core bolts;
    (b) the bearings;
    (c) the end-ring stress;
    (d) the stress in the teeth and wedges;
    (e) the critical speed;
    (f) short-circuit stresses between overhang conductors.
The details of these calculations will not be given, although it

must be emphasized that the mechanical design is of paramount importance. There is no unusual principle involved, however.

The static shaft deflection is 0·089 in. From the empirical expression

$$\text{critical speed} = 200/\sqrt{\text{deflection}}$$
$$= 200/\sqrt{0\cdot089} = 2\,250 \text{ r.p.m.}$$

This is 44 per cent above running speed.

7. Examples. Outline design figures are given below for three machines—

A. A standard *synchronous generator* for 1 250 kVA., 1 000 kW, at power factor 0·8 lagging, 6 600 V., 50-c/s., 3-phase, 20 poles.

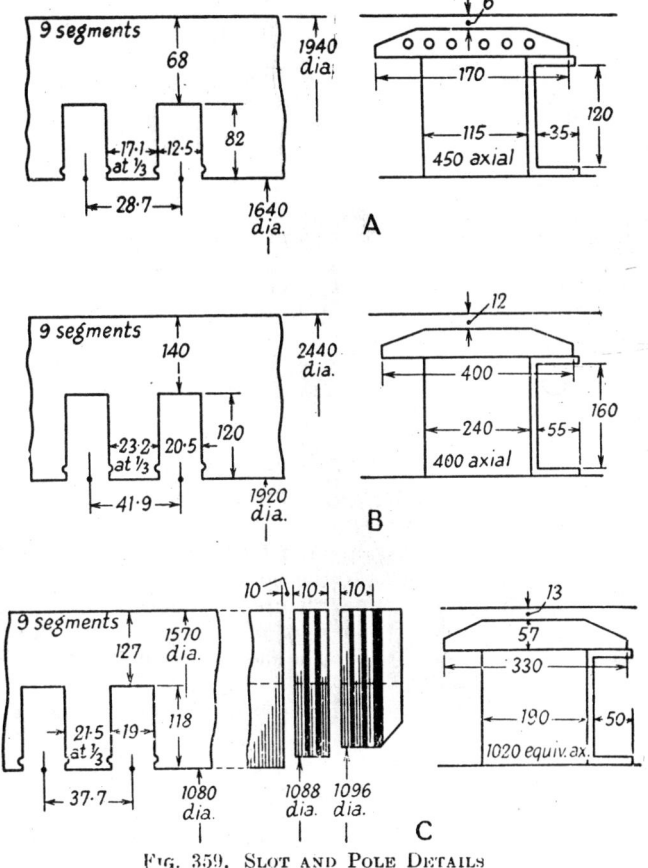

FIG. 359. SLOT AND POLE DETAILS
Dimensions in millimetres

300 r.p.m. Temperature rise, 40° C. The machine has 10-bladed fans, one on each side of the rotor.

B. A waterwheel *synchronous generator,* 3 750 kVA., 3 000 kW. at power factor 0·8 lagging, 10 000 V., 50-c/s., 3-phase, 10 poles, 600 r.p.m. The main dimensions of this machine are determined by the requirement of an 80 000 lb.-ft.² flywheel effect. The stator winding is set out in the example on page 220. Temperature rise, 65° C.

C. A *synchronous capacitor* for 2 500 kVAr. lagging/5 000 kVAr. leading, 11 000 V., 50-c/s., 3-phase, 6 poles, 1 000 r.p.m. Temperature rise, 85° C. Full-load loss, 125 kW. To reduce stray loss the stator core ends are stepped and the tooth stiffeners insulated from the core plates.

The chief dimensions are given in Fig. 359.

Machine			*A* Generator	*B* Generator	*C* Capacitor
Rating					
Full-load kVA. . . .	S		1 250	3 750	2 500/5 000
Full-load power, kW. .	P		1 000	3 000	0
Line voltage, V. . .	V_l		6 600	10 000	11 000
Phase voltage, V. .	V_{ph}		3 810	5 780	6 350
Power factor . .			0·8	0·8	0·0
Frequency, c.p.s. .	f		50	50	50
Speed, r.p.m. . .			300	600	1 000
No. of poles . .	$2p$		20	10	6
Main Dimensions					
Spec. mag. loading . .	$\bar{B}$		0·51	0·48	0·463
Spec. elec. loading . .	ac		38 200	51 500	55 000
Stator bore, cm. . .	D		164	192	108
Gross core length, cm. .	L		45	40	100
Radial ducts, no. . .	n_d		6	5	14
width, cm. . .	w_d		1	1	1
Slot length, cm. .	L_s		39	35	86
Iron length, cm. .	L_i		35·1	31·5	77·5
Pole-pitch, cm. . .	Y		25·8	60·5	56·6
Current per phase, A. .	I		109	217	262
Stator					
Winding . . .			Double-layer, diamond coils		
No. of circuits in parallel .	a		1	1	1
No. of slots . . .	S		180	144	90
Slots per pole per phase .	g'		3	4½	5
Conductors per slot .	z_s		2 × 5	2 × 5	2 × 4
Coil-pitch . . .			1–10	1–13	1–13
Chord ratio . .			1·0	12/14⅔	12/15
Conductor size, mm. .			5 × 5	2(10·5 × 3·8)	2(8·5 × 4·5)
area, mm.² .	a		25	80	76·5
Current density. A. per mm.²	δ		4·36	2·72	3·43
Mean conductor length, m.	L_{mc}		1·11	1·38	2·07
Resistance per ph. (hot), Ω	r		0·56	0·174	0·136
I^2R loss, kW. . .	P_c		20	24·5	28
Dissipation, W. per cm.²			0·091	0·0725	0·091
Weight of copper, lb. .	G_c		1 050	3 350	2 350
Rotor					
Type of pole			salient, laminated.	salient, laminated with cast-steel end-plates, bolted	salient, solid, screwed
Damper winding . . .			8 bars/pole, ⅞ in. dia.	none	none
Pole-arc, cm. . . .	b		17	40	33
Pole section, cm. . .			11·5 × 45	20 × 40	19 × 102
Pole length, cm. . .			13	20	22
Turns per pole . .			71½	99½	99½

Machine		A Generator	B Generator	C Capacitor
Rotor (contd.)				
Conductor size, mm.		35 × 1·4	55 × 1·3	50 × 1·75
area, mm.²	a	49	71·5	87·5
Full-load current, A.	I_f	135	207	270
Current density, A. per mm.'	δ	2·75	2·9	3·09
Mean turn, length, m.	L_{mt}	1·39	1·405	2·52
Resistance, hot, Ω	r	0·885	0·417	0·372
Volt drop, V.		120	86·5	100
I²R loss, kW.	P_c	16·1	18	27
Dissipation, W. per cm.²		0·185	0·298	0·36
Weight of copper, lb.	G_c	2 100	1 400	2 820
Peripheral speed, m. per sec.	u	25·5	60·5	11·15
Overspeed, %		0	80	0
Magnetization				
Flux per pole, Wb.	Φ_m	0·059	0·116	0·261
Core area, m.²		0·0238	0·0442	0·0098
density, Wb./m.²		1·24	1·31	1·33
Teeth area (⅓), total, m.²		1·08	1·055	1·50
pole-arc, m.²		0·036	0·070	0·150
density (⅓), Wb./m.²		1·61	1·67	1·74
Gap area, m.²		0·078	0·160	0·330
contraction coeff.	k_g	1·34	1·135	1·24
density, Wb./m.²		0·770	0·725	0·790
amp.-turns per pole		4 940	7 950	10 000
Pole area, m.²		0·005	0·0096	0·0194
density, Wb./m.²		1·20	1·23	1·40
Total AT. per pole, no load		7 500	8 950	10 350
full load	F_t	9 650	20 600	26 900
Armature AT. per pole	F_a	4 910	15 100	15 000
Efficiency				
Core loss, kW.		20	35	38
I²R loss, kW.		20	24·5	28
Stray (load) loss, kW.		7·4	12	5
Field loss, kW.		16·1	18	27
Fric. and windage loss, kW.		3·5	34·5*	23
Total loss, kW.		67	124	119
Output, kW.		1 000	3 000	0
Efficiency, %		93·8	96·0	0

* Includes friction and windage of Pelton wheel.

CHAPTER XXII

CONVERTING MACHINERY

1. Converting Plant. In spite of the increasing use of alternating current for all industrial purposes, there are nevertheless many cases in which the use of direct current is either necessary, or advantageous to a degree sufficient to make it preferable. For electric traction on suburban railways; for electrolysis, battery charging and chemical processes; and for conditions where extensive and economical speed control of motors is desired, direct current is employed. Since economy and the need for standardization have led to the use of alternating current for generation and transmission, some form of converting plant is necessary when bulk supplies of direct current are to be obtained from an a.c. network.

Since about 1920, conversion has become increasingly effected by rectification, and the present importance of new rotating converter plant is small. But since large numbers of converters are in use and will continue to be in service for a long time, they are still of interest, and in any case can still compete with rectifiers for low-voltage outputs.

Converting machines are of various design according to conditions, and fall into three classes—

1. Motor-generators.
2. Synchronous (or rotary) converters.
3. Motor-converters.

The *motor-generator* consists of an alternating-current motor (usually of the synchronous, but sometimes of the induction type) driving a direct-coupled d.c. generator. This is the simplest and most obvious converting unit. When a synchronous and a d.c. machine are combined, the resulting machine is termed a *synchronous converter*, a type that has been widely developed and used in Great Britain. The *motor-converter*, originally covered by the patents of J. L. Lacour, but now built if required by most manufacturers, is a combination of the induction motor-generator and the synchronous converter, the conversion of energy taking place partly mechanically through the shaft and partly electrically by the interconnection of the motor and converter windings.

A comparison between the three machines may most suitably be made here. The choice of a converting machine for a given case will be influenced primarily by economic considerations, and secondly by technical matters, although the two are, of course, mutually dependent to a great extent.

THE SYNCHRONOUS CONVERTER, in which the conversion occurs

580

in a purely electrical manner, has all the advantages and drawbacks of both the synchronous motor and the d.c. generator. It has the highest efficiency of the three and occupies the least space (not counting the transformer). Further, the cost, inclusive of the transformer, is the least. On account of its electrical arrangements, disturbances on the system to which one side of the converter is connected are transferred almost without diminution to the other side and its associated network. As a synchronous motor, it must be protected from hunting and falling out of step, and as a d.c. generator the liability to flash-over at the commutator must be taken into account, which is made more difficult since the number of poles and the brush spacing are controlled by the supply frequency and the peripheral speed.

THE MOTOR-GENERATOR, consisting of two separate machines with mechanical coupling only, is a robust unit, since each part can be designed to fit the conditions in the best manner without the compromise necessary in the case of the synchronous converter. The motor can now be constructed for large overload capacity, phase-advancing, self-starting, etc., while the synchronous-induction type can be applied if desired. The d.c. machine can be built purely as a generator, with the most economical number of poles, large overload capacity, and any desired regulation. For moderate a.c. supply voltages no transformer is necessary, and the a.c. machine can be connected directly to the supply. The efficiency is naturally lower than that of the synchronous converter, and more floor space is necessary. Where the required d.c. voltage regulation is wide, or regulation from zero voltage upwards is needed, a motor-generator is essential.

THE MOTOR-CONVERTER falls between the two foregoing types. The a.c. machine functions both as motor and transformer, while the d.c. machine is both generator and synchronous converter. The efficiency of the set is, consequently, better than that of the motor-generator, but not so high as that of the synchronous converter. The number of poles on the d.c. machine depends on the frequency, but presents more freedom than in the synchronous converter, for the frequency of the a.c. input to the converter machine is lower than that of the main supply. It is often one-half. A step-down transformer is in many cases unnecessary, since the stator and rotor of the a.c. motor acts as a voltage- as well as a frequency-changer. Stators can be wound directly for 10 000–15 000 V., except in small sizes.

A comparison is given in the Table on page 582.

2. **The Synchronous Motor-generator.** The most obvious method of obtaining d.c. from an a.c. supply is by connecting the two systems mechanically by means of a d.c. machine connected by direct coupling to a synchronous or induction motor. Consider the former combination. It has been seen that the behaviour of a

COMPARISON OF CONVERTING MACHINES

Machine	Advantages	Disadvantages
Induction Motor-generator	Self-starting Any frequency Operates direct on moderately high voltages Reliable and simple Wide regulation of d.c. voltage possible Free from reversal of polarity and flashover No limit to d.c. voltage	Low efficiency Low power factor unless compensated Not readily reversible (i.e. a.c. to d.c. only)
Synchronous Motor-generator	Any frequency Operates direct on moderately high voltages High power factor Wide d.c. voltage regulation Operation reversible Free from reversal of polarity and flashover No limit to d.c. voltage	Low efficiency Requires special starting gear Liable to fall out of step
Synchronous Converter	High efficiency High power factor Cheap	Requires step-down transformer D.C. voltage regulation limited Liable to flashover and reversal of polarity D.C. voltage 1 200–1 500 volts
Motor Converter	Self-starting Any frequency Operates direct on moderately high voltages Reliable and simple High power factor Wider regulation of d.c. voltage possible Relative freedom from reversal of polarity and flashover	Low speed, therefore expensive D.C. voltage 1 700–2 000 volts

synchronous motor connected to bus-bars of substantially constant voltage and frequency depends upon the load applied to its shaft and upon its excitation (Chapters XVII and XVIII). Thus on no load, which is the running condition when the induced e.m.f. of the d.c. machine is equal to the bus-bar voltage, the set takes no power (neglecting losses) on either side, but the synchronous machine will draw a purely reactive current from the line, lagging or leading according to the setting of its field regulator.

The *load* of the set depends entirely upon the setting of the field. regulator of the d.c. machine. If the e.m.f. of this machine is raised above the d.c. bus-bar voltage, it will generate, the magnitude of the generated current depending on the voltage difference. The d.c. machine operating as a generator throws a load on the synchronous machine which, by taking up a rotor displacement behind the neutral or no-load position (relative to its rotating magnetic field) is enabled to act as a motor and to draw an active component of current from the a.c. bus-bars. The active current component includes a part to cover all losses.

Just as under no-load conditions, an alteration of the excitation of the loaded synchronous machine alters only the power factor at which the machine works, and makes no difference (apart from minor secondary considerations) to the power component of the current. The flow of power in the circumstances described is from the alternating to the direct current side. To reverse the direction of power flow, the excitation of the d.c. machine is reduced until the induced e.m.f. is less than the bus-bar voltage, in which case the d.c. machine acts as a motor by urging the synchronous machine slightly ahead of its neutral position, and enables it to function now as a generator, the power factor being controlled by the excitation.

Assuming the excitation of each machine to be fixed, it will be seen that the set will run steadily at a constant load, provided that the two networks maintain rigidly constant voltage and frequency. This condition is not, however, exactly realizable. All systems possess a degree of regulation, or voltage and speed drop on load, and this will affect the operation of the motor-generator set. Assume, for example, that the d.c. network becomes heavily loaded ; its voltage will drop, resulting in a tendency for the d.c. machine to operate as a generator to deliver more power. If, on the other hand, the d.c. voltage rises or the a.c. frequency falls, the d.c. machine draws current as a motor, and drives the a.c. machine as a synchronous generator. The motor-generator set can thus in certain circumstances act as a reversible load-equalizer or interconnector set. In this connection it is clear that the set may become seriously overloaded if a fault occur on one of the systems. Should the d.c. network voltage become much reduced through accident, the d.c. machine will generate heavily, throwing a corresponding load on the synchronous machine. Protective gear is therefore necessary to provide for such emergencies.

It will be seen that the synchronous motor-generator is entirely reversible, while the power factor on the a.c. side can be adjusted at will. Furthermore, with a constant supply frequency, the speed of the set remains strictly constant.

OPERATION. As with the synchronous motor, the synchronous motor-generator is liable to *hunting* and requires a damping winding on its pole-faces. Auxiliary *starting* methods are necessary unless the d.c. side is available for this purpose, in which case the d.c. machine is used as a shunt motor. It is always necessary to *synchronize* the a.c. machine on to its supply. *Voltage regulation* is obtained in a precisely similar manner as for a normal d.c. generator, using an automatic regulator, a compounding winding, or hand control of the shunt rheostat.

3. Induction Motor-generator. Instead of a synchronous motor, an induction machine can be used for coupling to a d.c. machine. The excitation of the d.c. machine again decides the load, and on no-load the induction machine will operate with a small slip, taking a magnetizing current and a small loss component. If the excitation of the d.c. machine be increased, it will commence to generate, throwing a load back on to the induction motor, which will produce a driving torque. The torque of an induction motor running normally can only be increased if the slip increases, and this speed reduction will result also in a fall in the induced e.m.f. of the d.c. machine, so that it will reduce its generated load. Although the induction machine can be made to generate by lowering the excitation of the d.c. machine and thereby reversing the slip (i.e. running above the synchronous speed), this is seldom arranged for, because of the undesirably low leading power factor of the induction generator. While, therefore, the induction motor-generator presents a more flexible link between alternating and direct current networks than the synchronous motor-generator, and is less liable to excessive overloads through faults on the interconnected systems, it suffers from the disadvantages that power-factor compensation is not easily available, and reversibility is rarely feasible.

OPERATION. Since it requires no synchronizing, the *starting* of an induction motor-generator presents little difficulty. Where a slip-ring rotor is used, a set of starting rheostats connected across the slip-rings, together with suitable switchgear, is sufficient to enable the set to be run up to speed. Cage motors are usually started by applying reduced voltages to the stator by means of an auto-transformer. The d.c. voltage can be adjusted in the same way as with a normal direct current generator, although here a further complication is introduced in the change of speed with load. Thus, the induction motor-generator set is essentially one in which load fluctuations in the normal direction, alternating to direct current, are met by a change in voltage which has the effect of reducing the shocks on the supply system. The buffer action is still further enhanced by the inclusion of a certain amount of resistance in the rotor circuit of the induction motor, and by the use of a flywheel mounted on the common shaft. An increase of load—especially if sudden—is provided for by the released kinetic energy in the rotating mass as the set slows down.

CHAPTER XXIII

SYNCHRONOUS CONVERTERS

1. General Arrangement. The synchronous or rotary converter is used for changing polyphase alternating currents into direct currents, or in a few cases the reverse (inverted running). Since precisely similar field and armature designs can be used both for the synchronous a.c. and the d.c. machines in a motor-generator set, it is possible to combine the two units into a single machine, the armaturs of which is arranged with slip-rings at one end for the a.c. connectionc and a commutator at the other for connection to the d.c. side. Such a combination running in a normal salient-pole d.c. field system forms a synchronous converter. With the usual conditions a "motoring" current is fed to the a.c. slip-rings and a "generating" current taken from the commutator, so that in the same armature winding a considerable degree of neutralization occurs. The resultant torque is very nearly zero, a small amount only being needed to overcome no-load losses (in core, windage and friction) to keep the machine running at synchronous speed.

The converter is similar in construction to a d.c. generator, the chief differences being in the presence of slip-rings and damper windings, the absence of flange couplings for mechanical drive, and the provision in high-voltage machines (e.g. for traction) of flash barriers between the commutator brush-arms.

2. Voltage Ratios. Since the converter has a d.c. output, it must have a closed commutator winding, tapped at various points for the a.c. supply. From the a.c. viewpoint, therefore, the armature winding is an N-phase mesh-connected system. As developed in textbooks on d.c. machinery* the e.m.f. of a d.c. armature is

$$E_d = (p/a)nZ\Phi \text{ volts};$$

where
p = number of pole-pairs;
a = number of pairs of parallel armature paths;
n = armature speed in r.p.s.;
Z = total number of armature conductors;
Φ = flux per pole entering armature.

This, the e.m.f. between the commutator brushes, may be written—

$$E_d = 4pn(Z/4a)\Phi = 4fT_d\,\Phi \text{ volts} . \qquad . \qquad . \qquad . \qquad . \quad (178)$$

since $(Z/4a) = T_d$ is the number of turns in series between brushes, and $pn = f$ is the frequency of rotation, i.e. the frequency of the

* E.g. [A], p. 6.

induced e.m.f.'s in the individual coils. The maximum value of the e.m.f. per full-pitch turn for sinusoidal flux-distribution is

$$e_{tm} = 2\pi f \Phi \text{ volts.}$$

If a number T_{ph} of turns is tapped to form a *phase*, its e.m.f. will have the maximum value

$$e_{ph} = 2\pi f \cdot k_m T_{ph} \Phi \text{ volts,}$$

where k_m is the distribution factor, expressing the ratio of the vector sum of the individual coil e.m.f.'s to their arithmetic sum. For a uniformly distributed mesh winding tapped to N slip-rings to form

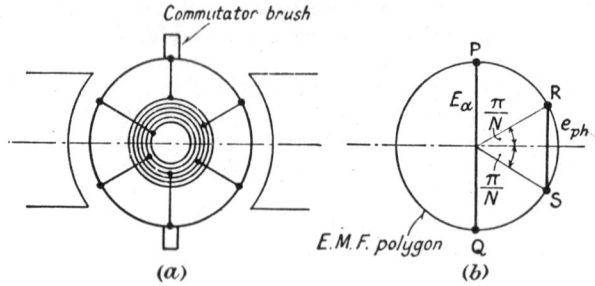

(a)   (b)

FIG. 360. ARMATURE OF THE SYNCHRONOUS CONVERTER

N phases, the number of turns per phase will be $T_{ph} = (2/N)T_d$, while $k_m = (\sin \frac{1}{2}\sigma)/\frac{1}{2}\sigma = (N/\pi) \sin (\pi/N)$ from eq. (72). Thus

$$e_{ph} = 2\pi f \cdot \frac{N}{\pi} \sin \left(\frac{\pi}{N} \right) \cdot \frac{2}{N} T_d \Phi$$

$$= 4fT_d \Phi \cdot \sin (\pi/N)$$

$$= E_d \cdot \sin (\pi/N) \text{ volts} \quad . \quad . \quad . \quad . \quad . \quad (179)$$

As has been seen (Chapter X, § 9), the value of k_m rises as the phase-spread $2\pi/N$ is made more narrow, and the winding from the view-point of output voltage is better utilized. Assuming sinusoidal variation of the a.c. voltage, the r.m.s. phase voltage from eq. (179) above is

$$E_a = e_{ph}/\sqrt{2} = (E_d/\sqrt{2}) \sin (\pi/N) \text{ volts} \quad . \quad . \quad . \quad (180)$$

A simple alternative method of finding the voltage ratio between the a.c. phase and d.c. brush e.m.f.'s is based on the fact that the coil-e.m.f. polygon in a closed symmetrical winding, uniformly distributed and rotating in a sinusoidal field, is a circle. Thus in Fig. 360 (a) the armature with its d.c. brushes in the geometrical neutral zone and its several phase tappings, carries a closed full-pitch winding of the commutator type. The e.m.f. of a coil at any instant depends on its relative position in the field. A coil spanning the vertical diameter (i.e. lying under the d.c. brushes) links maximum flux but has zero rate of flux-change: consequently it has zero e.m.f. On the other hand, a coil spanning the horizontal diameter has maximum e.m.f. The nearer the coils are to the pole-centres, the greater are their e.m.f.'s at the instant considered. Representing the e.m.f.'s by complexors (see Fig. 133A) their sum in series yields a vector polygon which approximates to a circle, Fig. 360 (b). The vertical diameter PQ is proportional to the sum of the coil e.m.f.'s in either half of the armature, and corresponds to the d.c. brush e.m.f., E_d. The latter is a constant voltage for, although the armature rotates, the d.c. brushes are fixed and connect between them always the same number of coils occupying the same position relative to the field system. When the phase of spread $2\pi/N$ lies symmetrically about the line of pole-centres, it has maximum e.m.f. e_{ph}, corresponding to the chord RS. The relationship between e_{ph} and E_a is clearly the ratio of chord to diameter, or sin (π/N). Whence

$$e_{ph} = E_d \sin (\pi/N)$$

and the r.m.s. value of the phase e.m.f. is

$$E_a = e_{ph}/\sqrt{2} = (E_d/\sqrt{2}) \sin (\pi/N). \qquad . \qquad . \qquad . \quad (181)$$

as before.

The value of the ratio E_a/E_d is given in the following Table—

SYNCHRONOUS CONVERTER: VOLTAGE RATIO

Number of rings, N . .	2	3	4	6	9	12
Phase spread, $2\pi/N$ . .	180°	120°	90°	60°	40°	30°
Distribution factor, k_m .	0·636	0·827	0·900	0·955	0·980	0·988
Ratio, $e_{ph}/E_d = $ sin (π/N) .	1·0	0·866	0·707	0·500	0·342	0·259
Ratio, $E_a/E_d = (1/\sqrt{2}) \sin (\pi/N)$	0·71	0·61	0·50	0·35	0·24	0·18

The most common number of rings is six. Although an increase in this number has technical advantages, both as regards a slight increase in output and a reduction of the heating, the amount is insufficient in most cases to offset the greater complication in construction of the converter and its transformer.

3. Current Ratios. The synchronous converter is a converting machine, developing no useful mechanical power. The electrical input will be equal to the output if losses be neglected. On this assumption, with that of unity power factor on the a.c. side, then

$$E_d I_d = N E_a I_a,$$

where I_d is the direct current output from the armature, and I_a is the phase current input in each of the N phases, under the normal a.c. to d.c. method of operation. Inserting the relation between E_a and E_d from eq. (180), the phase currents are

$$I_a = (\sqrt{2})I_d/[N \sin (\pi/N)] \quad . \quad . \quad . \quad . \quad . \quad (182)$$

The currents led to the slip-rings are those appropriate to the junction of successive phases. Considering, Fig. 361, any two successive symmetrical phases I and II, it is clear from Kirchhoff's first law that the line or slip-ring current I_l entering the junction of these two phases, must be the complex difference of the two phase

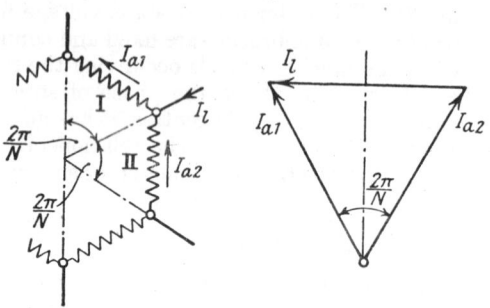

FIG. 361. RELATION OF LINE AND PHASE CURRENTS IN AN N-PHASE
MESH-CONNECTED WINDING

currents. Since the phase currents differ in time by $2\pi/N$, the relation between I_l and I_a is

$$I_l \doteq 2I_a \sin (\pi/N).$$

Putting in the value of I_a from eq. (182),

$$I_l = (2\sqrt{2})I_d \sin (\pi/N)/N \sin (\pi/N)$$
$$= (2\sqrt{2})I_d/N \quad . \quad . \quad . \quad . \quad . \quad (183)$$

The table below shows the current ratios for unity power factor and efficiency.

SYNCHRONOUS CONVERTER: CURRENT RATIOS

Number of Rings, N	2	3	4	6	9	12	∞
Ratio, E_a/E_d	0·71	0·61	0·50	0·35	0·24	0·18	0·0
Ratio, I_a/I_d	0·71	0·54	0·50	0·47	0·46	0·45	0·45
Ratio, I_l/I_d	1·41	0·94	0·71	0·47	0·31	0·24	0·0

In general the power factor will not be unity nor the losses negligible. The friction, windage and core losses will have to be supplied by an active current input to the a.c. side (under normal running conditions). The I^2R loss, on the other hand, entails an addition to the alternating voltage to overcome the internal IR drop. In most converters the latter is very small, and the voltage ratio can be considered unchanged. The currents can be assumed to be increased to take account of all the losses. Under this assumption the slip-ring current for an input power factor cos ϕ and efficiency η becomes

$$I_l = \frac{2\sqrt{2}}{N} \cdot \frac{I_d}{\eta \cos \phi} \qquad \qquad (184)$$

while the phase current is

$$I_a = (\sqrt{2})I_d/[\eta \cos \phi \ N \sin (\pi/N)]. \qquad (185)$$

4. Armature Heating. Since the power flow is usually from the a.c. to the d.c. side, the synchronous converter acts simultaneously as a synchronous motor and as a d.c. generator. The armature winding is common to both, and the two currents will generally be in opposing directions, and will neutralize to a considerable extent. The resultant I^2R loss may under favourable conditions be less than it would be if the machine were run *either* as a synchronous motor *or* as a d.c. generator alone. From the viewpoint of armature heating, therefore, the output of a synchronous converter of given size is usually somewhat greater than that of either of its corresponding constituent machines.

The diagrams in Fig. 362 illustrate the variations of current-distribution in the armature of a three-ring synchronous converter rotating under conditions of unity power factor and efficiency. At (a) is the two-pole armature of a d.c. generator producing an output current of $I_d = 100$ A. The two-circuit armature then carries 50 A. in each circuit. The r.m.s. phase current in a synchronous motor corresponding to this output is $I_a = (\sqrt{2}) \cdot 100/3 \sin (\pi/3)$ = 54 A., from eq. (182). The phase current fluctuates sinusoidally between the limits $\pm$ 54 . $\sqrt{2}$ A., and for the phases in the positions shown for the synchronous motor in Fig. 362 (b) will have the values 66·7 A., 66·7 A. and 0. Similarly, the corresponding slip-ring currents are 133·4, 66·7 and 66·7 A., respectively, and it will be seen that Kirchhoff's first law must be satisfied at the tapping points. If the d.c. generator and the synchronous motor armatures are superimposed in the converter, the current distribution is that given at (c).

Since the d.c. brushes form sliding contacts while the slip-rings are fixed contacts with the armature winding, and since the alternating currents pulsate whereas the direct currents are steady, the current-distribution in the armature will vary owing to the spatial alteration in the position of the tapping points with respect to the field system, coupled with the temporal alteration of the phase currents as they change sinusoidally. Fig. 362 (d) indicates a sequence of three successive instants with equal time intervals, showing progressive changes in the current-distribution.

The effective (heat-producing) current in a conductor is the difference between the a.c. and d.c. components in it, and varies in form both with time and with the position of the conductor with respect to the phase tapping point. Inspection of Fig. 362 (d) will show

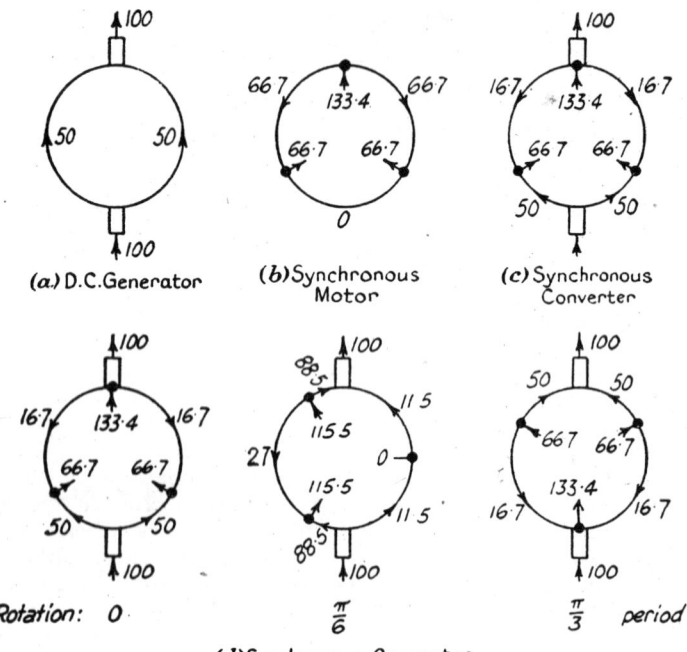

(a) D.C. Generator   (b) Synchronous Motor   (c) Synchronous Converter

(d) Synchronous Converter

Rotation: 0 $\frac{\pi}{6}$ $\frac{\pi}{3}$ period

FIG. 362. ARMATURE CURRENT DISTRIBUTION: 3-RING CONVERTER. UNITY POWER-FACTOR AND EFFICIENCY

that, whereas the middle conductor of a phase carries (in the case considered) a current not exceeding, and generally less than, 50 A., yet the conductor at a tapping point carries a maximum of nearly 116·7 A., which value will be further increased if the power factor is less than unity. Thus although the total or overall rate of heat development in the converter armature may be less than that of a corresponding d.c. generator, the distribution of heat-production is not uniform. The conductors at or near the tapping points carry the maximum currents and become the "hot-spots" in the armature.

The most commonly used converter has six slip-rings. Fig. 363 shows the distribution of the current in the armature of a two-pole six-ring converter with a commutator output current of 100 A., again assuming unity power factor and efficiency. The phase currents have a r.m.s. value of 47 A. and a maximum value of √2 . 47 = 66·7 A. In Fig. 363A it will be seen that the differences of the d.c. and a.c. armature components are small, so that the heating conditions will be more favourable than for the three-ring machine.

The wave-forms of the resultant current in two typical conductor positions are shown in Fig. 363B. For a coil midway between two tapping points, i.e. occupying the centre of a phase, the coil e.m.f. is a maximum at the same instant that the phase e.m.f. is a maximum— i.e. when the phase-centre (at which this coil is situated) passes

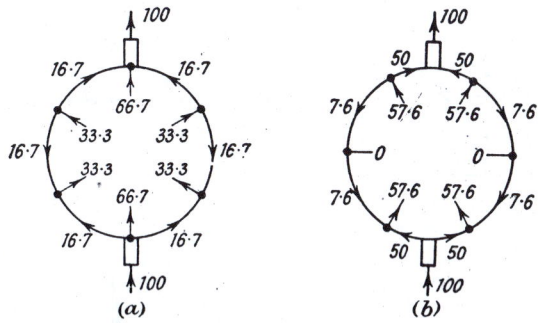

FIG. 363A. ARMATURE CURRENT DISTRIBUTION: 6-RING CONVERTER, UNITY POWER-FACTOR AND EFFICIENCY

the centre-line of a pole. The coil and phase e.m.f.'s are zero a quarter-period later, i.e. when the coil considered is passing under the d.c. brush. Thus the zero value of the centre-coil alternating current (for unity power factor) occurs at the same instant that its direct current is being commutated or reversed. The currents are thus in symmetrical opposition, as shown in (a). In the case of a coil adjacent to a slip-ring tapping, (b), the alternating component of current is displaced relatively by half the phase-spread, and the opposition is not now symmetrical. The resultant current has the differential wave-form shown, and taking the resistance as uniform, the varying rate of heat production, which depends on the square of the instantaneous current, is markedly different in the two cases.

In a multipolar machine the armature forms a mesh-connected *N*-phase system with *a* parallel circuits per phase, according to the type of winding. In most cases an ordinary lap winding (*a* = *p*) is used so that the number of parallel paths per phase is the same as the number of pole-pairs. Since all these parallel circuits must be connected to the *N* slip-rings, each ring will be connected to *a* tappings on the armature instead of only to one, as in the two-pole machine. The simple wave winding ($a = 1$) may be used for low output ratings if the voltage is suitably high.

The N phases of the armature must be symmetrical to avoid unbalanced slip-ring currents and commutation troubles. For this purpose a lap winding must have a number of slots per pole-pair divisible by the number of rings: i.e. S/Np must be integral. In a wave winding, S/N must be an integer.

CURRENT AND HEATING IN AN ARMATURE CONDUCTOR. The direct-current component in a conductor is $I_d/2a$, and the alternating

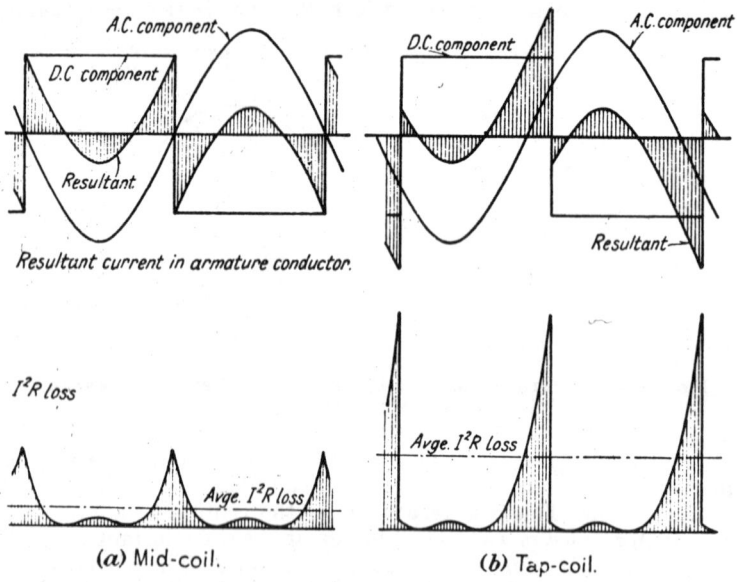

FIG. 363B. CURRENT AND LOSS VARIATION: 6-RING CONVERTER, UNITY POWER-FACTOR AND EFFICIENCY

The base in each diagram is one double pole-pitch in distance and one cycle in time

current is I_a/a, r.m.s. value. The bipolar case in Fig. 364 will serve equally well for a pole-pair in a multipolar machine. If a conductor situated centrally in the phase be considered, its current components attain their zero values together (Fig. 363B) provided that the alternating current component has unity power factor. Thus the a.c. component can be written as $(\sqrt{2}) (I_a/a) \sin \theta$, where θ is an angle swept out by the rotating armature and is reckoned as zero when the coil passes under a brush.

A conductor situated at an angle α to the phase centre, Fig. 364, will carry the same alternating current at the same instant as all the other conductors in the same phase: but if it is ahead of the mid-coil its d.c. component will be commutated α earlier. Again reckoning θ from the time of commutation of the coil considered, the a.c. is now relatively later, and has an instantaneous value given by $(\sqrt{2})\,(I_a/a)\sin(\theta - \alpha)$.

If the phase current has a lagging power factor of cos φ, then the a.c. in a conductor is φ later still, so that in general the alternating component is $(\sqrt{2})\,(I_a/a)\sin(\theta - \beta)$ where $\beta = \pm\,\alpha \pm \phi$: α is positive for coils ahead of the mid-coil in the direction of rotation, while φ is positive for lagging power factors.

The resultant current in a conductor is the difference of the components, or

$$j = (\sqrt{2})\,(I_a/a)\sin(\theta - \beta) - I_d/2a \qquad . \qquad . \qquad (186)$$

The squared current, on which the instantaneous heating depends, is

$$j^2 = 2(I_a/a)^2\sin^2(\theta - \beta) - (2\sqrt{2})\,(I_aI_d/2a^2)\sin(\theta - \beta) + (I_d/2a)^2.$$

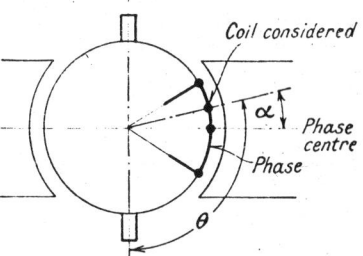

FIG. 364. PERTAINING TO THE CALCULATION OF ARMATURE HEATING

The mean square is the mean value of this expression. The average of $\sin^2\theta$ over π radians (one pole-pitch) is 0·5 regardless of the introduction of β, which is a constant for given conditions. The average of $\sin(\theta - \beta)$ is $(2/\pi)\cos\beta$. Thus the mean square of the conductor current is

$$J^2 = \left(\frac{I_a}{a}\right)^2 - \frac{2\sqrt{2}}{\pi}\cdot\frac{I_aI_d}{a^2}\cos\beta + \left(\frac{I_d}{2a}\right)^2$$

If the machine were a d.c. generator with the same commutator output, the current per conductor would be $I_d/2a$, and the heating would be proportional to the square of this. If k be the relative rate of mean heat production in the synchronous converter compared with the d.c. generator, then

$$k = \frac{J^2}{(I_d/2a)^2} = 4\left(\frac{I_a}{I_d}\right)^2 - \frac{8\sqrt{2}}{\pi}\left(\frac{I_a}{I_d}\right) + 1. \qquad (186)$$

With the substitution of the ratio I_a/I_d from eq. (185), the relative heating figure is

$$k = \frac{8}{\eta^2\cos^2\phi.N^2\sin^2(\pi/N)} - \frac{16\cos\beta}{\eta\cos\phi\pi N\sin(\pi/N)} + 1. \qquad (187)$$

This figure is for any specified coil. If it be integrated over the whole phase spread from $-(\pi/N)$ to $+(\pi/N)$ and averaged, then the relative heating figure over the whole phase (and therefore the whole armature of N identical phases) is

$$K = \frac{N}{2\pi} \int_{-\pi/N}^{+\pi/N} k \cdot d\alpha$$

$$= \frac{8}{\eta^2 \cos^2\phi \, N^2 \sin^2(\pi/N)} - \frac{16}{\pi^2 \eta} + 1 \qquad . \qquad . \qquad . \quad (188)$$

HEATING AND OUTPUT. The implications of the heating expressions in eqs. (187) and (188) are now discussed. The curves in Fig. 365 show the r.m.s. values of currents in conductors situated at

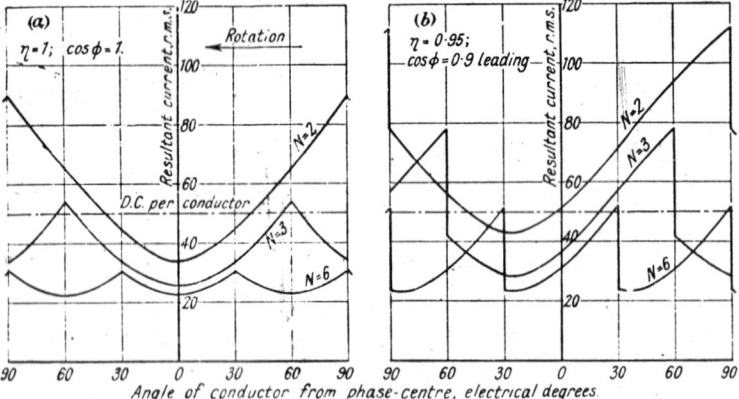

FIG. 365. ARMATURE CURRENT DISTRIBUTION: R.M.S. VALUES

various angles from the centre of a phase, per 100 A. output current per pair of parallel paths in the armature circuit of the machine as a d.c. generator. The curves at (a) are for unity power factor and efficiency; the curves at (b) for power factor 0·9 and efficiency 0·95. Consider curves (a). The r.m.s. resultant current in the mid-conductor of the single-phase case (two rings, $N = 2$) is seen to be considerably less than in a conductor of a d.c. generator of corresponding output, and that the farther the conductor is situated from the phase-centre (i.e. the nearer to a tapping point) the greater becomes the r.m.s. current. At the tapping point, 90° from the phase-centre, it is nearly double that of a d.c. generator. In the three-ring armature, the resultant currents are altogether lower, the tapping-point value at 60° from the phase-centre being about 54 A. The six- and twelve-ring armatures have a comparatively uniform current-distribution of about one-half that of the corresponding direct current. No great advantage of the twelve-ring over the six-ring armature is secured, and the latter is most commonly adopted. It is easily worked from a three-phase supply, and possesses great advantages compared with the three-ring machine.

The several curves in Fig. 365 (b) show the very great difference in r.m.s. current due to departure from unity power factor. The currents are now unsymmetrically distributed and, at the tapping points on one side of a phase, rise to considerably higher values. Since the heating of individual coils depends on the square of their r.m.s. currents, it is clear that with few rings the armature will tend to have "hot-spots" at which the rate of heat development is considerably greater than in the corresponding d.c. machine.

The lower *average* heating in the armatures of multi-ring converters suggests that these machines can be rated more highly on a temperature-rise basis than d.c. machines of corresponding dimensions. It must, however, be remembered (a) that local hot-spots tend to be produced even for small departures from unity power factor, and (b) that the design of synchronous converters is affected by the relationship between poles, speed and frequency, and will generally differ in any case from a d.c. machine of like rating, in which there is a comparatively free choice of poles.

The Tables below summarize some of the remarks already made. The average I^2R loss-ratio K, eq. (188), is given for a number of various conditions.

I^2R LOSS-RATIO K FOR WHOLE ARMATURE

No. of Rings N	Conditions of Load		
	$\eta = 1\cdot0$ $\cos\phi = 1\cdot0$	$\eta = 1\cdot0$ $\cos\phi = 0\cdot9$	$\eta = 0\cdot95$ $\cos\phi = 0\cdot9$
2	1·38	1·85	2·03
3	0·56	0·84	0·91
4	0·38	0·61	0·66
6	0·27	0·47	0·51
9	0·22	0·42	0·45
12	0·21	0·40	0·43
∞	0·19	0·38	0·40

The most common type of converter has six rings. The average armature heating and the heating of the tapped coil are given below for unity efficiency and various power factors—

I^2R LOSS-RATIOS FOR SIX-RING CONVERTER

Power Factor $\cos\phi$	Average for Whole Armature	Tapped Coil Only
1·0	0·27	0·42
0·95	0·36	0·79
0·9	0·47	1·04
0·8	0·77	1·55

The loading of a given armature cannot be determined solely on a basis of average heating, although it is not unusual to find the cooling good enough in an actual machine for no appreciable temperature-difference to be observed after a heat run : the conduction of the heat from hot to cooler parts and the efficient dissipation suffice to prevent undue hot-spots. In the design, it may be possible to include coils having maximum loss in the same slots as those with much lower loss, to equalize the temperature and loss-distribution.

The superiority of the six-ring over the three-ring converter is epitomized by the fact that for the former, 30 per cent leading current at usual efficiency produces a maximum tap-coil loss-ratio of about 0·8, compared with 2 for the three-ring machine.

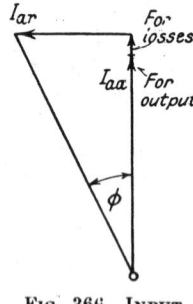

FIG. 366. INPUT CURRENT

On a basis of average armature heating, the output of a converter with a loss-ratio of K is $1/\sqrt{K}$ times as great as that of a d.c. generator of the same general design. This is because the dissipation of armature I^2R loss can be raised to the same value as for the d.c. machine by increasing the r.m.s. resultant current by $1/\sqrt{K}$. Thus a six-ring converter with an efficiency of 0·95 and a power factor of 0·9 will have an output of $1/\sqrt{0·51} = 1/0·714 = 1·4$ times the output of a corresponding d.c. generator. Actually the armature hot-spots will limit this to some smaller figure.

5. **Armature Reaction.** The consideration of armature reaction may be clarified by subdividing it into the following components—

(*a*) The d.c. reaction ;

(*b*) The a.c. reaction of the unity-power-factor current corresponding to (*a*) ;

(*c*) The a.c. reaction of the unity-power-factor loss-current required to drive the machine against rotational losses ;

(*d*) The a.c. reaction of the reactive component of the a.c. input.

In Fig. 366, the diagram shows the three current components in the a.c. input providing the reactions (*b*), (*c*) and (*d*) above, for an efficiency of 0·95 and power factor 0·0 loading. The combined a.c. reaction due to the active components in (*b*) and (*c*) is opposed to that of the direct-current output, since the former two are associated with the input plus losses, while the last is associated with the output. The two reactions are not, however, uniformly distributed over the pole-pitch, and while there is a general neutralization, local differences remain which affect commutation.

Consider the six-ring converter. The d.c. reaction F_d is due to the conductor current $I_d/2a$ flowing in the Z total armature conductors. The total ampere-conductors are $I_dZ/2a$, the ampere-turns $I_dZ/4a$, the ampere-turns per pole $I_dZ/8ap$, so that

$$F_d = I_dZ/8ap = 0.125I_dZ/ap \qquad . \quad . \quad . \quad . \quad (189)$$

The a.c. reaction is obtained from a consideration of the current-distribution in a three-phase narrow-spread winding, Chapter X, § 8. The reaction due to a current I_a per phase or I_a/a per phase-conductor is a changing value of fundamental amplitude $1.29g'z_s(I_a/a)$ ampere-turns per pole, the numerical coefficient varying actually between the limits 1·41 and 1·22 for the peaked and flat-topped distributions respectively (Fig. 139 (a) and (b)). Since $g'z_s/a = gz_s/3a$ $= 2pgz_s/6ap = Z/6ap$, the a.c. reaction is therefore

$$F_a = 1.29g'z_s(I_a/a) = 0.215I_aZ/ap \qquad . \quad . \quad . \quad . \quad (190)$$

It is most convenient to consider the components of this reaction. The cross-reaction, which will lie along the d.c. brush axis in direct opposition to the d.c. reaction, will be due to the active component of the alternating current I_{aa}, which at unity efficiency will be $0.47I_d$ from the Table on p. 588. The cross reaction will therefore be

$$F_{aq} = 0.215 . 0.47I_dZ/ap = 0.101I_dZ/ap,$$

for the fundamental, the numerical constant varying between 0·111 and 0·098 for the two limiting cases. Thus there is an unbalanced cross reaction varying between 0·014 and 0·027 of the amount I_dZ/ap, or 11 and 21·5 per cent of the d.c. reaction F_d.

The short-circuiting of certain armature coils by the d.c. brushes reduces F_d by a few per cent, to say $0.122I_dZ/ap$. Further, the divergence of the efficiency from unity increases F_{ac} by, for instance, 5 per cent. With these figures,

$$F_d \simeq 0.122I_dZ/ap \text{ and } F_{aq} \simeq 0.106I_dZ/ap,$$

the latter numerical constant varying between 0·103 and 0·117. The unbalanced reaction is now 4 to 15·5 per cent in the limiting cases, and about 11 per cent for the fundamental. This resultant is too small to cause any difficulty in commutation, and heavy overloads can usually be sustained without any adverse effect due to this cause alone. At the same time the local inequalities of the two m.m.f.'s produce harmonics of the reaction field acting in the interpolar zones, which tend to initiate voltage ripples detrimental to commutation. The m.m.f. distributions and their resultant are shown in Fig. 367.

The direct reaction F_{ad} is produced by the reactive component I_{ar} of the armature alternating current. Its amount depends on the power factor, i.e. upon the excitation of the field, in precisely the same way as in the synchronous motor (Chapter XVII).

6. Commutation. The small value of resultant armature m.m.f. in the interpolar commutating zone renders conditions of commutation relatively easy under normal operating conditions. Commutating poles require only to neutralize the reactance of the coils undergoing commutation, apart from the small resultant armature m.m.f. in the interpolar gap. In order to produce a rectilinear interpole field, i.e. one strictly proportional to the d.c. output, it is usual to employ a large interpole air gap. The poles may also be backed by plates of brass or other non-magnetic metal placed between the poles and the yoke to permit a "stiffer" interpole excitation to be used. This makes for stability on short-circuit.

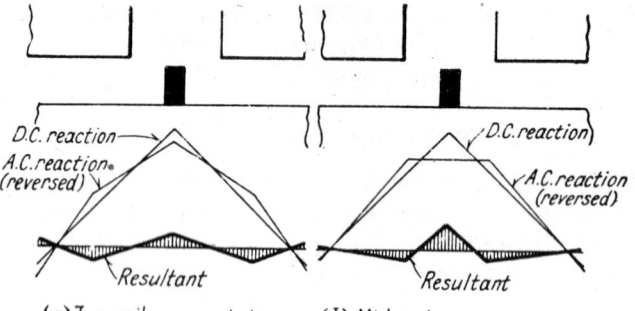

(a) Tap-coils commutating.   (b) Mid-coils commutating.

FIG. 367.   D.C. AND A.C. CROSS-REACTION M.M.F.'s

The reactive a.c. armature reaction, being a direct reaction on the *main* poles, does not normally affect the commutating zone.

Converters for traction, and similar services where overloads and heavy short circuits are common, require careful design and the employment of special equipment, more particularly as regards commutation. A sudden short circuit on the d.c. side of a converter results in a very great increase in the direct current. Owing to the reactance present in the transformers and lines on the a.c. side, the alternating current input to the converter cannot immediately attain a value corresponding to the short circuit. As a result, the balance of a.c. and d.c. ampere-turns becomes temporarily upset, and an armature m.m.f. appears in the commutating zone with which the interpoles are unable to deal. This, coupled with the greatly increased current to be commutated and the relatively slow increase of the interpole flux on account of eddy currents, etc., may produce sparking sufficient to form an arc between brush spindles of opposite polarity. The phenomenon is termed *flashover*. The means adopted to counteract or alleviate the effects of a short circuit are briefly as follows—

(a) *Magnetic Blow-out Brushgear.* This is designed to prevent flashover by blowing the arc magnetically away from the working surface of the commutator. The magnetic field is excited by means of a coil connected in series with the brush-arm in which it is incorporated.

(b) *Air Blast Fans* on the armature are arranged to sweep away the conducting vapour left round the commutator.

(c) *Mechanical Details* of design. Ample clearance is allowed as far as possible round the commutator. Bearing pedestals and base plates are insulated to prevent an arc between the commutator and

FIG. 368. 1 500-KW., 50-c/s., SELF-SYNCHRONIZING CONVERTER, 500 R.P.M.
With induction regulator and a.c. starting motor
(*British Thomson-Houston*)

earthed metal. The brush yoke does not overhang the commutator, being well insulated from it and from the brush holders. Flash barriers may be fitted between brush spindles with the minimum clearance from the commutator.

(d) *High-speed d.c. Circuit-breakers* have been developed to interrupt short circuits before they have had sufficient time to take dangerous effect.

7. **Voltage Limitations.** The voltage limitations of the synchronous converter are severely felt when high-voltage units for industrial frequencies (50 c/s.) are required. The commutator speed and the supply frequency fix the circumferential distance between adjacent brush spindles. Between these brush spindles, the number of commutator sectors is limited by the smallest width of sector which is

mechanically satisfactory. The mean voltage between commutator sectors obviously depends on the terminal voltage and the number of sectors per pole. The risk of flashover between adjacent brushes depends on the maximum voltage between commutator sectors, which, in turn, is related to the mean voltage between sectors.

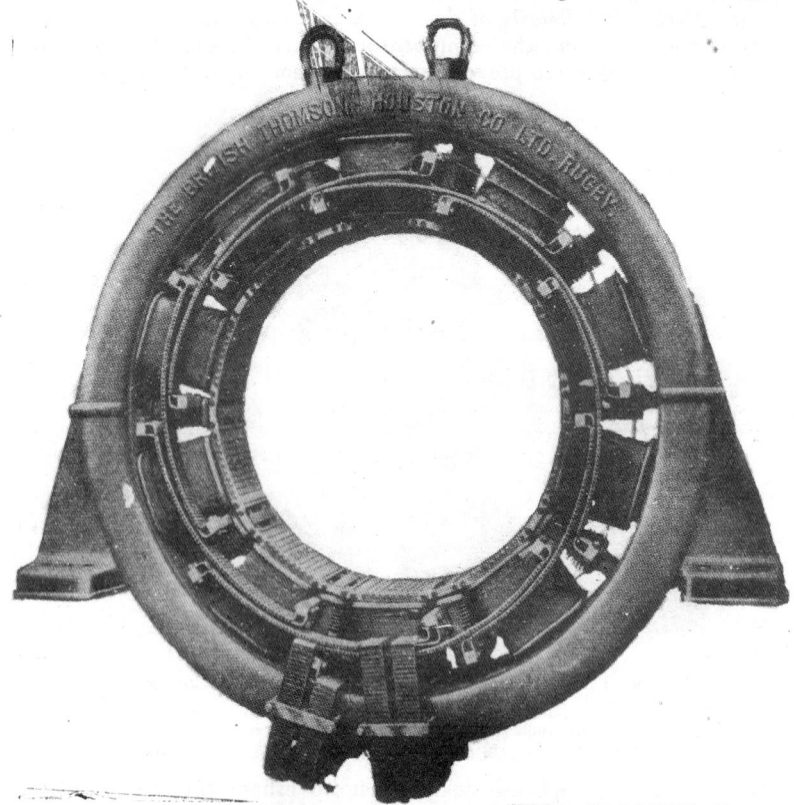

FIG. 369. CONVERTER MAGNET-FRAME
Showing main- and commutating-pole windings and damping grids
(*British Thomson-Houston*)

Consequently, the latter must be limited to reduce risk of flashover in case of sudden short circuit. Conditions become difficult when the mean voltage between sectors exceeds, say, 15 to 20 V. With a narrow-sector commutator running at high speed, wide-grooved micas between sectors are an aid to good working. To obtain high voltages such as 1 500 V. from a 50-c/s. supply, special attention has to be given to brush gear. Some manufacturers prefer two converters connected in series for such conditions.

The output $E_d I_d$ in watts of a d.c. machine can be expressed in terms of the peripheral speed u m. per sec. and the mean voltage $\bar{E}$ between sectors; for if

FIG. 370. 1 000 kW. SYNCHRONOUS CONVERTER FOR TRACTION SERVICE: 750 VOLTS, 750 R.P.M.
(*General Electric Co.*)

C = number of coils or commutator sectors;
Z = number of armature conductors;
p = number of pole-pairs;
a = number of pairs of parallel armature circuits;
D = armature diameter, m.;
n = armature speed, r.p.s.;
u = armature peripheral speed, m. per sec.;

then
$$E_d = \bar{E} \cdot C/2p$$

since there are $C/2p$ commutator bars between successive brushes, each with a mean voltage of $\bar{E}$ volts. The output current is

$$I_d = 2a\pi Dac/Z,$$

the specific electric loading *ac* depending on the heating. The output is

$$P = \bar{E}_d I_d = EC2a\pi Dac/2pZ.$$

In a large machine single-turn coils are almost inevitable, so that $Z = 2C$; and writing $u = \pi Dn$, the output becomes

$$P = \bar{E}au.ac/2pn \text{ watts.}$$

In converters it is usual to have $a = p$ with a lap winding. Further, since $pn = f$, the frequency of synchronous running, then

$$P = \bar{E}pu.ac/2f \text{ watts.}$$

If $\bar{E}$ is taken as 20 V., u as 45 m./sec., and ac as about 45000 amp.-cond. per m., then the output limit will be about 200 kW. per pole.

The output voltage E_d is $C\bar{E}/2p$. The quotient $C/2p$ can be written

$$C/2p = Cn/2f = Cyn/2fy = \pi D_c n/2fy = u_c/2fy$$

where y is the circumferential pitch of the commutator bars and $u_c = \pi D_c n$ is the commutator peripheral speed in m./sec. Taking as maximum figures $\bar{E} = 20$ V., $u_c = 40$ m./sec., and $y = 0.5$ cm. (i.e. the width of the commutator bar plus mica), then the limiting output voltage for 50-c/s. running is

$$E_d = 20 . 40/2 . 50 . 0.5 . 10^{-2} = 1\,600 \text{ V.}$$

With such values there is consequently a margin above 1 500 V. for 50-c/s. synchronous converters for 1 500 V. traction service. Such machines, however, demand very careful design to mitigate or avoid flashover.

8. **Construction.** Some examples of synchronous converter construction are shown in Figs. 368–70. A 1 500 kW., 50-c/s. converter running at 500 r.p.m. is illustrated in Fig. 368, from the a.c. end. At the side is an induction regulator for controlling the slip ring voltage, and the set is started by means of the induction motor on the end of the shaft. Fig. 369 shows a typical field system with main and commutating poles, and a damping winding. The parts of the latter that are embedded in the pole-faces are connected between poles by bridge pieces to form a continuous cage winding.

Fig. 370 shows a converter for traction service on a d.c. railway electrification. The machine develops 750 V. (two machines in

series supplying 1 500 V.) at 750 r.p.m. on a 50-c/s. supply. The
flash barriers are removed to show the arrangements for shielding
the pedestal bearing, the insulating plates behind the brush-gear,
and the special brush boxes. These together with open and well-
ventilated construction assist in reducing danger of flashover.

9. **Connections.** A schematic diagram of the connections of a six-
ring synchronous converter are given in Fig. 371. Large single-
phase converters are unsatisfactory as regards armature heating

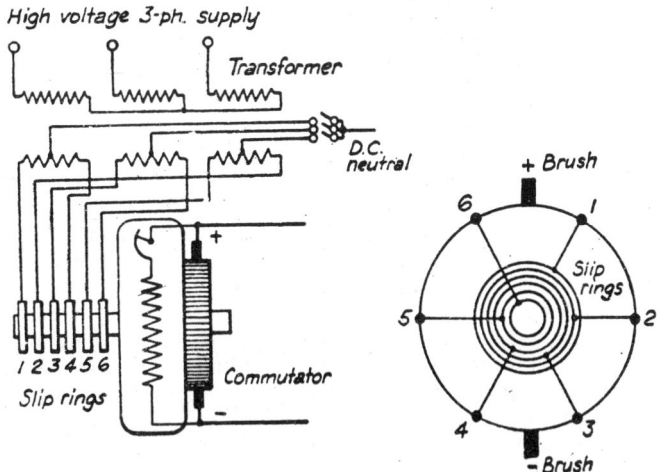

High voltage 3-ph. supply

Fig. 371. SIMPLIFIED DIAGRAM OF CONNECTIONS OF STANDARD
SIX-RING SYNCHRONOUS CONVERTER

and output, and have a large residual armature reaction pulsating
in the commutating zone at double supply frequency: they are
seldom used in any but miniature sizes. For a normal three-phase
supply it is common to employ six slip-rings on account of the
improved output of the six-phase armature, both in its induced
e.m.f. and its armature heating. For a multipolar armature there
will be more than one armature connection from each slip-ring,
unless the winding is of the wave type with only two circuits. The
connections shown in Fig. 371 are termed *diametral*, since each
secondary phase of the feeding transformer is connected to points
on the armature separated by 180 electrical degrees. Only one
secondary coil is needed per phase of the transformer, wherein lies
one of the advantages of the diametral connection. The method of
obtaining a neutral connection is shown—the mid-points of the
transformer secondary windings being tapped for this purpose.

If V_2 be the secondary voltage of the transformer with diametral connections, then the d.c. output voltage will be $V_d = (\sqrt{2})V_2$ since at two instants per cycle the d.c. brushes are connected directly through the slip-rings to the transformer secondary when at its maximum voltage. The transformer secondary current will be $P/3V_2\eta \cos\phi$, since each of the three phases contributes equally to the output power $\varGamma$.

Alternative connections are possible when the transformer carries more than one secondary coil per phase, but unless it is particularly desired to connect the secondary in mesh there is no advantage in the arrangement since it is more costly.

10. **Starting.** Since synchronous converters partake of the nature of synchronous motors, and as such have no inherent starting torque, they must be started and brought up to speed by external means or by the use of auxiliary gear, unless a d.c. supply is available. The common methods are starting with the machine acting as an induction motor, and starting by means of a small auxiliary a.c. motor. Methods of limited application include starting synchronously with an alternator and starting as a shunt motor from the d.c. side. In the latter case, a d.c. supply is needed, and the converter must be synchronized. Even where a source of direct current is available it is often the practice to use a.c. starting methods.

(a) STARTING AS AN INDUCTION MOTOR: "TAP-STARTING." Since it is general practice to provide the field system of a converter or synchronous motor with damping windings, these may be used for starting by operating the machine during the starting period as an induction motor with short-circuited secondary, the armature in this case having the function of the induction motor stator, while the damping winding acts as the cage rotor. The torque is naturally small, for the damping windings must be specially designed with low resistance since their primary function is to reduce hunting. During starting from rest, and at low speeds, the rotating field, produced by the three-phase currents fed into the armature at the slip-rings, induces in the main field windings mutual e.m.f.'s which may attain high values by reason of the large number of turns. The field winding must, therefore, be either split up (in which case the connections to the coils are brought out to an open-circuiting switch) or closed across the d.c. brushes, a two-pole reversing field switch being provided. In the latter case, the field winding, being in effect connected in parallel with the damping winding, reduces the starting torque by lowering the effective resistance. To start the machine, the field switch is closed in the normal starting direction, and a low voltage is applied to the slip-rings by means of tappings on the transformer secondary. The voltage applied is about one-third of the normal slip-ring voltage, being sufficient to produce the required torque, but avoiding an excessive current rush. The machine then develops a torque in the same way as an induction motor, and approaches synchronous speed. At speeds below the synchronous, of course, the e.m.f. across the commutator brushes is not direct, but pulsates with the frequency of slip. The exciting current from the commutator, therefore, pulsates also, and the remanent magnetism of the poles is destroyed. The voltage oscillations are shown on a centre-zero voltmeter connected across the d.c. brushes. At some instant when the slip is very small and the flux is sufficient, the machine will pull into step and become synchronized. The polarity of the d.c. brushes, however, will be quite arbitrary, and will depend

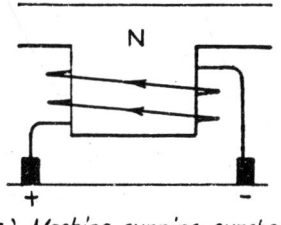

(a) Machine running synchronously with wrong polarity and with field switch in normal position.

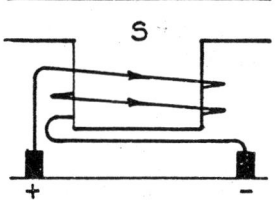

(b) Field switch reversed : torque reverses and armature begins to slip back.

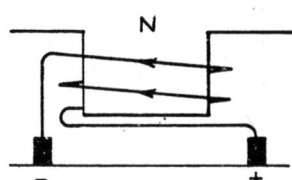

(c) Armature slips back by one pole-pitch ; torque still reversed tending to cause further slip.

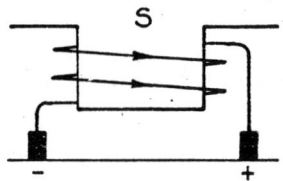

(d) Field switch returned to normal : correct polarity and torque.

FIG. 372. ILLUSTRATING THE EFFECT OF "SLIPPING A POLE"

on the position of the armature relative to the field system at the instant of synchronization. The polarity of the machine will be indicated by the pointer of a central-zero voltmeter. If this comes to rest on the wrong side, the field-reversing switch is used to correct the polarity by the following procedure, the effects of which are illustrated in Fig. 372. The machine is considered to be running in step, with the wrong brush polarity, in (a). The field switch is reversed, reversing the field flux and also the torque (b), causing the machine immediately to slow down. After one half-cycle (i.e. when the armature has slipped back by one pole-pitch) the field flux is still such as to produce a retarding torque (c), the field-reverse switch is returned to its normal position (d), establishing a driving torque with the brushes having correct polarity.

The next step is to increase the slip-ring voltage to normal, care being taken that the armature does not slip back while the transformer taps are being changed. Buffer resistances put momentarily into circuit may be used in large machines to prevent excessive current rushes when applying full voltage.

With large converters, the e.m.f.'s induced by the rotating field cutting the armature coils at slip frequency may cause injurious sparking, since the coils are short-circuited by the brushes during commutation. It becomes necessary, therefore, with units of 1 000 kW. or more in output, to provide brush-lifting gear to raise the d.c. brushes during starting. Two or four special hard-carbon brushes are left on to provide excitation of the field magnets. Such devices render the method complicated and expensive, and may make the use of a starting motor more desirable.

The remarks with reference to the e.m.f.'s induced in the converter field apply also to the field of the a.c. booster, if such be provided for voltage regulation. It is usual to short-circuit the booster field to prevent damage to its insulation.

(b) STARTING BY AUXILIARY A.C. MOTOR. Large converters are started by means of a small induction motor mounted on the end of the converter shaft (see Fig. 368). Since these motors are in service for a very short period, they are of robust and simple construction, and are highly rated. With an induction motor having the same number of poles as the converter and designed to work with a very small slip (by producing a strong field), the converter pulls into step when its field is energized.

As an alternative, a starting motor with two poles fewer than the converter, and designed for a large slip, may be employed. If a slip-ring rotor is used, the speed can be adjusted by rheostatic control: if—as is usual—a cage rotor, by varying the load on the motor by means of the shunt field regulator of the converter, which varies the excitation, and, therefore, the iron losses in the converter armature. Generally, it is not necessary to use even a cage winding on the rotor of the induction motor, as a solid cast-iron cylinder, sometimes with a steel sleeve, will serve the same purpose.

Motor starting is frequently combined with automatic synchronization. It has been seen in connection with tap-starting that a converter will pull into step when running with a small slip, but that the polarity is a matter of chance. If it were possible to leave the remanent magnetism of the field poles unreversed, there would be a definite tendency for the converter to pull into step with the correct polarity, since the flux (and, therefore, the synchronizing torque) would be greater when the armature polarity was such as to assist the remanence than when it opposed. Self-synchronizing methods depend upon this fact, the current flowing into the converter slip-rings being limited to a value insufficient to destroy the remanent magnetism. To this end, after one-half to two-thirds normal speed has been attained, the converter slip-rings are connected to the supply either through reactors or in series with the stator windings of the startor motor. The set then runs up to speed and synchronizes with correct polarity. Thereafter the current-limiting device, reactors or motor windings, is cut out of the armature circuit and full voltage applied to the converter slip-rings. Fig. 373 shows the simplified connections of one form of self-synchronizing

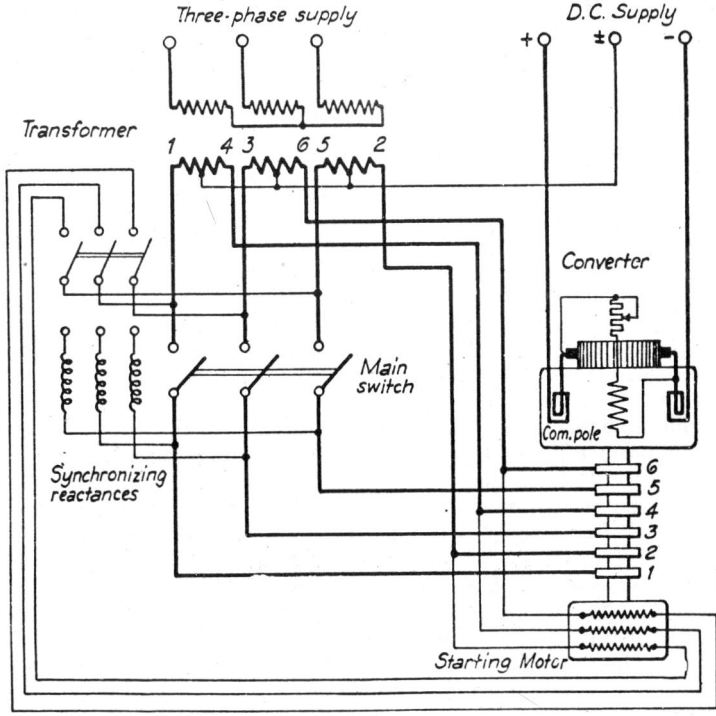

FIG. 373. CONNECTIONS OF A SELF-SYNCHRONIZING CONVERTER WITH STARTING MOTOR

converter set with startor motor. The latter is first connected across the transformer secondary windings alone, the converter itself being isolated. The set runs up to a speed approximating to synchronous speed. The starting switch is then thrown over, disconnecting the induction motor, and connecting the converter slip-rings to the transformer terminals through a separate synchronizing reactance. The converter then automatically synchronizes, since it has built up a field of correct polarity due to its remanent magnetism and develops a synchronizing torque. Finally, the main switch is closed, short-circuiting the synchronizing reactances.

11. Voltage Regulation. Variation of the excitation of a synchronous converter working normally has little effect on the direct voltage, since the machine operates on the a.c. side like a synchronous motor. Thus, an alteration of the excitation affects the power factor, while the effective flux is practically unaltered owing to the magnetizing or demagnetizing action of the reactive current component. To regulate the direct voltage it is therefore necessary either to control the alternating voltage applied to the converter armature, or to regulate the direct voltage at the commutator brushes by some additional apparatus. The former is the more practicable and common. The methods available for varying the alternating voltage applied to the slip-rings include (a) reactance control, (b) booster control, (c) transformer tappings, (d) induction regulator control.

(a) REACTANCE CONTROL. With this, a very common method of voltage regulation, use is made of the property of a synchronous machine whereby the power factor on constant load is dependent upon the excitation. When the machine is over-excited, the alternating supply current assumes a leading power factor, and *vice versa*. To obtain regulation, reactors may be included in the slip-ring leads or, more usually, a transformer used which has considerable reactance, augmented if necessary by the use of packets of transformer plates between the primary and secondary coils to increase the leakage flux. The effect in the two cases is the same, and can be better understood if a constant transformer secondary voltage V_2 be assumed, with reactors in the slip-ring connections. If I be the current, E_x the e.m.f. of self-inductance induced by I in the reactor, and V_c the voltage applied to the converter slip-rings: then with a lagging current as shown in Fig. 374 (a), the e.m.f. E_x is in lagging quadrature with the current and gives, as complexor sum with V_2, the converter voltage V_c. If now, by over-excitation of the converter field, the current I is made to lead, E_x will change its phase position accordingly, and the complexor resultant of V_2 and E_x will now be greater (Fig. 374 (b)). Thus, by control of the field regulator, the direct voltage with a constant alternating supply voltage can be raised or lowered within the limits imposed by the reactance of the reactor (or, in practice, of the transformer). Variations up to ± 6 per cent in the voltage can be obtained without difficulty. If a series winding be placed on the converter field, and connected in series with the d.c. brush lead, the regulation can be made to vary automatically with the load.

While reactance control is, from the view-point of operation in service, simple and widely employed, it has certain disadvantages. The power factor of the machine is now out of the control of the operator, being less than unity at both low and high loads. At one load only will the power factor be unity, although the leading power factor at higher loads tends to compensate for reactive line drop. This reflects on the field winding design. To obtain a current leading the supply voltage, the field winding must overcome the demagnetizing effect of that current. In addition to this, however, the flux must be increased since with leading currents the slip-ring voltage increases, requiring a corresponding increase of induced e.m.f., and flux. A feature of converters for reactance control is, therefore, a large amount of field copper, and the extra losses in both field and

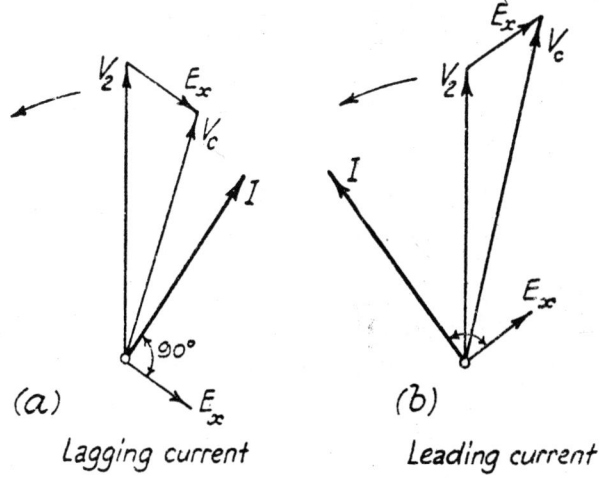

(*a*) Lagging current

(*b*) Leading current

Fig. 374. Illustrating Reactance Control

armature due to the departure from unity power factor lead to a drop in efficiency, and an increase in size and cost. The deleterious effect of low power factor has already been emphasized.

To obtain the necessary reactance, it is usual to employ transformers specially designed for considerable leakage, rather than auxiliary reactors. It may be necessary to use *reactive iron*, i.e. packets of sectionalized steel stampings placed in the ducts between the high- and low-voltage windings of the transformer to increase the leakage flux. This increases the core loss and size. To avoid saturation of the reactive iron, which tends to the production of voltage harmonics leading to the pitting of certain commutator sectors displaced by one-sixth of a pole-pitch, small air gaps are introduced.

Reactance control is widely used on account of its simplicity and relative cheapness.

(*b*) Booster Control. In this method, the slip-ring connectors lo not pass directly to the armature of the synchronous converter, but through the windings of a small a.c. generator mounted on the converter shaft, and having the same number of poles. An alternating voltage is thus introduced in series with the slip-ring supply voltage, its direction being dependent on the polarity of the booster field magnets, which may be shunt- or series-excited from the d.c. side of the set. When boosting up, the booster acts as a synchronous generator, being driven by the converter itself acting as a synchronous motor. When boosting down or bucking the voltage, the booster acts as a motor, driving the converter as a generator. This

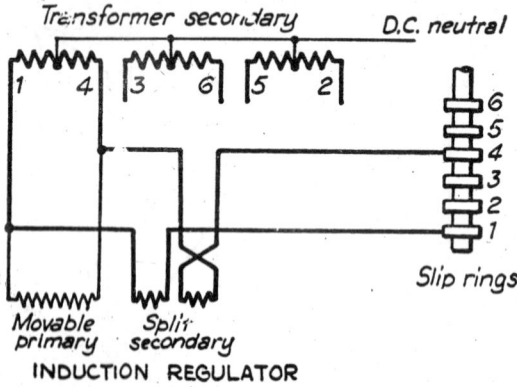

FIG. 375. CONNECTIONS OF INDUCTION REGULATOR IN
A SIX-RING CONVERTER SUPPLYING A THREE-WIRE
DIRECT-CURRENT SYSTEM

affects commutation since it upsets the balance of d.c. and active a.c. ampere-turns on the converter armature, necessitating more elaborate interpole circuit arrangements.

Booster control can be used for fairly large voltage variations, e.g. ± 20 per cent, and has the advantage that the power factor of the converter can be separately adjusted. Its cost, however, is relatively high.

(c) TRANSFORMER TAPPINGS. A common and easy method of voltage adjustment is provided by using a main transformer with on-load or off-load tap-changing. Remote control by push-button is readily arranged. The method has the advantages of comparative cheapness, and no interference with power factor control or commutation. On the other hand, *gradual* regulation is not possible.

(d) INDUCTION REGULATOR CONTROL. This is comparable in method with (c), with the additional advantage of providing gradual and continuous control. Remote operation and automatic working can be easily applied, no additional switches are necessary in the main circuit, and the construction is robust and simple. The induction regulator is very suitable for converters in which a greater voltage range than that available with reactance control is needed; also for converters running inverted. Fig. 375 gives a connection diagram for a six-ring converter and induction-regulator set supplying a d.c. three-wire system: only one phase is shown, for clearness. To balance the neutral point, which is taken from the phase centres of the transformer secondary winding, the regulator secondaries are split into two parts per phase, one part being connected into each line.

(e) BRUSH SHIFT. By rocking the d.c. brush-gear an appreciable variation of the d.c. voltage can be obtained, just as in a d.c. generator. This method impairs commutation, and is obviously unsuitable for interpole machines.

(f) SPLIT-POLE CONSTRUCTION. The same effect as with brush-shift, but without the necessity of shifting the brush-spindles, is obtained by making the main poles in two parts. According to the polarity to which each part is excited, the flux-axis can be shifted, and accordingly the commutator voltage changed. Commutation is difficult, and the interpoles must be specially designed and excited. There is not enough space to accommodate split-poles in restricted pole-pitches, so that 50-c/s. converters cannot as a rule be regulated in this way.

(g) D.C. BOOSTER. This is a d.c. machine connected in series with one of the d.c. output leads, and mounted (usually) on the converter shaft. Its obvious disadvantage is that its commutator has to carry the full main current, thus introducing further commutation difficulties. Two such machines are required for a three-wire supply. The operation of the machine is similar to that of the a.c. booster.

Where the voltage control in modern machines exceeds $\pm$ 6 per cent, induction regulators may be used, although in most cases on-load tap-changing transformers offer a cheap and adequate solution. Brush-shift, booster and split-pole voltage regulation are practically obsolete.

12. **Inverted Running.** Synchronous converters are occasionally required to run *inverted*, i.e. to convert d.c. to a.c. If working in parallel with other machines on the a.c. side the operation of the converter is quite satisfactory, since its speed is locked synchronously to the a.c. system. If working alone, on the other hand, excessive lagging current load may cause the machine to race by reason of a reduction of the main flux due to demagnetizing armature reaction. The use of a separate exciter and of a speed-limiting device is desirable for this case, while an a.c. booster or induction regulator may be necessary for the independent control of the alternating voltage output.

13. **Performance.** Although the converter has much of the nature of a synchronous motor, its load-angle is considerably less on account of the neutralization of the a.c. cross-reaction by the d.c. reaction. Consequently, while a synchronous motor will pull out of step at

loads above 2 or $2\frac{1}{2}$ times normal, a converter will keep in step with upwards of 10 times normal load *steadily* applied. Such loads cannot, however, be sustained on account of heating and commutation difficulties: with load-angles not abnormally high in synchronous motor operation, a converter would probably flash over on the d.c. side.

The effects of *harmonics* in the d.c. output voltage are sometimes troublesome. The chief causes are

(1) Slot ripples;

(2) Sixth harmonic of supply frequency.

Slot or tooth ripples (see p. 222) are avoided by skewing the slots or the pole-tips, and in modern machines will not be appreciable. The sixth-harmonic ripple is caused by the changing armature-reaction, Fig. 367. An intensification of this ripple, accompanied by blackening of certain commutator sectors, is due to third, fifth and seventh harmonics in the a.c. supply-voltage wave-form.

The full-load efficiency of a 1 000 kW. synchronous converter with transformer may be 2 per cent higher than that of a motor-converter, and 4 per cent higher than that of a synchronous motor-generator, when the two latter machines have stators connected to the supply without transformers. Where the supply voltage is too high to dispense with transformers, the superiority of the synchronous converter is more pronounced.

CHAPTER XXIV

MOTOR-CONVERTERS

1. Theory. Essentially, a motor-converter consists of an induction motor coupled both mechanically and electrically to a synchronous converter. The simplified diagram, Fig. 376, illustrates the method of connection. The stator of the induction motor is fed from the alternating current supply. This can be done without transformation unless the voltage exceeds about 11 000 V., above which it may be cheaper and simpler to construct a transformer for the high voltage and wind the stator for some lower voltage. The rotor of the induction motor is coupled mechanically to the armature of a converter, and is connected electrically, as shown, to the back of the converter armature winding.

The action of the machine is as follows—

Let f = frequency of supply, cycles per sec.;

f_2 = frequency of rotor e.m.f.;

f_r = speed frequency;

p_d = number of pole-pairs of converter:

p_a = number of pole-pairs of motor;

n = speed of set, r.p.s.

Then, since the rotor frequency is f_2,

$$f_r = f - f_2 = p_a n$$

where f_r is the speed frequency. But the rotor e.m.f.'s of frequency f_2 are to be applied to the converter armature, which must run synchronously at this frequency. Therefore.

$$f_2 = p_d n.$$

Consequently, $\qquad f = f_r + f_2 = p_a n + p_d n$

or speed $\qquad n = f/(p_a + p_d)$ . . . . . (191)

This is the only stable speed of the set, since there is a considerable synchronizing torque exerted by the converter to overcome any tendency to depart from this speed. When the numbers of poles on stator and field are the same, or $p_a = p_d$, then the normal speed n is one-half of the synchronous speed of the induction machine working on the supply frequency f.

2. Voltage and Current Ratios. *Alternating-current End.* The e.m.f. induced in a stator phase by the rotating field is

$$E_1 = 4.44 \, K_{w1} f T_1 \Phi \text{ volts.}$$

The speed of the set is, from eq. (191), $n = f/(p_a + p_d)$

so that the rotor frequency $f_2 = p_d n = p_a f/(p_a + p_d)$. The rotor e.m.f. per phase will therefore be

$$E_2 = 4.44\, K_{w2} \frac{p_a f}{p_a + p_d}\, T_2\, \Phi \text{ volts.}$$

From these equations it will be seen that

$$E_2 = \frac{K_{w2} \cdot T_2}{K_{w1} \cdot T_1} \cdot \frac{p_d}{(p_a + p_d)}\, E_1 = \frac{p_d}{r_e(p_a + p_d)}\, E_1$$

where r_e is the e.m.f. transformation ratio $K_{w1} \cdot T_1/K_{w2} \cdot T_2$.

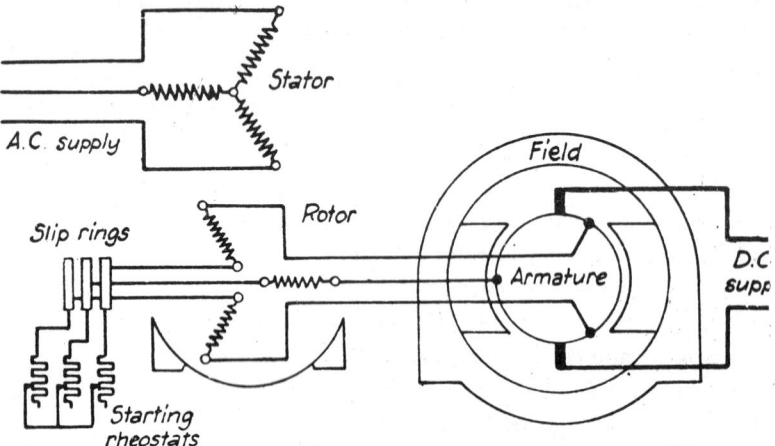

4. C. END   D. C. END
FIG. 376. SIMPLIFIED CONNECTION DIAGRAM OF MOTOR CONVERTER

Neglecting the magnetizing current, the ampere-turns on stator and rotor are identical,

$$K_{w1} N_1 T_1 I_1 = K_{w2} N_2 T_2 I_2,$$

where N represents numbers of phases. Hence,

$$I_2 = \frac{K_{w1} N_1 T_1}{K_{w2} N_2 T_2}\, I_1 = \frac{I_1}{r_i}.$$

where r_i is the ratio of current transformation, $K_{w2} N_2 T_2/K_{w1} N_1 T_1$.

The power factor can be adjusted to unity at the stator, so that the electrical power in the rotor is

$$\begin{aligned} P_2 &= N_2 E_2 I_2 = \frac{N_2}{r_e r_i}\, E_1 I_1 \frac{p_d}{p_a + p_d} \\ &= N_1 E_1 I_1 [p_d/(p_a + p_d)] \\ &= [p_d/(p_a + p_d)] P_1 \end{aligned}$$

where P_1 is the power supplied to the stator. The mechanical power at the rotor shaft is thus the remainder, i.e.

$$P_m = P_1 - P_2 = [p_a/(p_a + p_d)]P_1$$

Thus
$$P_1 : P_2 : P_m = 1 : p_d/(p_a + p_d) : p_a/(p_a + p_d)$$
$$= f : fp_d/(p_a + p_d) : fp_a/(p_a + p_d)$$
$$= f : \quad f_2 \quad : \quad f_r$$

The a.c. energy input to the stator is thus divided by the rotor into two parts, one, P_m, appearing as mechanical power at the shaft and proportional to f_r, and the other, P_2, as electrical energy in the rotor windings and proportional to f_2. The former drives the commutator machine as a d.c. generator, while the latter feeds it as a synchronous converter. Thus, each machine acts a dual role, the induction machine as a motor and transformer, the commutator machine as a converter and generator. In practice, the number of rotor and armature phases is not three, as shown for simplicity in Fig. 376. Twelve is a more usual number, allowing of a much better utilization of the converter armature. No slip-rings are required, so that no difficulty is found in providing for the large number of phases. The connectors are accommodated in a hollow shaft-coupling when a centre bearing is used, but some makers avoid this bearing and make connections directly between rotor and armature.

Direct-current End. The commutator machine operates simultaneously as a generator and as a synchronous converter. It receives the rotor phase currents

$$I_2 = P_2/N_2E_2 = (2\sqrt{2})P_2/N_2E_d$$

(where E_d is the commutator voltage) by means of the interconnectors between rotor and armature, and at the same time generates a further current by conversion of the mechanical power received through the common shaft.

3. **Action.** Since the speed is fixed by the numbers of poles and the frequency of supply, the rotor e.m.f. is also fixed. The armature induced e.m.f., on the contrary, can be controlled through a fairly wide range by means of the field regulator, since the induction machine possesses considerable inherent reactance: the action is analogous to that of voltage-control by reactance in synchronous converter practice. The difference between the inherent e.m.f. of the armature (i.e. that induced by rotation in its own field) and the rotor e.m.f. will cause a change of power factor in the current supplied by the rotor. The consequent reaction on the stator can be utilized to raise its power factor to unity, or to cause its current to lead. In other words, the magnetization of the induction machine can be provided from the d.c. end. In this respect the d.c. machine

acts as a phase-advancer to the induction motor. The air gap of the latter may be made larger than that of a normal induction motor.

The fact that the induction machine in the motor-converter set acts both as a voltage transformer and a frequency reducer, leads to a number of advantages. The a.c. slip-rings are small and few in number, and (except when a three-wire d.c. system is required) out of action during normal running. The speed is low, which enables good commutation to be obtained owing to the widely-spaced d.c. brush spindles even when the supply frequency is high, although the armature tends to be bigger and heavier than that of a corresponding synchronous converter.

4. **Operation.** (A) STARTING. As with the synchronous converter, the motor-converter may be started either from the direct current or from the alternating current end. The latter is more usual.

(*a*) *Starting from the Alternating Current Side.* Three phases are connected to a rotor starting resistance, the others being open-circuited. The d.c. side is self-excited and disconnected from the bus-bars. To start the supply is switched on to the stator, and the set begins to run up as a slip-ring induction motor. As the speed increases, the d.c. field builds up in the same manner as in an ordinary d.c. generator. The e.m.f.'s of rotor and armature are, therefore, successively in opposition and conjunction, so that the currents in the rotor starting resistance will vary in magnitude in a corresponding manner and with a periodicity which becomes progressively less as the speed approaches the normal running speed of the set. A synchronizing voltmeter connected across the starting resistance indicates this variation. When the oscillations are sufficiently slow and the voltage is nearly zero, the starting resistance is short-circuited, and, immediately after, a short-circuiting device connects all twelve phase-ends together, forming a star point. Usually, the starting resistance and shunt field regulator are adjusted so that no control is necessary during starting-up, the starting resistance being short-circuited when the synchronizing voltmeter indicates synchronous running.

The nature of the motor-converter makes it difficult to overload to the extent of dropping out of step when running normally.

Self-synchronization is accomplished by starting first on resistances, then changing over to reactors when normal speed is approached.

(*b*) *Starting from the Direct Current Side.* If a source of direct current supply is available, the direct current machine may be used as a motor to run the set up to speed, the alternating current rotor being first open-circuited to prevent a part of the starting current from being diverted through the rotor windings. Some form of synchronizing device for the a.c. end is necessary to indicate the approach to synchronous speed.

(B) VOLTAGE REGULATION. The requisite voltage variation can

be obtained by means of the field regulator. This affects the power factor on the a.c. side to a lesser extent than with the synchronous converter. As much as 20 per cent variation is obtainable without auxiliary apparatus.

(C) REVERSAL. Motor-converters are normally designed to convert from a.c. to d.c., but will reverse automatically if required.

CHAPTER XXV

THE E.M.F. OF THE GENERAL ELECTRICAL MACHINE

THE discussion on page 9 indicates the possibility of six cases for a general electrical machine, depending on whether the flux is constant or alternating, the coils in which the e.m.f. is induced are fixed or moving relative to the flux, and whether there are fixed tappings or commutator connections made to the coils.

GENERAL EQUATION FOR THE INDUCED E.M.F. IN A COIL. Considering a coil of T'_c turns linked completely with a magnetic flux of Φ webers—an arrangement to which all electromagnetic machines can be related for the purposes of comparison—then the number of flux-linkages concerned is $N = T_c\Phi$. If N change, an e.m.f. will be induced in the coil. The physical interpretation of the expression $e = -T_c(d\Phi/dt)$ volts is simplified by expressing the differential coefficient as

$$\frac{d\Phi}{dt} = \frac{\partial\Phi}{\partial\alpha} \cdot \frac{d\alpha}{dt} + \frac{\partial\Phi}{\partial t}$$

a form which allows for variation of linkage due to both causes occurring together. The first term refers to the change of flux linking a coil moving at angular speed ω, $= (d\alpha/dt)$ radians per sec. relative to the flux, the distribution of the flux at the instant considered being expressed as $\partial\Phi/\partial\alpha$. The second term refers to flux pulsation.

All modern electrical machines are constructed on the heteropolar principle, i.e. with alternate N and S poles: it is therefore possible to limit the generality implied in the partial derivative $\partial\Phi/\partial\alpha$. Fig. 377 shows a pair of poles, each with pole-pitch Y covering π electrical radians, and indicates some form of flux distribution over a pole-pitch. The shape of the flux-distribution curve, or B-curve, over a pole-pitch varies with different types, and, in any given type, also with the load. No-load conditions, however, are dealt with exclusively here, and magnetic balance is assumed—i.e. the amount of flux emerging from any north pole is the same as that entering the adjacent south pole. In Fig. 377, B denotes the flux density in Wh /m ^{2} in the air-gap. The flux Φ linking a concentrated full-pitch coil, such as that shown, will be represented by the shaded area, or

$$\Phi_a = \int_{-a}^{+a} BLR \, d\alpha$$

where L is the core-length in m. over which B extends, and R is

618

the radius of the armature. It is to be noted that α is measured in electrical radians, so that πR represents the pole-pitch Y whether the machine be of bipolar or of multipolar construction. For simplicity bipolar machines have been illustrated throughout.

It will be assumed that (*a*) the flux per pole pulsates sinusoidally with respect to time, and (*b*) the flux is distributed sinusoidally over

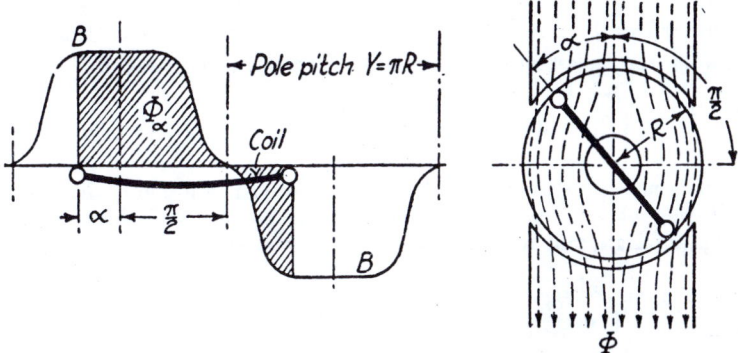

FIG. 377. FLUX DISTRIBUTION OVER A DOUBLE POLE-PITCH

the pole-pitch at any instant. Stated symbolically, the first assumption is that the total flux per pole at any instant is expressible by $\Phi = \Phi_m \sin \omega t$ where Φ_m is the time-maximum value of the flux per pole, and $\omega = 2\pi f$. This involves no unusual condition.

For the maximum induction density occurring at a point displaced by an angle α from the pole centres the second assumption gives the expression $B_\alpha = B_m \cos \alpha$. B_m is the maximum gap-flux density at the pole-centre and is thus the maximum in both space and time: Fig. 378. This assumption of harmonic variation may not, in practice, be completely justified, but the differences introduced can be dealt with without difficulty,

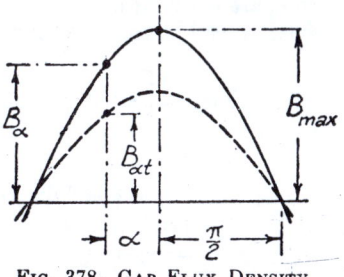

FIG. 378. GAP FLUX DENSITY AT A POINT

and are seldom important. The principle is not in the least affected by the assumption, while treatment for the present purpose is greatly simplified. Since the flux pulsates sinusoidally, the instantaneous value of gap density $B_{\alpha t}$ at any point in the gap can be expressed as $B_{\alpha t} = B_\alpha \sin \omega t$.

The value of Φ, the *instantaneous flux per pole*, is clearly the integral of the instantaneous flux density over the pole-pitch, or

$$\Phi = \Phi_m \sin \omega t = \int_{-\frac{1}{2}\pi}^{+\frac{1}{2}\pi} B_{\alpha t} LR d\alpha = \int_{-\frac{1}{2}\pi}^{+\frac{1}{2}\pi} B_m \sin \omega t . LR \cos \alpha . d\alpha$$

$$= 2LRB_m \sin \omega t = \frac{2}{\pi} LYB_m \sin \omega t$$

since $\pi R = Y$.

Now the *instantaneous flux linking the coil* shown in Fig. 377 is

$$\Phi_\alpha = \int_{-\alpha}^{+\alpha} B_{\alpha t} LR d\alpha = 2LRB_m \sin \omega t . \sin \alpha = \Phi_m \sin \omega t . \sin \alpha.$$

It is now possible to find the e.m.f. induced in a coil of T_c turns rotating at angular velocity ω_r radians per sec. Neglecting the negative sign, this will be—

$$e = T_c \left[\frac{\partial \Phi_\alpha}{\partial \alpha} \omega_r + \frac{\partial \Phi_\alpha}{\partial t} \right]$$

$$= \left[\omega_r T_c \Phi_m \sin \omega t . \cos \alpha + \omega T_c \Phi_m \cos \omega t . \sin \alpha \right].$$

Thus e is the instantaneous e.m.f. induced when the plane of the coil has reached an angle α from the pole centres. The total e.m.f. of the armature, as measured between phase tappings on an opened winding, or between brushes on a commutator winding, is the sum of the e.m.f.'s of the individual coils in series making up the phase or the part between brushes. If there be m of these coils occupying a phase-spread σ, then the phase e.m.f. will be—

$$\Sigma(e) = \left[\omega_r m T_c \Phi_m \sin \omega t . \frac{1}{\sigma} \int_{\beta - \sigma}^{\sigma} \cos \alpha . d\alpha \right.$$

$$\left. + \omega m T_c \Phi_m \cos \omega t . \frac{1}{\sigma} \int_{\beta - \sigma}^{\beta} \sin \alpha . d\alpha \right] . \qquad . \quad (192)$$

$$= e_r + e_p$$

Here β denotes the angle between the phase start and the pole-centre (Fig. 379); e_r is the first term involving ω_r, and is the e.m.f. induced by the movement of the coil in the flux, or briefly, the *e.m.f. of rotation*; and e_p is the e.m.f. induced by variation of the flux, or *e.m.f. of pulsation*. Both of these e.m.f.'s may be considered as appearing simultaneously in the same winding. Only the e.m.f. of rotation, e_r, represents the conversion of mechanical into electrical energy or vice versa, since e_r only is associated with the tangential force on the coil due to interaction of its current with the magnetic flux, the relative movement of the coil and pole being the condition resulting in the production or conversion of mechanical energy. The e.m.f. of pulsation, e_p, is the means whereby electrical energy is transferred from one electric system to another, as in the transformer and induction motor.

Clearly e_r will be a maximum when the phase-centre coincides with the pole-centres, while e_p is a maximum when the axes of the phase and poles are coincident.

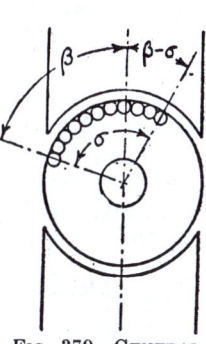

The expression in eq. (192) represents the general equation for the e.m.f. induced in a phase of any electromagnetic machine.

PARTICULAR CASES. There are six cases to be considered, according as the flux is constant or pulsating, the coil fixed or moving, and the armature connections made to tappings (e.g. slip-rings) or to a commutator and brushes; i.e. to points fixed relative to the field system or to the armature windings.

Fig. 379. GENERAL WINDING

The various combinations are listed in the Table on page 9.

Case A: Constant flux; moving coils; commutator connections (Fig. 380). The induced e.m.f. is obtained from eq. (192). Clearly the term e_p is zero, since the flux is constant, and the expression is thereby simplified to e_r. With the brushes in the neutral zone, $\beta = \pi/2$, and $\sigma = \pi$, so that the expression becomes—

$$\Sigma(e) = m\omega_r T_c \Phi \cdot \frac{1}{\pi} \int_{-\frac{1}{2}\pi}^{\frac{1}{2}\pi} \cos \alpha \cdot d\alpha = m\omega_r T_c \Phi \cdot \frac{2}{\pi}$$

assigning to the flux per pole the constant value Φ.

Since $\omega_r = 2\pi n$, and $m T_c = pZ/4a$ in a multipolar commutator armature, where n = armature speed in rev. per sec. and Z = total number of active conductors, then

$$\Sigma(e) = E_r = \frac{p}{a} nZ\Phi \text{ volts}$$

The flux Φ is constant: therefore E_r is constant and unidirectional since the brushes, fixed with respect to the field system, include between themselves a constant number of coils occupying always the *same position with respect to the flux*.

If the brushes stand on an axis displaced by an angle $\beta \neq \pi/2$ from the pole centres as in Fig. 380, the integration is performed between the limits β and $\beta - \pi$, resulting in the e.m.f.

$$E_r = \frac{p}{a} nZ\Phi \cdot \sin \beta \text{ volts.}$$

This case represents the common direct-current machine.

Case B: Constant flux; moving coils; tapping connections (Fig. 381). Considering a symmetrical armature winding divided into a number N of similar phases, the phase-spread σ will be $2\pi/N$, and the instantaneous e.m.f. of a phase will be given by

$$\Sigma(e) = m\omega_r T_c \Phi \frac{N}{2\pi} \int_{\beta - \frac{2\pi}{N}}^{\beta} \cos\alpha \,.\, d\alpha$$

$$= m\omega_r T_c \Phi \frac{N}{2\pi}\left[\sin\beta - \sin\left(\beta - \frac{2\pi}{N}\right)\right].$$

Since the flux is constant in value, no term in $d\Phi/dt$ will appear, i.e. the e.m.f. is entirely one of rotation. The expression will be a maximum when the integral is a maximum. It is clear from inspection that this occurs when $\beta = \pi/N$, i.e. when the phase-centre lies on the line of pole-centres. The maximum e.m.f. is then

$$\Sigma(e)_m = e_{ph} \,.\, {}_m = m\omega_r T_c \Phi \,.\, \frac{N}{2\pi} \,.\, 2\sin\frac{\pi}{N}$$

$$= m\omega_r T_c \Phi \times \frac{N}{\pi}\sin\frac{\pi}{N}.$$

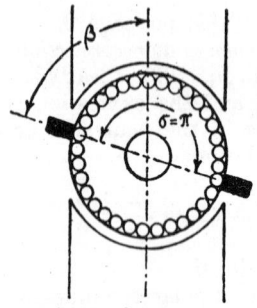

FIG. 380. COMMUTATOR MACHINE.

It is to be noted that $m\omega_r T_c \Phi$ represents the *arithmetic sum* of the maximum e.m.f.'s in the m coils, each of T_c turns, constituting a phase. The coils, being distributed over a spread $\sigma = 2\pi/N$ electrical radians, will not have their maximum e.m.f.'s occurring at the same instant, however, and the factor $(N/\pi)\sin(\pi/N)$ takes account of this: it is named the *distribution factor*, k_m. The values of k_m for various numbers of phases (N), each with spread $\sigma = 2\pi/N$, are shown in the Table below, corresponding to the figures obtainable from eq. (72)—

No. of Phases N	6	4	3	2
Phase Spread σ . .	$\frac{2\pi}{6} = 60°$	$\frac{2\pi}{4} = 90°$	$\frac{2\pi}{3} = 120°$	$\frac{2\pi}{2} = 180°$
Distribution Factor k_m	0·955	0·905	0·827	0·636

The e.m.f. of a phase pulsates sinusoidally, since the armature rotates at angular velocity $\omega_r = 2\pi f_r$. For this reason the factor $\sin\beta - \sin[\beta - (2\pi/N)]$ is expressible in terms of $\sin\omega_r t$, since the angular distance of the phase start at any instant t sec. from datum will be $\beta_0 + \omega_r t$, where β_0 represents the initial condition. For simplicity β_0 can be taken as π/N, so that the initial state at $t = 0$ is a coincidence of phase-centre and pole-centre. The phase e.m.f. at any instant then becomes

$$e_{ph} = m\omega_r T_c \Phi \frac{N}{2\pi}\left[\sin\left(\omega_r t + \frac{\pi}{N}\right) - \sin\left(\omega_r t - \frac{\pi}{N}\right)\right]$$

$$= m\omega_r T_c \Phi k_m \cos \omega_r t \text{ volts,}$$

showing that e_{ph} pulsates sinusoidally with frequency $f_r = \omega_r/2\pi$. The virtual value of the phase e.m.f., putting $mT_c = T$, the turns per phase, will be

$$E_{ph} = \sqrt{2}\pi f_r T\Phi \times k_m \text{ volts.}$$

This case represents that of the synchronous generator and motor.

Case C : Pulsating flux; fixed coils; commutator connections.
The only machine in this class is the "Transverter," constructed by Highfield and Calverley at the works of the English Electric Company. The "Transverter" consists essentially of a number of transformers, the phase-ends of which are connected to stationary commutators. Brushes are rotated round the commutators at synchronous speed, and thus are supplied with a continuous voltage. Suitable transformer connections to a 3-phase supply are employed to provide 36 secondary phases equally spaced in time, connected to 36- or $36p$-bar commutators, according to the number of pairs of brushes p. The d.c. brush voltage is identical with the maximum alternating voltage per phase when the brushes are

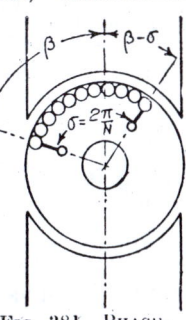

Fig. 381. Phase-wound Machine

arranged to connect with commutator bars at the instant of their maximum voltages.

Case D : Pulsating flux; fixed coils; tapping connections (Fig. 382). When the coils are stationary and linked with a pulsating flux, the absence of movement results in the absence of any conversion from electrical to mechanical energy, or vice versa: i.e. $e_r = 0$, and the e.m.f. in m coils is given by

$$\Sigma(e) = m\omega T_c \Phi_m \cos \omega t \frac{1}{\sigma}\int_{\beta-\sigma}^{\beta} \sin \alpha \cdot d\alpha.$$

This is the case of the induction regulator. The e.m.f. per phase then becomes, at any instant,

$$e_{ph} = 2\pi f T \Phi_m \cos \omega t \cdot \frac{N}{2\pi}\left[\cos \beta \cdot \cos \frac{2\pi}{N} + \sin \beta \cdot \sin\frac{2\pi}{N} - \cos \beta\right]$$

with $mT_c = T$ turns per phase, and $\sigma = 2\pi/N$. It is seen that e_{ph}

is an alternating voltage depending in amplitude upon the expression in brackets, i.e. upon the position of the phase in the field. In the (single-phase) induction regulator the magnitude of the induced e.m.f. is varied by altering the position of the "locked rotor" with respect to the flux axis.

Writing $\beta = (\pi/N) + \gamma$, then γ = angular displacement of the *phase-centre* from the pole-centres. The expression for e_{ph} then simplifies to

$$e_{ph} = 2\pi f T \Phi_m \cos \omega t \cdot \frac{N}{\pi} \sin \frac{\pi}{N} \cdot \sin \gamma$$

or, as virtual value

$$E_{ph} = \sqrt{2} \pi f T \Phi_m k_m \sin \gamma$$

since $(N/\pi) \sin (\pi/N) = k_m$, the distribution factor (see under case *B*).

The static transformer belongs also to the class of machine under consideration. The coils link the whole flux, so that the integral

$$\frac{1}{\sigma} \int_{\beta-\sigma}^{\beta} \sin \alpha \cdot d\alpha = 1$$

giving, for m coils of T_c turns (a total of T turns), the instantaneous value,

$$\Sigma(e) = e_{ph} = \omega T \Phi_m \cos \omega t$$
$$= 2\pi f T \Phi_m \cos \omega t$$

or the virtual value,

$$E_{ph} = \sqrt{2} \pi f T \Phi_m \text{ volts}$$

which is the well-known equation for the induced e.m.f. of a transformer. The factor $\cos \omega t$ in the expression for e_{ph} indicates that the e.m.f. lags by 90° upon the flux $\Phi = \Phi_m \sin \omega t$.

Case E: Pulsating flux; moving coils; commutator connections (Fig. 380). The a.c. commutator and the induction machines form the most generalized cases of e.m.f. induction, since both modes of linkage-change occur simultaneously in each. For the a.c. commutator machine, the case here dealt with, both terms of eq. (192) must be considered. With the brushes in the neutral axis, $\beta = \pi/N$, and $\sigma = \pi$. The brush or armature e.m.f. (single-phase) will then be

$$\Sigma(e) = e_a = \left[m\omega_r T_c \Phi_m \sin \omega t \cdot \frac{1}{\pi} \int_{-\frac{1}{2}\pi}^{\frac{1}{2}\pi} \cos \alpha \cdot d\alpha \right.$$
$$\left. + m\omega T_c \Phi_m \cos \omega t \cdot \frac{1}{\pi} \int_{-\frac{1}{2}\pi}^{+\frac{1}{2}\pi} \sin \alpha \cdot d\alpha \right]$$

But since the second integral is zero, the e.m.f. of pulsation, e_p, will not appear in the brush e.m.f. Putting $mT_c = pZ/4a$ as in case A, the brush e.m.f. at any instant is

$$e_a = e_r = \frac{p}{a}nZ\Phi_m \sin \omega t \text{ volts.}$$

which is an e.m.f. of fundamental (flux) frequency in phase with ⁻¹ flux.

Virtual value

$$E_a = E_r = \frac{p}{a}nZ\frac{\Phi_m}{\sqrt{2}} \text{ volts.}$$

Should the brushes stand at an angle $\beta \neq \pi/2$ from the pole-centres, the integration is performed between the limits β and $\beta - \pi$, giving

$$e_a = \left[\frac{p}{a}nZ\Phi_m \sin \omega t \cdot \sin \beta - \frac{p}{a}fZ\Phi_m \cos \omega t \cdot \cos \beta\right]$$

$$= \left[4f_rT\Phi_m \sin \omega t \cdot \sin \beta - 4fT\Phi_m \cos \omega t \cdot \cos \beta\right]$$

$$= e_r + e_p$$

showing that e_p now appears in the armature brush e.m.f. as a component lagging by 90° on e_r. The whole of the brush e.m.f. does not, therefore, represent "energy-converting" e.m.f., but includes a component of "transformer" e.m.f.

Case F: Pulsating flux; moving coils; tapping connections (Fig. 381). The final case is that of the single-phase induction machine. The e.m.f. in a phase is exactly as stated in eq. (192). The angle β can be expressed as $\beta = \omega_r t + (\pi/N)$, since the armature is rotating at angular speed ω_r, and can be assumed to start at $t = 0$ from a position such that the phase-centre is coincident with the pole-centre. The integration is now performed between the limits $\beta = \omega t + (\pi/N)$ and $\beta - \sigma = \omega_r t - (\pi/N)$. Then, with $mT_c = T$, and $(N/\pi) \sin (\pi/N) = k_m$,

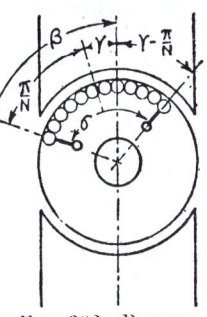

FIG. 382. PHASE-WOUND MACHINE

$$e_{ph} = \left[\omega_rT\Phi_m \sin \omega t \cdot k_m \cdot \cos \omega_r t\right.$$

$$\left. + \omega T\Phi_m \cos \omega t \cdot k_m \sin \omega_r t\right]$$

$$= k_mT\Phi_m\left[\omega_r \cdot \sin \omega t \cdot \cos \omega_r t + \omega \cdot \cos \omega t \cdot \sin \omega_r t\right]$$

It is usual, in the induction machine, to express ω_r, the rotational velocity, as $(\omega - \omega_2)$, where $\omega_2 = 2\pi f_2$, and f_2 is the "slip frequency." With this substitution,

$$e_{ph} = k_m T \Phi_{max} \left[\frac{2\omega - \omega_2}{2} \sin (2\omega - \omega_2)t - \frac{\omega_2}{2} \sin \omega_2 t \right]$$

$$= \frac{1}{2} k_m T \Phi_{max} \left[(2\omega - \omega_2) \sin (2\omega - \omega_2)t - \omega_2 \sin \omega_2 t \right].$$

The terms in the brackets involve sinusoidal variations at frequencies defined by $2\omega - \omega_2 = \omega + \omega_r$ and $\omega_2 = \omega - \omega_r$. The effect is as if the single *pulsating* flux Φ_m were divided into equal parts $\frac{1}{2}\Phi_m$ *rotating* in opposite directions at synchronous speed ω radians per sec. The armature runs at an angular speed ω_r differing by ω_2 from one of these rotating fluxes, and therefore by $2\omega - \omega_2$ from the other. The induced e.m.f. can be considered as the combination of the e.m.f.'s induced by the two fluxes acting separately. This equation forms the basis of the theory of the single-phase induction machine.

The e.m.f. equations found for the various types refer to single phases in every case. The arrangements of the various machines for two or more phases (generally three) requires phase-e.m.f.'s to be derived for the additional phases, with due regard to their relative geometrical spacing on the armature. In three-phase synchronous and induction machines the e.m.f.'s are simply phase-displaced by $2\pi/N$, a consequence of the constant-flux condition in these machines under balanced-load or no-load operation.

CHAPTER XXVI

THE CIRCUIT THEORY OF ELECTRICAL MACHINES

1. The Machine as a Circuit. The study of electrical machines is devoted (*a*) to understanding how they work, and (*b*) to predicting their performance. The former is the concern of the applications engineer and the student: the latter chiefly of the designer, who may use empirical methods not obviously based on clear analytical grounds.

Ideally, it ought to be possible to express the behaviour of electrical machines in terms of Maxwellian electromagnetic field equations. Such an approach is exceedingly difficult, and the normal method (used in previous Chapters) is through the simplified concepts of induction and interaction. Various techniques (e.g. complexor and locus diagrams, equivalent circuits), mostly of particular rather than of general application, have been employed. The disadvantage of this approach is its practical limitation to the steady state. But if one more step is taken from "reality" into the abstract, machines can be dealt with in terms of *circuits*.

So-called "equivalent" circuits are in common use, notably in the analysis of the transformer (e.g. Figs. 15 and 157) and the induction motor, but these are applicable only to particular conditions. A truly equivalent circuit will apply to its prototype machine under all conditions, steady and transient, balanced and unbalanced. To such a circuit can be applied all the powerful techniques of circuit analysis—circuit laws, network theorems, operational and matrix methods, network analysers and computers. The circuit can be used to obtain the three primary characteristics (voltage, current and torque) in advanced studies of generator faults, changes of load, or automatic control, all so important in modern engineering.

There are some sacrifices. Manipulation of the equations is only tractable on the assumption of linear parameters, Non-linear effects, such as saturation, must be the subject of subsequent adjustment in the light of experience by semi-empirical methods. Space-harmonics of flux, brush-contact phenomena and commutation effects must be ignored. These effects are far from negligible in practical machines; nevertheless to ignore them is to assume conditions differing only in degree from those of normal circuit theory. Finally, the circuits are still not truly equivalent because propagation phenomena are disregarded, so that surge-voltage effects need separate consideration.

2. Circuit Elements. When a voltage of instantaneous value v is applied to a series circuit comprising pure parameters of resistance

R, inductance L and capacitance C, a current i flows such that, by the energy-rate principle,

$$vi = (Ri)i + \left(L\frac{di}{dt}\right)i + \left(\frac{1}{C}\int i \cdot dt\right)i.$$

The terms in brackets denote the voltage drops across the parameters concerned. Eliminating i,

$$v = Ri + L\frac{di}{dt} + \frac{1}{C}\int i \cdot dt$$

or, using the operator p to represent d/dt,

$$v = Ri + Lpi + (1/Cp)i = [R + Lp + (1/Cp)]i.$$

With $p = 0$ this expression gives the steady-state d.c. condition. With $p = j\omega$ it corresponds to the steady-state sine condition at

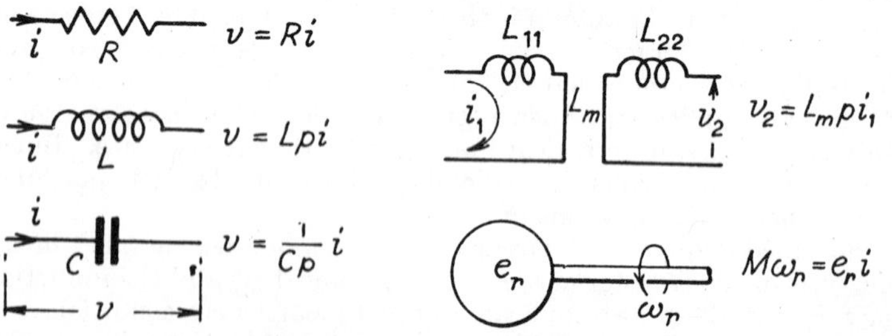

FIG. 383. BEHAVIOUR OF CIRCUIT ELEMENTS

frequency ω. As it stands, the expression can be used with operational calculus to give the current/time relation for an applied voltage stimulus v of *any* form applied at $t = 0$. The statement of circuit behaviour is thus general, and particular cases are readily derived from it.

A machine has windings with resistance and inductance, pairs of windings with mutual inductance and an armature in which rotational (and other) e.m.f.'s are induced. Certain machines have appreciable capacitance. The equations of behaviour for these elements are summarized in Fig. 383. The expressions hold for instantaneous values.

3. **Conventions.** All circuits have an applied voltage v derived from an external source; v can, on occasion, be zero. The current i has, when positive, the same direction as v, so that positive vi means power flow into the circuit from the external source. Thus for a resistor the relation is $vi = (Ri)i = Ri^2$, or $v = Ri$. For an inductive and dissipative circuit the relation is

$$v = Ri + L(di/dt) = (R + Lp)i$$

at every instant. In the case of a machine winding, additional terms may appear in order to represent voltage components required to overcome e.m.f.'s of mutual inductance and of rotation in a flux. When both v and i are positive at a given instant, and an active armature winding receives power from the external source, the machine of which it is a part is acting as a *motor*. A reversal of either v or i means *generator* action. Thus both generating and motoring conditions are included in the same equations.

In brief, the circuit analysis consists in setting up operational equations for comprehensive circuits representing the essential features of electrical machines and expressed in terms of the resistances, self-inductances and mutual-inductances of windings. Having obtained the equations of behaviour, they can be solved for any specified conditions.

TRANSFORMERS

4. Magnetically-coupled Coils. Consider two coils, Fig. 384 (a), having a partially common magnetic circuit. Current i_1 in the primary will produce a total flux Φ_{11} which, in the turns of the primary, develops a total linkage Ψ_{11}. Change of current i_1 with time t induces in the primary the e.m.f.

$$e_{11} = -\frac{d\Psi_{11}}{dt} = -\frac{d\Psi_{11}}{di_1} \cdot \frac{di_1}{dt} = -L_{11}\frac{di_1}{dt} = -L_{11}pi_1,$$

where $L_{11} = d\Psi_{11}/di_1$ is the total primary self-inductance. L_{11} depends on the geometry and number of turns of the primary; it is also a function of the permeability of the medium in which the coil is immersed, and if there are no saturation effects L_{11} is a constant. It is important to note that L_{11} is not the *leakage* inductance employed in transformer theory (e.g. Chapter III, §4), but the total or *magnetizing* inductance of the primary as a simple inductive reactor.

Of the flux Φ_{11} produced by i_1 a part, Φ_{12}, threads through the turns of the secondary winding producing therewith the linkage Ψ_{12}. Variation of the primary current will then induce in the secondary the e.m.f.

$$e_{12} = -\frac{d\Psi_{12}}{dt} = -\frac{d\Psi_{12}}{di_1} \cdot \frac{di_1}{dt} = -L_{12}\frac{di_1}{dt} = -L_{12}pi_1,$$

where $L_{12} = d\Psi_{12}/di_1$ is the mutual inductance of the primary with the secondary. L_{12} depends on the geometry, relative position and numbers of turns of the primary and secondary, and also on the magnetic properties of the ambient medium. Again if saturation is absent the mutual inductance L_{12} is a constant.

In like fashion a current i_2 in the secondary circuit will develop linkages Ψ_{22} in itself and Ψ_{21} in the primary, from which the secondary self-inductance L_{22} and mutual inductance L_{21} are defined.

Further, if there is no saturation, the two mutual inductances are equal: $L_{12} = L_{21} = L_m$.

If a voltage v_1 be applied to the primary, and the secondary circuit be closed through an impedance Z (expressed in operational form), then the following equations must be satisfied, Fig. 384 (b):

$$\left.\begin{array}{l} v_1 = r_1 i_1 + L_{11} p i_1 + L_m p i_2, \\ 0 = r_2 i_2 + L_{22} p i_2 + L_m p i_1 + Z i_2. \end{array}\right\} \quad (193)$$

The zero indicates the fact that no external source of energy is present

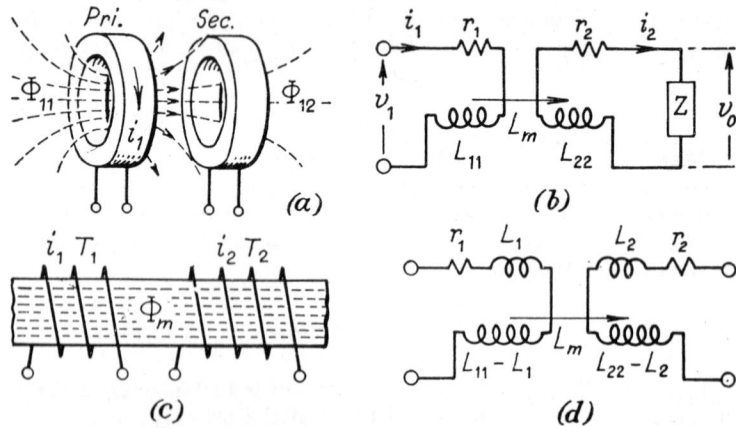

FIG. 384. TRANSFORMER: COILS WITH MUTUAL COUPLING

in the secondary circuit. The term $Z i_2$ corresponds to the load (or output) voltage v_0 of the secondary.

The discussion is quite general, and can be applied to any two-winding transformer with an ambient medium of constant permeability. When the coils are loose-coupled and their mutual inductance is low, the actual numbers of primary and secondary turns may have little significance. In the case of power transformers, in contrast, the coupling is tight and the secondary output and primary input voltages are closely associated with the turns-ratio. Unfortunately the use of a common magnetic core inevitably introduces saturation phenomena, so that the inductance coefficients become functions of current. The expressions in eq. (193) still apply, but they cannot be solved. The practical implications of this condition is discussed in Chapter V, § 13, in connection with transient currents.

IDEAL TRANSFORMER. The practical power transformer can be considered as a degraded form of ideal transformer. Such an ideal transformer, Fig. 384 (c),

(i) has no losses of any kind, so that the output and input volt-amperes are equal;

(ii) has perfect magnetic linkage between primary and secondary, so that any flux completely links both windings; and

(iii) has zero magnetic reluctance so that a flux requires a vanishingly small m.m.f. to excite it. All the inductances are now *infinite*.

Using the same method of definition as before, then

$$L_{11} = \Psi_{11}/i_1 = T_1\Phi_{11}/i_1; \qquad L_{22} = \Psi_{22}/i_2 = T_2\Phi_{22}/i_2;$$

and
$$L_{12} = L_{21} = L_m = T_1\Phi_{22}/i_2 = T_2\Phi_{11}/i_1;$$

whence $L_m = \sqrt{(L_{11}L_{22})}$, and the inductances are related to the turns by the expressions

$$L_{11} = L_m(T_1/T_2) \qquad \text{and} \qquad L_{22} = L_m(T_2/T_1).$$

When both coils carry current together there is a single mutual flux Φ_m and the primary ampere-turns exactly balance those of the secondary. The complexor diagram for steady state sine conditions is shown in Fig. 12, expressed in r.m.s. values. Here $E_1 = -j\omega L_m I_2$ and $E_2 = -j\omega L_m I_1$.

TRANSFORMER WITH LEAKAGE. A method of bringing leakage inductance into explicit account in the practical transformer is to write, for Fig. 384 (*d*),

$$L_m(T_1/T_2) = L_{11} - L_1 \qquad \text{and} \qquad L_m(T_2/T_1) = L_{22} - L_2,$$

where L_1 and L_2 are the *leakage inductances* used so commonly and having such important consequences in simple theory of power transformers. Then in place of $L_m = \sqrt{(L_{11}L_{22})}$ as in the ideal transformer, the relation is

$$L_m = \sqrt{[(L_{11} - L_1)(L_{22} - L_2)]} = k\sqrt{(L_{11}L_{22})},$$

introducing the coefficient of coupling k which will normally be of the order 0·95. The expression is true only for constant permeability, and therefore would seem to be inapplicable to a power transformer; but in fact while L_{11}, L_{22} and L_m vary with current, the leakage parts L_1 and L_2 remain substantially constant because they are concerned with flux in non-ferrous paths. This fact accounts for the successful use of leakage reactance in the general theory of power transformers to determine such important characteristics as regulation.

The effect of separating out L_1 and L_2 as leakage inductances, substantially constant, is to leave the remainders $(L_{11} - L_1)$ and $(L_{22} - L_2)$ again forming together an ideal transformer. Likewise the winding resistances, core losses and magnetizing inductance are introduced, as shown in Fig. 15, as defects separated from the idealized part of the transformer.

5. **Circuit Equations.** Eqs. (193) can be written

$$\left.\begin{aligned} v_1 &= (r_1 + L_{11}p)i_1 + L_m pi_2 &= Z_{11}i_1 + Z_{12}i_2 \\ 0 &= L_m pi_1 + (r_2 + L_{22}p + Z)i_2 &= Z_{21}i_1 + Z_{22}i_2 \end{aligned}\right\} \quad . \quad (194)$$

in which Z_{11} is the total impedance of the primary circuit, Z_{22} the total impedance (including the load impedance Z) of the secondary circuit and $Z_{12} = Z_{21} = L_m p$ is the mutual impedance. The secondary equation gives $i_2 = -i_1(Z_{21}/Z_{22})$ and the following results are readily obtained:

Current Ratio: $i_2/i_1 = -(Z_{21}/Z_{22})$.

Voltage Ratio: Calling the output voltage $v_0 = Zi_2$, then

$$v_0/v_1 = ZZ_{12}/(Z_{12}^2 - Z_{11}Z_{22}).$$

Effective Primary Impedance: This is the impedance presented by the primary to the supply, as affected by the secondary load:

$$Z_1' = v_1/i_1 = (Z_{11}Z_{22} - Z_{12}^2)/Z_{22}.$$

The implications of the above are seen more clearly if the case of

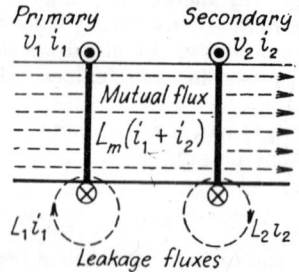

FIG. 385. TRANSFORMER IN PER-UNIT TERMS

an ideal transformer, loaded on to an impedance Z, is taken for steady-state sine conditions at angular frequency ω. Then

$$Z_{11} = j\omega L_{11}, \qquad Z_{22} = j\omega L_{22} + Z, \qquad Z_{12} = j\omega L_m, \qquad L_m^2 = L_{11}L_{22}.$$

The current ratio now becomes

$$i_2/i_1 = -j\omega L_m/(j\omega L_{22} + Z) = -L_m/L_{22} = -T_1/T_2,$$

because ωL_{22} is infinite, making any finite load impedance of no effect and preserving the identity of the current-ratio and inverse turn-ratio. The voltage ratio is

$$v_0/v_1 = -Z \cdot j\omega L_m/j\omega L_{11} \cdot Z = -L_m/L_{11} = -T_2/T_1,$$

because $L_m^2 - L_{11} \cdot L_{22} = 0$. The effective primary input impedance is

$$Z_1' = j\omega L_{11}Z/(j\omega L_{22} + Z) = Z(L_{11}/L_{22}) = Z(T_1/T_2)^2.$$

Thus the ideal transformer acts as a perfect converter of impedance, and can be used to "match" one impedance to another. Also its voltage and current ratios are fixed by the ratio of turns.

6. **Per-unit Transformer.** A power transformer, so far as its winding resistance and inductance are concerned, can be represented as in Fig. 385 by two single turns on a common core. The turns-ratio

is unity, and if all quantities are expressed in *per-unit* values, the circuit equations can be written

$$v_1 = r_1 i_1 + (L_m + L_1)p i_1 + L_m p i_2,$$
$$v_2 = r_2 i_2 + (L_m + L_2)p i_2 + L_m p i_1,$$ . . (195)

where v_1 and v_2 are the voltages *applied* to primary and secondary. This form of circuit is introduced because it is of the kind used in the analysis of rotating machines in the following paragraphs. L_m is so greatly affected by saturation, however, that the practical utility of the equations is very limited. In machines the air-gap provides a legitimate reason for neglecting saturation and assuming all inductance coefficients to be constant.

ROTATING MACHINES

7. The Basic Machine. The common features of all rotating electro-magnetic machines were discussed in Chapter VIII. Much unnecessary generalization is avoided by excluding reluctance machines that are salient on both sides of the air gap. The basic unit is then a 2-pole machine with one cylindrical and one salient element. The saliency leads naturally to the adoption of the two-reaction analysis.

The direct axis is chosen horizontal, the quadrature axis vertical and the rotation anti-clockwise. Because only relative motion is significant, the salient element is taken as stationary, so fixing the d- and q-axes. The moving element (armature) rotates around or within the fixed element.

The windings on each element are grouped or resolved so as to magnetize along one or other axis. Positive current in any winding then magnetizes in the positive axis-direction. The d- and q-axis coil groups are in space quadrature, so that there is no mutual inductance between any pair of coils in different groups. But rotational e.m.f.'s can be developed in *armature* coils of one axis due to movement in the flux of coils on the other axis.

Per-unit values are employed throughout. Unit power, voltage and current correspond to the rating. Unit speed is 1 rad/s. Unit torque corresponds to unit power, at rated speed expressed in electrical radians per sec. for a 2-pole unit.

ARMATURE E.M.F.'s. The machine of Fig. 386 (*a*) has a commutator armature with brushes in the q-axis, so that it produces a q-axis m.m.f. The winding is represented by the fixed coil Q, as shown in (*b*). Similarly the field windings are represented by coil F. When the armature rotates, e.m.f.'s are taken as generated in the coil Q, which must therefore be assigned the property of developing a rotational e.m.f. *even though it does not move.* An equally valid alternative arrangement is shown at (*c*).

Magnetically, F and Q can be taken as *single turns*, but the fluxes due to them are nevertheless assumed to be distributed sinusoidally over the pole-pitch, i.e. space harmonics are ignored.

Consider, in (b), the effect on the coil Q of the flux due to coil F. No mutually-induced pulsation (or "transformer") e.m.f. can be developed in Q by any variation of i_f, because the magnetic axes of Q and F are at right-angles. There will, however, be an e.m.f. of rotation in Q when the winding it represents rotates at angular speed ω_r, of the value

$$e_{rq} = \omega_r \psi_d,$$

where ψ_d is the mutual magnetic linkage in the armature winding due to a d-axis flux produced by F.

Any q-axis flux (including that due to its own current i_q) will produce in Q a pulsation e.m.f.

$$e_{pq} = -p\psi_q$$

where ψ_q is the linkage with the armature winding on the q-axis.

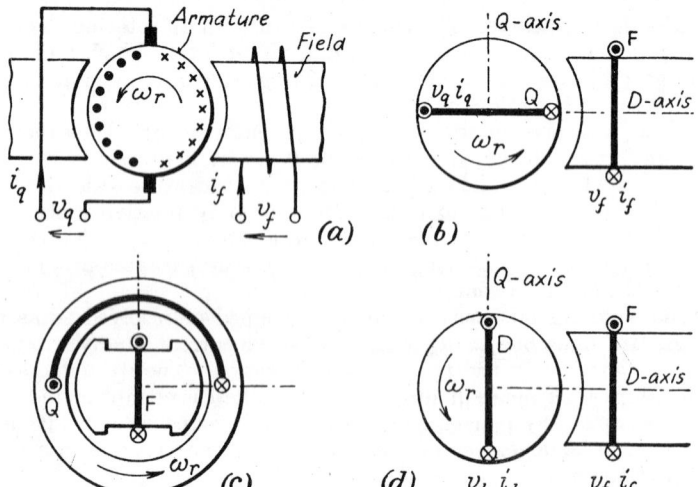

FIG. 386. DEVELOPMENT OF THE BASIC UNIT MACHINE

Suppose now that the armature winding has its brush-axis shifted by 90°, so that it can be represented by a d-axis coil D, Fig. 386 (d). There can be no rotational e.m.f. generated by the flux of F, but both coils are coaxial and any change of linkage ψ_d will produce in D the pulsation e.m.f.

$$e_{pd} = -p\psi_d.$$

If there were any q-axis flux with linkages ψ_q, a rotational e.m.f.

$$e_{rd} = -\omega_r \psi_q$$

would result. The negative sign is due to the fact that a d-axis coil with reference to a q-axis flux bears the opposite relation to that of a q-axis coil and a d-axis flux.

ARMATURE APPLIED VOLTAGES. Remembering that voltages are

considered as *applied* to the winding terminals and that a positive current flows in the corresponding direction, then such voltages must account for IR drops and leakage inductance effects, and overcome any e.m.f.'s resulting from mutual fluxes. Hence for armature coils D and Q,

$$v_d = r_d i_d + L_d p i_d + p\psi_d + \omega_r\psi_q$$

and $$v_q = r_q i_q + L_q p i_q + p\psi_q - \omega_r\psi_d$$

. (196)

Here r_d, r_q, L_d and L_q are the resistances and leakage inductances of coils D and Q.

AXIS FLUXES. The arrangement in Fig. 387 (a) consists of exclusively d-axis coils. The field winding F and armature winding D are,

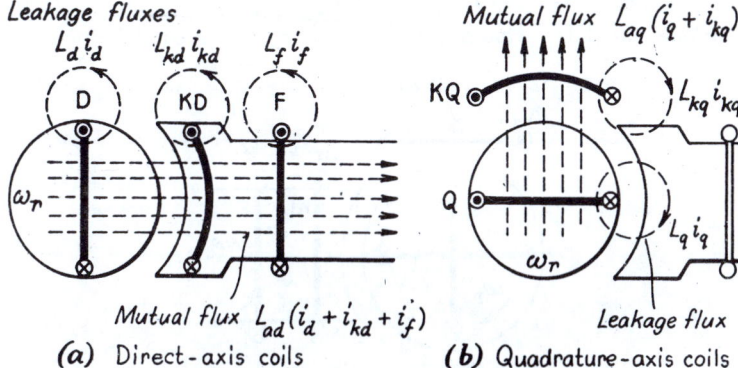

Leakage fluxes

$L_d i_d$ $L_{kd} i_{kd}$ $L_f i_f$

Mutual flux $L_{aq}(i_q + i_{kq})$

$L_{kq} i_{kq}$

$L_q i_q$

Mutual flux $L_{ad}(i_d + i_{kd} + i_f)$ Leakage flux

(a) Direct-axis coils   **(b)** Quadrature-axis coils

FIG. 387. MUTUAL AND LEAKAGE FLUXES OF AXIS COIL GROUPS

for generality, augmented by a third coil, KD, on the stationary salient pole. (KD represents, for example, a field collar or damping winding in a synchronous machine, or the equivalent of the eddy-current circuit in a solid pole-shoe.) The problem is to find the d-axis flux resulting from currents in all three coils.

There are three mutual inductances concerned, i.e. D with KD, KD with F, and F with D. A useful simplification is to take these as all equal to an inductance L_{ad}. Each coil will have a leakage inductance of its own, and the total inductances will consequently be

Coil D: $(L_{ad} + L_d)$ Coil KD: $(L_{ad} + L_{kd})$ Coil F: $(L_{ad} + L_f)$

The total flux due to a current i_f in F will thus be $(L_{ad} + L_f)i_f$, and so on. The mutual linkage with the armature coil D when there are currents in each coil will be

$$\psi_d = L_{ad}(i_f + i_{kd} + i_d) \qquad . \qquad . \qquad . \quad (197)$$

Considering the group of q-axis coils in Fig. 387 (b), and calling the mutual inductance L_{aq}, the mutual linkages will be

$$\psi_q = L_{aq}(i_q + i_{kq}) . \qquad . \qquad . \qquad . \quad (198)$$

If the leakage inductances are respectively L_q and L_{kq}, the total inductances will be $(L_{aq} + L_q)$ and $(L_{aq} + L_{kq})$ for Q and KQ respectively.

CIRCUIT EQUATIONS. It is now possible to write down the equations for all the circuits in the basic unit machine, Fig. 388.

Stator Field Circuit F: This has a resistance r_f and total inductance $(L_{aa} + L_f)$. Any other d-axis circuit will affect its flux, but no q-axis current can affect circuit F, nor do any rotational voltages appear; hence the applied voltage v_f will have the components

$$v_f = r_f i_f + (L_{aa} + L_f)p i_f + L_{ad}p i_{kd} + L_{ad}p i_d. \qquad . \text{(199)}$$

Stator Coil KD: This circuit differs in no way from F, so that

$$v_{kd} = r_{kd}i_{kd} + (L_{aa} + L_{kd})p i_{kd} + L_{ad}p i_f + L_{ad}p i_d \qquad . \text{(200)}$$

Stator Coil KQ: This has a resistance r_{kq} and a total inductance

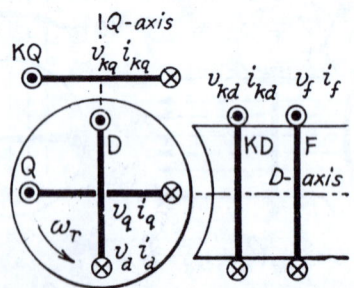

FIG. 388. UNIT MACHINE

$(L_{aq} + L_{kq})$. It has no rotational e.m.f.'s, but will be affected magnetically by any current i_q in armature coil Q: thus any applied voltage v_{kq} must satisfy the expression

$$v_{kq} = r_{kq}i_{kq} + (L_{aq} + L_{kq})p i_{kq} + L_{aq}p i_q. \qquad . \text{(201)}$$

Armature Coil D: The armature coils D and Q have the additional property of inducing e.m.f.'s of rotation. Their applied voltages are given in eq. (196), and using the expressions for ψ_d and ψ_q in eqs. (197) and (198).

$$\begin{aligned} v_d = r_d i_d + (L_{aa} + L_d)p i_d &+ L_{ad}p i_f + L_{ad}p i_{kd} \\ &+ L_{aq}\omega_r i_{kq} + (L_{aq} + L_q)\omega_r i_q \end{aligned} \qquad . \quad . \text{(202)}$$

Armature Coil Q: In a similar manner

$$\begin{aligned} v_q = r_q i_q + (L_{aq} + L_q)p i_q &+ L_{aq}p i_{kq} \\ &- L_{ad}\omega_r i_f - L_{ad}\omega_r i_{kd} - (L_{aa} + L_d)\omega_r i_d \end{aligned} \qquad . \text{(203)}$$

EQUATIONS IN MATRIX FORM. The above equations can be written down accurately and rapidly in matrix form with the aid of a few simple rules.* For the five coils F, KD, KQ, D and Q, a 5-row

* Gibbs, "Algebra of Electric Machine Analysis," *B.T.-H. Tech. Monograph*, TMS 757.

5-column matrix is set out as in Fig. 389. The rows are related to the voltages and the columns to the corresponding currents. Compartments in the main diagonal are emphasized: these contain the self-impedances. Thus r_f is entered only in row F and column F because only current i_f can produce a voltage drop in r_f, no resistance carrying more than one current. For the same reason the total self-inductances are also entered in the compartments of the main diagonal.

The remaining compartments contain the mutual impedances, where any exist. Thus in row F, the second compartment contains

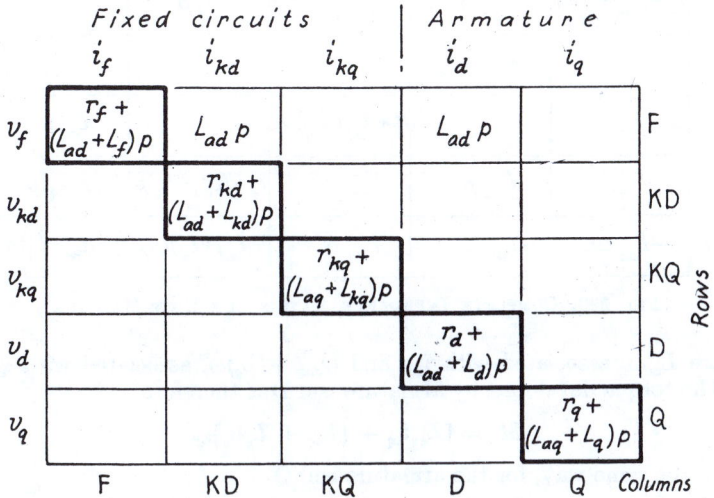

FIG. 389 CONSTRUCTION OF IMPEDANCE MATRIX FOR UNIT MACHINE

$L_{ad}p$, the third zero, the fourth $L_{ad}p$, and the fifth again zero. There will be no entries in row-F compartments lying in q-axis columns. Now, by inspection of the F-row, v_f is read as the sum of the products of the compartment impedances with their appropriate currents. This gives eq. (199).

The same procedure is employed to fill and read each row of the matrix. But in the case of the fourth and fifth rows (and only in these rows), rotational-e.m.f. terms will appear. Every coefficient of p in row Q appears again in row D in the same column, but with p replaced by $-\omega_r$. Similarly, every coefficient of p in row Q appears again in row D in the same column, but with p replaced by $+\omega_r$. The matrix is now complete, and if read off as already described it will give eqs. (199) to (203) directly.

Fig. 390 gives the completed matrix. If the machine has additional d- or q-axis coils, it is only necessary to introduce another column and row for each. If certain coils are absent, their columns and rows are simply omitted.

TORQUE. Torque is developed only in armature windings which, with the convention here adopted, move with respect to the fixed d- and q-axes. The torque produced by any winding is such that $M\omega_r = e_r i$, so that M is obtained from the rotational voltage multiplied by the current of the winding concerned, and divided by ω_r.

Consider row D of the matrix in Fig. 390. The rotational terms

	F	KD	KQ	D	Q
F	$r_f + (L_{ad} + L_f)p$	$L_{ad}\, p$		$L_{ad}\, p$	
KD	$L_{ad}\, p$	$r_{kd} + (L_{ad} + L_{kd})p$		$L_{ad}\, p$	
KQ			$r_{kq} + (L_{aq} + L_{kq})p$		$L_{aq}\, p$
D	$L_{ad}\, p$	$L_{ad}\, p$	$L_{aq}\, \omega_r$	$r_d + (L_{ad} + L_d)p$	$(L_{aq} + L_q)\omega_r$
Q	$-L_{ad}\, \omega_r$	$-L_{ad}\, \omega_r$	$L_{aq}\, p$	$-(L_{ad} + L_d)\omega_r$	$r_q + (L_{aq} + L_q)p$

FIG. 390. COMPLETE IMPEDANCE MATRIX FOR UNIT MACHINE

are $L_{aq}\omega_r$ associated with i_{kq}, and $(L_{aq} + L_q)\omega_r$ associated with i_q. The torque developed by armature coil D is therefore

$$M_d = [L_{aq}i_{kq} + (L_{aq} + L_q)i_q]i_d.$$

In the same way, for the armature coil Q.

$$M_q = [-L_{ad}i_f - L_{ad}i_{kd} - (L_{ad} + L_d)i_d]i_q.$$

The total armature torque is the sum of these two components, giving

$$M = L_{aq}(i_{kq} + i_q)i_d - L_{ad}(i_f + i_{kd} + i_d)i_q - (L_d - L_q)i_d i_q \quad (204)$$

The last term is due to the saliency, which makes $L_d > L_q$.

Mechanical torque is defined as that *applied* to the shaft, so that the torque developed by an electrical input becomes a mechanical output and is therefore reckoned as negative. The total torque of a machine will, of course, include both electrically-developed and mechanically-applied torques. These normally balance (i.e. they sum to zero), but if they do not, an acceleration torque $Jp\omega_r$ will occur to alter the speed ω_r of the total inertia J. It has been pointed out in Chapter VIII that an electrical machine couples an electric with a mechanical energy system so that, where speed changes occur, the mechanical characteristics are a prime factor.

8. **D.C. Machine.** This has a clear separation of its stator and rotor windings into d- and q-axis groups. Consider the elementary machine represented in Fig. 386 (*a*) and (*b*). Most of the flux produced by i_f in F crosses the gap and is mutual with the armature, while the remainder is field leakage. Call the former $L_{ad}i_f$ and the latter $L_f i_f$, where L_{ad} and L_f are per-unit inductance coefficients representing flux per unit current. Then the total self-inductance of F is $(L_{ad} + L_f)$. With a per-unit field-circuit resistance r_f the voltage equation for the field is

$$v_f = r_f i_f + (L_{ad} + L_f)pi_f.$$

The armature winding Q has a resistance r_q and a total self-inductance that can conveniently be called $(L_{aq} + L_q)$. It also has an e.m.f. generated because the armature winding it represents rotates at angular speed ω_r in the mutual part of the field flux $L_{ad}i_f$. Consequently

$$v_q = r_q i_q + (L_{aq} + L_q)pi_q - L_{ad}\omega_r i_f.$$

The negative sign arises from the convention of chosen positive directions. The equations for v_f and v_q could be obtained directly from the matrix of Fig. 390 by eliminating all columns and rows except those for coils F and Q, showing that the d.c. machine considered is a form of the general basic machine.

The power supplied to the armature circuit is

$$v_q i_q = r_q i_q{}^2 + (L_{aq} + L_q)i_q \cdot pi_q - L_{ad}\omega_r i_f i_q.$$

The first term on the right is the I^2R loss. The second is the rate of change of stored magnetic energy (see Chapter VIII). The third term is the rate at which electrical-mechanical energy conversion is taking place. The term indicates a negative input, i.e. a positive mechanical power output, and from it the torque is derived by dividing by ω_r to give

$$M = - L_{ad}i_f i_q \text{ per unit.}$$

For shunt and series machines it is necessary to take account of the connections. For a shunt machine $v_q = v_f$; while for a series machine $i_q = i_f$, and $v_q + v_f = v$, the total applied voltage.

The expressions obtained can be applied to transient conditions by use of operational calculus. If $j\omega$ is written for p, they hold for steady-state a.c. operation as for a commutator motor. With $p = 0$ the conditions become those of steady-state d.c. operation, i.e.

$$v_f = r_f i_f \quad \text{and} \quad v_q = r_q i_q - L_{ad}\omega_r i_f i_q.$$

The latter is of the well-known form $V = rI + E$.

OTHER VARIANTS. With two armature brush-pairs, one in the d- and the other in the q-axis, a cross-field machine results. Further, a coil KQ can be introduced to represent commutating poles, and KD as a compensating winding. In each case a restraint is put on the currents by the method of connection. These cases will not be considered further here.

9. Synchronous Machines. The commutator winding of the d.c. machine, with its stationary axis of m.m.f., is equivalent by nature to a fixed axis coil. A three-phase winding, however, rotates with respect to the d- and q-axes, and it is necessary to show (i) that it can be replaced by fixed D and Q armature coils, and (ii) the relation between the actual phase currents and their axis counterparts.

ARMATURE PHASE INDUCTANCE. Consider phase A of a three-phase winding. Its inductance will be a combination of its own self-inductance L_{aa} and its mutual inductances L_{ab} and L_{ac} with phases B and C. All three inductance coefficients vary with the relative

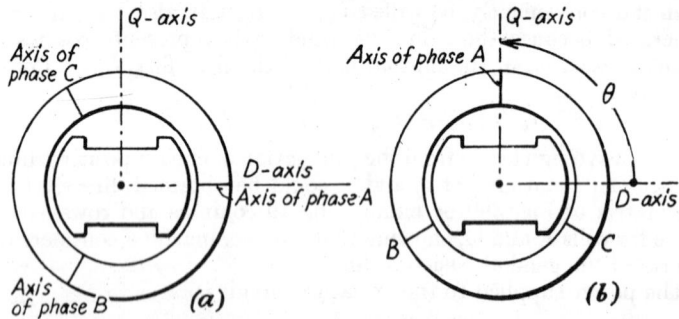

FIG. 391. EFFECT OF RELATIVE POSITION ON PHASE INDUCTANCE

position of stator and rotor because of the effect of saliency on the air gap. When the axis of phase A coincides with the direct (pole) axis, Fig. 391 (a), its self inductance will be a maximum: when at right-angles to the d-axis, as in (b), it will be a minimum. Thus L_{aa} fluctuates twice per revolution and can be approximately expressed as

$$L_{aa} = L_0 + L_2 \cos 2\theta.$$

Fig. 391 also shows that the mutual inductance of phases B and C is a maximum at (a) and minimum at (b). Again assuming a harmonic fluctuation, then

$$L_{bc} = L_{cb} = -L_{m0} + L_{m2} \cos 2\theta,$$

from which

$$L_{ab} = -L_{m0} + L_{m2} \cos (2\theta - 2\pi/3),$$

and

$$L_{ac} = -L_{m0} + L_{m2} \cos (2\theta + 2\pi/3).$$

With balanced phase currents and normal-speed rotation corresponding to the frequency, then for the three phases A, B and C:

$$i_a = i_m \sin (\theta - \gamma);$$
$$i_b = i_m \sin (\theta - \gamma - 2\pi/3);$$
$$i_c = i_m \sin (\theta - \gamma + 2\pi/3).$$

The position $\theta = 0$ is the coincidence of the axis of phase A with the pole-centre; and γ is the internal phase angle of the machine, i.e. the angle by which the current lags the induced e.m.f.

The total flux linkage of phase A contributing to its inductance is consequently

$$L_{aa}i_a + L_{ab}i_b + L_{ac}i_c = (L_0 + L_2 \cos 2\theta)i_m \sin (\theta - \gamma)$$
$$+ [- L_{m0} + L_{m2} \cos (2\theta - 2\pi/3)]i_m \sin (\theta - \gamma - 2\pi/3)$$
$$+ [- L_{m0} + L_{m2} \cos (2\theta + 2\pi/3)]i_m \sin (\theta - \gamma + 2\pi/3).$$

Simplifying this and neglecting the third harmonic gives the result

$$(L_0 + L_{m0}) \sin (\theta - \gamma) - (\tfrac{1}{2}L_2 + L_{m2}) \sin (\theta + \gamma).$$

Suppose $\gamma = 0$; then phase A has peak current when $\theta = \pi/2$, i.e. when its axis coincides with the q-axis. Its effective inductance is then

$$L_0 + L_{m0} - \tfrac{1}{2}L_2 - L_{m2} = L_{aq}.$$

If $\gamma = \pi/2$, then phase A has peak current when on the direct axis with $\theta = 0$. The effective inductance is

$$L_0 + L_{m0} + \tfrac{1}{2}L_2 + L_{m2} = L_{ad}.$$

Thus the flux per unit armature current (i.e. the armature reaction) is affected by saliency, and depends on how the armature m.m.f. is directed with respect to the d- and q-axes. The larger flux per unit armature current L_{ad} applies when the m.m.f. and d-axis coincide; the smaller, L_{aq}, when the armature m.m.f. is along the q-axis. These differ by $(L_2 + 2L_{m2})$, second-harmonic space-distribution coefficients that vanish for purely cylindrical geometry.

ARMATURE TO AXIS CURRENT TRANSFORMATION. It must be remembered that, under steady-state conditions, the armature m.m.f. is *stationary* with respect to the d- and q-axes. It can be resolved along these axes, and the two components then imagined to be produced by appropriate *direct currents* i_d and i_q in fixed axis coils D and Q. It is only necessary for the axis-coil currents together to produce the same m.m.f. in magnitude and direction as the actual currents i_a, i_b and i_c. Then i_d acts on L_{ad} and i_q on L_{aq}, determining at once the armature flux for any given conditions.

Just as for the d.c. machine, coils D and Q have the property of accommodating e.m.f.'s of rotation (as well as of pulsation) although they do not move.

The instantaneous axis currents are related to the instantaneous 3-phase armature currents and the instantaneous position θ of the axis of phase A by the expressions

$$\left.\begin{aligned} i_d &= \tfrac{2}{3}[i_a \cos \theta + i_b \cos (\theta - 2\pi/3) + i_c \cos (\theta - 4\pi/3)] \\ i_q &= \tfrac{2}{3}[i_a \sin \theta + i_b \sin (\theta - 2\pi/3) + i_c \sin (\theta - 4\pi/3)] \end{aligned}\right\} \quad . \quad (205)$$

The axis voltages, v_d and v_q, are related to the armature phase voltages v_a, v_b and v_c by expressions of identical form.

The reverse transformations are

$$\left.\begin{array}{l} i_a = i_d \cos \theta + i_q \sin \theta \\ i_b = i_d \cos (\theta - 2\pi/3) + i_q \sin (\theta - 2\pi/3) \\ i_c = i_d \cos (\theta - 4\pi/3) + i_q \sin (\theta - 4\pi/3) \end{array}\right\} \quad . \quad (206)$$

Again, for the phase voltages v_a, v_b and v_c in terms of v_d and v_q, expressions indentical in form to eq. (206) are employed.

If there are zero-sequence components, then the further equation $i_0 = \frac{1}{3}(i_a + i_b + i_c)$ is added to eqs. (205) and correspondingly for

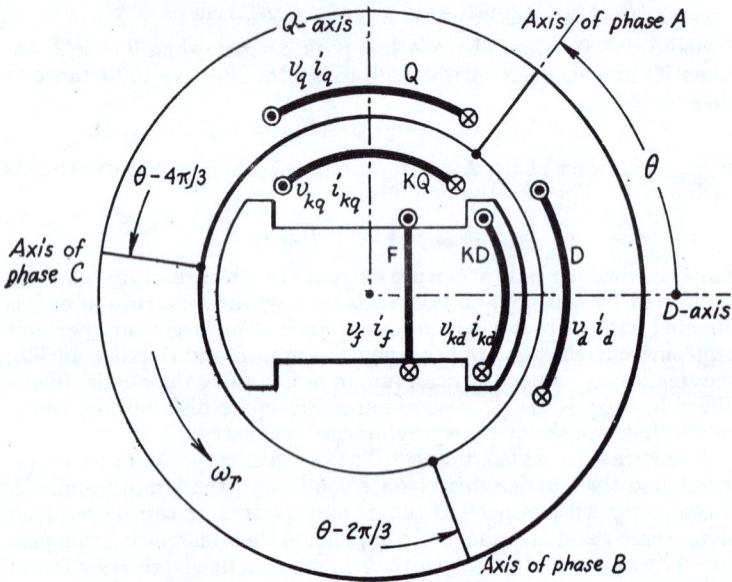

FIG. 392. UNIT SYNCHRONOUS MACHINE

v_0. The reverse transformations in eqs. (206) simply have the addition of i_0 to the right-hand side, and correspondingly the addition of v_0 for the voltages.

POWER. As noted in §7, unit power is that for the machine as a whole, so that with per-unit voltages and currents in a 3-phase machine the per unit power is $P = \frac{2}{3}(v_a i_a + v_b i_b + v_c i_c)$. The transformations in eqs. (205) and (206) then correctly give $P = \frac{1}{2}(v_d i_d + v_q i_q)$.

SUMMARY. The analysis in circuit form of a synchronous machine requires first the reduction of all windings to d- and q-axis groups. Fig. 392 shows the basic machine. Coil F is always a d-axis winding. The armature winding is transformed into the axis coils D and Q. The windings KD and KQ represent damping windings. They are single-turn equivalents of field collars, pole-shoe eddy-current paths and pole-face damper bars. Usually all such circuits are closed, and have no applied voltage so that v_{kd} and v_{kq} are zero. Strictly, several KD and KQ coils are required if there are several paths giving eddy-current damping, but it is sufficient here to assume one only in each axis. Then the matrix in Fig. 390 applies.

10. **Synchronous Generator in the Steady State.** Let the machine run on steady balanced load on busbars with symmetrical phase voltages v_a, v_b and v_c of angular frequency $\omega = 2\pi f$. The speed ω_r will then be equal to ω, and the angle θ will be ωt, reckoned from the instant that the axis of phase A coincides with the d-axis. The field current i_f is constant. Coils KD and KQ, permanently short-circuited, will under steady conditions have no induced e.m.f.'s or currents.

Let $v_a = v_m \cos(\omega t + \alpha)$ and $i_a = i_m \cos(\omega t + \beta)$ be the applied voltage and input current of phase A. Corresponding values for phases B and C are written by adding $-2\pi/3$ and $-4\pi/3$ respectively to the angles. Applying eq. (206) with $\theta = \omega t$ gives at once

$$i_d = i_m \cos \beta, \qquad v_d = v_m \cos \alpha,$$
$$i_q = -i_m \sin \beta, \qquad v_q = -v_m \sin \alpha.$$

These are all constant, so that the armature coils D and Q have direct applied voltages and carry constant direct currents. In eqs. (199) to (203), or in the matrix of Fig. 390, all terms with $p = d/dt$ will become zero. The equations therefore reduce to

$$v_f = r_f i_f$$
$$v_d = r_d i_d + (L_{aq} + L_q)\omega i_q \qquad = r_d i_d + x_{sq} i_q$$
$$v_q = r_q i_q - L_{ad}\omega i_f - (L_{ad} + L_d)\omega i_d = r_q i_q - x_{ad} i_f - x_{sd} i_d$$

Here $\omega L_{ad} = x_{ad}$, the armature-reaction reactance for the d-axis; $\omega(L_{ad} + L_d) = x_{ad} + x = x_{sd}$, the d-axis synchronous reactance; and $\omega(L_{aq} + L_q) = x_{aq} + x = x_{sq}$ is the q-axis synchronous reactance. In the synchronous machine the leakage inductances L_d and L_q are naturally equal and are both written as L, with $\omega L = x$ as the corresponding leakage reactance; also $r_d = r_q = r$.

These results can be used in a complexor diagram, it being remembered that the voltages are *applied*, and the currents are *input* currents. For a generator, therefore, the currents will have a resultant in phase-opposition to the resultant voltage, indicating electrical output. Using r.m.s. values, Fig. 393, then the applied voltage complexor for phase A is

$$V_1 = V_d + V_q = (v_m \cos \alpha + j v_m \sin \alpha)/\sqrt{2}$$
$$= (v_d - j v_q)/\sqrt{2},$$

and similarly

$$I = I_d + I_q = (i_m \cos \beta + j i_m \sin \beta)/\sqrt{2}$$
$$= (i_d - j i_q)/\sqrt{2}.$$

For no load and zero current the axis voltages are $v_d = 0$, $v_q = -x_{ad}i_f$. The no-load r.m.s. complexor is therefore

$$V_t = -E_t = jx_{ad}i_f/\sqrt{2},$$

so that for load conditions the applied voltage is

$$V_1 = rI + jx_{sd}I_d + jx_{sq}I_u - E_t$$

as shown in Fig. 393. The open-circuit applied voltage (i.e. the reverse of the open-circuit induced e.m.f. E_t) is a q-axis quantity, as will appear from the discussion of Fig. 387 (b).

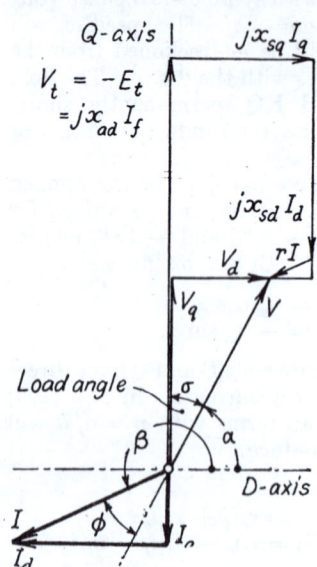

FIG. 393. COMPLEXOR DIAGRAM OF SYNCHRONOUS GENERATOR IN THE STEADY STATE

TORQUE. In per-unit terms, the inductance coefficients of eq. (204) can be expressed as their corresponding reactances. Only currents i_f, i_d and i_q are concerned, so that

$$M = -(x_{sd} - x_{sq})i_d i_q - x_{ad}i_f i_q.$$

The first term is the reluctance torque. In a cylindrical machine it would vanish, leaving only the torque due to the main flux $x_{ad}i_f$.

11. **Synchronous Generator on Sudden Short-Circuit.** Any fault condition can be handled by eqs. (199–203). The most straightforward case will be considered, namely a three-phase short circuit occurring on a machine initially on no load with zero armature current. The conditions obtaining subsequent to such a short-circuit have been discussed in Chapter XVIII, §5.

In dealing with transients it is convenient to have $t = 0$ at the instant of fault. Let the axis of phase A make the angle θ_0 with the d-axis at $t = 0$; then $\theta = \omega t + \theta_0$. Before $t = 0$ the armature axis voltages are $v_d = 0$ and $v_q = -x_{ad}i_{f0}$. The effect of a short circuit is therefore to reduce v_q to zero, which can be taken as equivalent to the sudden application of a step-function voltage $-v_q 1$ to the q-axis terminals. The currents that then flow in all the circuits are to be considered as superimposed on those existing prior to $t = 0$. In fact there are such currents only in the coil F, because the assumed initial conditions are of steady running on no load. Thus the solutions for i_d, i_q, i_{kd} and i_{kq} will be the actual currents in D, Q, KD and KQ, while the solution for i_f will be superimposed on the existing current i_{f0}.

The change of speed during short circuit can be neglected, and in consequence $\omega_r = \omega$ throughout. Eqs. (199–203) now apply.

GENERATOR WITHOUT DAMPERS OR RESISTANCE. This is a useful, though simplified, case of a machine represented by coils F, D and Q alone. With the assumption of negligible resistance, the principle of constant linkages is being applied, because during the short circuit all three circuits are closed. Only eqs. (199), (202) and (203) are concerned. With inductance coefficients changed for convenience to reactances, the expressions reduce to

$$0 = (1/\omega)(x_{ad} + x_f)pi_f + (1/\omega)x_{ad}pi_d \qquad \text{(i)}$$
$$0 = (1/\omega)(x_{ad} + x)pi_d + (1/\omega)x_{ad}pi_f + (x_{aq} + x)i_q \qquad \text{(ii)}$$
$$- v_q 1 = (1/\omega)(x_{aq} + x)pi_q - x_{ad}i_f - (x_{ad} + x)i_d \qquad \text{(iii)}$$

The current i_f, from (i), is $- i_d \cdot x_{ad}/(x_{ad} + x_f)$, and can be eliminated from (ii) and (iii). Then from (ii)—

$$i_q = - \frac{1}{x_{aq} + x}\left(\frac{x_{ad}x_f}{x_{ad} + x_f} + x\right)\frac{p}{\omega}\,i_d = - \frac{x_d{}'}{x_{sq}}\cdot\frac{p}{\omega}\cdot i_d,$$

where $x_d{}'$, the term in brackets, is called the *d-axis transient reactance*. Substituting for i_f and i_q in (iii) and simplifying then gives

$$i_d = \frac{\omega^2}{x_I{}'}\cdot\frac{1}{p^2 + \omega^2}\cdot v_q 1.$$

Applying the standard methods of operational calculus, by writing $(p^2 + \omega^2) = (p + j\omega)(p - j\omega)$, gives for the current in D:

$$i_d = (v_q/x_d{}')(1 - \cos \omega t).$$

Using the expression for i_q above, then in Q

$$i_q = - (v_q/x_{sq})\sin \omega t.$$

It remains to apply eq. (206) to obtain the actual current in phase A, with $\theta = \omega t + \theta_0$:

$$i_a = v_q\left[\frac{1}{x_d{}'}\cos(\omega t + \theta_0)\right.$$

normal-frequency short-circuit current

$$- \frac{x_{sq} + x_d{}'}{2x_d{}'x_{sq}}\cos\theta_0$$

constant asymmetric current

$$\left. - \frac{x_{sq} - x_d{}'}{2x_d{}'x_{sq}}\cos(2\omega t + \theta_0)\right]$$

double-frequency short-circuit current . (207)

This corresponds to the shape indicated in Fig. 307 for a short circuit at $\theta_0 = 0$. The currents i_b and i_c are found if θ_0 is replaced by $(\theta_0 - 2\pi/3)$ and $(\theta_0 - 4\pi/3)$ respectively.

The component of field current superposed on the initial current i_{f0} is closely related to i_d:

$$i_f = \frac{v_q \cdot x_{ad}}{x_{ad}x_f + x_{ad}x + x_f x} [1 - \cos \omega t]$$

which has the shape shown in Fig. 307. In the case considered, the absence of resistance implies that all components of current persist without decay.

EXAMPLE. A machine running on no load and excited to normal busbar voltage is dead-short-circuited at the instant that the axes of the field and phase A coincide. Find the currents in phase A and the field winding, neglecting all resistances. Per-unit reactances: $x_{ad} = 1\cdot50$, $x_{aq} = 0\cdot60$, $x_f = 0\cdot13$, $x = 0\cdot10$ p.u.

Here $v_q = 1\cdot0$ p.u., and $\theta_0 = 0$. The d-axis transient reactance is

$$x_d' = \frac{x_{ad}x_f}{x_{ad} + x_f} + x = 0\cdot22 \text{ p.u.,}$$

and $x_{sq} = x_{aq} + x = 0\cdot70$ p.u. Using the expressions found above,

$$i_a = 4\cdot5 \cos \omega t - 3\cdot0 - 1\cdot5 \cos 2\omega t,$$
$$i_f = 4\cdot5 (1 - \cos \omega t) + 1\cdot0 = 5\cdot5 - 4\cdot5 \cos \omega t.$$

These correspond closely to the waveforms in Fig. 307. The unity added to i_f is $i_{f0} = 1\cdot0$ p.u. for the initial field current.

GENERATOR WITHOUT DAMPERS BUT WITH FIELD RESISTANCE. The equations (i), (ii) and (iii) above are again applicable, except that (i) now includes the term $r_f i_f$ on the right-hand side. All transient components now decay at rates depending on their respective effective inductance/resistance ratios, i.e. their time-constants. The procedure is as before. From (i),

$$i_f = \frac{x_{ad}p}{\omega r_f + (x_{ad} + x_f)p} i_d.$$

This result is used in (ii), where the terms in i_d and i_f simplify to

$$(1/\omega)(x_{ad} + x)i_d + (1/\omega)x_{ad}i_f$$

$$= \frac{(x_{ad} + x)}{\omega} \left[\frac{1 + \dfrac{1}{\omega r_f}\left(\dfrac{x_{ad}x}{x_{ad} + x} + x_f\right) p}{1 + (1/\omega r_f)(x_{ad} + x_f)\, p} \right] i_d$$

$$= \frac{x_{sd}}{\omega} \left[\frac{1 + p\tau_d'}{1 + p\tau_{d0}'} \right] i_d,$$

where $\tau_d' = \dfrac{1}{\omega r_f}\left(\dfrac{x_{ad}x}{x_{ad} + x} + x_f\right)$ and $\tau_{d0}' = \dfrac{1}{\omega r_f}(x_{ad} + x_f)$

are the d-axis *transient* and *open-circuit* time-constants of current decay.

The final term in (ii) is $(x_{aq} + x)i_q \neq x_{sq}i_q$, so that i_q is obtained in terms of i_d. Eq. (iii) now gives

$$i_d = \frac{\omega^2}{x_{sd}} \cdot \frac{1}{p^2 + \omega^2} \cdot \frac{1 + p\tau_{d0}'}{1 + p\tau_d'} \cdot v_q 1.$$

The effect of field resistance is seen to be that the current i_d decays to v_q/x_{sd}, which is the normal steady-state condition as limited by the d-axis synchronous reactance x_{sd}.

If the resistance r of the D and Q coils had been included, the rates of decay would have been further influenced. In obtaining the solution it is desirable to ignore ri_d in (ii) when obtaining the expression for i_q in terms of i_d: this is legitimate because the per-unit value of r is very much smaller than those of the reactances. The term ri_q is included in (iii).

GENERATOR WITH DAMPING CIRCUITS. All four circuits, F, KD, KQ, D and Q are now concerned. Eqs. (199–202) are written with zero applied voltage, and (203) with $-v_q1$. The equivalent machine is that shown in Fig. 392.

The per-unit values of r_{kd} and r_{kq} are normally large, introducing terms with rapid decay corresponding to the *sub-transient*. The solution of five simultaneous operational equations is somewhat formidable unless approximations are made.* The solution for the current in phase A is

$$i_a = v_q \left[\frac{1}{x_{sd}} \cos (\omega t + \theta_0) \right.$$

final steady-state short-circuit current

$$+ \left(\frac{1}{x_d'} - \frac{1}{x_d} \right) \exp (-t/\tau_d') \cos (\omega t + \theta_0)$$

normal-frequency decaying transient current

$$+ \left(\frac{1}{x_d''} - \frac{1}{x_d'} \right) \exp (-t/\tau_d'') \cos (\omega t + \theta_0)$$

normal-frequency decaying sub-transient current

$$- \frac{x_d'' + x_q''}{2x_d'' x_q''} \exp (-t/\tau_a) \cos \theta_0$$

asymmetric d.c. decaying current

$$- \left. \frac{x_d'' - x_q''}{2x_d'' x_q''} \exp (-t/\tau_a) \cos (2\omega t + \theta_0) \right]$$

double-frequency decaying current   (208)

The normal-frequency components can be identified in Fig. 310, and

* Adkins, "Transient Theory of Synchronous Generators connected to Power Systems," *Proc.I.E.E.*, 98 (II), p. 510 (1951); also *General Theory of Electrical Machines* (Chapman & Hall).

the asymmetric component in Fig. 308 (b), in connection with which the various components are briefly discussed.

The quantities in eq. (208) are summarized below, all in per-unit terms:

Resistances and Leakage Inductances.

Armature, field, d-damper, q-damper $\qquad\qquad r, r_f, r_{kd}, r_{ku}$

$$L, L_f, L_{kd}, L_{kq}$$

Armature-reaction (Magnetizing) Inductances:

D-axis, Q-axis $\qquad\qquad\qquad\qquad\qquad\qquad\qquad\qquad L_{ad}, L_{aq}$

Reactances:

Corresponding to inductances $\qquad \omega L = x, \omega L_f = x_f \ldots \omega L_{aq} = x_{aq}$

Synchronous Reactances.

D-axis $\qquad\qquad\qquad\qquad\qquad \omega(L_{ad} + L) = x_{ad} + x = x_{sd}$

Q-axis $\qquad\qquad\qquad\qquad\qquad \omega(L_{aq} + L) = x_{aq} + x = x_{sq}$

Transient Reactance.

D-axis

$$x_d' = \frac{x_{ad} x_f}{x_{ad} + x_f} + x$$

Sub-transient Reactances:

D-axis

$$x_d'' = \frac{x_{ad} x_f x_{kd}}{x_{ad} x_f + x_f x_{kd} + x_{kd} x_{ad}} + x$$

Q-axis

$$x_d'' = \frac{x_{aq} x_{ku}}{x_{aq} + x_{kq}} + x$$

Time Constants

D-axis open-circuit transient

$$\tau_{d0}' = \frac{1}{\omega r_f} (x_{ad} + x_f)$$

D-axis short-circuit transient

$$\tau_d' = \frac{1}{\omega r_f} \left(\frac{x_{ad} x}{x_{ad} + x} + x_f \right)$$

D-axis o.-c. sub-transient

$$\tau_{d0}'' = \frac{1}{\omega r_{kd}} \left(\frac{x_{ad} x_f}{x_{ad} + x_f} + x_{kd} \right)$$

D-axis s.-c. sub-transient

$$\tau_d'' = \frac{1}{\omega r_{kd}} \left(\frac{x_{ad} x_f x}{x_{ad} x_f + x_f x + x_{ad} x} + x_{kd} \right)$$

Q-axis o. c. sub-transient

$$\tau_{q0}'' = \frac{1}{\omega r_{kq}} (x_{aq} + x_{kq})$$

Q-axis s.-c. sub-transient

$$\tau_q'' = \frac{1}{\omega r_{kq}} \left(\frac{x_{aq} x}{x_{aq} + x} + x_{kq} \right)$$

Armature short-circuit

$$\tau_a = \frac{1}{\omega r} \left(\frac{2 x_d'' x_q''}{x_d'' + x_q''} \right)$$

The symmetrical part of the short-circuit current takes the form shown in Fig. 310, where all the reactances and time-constants are d-axis quantities. It is important to note that the two-axis analysis is necessary even for cylindrical machines for short-circuit conditions, because the field winding is a d-axis circuit and introduces saliency effects during the flow of transient currents.

OTHER APPLICATIONS. The same method is applicable to the evaluation of torque during fault conditions. Strictly, the introduction of torque will result in speed changes and it is no longer true that $\omega_r = \omega$, and mechanical characteristics must be introduced to complete the electrodynamic system. With ω_r a variable the equations of operation are no longer linear.

Running-up, synchronizing, cyclic speed-variation and voltage regulation problems can be solved, although not so directly as in the simple short-circuit case.

12. **Induction Machines.** With both stator and rotor cylindrical and symmetrical, the d-axis is chosen arbitrarily as the axis of the stator phase A. The 3-phase winding is converted to a 2-phase one, so that the axis of the second stator phase becomes the q-axis. The two stator phases are then fixed axis coils, called 1D and 1Q respectively.

The rotor carries a slip-ring winding or a short-circuited cage. In the former case it is possible to convert the winding to fixed d- and q-axis coils, just as for the armature of the synchronous machine in §8. A cage rotor is rather more complex; but insofar as it can, and does, produce an armature reaction m.m.f. substantially identical (apart from space harmonics) with that of a slip-ring winding, it can also be represented by d- and q-axis coils. These latter are called 2D and 2Q.

Disregarding zero-sequence voltages and currents, the conversion of a 3-phase stator winding to a 2-phase axis winding is by

$$v_{1d} = \tfrac{2}{3}[v_a - \tfrac{1}{2}v_b - \tfrac{1}{2}v_c]$$

and
$$v_{1q} = \tfrac{2}{3}[-(\sqrt{3}/2)v_b + (\sqrt{3}/2)v_c]. \qquad . \ (209)$$

Identical coefficients relate the currents. For any zero-sequence voltage, $v_{10} = \tfrac{1}{3}(v_a + v_b + v_c)$. The reverse transformation is

$$v_a = v_{1d}; \qquad v_b = -\tfrac{1}{2}v_{1d} - (\sqrt{3}/2)v_{1q};$$
$$v_c = -\tfrac{1}{2}v_{1d} + (\sqrt{3}/2)v_{1q}; \qquad . \qquad . \qquad . \ (210)$$

and similarly for the currents. Any zero sequence voltage v_0 is added to each right-hand side.

For the rotor, the appropriate transformations are those given in eq. (205) and (206).

Using the same rules as for Fig. 390, the impedance matrix can be directly set up. It is only necessary to identify 1D with F, 1Q with

KQ, 2D with D, and 2Q with Q in Fig. 392, so that the unit induction machine becomes the arrangement of Fig. 394.

The stator and rotor resistances and reactances are r_1, r_2, L_1 and L_2 for each axis. 1D and 2D have the mutual inductance L_m, and similarly for 1Q and 2Q. Rotational e.m.f.'s appear only in the rotor coils. Thus 2D rotates in the mutual flux of 1Q, and also in the total flux of 2Q. The voltage equations are read from the matrix, Fig. 394, as before.

The torque is obtained directly from the rotational-voltage terms, i.e.

$$M = L_m(i_{1q}i_{2d} - i_{1d}i_{2q}).$$

STEADY LOAD. Here it is necessary only to replace p by $j\omega$, and ω_r by $\omega(1 - s)$, where s is the slip. Then ω can be combined with

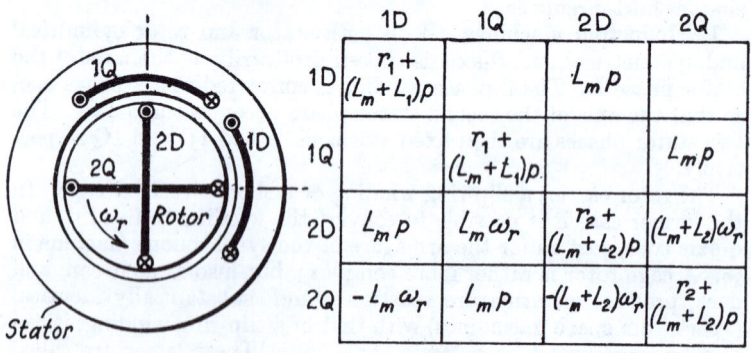

	1D	1Q	2D	2Q
1D	$r_1 +$ $(L_m+L_1)p$		$L_m p$	
1Q		$r_1 +$ $(L_m+L_1)p$		$L_m p$
2D	$L_m p$	$L_m \omega_r$	$r_2 +$ $(L_m+L_2)p$	$(L_m+L_2)\omega_r$
2Q	$-L_m \omega_r$	$L_m p$	$-(L_m+L_2)\omega_r$	$r_2 +$ $(L_m+L_2)p$

FIG. 394. CIRCUITS AND IMPEDANCE MATRIX FOR INDUCTION MOTOR

the inductances to form reactances. Reading from the matrix, the following can be written at sight, inserting the usual condition that $v_{2d} = v_{2q} = 0$ for a short-circuited rotor:

$$v_{1d} = [r_1 + j(x_m + x_1)]i_{1d} + jx_m i_{2d}$$

$$v_{1q} = [r_1 + j(x_m + x_1)]i_{1q} + jx_m i_{2q}$$

$$0 = jx_m i_{1d} + (1 - s)x_m i_{1q} \\ + [r_2 + j(x_m + x_2)]i_{2d} + (1 - s)(x_m + x_2)i_{2q}$$

$$0 = -(1 - s)x_m i_{1d} + jx_m i_{1q} \\ - (1 - s)(x_m + x_2)i_{2d} + [r_2 + j(x_m + x_2)]i_{2q}$$

There is no difference in the d- and q-phases except one of time, so that it would be possible to write

$$v_{1d} = v_1, \quad v_{1q} = -jv_1, \quad i_{1d} = i_1, \quad i_{1q} = -ji_1,$$
$$i_{2d} = i_2, \quad i_{2q} = -ji_2$$

and the d- and q-axis can be taken separately as each contributing one-half of the torque.

TRANSIENT STATE. There is little advantage in the circuit method compared with the usual rotating-field analysis for steady-state conditions. But for running on unbalanced voltage the method is advantageous, while for transients it is essential.

INDEX